COASTAL DYNAMICS '97

CONFERENCE PROCEEDINGS

EDITED BY
Edward B. Thornton

**American Society
of Civil Engineers**
1801 ALEXANDER BELL DRIVE
RESTON, VIRGINIA 20191–4400

Abstract: This proceedings, *Coastal Dynamics '97*, consists of papers presented at the third Coastal Dynamics Conference held in Plymouth, U.K., June 1997. The Coastal Dynamics Conferences provide a forum for researchers involved in the study of beach and nearshore dynamics. While the focus of these conferences is relatively narrow, dealing primarily with natural beach and nearshore environments and only with man-made structures to the extent that they influence the neighboring environment, the range of topics available for study is great. Some of the areas covered in this proceedings are: 1) sand suspension; 2) wave propagation in shallow water; 3) boundary layers and mixing; 4) sediment transport predictors; 5) breaking waves; 6) wave-current interaction; 7) beach profiles; 8) rip currents; 9) effects of coastal structures; and 10) wave-induced circulation.

ISBN: 0-7844-0321-X

PREFACE

The world's beaches are amongst the most dynamic environments of our planet, and yet are also experiencing the greatest pressure from man's activities. Proper management of beaches, for both development and preservation, demands a better understanding of the physical and sedimentological processes acting to mould and change the coastline, and the development of adequate predictive tools which can guide the manager. Important advances have been made by research into coastal dynamics over the past few decades, particularly with the development of instruments able to make measurements in the highly energetic beach environment, and with the rapid increase in computing power available for numerical modelling. However, this research has also shown the complexity of the processes and the need for a collaborative, interactive approach to solving the problems. Numerical models need field observations to test and extend them; field studies need theoretical advances to provide indications of possible processes and ways of interpreting measurements; laboratory observations provide valuable insights where the complexities of the natural environment can be treated in a controlled manner.

The *Coastal Dynamics* Conferences provide a forum for researchers involved in the study of beach and nearshore dynamics. The focus of the Conferences is a relatively narrow one, dealing primarily with natural beach and nearshore environments and only with man-made structures to the extent that they influence neighbouring natural enviroments. The range of approaches to this study area is wide however, and embraces field, laboratory, theoretical and numerical studies.

Coastal Dynamics '97 is the third Conference in the series, which was inaugurated in Barcelona, Spain, with *Coastal Dynamics '94*, and was followed by *Coastal Dynamics '95* in Gdansk, Poland. In common with the first two Conferences, *Coastal Dynamics '97* took as a sub-theme an emphasis on large scale experiments, an inclusive rather than exclusive phrase meant to reflect the growing importance of multi-institutional research projects such as those contained in the EC MAST programmes and the US field studies centred at Duck, and also to recognise the advances being made in our ability to study coastal processes over larger spatial and temporal scales, as for example in the Argus remote sensing project.

Judging by the positive response to this conference and the lively exchanges which characterised the technical sessions and informal gatherings, it is probably fair to conclude that the *Coastal Dynamics* series has now established itself as a very valuable and productive forum. Its relatively small size and clear focus allowed for very constructive and helpful interaction between the researchers. The papers contained in this Proceedings volume show the range, depth and liveliness of Coastal Dynamics research worldwide. After a break in two years time, when a *Coastal Sediments* Conference will be held instead, the next *Coastal Dynamics* Conference will be held in Copenhagen in the year 2001, and will provide opportunity to review the state of the art at the start of a new millenium.

The topics contained in these Proceedings were originally selected from more than twice as many abstracts submitted to the Paper Review Panel, which faced the difficult task of selecting papers to be presented at the Conference. Each of the papers included in this volume has been accepted for publication by the Proceedings Editors. All papers are eligible for discussion in the *Journal of Waterways, Port, Coastal and Ocean Engineering*, and all papers are eligible for ASCE awards.

We would like to acknowledge the support of those who made *Coastal Dynamics '97* possible. The EC MAST Office generously provided funds to allow seven European post-graduate students and eight researchers from eastern Europe and Turkey to attend the Conference. The City of Plymouth and the University of Plymouth (Institute of Marine Studies and Department of Civil and Structural Engineering) also provided valuable support. We also thank all those who contributed to the organisation of the Conference, particularly the International Steering Committee, the Paper Review Committee, the Local Organising Committee and especially the Conference Secretariat, Mrs Denise Horne, Mrs Lesley Pengelly and Mrs Penny Grove.

David A Huntley
University of Plymouth,
Institute of Marine Studies.
Plymouth, UK.

Edward B. Thornton,
Naval Postgraduate School,
Department of Oceanography.
Monterey, Ca., USA.

September 15th 1997.

Contents

Boundary Layers and Mixing

Sediment Transport Predictors

Breaking Waves

ix

Seabed Effects

Patterns in the Water: Patterns in the Sand?

A. J. Bowen[1]

Abstract

The nearshore zone is a fascinating region in which to study the interplay of theory and observation. Many of the processes of interest are of a wave-like nature, very amenable to analysis using classical wave and stability theory. From these theories rise a set of possible interactions, structure and patterns in the hydrodynamic flow and, with the inclusion of simplified sediment dynamics, in the seabed topography. Some of these interactions can be clearly demonstrated in simplified laboratory experiments. A few have been clearly seen in very particular environments in the real world.

However, this knowledge of what might happen is subject to rigorous examination in terms of the observations of what actually does happen. In reality, on most of the beaches that have been extensively studied, the nearshore flow seems to be the superposition of a surprisingly large number of different patterns. So much so, that the possible significance of some modes, notably shear waves and bar-trapped edge waves has only been recognised relatively recently, even though these modes seem well described by classical theory. As this complexity needs to be disected by measurements using a large array of instruments; such data is, as yet, available from only a few, particular locations. It remains to be seen whether general patterns of behaviour can be found which provide useful predictions of beach evolution. Recent long term observations of different environments using overwater video suggest a very wide range of complexity.

Introduction

One of the exciting aspects of nearshore research is the inherently multidisciplinary nature of the topic. However, as always, there is good news and bad news. The good news is the breadth of interests involved, the stimulation of working on problems of wide interest, and the new concepts that are imposed by other disciplines. The bad news is that while there may be some general ideas across these disciplines as to what the significant problems are,

---

[1]Department of Oceanography, Dalhousie University, Halifax, Nova Scotia, Canada, B3H 4J1.
E-mail: Tony.Bowen@Dal.Ca

there is much less agreement as to what constitutes an acceptable solution. An obvious example is the practical use of the CERC formula for longshore sand transport, a formula that contains only incident wave energy and direction as parameters. There is no dependence on sediment size, on beach topography (bar structure), groundwater dynamics, or anything else. The observed spatial and temporal patterns in the nearshore flow are ignored, even those that are essentially of first order, i.e. of the same order of magnitude as the incident waves and wave driven currents. Examples are

- spatial variability – the longshore current structure due to bar topography
  - rip currents
  - other flow patterns induced by beach cusps, longshore bars.

- temporal variability
  - wave grouping
  - low frequency waves, edge waves
  - shear waves
  - tidal effects

- swash and run-up, where low frequency waves may dominate the wave energy spectrum close to the shore.

These patterns are indicative of potentially significant processes that are operative in the nearshore zone that have been "integrated over" in the simple formulae. The expectation that a number of such processes do, in practice, play significant roles in longshore sand transport, and their importance can vary significantly in different beach environments, has been expressed in terms of a classification of beach types (Wright and Short, 1984), but this does not seem to have led to a simple, improved version of the CERC formula that has been widely accepted. Instead, increasingly complex 2-D and 3-D numerical models have been developed, using a variety of formulations and parameterizations of both the hydrodynamic and the sedimentary processes.

It is therefore timely to revisit some of the existing data, to try and define the processes which may be significant in reality and which should be reproduced in these complex models if they are to be sufficiently realistic analogues of the real world that they can be used to investigate, for example, the parameterization of high frequency processes. This is clearly necessary, as it is extremely unlikely that models with real time wave dynamics will be run out to decadal time scales needed to predict coastal evolution.

Long Waves, Edge Waves, and Shear Waves

> "edge-waves...the amplitude diminishes exponentially as the distance from shore increases. The wave velocity is less than waves of the same length in deep water. *It does not appear that the type of motion here referred to is very important.*"

Lamb. Hydrodynamics, p. 447

In looking at the nearshore environment, the dominance of the incident waves, particularly in storm conditions, is apparently obvious. The role of other waves is less clear and an awareness of the existence of other modes depends strongly on location. On the Oregon Coast, the run-up is clearly of very low frequency (Fig. 1) but the longshore structure is not obvious to the casual observer. The natural interpretation of Lamb's comments is that edge waves are not seen in the real world.

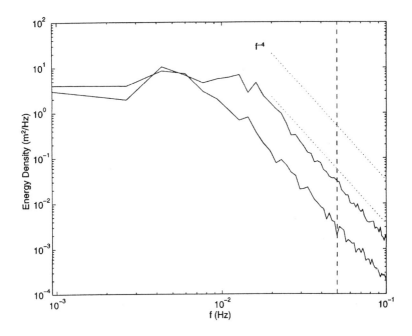

Figure 1: Shoreline run-up spectra for two locations on Agate Beach, Oregon, USA. The dashed line indicates the frequency of the incident waves (courtesy P. Ruggerio and R.A. Beach).

However, relatively simple measurements on a beach quickly reveal a spectral evolution with the tendency for low frequency energy to increase shorewards and, on all but very steep beaches, to dominate the run-up spectrum. While this indicates the distribution of wave energy, a rather sophisticated array of suitable instruments is required to identify leaky modes, edge waves or shear waves as specific modes in a frequency-wavenumber diagram. It was therefore not until the 1980s that the first definitive measurements of edge waves were made in the field (Huntley et al., 1981), and the number of such measurements that have been made is still not very large.

The earlier work tended to be focussed on the theoretical properties of individual modes of a specific frequency σ and longshore wavenumber k. Such modes seem to provide patterns of longshore variability that might be associated with rip currents, beach cusps and crescentic bars (Bowen and Inman, 1971; Huntley and Bowen, 1978; Huntley and Bowen, 1984). However, in considering how edge waves are forced it is apparent that a whole suite of triad interactions were possible (Fig. 2). The most effective way of forcing edge waves appears to be in the difference interaction of two incident waves (a3), as in Lippmann *et al.* (1997), but the subharmonic resonance (a2) is effective and readily observable in the laboratory (i.e. Guza and Bowen, 1976). Energetic edge waves can then exchange energy amongst themselves (a1), or to leaky modes (b1). A set of very special interactions occur when the difference frequency is zero and the resulting "mode" is a steady flow or a topographic feature (c1, c2, c3), an example would be Bragg scattering of edge waves by topography of longshore wavenumber $2k$ which reflect an edge wave (σ, k) as a mode $(\sigma, -k)$ as shown in (c2).

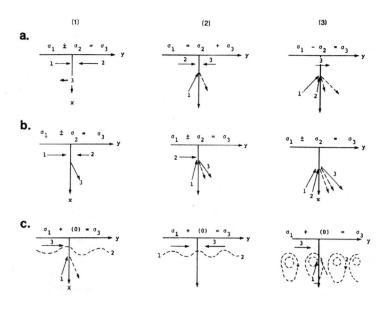

Figure 2: Wave-wave interaction diagrams for triad interactions in the nearshore zone. Edge waves are indicated by alongshore wave numbers, incident or leaky modes by vectors in other directions. Class (a) interactions lead to the trapping of energy in edge wave modes, (b) the release of energy seawards, and (c) the special cases of interactions between two waves of the same frequency.

While these interactions are considered individually, the important implication is that the different frequencies and wavenumbers in an incident wave spectrum can generate a wide range of wave modes (Bowen and Guza, 1978), modes that can then further interact and spread energy in frequency/wavenumber space.

Patterns of Flow

On a planar beach, the theoretical solutions for free long waves or edge waves have maximum values at the shoreline. In reality, wave breaking severely modifies the form of the incident wave and has potentially significant effects on other frequencies. As usual, there are many possibilities. The breaking of the incident waves forces long waves (Symonds *et al.*, 1982; Lippmann *et al.*, 1997) but also provides turbulent processes that can reduce or suppress long waves (Guza and Bowen, 1976). There is also the possibility that in some limit edge waves and long waves are limited by mutual breaking processes (arising from the summation of different modes), so local breaking occurs when some parameter value (i.e. for acceleration) is exceeded.

However, there are suggestions from the limited data base available, much of which has been obtained by arrays of instruments deployed in the Duck experiments, that the use of the theoretical model shapes to describe long waves in the nearshore environment is not unreasonable (Holland *et al.*, 1995; Herbers *et al.*, 1995). This is good news for modelling. The bad news is that the data suggests that energy is distributed by a very wide range of frequencies, each associated with a large number of possible modes! Lippmann et al (in press) have taken this to the limit in assuming that, over a broad frequency range (0.004-0.05 Hz), all possible modes exist and have equal amplitude at the shoreline. They show that, for high mode edge waves on a plane beach, the ratio between the variance of the currents and the pressure is given by a ratio R, where

$$R = 2(u^2 + v^2)h/p^2g = 2 \tag{1}$$

where u and v are the on/offshore and longshore components of velocity, p the pressure and h the water depth. R is then much larger for shear waves, which are almost non-divergent and p is small (Bowen and Holman, 1989). Data taken at Duck show a value of R frequently of about 2, particularly seaward of the surf line.

The increasing use of video in nearshore research has meant that the run-up at the shoreline is a frequently measured parameter. Such measurements tend to confirm that energy is distributed widely in frequency space. Run-up spectra are frequently rather featureless showing a broad spread of low frequency with a drop off to high frequencies approximately as f^{-4} (Huntley *et al.*, 1977; Holland *et al.*, 1995) as in Figure 1.

Patterns in the Seabed

For theoretical studies, a planar beach allows analytic solutions; in the laboratory, a plane beach is structurally simple. However, in the field beaches are rarely planar and often show structure in both horizontal dimensions. Nevertheless, early studies of the role of long waves in shaping seabed topography assume a quasi-planar beach and look for hydrodynamic forcing which might lead to preferential erosion and deposition. However, it was realised that as the topography becomes developed it may play a dynamic role in the interactions (Fig. 2) or significantly change the physical structure of the waves.

The assumption that longwaves are a set of modes of uniform shoreline amplitude leads to a simple result on a planar beach, where the energy then simply decreases seawards. If there are no particularly energetic modes at the shore, then there can be no preferred lengthscales offshore. The dynamics associated with each individual mode is averaged over sets of orthogonal modes.

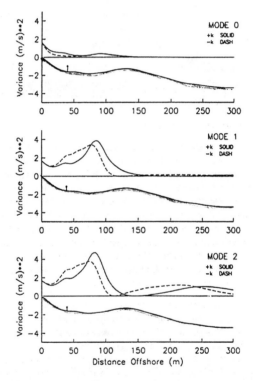

Figure 3: Theoretical edge wave solutions, in terms of the on-offshore velocity variance for three edge wave modes of frequency $f = 0.0175$Hz. Topography from Duck, 15 October 1986 (after Howd *et al.*, 1991). A weak longshore current leads to small differences in shape for waves propagating in the current direction $(+k)$ cf. those propagating against $(-k)$.

However, the calculations made for real barred topographies show that the shape of the long wave modes can be strongly influenced by topography. The fact that the edge wave solutions offshore can be larger is shown in Figure 3, from Howd *et al.* (1991). In Figure 3, the solutions depend partly on the depth profile but also on the propagation direction of the edge waves relative to the mean longshore current. Howd *et al.* (1992) showed that the edge wave equation for arbitrary depth profile $h(x)$, mean longshore current $V(x)$, can be transformed to the normal e-w equation with no currents.

The surface displacement, η, is given by

$$[H.\eta_x]_x + [\sigma^2/g - k^2.H].\eta = 0 \tag{2}$$

where σ is the wave frequency and k the longshore wavenumber. The phase speed is

$$C = \sigma/k \tag{3}$$

and the 'effective depth'

$$H(x) = h_{eff}(x) = h(x).[1 - V(x)/C]^{-2} \tag{4}$$

where both V and C are positive in the positive y direction (k positive), so that the 'effective' depth is reduced for waves and currents moving in the same direction.

If $V(x) = C$ somewhere, the solutions are shear waves.

Shonfeld (1995) showed a particularly impressive amplification of the edge wave solutions relative to the shoreline value in the presence of a strong longshore current. The fact that there is no real limit to the magnitude of this amplification, given the right bar scales for the particular mode, has been shown by Bryan and Bowen (1996). Edge waves can be almost totally trapped on any bar, the trapping being most effective when the phase speed of the wave $C \rightarrow (gh_{bar})^{1/2}$ where h_{bar} is the depth (or effective depth) at the bar crest. In a multiple bar situation each bar has its own set of trapped modes! Perhaps the most significant new result is that all the modes trapped on a bar have rather similar shape and similar phase velocity. The effect of adding up a large number of modes is no longer so clearly a cancellation. In addition, the assumption that all modes have equal shoreline amplitude would result in very large edge wave amplitudes over the bar.

While this theory is interesting, perhaps the most significant result of Bryan and Bowen (1996) is that, in a single bar system at Duck the energy in the longshore component of velocity lies preferentially along lines of a constant phase speed, consistent with the theoretical (Fig. 4). This is a little surprising as the bar system at Duck in 1990 was not particularly large and therefore the trapping effect for low frequencies was not particularly strong. Other locations with larger bar systems should trap low frequency, long waves much more effectively.

Conclusions

It seems likely that in saying that edge waves were not important, Lamb meant that they were a theoretical construct which had not been readily observed in practice. However, detailed observations using appropriate instrument arrays have now revealed a whole galaxy of long waves, edge waves and shear waves. The detailed behaviour varies from place to place, but these waves are energetic and ubiquitous.

In the early analysis of edge waves it was assumed that, as is the case for many wave motions, a few of the lowest modes would be dominant, and the theoretical calculations made at the time reflect this expectation (i.e. Bowen and Inman, 1971). This has also been shown to be wrong. Recent observations in relatively deep water have shown significant

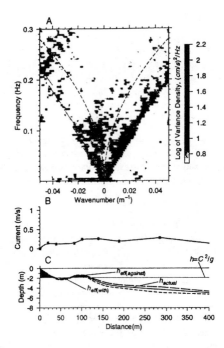

Figure 4: A. Frequency-wavenumber spectra of the longshore component of velociuty, 16 October 1990 at Duck. B. Profile of the mean longshore current, $V(x)$. C. Observed and effective depth. Note that although the energy lies along lines of slope $(gh_{bar})^{1/2}$, it is distributed over a wide range of frequencies (after Bryan, 1996).

energy propagation alongshore, implying the existence of energetic modes of high model number (Herbers *et al.*, 1995).

In laboratory experiments, the role of edge waves as a generation mechanism for a variety of three dimensional phenomena (beach cusps, crescentic bars, rip-currents) can be clearly demonstrated. There are observations that do suggest that sub-harmonic and lower frequency waves can be locally energetic and provide particular length scales (Bowen and Inman, 1971; Huntley and Bowen, 1978; Wright and Short, 1984; Aagaard, 1990). However, in many cases, and particularly in the Duck experiments, there seem to be far too many frequencies and modes co-existing to provide a preferred spatial scale. There is a rich field of investigation of the interaction between the waves and the sedimentary structures, both in terms of deterministic and rule-based modelling. However, the existence of wave breaking creates difficulties in assessing the predictions of sophisticated models of nearshore interactions in the absence of physical insight or observational guidance as to the dominant processes. As always, there are a lot of things that might happen, the requirement is to predict what does happen!

Simple models are very appealing, they appear to explain so much, particularly when there is an apparent match to a complex shape such as a crescentic bar. However, in practice, there seems to be an expanding range of processes of interest. Including shear waves and bar-trapped edge waves leads to "naming of names", a clearer identification of the possible processes. However, as more such processes are recognised, the real world appears more complex, and the complexity of nearshore processes and nearshore bathymetric changes is a fact that long term video observations are increasingly confirming (Lippmann and Holman, 1990).

Lamb thought that edge waves might not be important because they were not observed. We now know they are found everywhere, their patterns in the hydrodynamics have been established, but their other possible role, that of making patterns in the nearshore sediment, is still far from clear. We have not yet altogether disproved Lamb's contention!

References

Aagaard, T. (1990), Infragravity waves and nearshore bars in protected, storm dominated coastal environments. Mar. Geol. 94, 181-203.

Bowen, A.J., and R.T. Guza (1978). Edge waves and surf beat. J. Geophys. Res. 93, 1913-1920.

Bowen, A.J., and R.A. Holman (1989), Shear instabilities of the mean longshore current. I, Theory. J. Geophys. Res. 94, 18,023-18,030.

Bowen, A.J., and D.A. Huntley (1984). Waves, long waves and nearshore morphology. Mar. Geol. 60, 1-13.

Bowen, A.J. and D.L. Inman (1971), Edge waves and crescentic bars. J. Geophys. Res. 76, 8662-8671.

Bryan, K.R. (1996), Bar-trapped edge waves. Ph.D. thesis, Dalhousie University, Halifax, NS. 174 pp.

Bryan, K.R., and A.J. Bowen (1996), Edge wave trapping and amplification on barred beaches, J. Geophys. Res. 101, 6453-6552.

Guza, R.T., and A.J. Bowen (1976). Resonant interactions of waves breaking on a beach. Proc. 15th Coastal Engineering Conf. ASCE. 560-579.

Herbers, T.H.C., S. Elgar, R.T. Guza and W.C. O'Reilly (1995), Infragravity-frequency (0.005–0.5 Hz) motions on the shelf, Part II: Free waves. J. Phys. Oceanogr., 25, 1063-1079.

Holland, K.T., B. Raubenheimer, R.T. Guza and R.A. Holman (1995). Run-up kinematics on a natural beach. J. Geophys. Res. 100, 4895-4993.

Howd, P.A., A.J. Bowen, and R.A. Holman (1992), Edge waves in the presence of strong longshore currents. J. Geophys. Res. 97, 11,357-11,371.

Howd, P.A., A.J. Bowen, R.A. Holman, and J. Oltman-Shay (1991), Infragravity waves, longshore currents and linear sand bar formation. Proc. Coastal Sediments '91. ASCE. 72-84.

Huntley, D.A., and A.J. Bowen (1978). Beach cusps and edge waves. Proc. 16th Coastal Engineering Conference, ASCE. 1378-1393.

Huntley, D.A., R.T. Guza, and A.J. Bowen (1977). A universal form for shoreline run-up spectra? J. Geophys. Res. 82, 2577-2581.

Huntley, D.A., R.T. Guza, and E. B. Thornton (1981), Field observations of surf beat, Part I, Progressive edge waves. J. Geophys. Res. 86, 6451-6466.

Lippmann, T.C., T.H.C. Herbers, and E.B. Thornton (in press), Infragravity wave pressure and velocity variance in the nearshore. J. Geophys. Res.

Lippmann, T.C., and R.A. Holman (1990), The spatial and temporal variability of sand bar morphology. J. Geophys. Res. 95, 11,575-11,590.

Lippmann, T.C., R.A. Holman, and A.J. Bowen (1997). Generation of edge waves in shallow water. J. Geophys. Res. 102, 8663-8679.

Schonfeldt, H.J. (1995), On the modification of edge waves by longshore currents. Cont. Shelf Res. 15, 1213-1221.

Symonds, G., D.A. Huntley and A.J. Bowen (1982), Two dimensional surf beat: long wave generation by a time-varying break point. J. Geophys. Res. 87, 492-498.

Wright, L.D. (1982), Field observations of long-period surf-zone standing waves in relation to contrasting beach morphologies. Aust. J. Mar. Freshwater Res. 33, 181- .

Wright, L.D. and A.D. Short (1984), Morphodynamic variability of surf zones and beaches: a synthesis. Mar. Geol. 56, 93-118.

Field Observations of Nearshore, Wave-Seabed Interactions

Daniel M. Hanes, Member, American Society of Civil Engineers,
Chris D. Jette, Eric D. Thosteson, and Chris E. Vincent[1]

A series of field experiments have been conducted to investigate small scale sediment dynamics near the seabed in the nearshore region. The experiments took place at the U.S. Army Corps of Engineers Field Research Facility, in Duck, North Carolina, U.S.A., where the seabed typically consists of fine to medium sized sand. Using the Sensor Insertion System, which is cantilevered off of a massive pier, an array of instrumentation was deployed at various locations across the surf zone and nearshore region between depths ranging from 1 to 5 meters. The instruments measured the local hydrodynamics using a pressure sensor and 3 axis Acoustic Doppler Velocimeter (ADV). The suspended sediment concentration was measured with an Acoustic Concentration Profiler (ACP), and the local bedforms were measured with a Multi-Transducer Array (MTA). Surficial sediment samples were obtained at most sites, and sieved to determine the median sediment size. Water temperature, turbidity, and some video images (when the water wasn't too turbid) were also obtained during most runs. A schematic of the instrumentation is shown in Figure 1.

The bedform measurements were made with a commercially available 64 element SeaTek MTA, as shown in Figure 2. This 2.5 meter long array was composed of three smaller arrays. The center MTA was 50 cm long and consisted of 32, 5 MHz transducers with a spacing of 1.5 cm. This center section was very similar to the single MTA used in the SIS95 experiment. The center MTA was flanked on both

Department of Coastal and Oceanographic Engineering, University of Florida, Box
116590, Gainesville, FL 32611, U.S.A. email: hanes@coastal.ufl.edu
[1]School of Environmental Science, University of East Anglia, Norwich, UK

sides by 100 cm long arrays consisting of 16, 2 MHz, transducers with 6 cm spacing. The MTA recorded one bedform profile every 2 seconds. This system allowed for the measurement of small scale bedforms with the center MTA and large scale bedforms with lengths of over 1 meter.

The data were obtained over a variety of wave and current conditions, at several different cross-shore locations. Median sediment sizes ranged from 0.12 to 0.21 mm, depths ranged from 1.4 to 7 meters, and H_{mo} wave heights ranged from 0.32 to 1.2 meters. These conditions resulted in a range in mobility numbers from 26 to 256. This paper will focus primarily on the bedform geometry, and its temporal variability. Some examples of small scale ripples, larger mega-ripples, and superimposed small scale ripples and mega-ripples will be presented. We will also present evidence of ripple flattening and reformation on the time scales of a wave group.

Bedforms consisted mainly of two different types. Wave formed ripples were measured with heights up to 2 cm and lengths from 10 to 15 cm. Mega-ripples, with heights of 5 to 10 cm and lengths of 75 to 150 cm were also measured. Each of these bedform types sometimes occurred alone, and other times both types were superimposed. For example, Figure 3 illustrates the bedforms when both types were present. The center section of the MTA has higher resolution, and is blown up in the lower plot to show the smaller scale bedforms. The lines in this figure represent a sequence of scans (bedform measurement) separated in time by 1 minute; the lines have been displaced vertically by 1 mm in order to better see them. These measurements were obtained outside the surf zone in a depth of 3.9 m with a surficial sand size D_{50}=.19 mm, under conditions with H_{mo}=0.50 m, T_p=10.7 sec.

Interestingly, under very similar conditions, but closer to the breaker zone, the smaller scale ripples were not present, as seen in Figure 4. These mega-ripples were observed in a depth of 1.6 m, just outside the surf, with a surficial sand size D_{50}=.19 mm, under conditions with H_{mo}=0.53 m, T_p= 9.8 sec. It would appear that the threshold for sheet flow has probably been exceeded in this case.

An acoustic doppler velocimeter (ADV) and a pressure transducer were used to measure the current and wave conditions. For most of the data presented herein the near-bottom orbital velocity and orbital excursion are presented. These time series were derived from the ADV time series. A fast Fourier transform (FFT) was first performed on the cross-shore velocity time series. This frequency spectrum was then transposed to the bottom using the linear theory correction factor, k_c.

$$k_c = \frac{\cosh k(h + z)}{\sinh kh}$$

Field Instrument Array Setup
with single MTA

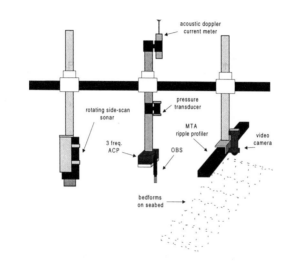

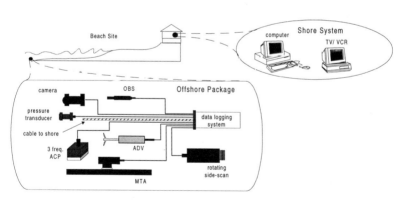

Figure 1: Instrumentation Schematic

An inverse fast Fourier transform (IFFT) was then performed, resulting in the depth corrected bottom time series for orbital velocity. This time series was then integrated to determine a near-bottom orbital excursion time series.

A time series of mobility number, defined as:

$$\Psi(t) = \frac{u(t)|u(t)|}{(s-1)gd_{50}}$$

can then be found from the time series of near-bottom orbital velocity, u(t), median sediment diameter, d_{50}, sediment specific gravity, s, and the acceleration of gravity, g. Mobility number has been shown by several authors to be an important parameter in bedform dynamics (Dingler and Inman, 1976, Nielsen 1981).

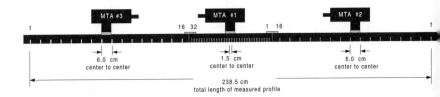

Figure 2. SeaTek 64 element MTA.

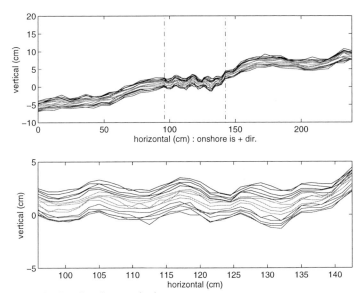

Figure 3: Small and mega-ripples

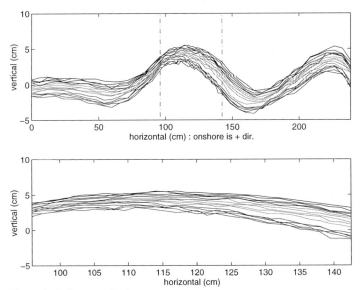

Figure 4: Only mega-ripples

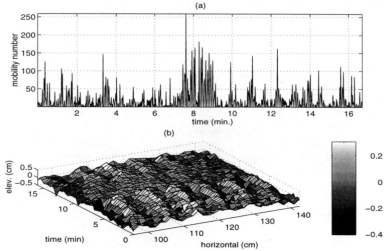

Figure 5: Time series of a) mobility number and b) ripple profiles.

There were periods during which it appeared that small scale ripples were flattened by wave groups, and then quickly reformed. Observations of bedform dynamics from the SIS96 experiment indicate that ripple flattening and reformation is related to mobility number. During periods of high peak mobility numbers, ripples tend to be flattened. The reformation of these ripples occurs within a range of mobility numbers that are large enough to move sediment but not so large as to cause flattening. For example, Figure 5a shows the mobility number (proportional to the square of the cross-shore velocity) for run 16, which lasted approximately 16 minutes. Figure 5b shows the corresponding ripples during the same period. It is apparent that one effect of the large wave group during minutes 7 through 9 is the flattening of the ripples. This is followed by a rapid rebuilding phase.

Conclusions

A 64 element multiple transducer array (MTA) is possible of high-resolution measurements of small and large scale bedform profiles. Temporal scales on the order of wave periods can be measured as well as the migration, flattening, and rebuilding of small-scale bedforms with heights of less than 1 cm.

Observations from the SIS96 experiment indicate that the flattening of small scale, vortex, ripples is a function of mobility number. Groups of waves with peak mobility numbers of over approximately 150 can potentially flatten ripples. Single waves with peak mobility numbers of larger than approximately 240, which also was suggested by Dingler and Inman (1976), can flatten a rippled bed. Ripple reformation can occur within a minute after flattening, and in certain conditions within a couple of wave periods. Rapid ripple reformation appears to occur when peak mobility numbers are between 50 and 150. Ripple reformation is slow when peak mobility numbers remain less than approximately 50. When mobility numbers reach values of over approximately 150 during a rebuilding period, ripple flattening can occur. Thus mobility numbers larger than 150 can impede the reformation process.

Ripples were observed in every run during the SIS96 field experiment. Seabed measurements were made at several cross-shore locations and under many different wave conditions. Many of the measurements contained both small and large scale ripples, while others contained only small or only large scale ripples. Even in some conditions when plane-bed conditions were anticipated, small scale ripples were observed to form in the periods between wave groups. These measurements indicate that in order to properly model sediment transport processes, ripple dynamics such as ripple migration, flattening, and reformation should be considered.

Acknowledgements:

This work was funded by the Coastal Sciences Program, U.S. Office of Naval Research.

References

Bagnold, R. A., Motion of waves in shallow water: Interaction of waves and sand bottoms, Proc Roy Soc Lond, A 187, 1-15, 1946.

Clifton, H, E., and J. R. Dingler, Wave-formed structures and paleoenvironmental reconstruction, Mar. Geol., 60, 165-198, 1984.

Dingler, J. R., and Inman D. L., Wave-formed ripples in near-shore sands, Proc. Fifteenth Conf. Coastal Engng., Amer. Soc. Civil Eng., 1976.

Fresdsøe J. (1978). Sedimentation of river navigation channels. Journal of the Hydraulics Division (ASCE), Vol. 104, No. HY2, pp.223-236.

Hay, A. E. and A. J. Bowen, 1993. Spatially correlated depth changes in the nearshore zone during autumn storms, J. Geophys. Res., 98: 12,387-12,404.

Meyer-Peter, E and R Muller (1948): Formulas for bed-load transport. Proc Int Ass Hydr Struct Res, Stockholm.

Miller, M. C., and P. D. Komar, Oscillation sand ripples generated by laboratory apparatus, J. Sediment. Petrol., 50, 173-182, 1980.

Nielsen, P (1992): Coastal bottom boundary layers and sediment transport. Advanced Series on Ocean Engineering - Vol. 4, pp.129-142. World Scientific, New Jersey.

Nielsen, P., Dynamics and geometry of wave-generated ripples, J. Geophys. Res., 86, 6467-6472,1981.

Osbourne, P. D. and C. E. Vincent (1993): Dynamics of large and small scale bedforms on a macrotidal shoreface under shoaling and breaking waves. Marine Geology, 115, 207-226, Elsevier Publ., Amsterdam.

Vincent C.E. & P.D.Osbourne (1993) Bedform dimensions and migration rates under shoaling and breaking waves. Continental Shelf Research, 13, 1267-1280

Vongvissessomjai, S. (1984): Oscillatory ripple geometry. Fournal of Hydraulic Engrg., ASCE, Vol. 110, No. 3.

Wiberg, P.L., and C. K. Harris, Ripple geometry in wave dominated environments, J. Geophys. Res., 99, 775-789, 1994.igure 5: Run 16: (a) Time series of mobility number ; (b) Time series of ripple profiles.

Mechanisms of Sand Suspending Under Non-Breaking and Breaking Irregular Waves

Pykhov N.V.*., Kuznetsov S.Yu*., Kunz H.**

Abstract

The mechanisms of sand suspension by irregular waves are discussed on the basis of two field experiments data. It is found that lee vortex ejection from the sand ripple crests is the basic mechanism of sand suspension under non-broken irregular waves outside the surf zone.

Vortex formation and its ejection from the bottom owing to instability of the bottom boundary layer during deceleration of water flow is probably the main mechanism of sand suspension in outer part of surf zone for spilling breakers. Inside the surf zone suspension events are caused by vortexes which are formed under breaking waves and penetrating to the bottom.

Introduction

From physical point of view the presence of instantaneous water flows from the bottom with a velocity exceeding the sand particle settling velocity is a necessary condition for the sediment suspension in wave and current flows.

Three basic mechanisms of sand suspending can be distinguished from the laboratory investigations with monochromatic waves.

The first mechanism is sand suspending by vortexes, which are formed behind the ripple crests. Lee vortexes eject the sand from the bottom twice per wave period at the moments of flow reversal for the case of 2-D vortex ripples. In this case, the suspended sand concentration has phase lag $\approx -\dfrac{\pi}{2}$ relatively the maxima of cross-shore velocity (Nielsen, 1979; 1991; Sleath 1982). For the present time there are no

----

* P.P.Shirshov Institute of oceanology, Russian Academy of Sciences, 23 Krasikov str., 117218 Moscow, Russia. Fax. +7 095 124 59 83. E-mail: pykhov@coast.msk.ru, kuznetsov@coast.msk.ru
** Coastal Research Station. NLOE-Forschungsstelle Kueste. An der Muehle 5, D-26548, Norderney, Germany. E-mail: kunz.crs@t-online. de

measurements of the temporal variation of suspended sand concentration in wave flow for 3-D ripples at the bottom. Such observations for steady flow have shown, that sand suspending are determined by the 3-D horse-shoe vortexes, which are formed behind the ripple crests and quasi-periodically ejected into the water column (Syunsuke, Asaeda, 1983).

The second mechanism is sand suspending by vortexes, which are formed due to instability of the turbulent bottom boundary layer over rough nearly plain bottom during the flow deceleration. Laboratory study of a turbulent structure of the bottom boundary layer in the wave flow above the rough bottom (Hino et al., 1983; Sleath,1987) has shown that a sharp increase of turbulent kinetic energy occurs in the time of the flow deceleration and its full damping at the time of velocity direction change. Visual observations with the help of special techniques have shown the presence of horse-shoe vortexes in these moments (Hino et al., 1983). Those vortexes were ejected from the bottom. According to experimental data of Hino et al., 1983, the phase lag between the burst of turbulence and maximum of cross-shore velocity above the bottom boundary layer is about $-\dfrac{\pi}{4}$. The possibility of realization of such a sand suspension mechanism has been discussed in the papers of Foster et al. (1994) and of Pykhov et al. (1995).

The third mechanism is sand suspending by large-scale vortexes with horizontal or vertical inclined axis, which are formed under the plunging and spilling breakers (Nadaoka, Kondoh, 1989; Zhang at al., 1994). Plunging breaker are characterized by a jet impinging on the oncoming trough and subsequent violent transition from irrotational to rotational motion with horizontal axis. Wave breaking in spilling breakers appear as a broken water spilling down the front face of the wave and fundamentally different from wave breaking in plunging breakers. In spilling breakers there are dominated vortexes with vertical inclined axis. Such vortexes penetrate to the bottom and catch the sand from the bottom like tornado (Zhang et al., 1994). For sediment transport modeling, it is significant to predict the spatial and temporal variations of turbulent kinetic energy (TKE) and suspended sediment concentration in order to model the rates and direction of sediment transport. Investigations of the temporal variability of turbulence in breakers show, that the turbulence decay time in the plunging breaker dies out between bores and it is large compared to the wave period in the spilling breaker (Ting, Kirby, 1994; 1995; 1996).

The reality of this mechanism under irregular waves in the field conditions is confirmed by some measurements of the macro turbulence in the surf zone (George et al., 1990; Rodriguez et al., 1995). The results of this research give a possibility to assess a time-mean value of the turbulent kinetic energy, the integral length scale, the turbulent viscosity. But they do not afford to trace the temporal variability of these parameters, which are necessary for the analysis of temporal fluctuations of the suspended sediment concentration and transport rate..

The hypothesis about sediment suspension by such vortexes is shared by the majority of investigators, but the confirmation of it by direct field measurements is poorly now. In the field conditions the presence of group structure of waves, infragravity waves and time-mean currents (undertow) can change the time scales of sand suspending in comparison with laboratory experiments for monochromatic waves.

In the present paper these mechanisms of sand suspending are examined on the basis of spectral and mutual spectral analyses of the data of synchronously measuring of suspended sediment concentration, cross-shore water velocity and free surface elevations for irregular waves on the basis of the field experiments «Novomichailovka'93» in the Black Sea and «Norderney'94» in the North Sea.

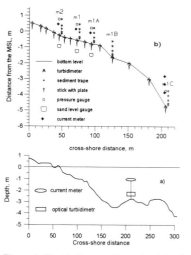

Figure 1. The Bottom profile and points of measurements.
a). «Novomichailovka'93»,
b) «Norderney'94».

Experiment

Field experiment "Novomichailovka'93" was carried out from the trestle, 300 m long near the Novomikhailovka site on the Black Sea (Krasnodar region) in December 1993 for non-tidal conditions. In the place of experiment isobaths in the nearshore zone are parallel to the coastline, and bottom sediments up to the depth of 10 m are represented by fine sand with the mean diameter of $0.21 \div 0.24$ mm. Bottom profile along the trestle and point of measurement are shown in Fig.1a. During this experiment the measurement were produced predominantly offshore of surf zone.

Experiment «Norderney'94» was carried out on the beach of Norderney Island, North Sea in October 1994 (Fig. 1b). Bottom profile and points of measurement are shown in Fig.1b. Amplitude of tidal water level oscillations with semi-diurnal period was 1.2 m. Measurements were produced during high water level predominantly in middle and outer part of surf zone. The mean diameter of bottom sand was about 0.23 mm.

The suspended sand concentration, cross-shore, longshore and vertical components of water velocity, free surface elevations were measured synchronously with sampling rate of 4.5 Hz over one hour and 18.2 Hz over 20 minutes.

The suspended sand concentration was measured by optical turbidimeter (Kos'yan et al., 1994) at the level from 4 to 8 cm above the bottom during «Novomichailovka'93» experiment and 4-20 cm during «Norderney'94» experiment.

Cross-, longshore and vertical velocities were measured by electromagnetic current meters at 10-25 cm above the bottom.

Free surface elevations were recorded by wire resistance gauge in the first experiment and by pressure gauge during second experiment.

The parameters of bottom ripples were measured by scuba divers for several runs during experiment at the Black Sea and estimated by formulae of Nielsen (1991) and by diagram of Kaneco (1981) for other runs, when observations by divers were impossible.

Spectral and mutual spectral analysis was performed by Uelch method with Nuttal spectral window (Marpl, Jr., 1987) with the spectral window width of 0.01 Hz. The number of the degrees of freedom was 70.

The envelope of free surface elevation and cross-shore velocity were calculated by Hilbert transform method (Bendat, Pirsol, 1981).

Spectral analysis show the turbulent fluctuations of velocity components at the frequencies more than 0.8 Hz. This is confirmed by absence of coherence between cross-shore velocity and free surface elevation at this frequencies, by changing of the velocity spectrum slope from -4 to ~ -2 at 0.8 Hz and by absence of slope changing in free surface elevation spectrum. Turbulent fluctuations of cross- and longshore velocities were extracted by digital filtering.

Results

The fragment of temporal variation of suspended sand concentration and cross-shore velocity, results of spectral calculations for the case of 2-D ripples with straight fronts, which were confirmed by diving observations, are shown in Fig.2. The narrow suspended sand events are coincide with passing of large waves in a group. The suspended sand concentration decrease to zero during low waves. In this case sand is suspended by lee vortex outburst from 2-D ripple crests. This is confirmed by the presence of local peaks at the frequency of spectrum maximum of the incident wind waves (0.2 Hz) and it first harmonics (0.4 Hz), by statistically significant value of coherence squared between concentration and cross-shore velocity at 0.2 Hz and by phase lag between these parameters $\sim -\dfrac{\pi}{2}$ at this frequency (Fig.2). By contrast to

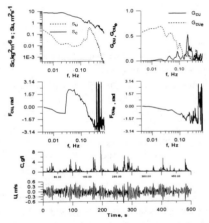

Figure 2. Suspended sand concentration spectra (S_c), of cross-shore water velocity (S_u), phase lag (F_{cu}) and coherence (G_{cu}) between C and U and between concentration and velocity envelope (F_{cu_e}, G_{cu_e}), temporal variation of cross-shore water velocity (u) and instantaneous values of suspended sediment concentration (C). «Novomichailovka'93», H_s = 0.48 m. H_s =0.48 m, T_p =4.5 s, h = 2.7 m.

regular laboratory waves the sand is suspended only by a groups of highest waves, that is confirmed by high values of coherence squared between suspended sand concentration and envelope of cross-shore velocity.

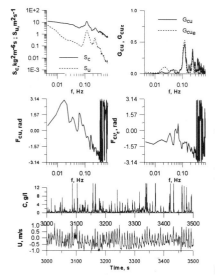

Figure .3. Suspended sand concentration spectra(S_c), of cross-shore water velocity (S_u), phase lag (F_{cu}) and coherence (G_{cu}) between C and U and between concentration and velocity envelope (F_{cu_e}, G_{cu_e}), temporal variation of cross-shore water velocity (u) and instantaneous values of suspended sediment concentration (C). «Novomichailovka'93», $H_s = 0.85$ m, $T_p = 8.3$ s, h = 2.7 m

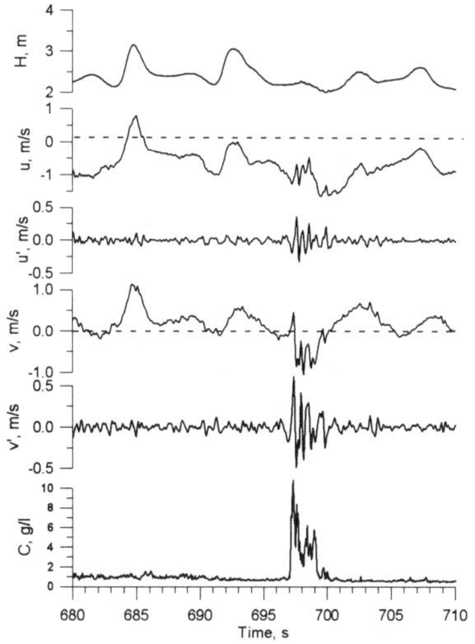

Figure 4. The example of single suspension event. . $H_S = 1.07$ m, $T_P = 8.7$ s, h =1.53 m, Spilling breaker. «Nordemey'94» experiment.

The example of the same characteristics for the case 3-D ripples at the bottom in the phase of its effacing is shown in Fig.3. During this run only the highest swell waves were breaking by spilling at the point of measurement. The narrow suspended sand events occur practically for each passing wave at the time of cross-shore velocity deceleration. This is the case of sand suspending by the mechanism of shear instability of bottom boundary layer. It is conformed by statistically significant coherence between concentration and cross-shore velocity at the main frequency of incident waves (0.11 Hz) and its first harmonic (0.22 Hz), by phase lag $\sim -\frac{\pi}{4}$ at 0.11 Hz and absence coherence between concentration and cross-shore velocity in the infragravity frequencies band.

The typical example of sand suspending by vortex in the surf zone is shown in Fig.4. Suspension event with nearly vertically forward front coincides well with strong fluctuations of cross- and longshore velocities. Temporal variation of 2-D turbulent kinetic energy and suspended sand concentration have a good coincidence at this time

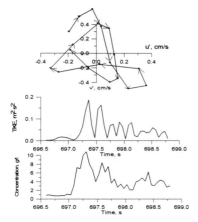

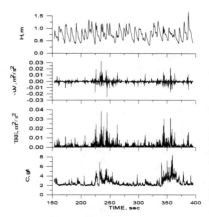

Figure 5. Temporal variation of concentration ,TKE and turbulent velocity vector diagram for the time interval of suspension event in Fig.4.

Figure 6. The example of temporal variation of suspended sand concentration, turbulent kinetic energy and shear stress in the surf zone.. «Nordeney'94» experiment.

$H_s = 0.90$ m, $T_p = 8.7$ s, $\overline{h} = 1.62$ m.

interval (Fig.5). In upper part of this figure it can be seen, that the vector of turbulent velocity in horizontal plane run few circles during this time period. It is probably, that few vortexes with vertical axis are passed across the measurement point. This is confirmed with laboratory data of Zhang et al. (1994).The results of our runs are situated in the region of three and more vortexes with inclined vertical axis on the vortex type diagram for surf zone, presented in this work.

The estimation by Teylor's frozen turbulence hypothesis for such situations give the spatial scale of vortexes from 0.3 to 1.5 m. A good coincidence between suspended sand concentration, horizontal component of shear stress and turbulent kinetic energy, which is shown in Fig.6, is a good confirmation of a mechanism of sand suspending by vortexes forming near the water surface under breakers and penetrating to the bottom. It is seen, that broad burst of turbulent kinetic energy, shear stress and suspended sand concentration are repeated with period of hundred seconds.

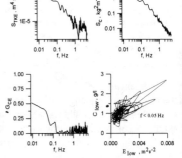

Figure 7. Spectra of TKE and suspended sand concentration (S_c), coherence between concentration and TKE (G_{cE}) and relation between low frequency fluctuations of the concentration and TKE in the surf zone. «Nordemay'94». $H_s = 1.07$ m, $T_p = 8.7$ s, $h = 2.3$ m.

A typical suspended sand concentration and TKE spectra, coherence between concentration and turbulent kinetic energy for surf zone are demonstrated in Fig. 7.

Statistically significant values of coherence at frequencies < 0.05 Hz and its absence at gravity band frequencies are observed. In the first approximation, the low frequencies values of concentration and TKE are connected nearly linearly (Fig.7). These results mean probably, that in field conditions only the vortexes with biggest values of TKE penetrate to the bottom and can suspend sand from it. Spectral densities of suspended sand concentration and TKE are increased monotonously with frequency decreasing.

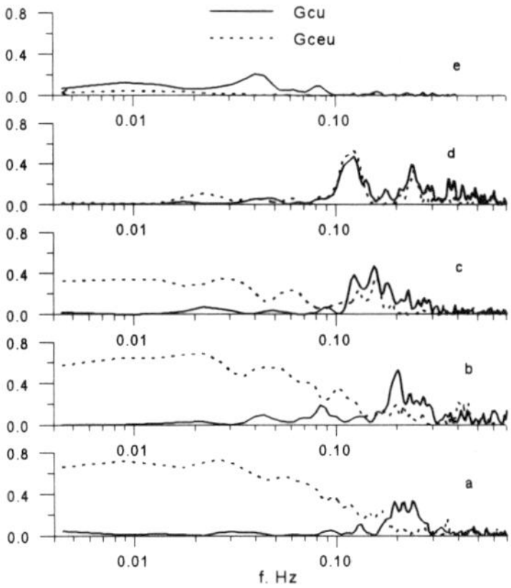

Figure 8. The coherence between suspended sand concentration and cross-shore velocity (Gcu) and its envelope (Gceu); a)non-deformed waves, b) slightly deformed, c) greatly deformed, d) 20-30% breaking by spilling, e) inner surf zone .

Frequency variability of coherence between concentration and cross-shore velocity and it envelope for the cases from non-deformed waves and 2-D ripples at the bottom to the plane bed in the surf zone are presented in Fig.8. It is seen, that coherence between concentration and envelope of cross-shore velocity at frequencies < 0.1 Hz decrease with increasing of wave deformation and changing of bottom ripples (Fig.8a-d). Inside the surf zone coherence is not statistically significant at all frequencies.

The phase lag between suspended sand concentration and cross-shore velocity at the frequency of spectrum peak of incident waves versus mobility number, which characterized the type of ripples is shown in Fig.9. Phase lag change from $-\dfrac{\pi}{2}$ for

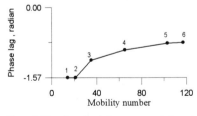

Figure 9. The phase lag between suspended sand concentration and cross-shore velocity at the frequency of spectrum peak of the incident waves versus mobility number. 1,2 -2D straight crests; 3 - 2D sinuous crests; 4,5 - 3D ripples; 6 - nearly flat bed. «Novomichailovka'93» experiment..

2-D straight crests ripples to $-\dfrac{\pi}{4}$ for 3-D ripples in the phase of its effacing.

In existing simple models for vertical distribution of suspended sand concentration the nearbottom reference concentration is determined usually through a square dependence on the nearbottom water velocity. In spite of our results such a dependence is questionable in the surf zone, where, as it has been shown above, fluctuations of suspended sediment concentration are not coherent to cross-shore velocity and it envelope. From physical point of view, the most effective way is the determination of relations between the suspended sediment concentration and turbulent kinetic energy, between which there are statistically significant values of coherence. To obtain such dependencies, the further research of spatial-temporal variability of large-scale turbulence and sand suspension under irregular waves in field conditions is necessary.

Conclusion

Lee vortex ejection from the sand ripple crests at the moments of flow reversal is the basic mechanism of sand suspension under non-broken irregular waves outside the surf zone.

Vortex formation and its ejection from the bottom owing instability of bottom boundary layer during deceleration of water flow is probably the main mechanism of sand suspension in outer part of surf zone for spilling breakers.

Inside the surf zone suspension events are caused by vortexes which are formed under breaking waves and penetrating to the bottom.

Acknowledgment

This research was supported by the grant 95-05-14343 from the Russian Foundation for Fundamental Research.

We wish to express our special thanks to the German Ministry for Science and Technology (BMFT/BMBF) for sponsor supporting of our joint investigations within the framework of the KFKI-research-program on Beach Nourishment ("Vorstrand- und Strandauffüllungen im Bereich von Buhnen-Deckwerks-Systemen").

Reference

Bendat J., Pirsol A.,1981. Applied analysis of incidental data." Mir" Publ., 540 pp. (in Russian).

Foster D.L., Holman R.A., and Beach R.A., 1994. Sediment suspension events and shear instabilities in to bottom boundary layer. In: Proc. Int. Conf. "Coastal Dynamics'94", Barcelona, pp. 712-716.

George K., Flick R.E., and Guza R.T., 1994. Observation of turbulence in surf zone. J. of Geoph. Res., Vol. 99 (C1): 801-810.

Hino M., Kashiwaynagi M., Nakayama A., 1983. Experiment on the turbulence statistics and the structure of a reciprocating oscillatory flow. J. Fluid Mech., 131: 363-400.

Kaneco A.,1981. Oscillation sand ripples in viscous fluids. Proc. Jap. Soc. Civ. Eng., 307: 113-124.

Kos'yan R.D., Kuznetsov S.Yu., Podymov I.S., Pykhov N.V., Pushkarev O.V., Grishin N.N., Harizomenov D.A., 1995. Optical device for measuring of suspended sediment concentration during a storm in the coastal zone. Oceanology, 35, N 2: 1-7. (in Russian).

Marple S.L., 1990. Numerical spectrum analysis and its application. "Mir" Publ., M., 584 pp. (in Russian).

Nadaoka K., Kondoh T., 1989. Turbulent flow field structure of breaking waves in the surf zone. J. Fluid Mech., 204: 359-387.

Nielsen P., 1979. Some basic concepts of wave sediment transport. Progr. Rept. Inst. Hydrodyn. and Hydraul. Eng. Techn. Univ. Denmark, 2, 160 pp.

Nielsen P., 1991. Coastal bottom boundary layers and sediment transport. Word Scientific, Singapore, New Jersey, London, Hong Kong, 324pp.

Pykhov N., Kos'yan R., Kuznetsov S., 1995. Time scales of sand suspending by irregular waves. In: Proc. of 2nd Int. Conf. "Medcoast-95", Tarragona, Spain, pp. 1073-1091.

Sleath J.F.A.,1982. The suspension of sand by waves. J. Hydraul. Res., 20: 439-452.

Sleath J.F.A.,1987. Turbulent oscillatory flow over rough beds. J. Fluid Mech., 182:369-400.

Syunsuke I., AsaedaT., 1983. Sediment suspension with rippled bed. J. Hydr. Eng., 109:409-423.

Zhang P., Sunamura T., Tanaka S., Yamamoto K., 1994. Laboratory experiment of longshore bars produced by breaker-induced vortex action,. In: Proc. Int. Conf. "Coastal Dynamics'94", Barcelona, pp. 29-43.

Ting F.C.K,, Kirby J.T., 1994. Observation of undertow and turbulence in a laboratory surf zone. Coastal Eng., 24: 51-80.

Ting F.C.K,, Kirby J.T., 1995. Dynamics of surf zone turbulence in spilling breaker. Coastal Eng., 27: 131-160.

Ting F.C.K,, Kirby J.T., 1996. Dynamics of surf zone turbulence in a strong plunging breaker.

CONVECTION AND DIFFUSION OF SEDIMENT UNDER LARGE WAVES.

A. H. Peet[1,2] and A. G. Davies[1]

[1]School of Ocean Sciences, University of Wales, Menai Bridge, Gwynedd, LL59 5EY, UK.
[2]Presently at: HR Wallingford Ltd., Howbery Park, Wallingford, Oxfordshire, OX10 8BA, UK.

Abstract
Reliable predictions of net sediment transport rates in the coastal zone are potentially limited by their failure to predict entrainment events at flow reversal. Even on plane beds these can significantly influence time-series of suspended sediment concentration during the wave cycle and, hence, the net suspended load transport rate. The present paper explores an entrainment process, involving the ejection of vortex pairs from the near bed layer just before the reversal of the free-stream velocity. Discrete vortices are formed from the velocity shear within the boundary layer. Interaction between vortices, formed in successive half cycles, gives rise to a vortex pair, which moves rapidly away from the bed and towards the edge of the boundary layer. This vortex pairing process is introduced as a perturbation to a classical one dimensional (1D) momentum/diffusion turbulent oscillatory boundary layer solution with prescribed vertical eddy viscosity. Both the sediment diffusivity and the bed shear stress are altered by the two dimensional (2D) entrainment processes, resulting in a qualitatively more realistic concentration field.

Introduction
One of the greatest causes of inaccuracy in morphological coastal area modelling is our continuing uncertainty about detailed sediment transport mechanisms. The understanding of small-scale sediment entrainment processes is essential for the accurate prediction of sediment movement on large scales. Existing sediment transport models appear to make reasonable transport estimates for coarse sediments above plane beds, where the majority of the transport occurs very close to the bottom. However, when the transport of finer grain sizes is modelled, transport rates do not compare well with observations.

In order to predict the wave-related sediment flux, accurate time-dependent predictions of concentration and velocity are required throughout the boundary layer. Recent advances in measurement techniques have allowed model results to be compared with accurate, high-resolution data. For example detailed time-series of suspended sediment concentration have been measured at various heights in the suspension layer above mobile plane beds by Murray et al. (1991) and Ribberink and Al-Salem (1995).

In both of the above mentioned studies, peaks in the concentration time-series were associated, as expected, with maxima in the magnitude of the time-varying bed shear stress. However, in addition to these 'diffusion' peaks, two concentration peaks occurred at times of flow reversal. In the wave-current experiments of Murray, where a fine grain size (diameter D=0.135mm) was used, these peaks were extremely abrupt, propagating upwards and attaining a greater relative importance than the 'diffusion' peaks with increasing height above the bed. In the data of Ribberink and Al-Salem (1995), presented in Figure (1) as a contour plot, directly above the sheet-flow layer the 'diffusion' peaks were dominant.

However, above heights of around 1cm, the 'flow reversal' peaks became dominant. The transition from the dominance of one peak to the other resulted in a substantial phase change in the concentration time-series. This greatly altered the wave-related, sediment flux, and hence the net transport. This flow reversal process cannot be modelled fully (see Li and Davies in these proceedings) by a standard turbulent diffusion approach since a completely different hydrodynamical process, involving convective transport of momentum and sediment, occurs around the time of flow reversal.

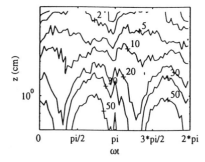

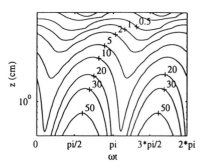

Figure (1) : Contour plot of the concentration field of Ribberink and Al-Salem (1995) experiment C10, (g/l).

Figure (2) : Contour plot of the concentration field calculated by the 1D model, (g/l).

Overview of model

Sarpkaya (1993) made observations of vortex pairs being ejected from the oscillatory boundary layer, using flow visualisation techniques. The flows were of high Reynolds numbers and had completely smooth boundaries. This implies that the existence of whatever process is responsible is not dependent on bottom roughness elements, and will therefore be present in oscillatory flow over plane beds. This paper describes a candidate model of vortex pair ejection from within the oscillatory boundary layer.

The present model is a combination of a standard, horizontally averaged, description of the boundary layer and a 2D discrete vortex model. Around the time of flow reversal the boundary layer can be subdivided into a lower and an upper layer in which the vorticity have opposite signs. The boundary layer is segmented into horizontal sections of length X (<< λ, the free-surface wavelength). During each model time-step a small proportion of the available circulation within the lower vorticity region is consolidated into a growing discrete vortex. The vortices then interact, grow and decay giving a 2D flow field. This flow field is then used to describe additional vertical mixing and entrainment of sediment.

Numerical studies have been made of the eruption of the near-wall region when a vortex is located close to the boundary. Doligalski and Walker (1984) have shown that a vortex near to a boundary induces an area of vorticity with the opposing sense of rotation. This locates the position of a new vortex, within the present model, which starts to grow at bottom flow reversal. This second vortex grows and interacts with the original vortex. The newly formed vortex is smaller in strength and therefore is advected with a greater velocity. Eventually the new vortex is advected upwards away from the vicinity of the bed and no longer grows in strength. At this stage if the two vortices are of similar strength they will behave like a vortex pair, inducing almost equal velocities on each other. If the line joining their two centres is not completely vertical, then they will have a vertical component to their advection velocity. In this fashion vortices of opposite signs, generated in consecutive half wave cycles, are able to propagate away from the bed through, and potentially out of, the boundary layer.

The moment that a vortex is advected out of the near bed vorticity region another is allowed to form. This vortex will only interact very weakly with the vortex pair higher up and will have little or no induced vertical velocity. It remains trapped within the near bed region rising slowly due to the thickening of the region. The vortex grows, governed by growth and decay rules, as the half wave cycle continues. At the bed, when the flow reverses, a new vortex of the opposite sign is initiated. This is located as before to the upstream side of the, now well developed, vortex. The entire procedure now repeats itself over the following half cycle with vortices of the opposite senses of rotation.

1D boundary layer model

The 1D momentum equation has been solved using a time-varying eddy viscosity. For symmetrical waves, this eddy viscosity varies at twice the fundamental frequency with peak values in phase with the shear stress maxima. The eddy viscosity is therefore approximately at its minimum at free-stream flow reversal. The eddy viscosity increases linearly with height away from the bed for the bottom sixth of the boundary layer and then remains constant above. The bottom boundary condition for sediment entrainment is driven by the bed shear stress and takes the form of a reference concentration. The concentration field produced by the 1D boundary layer model, for the test case C10 of Ribberink and Al Salem (1995), is given in Figure (2).

2D vortex model

The objective of the 2D model is to allow physical processes to occur which could not otherwise be described by the 1D model. However modelling additional processes, such as advective transport by coherent flow structures, is of most use if the final results can be incorporated in a 1D modelling scheme. This implies that output from the 2D model must be horizontally averaged to yield a 1D net response.

A 2D irrotational flow model describing detailed entrainment processes occurring within the boundary layer is coupled with a standard progressive wave 1D model. Due to the weakness of the horizontal velocity gradients they can be considered negligible within a small section, X, where $X \ll \lambda$ the surface wavelength, the phase can therefore be considered approximately constant within the 2D model at any instant.

The assumption underlying the present vortex modelling approach is that vorticity may be localised within the shear flow into areas of higher vorticity. These areas of high vorticity are considered as discrete vortices. Not all of the circulation of the shear flow contributes to the discrete vortices. In fact only a small proportion of the available circulation is used to drive an irrotational flow sub-model. Figure (3) demonstrates schematically how velocity shear within the oscillatory boundary layer is used to drive the discrete vortices.

The total circulation within the lower boundary layer vorticity region can be calculated within a finite, along-flow domain. The vorticity can be integrated in the x and z directions within defined limits. Under the assumption that $X \ll \lambda$ it is considered reasonable to assume no variation of velocity in the along-flow direction over the domain of width X. A small percentage, c_Γ, of the total instantaneous circulation is made available to growing discrete vortices, where $c_\Gamma = O(0.01)$. The contribution to a vortex, within the lower boundary layer region, over a small time increment Δt can be approximated as

$$\Gamma_n^+ \approx \Gamma dt = -c_\Gamma \; X \, u_2 \; \Delta t, \tag{1}$$

where u_2 is the velocity at the upper boundary of the region. This circulation contribution is added to the existing vortex circulation to give a new complex potential. The position of the vortex is altered by the contribution, Γ_n^+, depending on the height of the centre of vorticity, in an analogous fashion to a centre of mass correction to a large body with a small addition off centre. No alteration is made to the horizontal vortex location.

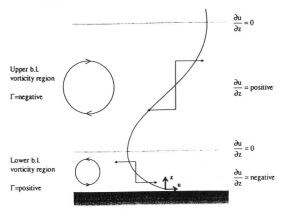

Figure (3) : Discrete vortex generation from regions of vorticity within a vertical velocity profile.

The velocity field around a vortex is given by the complex velocity. Instead of the vorticity being located at one singular point it is distributed uniformly over a small 'core', this avoids the unphysical velocities near its centre. The core radius grows in time, resulting in the energy within that vortex decreasing. The total circulation however remains constant, or is increased every time a vortex is created or increased in strength. In order to ensure a reduction of the circulation of a vortex in time an exponential decay rate is introduced of the form

$$\Gamma_n(t + \Delta t) = \Gamma_n(t) e^{-K_\Gamma \Delta t} \tag{2}$$

where K_Γ is a decay constant.

The two methods give growth of the vortex radius and decay of circulation. Together with the growth of vortex circulation described earlier, the entire lifetime of a vortex can be represented. The techniques above were however designed for use outside of boundary layers where the flow is laminar. Here vortex behaviour is modelled within the oscillatory boundary layer and therefore the radii growth will not depend on the kinematic viscosity but on the turbulent eddy viscosity, and the decay constant, K_Γ, will be significantly larger.

Irrotational flow theory states that each vortex is advected by the flow field generated by all the other vortices. The vortices at different heights will be advected to a different extent by the background flow, due to the vertical variation in horizontal velocity. This must therefore be accounted for each time step giving an altered vortex position.

The 2D model incorporates an interactive procedure for the sediment concentration being predicted by the 1D model. The alteration to the instantaneous b.l. originates in the velocity shear such that it is reduced by the factor $(1 - c_\Gamma)$. This loss of vorticity from the 1D solution is compensated by a gain of vorticity from the decay of the vortices. The total change to the velocity gradient at each height is

$$\left.\frac{\partial u}{\partial z}\right|^{new} = \left(1 - c_\Gamma\right)\left.\frac{\partial u}{\partial z}\right|^{old} + V \tag{3}$$

where V is the horizontally averaged vorticity arising from decaying vortices at each height. It is assumed that the circulation of a vortex comprises the vorticity spread in a Gaussian fashion about the vortex centre. The use of the radius as the standard deviation ensures that 95% of the circulation is restricted to an area of twice the solid core radius. The total decayed vorticity over each height interval per time step, Γ_i^-, over the along flow domain is the sum of all N vortex contributions. It follows that the total horizontally averaged, instantaneous decayed vorticity at each height is

$$V_i = \frac{\Gamma_i^-}{X \, \Delta z \, \Delta t}. \tag{4}$$

The transfer of vorticity between the 1D and 2D models produces no net change of circulation over a wave cycle, but is responsible for altering the time variation and vertical distribution of vorticity, such that the momentum equation may be written

$$\frac{\partial u}{\partial t} = \frac{\partial U_\infty}{\partial t} + \frac{\partial}{\partial z}\left[v_t\left(\left(1 - c_\Gamma\right)\frac{\partial u}{\partial z} + V\right)\right]. \tag{5}$$

Effective sediment diffusivity

The velocity field produced by the 1D b.l. solution has no vertical component and is assumed invariant in the along-flow direction. The 2D irrotational flow model involves both vertical velocities and along-flow variations caused by the presence of vortices. Upwards travelling fluid carries sediment from a region of high concentration to a region of low concentration, producing an upward sediment flux. Downward travelling fluid contains a lower concentration of sediment than the region into which it is moving. This also reduces the vertical concentration gradient and can therefore be considered as an upward sediment flux. A possible vertical advection process is associated with the fluid jet which flows between closely spaced vortices, as depicted in Figure (4).

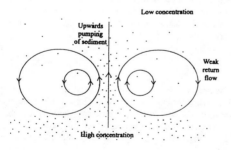

Figure (4) : Sediment pumping mechanism in the fluid jet between two vortices.

This mixing process can be considered on the basis of the 2D advection-diffusion equation for sediment motion:

$$\frac{\partial \overline{c}}{\partial t} + \overline{u}\frac{\partial \overline{c}}{\partial x} + (\overline{w} - w_s)\frac{\partial \overline{c}}{\partial z} = -\frac{\partial}{\partial x}(\overline{u'c'}) - \frac{\partial}{\partial z}(\overline{w'c'}), \tag{6}$$

$$\frac{\partial \overline{c}}{\partial t} = -\frac{\partial}{\partial x}\left[\overline{uc} + \overline{u'c'}\right] - \frac{\partial}{\partial z}\left[(\overline{w} - w_s)\overline{c} + \overline{w'c'}\right]. \tag{7}$$

An overbar denotes phase averaging and a prime denotes tubulent fluctuation, u is the along-flow velocity in the x direction, w is the vertical velocity in the z direction and w_s is the sediment settling velocity. The non-turbulent, phase averaged terms can be expressed as the sum of a horizontally averaged term, $<>$, and a horizontally fluctuating term, subscript p.

$$\bar{c} = <c> + c_p(x), \quad \bar{u} = <u> + u_p(x), \quad \bar{w} = <w> + w_p(x), \tag{8}$$

$$\overline{u'c'} = <\overline{u'c'}> + (\overline{u'c'})_p, \quad \overline{w'c'} = <\overline{w'c'}> + (\overline{w'c'})_p.$$

Equation (7) can therefore be rewritten

$$\frac{\partial \bar{c}}{\partial t} = -\frac{\partial}{\partial x}\Big[<u><c> + <u>c_p + u_p<c> + u_p c_p + <\overline{u'c'}> + (\overline{u'c'})_p\Big]$$
$$-\frac{\partial}{\partial y}\Big[-w_s(<c>+c_p) + <v><c> + <v>c_p + v_p<c> + v_p c_p + <\overline{v'c'}> + (\overline{v'c'})_p\Big]. \tag{9}$$

The horizontal average of this equation becomes

$$\frac{\partial <c>}{\partial t} = \frac{\partial}{\partial z}\Big[w_s<c> - <w_p c_p> - <\overline{w'c'}>\Big]. \tag{10}$$

This uses the result that $<w> = 0$, since there is no net vertical transport of mass. Equation (10) is similar to the 1D linearised advection-diffusion equation but with the addition of the horizontally averaged transport term $-<w_p c_p>$

Since

$$\frac{\partial c_p}{\partial t} = \frac{\partial \bar{c}}{\partial t} - \frac{\partial <c>}{\partial t}, \tag{11}$$

Equations (9) and (10) may be used to express Equation (11) as

$$\frac{\partial c_p}{\partial t} = -\frac{\partial}{\partial x}\Big[<u>c_p + u_p<c> + u_p c_p + (\overline{u'c'})_p\Big]$$
$$-\frac{\partial}{\partial y}\Big[-w_s c_p + v_p<c> + v_p c_p - <v_p c_p> + (\overline{v'c'})_p\Big] \tag{12}$$

Suppose initially that a horizontal layer has a homogeneous sediment distribution, $c_p = 0$. The change in c_p will then be

$$\frac{\partial c_p}{\partial t} = -\frac{\partial (\overline{u'c'})_p}{\partial x} - w_p\frac{\partial <c>}{\partial z} - \frac{\partial (\overline{w'c'})_p}{\partial z}. \tag{13}$$

In order for this equation to hold in the present situation, the background concentration must be assumed to redistribute instantaneously in the horizontal. This is justifiable because the problem is time-varying and therefore the steady equilibrium sediment distribution, produced by the instantaneous vortex positions, is not sought. It is assumed also that the turbulent diffusion is horizontally homogeneous and does not depend upon the phase averaged vortex velocities, such that

the gradient of along flow variations in eddy diffusivity can be approximated to zero. This reduces Equation (13) to

$$\frac{\partial c_p}{\partial t} = -w_p\frac{\partial <c>}{\partial z}. \tag{14}$$

Integrating this over one time step

$$c_p = \int_0^{\Delta t} w_p\frac{\partial <c>}{\partial z}dt. \tag{15}$$

and assuming further that neither the horizontally averaged vertical concentration gradient nor the vertical velocity vary significantly over a single time step, Equation (15) may be evaluated approximately as follows:

$$c_p = - v_p \frac{\partial <c>}{\partial y} \Delta t \ . \tag{16}$$

Hence the additional term in the advection-diffusion equation may be expressed

$$<w_p c_p> \ = \ - \ <w_p^2> \frac{\partial <c>}{\partial z} \ \Delta t . \tag{17}$$

Equation (10) can now be rewritten using the eddy diffusivity formulation for the turbulent transport of sediment and with the additional effective eddy diffusivity obtained from Equation (17)

$$\frac{\partial <\bar{c}>}{\partial t} \ = \ \frac{\partial}{\partial z}\left[w_s <\bar{c}> +\left(<w_p^2> \Delta t + \varepsilon_s \right)\frac{\partial <\bar{c}>}{\partial z} \right]. \tag{18}$$

The additional diffusivity, at any height, will be non-zero if there is any along-flow variation in the instantaneous vertical velocity, i.e. $w_p \neq 0$.

Separate analytical expressions can be found for all of the contributions to the horizontal integral of w^2. These are (i) the interaction between two vortices, $w_i w_j$, outside both core regions; (ii) the interaction between two vortices within cores; (iii) the contribution from each individual vortex, w_i^2, outside its core; (iv) the contribution from each individual vortex within its core; (v) the interaction between a vortex and its image, $w_i w_{i^*}$, outside of the core; and (vi) the interaction between a vortex and its image within the core. The horizontal average $< w^2 >$ can then be found by summing all of the contributions to the integral and dividing by the domain width, X.

A contour plot of the effective diffusivity, $< w^2 > \Delta t$, is presented in Figure (5). The most dominant feature of the time-series, at each height, is the double peak per wave cycle. This is primarily a result of the contribution from the strongest vortex, caused by interaction (iii) above. The time-series of circulation of the vortex with the largest strength is approximately in phase with the free-stream velocity, hence the maximum vortex circulation, and therefore maximum $< w^2 >$, occurs around peak velocity and not flow reversal. The double peak of effective diffusivity is qualitatively similar to the time-varying sediment diffusivity, ε_s, used within the 1D model. The effective diffusivity field is completely dominated by the contributions from individual vortices, i.e. processes (iii) and (iv) above. The contributions to Δt that are expected to give the sediment pumping mechanism are the cross-terms, (i) and (ii).

The presence of the vortices produces an effective diffusivity similar to that used to produce the background concentration field. Thus the major effect of is already implemented within the 1D b.l. sediment diffusion model. However there is no account of the much smaller cross-terms, resulting from the interaction of the vortex pair. The maximum value of $\Delta t \approx 300 \text{cm}^2\text{s}^{-1}$ in Figure (5), whereas the maximum contribution to this from the cross-terms is approximately $0.25 \text{cm}^2\text{s}^{-1}$, two orders of magnitude smaller. This term, shown as a contour plot in Figure (6), is not the sole diffusivity due to the vortex pair, but is indicative of the vortex pair interaction and hence sediment pumping capability. The cross-terms are therefore qualitatively, but not quantitatively, the desired diffusivity.

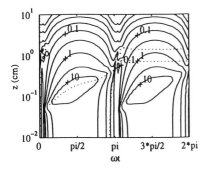 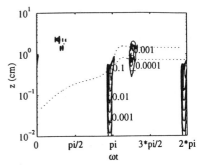

Figure (5) : Contour plot of the effective
sediment diffusivity, $< w^2 > \Delta t$ (cm^2s^{-1}).

Figure (6) : Contour plot of the cross-term,
$< w_i w_j > \Delta t$ (cm^2s^{-1}).

The dotted lines in both figures show the paths of two vortex cores.

The cross-term, , can be normalised by its maximum value to yield a shape function, I.

$$I(z, t) = -\frac{< w_i w_j >}{< w_i w_j >_{max}} \qquad (19)$$

where is the maximum value of the cross-term within the height/time domain. The effective
sediment diffusivity, applied at all phase instances, is proposed as

$$I < w^2 > \Delta t \qquad (20)$$

The total sediment diffusivity, which includes the addition from the vortex motions, is displayed as a
contour plot in Figure (7).

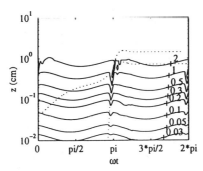

Figure (7) : Contour plot of sediment diffusivity, ε_s, in interactive convective event model, (cm^2s^{-1}).

The near bed sediment concentration has relatively low values at the times of flow reversal. There is
therefore not a large amount of sediment to be mixed vertically by the enhanced diffusivity, which
can be seen at these phases by a sharp minimum in the contour lines. Almost no noticeable change
occurs in the resulting concentration field. It can be concluded at this stage, that, although a
mecanism exists by which vertical sediment transport may be enhanced, insufficient sediment is
available for entraainment due to the nature of the bottom boundary condition.

Additional sediment entrainment

The stress at the bed, which is responsible for the sediment entrainment, is the tangential turbulent Reynolds stress, . In the 1D b.l. model flow is assumed to occur only in the along-flow direction, and hence the Reynolds stress is defined in terms of the eddy viscosity, v_t, and the mean flow velocity gradient, . Little alteration arises in this stress within the modified boundary layer because the velocity gradients are only slightly altered. In contrast, within the 2D modelling framework there exist vertical as well as horizontal velocities. This means that the Reynolds stress can be calculated directly from .

In the absence of vortices is always zero for horizontally uniform oscillatory flow. For a single vortex the flow close to the bed is parallel to the bed and there are no vertical velocities. When two vortices of opposing senses of rotation exist close to the bed then they induce vertical velocities in that region. The above two scenarios are depicted by the streamlines in Figure (8).

Single Vortex Above Bed   Vortex Pair Above Bed

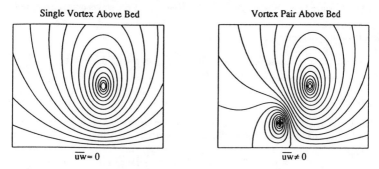

$\overline{uw} \approx 0$ $\overline{uw} \neq 0$

Figure (8) : Streamlines around near bed vortices showing regions of non-zero shear stress.

The additional entrainment is found by calculating the value of at all positions in the along flow domain at the height of one grain diameter. Only the values that are larger than the threshold of motion for the grains are then summed and divided by the total number of positions within the domain to give an additional bed shear stress.

Results of additional diffusivity and entrainment

The resulting concentration field, from the combined convection/diffusion model, is presented in the contour plot of Figure (9). The additional peak of sediment, that was not present in the earlier model concentration field, is now apparent after flow-reversal. This peak occurs slightly later than in the experimental data and does not grow in relative importance with increasing height as observed experimentally. The increase in sediment diffusivity is not sufficient within the interior of the boundary layer to produce the relative growth of the 'convetive' peak in comparison to the 'diffusive' peak.

The life cycle of individual vortices is central to their behaviour and interaction. For the vortex pairing mechanism to reach its required strength, the vortex initiating at flow reversal must attain greater vorticity before ejection. This may be possible by allowing it to be initiated earlier than flow reversal, due to the induced counter rotating, near bed vorticity formed by a single vortex close to a boundary.

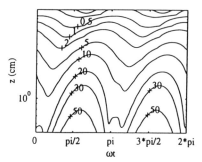

Figure (9) : Contour plot of concentration produced by combined diffusion/convection model. Comparison to Ribberink and Al-Salem (1995) experiment C10, (g/l).

It can be concluded that this candidate model is able to describe only qualitatively a physical mechanism by which additional sediment is advected from the bed and mixed into the boundary layer. The maximum mixing occurs at mid heights within the boundary layer and is therefore consistent with the increase in importance of the 'convective' peak of sediment at these heights. However the magnitude of the additional diffusivity is too small to yield the observed values of enhanced vertical transport.

References.

Doligalski, T.L. and J.D.A. Walker, 1984. The boundary layer induced by a convected two-dimensional vortex. *J. Fluid Mech.*, 139: 1-28.

Murray, P.B., A.G. Davies and R.L. Soulsby, 1991. Sediment transport by large waves combined with a steady current. HR report.

Ribberink J.S. and A.A. Al-Salem, 1995. Sheet flow and suspension of sand in oscillatory boundary layers. *Coastal Eng.*, 25: 205-225.

Sarpkaya, T., 1993. Coherent structures in oscillatory boundary layers. *J. Fluid Mech.*, 253: 105-140.

COASTAL SEDIMENT TRANSPORT ON SHINGLE BEACHES

Erik Van Wellen[1], Andrew J. Chadwick[1], Paul A. D. Bird[1],
Malcolm Bray[2], Mark Lee[3] and John Morfett[4]

Abstract

A large scale collaborative field based research programme aimed at improving the current knowledge and understanding of the coastal processes associated with shingle beach morphodynamics is presently underway in the South of the UK. The primary stage consisted of a field measurement campaign undertaken in the autumn of 1996 on an undisturbed shingle beach. The campaign and the data collection is described and some of the initial data collected during that campaign are presented.

Introduction

The management of beaches is an important and effective means for protecting coastal areas. Increasing research efforts in this field have been aimed mostly at trying to understand and quantify the elements which govern the morphodynamics of beaches over both long and short time scales in terms of the longshore and cross-shore movement of sediment. To date, predominantly sand beaches have received the bulk of the attention. The number of documented studies and available data on sand beaches ranging from analytical/numerical models, over laboratory tests, to large scale field experiments is therefore considerable. In strong contrast is the moderate attention that coarser grained (*i.e.* shingle) beaches have received. Studies of the processes governing shingle beach profiles have mainly been limited to empirical models of profile development based on laboratory studies (Pilarczyk & den Boer, 1983 and Powell, 1990). The study of longshore transport on shingle beaches has been mainly restricted to the use of variations of the CERC formula from the US

[1] School of Civil & Structural Engineering, University of Plymouth, Palace Court, Palace Street, Plymouth, PL1 2DE, Devon, UK, EVANWELLEN@plymouth.ac.uk
[2] Department of Geography, University of Portsmouth
[3] Department of Oceanography, University of Southampton
[4] Department of Civil Engineering, University of Brighton

Army Corps of Engineers (1984) combined with laboratory studies. Almost all models focusing on shingle beach processes have relied on laboratory data. Two exception are Chadwick's model (1991) which was validated from his own field study (Chadwick, 1989) and a longshore transport formula for bed load by Damgaard and Soulsby (1996) which was validated against field data from several sites including the work of Chadwick (1989). This scarcity of field data indicates that the best way in which the understanding of shingle beach morphodynamics could be furthered is through the execution of a large scale field experiment.

Such an experiment has now been designed by a partnership made up of HR Wallingford and the Universities of Portsmouth, Brighton, Southampton and Plymouth. The funding for the project is provided by the UK Ministry of Agriculture, Fisheries and Food (MAFF).

This paper seeks to describe the methods used and the measurements taken by the respective partners in the collaborative field programme aimed at determining and quantifying the governing factors which control shingle beach morphodynamics.

Aims and objectives

The general project aim is to develop improved techniques for the prediction of shingle beach transport and morphological development based upon large scale field experiments. The objectives to achieve this aim are to:

1. undertake comprehensive series of field measurements of hydrodynamic conditions and concurrent sediment transport.

2. use the field data to test and/or adapt existing shingle transport models

3. link transport processes with resulting beach morphology over a range of spatial and temporal scales and develop predictive techniques for use by beach managers.

4. compare the response of a natural beach and one managed by control structures.

The **first year field programme** (Autumn '96) involved measurements on a simple open beach at Shoreham, southern England. The main objectives were as follows:

• To co-ordinate and optimise a variety of measurement techniques so as to obtain concurrent data relating to hydrodynamics, shingle transport and beach morphodynamics

• To undertake electronic and aluminium tracing studies to obtain measurements of transport volumes.

• To undertake 3D beach surveys of the whole of the open Shoreham West Beach, at a spatial resolution sufficient for production of digital ground models and computation of volumetric changes between surveys.

- To undertake measurements at a variety of temporal resolutions ranging from one or two semi-diurnal tidal cycles (tracers, morphological surveys) to semi-continuous samples of individual events (wave recording, scour monitors).

- To undertake measurements during a variety of wave conditions, especially high energy conditions where there are few previous shingle beach measurements.

Complex situations in the presence of control structures and beach replenishment will be studied in the **second year field programme** (Autumn '97). Methods which proved successful and knowledge acquired from study of the natural portion of Shoreham Beach in year one will be applied to tackle research questions specific to a neighbouring rock groyne and shingle replenishment scheme. The objectives in this part of the study are:

- To investigate the effects of rock groynes in controlling the behaviour of the recharged beach.

- To study the effects of the scheme on the adjoining frontages.

- To use results to develop predictive models of future evolution since typical schemes have lifetimes of 30, or more years, and far in excess of the time available to this project for measurements.

Complementary studies of the nature and historical behaviour of the field site are also in progress to ascertain the representativeness of both the conditions and responses sampled during the field work in terms of the overall spectrum of shingle beach types, and of the magnitudes and frequencies of the events that control their development.

Field site

The first year experiment was undertaken on the West Beach, at Shoreham, southern England (fig. 1). Part of a series of shingle beaches protecting the lowlands of the West Sussex coastal plain, the beach comprises the western portion of a barrier type shingle spit which has, in the past, deflected the River Adur eastwards. Much of the coast to the west is controlled by extensive arrays of timber groynes and wooden revetments, but for a distance of 1.5km the beach is open and natural, being confined to the east by a harbour breakwater. The beach comprises a massive medium shingle storm berm resting upon a gently sloping sandy lower foreshore. The beach is macro-tidal with a maximum spring range of 6.5m and the shingle toe is well exposed at low water spring tides. The site is exposed fully to storm waves generated within the English Channel and the predicted annual maximum significant wave height offshore is about 4.0m for waves approaching from the south and south west, (Hague, 1992). Prevailing wave direction is from the SW and SSW and net beach drift is from the west to the east so that shingle accumulates against the breakwater at a mean rate of 15,000-20,000m^3/yr (Chadwick, 1989). The site was selected due to the simple morphology and bathymetry, unconfined transport and abundant shingle available for transport and natural profile adjustments. These qualities were considered essential for reliable measurements of hydrodynamic conditions and natural transport that would be used for model development and verification.

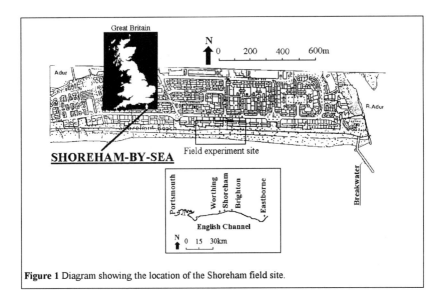

Figure 1 Diagram showing the location of the Shoreham field site.

The second year experiment will be undertaken immediately to the west of the open beach at Lancing where a scheme involving the construction of four rock groynes with shingle replenishment was completed in April 1997. This experiment design permits comparisons and application of knowledge from the open beach as the two sites are subject to similar hydrodynamic conditions, but differ significantly in degree of structural control.

Data collection and sample

The field programme was scheduled between mid-September and mid-November 1996 to permit sampling within known storm seasons, but at times when daylight hours were amenable to tracer and morphological measurements. Measurements were planned in terms of **core activities** vital to the success of the project and **trial activities** likely to yield valuable information, but in need of testing and evaluation before extensive adoption. Prioritisation was necessary due to the technical and practical difficulties of operating so many techniques simultaneously. Highest priority was to obtain a continuity of core measurements at high resolution, especially during high energy conditions.

Core activities
- Offshore wave recording (WRS)
- Inshore wave recording (IWCM)
- Sediment tracing
- Sediment sampling
- Beach elevation monitoring (Tell-Tails)
- Beach morphology survey (GPS)

Trial activities
- Current monitoring (EMCM)
- Mobile sediment trapping (NESSIE)
- Wave monitoring (X-band radar)

Offshore wave recording was conducted using the Wave Recording System (WRS), which is a solid-state data acquisition system capable of long term autonomous operation under water. The system comprises six pressure transducers which report the fluctuations in pressure at the seabed to the central data storage system where the information is stored until it is downloaded for further processing. For this project the WRS was programmed to take five measurements per tide of 17 minutes each, spread evenly over the top half of the tide. This measurement arrangement allowed for the WRS to tie in with the measuring scheme of the Inshore Wave Climate Monitor (see below). This ensured that both the instruments were recording the same sea state.

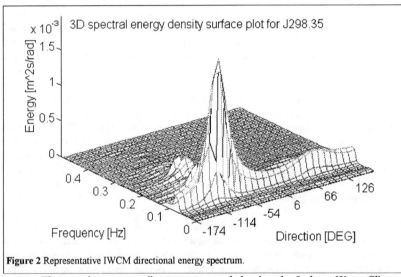

Figure 2 Representative IWCM directional energy spectrum.

The nearshore wave climate was recorded using the Inshore Wave Climate Monitor (IWCM). This device consists of a star array of four 6m resistive sensors mounted on a 6m sided triangular aluminium tubular frame. The IWCM performed well throughout the first 6 weeks of field work. However, the lower 1 metre section of all sensors was repeatedly damaged by shingle impact during high energy conditions with subsequent loss of water level measurements less than a metre from the beach surface. Towards the end of this field exercise the IWCM was torn from the beach during a Storm Force 10, the peak of which co-incided with high water, on the night of 28th October 1996 when a steel cable became entangled with the rig. A representative directional plot as obtained from data recorded using the IWCM can be seen in fig. 2 on which the southern direction is represented by -3 degrees with the positive numbers lying toward the south west.

Direct field measurements of sediment transport were obtained using tracers designed to mimic the indigenous pebbles. Tracers should not only behave exactly the same as indigenous pebbles, but also be recoverable in sufficient numbers to provide a representative sample of those injected. Unlike earlier methods of painting or marking indigenous materials, the aluminium and electronic tracer pebbles used in this project permit representative sub-surface recoveries. Artificial replica pebbles are cast in aluminium according to patterns produced from indigenous beach pebbles. The specific gravity of aluminium and flint are approximately the same (2.65 to 2.75). Pebbles are each stamped with an identification code. Recovery is undertaken by sweeping the beach with metal detectors. Tracers are dug to reveal their code and to measure burial depth. The depth detection range is 0.35 to 0.50m (Wright $et\ al.$, 1978 and Bray, 1990). Electronic pebbles comprise an electronic transmitter encapsulated within a loaded resin to mimic the density and shape of indigenous pebbles. The pebble circuit is powered by a watch-type battery with a life of over 2 years, and each emits low frequency magnetic pulses. Each pebble has an individual code, which is identified by an electronic detector within a depth range of up to 1.5 metres. Signals are picked up by a search coil and amplified and rectified to produce pulses audible within headphones. In a beach survey, the pebbles are first detected by a roving, wheeled multiple coil detector. The pebble is subsequently pin-pointed by a hand-held detector. Although the system is designed to determine burial depth remotely, pebbles were dug up during these experiments as an additional check. Details of the electronic pebble tracing system are provided in Workman $et\ al.$ (1994). Throughout the Autumn '96 field work a total of four aluminium tracer experiments and seven electronic tracer experiments were successfully conducted. Each tracer experiment followed a pre-set format and the longshore positions of the tracer centroid (centre of mass) and the depths of burial of the tracers (in addition to depth of disturbance data) obtained from these experiments were then used to calculate longshore drift rates following a method similar to that of Komar and Inman (1970).

Concurrent with the tracer experiments, shovelled sediment samples of approximately 70kg each were collected at each low water which occurred during a tracer study. Samples were taken from each of three sites adjacent to the injection sites. The samples are presently undergoing sieve analysis at half ϕ intervals.

Changes in beach elevation were monitored continuously using 5 vertical arrays of Tell-Tail scour monitors (Coates, 1997). Each array comprised of eight monitors set at 10cm intervals down a support pole connected to a self-contained data logger. The Tell-Tails oscillated when exposed to moving water, sending signals to the logger; burial of a Tell-Tail results in no signal. A change in the logger record from a signal to no signal, or vice versa, indicates a 10cm change in beach elevation at a time resolution of 10 minutes. The system is thus able to monitor changes in beach elevation continuously whilst the tide is in. The monitors were deployed initially with 4 or 5 sensors buried and 4 or 3 exposed to wave action. The 5 arrays covered a cross-shore distance of 24m with the landward array set at the approximate mean high water contour and the seaward array set at the approximate mean sea level contour with the intention of providing good spatial coverage within the main active area of the beach. Preliminary analysis of the data shows a number of periods of very rapid change in beach elevation that would not be evident from standard survey techniques. The significance of these changes to beach transport processes and particularly the depth of the mobile layer will be established when more of the information on waves, depth of tracer core disturbance and tracer distribution becomes available.

The method used for conducting 3D beach morphology surveys was GPS (Global Positioning System) surveying using a dual frequency GPS. From the GPS data digital terrain models and contour maps were produced for each survey. Using the SURFER software package, it was then possible to calculate volume levels for individual sections and to take horizontal slices through the beach surface. Calculations of the volume levels show that the early strong easterly winds in late September reduced the total sediment volume of the beach. This resulted in the adjustment of a typical summer profile, to a storm profile, where sediment was combed down onto the lower sections of the beach. As expected, the storm of the 28th October had the biggest impact upon the beach. Profile analysis suggests that large amounts of sediment were redistributed throughout the beach, being removed from the beach crest area and redeposited onto the lower parts of the beach.. The main crest, as defined by the five metre contour line, was cut back along the entire beach, especially in the central and easterly exposed section. The main area which was cut back by the storm receded between 2-11 metres, with the sediment being combed down onto the lower parts of the beach. Further volumetric analysis is required to demonstrate the extent to which different sections of the beach were affected by the storm. The height data contained in one Digital Elevation Model (DEM) can be subtracted from the DEM of a different survey. This will reveal areas of accretion and erosion. The results of this show that as expected from the profile analysis, most of the sediment was removed from the beach crest area and redeposited lower down on the beach (See fig. 3). This diagram also suggests, that sediment was removed from the lower section of the beach, suggesting it was either pushed up the beach, or, removed offshore. The volumetric analysis so far, suggests that the major storm resulted in a rearrangement of sediment within the beach system rather than significant quantities of shingle being removed from the beach. Further volumetric analysis is needed to confirm this observation. When the GPS equipment was not available, additional profile surveys were carried out using a Total Station. GPS surveys have been undertaken throughout the winter at the proposed site for the 1997-1998 field programme at Lancing to provide long-term development information and a baseline for establishing the other studies. During the Autumn '96 field work, a total of 22 full scale beach surveys were conducted from updrift of the eastern groynes to the Shoreham breakwater and therefore provide good volumetric data for establishing short and medium term drift volumes.

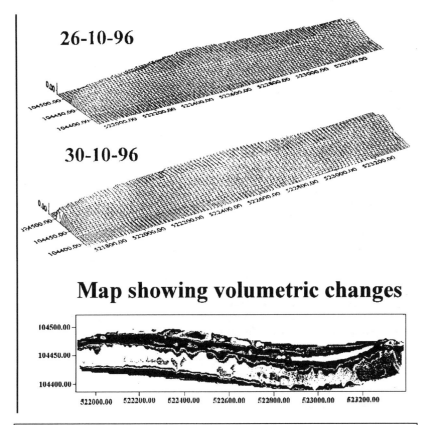

Map showing volumetric changes

The co-ordinates on the diagrams are given in m and refer to Ordnance Survey grid co-ordinates. The dark patches on the diagram indicate areas of erosion whilst the lighter areas represent accretion.

Figure 3 Beach survey results before and after a storm event.

Nearshore currents were measured using EMCM's. These have rarely been deployed on shingle beaches, due to their expense and comparative fragility. However, this makes any recorded data especially valuable, particularly when a vertical current profile can be obtained, which can be used to estimate shear stresses at the seabed. Up to 3 EM's were deployed at Shoreham as part of the TOSCA instrumentation suite (Voulgaris *et al.* 1995), modified for installation in the inter-tidal zone. All sensors were burst-sampled simultaneously at 4 Hz for 17.07 minutes every 20 minutes, throughout the period when the lowest sensor was covered by the tide. The EM's were deployed only in reasonably mild conditions during the 1996 fieldwork, but they yielded useful data and will be included in the 1997 fieldwork as a core activity.

The NEar Shore Sediment Investigation Equipment (NESSIE) consists of a small trap (mouth width 20cm) at the far end of a pole which is mounted on a carriage. Steel wires are used to stiffen the structure without adding significant weight. The trap can be orientated to face alongshore or across-shore. A pair of stop poles are driven into the beach to hold the carriage at a fixed position. For each measurement, the carriage is manoeuvred into position manually against the stop poles and the trap lowered and held securely onto the beach face to intercept the sediment. The trap is left in position for a short time (5-10 wave periods for the longshore direction, or wave-by-wave for the cross-shore direction) and then lifted and withdrawn shorewards. The sediment is weighed and collected for sieve analysis. By repeating the measurements as the tide rises and/or falls, sediment transport rates at different positions across the swash zone are determined. So far only the feasibility of deploying this apparatus has been investigated. For example on the 18th October 1996, during a high energy shore normal event, uprush and backwash cross-shore transport rates of 7.3 and 2.3 tonnes/m/hr respectively were measured just above the wave breaking zone. A net longshore transport rate of only 0.6t/m/hr was measured. In comparison, on a day with similar wave energy and a significant wave angle from shore normal, a net longshore transport rate of 2.0t/m/hr was found. More quantitative results are planned for the next phase of field work.

The X-band Wave Radar uses digitised radar backscatter from the sea surface to identify the position and orientation of wave crests for all wavelengths larger than 15m (Tenorio-Gonzalez, 1993). The system is capable of capturing a time series of images and hence can provide data on celerity, period and angle of approach of waves for individual wave frequencies. The system provides data over an area of $22km^2$ and can be used in severe weather conditions (wind speeds up to 20m/s).

Discussion

A large amount of high quality data has been collected and brought together. This is the first time that such an extensive database of concurrent field data has been collected on a shingle beach. Combined with earlier information gathered on this site (Chadwick, 1989) the research team can now also look at the evolution of the beach.

Analysis of the provisional sediment transport data appear to show a significant sensitivity to the wave angle of approach similar to that which can often be found in field data obtained from sandy beaches. The depth of disturbance data collected during the experiment is useful for the insight it gives into surface events, and will be useful for calibrating future computational models. Tracer studies could not be taken down to sub-tidal resolution and can only give an average bulk volume of sediment transport. In the next field work the NESSIE system is therefore intended to act as a check on the instantaneous rates and the distribution of the longshore transport.

A full report on the 1996 and 1997 (still to be undertaken) field work will be published by HR Wallingford and all data collected will be archived and made available after project completion.

Acknowledgements

The authors wish to acknowledge the support of MAFF and the Environment Agency in the realisation of the field work. The following people are also thanked for their vital contributions: D. Pope, B. Baily, T. Mason, M. Wilkin, J. Hague, T. Tapp, J. Davis, T Coates, B. Mcguire, Capt. G. Meadows, the Maritime Volunteer Service Shoreham Unit and S. Wallbridge.

References

Bray M.J., Geomorphological investigation of the South West Dorset Coast: Volume 2 Sediment Transport Report to Dorset County Council and West Dorset District Council, Department of Geography London School of Economics, 798p, 1990

Chadwick A.J., Field measurements and numerical model verification of coastal shingle transport BHRA, The Fluid Engineering Centre, Chapter 27, pp 381-402, 1989

Chadwick A.J., An unsteady flow bore model for sediment transport in broken waves. Part 1 & 2, Proc. Inst. Civ. Engrs, Part 2, pp 719-793, 1991

Coates T.T., Collaborative Shingle Beach Transport Project: Interim Report April 1996-March 1997, HR Report TR 29, pp 8-9, 1997

Damgaard J.S. & Soulsby R.L. Analytical longshore bed-load sediment transport formula, Proc. 25th Int. Conf. on Coastal Eng., Orlando, ASCE, pp 3614-3627, 1996

Tenorio-Gonzalez M.A., The uses of marine radar for coastal oceanographic studies, Unpublished MPhil/PhD Transfer Report, University of Southampton 115p, 1993

Hague, R.C., UK South Coast Shingle Study: Joint Probability Assessment. HR Wallingford Ltd. Report No. SR 315. 14p plus Tables, Figures and Appendices, 1992

Komar P.D. and Inman D.L., Longshore sand transport on beaches, Journal of Geophysical Research, 75, pp 5914-3927, 1970

Pilarczyk K.W. & den Boer K., Stability and profile development of coarse materials and their application in coastal engineering. Delft, Holland, Delft Hydraulics, publication No. 293, 1983

Powell K.A., Predicting short term profile response for shingle beaches, Report 219, HR Wallingford, February 1990

US Army Corps of Eng., Shore Protection Manual, Coastal Eng. Research Centre, Washington, 1984

Voulgaris G., Wilkin M. and Collins M. B., The in situ passive acoustic measurement of shingle under waves and currents: instruments (TOSCA) development and preliminary results, Continental Shelf Research, 15, pp 1195-1211, 1995

Workman M., Smith J., Boyce P., Collins M.B. and Coates T.T., . Development of the Electronic Pebble System. HR Report SR 405, HR Wallingford Ltd., 70p., 1994

Wright, P., Cross, J.S. and Webber, N.B., Aluminium pebbles: a new type of tracer for flint and chert pebble beaches. Marine Geology, 27, M9-M17., 1978

Hydrodynamics and Sediment Transport on Composite (Mixed Sand/Shingle) and
Sand Beaches: A Comparison

T. Mason[1], G. Voulgaris[2], D. J. Simmonds[3] & M. B. Collins[1]

Abstract

A comparison was made between wave reflection and beach groundwater
behaviour on a composite-type mixed beach and a sand beach, in order to assess the
influence of sediment composition on cross-shore morphodynamic response. Swell wave
reflection from the sand/shingle section of the mixed beach was higher than from a sand
beach, due to the steeper beach gradient. Wind wave reflection was unaffected by the
gradient increase. The hydraulic conductivity of the sand content was the controlling
factor for beach groundwater fluctuations. The main impact of the shingle fraction on
the mixed beach was via a steeper beach gradient, with subsequent influence on wave
reflection, rather than through energy dissipation through infiltration.

Introduction

Mixed beaches, though comparatively rare worldwide, are common around the
shores of the UK and other regions where the effects of glaciation provided abundant
supplies of sand and gravel for subsequent re-working by rising sea levels during the
Holocene. Accordingly, most research on mixed beaches emanates from New Zealand,
Ireland and Canada. However, the difficulty in deploying instrumentation in the
conditions when shingle is actually mobile, means that few hydrodynamic measurements
have been made on mixed beaches, despite their being "morphologically distinct from and
more complex than either sand or shingle beaches" (Kirk 1980).

---

[1] Department of Oceanography, University of Southampton, Southampton Oceanography
Centre, Southampton, Hampshire, SO14 3ZH, UK
[2] Woods Hole Oceanographic Institution, MS#22, Woods Hole, MA 02543, USA
[3] School of Civil and Structural Engineering, University of Plymouth, Palace Court,
Palace Street, Plymouth, Devon, PL1 2DE, UK

There are, in fact, two categories of mixed beach. The first is a broadly homogeneous mixture of sand and shingle[4], with some grading both cross- and longshore. Any sandy region to seaward is usually exposed only during low water spring tides. The second category is a composite-type beach, where a wide, sandy, inter-tidal terrace is flanked by a shingle ridge. The division between the two sediment types is often marked by an abrupt change of slope, so that the beach state changes suddenly from predominantly dissipative to reflective conditions. Both types of mixed beach may have a complex vertical sedimentary structure where, below the surface layer of cobbles and pebbles, layers of mixed pebble sizes are supported by a sand matrix and underlain by sand.

The objective of this paper is to examine two of the ways in which the distinctive sedimentary characteristics of a composite mixed beach interact with the hydrodynamic regime in comparison to the response of a sand beach.

The first aspect is the influence of the sand/shingle bank on wave reflection, since on a macro-tidal mixed beach, the sediment composition and gradient of the reflecting surface vary considerably through the period of tidal inundation. Both factors are first-order determinants of wave reflection. Reflection is reduced by percolation through a permeable layer (Kobayashi & Wurjanto 1992) but increases with increasing beach slope on natural beaches (*e.g.* Takezawa *et al.* 1990) permeable breakwaters (Bird *et al.* 1996) and laboratory structures (Benoit & Teisson 1994). It might be expected, therefore, that reflection from a mixed beach should be greater than from a shingle beach, if less energy can be dissipated through percolation, whilst also being greater than from an unrestrained sand beach, since a steeper beach gradient can be maintained.

The second aspect concerns the influence of the hydraulic conductivity of the sediment in controlling the translation of the seepage zone across the beachface. The presence and persistence of a seepage face is important for swash zone sediment transport through net infiltration losses as envisaged by Grant (1948) and for pressure gradient reversals with subsequent potential for fluidisation of the sediment (Baird *et al.* 1996). All recent beach groundwater modelling and, with only a few exceptions, all the field research has been concerned with sandy beaches, yet the hydraulic conductivity of a mixed beach sediment is far more variable both vertically and horizontally than on a sand or shingle beach.

Methods

Field data were obtained from three contrasting locations. The composite beach site was at Morfa Dyffryn, Meirionnydd, North Wales. The beach is macro-tidal with a series of ridges and runnels across the wide, dissipative inter-tidal zone. Reflection

---

[4] The term *shingle* is used to represent all sediment with D_{50} larger than 2 mm.

coefficients from the composite beachface at Morfa Dyffryn were compared with those from a sandy beach at Nieuwpoort-aan-Zee, Belgium, in order to assess the rôle of sediment composition and beach slope on wave reflection. The beach at Nieuwpoort is also macro-tidal, with an extensive series of 5 ridges and runnels across its 500 m wide inter-tidal zone (Voulgaris *et al.* 1997). Fluctuations of the beach groundwater table at Morfa Dyffryn were compared with those from a micro-tidal, sandy beach at Canford Cliffs, Dorset (Baird *et al.* 1996) to examine the influence of the hydraulic conductivity of the sediment. At Nieuwpoort and Morfa Dyffryn, waves, currents and suspended sediment concentration were measured using a co-located pressure sensor, electro-magnetic current meters and optical backscatter sensors. The elevation of the water table at Morfa Dyffryn and Canford Cliffs was measured with pressure sensors in a cross-shore transect of perforated dipwells.

Wave reflection

Reflection coefficients were calculated using a co-located EMCM and pressure sensor, according to the time-domain method of Guza *et al.* (1984). The time series of the sea surface elevation is decomposed into incoming and outgoing components based on linear shallow water theory:

$$\eta_{in}(t) \ = \ \frac{\eta(t) \ + \ \sqrt{\frac{h}{g}} \cdot u(t)}{2} \tag{1}$$

$$\eta_{out}(t) \ = \ \frac{\eta(t) \ - \ \sqrt{\frac{h}{g}} \cdot u(t)}{2} \tag{2}$$

where h (L) is mean water depth, $\eta(t)$ (L) and $u(t)$ (LT^{-1}) are the detrended sea surface elevation and cross-shore velocity time series respectively and g (LT^{-2}) is the acceleration due to gravity. The Frequency-Dependent Reflection Function (FDRF) is then derived from the square root of the ratio of the power spectra of the outgoing, η_{out} , and incoming, η_{in} , time series. The time domain method has been widely used but results can be artificially high if there is significant noise in either or both of the time series. The level of coherence between η_{in} and η_{out} is used to provide a value of the bias to the FRDF caused by signal noise which can then be corrected according to the method of Huntley *et al.* (1995). FDRF values were derived when the following criteria were fulfilled: (i) there was a distinctive spectral peak of the sea surface elevation with confidence values greater than 95%; (ii) coherence between both η_{in} and η_{out} and between the pressure and velocity time series was greater than 0.6; (iii) FDRF is less than 1.

Reflection coefficients were calculated at half hourly intervals over a tidal cycle at Morfa Dyffryn for the morning tide of Julian Day 277, 1994 (subsequently referred to as Tide 32). The spring tidal range of 4.7 m meant that the tide progressed across a variety of gradients and sediment types as shown in Table 1, where tidal stage is given in hours relative to local HW. Energy at frequencies $0.05 < f \leq 0.1$ Hz (10-20 s) is considered to represent swell waves and $f > 0.1$ Hz the wind waves.

Tidal Stage (hours)	H_s (m)	Surf/swash zone width (m)	Beach gradient		Sediment type
			tan β	deg.	
HW-1.5	0.19	10	0.02	1.2	Sand (inner ridge)
HW-1	0.19	8	0.08	4.5	Sand (beachface)
HW-0.5	0.16	7	0.09	5.0	Mixed sand/shingle
HW	0.15	5.8	0.10	5.7	Shingle
HW+0.5	0.15	7	0.09	5.0	Mixed sand/shingle
HW+1	0.15	8	0.08	4.5	Mixed sand/shingle
HW+1.5	0.16	10	0.02	1.0	Sand (inner ridge)

Table 1

The complex topography at Nieuwpoort meant that the variation in gradient of the reflecting surface at Nieuwpoort was not systematic during the tide, as at Morfa Dyffryn. The average gradient was 0.03 (~2°), but the steepest gradient was during the low water ebb (0.07, 4°). Reflection coefficients were derived for the afternoon tide of Julian Day 057, 1994 at Nieuwpoort, when wave conditions resembled most closely those at Morfa Dyffryn. T_z at both sites was around 4 s, with T_p of 8-10 and 10-12 s at Morfa Dyffryn and Nieuwpoort respectively. Tidal maximum H_s at Nieuwpoort was 0.3m, double that at Morfa Dyffryn. There was evidence of some low frequency energy at Nieuwpoort, but only swell and wind wave frequencies will be considered, since long wave reflection at Morfa Dyffryn was negligible during all the tidal cycles at Morfa Dyffryn used for reflection analysis.

The relationship between wave reflection and beach gradient at both sites is shown at Figure 1. Within each frequency group, maximum reflection at Morfa Dyffryn occurred around HW, although the increase was proportionally greater for swell waves than for incident wind waves. Hence, within HW ± 0.5 hours, about 85% of swell wave amplitude was reflected, whilst 30-40% of higher frequency waves were reflected. In contrast, reflection around low water was reasonably constant around 20-40% for all significant frequencies. FDRF values were derived for two further tidal cycles, with similar conclusions. In contrast, FDRF values for swell and wind wave frequencies at Nieuwpoort remained roughly constant at 10% throughout the entire tide. Similarly low reflection of incident waves has been reported from natural beaches elsewhere (*e.g.* Davidson *et al.* 1994, Miles *et al.* 1996) and during an earlier tide at Morfa Dyffryn, when H_s was about 0.3 m and the tide covered only the sandy ridge and runnel section of the beach, wind wave reflection coefficients were around 0.2.

There was no tidal signature in wave reflection at Nieuwpoort, suggesting that the tidal symmetry observed at Morfa Dyffryn is not a symptom of measuring a *local* reflection coefficient at one point *i.e.* the tidal variation is not purely a function of

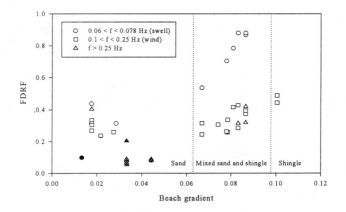

Figure 1 Variation in FDRF with beach gradient at Morfa Dyffryn Tides 31 and 32 (open symbols) and Nieuwpoort (filled symbols)

changing water depth, with accompanying changes in the relative proportion of energy dissipation. Possible reasons for the lack of tidally varying reflection at Nieuwpoort are that (i) the beach gradient has to be above a critical steepness, of at least 0.06, 3.4°, before frequency-selective reflection of swell waves occurs; or (ii) reflection from several surfaces masked any tidal signature in FDRF.

A third major determinant of reflection, in addition to the wave frequency and beach gradient already discussed, is wave amplitude or steepness. Therefore, in order to isolate the influence of beach gradient from wave amplitude and frequency, FDRF values are plotted against the non-dimensional Irribarren Number, Σ_b , which encompasses the three important parameters for wave reflection, beach gradient, wave steepness and frequency:

$$\Sigma_b = \frac{\tan\beta}{\sqrt{H_b/L_o}} \tag{4}$$

where $\tan\beta$ is beach gradient, H_b (L) is breaking wave height and L_o (L) the deepwater wavelength. Figure 2 shows reflection coefficients for Tides 31 to 33 at Morfa Dyffryn, when wave amplitude was reasonably constant at 0.08 m; hence for a given frequency, variation in Σ_b during these tidal cycles is due solely to beach gradient. Clearly, reflection of the swell waves (circles) is very sensitive to beach gradients in excess of 0.08 (4.5°), whilst reflection of wind wave frequencies (squares and triangles) is not.

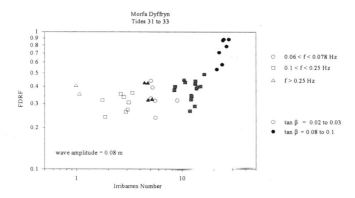

Figure 2. Irribarren Number vs. FDRF for constant wave amplitude and beach gradient of 0.02 to 0.03 (open symbols) and 0.08 to 0.1 (filled symbols)

In summary, the mixed sand/shingle section of the beach operated as a strong reflector of swell waves; swell wave reflection increased proportionally with beach gradient above 0.06 (3.5°), whilst reflection of incident wind waves did not and was below 40%. For gradients lower than 0.06, there was no relationship between gradient and reflection coefficients and reflection of both swell and wind waves remained less than about 40%. Two qualitative observations were noted which suggest that the effect on wave reflection of energy dissipation through percolation bears further investigation; (i) Bird *et al.*'s (1996) examination of an extensive field database of wave reflection from a porous rock island breakwater concluded that 60% reflection was the maximum possible for the 1 in 0.8 sloping structure, since the remaining wave energy was dissipated or transmitted through structure. The steepest part of the mixed beach profile at Morfa Dyffryn reflected over 80% of the swell wave amplitude and therefore a more permeable beach of similar gradient might be expected to have lower reflection characteristics; (ii) during the ebb tide, there was a tendency for reflection at a given frequency to be marginally higher than during the flood tide. A possible cause of the slight tidal asymmetry is wave/tidal current interaction and resultant modification of the wavelength, although the state of saturation of the sediment may be responsible.

Beach Groundwater

The response of the watertable to tidal forcing was modelled using the GRIST finite difference numerical model (Baird *et al.* 1996), based on the one-dimensional form of the Boussinesq equation:

$$\frac{\partial h}{\partial t} = \frac{1}{s} \frac{\partial}{\partial x} \left(Kh \frac{\partial h}{\partial x} \right) \tag{5}$$

where h (L) is the elevation of the water table above a given datum (*i.e.* the aquifer depth), K (LT^{-1}) is hydraulic conductivity, s (dimensionless) is specific yield, x (L) is distance and t is time. The model was run using hydraulic conductivity (K, cm s^{-1}) of 1, 0.5, 0.1, 0.05, 0.025, 0.01, 0.005 and 0.001 and specific yield, s, of 0.2, 0.25 and 0.3, which encompass typical values for coarse to fine sands. The inland boundary condition was a specified head of 0.44 m and the lower boundary was Chart Datum. Each simulation was run from HW -0.5 to HW +3.5. The predicted elevation of the water table was compared to that measured at half-hourly intervals from HW, examples of which are given in Figure 3. A combination of graphical and statistical methods of deriving the best-fit parameters found that K of 0.1 cm s^{-1} and s of 0.25 predicted a seepage zone seawards of the base of the sand/shingle slope at HW +1.5, but with no outcropping at the surface by HW +2. These parameters are typical of medium sand and represented well the measured water table.

Laboratory permeameter tests on sand/shingle mixtures demonstrated that the extremes of sand content make a considerable difference to hydraulic conductivity. The tests used fine beach shingle, $D_{50} = 4$ mm, with 10% incremental proportions of fine and medium sand and a number of tests for coarse sand also. An admixture of only 20% medium sand reduced the hydraulic conductivity of the shingle to around one third that of shingle alone and for fine sand the reduction was nearly 90%.

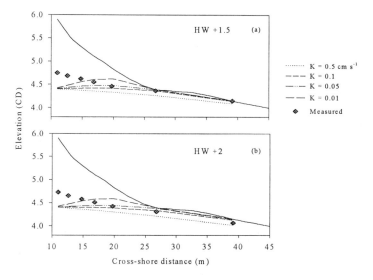

Figure 3 Measured and predicted water table elevation, Morfa Dyffryn Tide 32 at (a)
HW +1.5 and (b) HW +2. Specific yield = 0.25.

Implications of sediment composition for morphodynamic response

        The groundwater response of the mixed sand/shingle beach was not significantly
different to that of a sand beach, which suggests that the sub-surface sand content
determined the tidally-forced response of the water table. This, in turn, has a number of
interesting implications for the behaviour of beaches artificially replenished with shingle
e.g. Hayling Island, Hampshire and Seaford, E.Sussex. Such schemes are generally
designed as replenished shingle beaches, although the borrow is rarely pure shingle but
is dredged from offshore and contains a proportion of sands and fines. The rapid and
unforeseen "cliffing" (formation of near-vertical walls) observed on artificially
replenished beaches (McFarland et al. 1994) has been attributed to the combined effects
of compaction and cementation. However, it is possible that the additional sand and
fines content may be a controlling factor in the profile response of the replenished beach,
so that where a mixed sand/shingle layer exists at a depth below the surface which is
higher than the tidally-induced fluctuation of the water table, the shingle will not fulfill
its "rôle" in dissipating its energy through percolation (although energy will still be lost
through friction).

Although steep berms can form on shingle beaches, the higher hydraulic conductivity permits greater energy dissipation through infiltration and non-Darcian flow than on similarly steep mixed beaches, where even 20% sand content reduces the hydraulic conductivity by an order of magnitude. Yet the finding that the tidally-induced fluctuations of the groundwater table over such a complex sedimentary structure can be successfully modelled using the single value of hydraulic conductivity for medium sand suggests that the main impact of the shingle fraction is not on the groundwater behaviour, but via the beach gradient and subsequent influence on wave reflection. Ultimately then, the major significance of the sand and shingle mixture for the morphodynamic response of the ridge is through its ability to maintain a steeper slope than would be achieved by an unrestrained sand beach. Hence, the abrupt change in beach gradient has more impact on the hydrodynamics than does the sediment composition.

The steep gradient, which the shingle fraction is able to support, increases reflection of swell waves greatly in excess of that from unrestrained sand beaches. Thus, the mixed beach profile reflects more energy than a sand beach and dissipates less energy through infiltration than a shingle beach. Both these factors apply in the cross-shore direction, but the likely implication is that sediment mobility may be enhanced in comparison to either a purely sand or shingle beach, with net transport in the direction of the dominant mean currents.

Swell waves are usually thought of as "constructive" waves, but even during relatively calm conditions (H_s < 0.2 m), a considerable proportion of their energy can be reflected if the gradient is sufficiently steep. Accordingly, the consequences for sediment transport of swell waves on sand beaches may be different for mixed beaches, when the breakpoint of small plunging waves is very close to the shoreline. Swell waves have been held to account for the bulk of the annual net longshore transport rates along the mixed beach frontage at Hayling Island on the south coast of the UK *i.e.* storms were not the major contributor to the annual sediment transport rates (Whitcombe 1995). The waves encountered during the experimental period are not atypical of the average seasonal conditions for the whole of Cardigan Bay and since composite beaches form a considerable proportion of the coastline of the Bay, the influence of the shingle ridge may contribute significantly to the overall sediment dynamics of the nearshore region.

Conclusions

Reflection of swell waves increased with gradient once a threshold steepness of 0.06 (3.5°) was reached. Reflection of wind waves was not dependent on beach gradient. Higher reflection coefficients were observed than have been measured from porous structures, where energy is dissipated through wave breaking, friction, percolation and transmission. The reflection characteristics of swell waves from mixed sediment banks may be an important factor for long-term sediment transport patterns on short-fetched, macro-tidal mixed beaches. The tidally-induced fluctuations of the water table on a mixed beach are dominated by the sand content. The GRIST model for prediction of water table elevation in response to tides is applicable for mixed sediment beaches without calibration.

Acknowledgements

The authors acknowledge, with thanks, the Countryside Council for Wales, Snowdonia National Park Authority and Meirionnydd District Council for permission to carry out research at Morfa Dyffryn and for their field assistance. Our thanks also to Poole Borough Council for provision of facilities at Canford Cliffs. Field data from Nieuwpoort were obtained as part of the MAST II (CSTAB) programme, Contract No. MAS2-CT92-0024-C. Meteorological observations were kindly supplied by the Meteorological Office, Aberporth. The GRIST model was kindly made available by Dr A Baird, Department of Geography, University of Sheffield. The invaluable assistance of M P Wilkin, S Wallbridge and M Lee during the fieldwork was greatly appreciated. The first-named author was in receipt of NERC Grant No. GT4/93/250/P. Travelling funds for G. Voulgaris were provided from the Dept. Of Geology and Geophysics, WHOI and the Andrew W Mellon Foundation, through the Reinhart Coastal Research Center.

References

Baird A J, Mason T and Horn D P (1996). Mechanisms of beach ground water and swash interaction. *Proc. 25th Int. Conf. Coastal Eng.,* ASCE; 4120-4133.

Bird P A D, Davidson M A, Ilic S, Bullock G N, Chadwick A J, Axe P and Huntley D A (1966). Wave reflection, transformation and attenuation characteristics of rock island breakwaters. Advances in Coastal Structures and Breakwaters ed. J E Clifford, Thomas Telford; 93-109.

Grant U S (1948). Influence of the water table on beach aggradation and degradation. *Journal of Marine Research,* 7; 655-660.

Guza R T, Thornton E B & Holman R A (1984). Swash on steep and shallow beaches. *Proc. 19th Int. Conf. Coastal Eng.,* ASCE; 708-723.

Huntley, D A, Simmonds D J & Davidson M A (1995). Estimation of frequency-dependent reflection coefficients using current and elevation sensors. *Coastal Dynamics ,95,* ASCE; 57-68.

Kirk R M (1980). Mixed sand and gravel beaches: morphology, processes and sediments. *Progress in Physical Geography,* 4; 189-210.

Kobayashi N & Wurjanto A (1992). Irregular wave interaction with permeable slopes. *Proc. 23rd Int. Conf. Coastal Eng.,* ASCE; 1299-1312.

McFarland S, Whitcombe L J & Collins M B (1994). Recent shingle beach renourishment schemes in the UK: some preliminary observations. *Ocean and Coastal Management,* 25; 143-149.

Takezawa M, Kubota S & Nakamura N (1990). A field study of swash oscillation on a steep beach face. *4th Pacific Conf. Marine Science and Technology;* 282-289.

Voulgaris G, Michel D, Simmonds D, Howa H, Collins M B & Huntley D A (1997). Measuring and modelling sediment transport on a macrotidal ridge and runnel beach: an intercomparison. *Journal of Coastal Research* (in press).

Whitcombe L J (1995). Sediment transport processes, with particular reference to Hayling Island. Unpublished PhD thesis, Department of Oceanography, University of Southampton.

Grain-size influence on sand transport mechanisms

C. Marjolein Janssen[1], Wael N. Hassan[2], Remmelt v.d. Wal[3], Jan S. Ribberink[4]

Abstract

Little is known about the influence of the grain size on sand transport mechanisms in oscillatory sheet flow. Therefore experiments with fine sand (D_{50} = 0.13 mm) were carried out in the Large Oscillating Water Tunnel of Delft Hydraulics under the same flow conditions as in earlier experiments with unsieved dune sand (D_{50} = 0.21 mm; Katopodi et al., 1994). An intercomparison is presented between the measured time-dependent sediment concentrations and sediment fluxes of the two sands. In addition, a 1DV boundary layer model is used to analyze the data in more detail. Both methods will reveal information about the grain size influence on the transport mechanisms and their effect on the net transport rates.

Experiments

Fig.1 shows the general outline of the Large Oscillating Water Tunnel (LOWT) of Delft Hydraulics, which is equipped with a recirculation system, such that experiments with combined wave-current flow can be performed at full scale (1:1). The tunnel has the shape of a vertical U-tube with a rectangular horizontal test section and two cylindrical risers on either end. The desired oscillatory water motion inside the test section is imposed by a steel piston in one of the risers. The other riser is open to the atmosphere. The test section is 14 m long, 1.1 m high

---

1) Delft University of Technology, Hydraulics & Offshore Eng. Section, p/a: University of Twente, Department of Civil Engineering, PO Box 217, 7500 AE Enschede, The Netherlands. e-mail: c.m.dohmen-janssen@sms.utwente.nl
2) Hydraulics and Sediment Research Institute, Delta Barrage, PO Box 13621, Egypt. tel: 202-2188268, fax: 202-2189539
3) Delft University of Technology, Hydraulics & Offshore Engineering Section, PO Box 5048, 2600 GA Delft, The Netherlands.
4) University of Twente, Department of Civil Engineering, PO Box 217, 7500 AE Enschede, The Netherlands. e-mail: j.s.ribberink@sms.utwente.nl

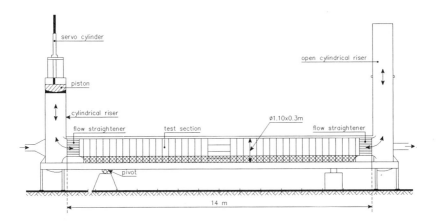

Figure 1:   General outline of the Large Oscillating Water Tunnel.

and 0.3 m wide. A 0.3 m thick sand bed can be brought into the test section, leaving 0.8 m height for the flow above the bed. The range of velocity amplitudes is 0.2 - 1.8 m/s and the range of oscillation periods is 4 - 15 s. The maximum superimposed current velocity in the test section of the tunnel is 0.5 m/s.

The sand used in the present fine sand experiments had the following characteristics: $D_{10} = 0.10$ mm, $D_{50} = 0.13$ mm and $D_{90} = 0.18$ mm, in which D_i is the grain size for which $i\%$ of the sand (by weight) is finer than D_i. The characteristics of the unsieved dune sand, used in earlier experiments are: $D_{10} = 0.15$ mm, $D_{50} = 0.21$ mm and $D_{90} = 0.32$ mm. The two flow conditions for which time-dependent velocities and concentrations were measured both lie in the sheet flow regime and are simulations of a sinusoidal wave superimposed on a net current. The values of the oscillation period T, the amplitude of the oscillatory velocity û and the net current velocity $<u>$ (at $z = 0.1$ m) for the two conditions are:

Condition	T (s)	û (m/s)	$<u>$ (m/s)
H6	7.2	1.50	0.25
H9	7.2	0.95	0.45

The present paper focuses on condition H6, because unsteady and mobile bed effects, described later, are most clearly visible for this condition.

Time-dependent flow velocities were measured using a 2D forward scatter Laser Doppler Anemometer (LDA) and a 3D Acoustic Doppler Velocitymeter (ADV). In the suspension layer, time-averaged concentrations were measured using a Transverse Suction System and time-dependent concentrations using an Optical Concentrationmeter (OpCon). Time-dependent concentrations in the sheet flow layer were measured using a Conductivity Concentration Meter (CCM).

In addition, net transport rates were measured for these two and 10 other flow conditions. The results of these tests are described in Janssen & Ribberink (1996).

Time-dependent sediment concentrations

Fig.2 shows the time-dependent ensemble-averaged concentrations in the sheet flow layer (measured by CCM) and in the suspension layer (measured by OpCon) for the 0.13 mm sand (left-hand-side panel) and for the 0.21 mm sand (right-hand-side panel). The velocity at 10 cm above the bed, i.e. outside the wave boundary layer, is given as a reference.

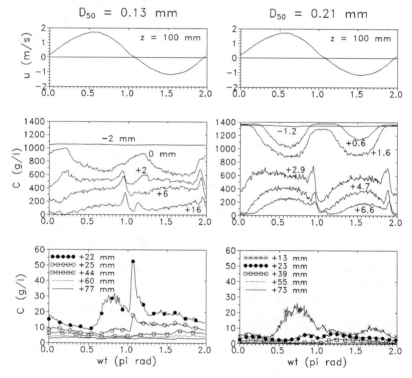

Figure 2:　　Ensemble-averaged concentrations in the sheet flow and the suspension layer for the 0.13 and the 0.21 mm sand.

The measurements show that at a certain level below the average still bed level ($z = 0$), the concentration remains constant throughout the wave cycle, indicating that no sediment is moving at these levels ($z = -2$ mm for the 0.13 mm sand; $z = -1.2$ mm for the 0.21 mm sand). The magnitude of the concentration is somewhat different for the two sands: around 1050 g/l for the 0.13 mm sand and about 1350 g/l for the 0.21 mm sand. These differences may be partly caused by the instrument which has an accuracy of about 10%. Also the upper layer of the 0.13 mm may be somewhat less packed than the 0.21 mm sand.

Just above these levels, sediment is being picked up during maximum (positive or negative) velocity, resulting in a decrease in concentration. It is clear that especially the increase in concentration due to the settlement of the particles back to the bed takes place quicker in the case of the coarser sand: in the case of the 0.21 mm sand the concentration at 1 and 2 mm above the bed returns to the still bed value around flow reversal, while for the fine sand the concentration at $z = 0$ mm remains smaller than the still bed value. Apparently, not all sediment returns to the bed. For the 0.21 mm sand the decrease in concentration is largest under the largest (positive) velocity. The situation is opposite for the 0.13 mm sand. This may be caused by the fact that the particles which are picked up under the positive half wave cycle do not all return to the bed, resulting in smaller concentrations just before the negative half wave cycle and therefore a larger decrease in concentration under the negative than under the positive half wave cycle.

In the upper sheet flow layer, above the pick-up layer, the behaviour is opposite: maximum concentrations occur under maximum positive or negative velocity. For both sands the largest (positive) velocity results in the largest concentration, although this is more distinct for the 0.21 mm sand. Again the coarse particles settle much quicker to the bed under decreasing velocities, as can be seen by the faster decrease in concentration after the maximum velocity. In this region the concentrations of the 0.13 mm sand are much higher than those of the 0.21 mm sand. Compare for example the concentrations at 6 mm for the 0.13 mm sand with the concentrations at 7 mm for the 0.21 mm sand.

The much larger concentrations for the finer sand are also observed in the suspension layer, as shown in the lower panels of Fig.2. (Note the difference in concentration scale between the plots for the sheet flow layer and those for the suspension layer). The concentration at 13 mm in the 0.21 mm sand is even smaller than the concentration at 22 mm in the 0.13 mm sand. The difference can also clearly be seen by comparing the concentrations at 22 mm in the fine sand with the concentrations at 23 mm in the 0.21 mm sand (lines with solid circles). The figure shows that the maximum in concentration occurs at a later moment for higher levels. For example in the 0.21 mm sand the maximum concentration under the negative half wave cycle occurs at $\omega t = 1.55\pi$ at $z = 3$ mm, at $\omega t = 1.6\pi$ at $z = 5$ mm and at $\omega t = 1.65\pi$ at $z = 7$ mm. This phase shift, together with the fact that the fine sand settles very slowly back to the bed may cause some sediment which is picked up under the positive half wave cycle to be transported by the succeeding negative velocity, resulting in decreasing *net* (i.e. time-averaged) sand transport rates.

Finally, the most striking phenomenon in the ensemble-averaged concentrations are the sharp peaks in concentration around flow reversal, i.e. around 0.95π and 1.9π in the sheet flow layer and around 1.05π and possibly 0π in the suspension layer. These peaks are not present in the pick-up layer. The peaks are especially large in the fine sand and even dominant over the maximum concentration under the maximum velocity. They occur just before, or at higher levels just after flow reversal. This means that the sand is mainly transported by the succeeding half wave cycle. Because the peaks are slightly larger after the positive than after the negative velocity, this may result in a decrease in the net sand transport rates, which can be significant for the fine sand.

Time-averaged concentration

The concentration can be averaged over the wave cycle, which of course results in larger values for the fine sand, because it was shown that the ensemble-averaged concentrations are larger for the fine sand. Fig.3 shows a detailed plot of the *time-averaged* concentration profile in the sheet flow layer. Like in the time-dependent measurements the concentrations below the average still bed level are somewhat lower for the fine sand. Additionally, the decrease in time-averaged concentration starts lower for the fine sand, which is probably caused by the slightly different behaviour in the pick-up layer, as described above.

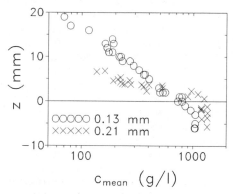

Figure 3:   Time-averaged concentration in the sheet flow layer for the fine and the coarser sand.

The coarser sand shows a much larger concentration gradient, with concentrations decreasing from about 1350 g/l at $z = 0$ to 200 g/l at $z = +5$ mm. Defining the thickness of the sheet flow layer δ_s, as the level where the time-averaged concentration equals 1 vol% = 26.5 g/l, the following results are found:

D_{50} (mm)	δ_s (mm)	δ_s/D_{50} (-)
0.13	20	154
0.21	9	43

The choice of 1 vol% for the concentration as a boundary between the sheet flow and the suspension layer is rather arbitrary. If a value of 8 vol% or 200 g/l was chosen, the thicknesses are smaller of course. Still, the sheet flow layer in the fine sand would be about twice as thick as in the coarse sand.

Time-averaged sediment flux and net transport rate

The net sediment flux $<\varphi>$ is the time-averaged value of the product of the time-dependent velocity and concentration:

$$\langle \varphi(z) \rangle = \langle u(z,t) * c(z,t) \rangle$$

$$= \langle u(z) \rangle * \langle c(z) \rangle \tag{1}$$

$$+ \langle u_{osc}(z,t) * c_{osc}(z,t) \rangle$$

$$= \varphi_c(z) + \varphi_w(z)$$

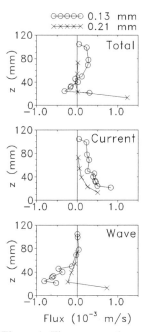

○○○○○ 0.13 mm
××××× 0.21 mm

Total

Current

Wave

Flux $(10^{-3}$ m/s)

Figure 4: Time-averaged suspended sediment flux profile.

The product of the time-averaged components of velocity and concentration, $<u(z)> * <c(z)>$ is the current-related flux, while the time-averaged value of the product of the oscillatory components, $<u_{osc}(z,t) * c_{osc}(z,t)>$ is the wave-related flux.

Fig.4 shows the total time-averaged *suspended* sediment flux and the current- and wave-related components. Only the suspended fluxes are presented, because it was impossible to measure velocities closer to the bed, due to the large sediment concentrations. The figure shows that the current-related flux, is much larger for the fine than for the coarser sand, due to the much larger time-averaged concentrations (the velocities are the same). However, the wave-related fluxes are also much larger but in opposite direction, due to the increased time-lag between velocity and concentration and the dominant concentration peaks at flow reversal. These unsteady effects are largest *after* the positive half wave cycle, resulting in a negative wave-related flux.

More important than the suspended sediment fluxes are the total net transport rates, defined as:

$$\langle q_s \rangle = \int_0^h \langle \varphi(z) \rangle \, dz \tag{2}$$

These net sand transport rates have been measured using a mass conservation technique (see Janssen and Ribberink, 1996). For the two sands the results are presented in Table 1, together with the values of the suspended net transport rate, derived from integration of the suspended sediment flux profiles. The difference between these two values must be the amount of sand transported in the sheet flow layer, i.e. between $z = 0$ and δ_s. Apparently, the unsteady effects do not

only occur in the suspension, but also in the sheet flow layer: unless the larger
concentrations for the fine sand, the *net* sand transport rates are smaller than for
the coarser sand.

Table 1: Net transport rates, divided between suspension and sheet flow layer

D_{50} (mm)	$<q_s>$ ($*10^{-6}$ m^2/s)		
	total	suspended	sheet flow
0.13	63.3	9.9	53.4
0.21	112.0	6.1	105.9

1DV boundary layer model

In order to further analyze the data, they are compared with calculations of
the 1DV boundary layer model of Ribberink & Al-Salem (1995), which consists
of the momentum equation for water and the mass balance of the sediment:

$$\frac{\partial u}{\partial t} = -\frac{1}{\rho}\frac{\partial p}{\partial x} + \frac{\partial}{\partial z}\left[v_t\frac{\partial u}{\partial z}\right]$$

$$\frac{\partial c}{\partial t} = \frac{\partial}{\partial z}\left[w_s c + \varepsilon_s\frac{\partial c}{\partial z}\right]$$

(3)

Here u is the horizontal velocity, t is time, ρ is the density of water, p is the
pressure, z is the level above the bed, v_t is the turbulent eddy viscosity, c is the
sediment concentration, w_s is the settling velocity and ε_s is the sediment
diffusivity. The turbulence is modelled according to Prandtl mixing length theory,
with the possibility of an extra reduction factor to the calculated eddy viscosity, ß:
$v_t = ß*v_{t,p}$, in which $v_{t,p}$ stands for the original Prandtl eddy viscosity, calculated
by $(\kappa z)^2 * \partial u/\partial z$. Here κ is the Von Karman constant (0.4). The sediment
diffusivity ε_s is equal to the eddy viscosity v_t. The boundary conditions are:

$u = 0$ at $z = z_0 = k_s/30$ $c = c_a$ at $z = z_a$
$u = u_\infty$ at $z = h$ $\partial c/\partial z = 0$ at $z = h$

Here z_0 is the level where the velocity becomes zero, k_s is the roughness height,
u_∞ is the free-stream velocity, i.e. above the wave boundary layer, h is some
arbitrary level above the wave boundary layer, c_a is the reference concentration at
the reference level z_a. c_a is calculated according to the formula of Engelund &
Fredsøe (1976) and the standard settings used in the calculations are: $k_s = 3D_{90}$, ß
$= 1.0$, $z_a = 2D_{50}$, h $= 0.5$ m, $u_\infty = <u> + $ûsin($2\pi t/T$), with $<u> = 0.34$
m/s, û $= 1.47$ m/s and T $= 7.2$ s. The input conditions at 0.5 m above the bed
were chosen such that the measured velocities at 0.1 m above the bed were
reproduced correctly by the model.

Fig.5 shows a comparison between the measurements and the results of the
model with standard settings. The velocities, shown in the upper panels, are

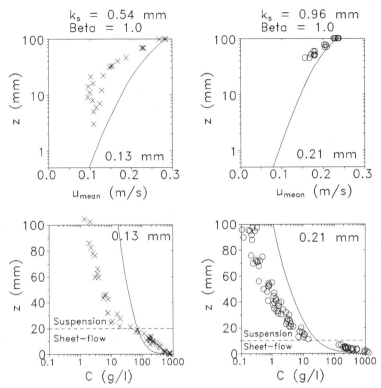

Figure 5:   Comparison between measurements (symbols) and results of the standard model (lines) for the 0.13 and the 0.21 mm sand.

measured by ADV for the 0.13 mm sand (left-hand-side panel) and by LDA for the 0.21 mm sand (right-hand-side panel). The LDA can only measure velocities further away from the bed, where sediment concentrations are small. The ADV can measure velocities in much larger concentrations and thus closer to the bed. Still, it seems that for very large concentrations ($<c> > 100$ g/l) the ADV does not measure velocities correctly, as shown by the constant velocity below $z = 20$ mm, which does not seem very realistic. The measured concentrations, shown in the lower panels, are obtained from CCM, OpCon and Suction measurements.

Both for the fine and the coarser sand the velocities and the concentrations are largely overpredicted. This may be caused by the sediment-flow interaction, which is an important aspect in sheet flow that is not included in the model: The sheet flow layer acts as an increased roughness to the flow above it resulting in a decrease in velocity gradient close to the bed, and an increased velocity gradient further away. In addition, the sheet flow layer may cause turbulence damping, (decreased mixing), which reduces the sediment concentration.

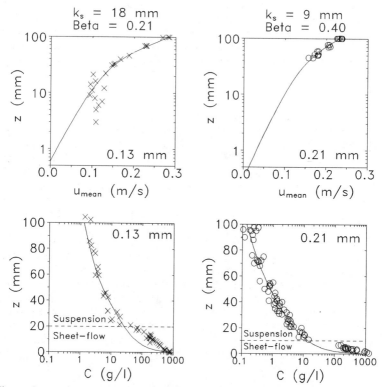

Figure 6:   Comparison between measurements (symbols) and results of the
                tuned model (lines) for the 0.13 and the 0.21 mm sand.

In order to take account of these effects, two parameters are used to tune the
model: the roughness height k_s and the eddy viscosity reduction factor ß. Fig.6
shows that in the suspension layer good agreement between the measurements and
the model results can be found by using the thickness of the sheet flow layer as
the roughness height, provided that the eddy viscosity is reduced stronger if the
sheet flow layer is thicker. This seems reasonable, because the thicker the sheet
flow layer the larger the mobile bed effects will become. The results can be
summarized as follows:

D_{50} (mm)	measured δ_s (mm)	applied k_s (mm)	required ß (-)
0.13	20	18	0.21
0.21	9	9	0.40

Apparently, the influence of the mobile bed effects on the velocity and concentration in the suspension layer, caused by the presence of the sheet flow layer, can be simulated relatively simple. However, it is clear that within the sheet flow layer there is still a large discrepancy between the measurements and the predictions. This is caused by the different transport mechanisms in the sheet flow layer, like grain-grain and grain-fluid interactions, which are not included in the model.

Conclusions

Comparing the measurements of the 0.13 mm and the 0.21 mm sand shows that for the fine sand unsteady effects are very important: The time-dependent concentrations show that some of the fine sand particles, which are entrained under the positive half wave cycle, do not return to the bed and are therefore transported by the succeeding negative half wave cycle, in opposite direction. In addition, the dominant concentration peaks at flow reversal, which are largest after the largest positive velocity, result in an extra amount of sand transported in negative direction and thus decreasing the *net* sand transport rate. This was also shown by the large negative wave-related sediment fluxes for the fine sand and indeed by smaller measured net sand transport rates than for the coarser sand.

Comparing the measurements with the results of a 1DV boundary layer model, indicates the mobile bed effects are larger if the sheet flow layer is thicker: In the suspension layer good agreement can be found by using the measured thickness of the sheet flow layer as the roughness height, and reducing the eddy viscosity stronger if the sheet flow layer is thicker. Despite the possibility to model the influence of the sheet flow layer on the velocity and concentration in the suspension layer, still a large discrepancy exists between the measured and predicted concentration within the sheet flow layer. This is caused by the different transport mechanisms, like grain-grain and grain-fluid interaction.

Acknowledgements

The authors want to thank the Dutch Ministry of Transport and Public Works (RIKZ), for the financial support to perform the experiments. The first author also wants to thank the Technology Foundation (STW) for their financial support during the complete research project.

References

Engelund, F. and J. Fredsøe (1976). A sediment transport model for straight alluvial channels. Nordic Hydrol., 7, pp. 293-306.

Janssen, C.M and J.S. Ribberink (1996). Grain-size influence on sand transport in oscillatory sheet flow. Proc.25th ICCE, Orlando, pp. 4779-4792.

Katopodi, I., J.S. Ribberink, P. Ruol and C. Lodahl (1994). Sediment transport measurements in combined wave-current flows. Proc. Coast. Dyn.'94, Spain.

Ribberink, J.S. and A.A. Al-Salem (1995). Sheet flow and suspension in oscillatory boundary layers. Coast. Eng., Vol. 25, pp. 205-225.

Fractional Threshold of Motion for Mixed Sand Beds and Implications for Sediment Transport Modelling

S. Wallbridge[1] and G. Voulgaris[2]

Abstract

Many models consider sediment transport to be a function of the excess bed shear stress over the critical (threshold) value for initial grain motion *i.e.* transport$=f(\theta-\theta_c)$, where $\theta=$dimensionless shear stress and $\theta_c=$critical value. Smith and McLean (1980) also relates suspended sediment concentration to bed shear using $(\theta-\theta_c)$ in a sediment pick-up function. Mixed bed threshold models (*e.g.* Wiberg and Smith, 1987) show that different grain sizes within a mixed-sized sediment have different θ_c values than the uniform sediment. This study verifies by laboratory experiment a mixed bed threshold model and applies the results of this model to bedload and suspension formulae, for comparison with results obtained using single threshold values only. The multi-fraction approach is found to have most effect in suspended transport calculations, primarily as a result of the introduction of fractional settling velocities. Entrainment threshold is found to be a secondary factor only and we consider it is likely that the theoretical model becomes irrelevant as bed conditions change under higher flows.

Introduction

The majority of sediment transport models used in morphodynamical and coastal evolution studies use a threshold term. Bedload transport models similar to that of Meyer-Peter and Müller (1948) express the amount of sediment in transport as function of sediment size (D) and the excess shear stress $(\tau-\tau_c)$, where τ_c is the critical stress for initiation of sediment movement. Suspended sediment transport models are based on the coupling of the vertical distribution of suspended sediment concentration and current velocity. The former distribution is usually scaled with a reference

---

[1] Department of Oceanography, University of Southampton, Southampton Oceanography Centre, European Way, Southampton, SO14 3ZH, UK
[2] Woods Hole Oceanographic Institution, Woods Hole, MA 02543, USA

concentration (C_r) that is defined at a specific reference level above the sea bed. Pick-up functions are used to estimate this reference concentration by relating the amount of sediment available for suspension to a resuspension coefficient (γ) and the excess shear stress. The excess shear stress depends on the sediment threshold conditions, so is directly related to the seabed sediment. Threshold, as expressed by hydrodynamic force balance, may be applied to the case of a single grain on a sediment bed. An average value can be assigned for the entire sediment population or for a particular size fraction of it.

In most conventional studies of sediment transport, the bed is represented by the size of the mean fraction and the threshold is derived for this fraction using Shields (1936) threshold curve, which assumes a mono-sized bed and does not account for the heterogeneity of natural beds. Wiberg and Smith (1987) and Bridge and Bennet (1992) showed analytically that, for the case of steady flow, the grain threshold of individual fractions within a bed vary as a function of the ratio of grain size to bed roughness (D/k_s, where k_s is frequently taken as the mean bed grain size K).

The aim of the present work is to examine (i) the threshold of fractions of mixed beds with specific reference to an analytical model and for oscillatory conditions; and (ii) the implications of such a fractional threshold definition for bed- and suspended load transport modelling. In this contribution preliminary results of experimental data on the threshold for wave and current conditions of individual fractions of a mixed bed are presented. Further, sediment transport is modelled with and without a fractional threshold criterion to assess the importance of including this parameter in transport studies.

THRESHOLD OF MOTION

The fractional threshold model used here is that of Wiberg and Smith (1987). The model is based upon the force balance acting on a single grain at the bed surface under the action of a steady current. Threshold stress is given by

$$\theta_c = \frac{2}{(C_D)_c \alpha} \frac{1}{<f^2(z/z_0)>} \frac{(\tan\phi_0 \cos\beta - \sin\beta)}{[1 + (F_L/F_D)_c \tan\phi_0]} \tag{1}$$

where C_D is the drag coefficient, F_L and F_D are the fluid lift and drag forces, $f^2<z/z_0>$ is the square of the cross sectionally averaged velocity profile function and α is a grain shape coefficient. The angles β and ϕ_0 are the bed slope and pivoting angles respectively.

Experiments

Thresholds of four different grain sizes were studied. $\frac{1}{2}\phi$ sieve fractions of mean diameter 215, 295, 420 and 595μm were used, within log-normally distributed mixed

beds composed of seven ½ɸ fractions. Each size is studied acting as the mean size of
the mixture, as well as both one fraction finer than and one fraction coarser than the
mean size. For each experiment, the fraction under study was 22% of the total bed by
weight. Threshold was defined using the statistical Yalin (1972) criterion, counting
only the numbers of dyed grain movements.

Thresholds under combined unidirectional and oscillatory flows were measured in a
laboratory flume (Wallbridge *et al.*, *in prep*). The channel dimensions were 5m x
30cm x 60cm (height), with a plate in the base of the flume that could be oscillated
to simulate wave orbital motion. A central recessed area of the plate contained the
sediment sample, flattened and levelled with the surrounding plate (roughened with
a fixed layer of identical grain size distribution). The plate oscillated at a given
frequency and orbital velocity and the unidirectional flow was steadily increased until
threshold was satisfied. Flow conditions were then recorded (in the absence of the
wave motion) using a Laser Doppler Anemometer at 5mm above the bed. This method
was repeated four times for each wave and bed combination, with waves of 5 and 10s
period and orbital velocities ranging from 0.02-0.25ms^{-1}. The results presented here
are mean values from all the experiments.

Results

The coarser grains have lower thresholds than fine grains in a mixture (fig 1), as
reported elsewhere (*e.g.*Kirchner *et al*, 1990). The values of the ratio D/K of the
Wiberg and Smith (1987) curves shown (0.75, 1 and 1.5) compares with the actual
values of 0.7, 1 and 1.41 for the grains studied. The results of these experiments confirms that the model of Wiberg and Smith (1987), although developed for steady flows, is accurate under combined flows.

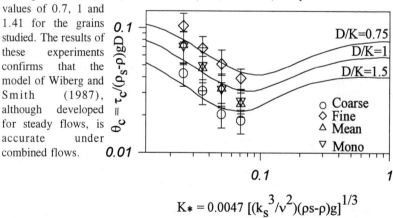

Figure 1. Combined flow threshold data plotted on the
model of Wiberg & Smith (1987).

SEDIMENT TRANSPORT MODELLING

In order to examine the implications for sediment transport modelling of fractional threshold conditions, formulae for bed- and suspended load transport are applied to 2 sediment beds under given wave-current conditions and compared for several threshold definitions. The analytical transport model of Wikramanayake and Madsen (1994) is used, which splits both the flow and suspended sediment concentration into a mean and an oscillatory component.

A three layer, time-invariant eddy viscosity (v_t) model (Madsen and Wikramanayake, 1991) is used to solve the governing equations for the mean current (u_c):

$$v_t \frac{du_c}{dz} = \frac{\tau_c}{\rho} \tag{2}$$

the time-varying oscillatory current (u_w):

$$\frac{\partial u_w}{\partial t} = -\frac{1}{\rho}\frac{\partial p_w}{\partial x} + \frac{\partial}{\partial z}[v_t \frac{\partial u_w}{\partial z}] \tag{3}$$

and the time-varying suspended sediment concentration (c)

$$\frac{\partial c}{\partial t} = w_s \frac{\partial c}{\partial z} + \frac{\partial}{\partial z}[v_t \frac{\partial c}{\partial z}] \tag{4}$$

where w_s is the sediment settling velocity, p is the time-varying component of pressure, ρ is the density of water, x is the direction of wave motion, z the vertical direction. The pick-up function used to relate the instantaneous reference concentration C_r to the instantaneous excess skin friction shear stress is:

$$C_r(t) = C_b \cdot \gamma \cdot \frac{\tau_b(t) - \tau_c}{\tau_c} \tag{5}$$

where C_b is the volumetric bed concentration, γ is the resuspension coefficient and τ_b is the instantaneous skin friction bed shear stress. Suspended sediment transport is the sum of the fluxes due to the time-varying and the mean components of velocity and concentration integrated over the whole water column.

The instantaneous bedload transport is derived using Madsen's (1991) general form of the Meyer-Peter and Müller (1948) formula, given by

$$Q_b(t) = \sqrt{(s-1)gD^3} \cdot \frac{8}{1 + \frac{\tan\beta(t)}{\tan\phi_m}} \cdot (\theta(t) - \theta_c)^{3/2} \cdot \frac{\tau_b(t)}{|\tau_b(t)|} \tag{6}$$

where θ is the Shield parameter and τ_b is the instantaneous skin friction bottom shear stress. Bedload is calculated when $(\theta-\theta_c)>0$ (*i.e.* bed stress exceeds bed threshold). The angles $\beta(t)$ and ϕ_m are the slope of the bed in the direction of $\not{F}$ (t) and the friction angle of the sediment respectively. In the current analysis flat bed was assumed.

Model Runs

The model was run for two sediment populations, both with the same mean grain size (=0.183mm), though one is well sorted while the second sample is poorly sorted (with an additional coarse grain content) (fig 2). Sediment transport rates were estimated using three different approaches: (i) assuming the D_{50} (mean) grain size represents the whole bed, with a single threshold value θ_{D50} obtained from the Shields curve; (ii) using the analytical model of Wiberg and Smith (1987) to define separate thresholds for each grain size fraction within the beds (θ_{ci}), with the skin friction roughness defined by the D_{50}; (iii) similar to case (ii) but with the threshold criterion for each fraction *i* derived from the Shields curve (*i.e.* no hiding and protrusion effects). For cases (ii) and (iii), the model is run separately for each fraction *i* and total transport was obtained as:

$$Q=\sum_{i=1}^{N} f_i \cdot Q_i \qquad (7)$$

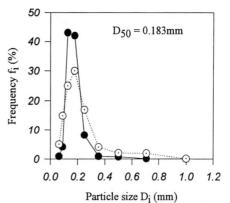

Figure 2. Grain size distributions of the sediment beds, with equal mean size and well sorted (filled symbols) and poorly sorted (open symbols) sediment

where Q_i is the transport rate calculated by the model for the fraction *i* with particle size D_i and f_i is the percentage of occurrence of the particular fraction. The above listed three cases are referred to hereafter as: (i) 'mean size, Shield's criterion' (MSSC); (ii) "fractional approach, differential threshold criterion" (FADC) and (iii) "fractional approach Shields criterion" (FASC), respectively.

Modelled flow conditions approximate coastal conditions (mean current (0.37ms⁻¹) approximately perpendicular to the waves). Shoaling (asymmetric) waves are obtained by the addition of two sinusoidal components, one being the subharmonic of the other (U_{rms} 0.63 and 0.15 ms⁻¹, respectively, wave period 5.5s). The model calculates the bedload and suspended load transport rates for each component of the flow separately *i.e.* mean flow and each of the wave components, totalling them to obtain overall transport rate.

Model Results

Vertical Distribution of Suspended Sediment Concentration

Volumetric sediment concentrations due to the mean flow component are several orders of magnitude larger than the oscillatory components except very near to the bed (in the wave boundary layer) (fig 3). The profile of suspended sediment concentration estimated by MSSC is smaller by an order of magnitude than the FA methods. The differences between the FA profiles are much lesser. A similar effect is found for the oscillatory components of the suspended sediment concentrations.

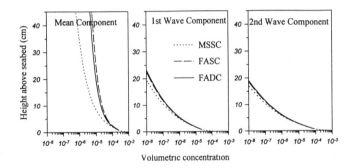

Figure 3. Vertical profiles of suspended sediment concentration due to each component of the flow (for the well-sorted sediment).

Sediment Transport Rates

Figure 4 displays the absolute values of the transport rates and reference concentration calculated by the model, unscaled by the bed distribution. Bedload transport rate calculated using the FASC decreases with increasing grain size as larger, heavier grains move less easily than the fine grains. The FADC compensates for this inertial effect by the increased protrusion (larger D_i/K) and predicts approximately equal transport rates for each fraction, even increasing slightly with grain size. Suspended load follows a similar pattern, with a consistent decline in transport rate with size for the FASC. Using the FADC threshold values, suspended load transport actually begins to increase for grains larger than $500\mu m$. This is clearly reflected in the

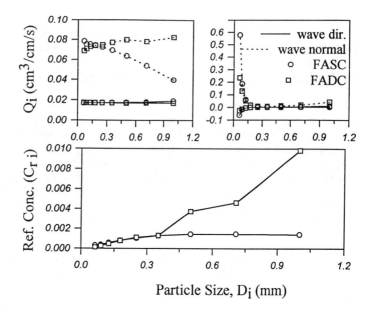

Figure 4. Absolute bed- and suspended load transport rates calculated using each fractional threshold model for the well-sorted bed, along with reference concentration.

concentration, which rises to very high levels at the coarsest grain sizes.

DISCUSSION

Evaluation of threshold approaches

The model indicates that suspended transport due to the mean current is the most important mechanism. Fractional methods result in much higher transport rates, due to the inclusion of separate settling velocities for each grain size and a more accurate calculation of suspended grain concentrations. The FADC implies lower mobility of fine grains due to sheltering in the bed and this is carried directly into transport rate calculations. Transport of fine grains is reduced as a result, thus the FADC predicts lower transport rates than the FASC.

Table 1 compares the total transport rates calculated for the well sorted bed with each of the three threshold models. Table 1 also gives the same data for the poorly sorted beds. Transport rates for each bed are identical when using the MSSC, since θ_{D50} is the same for both beds and the tails of the distribution are not taken into account. Bedload transport rates for the fractional methods are very similar to the mean value for both the well and poorly sorted beds with only very small differences in the transport due to the additional coarse material mobilised in the FADC model.

	Well sorted sediment			Poorly sorted sediment		
	Q MSSC	$\Sigma f_i Q_i$ FASC	$\Sigma f_i Q_i$ FADC	Q MSSC	$\Sigma f_i Q_i$ FASC	$\Sigma f_i Q_i$ FADC
Bedload						
Wave dir	0.0172	0.0172	0.0172	0.0172	0.0172	0.0172
Wave norm	0.074	0.0745	0.0736	0.074	0.0738	0.0735
Suspended						
Mean mode	-0.0017	-0.005	-0.0041	-0.0017	-0.0085	-0.0055
Osc. mode	0.0019	0.0017	0.0016	0.0019	0.0016	0.0018
Σ wave dir	0.0002	-0.0034	-0.0024	0.0002	-0.0068	-0.0038
Wave norm	0.00157	0.0472	0.038	0.0157	0.0792	0.0520
Total						
Wave dir	0.0174	0.0138	0.0148	0.0174	0.0104	0.0134
Wave norm	0.0897	0.1217	0.1116	0.0897	0.1530	0.1255

Table 1. Calculated sediment transport rates (Q, cm³/cm/s) for each method and each bed, in both wave normal and wave directions.

The greatest differences occur in the calculation of suspended transport. The effect of the oscillatory motion is confined close to the bed where all runs predict very similar sediment concentrations and hence there is little effect of using different threshold criteria. The fractional methods, which increase the suspension concentrations at height, increase transport due to the mean flow. In the present case, this produces the result that net suspended transport in the wave orbital plane actually reverses direction when using fractional methods rather than θ_{D50}. Total transport in the wave direction with the reference FADC is actually larger than with the FASC. For the fine grains, $\theta_{ci} > \theta_{si}$ and so suspension of these fine grains is reduced. Bedload and suspended load transport are in opposite directions in this plane, and so net transport is higher for the fractional method. In the wave normal direction, transport is much greater for the FASC model.

Application to transport modelling

Intuitively, the fractional approach to sediment transport modelling is preferable. The suspended transport rates are a strong function of two parameters, threshold θ_c and settling velocity W_s. The primary factor is W_s, which exerts most control on the vertical distribution of grains. The effect of θ_c is secondary, affecting reference concentrations and so slightly altering the vertical concentration profiles. The calculation of W_s is unaffected by which fractional threshold method is used.

FADC decreases the suspension potential of the fine grains and increases that of the coarse grains. With poorer sorting, this would lead to a coarser suspended load population, concentrated much lower in the flow, with a reduction in overall transport rate. A decrease in sorting makes it more important to choose the correct threshold definition. The relative merits of the FASC and FADC models are unclear. The experiments presented have clearly shown that fractional threshold is a real phenomenon. However, conventional wisdom finds that suspended sediment concentration should decrease with size rather than increase and that the finest grains should be more easily suspended. Use of the FADC (using θ_{ci}) contains the opposite implication. We question whether thresholds relevant to grain transport vary as a function of D_i/k_s or simply of D_i, which would govern the application of fractional threshold models to sediment transport modelling.

Conclusions

(i) Experiments indicate that grain threshold in mixed beds varies as a function of D_i/K. The mixed bed threshold model of Wiberg and Smith (1987) adequately describes the threshold of size fractions within mixed sand beds, even under the influence of waves, provided the bed pivoting angle is properly defined.

(ii) Mixed-bed transport modelling using only a single grain size is inadequate and becomes increasingly less accurate as the sorting of the bed decreases. Modelling of suspended sediment transport should be undertaken using all fractions of a heterogeneous bed, since the inclusion of separate settling velocities for each fraction develops a more accurate suspended sediment profile.

(iii) Inclusion of differential thresholds in the fractional approach is less significant in transport rate calculation. In highly energetic conditions it is believed that the effect of bed roughness on grain transport is insignificant. However, the effect may be of importance for the natural sorting of the bed under near-threshold conditions.

Acknowledgements

During this study, G. Voulgaris was the recipient of a Woods Hole Oceanographic Institution Post-doctoral Scholar Award funded by the Andrew W Mellon Foundation. Travelling funds were povided from the Dept. of Geology and Geophysics, Woods Hole Oceanographic Institution and the Andrew W Mellon Foundation, through the Rinehart Coastal Research Centre.

References

Bridge, J. and Bennet, S. (1992) A model for the entrainment and transport of
    sediment grains of mixed sizes, shapes and densities. *Water Resources
    Research* **28**(2) p337-63
Kirchner, J., Dietrich, W., Iseya, F. and Ikeda, H. (1990) The variability of critical
    shear stress, friction angle and grain protrusion in water-worked sediments.
    Sedimentology **37** p647-72
Madsen, O. (1991) Mechanics of Cohesionless sediment transport in Coastal Waters,
    Coastal Sediments '91, ASCE: p15-27.
Madsen, O. and Wikramanayake, P. (1991) Simple Models for Turbulent
    wave-current bottom boundary layer flow. U.S Army Corps of Engineers,
    CERC, CR DRP-91-1.
Meyer-Peter, E. and Müller R. (1948) Formulas for bedload transport. *Report of 2nd
    meeting of the International Association for Hydraulic Structures Research*,
    Stockholm, p39-64
Miller, R. and Byrne, R. (1965) The angle of repose for single grains on a fixed
    rough bed. *Sedimentology* **6** p303-14
Shields, A. (1936) Application of similarity principles and turbulence research to
    bedload movement. Ott, W. and Van Uchelen, J. (translators), California
    Institute of Technology, W.M. Keck Laboratory of Hydraulics and Water
    Resources, 26pp.
Smith, J. And McLean, S. (1977) Spatially averaged flow over a wavy surface.
    Journal of Geophysical Research **82**, p1735-46
Wallbridge, S., Voulgaris, G., Tomlinson, B. and Collins, M. Initial motion and
    pivoting characteristics of sand particles in uniform and heterogeneous
    sediments: experiments and modelling. Submitted *Sedimentology, Aug 1997*
Wiberg, P. and Smith, J. (1987) Calculations of the critical shear stress for motion
    of uniform and heterogeneous sediments. *Water Resources Research* **23**(8),
    p1471-80
Wikramanayake, P. and Madsen, O. (1994) Calculation of suspended sediment
    transport by combined wave current flows. U.S. Army Corps of Engineers,
    CERC, CR DRP-94-7, 148pp plus Appendices.
Yalin, M.S. (1972) Mechanics of sediment transport. Pergamon Press, New York.
    290pp

Selective Sediment Transport in the Nearshore Zone: Field Observations and Potential Mechanisms

Piet Hoekstra[1] and Klaas T. Houwman[1]

Abstract

Field observations of selective sediment transport in the nearshore zone are reported from Terschelling, The Netherlands. The results are based on the developments prior to and shortly after the implementation of a shoreface nourishment. Selective sediment transport mechanisms are discussed and appear to be related to combinations of all major sediment transport components.

Introduction

Sediment properties have since long been recognized as being an important factor in sediment transport processes and, as such, may play an important role in coastal morphodynamics. Especially the sediment fall velocity or equivalent grain size is a relevant parameter in this respect. From studies in the past, it is a well known fact that the order of magnitude of sediment transport substantially varies for different (median) grain sizes. A coarsening of the sediment, for example, results in a decrease of the mobility of the sand grains in relation to the criterion for the initiation of motion and, in case of highly graded sediments, may also lead to a certain degree of armouring of the bed. In addition, not only the order of magnitude but also the (net) direction of the transport seems to change, e.g. due to a different relative contribution of (wave-dominated) bedload versus (longshore current-dominated) suspended load transport (Van Rijn, 1995).

One has also noticed that there may be a differential behaviour of grain size fractions under the same hydrodynamic conditions. This differential behaviour is observed in response to either a natural range in overall settling velocities of the sediment (e.g. Guillen and Hoekstra, 1996) or, more specifically may result from a variability in the specific density of the sediment due to mixtures of light and heavy minerals (Tanczos, 1996).

Research Objectives

In the following section a brief description of the main results of the NOURTEC field experiment of Terschelling (The Netherlands) is given, demonstrating the different responses of various grain size fractions in the nearshore zone after the introduction of a perturbation in the sediment grain size composition due to the implementation of a shoreface nourishment. The main objective of the present study is to evaluate and

[1]   Institute for Marine and Atmospheric Research, Utrecht University, P.O.Box 80.115, 3508 TC Utrecht, the Netherlands, e-mail: p.hoekstra@frw.ruu.nl

understand the principal sediment transport mechanisms that are expected to be responsible for the observed selective sediment transport in the nearshore zone.

NOURTEC Observations of Selective Sediment Transport

In the spring and summer of 1993, a shoreface nourishment was carried out along the barrier island coast of Terschelling, the Netherlands (Hoekstra et al. 1997). The shoreface nourishment formed part of a larger international (EC-MAST2) experimental programme, called NOURTEC, to study the feasibility, effectiveness and optimum design characteristics of shoreface nourishment techniques in different marine environments. The Terschelling shoreface nourishment was located offshore, in the depth interval between -5 and -7 m below NAP (Dutch Ordnance Datum), filling up the trough between two nearshore and (almost) shore-parallel breaker bars. Sediment sampling was carried out in various campaigns prior to (March 1993 - T0) and after the nourishment (November 1993 - T1, April and October 1994 - T2 and T3, respectively). In total, more than 1000 samples were collected and elaborated. Sediment sampling procedures, grain size and data analysis are presented by Guillen and Hoekstra (1996).

Prior to - but also after - the nourishment a clear overall trend was found in the cross-shore distribution of median grain sizes (Fig. 1). The median grain size reaches a minimum (150-160 micron) at about 6 to 8 m depth and shows a coarsening in both a landward and seaward direction. The coarsest sediment is located on the intertidal beach and in the swash zone and the median grain size of this sediment locally varies from

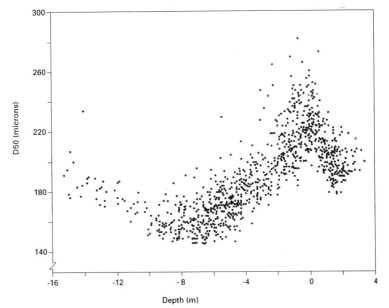

Figure 1.   Cross-shore variation of median grain size with depth for all samples taken in the Terschelling study area (based on 4 sampling campaigns).

200-260 micron; on the beach the D_{50} reduces again towards the dune foot. The general patterns that emerge are the result of the underlying, overall distribution in grain size fractions in the nearshore zone. As a matter of fact, also the cross-shore distribution of grain size fractions shows a clear relation with depth (Guillen and Hoekstra, 1996). For example the finest fractions (100-150 and 150-200 micron, respectively) display a maximum in the relative distribution at 6-8 m depth; in general almost 50-60 % of these fractions, present in an individual cross-shore profile, is found in this zone.

The nourished sand differed in two ways from the native, local deposits at the nourished site (Guillen and Hoekstra, in press):
    the nourished sand was in general coarser than the native deposits (Fig. 2); consequently, the nourished area was characterized by an overall coarsening of the sediment with more than 20 micron (Fig. 2) in comparison to the original deposits (the actual range in coarsening varied locally from about 15 to 60 micron);
    the nourished sediment was more poorly sorted than the native deposits.

In general, there is a very rapid response of the grain size characteristics immediately after the implementation of the nourishment. About half a year after the nourishment an overall fining is observed in the area of the nourishment (Fig. 2). Most of the sediment is about 15 micron or more smaller than in the previous post-nourishment situation and grain sizes are comparable again with those observed during the pre-nourishment conditions (Fig. 2). This fining is mainly related to a loss of the coarser fractions (> 200 micron). Meanwhile a coarsening is observed closer to the beach, especially due to the increase of the fraction 200-250 micron. However, whether this material is actually originating from the nourishment or just another zone, is unknown. A longshore origin is not expected, given the longshore uniformity in sediments (next sections). The differences in median grain size between the T2 (April 1994) and T3 (October 1994) campaigns are only marginal (Fig. 2) and a (seasonal ?) fining is observed in a longshore-oriented zone close to the beach, in an area West, East and landward from the nourishment. The finest fractions of the nourishment (100-150 micron) though tend to be dispersed into an offshore and Eastward direction (Guillen and Hoekstra, in press); the latter direction is corresponding with the dominant wave- and wind-driven longshore drift (Hoekstra et al., 1997).

Summarizing the results, the natural (cross-shore) patterns in grain size characteristics, the onshore coarsening of the sediment and the offshore and longshore dispersion of fine-grained deposits (100-150 micron) after the nourishment demonstrate that fine and coarse grain size fractions behave differently under the same hydrodynamic conditions. The natural distribution of median grain sizes and the related grain size fractions, as well as the rather quick "recovery" of the former pre-nourishment grain size distribution indicate that there is an optimum, depth-dependent interval for the presence of specific grain size classes (and, logically the associated D_{50}). The time-averaged and remarkably stable character of this distribution directly implies that selective sediment transport mechanisms have to be operating in the nearshore zone.

A further analysis of wave- and current-driven sediment transport processes is carried out in order to identify the (potential) processes and mechanisms that are responsible for selective sediment transport across the (upper) shoreface. In a first approach, only cross-shore sediment transport processes are taken into account, since the cross-shore grain size distribution along the barrier island coast of Terschelling shows a great longshore uniformity (Guillen and Hoekstra, in press; see also partly Fig. 1).

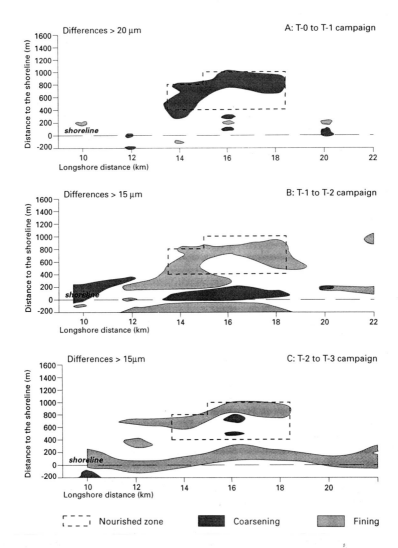

Figure 2. Coarsening and fining inside and outside the nourished area of Terschelling.

Selective Sediment Transport Mechanisms

The total cross-shore sediment transport in the nearshore zone is commonly
decomposed into bedload and suspended load components. Suspended load components
may be further subdivided in current-dominated fluxes due to mean flows (such as a
cross-shore directed undertow or a rip current) and oscillatory fluxes related to high- and
low-frequency waves. As a result one can easily distinguish a number of the most
obvious combinations of sediment transport mechanisms that can lead to selective
sediment transport and a segregation of grain size fractions for graded sediments:

1:    *Onshore bedload transport* due to wave asymmetry and streaming versus *offshore
      suspended load transport* due to mean currents;

2:    *Onshore directed suspended load* transport due to *wave asymmetry*, in combi-
      nation with *offshore directed suspended load* transport due to mean currents
      (*undertow*);

3:    Alternating *onshore* and *offshore* directed *oscillating suspended load* transport.

Mechanism 1. This is one of the mechanisms emerging from the spatially varying
pattern of grain size characteristics prior to and shortly after the execution of the
shoreface nourishment. The field observations at the NOURTEC site suggest that the
greater part of the finer sediments has predominantly been transported as suspended load
in both an offshore and longshore direction. The offshore transport is expected to be the
result of undertow, probably in combination with suspension fall-out from rip currents.
Meanwhile, according to the morphological evolution since the nourishment, a substantial amount of nourished sediment has also moved in an onshore direction (Hoekstra et al.,
1997). This material may partly represent the coarser bedload fractions of the nourished
sediment which, in combination with oscillatory suspended load, is thought to be
transported onshore in response to a landward-directed wave-asymmetry (and streaming).

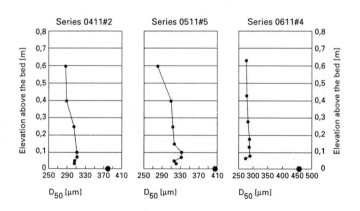

Figure 3.   Grain size variation as a function of height above the bed for bedload and
                 suspended load (time-averaged samples).

Therefore, for the proposed mechanism there basically are two requirements to generate selective sediment transport: 1) *opposing* flow/transport directions for bedload and suspended load, and 2) natural *vertical* gradient in grain size for suspended load relative to bedload. The first condition is easily satisfied for irregular and natural breaking waves in the surfzone. A mixture of breaking and non-breaking waves leads to a simultaneously existing breaker-induced undertow and the presence of highly asymmetric waves. Hydrodynamic and sediment transport measurements at the NOURTEC - site clearly confirm the co-existence of both processes (Houwman and Ruessink, 1997; Hoekstra et al., 1997; see also next section on mechanism 2). For the second condition though, no data is available from the NOURTEC site and information is based on field observations at the Egmond aan Zee experimental site, along the West coast of the Netherlands (courtesy of Wolf, 1997). A comparison of the grain size characteristics of time-averaged suspended load samples and bedload samples, taken in the inner

nearshore zone of Egmond aan Zee for conditions with breaking waves, shows a considerable range in median grain size (Fig. 3). On average, the D_{50} of the bedload is more than 80 micron coarser than the suspended load. This rather extreme difference between bedload and suspended load may be due to the fact that the samples were taken in an area relatively close to the beach. In conclusion, the processes of onshore bedload transport due to wave asymmetry (and streaming) versus offshore suspended load transport due to mean currents will certainly contribute to the effect of selective sediment transport.

Mechanism 2. In the surfzone, for conditions with breaking waves, suspended load transport is commonly dominant and bedload usually represents only a minor part of the total sediment transport (Van Rijn, 1995). However, as demonstrated by Houwman and Ruessink (1997), this bedload component may still be significant in determining the net sediment transport magnitude and direction. Their results show that the onshore directed oscillating suspended load transport is more or less equal to and balanced by the offshore directed mean suspended load transport. Therefore, the contribution of a minor transport component may still be relevant in these conditions. Part of the approach of Houwman and Ruessink (1997) is adopted in this paper to evaluate the impact of grain size characteristics on both suspended load components. A sensitivity analysis is carried out by applying the *modified* Van Rijn/Ribberink model (Van Rijn, 1993) for sediment transport computations. Since this approach does not include the oscillating suspended load transport - the suspended sediment is only transported by mean currents - the model has been modified to incorporate the oscillatory fluxes (Houwman and Ruessink, 1997). In addition, the original model required the input of a depth-averaged mean velocity and a wave height. Instead of this, a second modification was implemented and the model is now run with timeseries of measured near-bed velocities - so including measured wave-orbital velocities - and water levels. As a result, the modified sediment transport model computes the bedload, transported by both waves and currents, the oscillating suspended load transport and the mean suspended load transport.

During the NOURTEC experiment of Terschelling a number of process-oriented measuring campaigns have been carried out. Each campaign lasted for a period of about 6 weeks (Hoekstra et al., 1997). Data from these measurements are used in the present analysis. As an illustration of the procedure, data are taken from a tripod (F3), located at the seaward side of the outer bar at an average depth of about 5-6 m.

The following procedure is applied:

1) Sediment transport computations are performed with the locally measured time series of waves and currents and by using the local native, median grain size (here: 165 micron);

2) Sediment transport fluxes are related to the ratio of local wave height and local water depth (Hm0/h): the transport fluxes are represented and classified as a function of this relative wave height Hm0/h (see Fig. 4);

3) The long-term or annual *Probability density function (Pdf)* of the relative wave height (P{Hm0/h}) is determined; the long-term contribution of the sediment transport fluxes is calculated for each class of Hm0/h by multi-plying the corresponding sediment transport fluxes with the related *Pdf*;

4) The final result is a diagram representing the magnitude and direction of the various sediment transport components as a function of relative wave height and frequency of occurrence, the "yearly mean transport rates" (Houwman and Ruessink, 1997; Fig. 4, right panel; positive transport means onshore, negative is offshore).

The results depicted in Fig. 4 first of all indicate that both suspended load components indeed largely exceed the bedload component. Secondly, the mean suspended load component is offshore directed (undertow), whereas the oscillating component is entirely onshore directed (wave asymmetry; the flow measurements do not include streaming effects). Finally, the mean suspended load component is in general larger than the oscillating component. It means that for this particular example, the ratio of offshore directed and onshore directed suspended load components is larger than one. As a next step in the analysis, the sediment transport computations are repeated for different local median grain sizes. The left panel in Fig. 4 shows similar computations at location F3, but now for a median grain size of 100 micron. The result is a considerable increase in sediment transport rates for the different components as well as a changing relative contribution of offshore versus onshore directed suspended load fluxes.

This can be further illustrated by normalizing the ratio for "equilibrium" conditions with a median grain size of 165 micron. By definition, in these conditions (D_{50} = 165 micron) the ratio is made equal to unity, corresponding with no net transport (Fig. 5). Likewise, for another median grain size the ratio of offshore versus onshore directed suspended load fluxes is normalized by using the ratio in "equilibrium conditions". The results are presented in Fig. 5; position F2 is located on the shoreface, at an average depth of about 9 m. For both positions F2 and F3 the general trend is quite obvious: if sediments are smaller than the average local grain size the offshore directed transport processes are dominating. A coarsening of the sediment results in the opposite effect and onshore directed transport processes are prevailing. The main reason for this effect is the changing relative contribution of both suspended transport components with height above the bed. The lower part of the suspension vertical is dominated by the oscillating suspended transport whereas at greater heights above the bed, the mean suspended transport is the dominant component. A decrease in local grain size leads to increasing suspended sediment concentrations, in particular at greater heights above the bed. Consequently, both the oscillating and mean suspended transport will increase. In reponse to the vertical distribution though, the latter transport component will increase more rapidly than the former one. Therefore, a local coarsening is in favour of onshore transport whereas a local fining promotes the offshore transport. This is in general in line with the results of the NOURTEC grain size study.

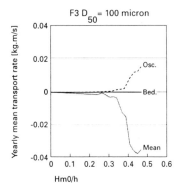
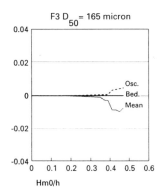

Figure 4.   The effect of median grain size (D_{50}) on the computed cross-shore suspended load components at location F3 for native deposits (D_{50} = 165 micron; right panel) and finer sediments (D_{50} = 100 micron; left panel).

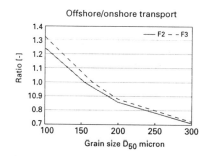

Figure 5.   Ratio of onshore and offshore directed yearly suspended load transport at locations F2 and F3 as a function of local median grain size (for computational procedures see text).

Mechanism 3. Another potential mechanism for the segregation of grain size fractions and selective sediment transport is the net result of the oscillating suspended load transport. In general the net result of the alternating onshore and offshore directed oscillating suspended load components greatly depends on the degree of wave asymmetry and the existence of time lag effects between the high-frequency velocity and concentration signals. An impression of these time lag effects can be obtained by plotting the mean (high-frequency) shape of the orbital velocities and concentrations, showing the mutual phase relationships (procedure of ensemble-averaged signals). Shortly after the nourishment, during the T2 measuring campaign (Spring 1994) and as observed for only a restricted number of measurements, there appears to be a considerable time lag between orbital velocities and concentrations. Onshore orbital velocities coincide with

low concentrations (Fig. 6) whereas the transition from onshore to offshore orbital flow
marks the development of a sharp concentration peak. Surprisingly enough, a level of
higher concentrations is maintained during the offshore directed orbital flows (Fig. 6).
As a result a net offshore directed oscillating flux will be observed. In the past, a
significant time lag effect is frequently explained by the presence of bedforms. A ripple-
induced vortex is developed during onshore flow and, together with suspended sediment,
is being lifted from the bed during flow reversal. During the offshore directed flow, the
associated sediment is partly being moved and gradually settles (e.g. Horikawa, 1988).

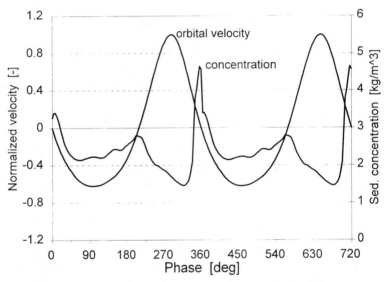

Figure 6.   Average wave shape and concentration showing time lag effects for suspended
                 sediment concentrations relative to wave orbital velocities; relatively high
                 concentrations are observed during offshore directed orbital velocity compo-
                 nents.

However, another possible explanation for the features observed in Fig. 6 could be
the relative abundance, at that stage in the post-nourishment situation, of relative fine
sediment fractions. The extreme peak in concentration during flow reversal may be
associated with a wave bottom boundary layer shear instability in relation to this flow
reversal, as described by e.g. Foster et al. (1994). The finer suspended sediments are
distributed across the vertical and start to settle again but as long as the settling time is
larger than the average half wave period (0.5T), the concentration just gradually diminishes. This may cause a dominantly seaward directed, oscillating flux of fine grained
sediments. Given the present lack of information concerning the grain size characteristics
of instantaneously suspended sediments and the (unknown) origin of concentration peaks
around flow reversal, for the time being, the latter explanation can only be considered as
a hypothesis.

Conclusions

The result of the NOURTEC field study leads to the inevitable conclusion that selective sediment transport processes are operating in the nearshore zone. Consequently, for naturally graded coastal sediments, the use of a single grain size parameter to represent the sediments - for example, as applied in sediment transport formulae - becomes less meaningful or even questionable. Without any doubt, the mechanism of onshore bedload transport due to wave asymmetry and streaming versus offshore suspended load transport due to mean currents (undertow) is an important mechanism responsible for this selective transport. Based on a sensitivity analysis with the modified Van Rijn/Ribberink model, the combination of onshore directed suspended load transport due to wave asymmetry and offshore directed suspended load transport due to mean currents may be a second important mechanism. Alternating onshore and offshore directed oscillating fluxes, in particular in relation to fine-grained sediments and the occurrence of phase lag effects between velocities and concentrations, represent a third potential mechanism although this effect is extremely difficult to prove.

References

Foster, D.L., Holman, R.A. and Beach, R.A. , 1997. Sediment suspension events and shear instabilities in the bottom boundary layer. Proc. Coastal Dynamics '94, ASCE New York, 712-726.

Guillen, J. and Hoekstra, P., 1996. The "equilibrium" distribution of grain size fractions and its implications for cross-shore sediment transport: a conceptual model. Marine Geology 135, 15-33.

Guillen, J. and Hoekstra, P., in press. Sediment distribution in the nearshore zone: grain size evolution in response to shoreface nourishment (Island of Terschelling, the Netherlands). Estuarine, Coastal and Shelf Science.

Hoekstra, P., Houwman, K.T., Kroon, A., Ruessink, B.G., Roelvink, J.A. and Spanhoff, R., 1997. Morphological development of the Terschelling shoreface nourishment in response to hydrodynamic and sediment transport processes. Proc. 25th Int. Conf. on Coastal Eng., ASCE New York, 2897-2910.

Horikawa, K., 1988. Nearshore Dynamics and Coastal Processes. Univ. of Tokyo Press, 522 pp.

Houwman, K.T. and Ruessink, B.G., 1997. Cross-shore sediment transport mechanisms in the surfzone on a timescale of months to years. Proc. 25th Int. Conf. on Coastal Eng., ASCE New York, 4793-4806.

Tanczos, I.C., 1996. Selective Transport Phenomena in Coastal Sands. Thesis University Groningen, 184 pp.

Van Rijn, L.C., 1993. Principles of sediment transport in rivers, estuaries and coastal seas. Aqua Publ., Amsterdam.

Van Rijn, L.C., 1995. Yearly-averaged sand transport at the -20 m and -8 m NAP depth contours of JARKUS profiles 14, 40, 76 and 103. Report H1887, Delft Hydraulics, 98 pp.

Wolf, F.C.J., 1997. Hydrodynamics, Sediment Transport and Daily Morphological Developments of a Bar-Beach System. Thesis Utrecht University.

CROSS-SHORE SAND TRANSPORT AND BED COMPOSITION

Leo C. van Rijn[1]

1. INTRODUCTION

The sediment bed of the coastal zone usually exhibits a large variation of sediment sizes. Local variations related to the presence of bed forms (differences in size at the top and in the trough) may occur, but cross-shore sorting due to selective transport processes is a more important process in nature (fining in seaward direction). These effects can only be represented by taking the full size composition of the bed material, which may vary across the profile, into account.

2. DESCRIPTION OF MODELS

2.1 Wave propagation and longshore currents

The propagation and transformation of individual waves (wave by wave approach) is described by a probabilistic model (Van Rijn and Wijnberg, 1996). The individual waves shoal until an empirical criterion for breaking is satisfied. Wave height decay after breaking is modelled by using an energy dissipation method. Wave-induced set-up and set-down and breaking-associated longshore currents are also modelled. The near-bed orbital velocities of the high-frequency waves (low-frequency effects are neglected) are described by second order Stokes theory and by linear wave theory in combination with an empirical correction factor. The depth-averaged return current (u_r) under the wave trough of each individual wave (summation over wave classes) is derived from linear mass transport and the water depth (h_t) under the trough. Streaming in the wave boundary layer due to viscous and turbulent diffusion of fluid momentum is taken into account. The streaming (u_b) in the wave boundary layer is of the order of 5% of the orbital peak velocity and generally onshore-directed in deeper water (symmetric waves). In shoaling waves in shallow water, the streaming in the boundary layer may be offshore-directed.

2.2. Sand transport; single fraction method (SF-method)

The sand transport rate for each wave (or wave class) is based on the computed wave

---

[1] Delft Hydraulics and Dep. of Phys. Geography, Univ. of Utrecht, P.O. Box 177, 2600MH, Delft, The Netherlands

height, depth-averaged cross-shore and longshore velocities, orbital velocities, friction factors and sediment parameters..

The net (averaged over the wave period) total sediment transport is obtained as the sum of net the bed load (q_b) and net suspended load (q_s) transport rates.

The net bed-load transport rate is obtained by time-averaging (over the wave period) of the instantaneous transport rate using a formula-type of approach.

The net suspended load transport is obtained as the sum $(q_s = q_{s,c} + q_{s,w})$ of the net current-related and the wave-related transport components (see, Van Rijn, 1993). The current-related suspended load transport $(q_{s,c})$ is defined as the transport of sediment particles by the time-averaged (mean) current velocities (longshore currents, rip currents, undertow currents). The wave-related suspended sediment transport $(q_{s,w})$ is defined as the transport of sediment particles by the oscillating fluid components (cross-shore orbital motion). The $q_{s,w}$-component includes a calibration coefficient γ in the range between 0.3 and 0.7. Computation of the wave-related and current-related suspended load transport components requires information of the time-averaged current velocity profile and sediment concentration profile.

The current velocity profile is represented as a two-layer system to account for the wave effects in the near-bed layer (Van Rijn and Kroon, 1992; Van Rijn, 1993). The convection-diffusion equation is applied to compute the equilibrium time-averaged sediment concentration profile for current-related and wave-related mixing. The effect of the local cross-shore bed slope on the transport rate is taken into account (see Van Rijn, 1993). Details are given by Van Rijn and Kroon (1992); Van Rijn (1993, 1997).

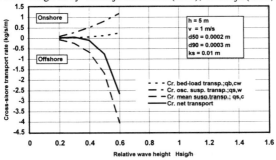

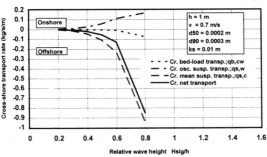

Figure 1   *Cross-shore transport components as a function of relative wave height*

To get a better understanding of the relative importance of the three cross-shore transport components $q_{b,cw}$ =mean and oscillating bed-load transport , $q_{s,w}$ = oscillating suspended transport (only high-frequency) and $q_{s,c}$ = mean suspended transport; some sensitivity computations using the single wave method and the single fraction method have been made for two water depths. The γ-factor of the $q_{s,w}$ transport component is taken as 0.3 to obtain values of the right order of magnitude compared to measured data (Kroon, 1994; Houwman and Ruessink, 1996 and Wolf, 1997). The computations have been made for uniform water depths of 1 and 5 m. The wave period is 7 s. The waves are normal to the coast. The longshore (tide- or wind-driven) depth-averaged current velocities are taken as 0.7 and 1 m/s. The bed roughness is $k_{s,c}=k_{s,w}=0.01$ m for all conditions. The water temperature is 15 °C, the salinity is 30 promille.

Figure 1 shows computed values of the three cross-shore transport components and the net transport as a function of relative wave height H_s/h for bed material of 200 μm. For a depth of 5 m the net total transport is offshore-directed for $H_s/h > 0.3$. For a depth of 1 m the net total transport is offshore-directed for all conditions. Since, the net total transport in cross-shore direction is a delicate balance of all cross-shore transport components involved, the γ-factor of the osc. suspended transport has a great influence on the net transport. Additional calibration of the γ-factor remains necessary. Furthermore, also the low-frequency transport component (neglected in this study) may have some effect on the cross-shore transport balance.

2.3   Multi-fraction method (MF-method)

Generally, the approach is to divide the bed material in a number of size fractions and to compute the sand transport rate of each size fraction using an existing single fraction method (replacing the median diameter of the bed material by the mean diameter of each fraction) with a correction factor (ξ_i) to account for the non-uniformity effects. Herein, the correction factor of Egiazaroff (1965) is used. The total sand transport rate for all size fractions can be obtained by summation of the transport rates per fraction taking the probability of occurrence of each size fraction into account. The multi-fraction method incorporates two effects:

- transport is non-linearly dependent on particle diameter, for example: suspended load transport is inversely proportional to size; $q_s \approx d^m$ with m between -0.5 and -2,
- hiding effect of smaller particles between the larger particles.

The influence of the hiding factor is a reduction of the transport rate in the case of dominant suspended load transport, because the critical bed-shear stress of the finer particles (hiding between the coarser particles) is enlarged.

2.4   Bed Level changes and sediment composition

Bed level changes per fraction i are described by:

$$\rho_s(1-e)\partial z_{b,i}/\partial t + \partial(p_i q_{t,i})/\partial x = 0 \qquad (1)$$

with: z_b = bed level to datum, $q_{t,i} = q_{b,i} + q_{s,i}$ = volumetric total load (bed load plus suspended load) transport per fraction i, p_i = value of fraction i, ρ_s = sediment density, e = porosity factor.

The total bed level change is obtained by summation of fractional bed level changes over all N-fractions: $\Delta z_{b,x,t} = \Sigma \Delta z_{b,i,x,t}$

The bed material is computed in a thin (order of 0.1 m) surface mixing layer of thickness δ applying a one-layer approach. This approach was introduced by Hirano (1971) and later extended by Ribberink (1987). The thickness of the surface layer is herein assumed to be constant in space and time and is moving in vertical direction with the bed surface in response to bed level changes (deposition upwards and erosion downwards). Thus, the surface layer is always at the top of the bed. The mixing of sediment within the surface layer is assumed to be effectuated within each time step (instantaneous mixing) through small-scale bed form migration processes in the lower regime or by wave-induced vortices in the sheet flow regime.

At present stage of research the bed material composition of the subsoil below the surface layer is assumed to be uniform (no layered structure) and equal to the initially specified fraction values ($p_{0,i,x}$).

The bed material composition is computed according to the following procedure:

- the sediment mass M of the surface layer at t = 0 is subdivided in masses $M_{x,i}$ based on the initial fraction values $p_{o,x,i}$, as follows: $M_{x,i} = p_{o,x,i} M$,

- the sediment mass $M_{x,i}$ of fraction i changes due to sediment deposition or erosion at the surface of the bed; in case of deposition the mixing layer will move upward at a rate equal to the deposition rate, while an equal amount of sediment with the composition of the mixing layer will be lost at the bottom (exchange at base) of the mixing layer; in case of erosion the opposite process will take place and the mixing layer will move downward eroding itself into the subsoil, hence sediment with the composition of the subsoil will be absorbed by the mixing layer;

- the new composition of the bed material is given by: $p_{x,i,t+\Delta t} = M_{x,i,t+\Delta t}/M$

3.   SENSITIVITY COMPUTATIONS SAND TRANSPORT

3.1   Effect of single and multi-wave approach on sediment transport in uniform conditions (constant depth)

To analyze the effect of single and multi-wave modelling on the current-related (longshore) sand transport rates, sensitivity computations have been made for a specific case with a constant water depth h = 3m. The bed material is represented as single fraction material with $d_{50} = 0.0002$ m and $d_{90} = 0.0003$ m and the bed roughness is taken to be $k_{s,c} = k_{s,w} = 0.01$ m. The irregular waves (normal to the coast) are assumed to have a Rayleigh-distribution with $H_{rms} = 0.71$ m, $H_{1/3} = 1$ m and $T_p = 7$ s. The longshore current velocities have been varied in the range between 0.2 and 2 m/s. The water temperature is taken as 15 °C and the salinity as 30 promille.

The current-related sand transport rates have been computed in three ways:

- based on a single representative wave height equal to H_{rms},
- based on a single representative wave height equal to $H_{1/3}$,
- based on multi-wave approach, schematizing the Rayleigh-distribution to 8 wave height classes between 0.2 and 1.6 m and wave periods between 6.2 and 7.6 s.

The results are presented in Figure 2, showing the ratio of transport rates based on the single wave approach and the multi-wave approach. Application of $H_{1/3}$ as the representative wave height yields transport rates which are considerably larger than those of the multi-wave approach (factor 2 for low velocities). Application of the H_{rms} leads to transport rates, which are much smaller (factor 2 for low velocities). The discrepancies are relatively large for low current velocities, when the waves are the dominant factor in the transport process. For high current velocities the proper representation of the wave field becomes less important with respect to the transport process.

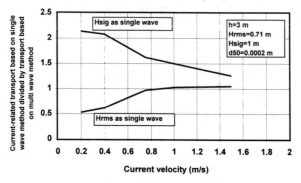

Figure 2   *Influence of single and multi-wave approach on sand transport*

3.2   Comparison of single and multi-fraction method for sediment transport in uniform conditions (constant depth)

To compare the results of both methods, computations have been made for three types of sediment material ($d_{50}=0.2$, 0.4 and 0.8 mm) with a symmetrical size distribution using seven fractions ($N=7$). The water depth was taken as $h=3$ m. The current velocity was varied in the range of 0.3 to 2 m/s. The wave heights were taken as $H_s=0$ m and $H_s=2$ m with a period of $T_p=7$ s. The angle between the wave and current directions was $90°$. The current-related and wave-related bed roughness heights were assumed to be equal and varied in the range of 0.05 to 0.01 m, depending on the hydraulic conditions (smaller values at higher bed-shear stresses). The water temperature was taken to be 15 °C and the salinity 30 promille.

Computation of the suspended load transport using the single fraction method requires information of the size (d_s) of the suspended sediment. Two series of computations have been made for $N=1$; $d_s=d_{50,bed}$ for all conditions and $d_s=0.7$ $d_{50,bed}$ at small bed-shear stresses increasing to $d_s=d_{50,bed}$ at high values of the bed-shear stress.

Figures 3 and 4 show the ratio of the transport rates according to both methods for sediment of 0.2 and 0.4 mm. The bed-load transport includes the wave and current-related transport components; the suspended transport only includes the current-related transport component (wave-related has been neglected).

Bed-load transport ratios ($q_{b,N=1}/q_{b,N=7}$) as well as suspended load transport ratios ($q_{s,N=1}/q_{s,N=7}$) are presented. $N=1$ refers to the single fraction method; $N=7$ refers to the multi-fraction method.

Generally, the deviations between the MF and SF-method are largest in the lower transport regime (bed-shear stress slightly greater than critical shear stress) and smallest in the higher transport regime when the bed-shear stress is much larger than the critical bed-shear stress. For all cases (see Figs. 3 and 4) the application of the multi-fraction method results in substantially smaller transport rates, especially at lower velocities. This is caused by the effect of the hiding factor on the reference concentration.

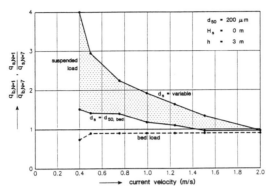

Figure 3   *Ratio of transport rates according to the SF and MF methods (0.2 mm)*

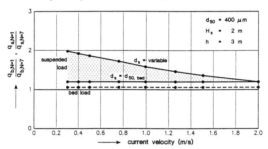

Figure 4   *Ratio of transport rates according to the SF and MF methods (0.4 mm)*

3.3   Comparison of sand transport rates across the surf zone using the single/multi-wave and the single/multi-sand fraction methods

3.3.1 Input data and boundary conditions

Computations have been made for a cross-shore profile of the barrier island coast of Terschelling, The Netherlands (see Van Rijn and Wijnberg, 1996).
Wave conditions: $H_{rms,0}$= 2.8 m T_p= 9 s
Three bed material types: d_{50}= 0.00023 m, d_{50}= 0.0004 m, d_{50}= 0.0008 m
Both the number (N) of sand fractions and the number (NW) of wave height classes have been varied. As regards the sand fractions, N=7 and N=1 have been used. As regards the wave classes, NW=13 (multi-wave approach), NW=5 and NW=1 (single wave approach) have been used. In the latter case the rms-wave height was used as the representative wave height; the wave aproach angle was constant (= 30°) for all NW values. The wave-related suspended transport was not taken into account (γ=0). The longshore velocity (x=0 m) is v_0= 0.7 m/s. Bed roughness is $k_{s,w}$= 0.01 m and $k_{s,c}$= 0.05 m, water temperature Te= 15 degrees C, salinity Sa= 30 promille.

3.3.2 Computed results; influence of number of wave classes (NW=1, 5, 13)

Case 1: storm event $H_{rms,0}$ = 2.8 m, d_{50} = 0.00023 m (N=1)
The basic features of the computational results are:

- wave height decreases by bottom friction seaward of the outer bar and by wave breaking at the outer bar to a value of about 2 m; the wave heights are about 15% lower using 13 classes (NW=13) in stead of 1 class (NW=1 based on H_{rms}); the wave heights are within 3% for NW=5 and NW=13 (not shown);
- longshore current is maximum (about 1.8 to 2.4 m/s) just landward of the outer bar; using NW=13 the maximum longshore current is about 30% (1.8 m/s in stead of 2.4 m/s) smaller than for NW=1;for NW=5 and NW=13 the longshore current velocities are within 3% in the bar zone and within 10% near the shoreline;
- offshore current (undertow) is maximum at the crest (about 0.6 m/s); using NW=13 yields smaller (30%) current velocities seaward of the outer bar; for NW=5 and NW=13 the offshore current velocities are within 3% in the bar zone and within 10% near the shoreline;
- the longshore transport rate (sum of bed load and suspended load) shows maximum values just landward of the bar crests (about 10 to 35 kg/s/m) and near the shoreline;
  using NW=1 results in an increase of the longshore transport by a factor 4 compared with the values for NW=13; thus the schematization of the wave spectrum of a storm event has a very pronounced effect on the computed transport; NW=1 (1 wave class) leads to somewhat higher wave heights (5%), but significantly larger longshore current velocities (30%) and hence significantly larger values of the longshore transport rate;
  for NW=5 and NW=13 the longshore transport rates are within 10% in the bar zone and within 30% near the shoreline;
- the cross-shore transport is onshore-directed seaward of the outer bar and in the trough between the bars due to the dominating effect of the bed-load transport, but is offshore-directed in the bar crest zones where the suspended transport is dominant; using NW=1 results in an increase of the cross-shore transport by a factor 4 compared with the values for NW=13, mainly because of an increase of the undertow velocity;
- the cross-shore transport rates in the bar crest zone generally are a factor 10 to 20! smaller than the longshore transport rates;

Computational results for d_{50} = 0.0004 m and d_{50} = 0.0008 m using NW=1 and NW=13 showed similar effects (increase of transport rates at bar crests by factor 4).

3.3.3 Computed results; influence of number of size fractions (multi fraction approach)

Case 1: storm event $H_{rms,0}$ = 2.8 m, d_{50} = 0.00023 m
The effects of the number of sediment fractions on the computed longshore and cross-shore total transport rates along the cross-shore profile are presented in Figure 5. The number of wave classes is NW=13.

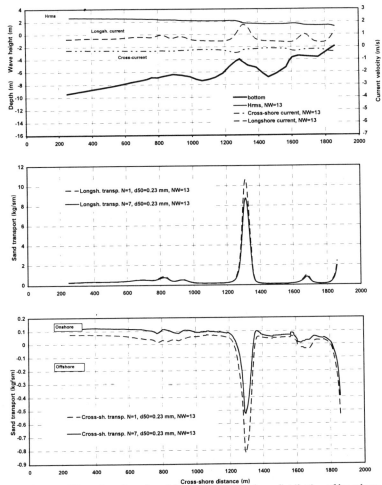

Figure 5 *Effect of number of sand fractions on cross-shore distribution of longshore and cross-shore transport for 230 μm-sediment*

The basic features of the computational results are:

- the longshore transport rate (sum of bed load and suspended load) shows maximum values just landward of the bar crests (about 8 to 10 kg/s/m) and near the shoreline (about 2 kg/s/m);
  the number of sand fractions (N=1 or 7) has little influence on the longshore transport at the crest of the outer bar (10 to 20% difference);
- the cross-shore transport is onshore-directed seaward of the outer bar and in the trough between the bars due to the dominating effect of the bed-load transport, but

is offshore-directed at the bar crest zones where the suspended transport is dominant;

- the cross-shore transport rates in the bar crest zone generally are a factor 10 to 20! smaller than the longshore transport rates;
- the onshore transport seaward of the outer bar shows a large increase (factor 2 to 4) using $N=7$ sand fractions in stead of $N=1$; this can be be explained by comparing the data at $x=804$ m, which are:

$N=7$	$q_{b,cross}=$	0.204	kg/s/m
	$q_{s,cross}=$	-0.089	kg/s/m
	$q_{t,cross}=$	0.114	kg/s/m
$N=1$	$q_{b,cross}=$	0.155	kg/s/m
	$q_{s,cross}=$	-0.124	kg/s/m
	$q_{t,cross}=$	0.031	kg/s/m;

using $N=7$ yields a larger bed-load transport (about 30%),but a smaller suspended transport (about 30%);

due to the opposite sign of the cross-shore bed load and suspended transport, small differences in each of these terms may result in a rather large difference of the total cross-shore transport (almost factor 4!);

the offshore transport at the bar crest is only weakly affected by the number of sand fractions ($N=7$ yields smaller transport rates because of the dominant influence of the suspended transport at the crest);

the sand transport gradients do not seem to be much affected by the number of size fractions.

Computational results for $d_{50}=0.0004$ m and $d_{50}=0.0008$ m showed similar effects.

4.   SENSITIVITY COMPUTATIONS BED COMPOSITION AND BED LEVEL CHANGES;   Test case symmetrical bar

To study the effect of selective transport proceses on cross-shore bed composition and bed level changes, a test case with low waves over a symmetrical bar is considered (see Figure 6). The toe of the bar is situated at -20 m and the crest of the bar at -3 m (below MSL). The bar slopes are 1 to 20. As the depth at both ends of the bar is relatively large ($h=20$ m), the sand transport rate is zero under the present conditions with low waves. Hence, no sediment can enter or leave the computational domain. Sediment of a certain size fraction can only leave or enter the mixing layer via the base of this layer. The input data are, as follows:

$H_{1/3,0}=1$ m, $h_0=20$ m, $\theta=20°$, $v_0=0$ m/s, $T_p=8$ s, $k_{s,c}=0.05$ m, $k_{s,w}=0.05$ m, $\gamma=0.7$, $d_{50}=0.2$ mm and $N=4$ ($d_i=0.1$; 0.2; 0.3; 0.5 mm), ($p_i=0.2$; 0.3; 0.3; 0.2).

Computed results after 1 month for various values of the mixing layer thickness (DL=0.05 to 0.25 m) and the hiding factor (h.f.=variable and 1) are given in Figures 6 and 7. The initial bed composition is uniform along the profile ($d_{50}=0.2$ mm). Deep water (offshore) is situated at $x=0$.

The most important results based on $N=4$ fractions are:

Effect of mixing layer thickness

- seaward of $x=375$ m the median particle size d_{50} decreases for all values of the mixing layer thickness DL, because the coarser particles are transported onshore in larger quantities than the finer particles (bed load transport is dominant);

- landward of $x=375$ m the median particle size d_{50} increases for all values of DL, because the coarser particles are deposited in this zone; furthermore coarser material is lost to the subsoil in the deposition zone;
- the effect of selective transport processes on the particle size distribution decreases with increasing value of DL;
- the bar crest migrates onshore for all values of DL (Figure 7); the migration rate of the crest is smallest for DL=0.05 m;
- the computed bed level after 1 month (Figure 7) is not much affected for DL between 0.1 and 0.25 m;
- the model yields stable results for all values of DL, provided that the time step is sufficiently small; DL=0.05 m requires a relatively small time step;

Effect of hiding factor
- the hiding factor was set to 1 for DL=0.15 m leading to a significant decrease of the fining process seaward of $x=375$ m (in the crest zone);
- a hiding factor of 1 leads to a reduction of the migration rate of the crest, which is caused by the increase (factor 2) of the offshore-directed suspended transport.

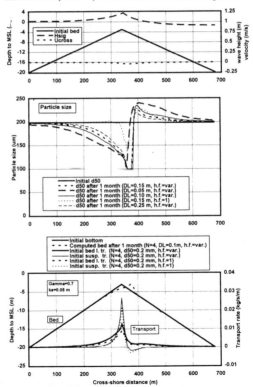

Figure 6   *Effect of mixing layer and hiding factor on computed bed composition and bed profile along symmetrical bar*

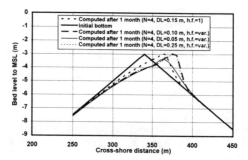

Figure 7   *Effect of mixing layer and hiding factor on computed bed profiles along symmetrical bar*

REFERENCES

Egiazaroff, P.I., 1965. *Calculation of non-uniform sediment concentrations, Journal of the Hydraulics Division, ASCE, Vol. 91, HY 4*

Hirano, M., 1971. *River bed degradation with armouring. Trans. JSCE, Vol. 3, part 2*

Houwman, K.T. and Ruessink, G., 1996. *Cross-shore sediment transport mechanisms in the surf zone. ICCE, Orlando, USA*

Kroon, A., 1994. *Sediment transport and morphodynamics of the beach and nearshore zone near Egmond, The Netherlands. Doc. Thesis, Dept. of Physical Geography, Univ. of Utrecht, The Netherlands*

Ribberink, J.S., 1987. *Mathematical modelling of one-dimensional morphological changes in rivers with non-uniform sediment. Dissertation, Civil Eng. Dep., Delft Univ. of Technology, Delft, The Netherlands*

Van Rijn, L.C., 1993. *Principles of sediment transport in rivers, estuaries and coastal seas. Aqua Publications, Amsterdam, The Netherlands*

Van Rijn, L.C., 1997. *Cross-shore modelling of graded sediments. Delft Hydraulics, Report Z2181, Delft, The Netherlands*

Van Rijn, L.C. and Kroon, A., 1992. *Sediment transport by current and waves, p.2613-2628. 23rd ICCE, Venice, Italy*

Van Rijn, L.C. and Wijnberg, K.M., 1996. *One-dimensional modelling of individual waves and wave-induced longshore curents in the surf zone. Coastal Engineering, Vol. 28, p. 121-145*

Wolf, F., 1997. *Morphology of bars in inner surf zone near Egmond, The Netherlands.Dept. of Physical Geography, Univ. of Utrecht, The Netherlands*

A NONSTATIONARY, PARAMETRIC
COASTAL WAVE MODEL

N. Booij,[1] L.H. Holthuijsen[1] and R. Padilla-Hernandez[1]

ABSTRACT

The HISWA model is a parameterized spectral wave model that is used to compute irregular, short-crested wave fields in the coastal zone. The Cycle 1 version of this model is presently used in a large number of institutes where it generally provides good results. However, there are a number of limitations that are operationally inconvenient. This Cycle 1 version is (a) stationary and (b) it employs an explicit forward marching technique in geographic space, which limits wave propagation to a directional sector of 60° to either side of the main orientation of the computational grid. A new version, HISWA Cycle 2 is described here that overcomes these limitations by using a nonstationary, four-sweep, forward marching scheme that is implicit, thus allowing waves to propagate in all directions in nonstationary conditions with relative large time steps. The computed significant wave height agrees well with observations in a complex field case with barrier islands. The wave directions that are computed with the Cycle 2 version of the model are more realistic than those computed with the Cycle 1 version.

INTRODUCTION

The classical approach to compute waves in coastal regions is to transport the wave energy along wave rays to the shore. The energy balance of these waves can be computed along these wave rays but this is numerically very inefficient because the computation of nonlinear processes requires the wave conditions at each point along the wave rays. This can be obtained only from other wave rays which are usually not systematically available in the computational area. Moreover, the numerical inclusion of nonstationary current-induced refraction is not readily achieved. The most obvious alternative to this Lagrangian approach is to formulate the evolution of the waves with a Eulerian energy balance on a regular grid. Such formulation is numerically efficient because (a) all wave information is available in each grid point, (b) the random, short-crested nature of the waves is readily

---

[1]   Delft University of Technology, Department of Civil Engineering, P.O. Box 5048, 2600 GA Delft, Netherlands.

accounted for by using a spectral representation of the energy balance equation and (c) the effects of currents are readily included. This Eulerian approach can be implemented in several ways, depending on the available computer capacity. For very limited computer capacity, a parametric version was developed by Holthuijsen et al. (1989). It employs an explicit forward marching technique in space, with a computational grid that must be oriented roughly parallel to the main wave propagation direction. This marching technique limits the sector of spectral directions to about 60° to either side of the main axis of the computational grid. The model is used in a large number of institutes where it generally provides good results. However, there are a number of situations in which it is operationally inefficient or in which the actual wave directions do not fall in the directional limitations of the model or in which the situation is not stationary. For instance behind a barrier island the change in wave direction is larger than allowed by the directional limitation and a combination of swell and local wind sea from different directions cannot be accommodated. Moreover, wind and current conditions may change too rapidly in time in view of the stationarity of the model. To remedy these limitations and to further improve the wave model, several modifications and additions are possible. Firstly, the explicit (stationary) numerical propagation scheme can be replaced by an implicit (nonstationary) scheme. Secondly, the parametric description of the waves and of the physical processes can be replaced by a discrete spectral representation. The first development is described in the present paper. The second development is described in an accompanying paper (Holthuijsen et al., 1997).

THE HISWA MODEL

The above mentioned parametric coastal wave model of Holthuijsen et al. (1989; HISWA Cycle 1) is derived from the spectral energy balance of random, short-crested waves which for Cartesian coordinates is (e.g., Whitham, 1974; Mei, 1983; Hasselmann et al., 1973):

$$\frac{\partial}{\partial t} N + \frac{\partial}{\partial x} c_x N + \frac{\partial}{\partial y} c_y N + \frac{\partial}{\partial \omega} c_\omega N + \frac{\partial}{\partial \theta} c_\theta N = \frac{S}{\sigma} \tag{1}$$

where $N = N(\omega, \theta)$ is the spectral action density (equal to the energy density divided by absolute frequency) as a function of absolute frequency and spectral wave direction. The first term in the left-hand side of this equation represents the local rate of change of action density in time, the second and third term represent propagation of action in geographical space (with propagation velocities in x-and y-space, c_x and c_y respectively). The fourth term represents shifting of the relative frequency due to variations in depths and currents (with propagation velocity in ω-space, c_ω). The fifth term represents depth-induced and current-induced refraction (with propagation velocity in θ-space, c_θ). The expressions for these propagation speeds are taken from linear wave theory (e.g., Whitham, 1974; Mei, 1983). The term S $(= S(\sigma, \theta))$ at the right hand side of the action balance equation is the source term in terms of energy density representing the effects of generation, dissipation and nonlinear wave-wave interactions.

The numerical implementation of the complete action balance equation of Eq. (1), would result in a computer code that would require considerable computer capacity. The prototype model that was developed for deep water and intermediate water depths is the WAM model (the WAMDI group, 1988, also Komen et al., 1994). To reduce these computer requirements, the basic balance equation can be parameterized. To retain the short-crested character of the waves (cross-seas due to refraction tend to be important in shallow water), the parameterization has been formulated in terms of the zero-th (m_0) and first (m_1) order moment of the spectrum in each spectral direction:

$$m_n(\theta) = \int_0^\infty \omega^n N(\omega, \theta) \, d\omega \tag{2}$$

Applying this definition operator to the spectral action balance equation results in the two evolution equations on which the HISWA model is based:
- an evolution equation for the zero-order moment of the action density spectrum for each spectral direction,

$$\frac{\partial}{\partial t}(m_0) + \frac{\partial}{\partial x}(c_x^* m_0) + \frac{\partial}{\partial y}(c_y^* m_0) + \frac{\partial}{\partial \theta}(c_\theta^* m_0) = T_0 \tag{3}$$

- an evolution equation for the first-order moment of the action density spectrum for each spectral direction,

$$\frac{\partial}{\partial t}(m_1) + \frac{\partial}{\partial x}(c_x^{**} m_1) + \frac{\partial}{\partial y}(c_y^{**} m_1) + \frac{\partial}{\partial \theta}(c_\theta^{**} m_1) = c_\omega^* m_0 + T_1 \tag{4}$$

The left-hand side of these equations represent propagation of the waves where c_x^*, c_y^* and c_θ^* are the propagation speeds of the zero-th moment in x-, y- and θ-space respectively. Similarly, c_x^{**}, c_y^{**} and c_θ^{**} are the propagation speeds of the first moment in x-, y- and θ- space respectively. Both are approximated with the corresponding speeds of the mean frequency. The propagation speeds c_x and c_y (in both equations) represent rectilinear propagation (including shoaling) whereas c_θ (in both equations) represents refraction. The effect of time variations in currents and depth on the mean frequency is represented by the term $c_\omega^* m_0$. Generation and dissipation of m_0 and m_1 are represented by the source terms T_0 and T_1 respectively. The physical phenomena which are accounted for in these equations are: (a) refractive propagation, (b) wind growth, (c) bottom dissipation, (d) surf dissipation and (e) current dissipation (blocking).

To further reduce the computer requirements, the wave conditions can be reduced to stationary conditions (removing the first and fourth term in Eqs. 3 and 4). This has been done for the original HISWA Cycle 1 model but not for the HISWA Cycle 2 model. The above Eqs. 3 and 4 basically describe the evolution of a superposition of wave groups in all spectral directions, each characterized by a total energy and a mean frequency per group. The propagation of the wave groups is computed with the linear theory of surface gravity waves. The generation and dissipation of the waves is described with simple evolution expressions of the wave energy and the mean frequency. For the energy these are: the fetch-limited expression of the Shore Protection Manual (CERC, 1973) for wind generation, a quadratic friction law (Putnam and Johnson, 1949) for bottom friction, the bore-based model of Battjes and Janssen (1978) for depth-induced wave breaking and instantaneous dissipation of energy above the critical frequency for current-induced wave blocking. The evolution of the mean frequency is coupled to these mechanisms (per direction) as follows. The wind-induced evolution of the mean frequency is based on an assumed universal relationship between total wave energy and mean frequency (deviations due to other processes are accounted for with a relaxation model). For the effect of dissipation on the mean frequency, a uni-modal k^{-3}-spectrum is assumed (k is wave number) where bottom dissipation and depth-induced breaking are concentrated in the lowest frequencies and current-induced wave blocking in the highest frequencies. For further details reference is made to the original HISWA publication (Holthuijsen et al., 1989).

COMPUTATIONAL TECHNIQUE

The computational technique of the HISWA Cycle 1 model is essentially a one-sweep, forward marching, first-order upwind scheme. It is an explicit scheme which limits wave propagation to a sector of $\arctan(\Delta x / \Delta y)$ (typically 60°) on either side of the direction of the x-axis (where Δx and Δy are the grid spacings in x- and y-direction). This limitation prohibits the calculation of cross seas (which extend over a wider range of directions). This is an operational problem in the sense that such cases cannot be computed or that such cases need to be composed of (at least) two separate computations (the results to be superimposed without potential nonlinear interactions). It is also operationally inconvenient that the computational grid has to be rotated towards the mean wave direction of the incident waves. It forces the user to change computational grid orientation for every computation with a different incident wave direction (sometimes in large sequences, e.g., when HISWA Cycle 1 is used as a subroutine in morphodynamic models). To avoid these directional limitations and to add nonstationarity to the model, the numerical scheme has been modified to an implicit, non-stationary, four-sweep version.

Since the nature of the basic equation is such that the state in a grid point is determined by the state in the up-wind grid points, the most robust scheme would be an implicit, up-wind difference scheme. Such a scheme would also be economic in the sense that implicit schemes are unconditionally stable. They therefore permit relatively large time increments in the computations (much larger than for explicit schemes in shallow water). The adjective "implicit" is used to indicate that the independent variable in the numerical scheme at integration level n is expressed in terms of that variable at that level n and, for the time derivative only, at a previous level n-1. Several years of experience in using the HISWA Cycle 1 model has shown that for coastal applications a first-order up-wind difference scheme is usually accurate enough in geographic space. However, this experience has also shown that a higher accuracy is desirable in directional space. This has been achieved (in Cycle 1 and 2) by supplementing the first-order upwind scheme in directional space with a central scheme (more economic than a higher-order up-wind scheme). For HISWA Cycle 2 therefore, fully implicit up-wind schemes in both geographic and directional space have been chosen, supplemented with a central scheme in directional space.

As for the new up-wind scheme in geographic space, the state in a grid point is determined by the state in the up-wind grid points (as defined by the direction of propagation). For economic reasons it is therefore obvious to decompose the directional space into four quadrants. In each of these quadrants the computations can be carried out independently from the other quadrants except for the boundary conditions between the quadrants. The wave components are correspondingly propagated in the HISWA Cycle 2 model with a first-order up-wind scheme in a sequence of four forward-marching sweeps (one per quadrant). To account for the boundary conditions between the quadrants, these sweeps are repeated until some convergence criterion is reached. Since all directions are thus accounted for in the model, the orientation of the computational grid is independent of the wave direction and the wave propagation is no longer limited to a finite directional sector, thus permitting cross-seas in the area. For each geographic grid point individually the geographic propagation is readily combined with the propagation in directional space (refraction). The integration in time is based on a simple backward finite difference scheme. A curvi-linear version has recently been implemented but it will not be addressed here (see the accompanying paper of Holthuijsen et al., 1997).

PROTOTYPE TEST

Both HISWA Cycle 1 and Cycle 2 are applied here to a realistic field case and the results are compared with measurements. The application area is the Friesche Zeegat, a tidal inlet with a complicated bathymetry and with current velocities up to 1.5 m/s, between the barrier islands of Ameland and Schiermonnikoog in the northern part of the Netherlands (see Fig. 1, with six buoy locations).

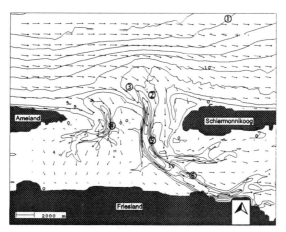

Fig. 1 The water depth and currents of the Friesche Zeegat during the computations. Buoy observations at locations # 1 to 6.

A system of long, narrow channels runs from the North Sea into an area of tidal flats with shoals and channels crossing towards the mainland. The Ministry of Transport, Public Works and Water Management in the Netherlands carried out a well documented field campaign in 1992 (e.g., Dunsbergen, 1995) to monitor the geophysical changes in this area after the closure of the Lauwers Zee in 1969 (located in the lower right-hand corner of Fig. 1).

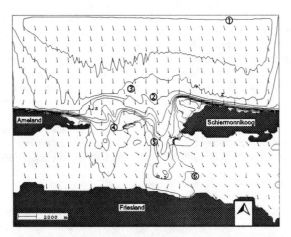

Fig. 2 The computed (HISWA Cycle 1) significant wave height and mean wave direction of the Friesche Zeegat field case.

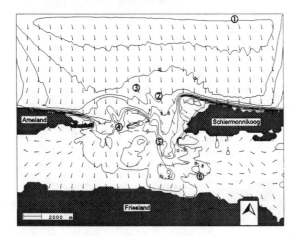

Fig. 3 The computed (HISWA Cycle 2) significant wave height and mean wave direction of the Friesche Zeegat field case.

The wind speed (11.5 m/s) and direction (320°) in the computations are assumed to be uniform over the area. The incident significant wave height, mean wave period (based on the first-order moment of the spectrum) and mean wave direction are 2.24 m, 5.6 s and 328° respectively. The current velocities and water depth that are used in the computations have been obtained with a hydrodynamic circulation model and are shown in Fig. 1.

The pattern of the significant wave height and of the mean wave direction as computed with both Cycles of the HISWA model are shown in Figs. 2 and 3. The main differences between the two patterns are in the mean wave directions and in the corresponding penetration pattern of the waves. The mean wave directions in the Cycle 1 results are clearly restricted to a limited sector (more nearly aligned with the incident wave direction) whereas the directions in the Cycle 2 results permit the waves to turn 180^0 behind the islands due to refraction. Overall, the computed patterns are consistent with the pattern of the observations: the wave height gradually decreases between the deep water boundary and the entrance of the tidal inlet. Results of repeated computations with and without wind show that the wind generates more wave energy in the deeper eastern entrance than in the shallower western entrance of the tidal inlet (thus erroneously suggesting that the waves penetrate deeper into the eastern channel than into the western channel). After the waves travel through the tidal gap between the two barrier islands they refract laterally to the shallower parts of the inlet. They completely reverse direction behind the two islands (in the Cycle 2 results). The larger directional freedom of the waves in the Cycle 2 computations also results in a more diffuse, but probably also more realistic penetration of the waves towards the mainland.

The computed significant wave height and mean period at the six observation stations are given in Fig. 4 and 5. The results of the two Cycles are almost identical, indicating that the waves at the buoy locations are not very sensitive to the change in wave directions. The agreement with the observed significant wave heights is very reasonable, but behind the islands the mean wave period is underestimated by approximately 1 s.

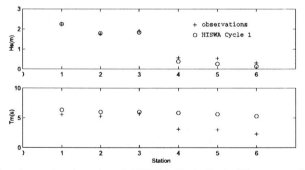

Fig. 4 The observed and computed (HISWA Cycle 1) significant wave height and mean wave periods at the six buoy stations in the Friesche Zeegat.

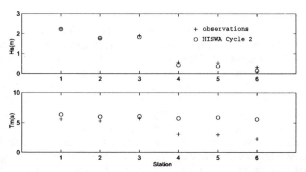

Fig. 5 The observed and computed (HISWA Cycle 2) significant wave height and
      mean wave periods at the six buoy stations in the Friesche Zeegat.

CONCLUSIONS

The one-sweep (120°) propagation scheme of the parametric spectral model
HISWA has been replaced by a four-sweep (4x90°=360°) scheme. The new scheme
is a nonstationary, implicit scheme which is unconditionally stable and therefore
allows time steps that are one or two orders of magnitude larger than in the more
conventional explicit schemes. More realistic wave directions are thus obtained in
complex conditions and the operational convenience of the model has improved. The
computed significant wave heights agree well with observations in a complex field
case with barrier islands but the mean wave period is poorly predicted behind the
islands. This is ascribed to the rather primitive formulation of the physical processes
in the model.

ACKNOWLEDGEMENTS

We were very fortunate to have access to the original data of the test case. We
thank John de Ronde, Joska Andorka Gal, Daan Dunsbergen, Miriam van Endt and
their co-workers at the Ministry of Transport, Public Works and Water Management
(the Netherlands) for providing us with the data. Our special thanks go to Bas Les
who carried out the hydrodynamic calculations to determine the currents and water
levels for the Friesche Zeegat case in the framework of his M.Sc. thesis at the Delft
University of Technology.

REFERENCES

Battjes, J.A. and J.P.F.M. Janssen, 1978: Energy loss and set-up due to breaking
      of random waves, *Proc. 16th Int. Conf. Coastal Engineering*, ASCE, 569-587
CERC, 1973: *Shore Protection Manual*, U.S. Army Corps of Engineers, Techn.
      Rep. No. 4, Vol. I
Dunsbergen, D.W., 1995: Verification set Friesche Zeegat -October 1, 1992-
      November 17, 1992-, Rep. RIKZ-95.035, Ministry of Transport, Public Works
      and Water Management, Den Haag, The Netherlands
Holthuijsen, L.H., Booij, N. and T.H.C. Herbers, 1989: A prediction model for
      stationary, short-crested waves in shallow water with ambient currents, *Coastal*

Engineering, **13**, 23-54

Holthuijsen, L.H., N.Booij and R. Padilla Hernandez, 1997: Wave propagation on a curvi-linear grid in a third-generation shallow water wave model, Proc. Coastal Dynamics '97, in press

Hasselmann, K., T.P. Barnett, E. Bouws, H. Carlson, D.E. Cartwright, K. Enke, J.A. Ewing, H. Gienapp, D.E. Hasselmann, P. Kruseman, A. Meerburg, P. Müller, D.J. Olbers, K. Richter, W. Sell and H. Walden, 1973: Measurements of wind-wave growth and swell decay during the Joint North Sea Wave Project (JONSWAP), *Dtsch. Hydrogr. Z. Suppl.*, **12**, A8

Komen, G.J., Cavaleri, L., Donelan, M., Hasselmann, K., Hasselmann, S. and P.A.E.M. Janssen, 1994: *Dynamics and Modelling of Ocean Waves*, Cambridge University Press, 532 p.

Mei, C.C., 1983: *The applied dynamics of ocean surface waves*, Wiley, New York, 740 p.

Putnam, J.A. and J.W. Johnson, 1949: The dissipation of wave energy by bottom friction, *Trans. Am. Geoph. Union*, **30**, 67-74

WAMDI group, 1988: The WAM model - a third generation ocean wave prediction model, *J. Phys. Oceanogr.*, **18**, 1775-1810

Whitham, G.B., 1974: *Linear and nonlinear waves*, Wiley, New York, 636 p.

A SIMPLE PC BASED SIGNIFICANT-WAVE MODEL

Tony Butt[1], Paul E. Russell[1] and David A. Huntley[1]

Abstract

Modifications are made to the wave prediction method of the U.S. Army Corps of Engineers *Shore Protection Manual* (1984), to allow for spatially and temporally heterogeneous windfields, and to predict a time-series of outputs of end-of-fetch significant wave height and period. The model will run on a P.C. and is designed to provide a quicker and more versatile method of 'significant wave' forecasting than the manual methods. Inputs are derived from surface atmospheric pressure charts. A hindcast is performed in the North-Atlantic to test end-of-fetch values against those measured by a wave-buoy. Predicted values of H_S are within 0.5m of those recorded, and maximum values occur within 6hrs. Maximum values of T_S are to within 1.5s, and also occur within 6hrs. Possible error sources include interpolation of the inputs and inherent overprediction of the *Shore Protection Manual* method.

Introduction:

Traditional 'manual' methods of wave forecasting are still widely used by engineers, to obtain estimates of significant height and period at single target locations, without having to gain access to large models which require significant amounts of computing power (e.g. WAMDI Group, 1988, Komen *et al*, 1994). In this study, two 'manual' methods are combined, and a simple model is developed which (a) gives a *time series* of outputs, rather than just a single value of significant height and period; and (b) allows for more realistic windfields to be considered, i.e. those which are not necessarily spatially and temporally homogeneous. The program is simple and may be run on a P.C., whilst also offering greater versatility and speed than would be possible applying the methods manually. To test the performance of the model, a hindcast is performed against data obtained from a wave buoy in the North Atlantic.

----

[1] Institute of Marine Studies, University of Plymouth, Drake Circus, Plymouth, UK

Shore Protection Manual model

In the 1940s wave prediction techniques were first developed to aid war-time landing operations. Empirical formulae were developed relating windspeed, fetch and duration, to wave height (Sverdrup and Munk, 1947). Improvements were made by Bretschneider (1952, 1958) and hence the well known, SMB method emerged.

The U.S. Dept. of the Army *Shore Protection Manual* (1984) (*SPM*) method is based upon a simplification of the parametric equations used by Hassleman *et al* (1976), to predict significant height and peak period, according to the JONSWAP spectrum. A nomogram similar to the 'SMB chart' (Bretschneider, 1958) is used, and the user reads off significant height and peak period, according to wind speed, fetch and duration.

Wilson's model

Wilson (1961, 1965) developed a model based on the empirical relations in the SMB model, but applicable to spatially and temporarily varying wind fields. Wilson differentiated the equations: $H_S = f(U, x)$ and $T_S = f(U, x)$ to obtain:

$$\partial H_S / \partial x = f'(U, x) \tag{1}$$

$$\text{and} \quad \partial C_g / \partial x = f'(U, x) \tag{2}$$

where H_S is significant wave height, C_g is the group speed, x is the fetch, and U is surface wind speed.

He introduced the concept of a *significant wave propagation path*, at Speed C_g across a space-time grid. Each point on the grid has a different wind speed U(x, t). Equations 1 and 2 are numerically integrated as the 'significant wave' propagates from a starting point to its target destination. This method has been used, with reasonable success, to hindcast hurricane waves (Wilson, 1965).

The model described in this paper

The equations of the *SPM* method give improved results for a 'significant wave' prediction technique over the traditional SMB method. However, the major disadvantage of this method is that it assumes spatially and temporally homogeneous windfields. The method of Wilson (1961) provided one way of allowing for varying windfields, but the windsea was always assumed to be fetch-limited. This works well when used to predict waves generated by very short, fast moving fetches, such as hurricanes, but is not easily applicable to mid-latitude depressions.

The method used here is one combining the techniques of the *SPM* with the ideas of Wilson. A set of differential equations are derived from the fetch and duration limited equations in the *SPM,* and these are then numerically integrated over time and space to allow different wind inputs at each grid point, during the integration process, thus computing significant height and period at the end of the fetch. Values for a fully arisen sea (FAS) are also calculated, thus allowing for the three possibilities of either fetch limited, duration limited or, fully arisen sea.

The *SPM* deep water wave generation formulae

The formulae given are based on the JONSWAP results (Hassleman *et al,* 1973), and the parametric model proposed by Hassleman *et al* (1976).   They predict H_S (significant height) and T_P (peak spectral period), for (i) fetch limited, (ii) duration limited, and (iii) fully arisen sea conditions.   Five basic formulae are given in dimensionless form, and the following are obtained from conversion into SI units:

Fetch-limited:

$$H_S = 5.112 \times 10^{-4} U_A x^{1/2} \tag{3}$$

$$T_P = 6.238 \times 10^{-2} (U_A x)^{1/3} \tag{4}$$

Fully arisen sea:

$$H_S = 2 \cdot 48 \times 10^{-2} \ U_A^2 \tag{5}$$

$$T_P = 0 \cdot 83 \ U_A \tag{6}$$

Duration limited:

$$H_S = 3.786 \times 10^{-5} (U_A^5 \ t^3)^{1/4} \tag{7}$$

$$T_P = 0.011 (U_A t)^{1/2} \tag{8}$$

where x is the fetch length, t is the duration, and U_A is the adjusted wind speed.

A simple procedure for deciding whether the sea is fetch-limited (FL), duration limited (DL) or FAS, would be to obtain H_S and T_P for all three conditions, and output the smallest values, as these would reflect the point at which the sea reached its limit.

Application of the equations to heterogeneous windfields

If the equations were applied directly to a uniform windfield, then the values of H_S and T_P for all three situations could be computed, and the smallest could be taken. This procedure could be performed, keeping x and U_A constant, but incrementing t, which would give 'running' values of H_S and T_P as the windsea progressed.

Now, H_S & T_P must be calculated in a similar way to above, but with U_A varying as a function of x and t.   A good starting point is to take U_A as varying with x only, and devise a method of putting different values of U_A into the fetch-limited equations, as x is incremented from zero to the fetch-length.   This is done by differentiating the equations to obtain the rate of change of H_S & T_P with x; and then integrating them numerically from zero to the fetch-length, whilst inserting the appropriate value of U_A at each step.

The method must also be applied to DL and FAS situations.   To compute DL values, the RMS value of U_A (U_{RMS}) over the fetch is taken for each time step.   The duration limited equations are then differentiated with respect to time.   This allows them to be integrated numerically over the duration, with different values of U_{RMS} to be inserted at each step.   This integration process is performed from zero to the duration up to that time-step. This gives 'running' values of H_S & T_P, as the windsea continued to exist.

To compute H_S and T_P for a FAS, the RMS value (over the duration) of U_{RMS} is taken, and inserted into the FAS equations. Note that the FAS equations yield H_S and T_P as a function of U_A only. The scheme is illustrated in Figure 1.

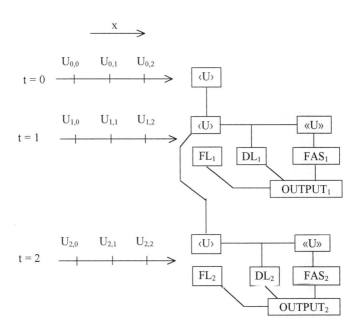

Fig 1: The computation scheme: Calculated at each time are values of the rms windspeed over the fetch (⟨U⟩), *the duration-rms of the fetch-rms windspeed* (⟪U⟫), *and the fetch-limited, duration-limited and FAS values of H_S and T_P . The smallest of the three is taken as the output at that time step.*

Differentiation

Differentiation with respect to x of the fetch limited equations (3 and 4), yields:

$$\partial H_S / \partial x = 2.556 \times 10^{-4} \, U_A \, x^{-1/2} \tag{9}$$

$$\text{and} \quad \partial T_P / \partial x = 2.079 \times 10^{-2} \, U_A^{1/3} \, x^{-2/3} \tag{10}$$

Differentiation with respect to t of the duration limited equations (7 and 8) yields:

$$\partial H_S / \partial t = 2.84 \times 10^{-5} U_A^{5/4} t^{-1/4} \tag{11}$$

and $\quad\quad \partial T_P / \partial t = 5.5 \times 10^{-3} U_A^{1/2} t^{-1/2} \tag{12}$

Numerical Integration

The integration scheme is of the simplest type, using a truncated form of the Taylor series. Applying this method to equation 9, and taking the step-number as n, we obtain, for fetch-limited height:

$$H_S (n + 1) = H_S (n) + 2.556 \times 10^{-4} \Delta x\, U_A(n + \tfrac{1}{2})\, [\, x\, (n + \tfrac{1}{2})\,]^{1/2} \tag{13}$$

This equation is computed say p times, incrementing n each time. The number of steps overall is p, which is the fetch-length divided by Δx.

For fetch-limited period:

$$T_P (n + 1) = T_P (n) + 2.079 \times 10^{-2} \Delta x\, [U_A(n + \tfrac{1}{2})\,]^{1/3}\, [\, x\, (n + \tfrac{1}{2})\,]^{-2/3} \tag{14}$$

For duration limited height:

$$H_S (m + 1) = H_S (m) + 2.84 \times 10^{-5} \Delta t\, [U_A(m + \tfrac{1}{2})\,]^{5/4}\, [t\, (m + \tfrac{1}{2})\,]^{-1/4} \tag{15}$$

and for the duration-limited period:

$$T_P (m + 1) = T_P (m) + 5.5 \times 10^{-3} \Delta t\, [U_A(m + \tfrac{1}{2})\,]^{1/2}\, [\, t\, (m + \tfrac{1}{2})\,]^{-1/2} \tag{16}$$

The duration limited equations are computed q times, incrementing m each time. In this case q is the duration up to that particular point, divided by Δt.

Inputs to the model

The fetch-line

Although, in reality, waves are produced over a two-dimensional spatial fetch field, here only a one-dimensional fetch line is considered. The fetch line is drawn as a great-circle route through the strongest part of the wind field, towards the target. For short distances, depending on the projection of the chart, this may be a straight line.

Grid inputs

The fetch line is split up into a number of points which act as grid points in the space-time domain. The model requires certain parameters to be input at each of these points. All of these are input at 100NM intervals along the fetch line, and for 12 hour intervals in time; which may easily be changed according to the scale of the application. The parameters are: (a): Isobar spacing: This is measured by hand from the chart, at each increment along the fetch line; (b): Latitude: To obtain windspeed values, the program requires the latitude of each grid point to be input; a single, average, latitude is unsatisfactory, especially for fetches which have a very 'north-south' orientation; (c): Angle between the fetch-line and the wind direction: This is required to obtain the component of the windspeed in the same direction as the windsea. A 15° friction modification is also applied (Barry & Chorley, 1992).

<u>Conversion of inputs to values directly applicable to the wave generation equations</u>

The SWAMP study (1985) concluded that many models overlooked the importance of accurate windfield specification, and large errors could occur if the windspeed inputs to the model were incorrect in the first place. Here, although some approximations must be made, care is taken so that the wind speed parameter is as accurate as possible. From the inputs, the model initially calculates the geostrophic windspeed at each grid point, then applies various algorithms to arrive at the surface windspeed values, for input to the equations. The following steps must be applied: (a): Obtain windspeed component in the direction of the fetch line; (b): Convert from geostrophic wind to 10m wind (*SPM, 1984*); (c): Correct for air-sea temperature difference (Resio and Vincent, 1977); (d): Compensate for the non-linear relationship between surface stress and windspeed (Bishop, 1983).

<u>Outputs</u>

The model produces a series of values of significant height and period, at 2hr. intervals, at the downwind end of the fetch-line. These values could then be propagated away as oceanic swell or input to a spectral parameterisation model.

<u>Hindcast</u>

A hindcast was performed to test the model's output against real data obtained from a waverider buoy in deep water in the North Atlantic, with inputs to the model taken from surface pressure charts at 12 hour intervals.

<u>Data collection</u>

It was found that waverider buoy measurements could be obtained on an hourly basis through the Internet, together with sea surface temperatures (Brugge, 1995). Data was collected throughout November and December 1995, and a suitable windsea growth and decay pattern was chosen. The buoy is located at 53.6°N, 15.5°W, about 550km west of Ireland. The windsea chosen was between 1200 GMT, 24 November to 0900 GMT, 26 November 1995, during which time the value of H_S was recorded to reach a maximum of 9.5m.

The value of H_S and T_S recorded by the buoy were given to the nearest 0.5m and 1s respectively, therefore errors were plotted as ± 0.25m and ± 0.5s respectively. To allow direct comparison with the model output, the conversion factor of $T_S = 0.815$ T_P (Sylvester, 1974) is used.

Running the model

Surface pressure charts were obtained from the British Meteorological Office, covering 0000 GMT, 24 November 1995, to 1200 GMT, 26 November 1995 at 12hr intervals. Inspection of the charts revealed that the windsea was being generated by a low pressure centred approximately over Ireland. The depression remained fairly stationary over the period, but the windfield on its NW flank could be seen to vary in intensity. A fetch-line was drawn on the charts ending at the buoy itself.

Results

Comparisons of recorded and predicted values of end-of-fetch H_S and T_S are shown in Figures 2 and 3 respectively. The time is normalised to $t0 = 0000$ GMT, 24 November 1995. Since there was already an appreciable amount of windsea, the initial 22hrs of the model run, representing the duration-limited 'build-up time', may be ignored.

Ignoring the first 22hrs, both the recorded and predicted values of significant height show a definite growth and decay pattern. The maximum values of H_S are 10m predicted, and 9.25m to 9.75m recorded. Maximum values occur within about 6hrs of each other. The decay of the windsea shows good agreement between recorded and predicted characteristics, although the recorded values occur about 1 to 4 hrs later than predicted. The growth part of the curve also occurs later than predicted by about 5 to 7 hrs.

For significant period, the late arrival of the peak compared with that predicted is consistent with the H_S values (above), since the equations on which the model is based predict the highest waves to have the lowest frequency in the generating area. The predicted and recorded maxima are 13.5s and 11.5s to 12.5s respectively.

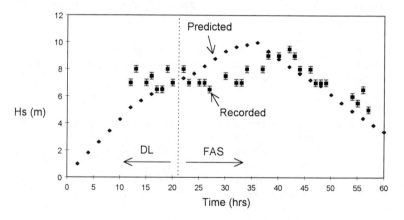

Fig. 2: recorded v. predicted H_S.

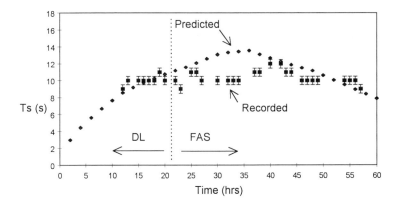

Fig. 3:   recorded v. predicted T_S

Discussion

To obtain a direct comparison between predicted and recorded values during the growth and decay of the windsea, the results are re-plotted with the recorded values shifted back 6hrs (figures 4 and 5).   Here it can be seen that H_S shows good agreement, although slightly over predicting throughout.   The values of T_S show poorer agreement, with considerable over-prediction during the growth and decay.   Note that the best agreement is at t = 24, 36 and 48hrs, when the inputs from the surface pressure charts were obtained.   This suggests there may be some error associated with interpolation of the windfield inputs.   After conversion to surface windspeed values, the model performs a linear interpolation to provide inputs to the equations at 2hr intervals.

The overall slight overprediction of both H_S and T_S may be attributed to an inherent overprediction in the *SPM* method. A comparison of the various 'manual' methods is given by Bishop (1983), and one conclusion is that the *SPM* overpredicts values of H_S and T_S. This is based on the conversion of the 10m windspeed to the *adjusted surface windspeed* using the equation: $U_S = 0.71(U_T)^{1.23}$.   Bishop sees no justification for using this relation, although it was used in this model.

The model predicts the peak of the windsea to arrive at the end of the fetch, about 6hrs too early. One possible reason for this is that the state of the windsea at the end of the fetch is assumed to respond instantaneously to any changes throughout the fetch area. Clearly, it would take a certain time for any larger waves generated at the 'back' of the fetch to propagate to the 'front', and the method used here to accommodate for heterogeneities in the windfield does not consider this.

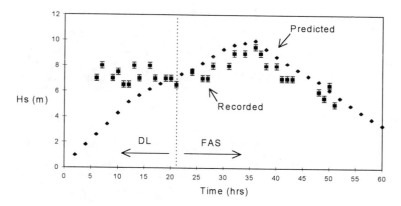

Fig. 4: recorded v. predicted H_S with recorded values shifted back 6hrs.

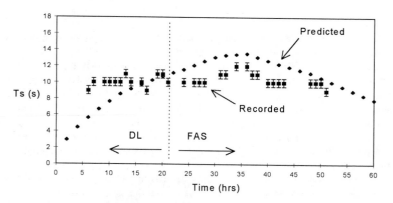

Fig. 5: recorded v. predicted T_S with recorded values shifted back 6hrs.

Conclusions:

- A wave prediction model has been developed which represents an advance over traditional significant wave models principally by (a): allowing for heterogeneous windfields in the fetch area; and (b): automatically producing a time-series of values (without having to be run several times).

- Values of significant wave height and period, including the growth and decay characteristics, at the end of the fetch, of an accuracy which is sufficient for many purposes, may be obtained quickly and easily from a program which will run on a PC; avoiding the laborious calculations involved in traditional significant-wave models.

- A hindcast has been performed, in the North Atlantic, to test the end-of-fetch outputs of significant height and period, against data provided by a waverider buoy. Values at the peak of the growth and decay of the windsea are within 0.5m and 1.5s for H_S and T_S respectively.
- Possible error sources are identified as (a) interpolation of inputs, (b) inaccuracies of the original model of the *SPM*, and (c) the inevitable non-inclusion of physical processes in such an empirical model.

References

Bishop, C. 1983. Comparison of manual wave prediction models. *J. Waterway, Port, Coastal Ocean Engineering.* **109**: 1-17.

Bretschneider, C. L. 1952. The generation and decay of wind waves in deep water. *Trans. Am. Geophys.Union,* **33** (3): 381-89.

Bretschneider, C. L. 1958. Revisions in wave forecasting: deep and shallow water. *Proc. 6th. Coast. Eng.* pp. 30-67.

Brugge, R. 1995. Meteorology and the Internet - current surface weather information. *Weather* **50** (11): 387-388

Hassleman, K., D. Ross, P. Muller, and W. Sell. 1976. A parametric wave prediction model. *J. Physical Oceanography.* **6**: 200-228

Hassleman, K., T. Barnet, E. Bouws, H. Carlson, D. Cartwright, K. Enke, J. Ewing, H. Gienapp, D. Hasslemann, P. Kruseman, A. Meerburg, P. Muller, D. Olbers, K. Richter, W. Sell, H. Walden. 1973. Measurement of wind-wave growth and swell decay during the Joint North Sea Wave Project (JONSWAP). *Deutsche Hydrogaphische Zeitschrift Suppliment A8.*

Komen, G. K., L. Cavaleri, M. Donelan, K. Hassleman, S. Hassleman, P. Janssen. 1994. *Dynamics and Modelling of Ocean Waves.* Cambridge University Press.

Resio, D. and C. Vincent. 1977. Estimation of winds over the Great Lakes. *J. Waterways and Harbors Division,* ASCE. **103** (WW2): 265-281.

Sylvester, R. 1974. *Coastal Engineering, vol 1.* Elsevier Scientific Publishing.

Sverdrup, H. and W. Munk. 1947. *Wind, sea and swell theory of relationships in forecasting.* U.S. Dept. of the Navy Hydrog. Office Publ. No. 601.

Swamp group. (J. Allender *et al*). 1985. *Ocean Wave Modelling.* Plenum Press.

U.S. Army Corps of Engineers. 1984. *Shore Protection Manual, 4th. ed.* Coastal Engineering Research Centre, Waterways Experimental Station, Corps of Engineers.

Wamdi Group (K. Hassleman *et al.*). 1988. The WAM model - a third generation ocean wave prediction model. *J. Physical Oceanography.* **18**: 1775-1810.

Wilson, B. 1961. Deep water wave generations by moving wind systems. *J. Waterways and Harbors Division,* ASCE. **87** (WW2): 113-141.

Wilson, B. 1965. Numerical prediction of ocean waves in the North Atlantic for December 1959. *Deutsche Hydrogaphische Zeitschrift.* **18** (3): 114-130.

Finite Volume Solutions to Unsteady Free Surface Flow with application to gravity waves

Stefan Mayer [1] Antoine Garapon [1] and Lars Sørensen [1]

Abstract

A fractional step method is developed for solving the time dependent two dimensional Euler equations with full non-linear free surface boundary conditions. The geometry of the free surface is described by a height function and the fluid domain is discretized by a time-varying curvilinear finite volume grid which is adapted to all boundaries, including the free surface. The method is applied to non-linear standing waves, testing for compliance with mass and energy conservation. Further, nonlinear travelling waves are simulated on constant depth and on varying topographies in order to demonstrate the methods capability to describe nonlinear wave-wave interaction. Finally, wave-current interaction is treated, including simulation of wave blocking. Generally, in all test cases excellent agreement is seen with theoretical and experimental results. The method is intended to be a first step towards a full description of wave dynamics interaction with structures and currents.

Introduction

Many attempts have been made in the past to solve the unsteady incompressible Euler or Navier Stokes equations with a full nonlinear free surface, without the assumption of irrotational flow and without the use of the Bernoulli type dynamic free surface boundary condition. Unfortunately, it seems that these methods usually have suffered from relatively large numerical errors and especially numerical damping, which did not allow accurate long term simulations of travelling gravity waves, see e.g. Tsai and Yue (1996) [7]. Here, an attempt is made to extend the application range of free surface Euler equations, so that unsteady wave problems can be simulated with accuracies, which are at least comparable to those of potential flow methods.

---

[1]International Research Centre for Computational Hydrodynamics (ICCH), Agern Allé 5, DK-2970 Hørsholm, Denmark, e-mail: icch@dhi.dk

Method

The two-dimensional Euler equations,

$$\nabla \cdot \boldsymbol{u} = 0, \tag{1}$$

$$\frac{\partial \boldsymbol{u}}{\partial t} + \nabla(\boldsymbol{uu}) = -\nabla p, \tag{2}$$

are solved for the Cartesian velocity $\boldsymbol{u} = (u_x, u_y)$ and dynamic pressure p. The usual inviscid dynamic boundary conditions are imposed,

$$p = g\eta \,, \quad \nabla \boldsymbol{u} \cdot \boldsymbol{n} = 0, \tag{3}$$

η denoting the free surface elevation and $\boldsymbol{n} = (n_x, n_y)$ being the surface normal vector. The kinematic boundary condition is expressed in terms of the local volume flux $f = \boldsymbol{u} \cdot \boldsymbol{n}$ through the free surface

$$\frac{d\eta}{dt} = \frac{df}{dn_y}. \tag{4}$$

A finite volume method employing a cell-centred variable layout is set up to express the discretized mass and momentum balance equations on a time-varying, curvilinear grid in a conservative manner [4]. The elevation η is discretized at every grid cell face along the free surface, and (4) is then integrated in time by an explicit Adam Moulton third order multi-step method. An adaptive curvilinear grid is generated algebraically and the time integration of the fluid motion is performed by a second order fractional step method. The velocity field $\boldsymbol{u}^{n+1}$ at time t^{n+1} is split into a predictor velocity field $\boldsymbol{u}^*$ and an irrotational correction $\nabla\phi$,

$$\boldsymbol{u}^{n+1} = \boldsymbol{u}^* + \nabla\phi. \tag{5}$$

$\boldsymbol{u}^*$ is updated by time-integrating to second order the momentum equations (2)

$$
\begin{aligned}
\frac{\boldsymbol{u}^* - \boldsymbol{u}^n}{\Delta t} \quad + \quad & \frac{1}{2}(\nabla(\boldsymbol{u}_g\boldsymbol{u}^*) + \nabla(\boldsymbol{u}_g\boldsymbol{u}^n)) \\
= \quad & \frac{3}{2}\nabla p^n - \frac{1}{2}\nabla p^{n-1} \\
- \quad & \frac{3}{2}(\nabla(\boldsymbol{uu}))^n - \frac{1}{2}(\nabla(\boldsymbol{uu}))^{n-1},
\end{aligned} \tag{6}
$$

employing QUICK interpolation for the convective terms. The grid velocity $\boldsymbol{u}_g$ reflects the motion of grid lines from time step n to $n+1$. The correction $\nabla\phi$ is determined by restricting $\boldsymbol{u}^{n+1}$ to satisfy the continuity equation (1), hence

$$\nabla^2\phi = -\nabla \boldsymbol{u}^*. \tag{7}$$

The pressure is computed by the divergence of the momentum equation (2)

$$\nabla^2 p^{n+1} = -\nabla \cdot (\boldsymbol{u} \cdot \nabla\boldsymbol{u})^{n+1}, \tag{8}$$

together with boundary condition (3). The Poisson equations (7) and (8) are solved by a standard multi-grid method.

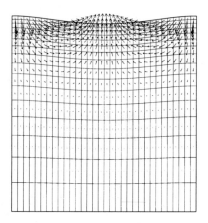

Figure 1: Standing wave of steepness $ka = 0.1403$. Fluid domain discretized by 32×16 cells, with equidistant horizontal resolution and stretched vertical discretization. Velocity vectors are shown at cell centres.

Wave generation and absorption

Travelling waves are generated by imposing at one side of the fluid domain second order Stokes velocity profiles of the form

$$u_x = (U_1(y)\sin(\varphi) + U_2(y)\sin(2\varphi)), \tag{9}$$

with $\varphi = \omega t + \theta$, ω and θ denoting the wave's angular frequency and phase angle, respectively. At the opposite end both the elevation η and the fluid velocity $\boldsymbol{u}$ are relaxed at every time step towards prescribed values, η_p and $\boldsymbol{u}_p$, respectively, by the following procedure

$$\eta = (1 - \alpha)\eta + \alpha\eta_p \,, \quad \boldsymbol{u} = (1 - \alpha)\boldsymbol{u} + \alpha\boldsymbol{u}_p. \tag{10}$$

α denotes a relaxation parameter, which increases softly as the waves are entering into the numerical sponge layer.

Results

Standing wave

The method is tested for a nonlinear deep water standing wave with steepness $ka = 0.1403$ and with a wave period which has been very accurately computed by

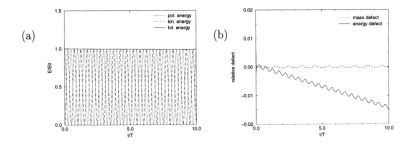

Figure 2: Standing wave of steepness $ka = 0.1403$ discretized in space by 64×32 cells and in time with time step $\Delta t/T_l = 0.009$. (a) Energy history, (b) Relative volume defect $(V_{tot}(t) - V_{tot}(0))/(LH)$ and relative energy defect $(E_{tot}(t) - E_{tot}(0))/E_0$ as function of time.

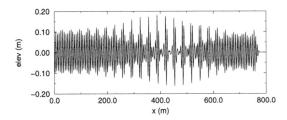

Figure 3: Propagation of modulated deep water wave (carrier wave: $kh = 2\pi$ and $ka = 0.10$, sideband perturbations: $\beta = 0.05$ and $\delta = 0.10$). Fully developed elevation profile along channel. Discretization by $L/\Delta x \approx T/\Delta t \approx 50$.

Agnon and Glozman (1996) [1]. At $t = 0$, in a square container with side length L (being the wavelength), the free surface elevation is prescribed including the first eight harmonic components according to [1], and fluid velocity is initially set to zero. Grids are generated with equidistant horizontal discretization and with vertical stretching towards the free surface boundary, see Fig. 1.

Due to the fully conservative discretization, errors in mass conservations are negligible in all cases. For different space and time discretizations, the wave period and the energy loss is computed. For numerical resolutions in time and space of $L/\Delta x >= 50$ and $T/\Delta t >= 100$, respectively, the dispersion relation is fulfilled to an accuracy better than 0.5 % and the wave is damped about 0.1 % during every period due to numerical dissipation, see Fig. 2.

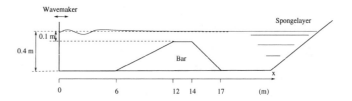

Figure 4: Wave tank with submerged bar on horizontal bottom.

Benjamin-Feir instability

The method is applied to the simulation of weakly non-linear deep water waves with superposed sideband perturbations, where the modulations gain energy by quartet wave-wave interactions and the wave train evolves into massive wave pulses [2]. In case no wave breaking occurs, those pulses then again recur back to an almost initial state. Numerical simulations of the Benjamin - Feir instability are quite demanding as the computational domain has to cover in the order of $O(100)$ wave length, hence requiring high computational efficiency with present computer technology at hand.

   In the present work, waves are generated at one end of a channel, having length and deepness of $l_{channel} = 130L$ and $h = L$, respectively, L being the wavelength. The carrier wave of steepness $ka = 0.1$ is generated by imposing second order Stokes velocity profile at one end of the channel. The side band waves with amplitude βa, $\beta = 0.05$ and frequency shift $\delta = \pm ka = \pm 0.1$ are generated by superposing first order Stokes profiles.

   Fig. 3 shows the fully developed elevation profile, as it appears after 350 wave periods. The growth of the sideband amplitudes is compared with the theoretical estimates of Crawford et al. (1981), showing good agreement. However, the computed recurrence length of about 800 m is larger than the value of 680 m according to the theoretical results of Stiassnie and Kroszynski (1982) [6]. Furthermore, the waves approaching the right end of the channel are damped to about 80 % of their initial height. Despite these considerable discrepancies, we consider the results to be excellent, since to our knowledge no comparable result has been achieved before with a similar free surface Euler code.

Shoaling wave over submerged bar

The method is used to study the propagation of regular incident waves with period $T = 2.02$ s and height $H = 0.02$ m over a submerged bar on a horizontal bottom, see Fig. 4, this test being investigated experimentally by Luth et al. (1994) [3]. On the upward slope the incoming waves are shoaling, nonlinearity hereby generating bound higher harmonics, which travel phase-locked to the primary wave. On the downward slope these harmonics are released as free waves resulting

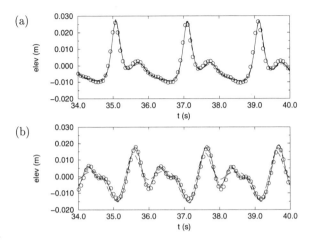

Figure 5: Propagation of regular waves with period $T = 2.02$ s and initial height $H = 0.02$ m over a submerged bar, see Fig. 4. Time series of surface elevations at (a) a location on top of the bar, $x = 13.5$ m, and (b) behind the bar, $x = 21$ m. (○) meas. by Luth et al. (1994) [3], (———) comp., $\Delta x = 0.015$ m, $\Delta t = 0.01$ s, (- - - - -) comp., $\Delta x = 0.03$ m, $\Delta t = 0.02$ s, (− − −) comp., $\Delta x = 0.06$ m, $\Delta t = 0.04$ s.

in an irregular wave pattern.

In the simulation, waves are generated for 80 s in fluid initially being at rest. The computed elevation history and its Fourier transform is compared to measurements at selected locations, see Figs. 5 and 6. On the upward slope excellent agreement with measurements is found, the nonlinear shoaling process being well described even with rather coarse discretization. On the lee side of the bar fine spatial discretization is required to resolve the released higher harmonic waves, which are otherwise damped by numerical dissipation. However, given sufficient discretization, the model describes phase-accurately the resulting irregular wave train behind the bar.

Wave-current interaction

In a channel with a submerged bar a opposing steady current is introduced by modifying the sponge layer and the wave-generating boundary condition, see Fig. 7. The local Froude number on top of the bar takes values of about $Fr \approx 0.5$. Waves of height $H = 0.005$ m and with periods $T = 2$ s and $T = 1$ s, respectively, are generated during 100 s and propagate against the mean flow direction. In the case, $T = 2$ s waves travel in close agreement with linear theory with regard to amplitude and wave number, the wave almost recovering to its initial shape after the bar, see Fig. 8.

In the case of $T = 1$ s, however, waves are blocked on the upward slope,

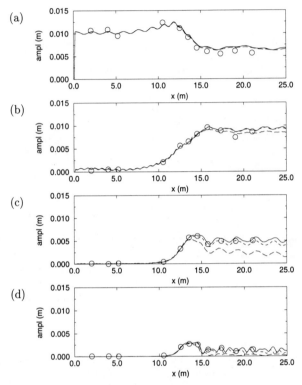

Figure 6: Propagation of regular waves with period $T = 2.02$ s and initial height $H = 0.02$ m over a submerged bar, see Fig. 4. (a-d) Amplitudes of 1st to 4th harmonic, respectively, as function of x-location along the channel. (o) measurements by Luth et al. (1994) [3], (——) comp., $\Delta x = 0.015$ m, $\Delta t = 0.01$ s, (- - - - -) comp. $\Delta x = 0.03$ m, $\Delta t = 0.02$ s, (− − −) comp. $\Delta x = 0.06$ m, $\Delta t = 0.04$ s.

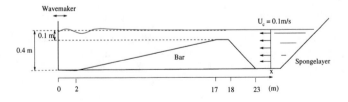

Figure 7: Wave tank with submerged bar on horizontal bottom for studying wave current interaction.

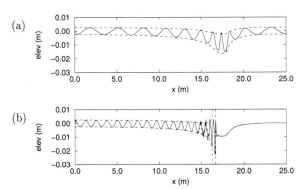

Figure 8: Propagation of waves over a submerged bar with opposing current, see Fig. 7. Elevation profiles at $t = 95T$. Initial wave height, $H = 0.005$ m, and current $U_o = 0.1$ m/s at 0.4 m depth. (a) Wave period $T = 2$ s. (b) Wave period $T = 1$ s. (- - - - -) envelope according to linear wave theory.

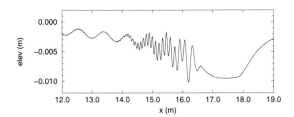

Figure 9: Elevation profile of wave with period $T = 1$ s at $t = 95T$ and initial height $H = 0.001$ m propagating against a steady current, see Fig. 7

since the current velocity exceeds the group velocity of the wave. The blocking
point agrees precisely to the location estimated by linear wave theory. In order to
resolve more closely the blocking process, discretization is refined to $\Delta t = 0.005$ s
and $\Delta x \approx 0.01$ m around the blocking point and waves with initial heights of
$H = 0.001$ m are generated, see Fig. 9. The elevation profile clearly shows the
incoming wave to be superposed by short reflected waves, the wave length of
which is shortened with increasing depth and damped by numerical dissipation.
The estimated wave numbers of the reflected wave are found to agree quite closely
to the theoretical estimate using linear wave action theory, e.g. [5].

Conclusion

A finite volume code employing a modified fractional step method has been applied to unsteady, incompressible free-surface flow. In contrast to similar attempts
in the literature, the accuracy of the present method is sufficiently good to allow
phase-accurate simulation of fully nonlinear wave trains within spatial dimensions of $O(50)$ wave lengths. This accuracy may not be as good as with free
surface potential flow methods. However, in the present formulation both steady
and unsteady rotational currents can be introduced directly. Further, it is easily
extended to viscous flow, as well as the incorporation of turbulence models is
straight forward.

In conclusion, the method is intended to be a supplement for investigation of
interaction of wave motion with other processes as wave-current interaction, interaction with laminar and turbulent bottom boundary layers. Further, we believe
that the present method is a good basis on which models describing the effect of
wave breaking on wave driven currents, turbulence, etc., can be developed.

Acknowledgement

This work was financed by the Danish National Research Foundation. Their
support is greatly appreciated.

References

[1] Y. Agnon and M. Glozman. Periodic solutions for a complex Hamiltonian
    system: New standing water waves. *Wave motion*, 769:1–12, 1996.

[2] T. B. Benjamin and J. E. Feir. The desintegration of wave trains on deep
    water, part 1. *J. Fluid Mech.*, 27:417–430, 1967.

[3] H. R. Luth, G. Klopman, and N. Kitou. Projects 13G: Kinematics of waves
    breaking partially on an offshore bar; LDV measurements for waves with and
    without a net onshore current. Technical Report H1573, Delft Hydraulics,
    1994.

[4] S. Mayer, A. Garapon, and L. Sørensen. A fractional step method for unsteady free surface flow with applications to non-linear wave dynamics. *submitted to Int. J. Num. Meth. Fluids*, 1997.

[5] D. H. Peregrine. Interaction of water waves and current. *Adv. in Appl. Mech.*, 16:10–117, 1976.

[6] M. Stiassnie and U. I. Kroszynski. Long-term evolution of unstable water wave train. *J. Fluid. Mech.*, 116:207–225, 1982.

[7] W. Tsai and D. K. P. Yue. Computations of nonlinear free-surface flows. *Ann. Rev. Fluid. Mech.*, 28:249–78, 1996.

A CURVILINEAR, THIRD-GENERATION COASTAL WAVE MODEL

L.H. Holthuijsen[1], N. Booij[1] and R. Padilla-Hernandez[1]

ABSTRACT

The SWAN model is a discrete spectral wave model of the third-generation to compute irregular, short-crested wave fields in the coastal zone. It is seen as a shallow water version of the WAM model (e.g. the WAMDI group, 1988, for deep water and intermediate-depth waters). It correspondingly represents all processes of generation, dissipation and nonlinear wave-wave interactions explicitly, without a priori assumptions about the shape of the spectrum. The SWAN model employs numerical propagation schemes that are implicit (both in geographic space and in spectral space). In Cycle 1 of the model the computational grid in geographic space is rectilinear, as usual for such models. In Cycle 2 of the model, this grid is replaced by a curvilinear grid. Test computations on a curvilinear grid in a complex region off the Dutch coast with barrier islands and tidal flats show good agreement with computations on a rectilinear grid and with buoy observations.

INTRODUCTION

Over the last decade, the traditional wave ray models in coastal engineering to compute waves in near-shore areas are being replaced by models that formulate the wave evolution in terms of a spectral energy balance on a regular grid. The prototypical implementation of this is the WAM model (the WAMDI group, 1988). A parametrical implementation for shallow water is the HISWA model which has been developed by Holthuijsen et al. (1989). A numerically improved version of the HISWA model is described in an accompanying paper (Booij et al., 1997). A further upgrading of this model, resulting in the SWAN model, is described here. The first upgrade makes the model a fully discrete spectral model of the third-generation. This implies that all processes of wave generation and dissipation and the wave-wave interactions are modelled explicitly, allowing the wave spectrum to develop without a priori constraints such as an assumed universal shape. The second upgrade formulates the model on a curvilinear grid rather than the conventional rectilinear grid. A rectilinear grid is often inconvenient in regions with highly variable scales

---

[1]   Delft University of Technology, Department of Civil Engineering, P.O. Box 5048, 2600 GA Delft, Netherlands.

such as tidal inlets, tidal flats and estuaries. A variable resolution grid (as in a curvilinear grid) would avoid the conventional nesting (with its extra computations), in particular if the grid would conform to the topography of the region.

THE SWAN WAVE MODEL

The SWAN wave model (Simulating Waves Nearshore) is a fully discrete spectral model based on the spectral action balance equation which for Cartesian coordinates is (e.g., Mei, 1983; Hasselmann et al., 1973):

$$\frac{\partial}{\partial t} N + \frac{\partial}{\partial x} c_x N + \frac{\partial}{\partial y} c_y N + \frac{\partial}{\partial \sigma} c_\sigma N + \frac{\partial}{\partial \theta} c_\theta N = \frac{S}{\sigma} \tag{1}$$

in which $N(\sigma, \theta, x, y, t)$ is the action density as a function of intrinsic frequency σ, direction θ, horizontal coordinates x and y and time t. The first term in the left-hand side of this equation represents the local rate of change of action density in time, the second and third term represent propagation of action in geographical space (with propagation velocities in x- and y-space, c_x and c_y respectively). The fourth term represents shifting of the intrinsic frequency due to variations in depths and currents (with propagation velocity in σ-space, c_σ). The fifth term represents depth-induced and current-induced refraction (with propagation velocity in θ-space, c_θ). The expressions for these propagation speeds are taken from linear wave theory (e.g., Mei, 1983). The term S ($= S(\sigma, \theta)$) at the right hand side of the action balance equation is the source term in terms of energy density representing the effects of generation, dissipation and nonlinear wave-wave interactions.

Wind input

Transfer of wind energy to the waves is described in the SWAN model with the resonance mechanism of Phillips (1957) and the feed-back mechanism of Miles (1957). The corresponding source term for these mechanisms is commonly described by the sum of linear and exponential growth:

$$S_{in}(\sigma, \theta) = A + B E(\sigma, \theta) \tag{2}$$

in which A and B depend on wave frequency and direction and wind speed and wind direction. The effects of currents are accounted for by using the apparent local wind speed and direction. The expression for the term A is due to Cavaleri and Malanotte-Rizzoli (1981) supplemented with a filter to avoid growth at frequencies lower than the Pierson-Moskowitz frequency (Tolman, 1992). In the present study this linear growth is ignored (the computations converge much faster with a reasonable first guess of the wave field that is used instead). Two optional expressions for the coefficient B are used in the model. The first (which is used in the present study) is due to Snyder et al. (1981), rescaled in terms of friction velocity U_* by Komen et al. (1984). The second is due to Janssen (1991).

Dissipation

Whitecapping is primarily controlled by the steepness of the waves (e.g., the WAMDI group, 1988):

$$S_{ds,w}(\sigma, \theta) = -\Gamma \, \bar{\sigma} \, \frac{k}{\bar{k}} \, E(\sigma, \theta) \tag{9}$$

where Γ is a steepness dependent coefficient, and $\bar{\sigma}$ and $\bar{k}$ denote the mean frequency and the mean wave number. The value of Γ depends on the wind input formulation that is used. Since two expressions are used for the wind input in the SWAN model, also two values for Γ are used. The first is due to Komen et al. (1984). The second expression is an adaptation of this expression based on Janssen (1991). They are used when the wind input according to Komen et al. (1984) or Janssen (1991) respectively are used.

Bottom friction can generally be represented as:

$$S_{ds,b}(\sigma, \theta) = -C_{bottom} \, \frac{\sigma^2}{g^2 \sinh^2(kd)} \, E(\sigma, \theta) \tag{10}$$

in which C_{bottom} is a bottom friction coefficient. The empirical JONSWAP model of Hasselmann et al. (1973), the drag law model of Collins (1972) and the eddy-viscosity model of Madsen et al. (1988) are available in the SWAN model. The JONSWAP approach (with the coefficient of Bouws and Komen, 1983) has been used in the present study.

Depth-induced wave breaking can be modelled with the random-bore model of Battjes and Janssen (1978). Laboratory observations (e.g., Battjes and Beji, 1992; Vincent et al. 1994) show that the shape of initially uni-modal spectra propagating across simple (barred) beach profiles, is fairly insensitive to depth-induced breaking. This has led to a simple spectral version of the energy dissipation:

$$S_{ds,br}(\sigma, \theta) = -\frac{D_{tot}}{E_{tot}} E(\sigma, \theta) \tag{11}$$

in which D_{tot} is the rate of total energy dissipation due to wave breaking according to Battjes and Janssen (1978). The maximum wave height-to-depth ratio which is a parameter in this model of Battjes and Janssen is $\gamma = 0.73$ in the present study.

Nonlinear wave-wave interactions

In deep water, quadruplet wave-wave interactions dominate the evolution of the spectrum. They transfer wave energy from the peak frequency to lower frequencies (thus moving the peak frequency to lower values) and to higher frequencies (where the energy is dissipated by whitecapping). In very shallow water, triad wave-wave interactions take over the role of the quadruplet wave-wave interactions. They transfer energy from lower frequencies to higher frequencies, often resulting in a secondary high-frequency peak in the spectrum (low-frequency energy generation by triad wave-wave interactions is not considered here). In the SWAN model the computations of the quadruplet interactions are carried out with the Discrete Interaction Approximation (DIA) of Hasselmann et al. (1985). The computations of the triad interactions are carried out with the Lumped Triad Approximation (LTA) of Eldeberky (1996).

Numerical schemes

The numerical schemes that are used in the SWAN model (Cycle 1, e.g. Ris et al., 1994; Holthuijsen et al., 1996) are the same implicit schemes as used in the HISWA Cycle 2 model. They have been described in an accompanying paper (Booij et al., 1997). The only differences with these schemes are that the SWAN model is fully spectral and that energy (or action) is transported along the frequencies due to time-variations in depth and currents (see Eq.1). This transport is modelled in the SWAN model with the same scheme as for transport along the spectral directions (in other words: the Doppler-type frequency shifting is based on the same scheme as the depth- and current induced refraction). The numerical schemes are further upgraded in SWAN Cycle 2 in the sense that the propagation scheme for geographic space is formulated on a curvilinear geographic grid (irregular, quadrangular, not necessarily orthogonal) rather than the rectilinear grid of SWAN Cycle 1. This modification is based on approximating the geographic distribution of the energy (action) density between each three neighbouring grid points with a flat triangle. The gradient in each grid point at location (x_i, y_j) is then readily approximated from the up-wind gridpoints. For the x-direction this is for grid point i, j (the grid points are ordered in x, y-space):

$$\frac{\partial C_x N}{\partial x} \simeq \left[\frac{[c_x N]_{i,j} - [c_x N]_{i-1,j}}{\Delta \tilde{x}_1} \right] + \left[\frac{[c_x N]_{i,j} - [c_x N]_{i,j-1}}{\Delta \tilde{x}_2} \right] \quad (12)$$

where $\Delta \tilde{x}_1 = \Delta x_1 - (\Delta y_1 / \Delta y_2) \Delta x_2$, $\Delta \tilde{x}_2 = \Delta x_2 - (\Delta y_2 / \Delta y_1) \Delta x_1$. The increments are $\Delta x_1 = x_{i,j} - x_{i-1,j}$, $\Delta x_2 = x_{i,j} - x_{i,j-1}$, $\Delta y_1 = y_{i,j} - y_{i-1,j}$ and $\Delta y_2 = y_{i,j} - y_{i,j-1}$. The gradient in y-direction is similarly estimated.

PROTOTYPE TEST

The SWAN model is applied here to a tidal inlet with a complicated bathymetry and current pattern. It is the Friesche Zeegat (the same situation as used in the accompanying paper of Booij et al., 1997) between the barrier islands of Ameland and Schiermonnikoog in the northern part of the Netherlands (see Fig. 1, with six buoy locations).

The rectilinear grid that is used for the computations has a resolution of 260 m x 230 m and it is oriented along the sides of the map shown in Fig. 1. It is visible in Fig. 1 as the initiation point of each current vector (diluted with a factor 5 in each direction). The curvilinear grid for the computations has been chosen mainly for the purpose of demonstration. It has not been generated by some objective criterion by some automated process. In fact, a skeleton grid was drawn by hand roughly along the coast line and along the main channels (it need not be orthogonal nor boundary fitting; in fact, the grid crosses the shore line and contains land points). The resolution is higher near the expected steepest gradients in significant wave height (edges of channels and shoals and shore line) but it is rather coarse. A finer grid has then been generated by linearly interpolating the coordinates of the skeleton grid (resolution 10 x finer in each direction, Fig. 2).

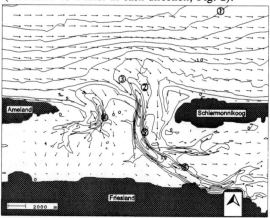

Fig. 1 The water depth and currents of the Friesche Zeegat during the computations. Buoy observations at locations # 1 to 6.

The wind speed (11.5 m/s) and direction (320°) in the computations are assumed to be uniform over the area. The incident significant wave height, mean wave period (based on the first-order moment of the spectrum) and mean wave direction are 2.24 m, 5.6 s and 328° respectively. The current velocities and water levels that are used in the computations have been obtained with a hydrodynamic circulation model and are shown in Fig. 1.

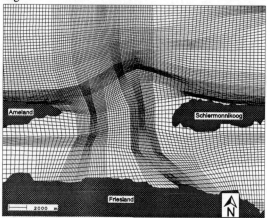

Fig. 2 The curvilinear computational grid for the Friesche Zeegat.

The pattern of the significant wave height and of the mean wave direction as computed with both the rectilinear grid and the curvilinear grid are shown in Figs. 3 and 4. The results are very similar; only near the coastline does the computation on the curvilinear grid show better representation of the coast than in the computation on the rectilinear grid (in particular near the two headlands of the islands).

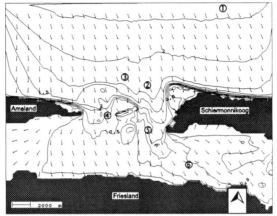

Fig. 3 The computed significant wave height and mean wave direction of the Friesche Zeegat field case (rectilinear computational grid).

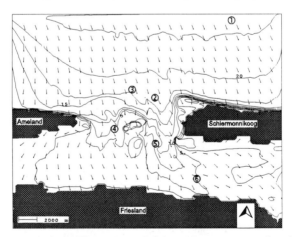

Fig. 4 The computed significant wave height and mean wave direction of the Friesche Zeegat field case (curvilinear computational grid).

Overall, the computed patterns are consistent with the pattern of the observations: the wave height gradually decreases between the deep water boundary and the entrance of the tidal inlet. Results of repeated computations with and without wind show that the wind generates more wave energy in the deeper eastern entrance than in the shallower western entrance of the tidal inlet (thus erroneously suggesting that the waves penetrate deeper into the eastern channel than into the western channel). After the waves travel through the tidal gap between the two barrier islands they refract laterally to the shallower parts of the inlet. They completely reverse direction behind the two islands. The computed significant wave height and mean period at the six observation stations are given in Figs. 5 and 6. The agreement with the observations is very reasonable, indicating a fundamental improvement over the parametric HISWA model (for a comparison with this model, reference is made to the accompanying paper of Booij et al., 1997).

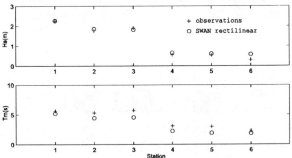

Fig. 5 The observed and computed significant wave height and mean wave period at the six buoy stations in the Friesche Zeegat (rectilinear computational grid).

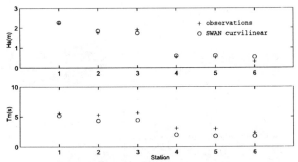

Fig. 6 The observed and computed significant wave height and mean wave period at the six buoy stations in the Friesche Zeegat (curvilinear computational grid).

CONCLUSIONS

A numerical technique has been developed that allows waves to propagate across a curvilinear grid (quadrangular, non-orthogonal) with a highly variable spatial resolution, thus avoiding the need of nested computations. It has been implemented in the third-generation wave model SWAN for shallow water with ambient currents. The agreement between computations with the new technique and observations in a rather complex field situation is very reasonable.

ACKNOWLEDGEMENTS

We were very fortunate to have access to the original data of the test case. We thank John de Ronde, Joska Andorka Gal, Daan Dunsbergen, Miriam van Endt and their co-workers at the Ministry of Transport, Public Works and Water Management (the Netherlands) for providing us with the data. Our special thanks go to Bas Les who carried out the hydrodynamic calculations to determine the currents and water levels for the Friesche Zeegat case in the framework of his M.Sc. thesis at the Delft University of Technology.

REFERENCES

Battjes, J.A. and J.P.F.M. Janssen, 1978: Energy loss and set-up due to breaking of random waves, *Proc. 16th Int. Conf. Coastal Engineering*, ASCE, 569-587

Battjes, J.A. and S. Beji, 1992: Breaking waves propagating over a shoal, *Proc. 23rd Int. Conf. Coastal Engineering*, ASCE, 42-50

Booij, N, L.H. Holthuijsen and R. Padilla Hernandez, 1997: A nonstationary, parametric coastal wave model, Proc. Coastal Dynamics '97, in press

Bouws, E. and G.J. Komen, 1983: On the balance between growth and dissipation in an extreme, depth-limited wind-sea in the southern North Sea, *J. Phys. Oceanogr.*, **13**, 1653-1658

Cavaleri, L. and P. Malanotte-Rizzoli, 1981: Wind wave prediction in shallow water: Theory and applications. *J. Geophys. Res.*, **86**, No. C11, 10,961-10,97

Collins, J.I., 1972: Prediction of shallow water spectra, *J. Geophys. Res.*, **77**, No. 15, 2693-2707

Eldeberky, Y., 1996: Nonlinear transformation of wave spectra in the nearshore zone, Ph.D. thesis, Delft University of Technology, Department of Civil Engineering, The Netherlands

Hasselmann, K., T.P. Barnett, E. Bouws, H. Carlson, D.E. Cartwright, K. Enke, J.A. Ewing, H. Gienapp, D.E. Hasselmann, P. Kruseman, A. Meerburg, P. Müller, D.J. Olbers, K. Richter, W. Sell and H. Walden, 1973: Measurements of wind-wave growth and swell decay during the Joint North Sea Wave Project (JONSWAP), *Dtsch. Hydrogr. Z. Suppl.*, **12**, A8

Hasselmann, S., K. Hasselmann, J.H. Allender and T.P. Barnett, 1985: Computations and parameterizations of the linear energy transfer in a gravity wave spectrum. Part II: Parameterizations of the nonlinear transfer for application in wave models, *J. Phys. Oceanogr.*, **15**, 11, 1378-1391

Holthuijsen, L.H., Booij, N. and T.H.C. Herbers, 1989: A prediction model for stationary, short-crested waves in shallow water with ambient currents, *Coastal*

Engineering, **13**, 23-54

Holthuijsen, L.H., R.C. Ris and N. Booij, 1996: A third-generation model for near-shore waves with ambient currents, 25th Int. Conf. Coastal Engng., in press

Janssen, P.A.E.M., 1989: Wave induced stress and the drag of air flow over sea waves, *J. Phys. Oceanogr.*, **19**, 745-754

Janssen, P.A.E.M., 1991: Quasi-linear theory of wind-wave generation applied to wave forecasting, *J. Phys. Oceanogr.*, **21**, 1631-1642

Komen, G.J., S. Hasselmann, and K. Hasselmann, 1984: On the existence of a fully developed wind-sea spectrum, *J. Phys. Oceanogr.*, **14**, 1271-1285

Komen, G.J., Cavaleri, L., Donelan, M., Hasselmann, K., Hasselmann, S. and P.A.E.M. Janssen, 1994: *Dynamics and Modelling of Ocean Waves*, Cambridge University Press, 532 p.

Madsen, O.S., Y.-K. Poon and H.C. Graber, 1988: Spectral wave attenuation by bottom friction: Theory, *Proc. 21th Int. Conf. Coastal Engineering*, ASCE, 492-504

Mei, C.C., 1983: *The applied dynamics of ocean surface waves*, Wiley, New York, 740 p.

Miles, J.W., 1957: On the generation of surface waves by shear flows, *J. Fluid Mech.*, **3**, 185-204

Phillips, O.M., 1957: On the generation of waves by turbulent wind, *J. Fluid Mech.*, **2**, 417-445

Ris, R.C., L.H. Holthuijsen and N. Booij, 1994: A spectral model for waves in the near shore zone, Proc. 24th Int. Conf. Coastal Engng, Kobe, Oct. 1994, Japan, pp. 68-78

Snyder, R.L., Dobson, F.W., Elliott, J.A. and R.B. Long, 1981: Array measurement of atmospheric pressure fluctuations above surface gravity waves, *J. Fluid Mech.*, **102**, 1-59

Tolman, H.J., 1992: Effects of numerics on the physics in a third-generation wind-wave model, *J. Phys. Oceanogr.*, **22**, 10, 1095-1111

Vincent, C.L., J.M. Smith and J. Davis, 1994: Parameterization of wave breaking in models, *Proc. of Int. Symp.: Waves - Physical and Numerical Modelling*, Univ. of British Columbia, Vancouver, Canada, M. Isaacson and M. Quick (Eds.), Vol. II, 753-762

WAMDI group, 1988: The WAM model - a third generation ocean wave prediction model, *J. Phys. Oceanogr.*, **18**, 1775-1810

VALIDATION OF THREE-DIMENSIONAL MULTIGRID WAVE MODEL AGAINST EXPERIMENTAL DATA

B. LI[1]   and   C.A. FLEMING[1]

Abstract

In this paper the results of a three dimensional multigrid model are compared against the experimental data. The model is valid for both linear and non-linear water waves. The Laplace equation is transformed from a irregular calculation domain to a regular one and the boundary conditions on water surface and sea bottom can be implemented precisely. The Multigrid Method is used to solve the governing equation and the requirement of the computer storage is very small. The present model is valid over the complete range of water depths. For fully nonlinear water wave problems the present model is particularly efficient.

1. Introduction

In the past decade the most widely used wave model in industry is the mild-slope equation. The mild-slope equation was derived by Berkhoff (1972) for monochromatic waves, allowing both wave refraction and diffraction have been taken into account through the single linear wave equation. Random waves can also be handled by the mild-slope equation (Li, Reeve and Fleming, 1993). The original mild-slope equation was in elliptical form and the latest numerical method for solving it is the generalized conjugate gradient technique (Li, 1994a). In order to solve the mild slope equation more efficiently the parabolic model was derived by Radder (1979) and the parabolic model without angle restrictions was derived by Li (1997). The hyperbolic version of the model was derived by Booij (1981), and the time dependent parabolic version (evolution equation) was derived by Li (1994b). Because the evolution equation model (Li, 1994b) is unconditionally stable, it has a fast convergent property and can be used for very large area, it has been extensively and successfully applied for tens of projects.

Three Types of The Mild Slope Equation are:
1). Elliptic Type: (Berkhoff, 1972)

$$\nabla \cdot (CC_g \nabla \Phi) \ + \ CC_g k^2 \Phi \ = \ 0$$

2). Hyperbolic Type: (Booij, 1981)

---

[1]SIR WILLIAM HALCROW & PARTNERS LTD, BURDEROP PARK, SWINDON, WILTSHIRE, SN4 0QD, UK

$$\frac{\partial^2 \Phi}{\partial t^2} - \nabla(CC_g \nabla \Phi) + (\omega^2 - k^2 CC_g)\Phi = 0$$

or (Copeland, 1985)

$$\frac{C_g}{C}\frac{\partial^2 \Phi}{\partial t^2} - \nabla \cdot (CC_g \nabla \Phi) = 0$$

3). Parabolic Type: (Li, 1994b)

$$2\omega i \frac{\partial \Phi}{\partial t} + \nabla \cdot (CC_g \nabla \Phi) + CC_g k^2 \Phi = 0$$

or in form of wave amplitude and wave number:

$$2\omega \frac{\partial A}{\partial t} + \frac{1}{A}\frac{\partial}{\partial x_i}(A^2 CC_g K_i) = 0$$

$$\frac{\partial K_i}{\partial t} + \frac{C_g}{k}K_j\frac{\partial K_i}{\partial x_j} - C_g\frac{\partial k}{\partial x_i} + \frac{1}{2\omega}\frac{\partial CC_g}{\partial x_i}(K_1^2 + K_2^2 - k^2) - \frac{1}{2\omega}\frac{\partial}{\partial x_i}(\frac{1}{A}\frac{\partial}{\partial x_j}(CC_g\frac{\partial A}{\partial x_j})) = 0$$

However, nonlinear wave effects are ignored in the mild slope equation. More recently the Boussinesq equation has been the focus of research focus and much effort has been made to extend the applicability of the equation to deep water, but for the time being two dimensional nonlinear equations which valid over all ranges of the water depth have not been derived.

It is proposed that the best way to transform the offshore wave data to inshore is to solve the original three dimensional Laplace equation with corresponding boundary conditions. The wave refraction, diffraction and nonlinear wave effects can then all be taken into account automatically.

Model	Mild Slope Equations	Boussinesq Equations	3-D Multigrid Model
Monochromatic waves	Yes	Yes	Yes
Irregular waves	No	yes	Yes
Weakly Nonlinear wave-wave interaction	No	Yes	Yes
Strong Nonlinear wave-wave interaction	No	No	Yes
Shallow water	Yes	Yes	Yes
Deep water	Yes	No	Yes

We can also compare the three dimensional multigrid method with boundary element method for solving the Laplace eqution for water wave propagation. For a case of wave propagation over a typical region of ten wave length by ten wave length, 100 grid points should be used in X and Y directions, respectively. 10 grid points should be used in the Z direction. The total grid points are 100,000.

	BOUNDARY ELEMENT METHOD	MULTIGRID METHOD
ORDER OF VARIABLES TO BE SOLVED	20,000	100,000
COMPUTER OPERATIONS	20,000×20,000= 4×10^8	50×100,000= 5×10^6
COEFFICIENTS STORED	20,000×20,000= 4×10^8	10×100,000= 1×10^6

It is revealed from the table that the 3D multigrid model is better than the boundary element model in respect of computer oprations and storage requirement.

2. Governing Equations

Three dimensional Laplace equation is

$$\frac{\partial^2\Phi}{\partial x^2} + \frac{\partial^2\Phi}{\partial y^2} + \frac{\partial^2\Phi}{\partial z^2} = 0$$

with the kinematic boundary condition

$$\frac{\partial\eta}{\partial t} + \frac{\partial\Phi}{\partial x}\frac{\partial\eta}{\partial x} + \frac{\partial\Phi}{\partial y}\frac{\partial\eta}{\partial y} = \frac{\partial\Phi}{\partial z} \qquad at\ z = \eta$$

and the dynamic boundary condition

$$\frac{\partial\Phi}{\partial t} + \frac{1}{2}\left[(\frac{\partial\Phi}{\partial x})^2 + (\frac{\partial\Phi}{\partial y})^2 + (\frac{\partial\Phi}{\partial z})^2\right] + g\eta = 0 \quad at\ z = \eta$$

and on the sea bottom

$$\frac{\partial\Phi}{\partial z} + \frac{\partial h}{\partial x}\frac{\partial\Phi}{\partial x} + \frac{\partial h}{\partial y}\frac{\partial\Phi}{\partial y} = 0 \qquad at\ z = -h$$

where Φ is the velocity potential, η is the water surface variable, g is the gravitational acceleration and h is the water depth. The coordinate system is located on undisturbed water surface with z direction being upward.
Using σ Transformation,

$$\sigma = \frac{z - z_b(x,y)}{\eta(t,x,y) - z_b(x,y)}$$

we can transform the Laplace equation from the physical domain (x,y,z) to calculational domain (x,y,σ):

$$\frac{\partial^2\Phi}{\partial x^2} + \frac{\partial^2\sigma}{\partial x^2}\frac{\partial\Phi}{\partial\sigma} + 2\frac{\partial\sigma}{\partial x}\frac{\partial^2\Phi}{\partial x\partial\sigma} + (\frac{\partial\sigma}{\partial x})^2\frac{\partial^2\Phi}{\partial\sigma^2} +$$

$$\frac{\partial^2 \Phi}{\partial y^2} + \frac{\partial^2 \sigma}{\partial y^2}\frac{\partial \Phi}{\partial \sigma} + 2\frac{\partial \sigma}{\partial y}\frac{\partial^2 \Phi}{\partial y \partial \sigma} + (\frac{\partial \sigma}{\partial y})^2\frac{\partial^2 \Phi}{\partial \sigma^2} +$$

$$+ (\frac{\partial \sigma}{\partial z})^2\frac{\partial^2 \Phi}{\partial \sigma^2} = 0$$

Where

$$\frac{\partial \sigma}{\partial x} = (\sigma - 1)(\frac{1}{h}\frac{\partial z_b}{\partial x}) - \frac{\sigma}{h}\frac{\partial \eta}{\partial x}$$

$$\frac{\partial \sigma}{\partial y} = (\sigma - 1)(\frac{1}{h}\frac{\partial z_b}{\partial y}) - \frac{\sigma}{h}\frac{\partial \eta}{\partial y}$$

$$\frac{\partial^2 \sigma}{\partial x^2} = (\sigma-1)\frac{\partial}{\partial x}(\frac{1}{h}\frac{\partial z_b}{\partial x}) + \frac{\partial \sigma}{\partial x}(\frac{1}{h}\frac{\partial z_b}{\partial x}) - \frac{\partial \sigma}{\partial x}(\frac{1}{h}\frac{\partial \eta}{\partial x}) - \sigma\frac{\partial}{\partial x}(\frac{1}{h}\frac{\partial \eta}{\partial x})$$

$$\frac{\partial^2 \sigma}{\partial y^2} = (\sigma-1)\frac{\partial}{\partial y}(\frac{1}{h}\frac{\partial z_b}{\partial y}) + \frac{\partial \sigma}{\partial y}(\frac{1}{h}\frac{\partial z_b}{\partial y}) - \frac{\partial \sigma}{\partial y}(\frac{1}{h}\frac{\partial \eta}{\partial y}) - \sigma\frac{\partial}{\partial y}(\frac{1}{h}\frac{\partial \eta}{\partial y})$$

$$\frac{\partial \sigma}{\partial z} = \frac{1}{h}$$

and h is the water depth

$$h(t,x,y) = \eta(t,x,y) - z_b(x,y)$$

The surface boundary conditions after coordinate transformation are:

$$\frac{\partial \eta}{\partial t} + (\frac{\partial \Phi}{\partial x} + \frac{\partial \Phi}{\partial \sigma}\frac{\partial \sigma}{\partial x})\frac{\partial \eta}{\partial x} + (\frac{\partial \Phi}{\partial y} + \frac{\partial \Phi}{\partial \sigma}\frac{\partial \sigma}{\partial y})\frac{\partial \eta}{\partial y} = \frac{\partial \Phi}{\partial \sigma}\frac{\partial \sigma}{\partial z} \quad at \ \sigma = 1$$

and

$$\frac{\partial \Phi}{\partial t} - \frac{1}{(\eta - z_b)}\frac{\partial \Phi}{\partial \sigma}\frac{\partial \eta}{\partial t} + \frac{1}{2}[(\frac{\partial \Phi}{\partial x} + \frac{\partial \Phi}{\partial \sigma}\frac{\partial \sigma}{\partial x})^2 + (\frac{\partial \Phi}{\partial y} + \frac{\partial \Phi}{\partial \sigma}\frac{\partial \sigma}{\partial y})^2$$
$$+ (\frac{\partial \Phi}{\partial \sigma}\frac{\partial \sigma}{\partial z})^2] + g\eta = 0 \quad at \ \sigma = 1$$

The bottom boundary condition is

$$\frac{\partial \Phi}{\partial \sigma} \frac{\partial \sigma}{\partial z} + \frac{\partial h}{\partial x} \left(\frac{\partial \Phi}{\partial x} + \frac{\partial \Phi}{\partial \sigma} \frac{\partial \sigma}{\partial x} \right) + \frac{\partial h}{\partial y} \left(\frac{\partial \Phi}{\partial y} + \frac{\partial \Phi}{\partial \sigma} \frac{\partial \sigma}{\partial y} \right) = 0 \qquad at \ \sigma = 0$$

3. Numerical Model

For the three dimensional fully nonlinear water wave problem the key issue is how to efficiently solve the Laplace equation. The transformed Laplace equation, Eq. (6), can be solved by the conventional iterative methods such as Jacobi, Gauss-Seidel or Simultaneous Over-Relaxation method, if proper boundary conditions are applied. It is clear that a direct method such as the Gauss Elimination method cannot be used for the time being, because of its huge requirement for computer storage and running time. For a three dimensional wave propagation problem the usual number of unknown variables is 100,000. Thus a fast numerical solver is essential to reduce the computer demands and ideally should also require very small computer storage.
The multigrid method which was first developed by Brandt (1977) is one of the fastest numerical solver for elliptical type of the partial differential equations. To solve a system of n discrete equations only $O(n)$ operations are required. Normally other numerical solvers such as Jacobi, Gauss-Seidel or Simultaneous Over-Relaxation method are much slower. For the wave propagation problem presented here the Gauss-Seidel method is ten times slower than the multigrid method. An excellent explanation and comparison of the multigrid method and the other iterative method have been given by Press et al (1992).
It should be noted that the Laplace equation in shallow water region can be written in a dimensionless form as

$$\mu^2 \left[\frac{\partial^2 \Phi'}{\partial x^2} + \frac{\partial^2 \Phi'}{\partial y^2} \right] + \frac{\partial^2 \Phi'}{\partial z^2} = 0$$

where $\mu=kh$ presents relative water depth, k is the wave number and h is the water depth. When $\mu<<1$ the Laplace equation becomes a degenerate elliptical equation. For such degenerate equations the Gauss-Seidel relaxation method can only efficiently smooth error oscillations in z direction, and not in x and y directions. It appears that the best scheme for the degenerate equation is the line Red-Black relaxation method. The line relaxation method involves solving simultaneously all equations on the same grid line parallel to the z direction. On each line of z direction a tridiagonol matrix can be generated and solved very easily by the Gauss Elimination Method.
After implementation of the Red-Black line relaxation scheme the speed of the convergence of the multigrid model in shallow water region is recovered and it is essentially the same as the convergence speed in the deep water region.

The implicit scheme for equation the kinematic boundary condition is

$$\frac{\eta^{t+\Delta t} - \eta^t}{\Delta t} + \frac{1}{2} \left\{ \left(\frac{\partial \Phi}{\partial x} + \frac{\partial \Phi}{\partial \sigma} \frac{\partial \sigma}{\partial x} \right) \frac{\partial \eta}{\partial x} + \left(\frac{\partial \Phi}{\partial y} + \frac{\partial \Phi}{\partial \sigma} \frac{\partial \sigma}{\partial y} \right) \frac{\partial \eta}{\partial y} - \frac{\partial \Phi}{\partial \sigma} \frac{\partial \sigma}{\partial z} \right\}^{t+\Delta t}$$

$$+ \frac{1}{2} \left\{ \left(\frac{\partial \Phi}{\partial x} + \frac{\partial \Phi}{\partial \sigma} \frac{\partial \sigma}{\partial x} \right) \frac{\partial \eta}{\partial x} + \left(\frac{\partial \Phi}{\partial y} + \frac{\partial \Phi}{\partial \sigma} \frac{\partial \sigma}{\partial y} \right) \frac{\partial \eta}{\partial y} - \frac{\partial \Phi}{\partial \sigma} \frac{\partial \sigma}{\partial z} \right\}^{t} = 0$$

and the dynamic boundary condition is

$$\frac{\Phi^{t+\Delta t}-\Phi^t}{\Delta t} - \frac{1}{2}\{[\frac{1}{(\eta-z_b)}\frac{\partial\Phi}{\partial\sigma}]^{t+\Delta t} + [\frac{1}{(\eta-z_b)}\frac{\partial\Phi}{\partial\sigma}]^t\}\frac{(\eta^{t+\Delta t}-\eta^t)}{\Delta t} +$$

$$+\frac{1}{2}\{\frac{1}{2}[(\frac{\partial\Phi}{\partial x}+\frac{\partial\Phi}{\partial\sigma}\frac{\partial\sigma}{\partial x})^2 + (\frac{\partial\Phi}{\partial y}+\frac{\partial\Phi}{\partial\sigma}\frac{\partial\sigma}{\partial y})^2 + (\frac{\partial\Phi}{\partial\sigma}\frac{\partial\sigma}{\partial z})^2] +g\eta\}^{t+\Delta t} +$$

$$+\frac{1}{2}\{\frac{1}{2}[(\frac{\partial\Phi}{\partial x}+\frac{\partial\Phi}{\partial\sigma}\frac{\partial\sigma}{\partial x})^2 + (\frac{\partial\Phi}{\partial y}+\frac{\partial\Phi}{\partial\sigma}\frac{\partial\sigma}{\partial y})^2 + (\frac{\partial\Phi}{\partial\sigma}\frac{\partial\sigma}{\partial z})^2] +g\eta\}^t = 0$$

When these two boundary conditions are used, initialization is made by using explicit schemes.

Along the offshore driving boundary, the incident wave field is assumed known.

Outgoing boundary condition is

$$\frac{\partial\Phi}{\partial t} + c\cos(\alpha)\frac{\partial\Phi}{\partial x} + c\sin(\alpha)\frac{\partial\Phi}{\partial y} = 0$$

where α is the angle between normal direction of the boundary and the direction of the outgoing waves, c is the wave phase velocity.

4. Solution Procedure

a. initialize the boundary conditions on the water surface and the incident boundary;
b. transform the Laplace equation and the boundary conditions from the physical domain to the calculation domain;
c. solve the Laplace equation by the multigrid method;
d. renew the boundary conditions on the water surface with updated value of the velocity potential;
e. if convergence has not been achieved go to b;
f. progress with a time step and go back to a.

Performance of the multigrid model

Loop	First	Second	Third	Fourth	Fifth	Sixth
Number of Iteration	45	30	10	2	1	1

5. Validation of Three-dimensional Multigrid Wave Model Against Experimental Data

1). Elliptical Shoal Case

The wave model is first verified against the experimental data of Berkhoff et al.(1982).Time Step is 0.025 second, the grid spacing is $\Delta x = \Delta y = 0.125$ m, $\Delta z = h/7$. The grid points in X direction are 145, in Y direction are 129 and in Z direction are 7. Total grid points are 130,935. Incoming wave height is 0.0464 metre and incoming wave period is 1 second. The area of numerical model covered: -8 metres to +8 metres in x direction and from -13 metres to +5 metres in y direction. The test case of the elliptic shoal was carried out on a desk top Pentium Pro 200MHZ PC with 64 MB RAM. It took 3 hours to simulate 30 seconds of wave propagation. Figure 1 shows the results.

2). Whalin's Experimental Case

Whalin (1971) conducted a series of laboratory experiments concerning wave convergence over a bottom topography that acts as a focusing lens.
The wave tank used in the experiments has the horizontal dimensions 25.603m×6.096m. The equations approximating the topography are given as follows (Whalin, 1971):

$$h(x,y) = \begin{cases} 0.4572, & (0 \le x \le 10.67 - G) \\ 0.4572 + \dfrac{1}{25}(10.67 - G - x), & (10.67 - G \le x \le 18.28 - G) \\ 0.1524, & (18.29 - G \le x \le 21.34) \end{cases}$$

$$G(y) = [y(6.096 - y)]^{\frac{1}{2}} \qquad (0 \le y \le 6.096)$$

where the length variables x and y are measured in metres. Whalin conducted three sets of experiments by generating waves in the deeper part of the model with periods of 1, 2 and 3 seconds. At the wave maker the waves are linear but after the focusing on the shoal higher harmonics become significant due to nonlinear effects. The summary of the experimental conditions are

Experimental information at water depth h=0.4572m

Wave period T(s)	$k_0 h_0$	Wave amplitude a_0(m)	$k_0 a_0$	Ursell number Ur
1.0	1.922	0.0097	0.041	0.0057
1.0		0.0195	0.082	0.0115
2.0	0.735	0.0075	0.012	0.0303
2.0		0.0106	0.017	0.0429
2.0		0.0149	0.024	0.0603
3.0	0.468	0.0068	0.007	0.0678
3.0		0.0098	0.010	0.0977
3.0		0.0146	0.015	0.1456

The 3D multigrid model was tested against the experimental results, which included 1, 2 and 3 seconds cases. Figure 2 shows part of the results.

6. Conclusion

Propagation of fully nonlinear water wave has been successfully simulated by present 3-D multigrid model. Numerical results agree quite well with experimental data. Computer time costs for running present three-dimensional wave model on a PC is about the same as computer time costs five years ago for running a two-dimensional model based on elliptical mild-slope equation. Treatment of boundary conditions on outgoing boundary should be further improved.

References

Berkhoff, J.C.W., 1972. 'Computation of combined refraction-diffraction', Proc. 13th Coastal Eng. Conf., Vancouver, Canada, A.S.C.E. 1 (1973), 471-490.

Berkhoff, J.C.W., Booij, N. and Radder, A.C., 1982. 'Verification of numerical wave propagation models for simple harmonic water waves', Coastal Eng., 6: 255-279.

Booij, N., 1981. 'Gravity waves on water with non-uniform depth and current', Report No. 81-1, Delft University of Tech., Dept. Civil Eng.

Brandt, A. 1977. 'Multi-level adaptive solutions to boundary-value problems'. Mathematics of Computation, Vol.31, No. 138, p333-390.

Copeland, G.J.M., 1985. 'A practical alternative to the "mild-slope" wave equation', Coastal Eng., 9: 125-149.

Li, B., Reeve, D. E. and Fleming, C. A., 1993. Numerical Solution of the Elliptic Mild-Slope Equation for Irregular Wave Propagation. Coastal Engineering, Vol 20:85-100.

Li, B., 1994a. A Generalised Conjugate Gradient Model for the Mild-Slope Equation. Coastal Engineering, Vol 23: 215-225.

Li, B., 1994b. An Evolution Equation for Water Waves. Coastal Engineering, Vol. 23: 227-242.

Li, B., 1997. Parabolic Model For Water Waves. Journal of Waterway,Port,Coastal & Ocean Eng., ASCE, JUL./AUG. 1997, VOL. 123 NO.4: 192-199.

Li, B. and Fleming, C.A., 1997. A Three Dimensional Multigrid Model For Fully Nonlinear Water Waves. Coastal Engineering, Vol 30: 235-258.

Press, W.H., Flannery, B.P., Teukolsky, S.A. and Vetterling, W.T., 1992. Numerical Recipes, The Art of Scientific Computing, Second Edition. Cambridge University Press.

Radder, A.C. 1979. 'On the parabolic equation method for water-wave propagation'. J. Fluid Mech. 95, 159-176

Rygg, O.B. 1988. Nonlinear refraction-diffraction of surface waves in intermediate and shallow water, Coastal Eng., 12, 191-211.

Whalin, R.W. 1971. The limit of applicability of linear wave refraction theory in a convergence zone. Res. Rep. H-71-3, U.S. Army Corps of Engineers, Waterways Expt. Station, Vicksburg, MS.

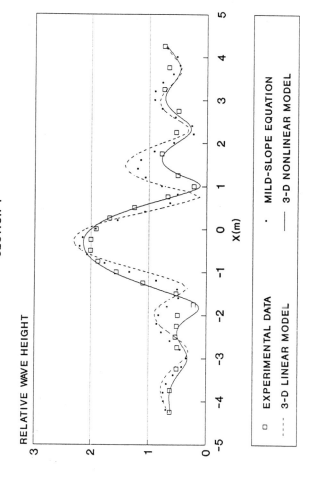

Figure 1

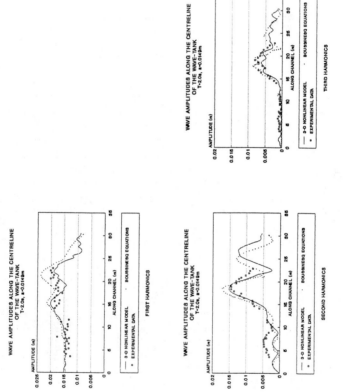

Figure 2

Numerical Simulations of Wave Transformation in the Nearshore Zone

Y. Eldeberky[1], O.R. Sørensen[1], and J.A. Battjes[2]

Abstract
A recent detailed set of observations of water level and wave characteristics in a cross-shore array is utilised to verify two different modeling approaches for wave transformation in the nearshore regions, namely, a phase-resolving approach and a phase-averaged approach. The phase-resolving model is based on time-domain Boussinesq equations. The model is used to simulate the cross-shore wave transformation, and wave runup on the dike. The results show the model capabilities in producing features of surf-zone dynamics including the shoreline movement. The phase-averaged model is based on an energy balance formulation, supplemented with simplified source terms representing the effects of wave breaking and generation of higher harmonics due to triad wave interactions. The model results are consistent with the observations, and indicate a decrease in the mean wave period over the offshore bar due to the generation of higher harmonics. The predicted wave heights are generally underestimated over the steep slopes along the cross-shore profile. As a remedy, a larger value of the breaker height-to-depth ratio could be used, but the present data do not allow a parameterization to include bottom slope effects in a generalised manner.

1. Introduction
Knowledge of wave characteristics in the shoaling region and the surf-zone, shoreline motion and wave runup are important to coastal engineering. In the nearshore regions, linear and nonlinear dynamical processes, such as refraction, shoaling, nonlinear interactions and breaking, change the characteristics of the wavefield. Nonlinear processes in shallow water significantly influence the shoreline oscillations and the wave runup. The understanding of these physical processes has been improved in recent years and a number of attempts have been made to incorporate this knowledge in numerical models for wave transformation in the nearshore regions.

[1]International Research Centre for Computational Hydrodynamics (ICCH), Danish Hydraulic Institute, Agern Allé 5, DK-2970 Hørsholm, Denmark
[2] Faculty of Civil Engineering, Delft University of Technology, P.O. Box 5048, 2600 GA Delft, The Netherlands

147

In this study, a recent detailed set of observations of water level and wave characteristics in a cross-shore array is utilised to verify numerical models for wave transformation in the near shore. Two different modeling approaches for wave transformation in the nearshore regions are utilised. These are a phase-averaged approach and a phase-resolving approach. The purpose of this paper is to examine the capabilities of both model-types in representing some specific physical processes during the wave transformation from deep water to the shoreline.

The arrangement of this paper is as follows. In section 2, the observation site and the measurements are briefly presented. The phase-resolving model and simulation-results are described in section 3. In section 4, the phase-averaged model is briefly described followed by the model results. To reduce excessive dissipation in the model calculation over steep slopes, a new breaking criterion is introduced in section 5. Finally a summary and conclusions are given in section 6.

2. Field Observations

An extensive measurement campaign has been conducted by the Dutch Department of Public Works in a cross-shore profile near Petten at the coast of The Netherlands. A detailed description of the measurement campaign is given in Andorka Gal (1996). This cross-shore profile is characterised by almost parallel depth contours with a bar and trough bounded by a steep (slope 1:4) high dike. The bottom topography and the measurement locations are illustrated in Fig. 1.

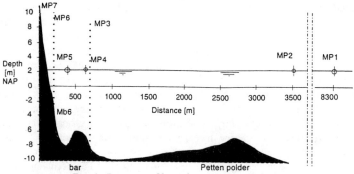

Fig. 1 Petten profile and wave sensor locations

The measurements were conducted during the storm season from October 1994 till April 1995. The observations contain two storm periods of similar intensity with 20 m/s wind speed. During the storms, all stations measured continuously the time series of the surface elevation. The water level was measured at station MP3 (distance to dike top 635 m). The water level varied for both storms between -0.5 m (low water) and

2.25 m (high water) measured with respect to normal Amsterdam level (NAP). The wave runup over the dike slope is measured using a runup gauge fixed at station MP7.

Numerical simulations have been carried out in this study, using both a phase-averaged and a phase resolving model, to simulate the cross-shore wave transformation during one event at 16 hour, January 1, 1995. This event has been chosen to have manifestation of both linear and nonlinear physical processes. The wind speed, during this event, approaches 20 m/s, and the wind and wave directions are nearly the same and perpendicular to the shore. The wavefield is directional with a spreading angle equal to 30°. The water level reaches 2 m (+NAP) due to the storm surge. The measured significant wave height at MP1 (8.3 km offshore) equals 4.8 m. The observed wave runup, defined by the level that 2% of the wave runup exceeds in a period of 20 minuets, indicates +7.2 m with respect to NAP.

The hydrodynamic and topographic conditions in the nearshore region of Petten induce the manifestation of various physical mechanisms. Among these are energy dissipation by wave breaking, generation of sub- and super-harmonics due to triad wave interactions, generation of surf beats and low-frequency waves induced by the short-wave groups, all of these over a multiple-bar profile, wave runup over the dike slope, and finally possible wave reflection.

3. Phase-resolving model and simulations

The numerical phase-resolving model used in this study is based on time-domain Boussinesq equations with improved dispersion characteristics (Madsen and Sørensen, 1992). The effect of energy dissipation due to wave breaking is incorporated using the surface roller concept (Schäffer et al., 1993). The moving boundary at the shoreline is treated numerically by replacing the solid beach by permeable beach characterised by an extremely small porosity. Previous calculations have included the transformation and breaking of wave groups and irregular waves over bars and constant sloping beaches with emphasis on the shoreline motion and the surf beat (Madsen et al., 1997). In this study the results of one-dimensional model simulations are compared against the Petten observation in the surf-zone, with emphasis on the runup predictions, to further test the model's capability to represent this phenomenon.

One-dimensional Boussinesq calculations are carried out along the cross-shore profile. The model boundary condition (at the sea side) is generated using linear theory according to the measured spectrum at MP3 (635m from the shoreline), with peak frequency of 0.1 Hz and significant wave height of 4.1 m. At the position of MP3 the waves are generated internally and re-reflection from this boundary is avoided by using a sponge layer offshore from the line of generation. In order to resolve the wave motion, a grid step of 0.2 m and a time step of 0.015 s are used in the simulations. The length of the simulation period is extended to 20 minuets in order to obtain phase-averaged parameters of the wavefield along the cross-shore profile. The numerical simulation was done on an IBM-type computer. The CPU time required for the simulation is 5740 s.

The model calculations are extended through the surf zone and the swash zone to include the shoreline movement. The predicted surface elevations by the Boussinesq model at MP6 (130 m from the shore-line) and at 90 m offshore as well as the shoreline movement are plotted in Fig. 2. The computed surface elevations show typical shallow-water wave profiles with sharp crests and flat troughs. The maximum wave height in the simulated surface elevation approaches 4 m at the location of MP6. The presence of these high waves in 4.2 m water-depth is consistent with the maximum wave height (3.9 m) observed in the measurements. The predicted shoreline motion converted into vertical displacement shows a maximum runup level of about 6 m with respect to NAP. The measurements indicate that the maximum runup level is slightly over 7 m. It can be seen from the model results that the predicted shoreline motion seems to be affected by low-frequency oscillations. These results show the model capabilities in describing the shoreline motion and the maximum wave height in the surf-zone. Detailed model-data comparisons are not meaningful due to the assumption of unidirectional incident waves and of linear boundary conditions.

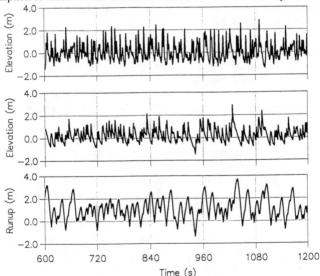

Fig. 2 Computational results of the Boussinesq model. Upper panel: surface elevation at 130m offshore, mid panel: surface elevation at 90m offshore, lower panel: shoreline movement converted into vertical displacement.

The variations of the significant wave height and skewness over the nearshore stretch of the Petten profile are plotted in Fig. 3. Over the nearshore bar, nonlinearity increases leading to an increase in the skewness, and wave breaking takes place resulting in a wave height decay. Wave breaking over the bar and decreasing nonlinearity beyond the bar result in a reduction in the skewness. Nonlinear shoaling evolution over the sloping beach leads to an increase in the skewness to a maximum just prior to wave breaking, which results in a wave height decay.

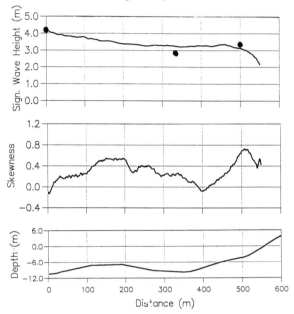

Fig. 3 The spatial variations of the significant wave height and the skewness computed by the Boussinesq model over the Petten profile. Circles denote observations

The predicted energy spectra are compared against the measurements in Fig. 4. The overall features of the higher harmonics are reproduced by the model. The model results show generation of low-frequency motion that is not present in the observations. The presence of this low frequency energy in the one-dimensional time-domain Boussinesq-calculation is also found in calculations performed by a one-dimensional deterministic spectral model (Eldeberky and Battjes, 1996). A possible explanation of the presence of low-frequency motion in the calculation is the assumption of unidirectional incident waves. Previous studies (Elgar et al., 1993, and many others) have indicated that the generation of higher harmonics (due to the sum triad interaction) is not sensitive to the directional characteristics of the wavefield, whereas the generation of lower harmonics (due to the difference triad interactions) is strongly influenced by the directional characteristics of the wavefield. Directional spreading causes a detuning from resonance (similar to the effect of weak dispersion) that reduces the transfer to the low frequency components.

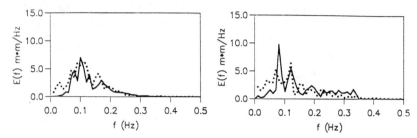

Fig. 4 Computed energy spectra based on Boussinesq calculations at MP5 (left panel) and MP6 (right panel).

4. Phase-averaged model and simulations

The numerical phase-averaged model is based on a one-dimensional spectral energy balance equation, with source/sink terms representing energy dissipation due to wave breaking and energy transfers to higher harmonics due to triad interactions. The average effect of triad wave interactions is modelled using a low-cost algorithm based on a lumped presentation of the interaction integral supplemented with a parameterization for the biphase evolution (Eldeberky and Battjes, 1997; and Eldeberky, 1996). The influence of depth-induced wave breaking is included by distributing the total energy dissipation, according to Battjes and Janssen (1978), in proportion to the spectral levels as in Eldeberky and Battjes (1996). The breaking criterion is based on waveheight-to-waterdepth ratio H_b/d at breaking, i.e., breaker index γ, equal to 0.74. This value represents an average for the breaker index in Battjes and Stive (1985) for different types of bathymetries and wave conditions. Numerical simulations of the Petten observations are carried out in particular to check the parametrization of the triad interactions.

One-dimensional spectral computations are carried out along the cross-shore profile. The model boundary condition is taken from the observations at 3500 m offshore. The computations are performed on a 486-DX2 personal computer. In the numerical discretisation, the frequency resolution is 0.01Hz (50 frequencies in the range 0.01-0.50 Hz), the spatial resolution is 2.5 m. The CPU time required for the simulation is 45 s.

The predicted energy spectra are compared against the observed spectra in Fig. 5. The comparisons show that the generation of higher harmonics is reasonably predicted by the triad parameterization. The predicted mean wave period and significant wave height are compared against the observed values in Figs. 6 and 7, respectively. The spatial variation of the mean wave period indicates a significant reduction from 8.5 s to about 6 s over the offshore bar as a result of energy transfers to higher harmonics due triad wave interactions. These results are consistent with the observed shift in the mean wave period. Without the inclusion of the triad parametrization, the variation in the mean wave period shows an increase to about 10 s.

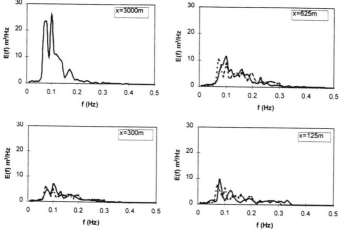

Fig. 5 Computed energy spectra at various location using the spectral model.

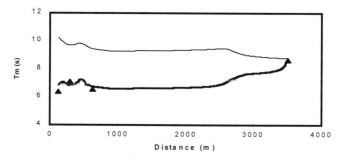

Fig. 6   The spatial variations of the mean wave period T_m. Thin line denotes computations using the spectral model without the inclusion of triad wave interactions, thick line denotes computations with triad interactions, and the triangles denote the observations.

The spatial variation of the significant wave height H_{m0} indicates a significant decay over the offshore bar due to wave breaking, and a nearly constant value beyond the bar. Close to the shore, the waves continue to break over the nearshore bar and farther over the steep beach. The calculated wave height at MP6 (2.6 m) is under-predicted compared to the observation (3.1 m). This suggest that the nominal breaking waveheight-to-waterdepth ratio (the breaking parameter γ in the Battjes and Janssen model) which was used in the calculations (0.74) was too low.

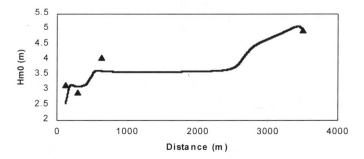

Fig. 7   The spatial variations of the significant wave height H_{m0} computed using the spectral model with γ=0.74. The triangles denote the observations.

A variety of empirical relations for the breaker index ($\gamma = H_b/d$) exists in literature. These relations are deduced based on analysis of wave breaking observations for various wave conditions and bathymetries. In spite of the differences between all these relations, they all seem to agree on the dependence of the breaker index (γ) on the beach slope α and the wave steepness (H/L), where L is the wave length. Battjes (1974) has combined these independent variables, i.e., the deep-water wave steepness H/L_o and the bottom slope α, and presented a plot of experimental results relating the breaker index to the surf similarity parameter $\xi = \alpha/\sqrt{H/L_o}$. This plot is reproduced in Fig. 8. Values of breaker index γ generally range between 0.7 and 1.2. For decreasing wave steepness and/or increasing beach slope, the breaker height-to-depth ratio increases and may exceed 1. This ratio approaches a value of 0.65 in cases with very gentle or flat slope and/or waves with low steepness. In fact, Nelson (1994) has suggested the value of 0.55 as the largest wave height-to-depth ratio for waves propagating over horizontal depth. Kamphuis (1991) has presented a simple relation for the significant wave height at incipient breaking which incorporates wave period (or wave length) in addition to effective beach slope and breaking water depth. Raubenheimer et al. (1996) found that the ratio of significant wave height at breaking to the water depth increases with increasing beach slope α and decreasing normalized (by a characteristic wavenumber k) water depth kh and is well correlated with the ratio α/kh. They have also shown that γ does not vary systematically with the offshore wave steepness.

The breaking criterion used in this study is based on a constant value of nominal breaking waveheight-to-waterdepth ratio (0.74). Over steep slopes, this value is relatively low and results in excessive dissipation and under-prediction of the wave height in the surf-zone. As a remedy, the influence of bottom slope on the process of wave breaking could be introduced in the breaker index γ. However, our present data set is too limited to allow a parameterization of bottom slope effects.

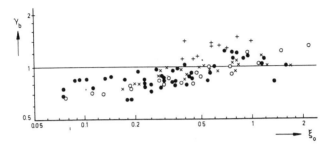

Fig. 8   The breaker height-to depth-ratio versus the surf similarity parameter
(Battjes, 1974)

6. Summary and Conclusions

One-dimensional numerical simulations of wave transformation in the nearshore are carried out using two different modeling approach, namely a phase-resolving approach and a phase-averaged approach. The phase-resolving model is based on time-domain Boussinesq equations. The model is used to simulate the cross-shore wave transformation, and wave runup on the dike. The results show the model capabilities in producing features of surf zone dynamics including the shoreline movement. The model prediction of the low-frequency motion is expected to be inaccurate due to the assumption of unidirectional incident waves. For accurate prediction of low frequency motion, two-dimensional simulations are needed.

The phase-averaged model is based on an energy balance formulation, supplemented with simplified source terms representing the effect of wave breaking and generation of higher harmonics due to triad wave interactions. The predicted energy spectra and mean wave period are in good agreement with the observations. The model predictions of the significant wave height along the cross-shore profile are generally underestimated. As a remedy, a larger value of the breaker height-to-depth ratio could be used, but the present data do not allow a parameterization to include bottom slope effects in a generalised manner.

A concluding remark that generally holds for both model-families is the increasing complexity, in the sense of more physics (e.g. triad interactions in phase-averaged models; wave breaking in phase-resolving models), extended range of validity, greater resolution and computing efficiency.

Acknowledgements
The work presented here is part of a project sponsored by the Dutch Department of
Public Works (RWS). The first author was supported jointly by the Danish National
Research Foundation and the Dutch Department of Public Works (RWS). Their
support is greatly appreciated. We thank J. de Ronde and J. Andorka Gal, of the
Dutch National Institute for Coastal and Marine Management (RIKZ), for providing
the data and for useful discussions.

References
Andorka Gal, J.H., 1996, Verification set Petten, *Technical report RIKZ/OS-96.137x*
    Dutch Ministry of Public Works (RWS).
Battjes, J.A., 1974, Computation of set-up, longshore currents, run-up and
    overtopping due to wind-generated waves, *Ph.D thesis*, Delft University of
    Technology.
Battjes, J.A., and J.P.F.M. Janssen, 1978, Energy loss and set-up due to breaking of
    random waves, *Proc. Coastal Eng. Conf*, 16th, pp. 569-587.
Battjes, J.A., and M.J.F. Stive, 1985, Calibration and verification of a dissipation
    model from random breaking waves, *J. Geophys. Res.,* 90(C5), pp. 9159-9167.
Eldeberky, Y., 1996, Nonlinear transformation of wave spectra in the nearshore
    zone, *Ph. D thesis,* Delft University of Technology.
Eldeberky, Y., and J.A. Battjes, 1996, Spectral modeling of wave breaking:
    Application to Boussinesq equations, *J. Geophys. Res.* 101(C1), pp 1253-1264.
Eldeberky, Y., and J.A. Battjes, 1997, Parameterized energy formulation for triad
    wave interactions in wave energy models, in preparation.
Elgar, S., R.T. Guza, and M.H. Freilich, 1993, Observation of nonlinear interactions
    in directionally spread shoaling surface gravity waves. *J. Geophysical Res.*, 98,
    20299-20305
Kamphuis, J.W., 1991, Incipient wave breaking, *Coastal Engineering,* 15, pp. 185-
    203.
Madsen, P.A., and O.R. Sørensen, 1992, A new form of the Boussinesq equation
    with improved linear dispersion characteristics, Part 2: a slowly-varying
    bathymetry, *Coastal Eng.*, 18, pp. 183-205.
Madsen, P.A., O.R. Sørensen and Schäffer, H.A., 1997, Surf zone dynamics
    simulated by a Boussinesq type model, Part II: Irregular waves and Surf Beat,
    Submitted to *J. Coastal Engineering.*
Nelson, R.C., 1994, Depth limited wave height in very flat regions, *Coastal
    Engineering*, 23, 43-59.
Raubenheimer B., R.T. Guza, and S. Elgar, 1996, Wave transformation across the
    inner surf zone, Vol. 101, No. C10, pp. 25,589-25,597.
Schäffer, H.A., R. Deigaard, and Madsen, 1993, A Boussinesq model for wave
    breaking in shallow water, *Coastal Eng.*, 20, pp. 185-202.

Prediction of Velocity Field under Waves over Varying Depth

A.K. Otta[1], G. Klopman[2], M.W. Dingemans[3]

Abstract

In this paper, prediction of velocity field due to waves over varying depth is addressed. We have considered the application of a Hamiltonian-based nonlinear model to obtain the velocity field due to waves propagating over finite depth. Comparisons with experimental measurements show that asymmetry and skewness of both the surface elevation and velocity field can be reproduced by the applied model for non-breaking waves. Agreement with the measurement is much better than that can be obtained based on the method of a 'local Fourier' approximation (LF) using an available time-record of the surface elevation. However, caution should be exercised in generalizing this observation. Some aspects of linear theory in calculating the velocity profile from an available time-record of surface elevation are also discussed.

1 Introduction

Several approaches are currently in practice in order to obtain the velocity field corresponding to a measured surface elevation record. Some of these theories have been inter-compared and in limited cases compared against measured data by Dean and Perlin (1986). Among the nonlinear formulations used to obtain wave kinematics under the assumption of steady propagation the Stream Function theory (Dean, 1965; Rienecker and Fenton, 1981) has highest analytical validity upto the highest waves without any constraint on the water depth. Modifications of the steady Stream Function theory have been proposed by several authors in order to improve their applicability to irregular, asymmetric waves (*e.g.*, Sobey, 1992b) or the effect of a linear shear current (Dalrymple, 1974). However, linear theory (see Sobey (1992a) for an useful discussion of alternative forms of applying the linear theory) is perhaps 'most commonly used' in normal practices primarily because of its simplicity. This is helped by the fact that field comparisons have not

---

[1]Corresponding author, International Research Centre for Computational Hydrodynamics (ICCH), Danish Hydrulic Institute, Agern Alle 5, DK-2970 Hørsholm, Denmark

[2]Netherlands Centre for Coastal Research, Delft University of Technology, The Netherlands

[3]Delft Hydraulics, The Netherlands

always pointed in favour of using a nonlinear theory such as the Stream Function theory over the linear theory. Whereas this apparent contradiction of analytical validity is perhaps due to the field conditions introducing features beyond the scopes of the applied theories, investigations are clearly needed to clarify such discrepancies.

For propagations over varying depth, wave steepness and nonlinearity-induced features become more prominent as waves go over a shallower area. In particular, the generated asymmetry and skewness are rather significant from the point of forces on a pipeline or sediment transport gradient. Some empirical formulations like that of Isobe & Horikawa (1982) (see also Hamm, 1996) exist to account for the asymmetry and skewness of a wave profile. However, such parametrizations can be in serious doubt to work reasonably in different conditions beyond a limited range on which they are based. For example, the formulations clearly break down to predict the change of asymmetry and skewness on a uniform depth due to a train of (self) interacting waves. This direction is not pursued here any further. In recent times, there has been considerable progress in the application of weakly nonlinear models to study the propagation of waves over varying depth and such computations are now being increasingly undertaken. Though much attention has been paid to verify the elevation-records computed from such models with experimental data, analysis of the associated velocity field is largely lacking. Therefore, one of the primary issues addressed in this article is the prediction of the velocity field using a nonlinear model. The model to be applied accounts for generation of higher harmonics, both bound and free, for moderately high waves and is discussed in the following section. Secondly, application of a Local Fourier approximation (also referred to as 'LF' method here) as proposed by Sobey (1992b) is considered to extract the vertical field of wave kinematics at a location from an available time-record of surface elevation. Two important drawbacks of the LF are that the presence of long waves is not accounted for within the local window selected for analysis and that all Fourier modes are assumed to be bound. The advantage of the method is that it accounts for full-nonlinearity in computing the kinematics from an available surface-record. For waves propagating over varying depth like a submerged bar wave shoaling, steepening and also the generation of free harmonics are significant. It is therefore expected that the comparison of the velocity field computed from both the methods will lead to some insight when compared against the experimental data.

2   Solution of the Wave Field through a Nonlinear Model

In this section we outline a nonlinear model derived from an explicit approximation of the Hamiltonian (sum of the potential and kinetic energy of the fluid). The theoretical basis of the formulations can be found in Radder (1992). Further elaborations and numerical implementation of the one-dimensional wave-motion are presented in Otta and Dingemans (1994) and Otta et al. (1996). An approximate Hamiltonian $\mathcal{H}$ satisfying the exact dispersion realtionship in the linearized case and guaranteeing a positive definite expression for kinetic-energy density is given by

$$\frac{\mathcal{H}}{\rho} = \frac{1}{2} \int_{-\infty}^{\infty} dx \, g\zeta^2 - \frac{1}{2\pi} \int_{-\infty}^{\infty} dx \int_{-\infty}^{\infty} dx' \varphi_x \varphi_{x'} \log \tanh \frac{\pi}{4} \left| \int_x^{x'} \frac{dx''}{\eta} \right| \quad . \tag{1}$$

where $\zeta(x,t)$, $\eta(x,t)$ and $\varphi(x,t)$ denote the free-surface elevation, total water depth $(h(x) + \zeta(x,t))$ and potential. Nonlinear evolution of an initial wave field is determined from the corresponding evolution equations which follow from the relation $\rho\zeta_t = (\delta\mathcal{H})/(\delta\varphi)$, $\rho\varphi_t = -(\delta\mathcal{H})/(\delta\zeta)$ to be

$$\frac{\partial \zeta}{\partial t} = \frac{1}{\pi} \frac{\partial}{\partial x} \int_{-\infty}^{\infty} dx' \varphi_{x'} \log \tanh \frac{\pi}{4} \left| \int_x^{x'} \frac{dr}{\eta} \right| \tag{2}$$

$$\frac{\partial \varphi}{\partial t} = -g\zeta - \frac{1}{2\eta^2} \int_{-\infty}^{x} dx' \int_x^{\infty} dx'' \frac{\varphi_{x'} \varphi_{x''}}{\sinh\left(\frac{\pi}{2} \int_{x'}^{x''} \frac{dr}{\eta}\right)} \tag{3}$$

The equations (2) and (3) are manipulated further for numerical implementation. Those details are described in Otta & Dingemans (1994) and are skipped here for brevity.

2.1 Velocity Computation

The basic principle to the velocity computation is that the instantaneous velocity potential $\Phi(x,z,t)$ in the entire wave field is determined by the two surface variables $\zeta(x,t)$ and $\varphi(x,t) = \Phi(x,\zeta,t)$ obtained from the aforementioned nonlinear evolution. It is easier to derive this velocity field in the transformed strip $(-\infty < \chi < \infty, \ 0 \le \xi \le 1)$ which is related to the physical space through the Woods transformation (see Radder 1992). Denoting the Fourier transform with respect to χ by the superscript 'top hat', it follows from the Laplace equation in the (χ, ξ) variables, i.e.,

$$-k^2\hat{\Phi} + \frac{\partial^2 \hat{\Phi}}{\partial \xi^2} = 0, \quad (-\infty < \chi < \infty), \ (0 \le \xi \le 1) \tag{4}$$

that the field variation of the potential is given by

$$\hat{\Phi}(k,\xi) = \hat{\Phi}(k,1) \frac{\cosh(k,\xi)}{\cosh(k)} \quad . \tag{5}$$

The velocity field is accordingly expressed by

$$\hat{u}(k,\xi) = (\imath k)\hat{\Phi}(k,1)\frac{\cosh(k,\xi)}{\cosh(k)}, \quad \hat{w}(k,\xi) = k\hat{\Phi}(k,1)\frac{\sinh(k,\xi)}{\cosh(k)} \quad . \tag{6}$$

It is evident from (6) that the linearized form of the equations yield a field which is identical to the classical linear theory for all wave numbers. In the nonlinear form the free-surface potential implicitly describes both the bound and the free modes at any instant, which are then exhibited in the computed velocity field. In this article, we shall focuss on near-bed ($\xi = 0$) tangential velocity u_b which upon using the convolution theorem on (6) is expressed in the physical variables by

$$u_b(x,t) = \frac{\pi}{2(h+\zeta)} \int_{-\infty}^{\infty} dx' \frac{\partial \varphi}{\partial x'} \operatorname{sech} \frac{\pi}{2} \left| \int_x^{x'} \frac{dr}{h+\zeta} \right| \quad . \tag{7}$$

3   Local Fourier Approximation Method

Inclusion of both cosine and sine series in the global representation of an assymmetric wave form through Stream Function Theory has been proposed by Dean (1965). It is conjectured by Sobey (1992b) that higher analytical fidelity may be achieved by assuming the waves to be locally steady instead. This argument is the primary motivation for adopting here the procedure of Sobey (1992b) for deriving the velocity field from a time-record of surface elevation due to unsteady, asymmetric waves. A brief account of the formulations is as follows. The velocity potential $\Phi(x, z, t)$ is assumed to be given by

$$\Phi(x,z,t) = C_E x + \sum_j A_j \frac{\cosh jk(h+z)}{\cosh(jkh)} \sin j(kx - \omega t) \tag{8}$$

where C_E represents a net Eulerian current assumed to be known. The form (8) for $\Phi(x, z, t)$ satisfies the Laplace equation over a locally flat bottom with $(3+M)$ unknowns, namely ω, k, kx and A_j's, for $j = 1, 2, \cdots M$. These unknowns are determined by an optimization of satisfying the free surface conditions

$$\frac{\partial \zeta}{\partial t} + u \frac{\partial \zeta}{\partial x} = w, \quad \frac{\partial \Phi}{\partial t} + \frac{1}{2} |\nabla.\Phi|^2 + g\zeta = B \tag{9}$$

for an available time-record of the elevation $\zeta(t)$ at the desired location. The spatial derivative $\partial \zeta / \partial x$ in the kinematic condition is replaced by the temporal derivative using the hypothesis of waves being locally steady; i.e., $\zeta_x = -\zeta_t/C$ where $C = \omega/k$ is the the phase velocity. This assumption of waves being locally steady is applied within each local window τ. Mathematically, the only requirement is that the number of observation points I within each local window need to be larger than $(3 + M)$. There are, however, several difficulties associated with the solution, specially for irregular profiles and the initial estimates may be crucial. Sobey (1992b) has presented a description of the nature of difficulties. No effort is made to investigate these aspects in this work. With the unknowns determined, the velocity field [u,w] is given by the expressions

$$u(z,t) = C_E + \sum_j (jk) A_j \frac{\cosh jk(h+z)}{\cosh(jkh)} \cos j(kx - \omega t) \tag{10}$$

$$w(z,t) = \sum_j (jk) A_j \frac{\sinh jk(h+z)}{\cosh jkh} \sin j(kx - \omega t) \quad . \tag{11}$$

4 Comparison with Experimental Data

Computed velocities from both methods are compared with experimental measurements due to waves propagating over a submerged bar (Luth *et al.*, 1994). The experimental setup is shown in Fig. 1 with the measurement locations for surface elevation and the velocity given in table 1. Distance shown is measured

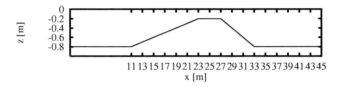

Figure 1: Geometry of the experimental setup

Elevation	3.04 7.04 9.44 20.04 24.04 26.04 28.04 30.44 33.64 37.04 41.04
velocity	23.84

Table 1: Locations of measurements in meters from the wavemaker.

with respect to the wavemaker. Three monochromatic wave conditions were used (Table 2). Waves were absorbed with the help of an active wave absorber located at 45 m at the right end of the wave flume. Waves were generated using

Case	h_0 (cm)	H_0 (cm)	T (s)	$T\sqrt{(g/h_0)}$	Ur	Remarks
A	80	4	2.86	10.0	0.11	Non-breaking
B	80	5.6	3.57	12.5	0.26	Mildly spilling breaker
C	80	8.2	1.43	5.0	0.04	Non-breaking

Table 2: Wave conditions used in the experiments. Ursell parameter, denoted by Ur, is defined as $(H_0/h_0)(kh_0)^{-2}$.

a second-order theory. In all three cases, however, the second-order component in the offshore boundary is very small and the waves are almost sinusoidal. To compare the computed time-records with those from the experiments, a shift of the origin was necessary for synchronisation. This was done as follows. The experimental data were collected in two series. For each series, the time-origin of the numerical record was shifted so that the leading peaks of the oncoming waves from both experiments and computations at a given location nearly coincided. Comparisons of time-records of surface elevation showed that both numerical and experimental records matched satisfactorily at all stations along the flume. Fig.

2 shows the computed and measured surface elevation and velocity for the three cases from Table 2. Because of the finite grid size, the exact location of measurement did not always correspond to a discrete grid location. Instead of using interpolation between the grids the time-record at the grid next to the measurement location is compared with that from the experiment.

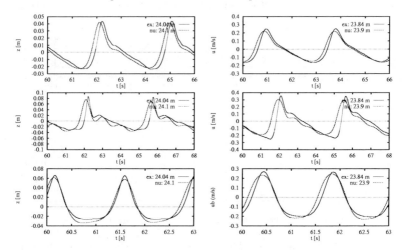

Figure 2: Surface elevation (left side) and near-bed horizontal velocity (right side) on top of the bar. Top: case A, middle: case B and bottom: case C from table 2. Measurements ('ex') and numerical prediction ('nu') are shown respectively by the solid and dashed line.

In Fig. 3, velocity field obtained from the LF method using three Fourier modes is compared against the measurements and the nonlinear Hamiltonian model (Otta & Dingemans, 1994). Since no surface elevation measurement is

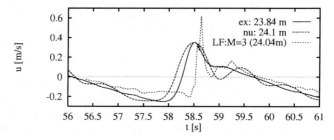

Figure 3: Horizontal velocity for case B (table 2). solid line: measurement at 23.84 m, dotted line: LF with three Fourier modes at 24.04 m, dashed line: Hamiltonian model at 24.1 m.

available at the location of the velocity measurement, measured surface elevation from the closest station at 24.04 m is used as input to the LF method. It is seen that the deviations of the LF prediction from the actual measurement are larger than that obtained from the solution to the complete wave field through the Hamiltonian model. Comparisons of LF predictions with different local windows show only minor variations without really improving the agreement with the measurement.

5   Discussion of the results

The comparisons presented in Figs. 2 and 3 show that the nonlinear wave model reproduces the skewness and asymmetry in all the three cases starting from the offshore wave conditions. Application of the local Fourier approximation (LF) as adopted in this work yields less good results. A further analysis is therefore presented to gain some insight. Table 3 shows a comparison of some properties of the wave profiles on top of the bar corresponding to the two cases B and C as computed from the steady Stream Function theory (Rienecker and Fenton, 1981) against those from the measurements. Water depth, wave height (from trough to crest, as measured) and period form the input for the Stream Function Theory. For the nearly symmetric, though skewed, profile of case C near 24 m,

case		a^C	a^T	u^c	u^T	u^1	u^2	u^3
B	Expt.	0.085	-0.029	0.353	-0.254	0.191	0.089	0.066
	Str. Fnc.	0.102	-0.012	0.461	-0.076	0.143	0.117	0.086
C	Expt.	0.065	-0.027	0.269	-0.200	0.222	0.075	0.016
	Str. Fnc.	0.066	-0.026	0.297	-0.170	0.233	0.075	0.017

Table 3: Crest 'c' and trough 'T' values of surface elevation and near-bed horizontal velocity from Stream Function theory (Str. Fnc.) and the measurement at 24.04 m. u^1, u^2 and u^3 are the first three harmonics of bed velocity u.

the calculated values of crest and trough elevation and velocity from the Stream Function theory are in excellent agreement with the experiment. The differences, however, are clear for case B. The reason for this is that the generated free harmonics are much stronger in case B, the wave period being much longer. Both the Stream Function theory for waves of permanent form and the local steady Fourier approximation assume all the modes to be bound and are therefore likely do show disagreement with observations at locations where significant free modes may be present.

It is useful to discuss the linear theory in calculating the velocity field. Linear theory is, of course, limited giving only the sinusoidal profile if only depth, wave height and period are known. For an available time-record of the surface elevation at a location, linear theory can be used to give a better representation of the velocity field. We use the following procedure here to compute the wave kinematics from an available time-record using the linear theory. Expressing the time-record of surface elevation in a Fourier series, i.e., $\eta(t) = \sum_j a_j \cos(j\omega t + \delta_j), t_1 \leq t \leq t_2$

where a_j's and δ_j's are the amplitudes and phases of the j-th harmonic in a Fourier anlysis of the surface-record on a wave-by-wave basis; with t_1 and t_2 denoting two consecutive zero-upcrossing points the velocity potential is given by

$$\Phi(x, z, t) = \sum_j a_j \left(\frac{g}{j\omega} \right) \frac{\cosh k_j(h + z)}{\cosh k_j h} \sin(k_j x - j\omega t - \delta_j) \,. \tag{12}$$

The wavenumber k_j and frequency $j\omega \, (= 2\pi j/(t_2 - t_1))$ are related through the linear dispersion relation $(j\omega)^2 = g k_j \tanh(k_j h)$. An external current, if known, can be included in (12) easily with an appropriate modification of the dispersion relation. The velocity field is determined by taking the gradient of Φ given by (12).

Using this procedure based on the linear theory, we return to the prediction of near-bed velocity for the wave cases shown in table 2. Measured surface elevation is used as input at each of the locations and the derived amplitudes of the first and second harmonics of horizontal velocity are compared against those from the nonlinear wave model (Fig. 4) which has been described earlier. The

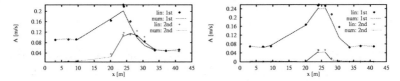

Figure 4: First and second harmonic amplitudes of horizontal velocity from linear theory using measured surface elevation (points) and nonlinear wave model (line). Left: case B and right: case C.

strikingly good agreement between the predictions from the linear theory and the nonlinear wave model in this case can be explained as follows. In general, the wave field comprises of both bound and free superharmonics. Whereas the Fourier approximation, satisfying the nonlinear free-surface conditions, assumes all modes to be bound as in (8); the procedure using linear theory assumes all modes to be free as in (12). Under certain conditions this assumption of all modes being free in the linear theory can yield realistic results. Fig. 5 illustrates this for monochromatic waves. The two curves show the ratio of the velocity associated with a second order bound wave to the calculated value if the bound superharmonic is interpreted to be a free superharmonic as in (12). For $kh \leq 0.5$ the assumption of the superharmonic being free yields no difference. For larger kh, this assumption underpredicts the horizonal velocity at the bottom and overpredicts the same closer to the free surface. For the wave conditions shown in Fig. 4 the bound superharmonics are stronger only over the shallow region of the bar. As the waves propagate past the top of the bar the bound modes are released and the free modes dominate the superharmonics. These test cases therefore present a situation where the linear theory predictions through (12) are likely to be good since the kh value is small (< 0.7 on top of the bar for case C) in the regions where the bound modes are most prominent.

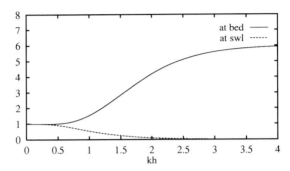

Figure 5: Ratio of the actual second-harmonic amplitude of the horizontal velocity to the calculated value if the second-harmonic bound component of the surface elevation is interpreted to be a free mode.

6 Summary & Conclusion

The limited comparisons presented here for analysing the performance of the prediction methods do not cover the wide-ranging situations that arise in the field. Notably absent factors are the directional spread, an external current and wave-breaking. The conditions studied are such that skewness and asymmetry develop during propagations of an initially sinusoidal train of waves. The present investigations can be summed up as follows:

We have discussed the use of three procedures in predicting velocity field under waves over varying depth. Of these three, the application of linear theory and the local Fourier approximation method require the availability of the time-records of the surface elevation. For a specified wave height and period, the velocity field under symmetric (about the crest), but skewed waves seem to be well represented by the steady Stream Function theory. Application of the local Fourier approximation (Sobey, 1992) to a skewed and asymmetric surface record shows that significant differences exist between the measured and calculated velocities. Two likely factors for this are the assumption of all modes being bound and the difficult nature of the solution over each local window. Performance of the LF method with respect to these two aspects need to be further investigated. In fact, the conjecture of LF method providing a better analytical fit than the global (over a wave period defined by two consecutive zero-upcrossigs) Fourier approximation (including both cosine and sine series) to an asymmetric surface record remains to be established. The good performance of the linear theory for the test cases presented here is discussed and explained in the preceding section using the concept of second-harmonic bound and free components of the surface elevation and their influence on the associated velocity field. Accordingly, the good performance of the linear theory observed in the present test cases may not be expected to be valid at higher value of kh for realistic waves. Application of the nonlinear wave model (Otta & Dingemans, 1994) to the three cases of non-breaking regular waves shows good agreement between the predicted and measured skewness and asymmetry of both the surface elevation and the velocity field due to waves transforming over varying depth.

Acknowledgements

This work was initiated during the first author's employment at Delft Hydraulics and later financed by the Danish National Research Foundation. Their support is greatly appreciated. The measurements reported here were carried out in Delft Hydraulics under the sponsorship of MAST-G8 [EC, MAS2-CT92-0027] and Coastal Genesis [Dutch Public Works Department, DG-469] programmes.

References

Dalrymple, R.A. (1974). A finite amplitude wave on a linear shear current, *J. Geophys. Res.*, (79): 4498-4504.

Dean, R.G. (1965). Stream Function represntation of nonlinear ocean waves, *J. Geophys. Res.*, (70): 4561-4572.

Dean, R.G. and M. Perlin (1986). Intercomparison of near-bottom kinematics by several wave theories and field and laboratory data, *Coastal Engg.*,(9): 399-437.

Hamm, L. (1996). Computation of the near-bottom kinematics of shoaling waves, in *Proc. 25th Int. Conf. on Coastal Engineering*, ASCE, 537-550.

Isobe, M. and H. Horikawa (1982). Study on water particle velocities of shoaling and breaking waves, *Coastal Engineering in Japan*, (25) 109-123.

Luth, H.R., G. Klopman and N. Kitou (1994). Project 13G, kinematics of waves breaking partially on an offshore bar, Delft Hydraulics report, H1573, 40pp.

Otta, A.K. and M.W. Dingemans (1994). Hamiltonian formulation of water waves; formulation, numerical evaluations and examples, Delft Hydraulics, H782, 70pp.

Otta, A.K., M.W. Dingemans and A.C. Radder (1996). A Hamiltonian model for nonlinear water waves and its applications, in *Proc. 25th Int. Conf. on Coastal Engineering*, ASCE, 1156-1167.

Radder, A.C. (1992). An explicit Hamiltonian formulation of surface waves in water of finite depth, *J. Fluid Mechanics*, (237) 435-455.

Rienecker, M.M. and J.D. Fenton (1981). A Fourier approximation method for steady water waves, *J. Fluid Mechanics*, (104) 119-137.

Sobey, R.J. (1992a). Estimation of irregular wave kinematics from a measured record, *Proc. 23rd Int. Conf. on Coastal Engineering*, ASCE, 644-657.

Sobey, R.J. (1992b). A local Fourier approximation method for irregular wave kinematics, *Applied Ocean Research*, vol. 14(2): 93-106.

LABORATORY OBSERVATIONS AND NUMERICAL SIMULATIONS
OF SHOALING SURFACE GRAVITY WAVES

Françoise BECQ [1], Michel BENOIT [2], Philippe FORGET [3]

Abstract

The present study aims at investigating some of the processes affecting shoaling waves in shallow water, with particular attention paid to the non-linear interactions between triplets of waves (triad interactions) and depth-induced breaking. Four monodimensional non-linear wave models (two phase-resolving and two phase-averaged spectral models) have been implemented and compared to laboratory experiments performed in a wave flume of the Laboratoire National d'Hydraulique (LNH). Tests were realised in breaking and non-breaking wave conditions. The non-linear mechanisms associated with the models are found to satisfactorily reproduce, both qualitatively and quantitatively, the wave spectra evolution along the bathymetric profile. The non-linear coupling effects are characterised by strong energy transfers between the interacting components of the wave field, and for some test-cases, by an important decay of wave energy.

1. INTRODUCTION

In the nearshore zone and in the coastal domain, surface gravity waves coming from offshore deep water undergo significant changes under the effects of various processes such as refraction, shoaling, surf-breaking, bottom friction and specific non-linear interactions. The wave spectrum can be profoundly modified, not only in its averaged integral parameters and moments, but also in its spectral shape and in the relationship between the phases of the various wave components. These modifications are of major concern for the hydrodynamics of the coastal zone and, in particular, for the design of coastal structures and for the modelling of morphodynamical evolutions.

The present study aims at investigating the influence of the non-linear interactions between triplets of waves on the energy density spectrum in shallow water. Triad interactions are usually considered as the main exchange mechanism for wave energy of an irregular wave train towards sub- or super-harmonics (e.g. Freilich and Guza,

---

[1] PhD Student — Maritime Hydraulics Section
[2] Research Engineer — Maritime Hydraulics Section
    EDF - Laboratoire National d'Hydraulique, 6, quai Watier 78400 CHATOU, FRANCE
[3] Research scientist — LSEET
    Université de Toulon et du Var, BP 132, 83957 LA GARDE, FRANCE

1984 and subsequent papers). The strength of the interactions is governed by the phase mismatch between bound and free waves. Since waves become non dispersive as water depth decreases, the phenomenon is enhanced towards the shore. For this reason, the domain of study has been extended up to the surf zone and the depth-induced wave breaking process was also considered.

The present work consists in experimental observations in a wave flume and numerical simulations using non-linear models of wave propagation in shallow water. We considered two kinds of models : phase-resolving and phase-averaged models. Phase-resolving models are commonly used to simulate the non-linear evolution of a wave field in shallow water. Many authors have been working on the derivation of different sets of evolution equations starting from the Boussinesq equations (Freilich and Guza, 1984), the Korteweg De Vries (KDV) equation (Mase and Kirby, 1992) or also the exact Euler equations. These models usually give an accurate description of the mechanisms involved in weakly non-linear situations. The drawback is the important computational cost associated with the method. Two such phase-resolving models have been tested in the present study. The first one was proposed by Madsen and Sorensen (1993) and is based on the Boussinesq equations. In order to extend the validity domain towards deep water, Agnon et al. (1993) proposed a model based on the Laplace equation. The phase-averaged spectral models offer the advantage of a lower computational cost. Abreu et al. (1992) first presented a non-linear spectral wave transformation model to simulate combined shoaling and triad-interactions effects. Their model assumes colinear resonant interactions implying a one way transfer towards higher frequencies. As these high frequencies do not satisfy the shallow water criterion, the model leads to an unexpected behaviour of the high frequencies part of the spectrum. To include off-resonant triad-interactions, Eldeberky et al. (1996) proposed a model based on the so-called Zakharov equation (Zakharov et al., 1992). Their model has been compared to the parametrized LTA model proposed by Elbeberky and Battjes (1995), which is characterized by the restriction to self-self triplet interactions. All the models have here been extended to the surf zone by the inclusion of a damping term in the equations, based on the parametrical model of Battjes and Janssen (1978).

After a brief presentation of the different models, a description of the laboratory experiments is made, followed by the comparison between measurements and simulations for non-breaking and breaking situations. As an outcome of this work, it is expected, on one hand, to increase our knowledge about the physical processes affecting shoaling waves in shallow water and, on the other hand, to develop efficient numerical formulations of these processes to be included in the spectral wave model TOMAWAC developed at LNH (Benoit *et al.*, 1996).

2 . PRESENTATION OF THE MODELS.

2.1. Phase-resolving models.

2.1.1. Model based on Boussinesq equations.

The Boussinesq equations are often used to study non-linear wave propagation in shallow water. They provide reasonably accurate descriptions of the wave field, by including the effects of weak dispersion and non-linearity under the assumptions that $\mu^2=(kh)^2<<1$, $\varepsilon=(a/h)<<1$ and $O(\mu^2)=O(\varepsilon)$, where k, h and a represent the wave number, the water depth and the wave amplitude respectively. In order to extend the Boussinesq type models towards intermediate water depths, Madsen and Sorensen (1992) improved their linear dispersion relationship and shoaling term by including a linear dispersion parameter B in the equations. It is more efficient to solve the equations in the frequency domain than in the time domain. To that end, starting from the extended Boussinesq equations (time domain) and expanding the free-surface elevation η(t) and the depth-integrated velocity P(t) in Fourier series, Madsen and Sorensen (1993) derived deterministic evolution equations for the amplitudes and phases of waves propagating over a mildly sloping bottom. The evolution equation for the complex Fourier amplitudes A_p reads:

(1) $$\frac{dA_p}{dx} = -C_{shoal}\frac{h_x}{h}A_p - 2ig\left[F_p^+ + F_p^-\right]$$

The first term on the right hand side represents linear shoaling whose coefficient C_{shoal} depends on the relative depth (μ) and the linear dispersion parameter (B). The dispersion relationship for component p, of angular frequency ω_p, is expressed by:

(2) $$\frac{\omega_p^2 h}{g} = \frac{\left(k_p h\right)^2 + B\left(k_p h\right)^4}{1 + (B + 1/3)\left(k_p h\right)^2}$$

The second term of (1) models the triad interactions. F_p^- and F_p^+ are coupling coefficients for the exchange of energy towards sub- and super-harmonics respectively. They depend on the amplitudes and phase mismatch of the interacting components. Their expressions can be found in Madsen and Sorensen (1993).

The standard form of the Boussinesq equations would lead to a similar equation as (1) with B=0.

2.1.2. Model based on the Laplace equation.

The Boussinesq models are restricted to weakly dispersive situations. To study the evolution of shoaling surface gravity wave fields of arbitrary spectra over bathymetric profiles from deep to shallow water, Agnon et al. (1993) derived an evolution equation in the frequency domain starting from the Laplace equation. Their model was derived using continuous Fourier integrals but, after discretization, the resulting evolution equation takes a form similar to (1). Differences mainly occur on the linear shoaling term, where full dispersion was taken into account, and on the coupling coefficients for triad interactions. Full expression of those terms can be found in Agnon et al. (1993) or in Kaihatu and Kirby (1995) (their non-linear parabolic mild-slope model is essentially a two-dimensional extension of the Agnon et al. (1993) model).

2.1.3. Inclusion of a breaking dissipation term.

The evolution equation for complex Fourier amplitudes has been extended to the surf zone by Eldeberky and Battjes (1996) with the inclusion of a damping term due to depth-induced breaking. Their model assumes that breaking dissipation, d_{br}, is proportional to the spectral energy of each frequency bin and reads:

(3) $$d_{br} = -\frac{1}{2}\frac{D_{tot}}{F_{tot}}A_p$$

F_{tot} is the total energy flux per unit width and D_{tot}, the total energy dissipation due to depth-induced breaking, can be obtained from several parametric models (Battjes and Janssen, 1978; Thornton and Guza, 1983). The evolution equation for complex Fourier amplitudes can be integrated with a fourth-order Runge-Kutta method allowing for the calculation of sea surface elevations or raw values of spectral energy densities along a bathymetric profile (Eldeberky and Battjes, 1996). The methods used for the calculation of the evolution of energy spectra and based on the Madsen and Sorensen (1993) and Agnon et al. (1993) models will be referred hereafter as SB (Spectral Boussinesq) and SL (Spectral Laplace) methods respectively.

2.2. Phase averaged models.

2.2.1. Model based on Zakharov equation for finite water depth.

Near-resonant triad interactions were formulated by Eldeberky et al. (1996) as a source term in phase-averaged spectral models. The formulation is based on the so-called Zakharov equation for resonant triplet interactions, which reads (Zakharov et al., 1992):

$$(4) \qquad \frac{dn_{\vec{k}}}{dt} = 4 \int \int d\vec{k}_1 d\vec{k}_2 \left[V_{k12}^2 N_{k12} \mu_{\omega-1-2} \delta_{\vec{k}-1-2} - 2 V_{1k2}^2 N_{1k2} \mu_{\omega-1+2} \delta_{\vec{k}-1+2} \right] \quad \text{where}$$

$$\mu_{\omega-1-2} = \frac{\Omega}{\left(\omega_k - \omega_1 - \omega_2 \right)^2 + \Omega^2}$$

N_{k12} depends on the spectral action density functions (n_i) of the interacting waves. V_{k12} is a coupling coefficient depending on the physics of the interacting waves. For surface gravity waves in finite depth, its expression has been given by Stiassnie and Shemer (1984). μ is a filter function of an unknown frequency parameter : Ω. For the resonant conditions $\vec{k} - \vec{k}_1 - \vec{k}_2 = 0$ and $\omega_k - \omega_1 - \omega_2 = 0$, Zakharov et al. (1992) imposed that $\Omega \to 0$, leading to $\mu_{k-1-2} = \pi \delta \left(\omega_k - \omega_1 - \omega_2 \right)$. To include the off-resonant triad interactions, Eldeberky et al. (1996), proposed to use a small (but finite) value of Ω.

The derivation of the energy source term for monodimensional situations was adapted from Eldeberky et al. (1996) model. We obtained:

$$(5) \qquad S_{tr}(\omega_k) = 16 \pi^2 g \int_0^\infty d\omega_1 \frac{C_{g,2}}{\omega_1 \omega_2} \left(T_{k12}^+ - 2 T_{1k2}^- \right) \mu_{k12} \quad \text{where}$$

$$T_{k12}^+ = V_{k12}^2 \left[\frac{\omega_k}{C_{g,k}} E_1 E_2 - \frac{\omega_1}{C_{g,1}} E_k E_2 - \frac{\omega_2}{C_{g,2}} E_1 E_k \right] \quad \text{and } T_{1k2}^- \text{ is derived from } T_{k12}^+$$

by the permutation of the interacting components. Variations of total energy can occur due to the non-resonant situation. However, the energy transfer is not rescaled to be equal to zero over the whole spectrum, as it is done in Eldeberky et al. (1996). The model will be referred hereafter as SZ (Spectral Zakharov) model.

2.2.2. Lumped Triad Approximation (LTA).

A parametrized approach to model the phase-averaged effects of triad wave interactions was proposed by Eldeberky and Battjes (1995). The LTA model is based on the deterministic model of Madsen and Sorensen (1993) (cf § 2.1.1). As the energy $E(f_p)$ is proportional to the square of the wave amplitude A_p, the evolution equation of the discrete spectral energy $E_p = \langle A_p A_p^* \rangle$ can be obtained from the evolution equation governing the individual complex Fourier amplitudes. The new equation depends on the so-called bispectrum defined as the third-order average amplitude product : $B_{m,p-m} = \langle A_m A_{p-m} A_p^* \rangle$. A quasi-gaussian hypothesis allows to express the fourth-order averages in terms of second-order averages and to close the system. To reduce the computational cost of this kind of model, Eldeberky and Battjes (1995) restricted the triad interactions phenomenon to self-self interactions. This considerably simplifies the net source term due to triad interactions for application in an energy balance equation:

$$S_{nl}(f_p) = S_{nl}^+(f_p) + S_{nl}^-(f_p)$$

(6) $$S_{nl}^+(f_p) = \alpha\, c_p c_{g,p} R_{(p/2,p/2)}^2 \sin\cdot\left|\beta_{p/2,p/2}\right|\left[E^2(f_{p/2}) - 2E(f_p)E(f_{p/2})\right]$$

$$S_{nl}^-(f_p) = -2\, S_{nl}^+(f_p)$$

The two terms of the source function represent respectively the sum and difference interactions between the three waves. c_p and $c_{g,p}$ are the phase and group velocities for the p harmonic, R is an interaction coefficient derived from the evolution equation (1) and α is a tuning parameter. A parametrical formulation for β, the biphase, has been proposed by Eldeberky and Battjes (1995) from laboratory observations.

2.2.3. Inclusion of a breaking dissipation term.

The spectral source terms associated to non-linear wave-wave interactions presented previously have been included in an energy balance equation. In order to extend the models to the surf zone, a breaking dissipation term was also included in the equation. The parametric model of Battjes and Battjes (1978) was chosen. As for the phase-resolving models, it was assumed that the dissipation was frequency independent.

3. EXPERIMENTAL INVESTIGATIONS IN A WAVE FLUME.

The non-linear models have been compared to laboratory measurements performed at the LNH. Experiments were made for non-breaking waves and for breaking waves.

3.1 Description of experimental set-up and test conditions.

The experiments were carried out in a wave flume (length: 45 m ; width: 0.60 m), equipped with a random piston-type wave-maker (figure 3.1.). Random wave trains were generated by a method of linear superposition of numerous sine components. These components have a random distribution of phases at the wave-maker and their amplitudes were computed from the target energy spectrum.

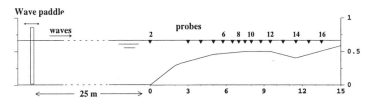

Figure 3.1. : Description of the wave flume.

Time series of free surface elevations were recorded by means of resistive probes at various locations in the wave flume, and in particular in the shallower part of the sloping bottom, where non-linear interactions and breaking are expected to occur in a significant way. The recording duration was chosen quite long in order to increase the number of degrees of freedom of the estimated spectra. Bottom friction dissipation and reflection from the beach were considered negligible in the experiments. Wave-by-wave (statistical) and spectral analysis were performed at each probe.

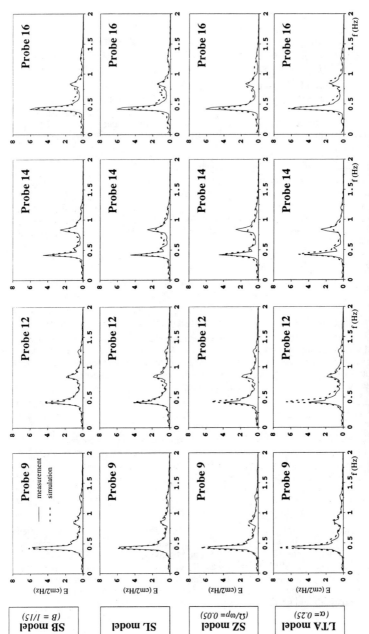

Figure 3.3 : Evolution of energy spectra along the bathymetric profile.
Comparison between measurements and simulations for a non-breaking test-case.

3.2. Results for non-breaking waves.

3.2.1. Presentation of the test-case.

The first test-case used for comparison of the models was conducted for non-breaking waves. The target spectrum at the wave generator is a JONSWAP spectrum with a peak frequency of 0.435 Hz and a significant wave height of 3.4 cm. We used as reference for numerical simulations the spectrum obtained at probe 2 (figure 3.2.). The water depth at the wave generator is 0.65 m.

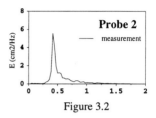

Figure 3.2

3.2.2. Comparison between simulations and measurements (figure 3.3).

The evolution of the energy spectrum along the bathymetric profile shows that, over the shoaling region, strong energy tranfers take place from the primary spectral peak towards higher harmonics (figure 3.3). These transfers are characterized by the generation of new spectral peaks and result only from the non-linear triad interactions. In a general manner, the four tested models are able to reproduce the generation of a secondary spectral peak. The phase-resolving models are characterized by an important computational cost compared to the spectral ones and the LTA model is the most efficient in CPU time due to its restriction to self-self interactions.

The Boussinesq deterministic approach particularly well predicts the strength of the energy transfers. Indeed, the measured energy spectra are well reproduced by the SB model, even very close to the shore at probe 16.

The SL model correctly reproduces the generation of the new spectral peaks even if the main spectral peak is slightly underestimated along the bathymetric profile.

The SZ model gives a good representation of the evolution of the energy spectra along the bathymetric profile. The new spectral peaks are shifted towards lower frequencies compared to the measurements. This behaviour results from the negative curvature of the dispersion relation which implies that if $\vec{k} = \vec{k}_1 + \vec{k}_2$ then $\omega_k < \omega_1 + \omega_2$ (Eldeberky et al., 1996) Compared to the formulation proposed by Eldeberky et al. (1996) for 1D situations, the expression implemented here does not generate the unwanted increase of energy towards high frequencies, even with important values for the frequency parameter Ω. Nevertheless, low values of this parameter lead also to an underestimation of the energy transfer towards bound higher harmonics.

The LTA model proposed by Elbeberky and Battjes (1995) overestimates the energy of the main spectral peak at probe 9 and 12. It can generate a secondary spectral peak, but, due to the restriction to self-self interactions, it is not able to create a third peak. Different values for α (betwen 0 and 1) have been tested. On this test-case the strength of the interactions, increasing with α, was only slightly modified.

3.3. Results for breaking waves.

3.3.1. Presentation of the test-case.

The test-case was characterized by strongly breaking waves. We used the same JONSWAP spectrum as before, but with a significant wave height of 8.0 cm (figure 3.4). The water depth at the wave generator was 0.61 m.

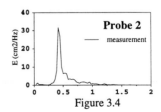

Figure 3.4

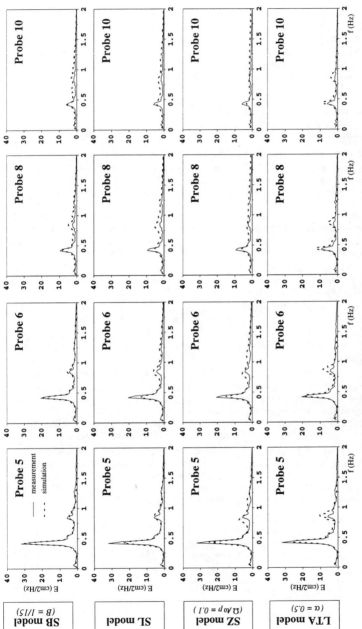

**Figure 3.5. : Evolution of energy spectra along the bathymetric profile.
Comparison between measurements and simulations for a breaking test-case.**

3.3.2. Comparison between simulations and measurements (figure 3.5).

The evolution of the energy spectra along the bathymetric shows that strong depth-induced wave breaking occurs in the shallower region (after probe 5) where dissipation is very important. The comparison between figure 3.4 and figures 3.5 shows that the energy transfer due to non-linear triplet interactions is important too and allows for the generation of new spectral peaks. In a general manner, all the models are able to reproduce the generation of a secondary peak. The transfer of energy towards higher frequencies is in general overestimated, but the results are quite good for the four models.

The phase-resolving models (and particularly the SB model) correctly simulate the laboratory measurements along the bathymetric profile. The SL model shows an important transfer of energy towards high frequencies and the energy of the main spectral peak is slightly underestimated. As proposed by Mase and Kirby (1992) or Chen and Guza (1997), we tried a breaking dissipation term proportional to the square of the frequency. The results were slightly improved for high frequencies where energy levels were found smaller. Such a refinement is not useful for the phase-averaged models tested in this study. They depend on tuning parameters governing the strength of the interactions and the evolution of wave spectra along the bathymetric profile is mainly governed by the choice of those parameters.

The spectral model based on the Zakharov equation generates the new spectral peaks at lower frequencies compared to the other models. Nevertheless, the wave spectra evolution is quite well reproduced along the bathymetric profile.

The LTA model gives a good representation of the spectral evolution of the wave field, even if the secondary peak is rather overestimated.

4 . CONCLUSIONS — FUTURE WORK

The evolution of measured energy spectra along a bathymetric profile shows that, over the shoaling region, strong energy tranfers take place from the primary spectral peak towards higher harmonics, leading to the generation of new spectral peaks. These energy transfers cannot be simulated with linear wave models and result from the triplet interactions process.

Four non-linear wave models were considered in the study. Comparisons between measurements and numerical simulations show that all the models are able to correctly reproduce the non-linear triad interactions phenomenon. The phase-resolving models are characterized by important computational costs but are able to fit very well the laboratory observations, especially for the Boussinesq type model proposed by Madsen and Sorensen (1993). When comparing the phase-resolving models, attention must be paid to the fact that the presented test-cases are representative of weakly dispersive situations. So, in deeper regions, the model proposed by Agnon et al. (1993) should lead to better results.

The phase-averaged spectral models offer the advantage of a lower computational cost, especially for the parametrized LTA model proposed by Eldeberky and Battjes (1995) which is restricted to self-self interactions. The two models correctly reproduce the evolution of the energy spectra along the bathymetric profile and they are being implemented in the wave model TOMAWAC (Benoit et al., 1996). The LTA model is very efficient for CPU time, but the model based on the Zakharov equation offers the advantage to be directionally coupled, allowing for the calculation of both colinear and non-colinear interactions. This later model is however characterized by a shift of the new spectral peaks towards lower frequencies compared to the measurements. This misbehaviour is induced by the negative curvature of the dispersion relation (Eldeberky et al., 1996).

The inclusion of a damping term due to depth-induced breaking in the equations allows to correctly simulate the evolution of the spectral shape for breaking waves. However, the results obtained with the phase-resolving models can be slightly improved by a repartition of the total breaking dissipation term proportionally to the square of the frequency (Chen and Guza, 1997).

5. REFERENCES

ABREU M., LARRAZA A. and THORNTON E. (1992). Nonlinear transformation of directional wave spectra in shallow water. *J. Geophys. Res.*, Vol 97, pp. 15579-15589.

AGNON Y., SHEREMET A., GONSALVES J. and STIASSNIE M. (1993). Nonlinear evolution of a unidirectional shoaling wave field. *Coastal Eng.*, 20, pp. 29-58.

BATTJES J.A., JANSSEN P.A.E.M. (1978) : Energy loss and set-up due to breaking of random waves. *Proc. 16th Int. Conf. on Coastal Eng.*, Vol 1, *pp 569-587.*

BENOIT M., MARCOS F., BECQ F. (1996) : Development of a third generation shallow-water wave model with unstructured spatial meshing. *Proc. 25th Int. Conf. on Coastal Eng. (ASCE)*, pp 465-478.

CHEN Y.H. and GUZA R.T. (1997). Modeling spectra of breaking surface waves in shallow water. *In press in J. Geophys. Res.*

ELDEBERKY Y. and BATTJES J.A. (1995). Parameterisation of triads interactions in wave energy models. *Proc. Int. Coastal Dynamics'95*, pp 140-148.

ELDEBERKY Y., BATTJES J.A. (1996) : Spectral modelling of wave breaking : application to Boussinesq equations. *J. Geophys. Res.*, 101 (C1), pp 1253-1264.

ELDEBERKY Y., POLNIKOV V. and BATTJES J.A. (1996). A statistical approach for modeling triad interactions in dispersive waves. *Proc. 25th Int. Conf. Coastal Eng.*, pp 1088-1101.

FREILICH M.H., GUZA R.T. (1984) : Non-linear effects on shoaling surface gravity waves. *Philos. Trans. Royal Soc. London*, A311, pp 1-41.

KAIHATU J.M. and KIRBY J.T. (1995). Nonlinear transformation of waves in finite water depth. *Phys. Fluids.*, 7 (8), pp. 1903-1914.

MADSEN P.A. and SORENSEN O.R. (1992). A new form of the Boussinesq equations with improved linear dispersion characteristics. Part 2: a slowly-varying bathymetry. *Coastal Eng.*, 18, pp. 183-205.

MADSEN P.A. and SORENSEN O.R. (1993). Bound waves and triads interactions in shallow water. *Ocean Eng.*, 20 (4), pp. 359-388.

MASE H. and KIRBY J.T. (1992). Hybrid frequency-domain KdV equation for random wave transformation. *Proc. of the 19th Int. Conf. on Coastal Eng.*, pp 474-487.

STIASSNIE M. and SHEMER L. (1984). On modifications of the Zakharov equation for surface gravity waves. *J. Fluid Mech.*, Vol. 143, pp. 47-67.

THORNTON, E.B. and R.T. GUZA (1983), Transformation of wave height distribution, *J. Geophys. Res.*, Vol 88, No. C10, pp 5925-5938.

ZAKHAROV V.E., L'VOV V.S and FALKOVICH G (1992). Kolmogorov spectra of turbulence, I. Wave turbulence, Springer Verlag.

High Resolution Measurements of Turbulent Fluxes and
Dissipation Rates in the Benthic Boundary Layer

George Voulgaris[1], John H. Trowbridge, William J. Shaw & Albert J. Williams III

Abstract

Two new techniques for estimating turbulent fluxes and dissipation rates in the benthic boundary layer are presented. Acoustic travel time instruments are used to obtain simultaneous measurements of sound speed fluctuations and 3-D flow velocities. The sound speed time-series are converted to temperature fluctuations and heat fluxes are estimated. Near the bed, Acoustic Doppler Velocimeters (ADV) are used to directly measure the 3-D velocity vector and estimate the momentum flux. A new technique using 2 ADV sensors, horizontally displaced, is utilized to calculate momentum fluxes in the presence of surface waves. This method removes the effect of wave contamination due to sensor misalignment. Examples of both techniques are presented from a field deployment of an 8m tripod.

Introduction

Understanding the mechanisms and implications of turbulent mixing in the bottom boundary layer of the coastal ocean requires high-resolution (in time and space) measurements of velocity, temperature and salinity. To date studies of bottom boundary layer dynamics have been limited by a number of technical problems. In particular, direct measurements of heat flux are limited by a lack of co-located, high frequency time-series of temperature and 3-D velocity while direct estimates of momentum flux in the presence of surface waves are subject to large errors due to sensor misalignment.

In order to overcome the above limitations scientists and engineers need to have in their disposal both accurate sensors and good methods. In this paper two major technological innovations are introduced:

(i) The development of an acoustic travel-time sensor for simultaneous high frequency measurements of fluid velocity and sound speed fluctuations; and

[1]Woods Hole Oceanographic Institution, Woods Hole, MA 02543

(ii) the use of horizontally separated Acoustic Doppler Velocimeters (ADV) for the estimation of near-bed turbulent momentum flux in the presence of waves.

Technological Innovations

Acoustic Thermometer.

Acoustic travel time instrumentation has been used extensively for current measurements (Williams et al., 1987). The travel times (t_1, t_2) of narrow beam acoustic pulses, propagating in opposite directions, depend on sound speed (c) and the component of water velocity (v) in the direction of the acoustic path of length l: $t_1 = l/(c+v)$ and $t_2 = l/(c-v)$. Knowledge of the difference of the arrival times (t_1-t_2) and the assumption that $c \gg v$ are adequate for the calculation of the water speed. Use of at least 3 pairs of transducers with different orientation of the acoustic axis are used to estimate the 3-D velocity field (Williams et al., 1987).

Previously, the acoustic travel time current meters were able to accurately measure the difference (t_1-t_2) in travel time and thus were able to provide information on current velocity only. A new generation of Bottom Acoustic Shear Stress (BASS) current meters has been developed (Trivett, 1991; Shaw et al., 1996) that, in addition to the time difference, measures the absolute times t_1 and t_2. These values can then be used to calculate the instantaneous value of sound speed:

$$c = \frac{2 \cdot l}{t_1 + t_2}, \qquad for \quad c \gg v \tag{1}$$

In principle, with knowledge of the local T-S relationship and use of an equation of state for seawater (e.g., McKenzie, 1981), sound speed can be related to temperature. Calculated T fluctuations correlated with co-located and simultaneous records of velocity fluctuations can provide information on heat fluxes and the structure of the benthic boundary layer during periods of neutral or well stratified conditions.

Differencing Technique for Reynolds Stress Estimates

Shear stress estimates in the constant stress layer can be obtained using the 'law of the wall', the inertial dissipation, or the Reynolds stress technique. The latter method is the most attractive; it provides a direct estimate of shear stress. However, it requires concurrent measurements of the 3-D flow field and, when waves are present, is sensitive to alignment of the sensor in relation to the local bed. The Acoustic Doppler Velocimeter (ADV) is a new coherent-pulse-to-pulse sensor that can provide accurate information on the 3-D flow field. Extensive laboratory work (Voulgaris and Trowbridge, in press) showed that a well calibrated ADV is capable of measuring Reynolds stresses, under steady flow conditions, within 1%. However, when waves are present, any error in alignment produces large errors in estimates of Reynolds stress. In such conditions, the fluctuating velocities (u, w) consist of a turbulent (u', w') and an oscillatory component (u_w, w_w). Assuming the two components are uncorrelated, it can be easily shown that the Reynolds stress estimate calculated from the measured values is:

$$<uw> = <u'w'> + <u_w w_w> \tag{2}$$

where the term $<u_w w_w>$ represents the error due to wave contamination.

Trowbridge (in press) demonstrated that the wave contamination can be removed using two sensors displaced horizontally (Fig. 1). Assuming that the horizontal separation (x) is: (i) much smaller than the wavelength (L) of the surface waves; and (ii) larger than the height above the bed (z), the wave components at the locations of the two sensors are highly correlated. On the other hand, the turbulent components, since turbulence is scaled by the elevation above the bed, are decorrelated at the two locations. Thus, the real Reynolds stress can be obtained by simply differencing the instantaneous velocity records obtained at the two displaced sensors:

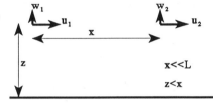

Figure 1. Schematic diagram of differencing technique.

$$<u'w'> = \frac{1}{2}<\Delta u \Delta w> = \frac{1}{2}<(u_1 - u_2)\cdot(w_1 - w_2)> \tag{3}$$

where the subscripts 1 and 2 denote instantaneous velocity values measured by two horizontally separated sensors (*see* Fig. 1). Using this technique, the error due to the wave contamination is $O[(kx)^2 <u_w w_w>]$ (Trowbridge, in press). In the case of a surface wave with wavelength of 100m and an horizontal separation of 1m, the error in Reynolds stress estimates using the differencing technique is reduced by a factor of approximately 3×10^{-3}.

Experimental Design

The experimental design presented here was developed for the Coastal Mixing and Optics Program of the Office of Naval Research. The main objective of the program is to quantify and understand the role of vertical mixing processes in determining the mid-shelf vertical structure of hydrography, optical properties of the seawater and particulate matter. Part of the program is concerned with the internal vertical mixing and in particular trying to identify the role of near-inertial and higher frequency shear and internal tides in enhancing mixing and also on the role of the surface and bottom generated turbulence.

In an attempt to ensure that the turbulence characteristics were measured at a substantial fraction of the bottom boundary layer, an 8m high tripod was constructed. The instrumentation suite installed on the tripod consisted of a vertical array of 7 new BASS sensors, measuring 3-D flow velocities and sound speed fluctuations at 35, 70, 105, 215, 320, 425 and 585 cm above the sea bed, two Seabird temperature / salinity gauges, at 35 and 585 cm above the seabed and a pressure sensor. In addition, turbulence characteristics near the seabed were measured using 3 ADV sensors. The ADV's were installed so that their sample volume was located at 35cm above the seabed. They were arranged in a triangular configuration with a separation of 220cm, so that the differencing technique could be applied to the 3 pairs of sensors, yielding three Reynolds stress estimates. The ADV output signal was recorded at 25Hz for 9.6 minute bursts, every hour, while the remaining sensors were sampled at 1.2Hz for 90 minute bursts every two hours.

The tripod was deployed south of Martha's Vineyard, in the mid-Atlantic Bight, at a mean water depth of 70m within the "Mud Patch". The seabed is characterized by a flat bottom and it consists of 53.7% sand, 38.8 silt and 7.5% clay. Examples of data from the first deployment period (August - September, 1996) are presented here.

<u>Results</u>

<u>Heat Fluxes</u>

Figure 2a shows the velocity variance during the deployment period as measured with the travel time (BASS) sensors. There are two significant peaks that occurred on 2nd and 14th September. These correspond to the passage of hurricanes Edouard and Hortense in the vicinity of the deployment area. Sound speed calculated from absolute travel time records at elevations of 35, 215 and 585 cm above the bed are shown in Figure 2b. It is worth noting that, although the sound speed varies with time, is constant with height during the first half of the displayed data. However, after 7th September, the sound speed at 585cm is greater than the sound speed at 35cm while the sound speed at 215cm fluctuates between the values of the lower and upper sensor. This exact pattern is reflected also in the temperature measurements obtained using the Seabird gauges (Fig. 2c). The high correlation between Seabird temperature and BASS sound speed records instills confidence in the latter. Stratification seems to be a seasonal phenomenon that is also affected by the storm-induced mixing. In particular, the water column is well mixed at the end of August despite the lack of storm activity. Hurricane Edouard assisted in maintaining these conditions. After the 8th September stratification starts developing and intensifies around the 10th September. After this, a well mixed period is observed that lasts until before hurricane Hortense. A new period of stratification was present during hurricane Hortense. The

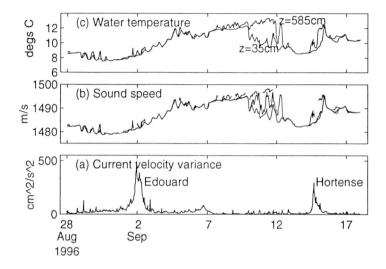

Figure 2. Time-series of burst-averaged (a) near-bed current velocity variance; (b) sound speed measured with BASS at 35, 215 and 585 cm; and (c) water temperature measured with the Seabird gauges at 35 and 585 cm above the sea bed.

persistence of the stratification throughout the hurricane period demonstrates the complexity of the system regarding the driving forces of both stratification and mixing. However, the high spatial and temporal resolution of the collected data will assist in further studying and understanding these processes and the fluxes associated with them.

Sound speed data for a particular burst corresponding to a period of well mixed conditions (31st August), were converted to temperature fluctuations assuming a constant salinity for the period of the burst (1.5 hours). The spectral characteristics of the temperature fluctuations are shown in Figure 3 together with the energy spectrum of the BASS measured vertical velocity. The noise floor of the temperature spectrum indicates that

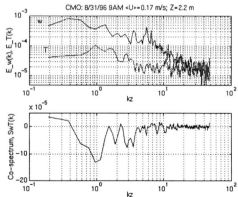

Figure 3. Top: vertical velocity (w, in m^3/s^2) and temperature (T, in deg^{2}m) wavenumber spectra of a burst time-series. Bottom: Co-spectrum of w and T for the same time-series.

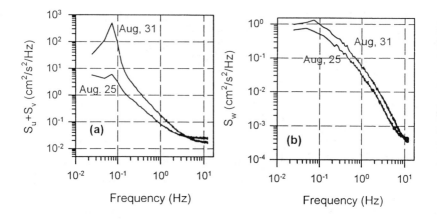

Figure 4. 24hour-averaged spectra of (a) horizontal and (b) vertical velocities for the periods 25th and 31st August, respectively.

this method can be used to estimate temperature fluctuations with accuracy of 10mC. The co-spectrum of w and T is shown on the bottom panel of Figure 3. For this particular period the co-spectrum indicates a downward heat flux. Co-spectra from other well mixed periods have indicated that upward heat fluxes can occur equally often. In contrast, data from periods of stable stratification have shown a predominantly downward heat flux.

Reynolds Stress Estimations

During the course of the experiment, periods of steady flow were identified as well as periods with oscillatory flow caused by large surface waves. Mean water speeds varied from 0 to 20cm/s. Direct inter-comparison of the mean flow measured by the three ADV sensors revealed differences in mean velocity up to 20%. Extensive laboratory work on the accuracy of the ADV sensors (Voulgaris and Trowbridge, in press) has shown that the sensor's accuracy in measuring mean flow is 1%, suggesting that the observed differences are true and not an artifact of measurement errors. A strong correlation of the velocity defect with the direction of the mean flow was found. The flow directions with the largest defects were associated with the tripod legs. This was a surprising result since the ADV's sample volumes were offset from the legs a distance 27 times the diameter of the legs. This is larger that the 20 times diameter usually recommended to obtain wake-free measurements. Despite this, flow statistics derived by averaging the readings of all 3 sensors are statistically significant as shown later (see Fig. 8).

Two 24-hour cycles were selected for presentation here. One data set was collected on 25th August and the second data set was collected on 31st August 1996, during the period hurricane Edouard was approaching the experimental site. Thus, the data from 25th August are characteristic of steady flow conditions whilst the data collected on the 31st August represent conditions of combined steady and oscillatory wave conditions. Mean current speeds varied between 2 and 16 cm/s during each of the days under consideration. The 24 hour averaged water speed was 8.2 and 7.2 cm/s whilst the wave-induced orbital velocity ranged between 0 and 1.5 (average 0.6cm/s) and 4 and 11cm /s (average 7.6cm/s) for the 25th and 31st August, respectively. 24-hour averaged spectra of the horizontal and vertical velocities are shown in Fig. 4.

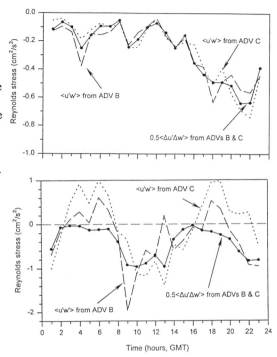

Figure 5. Reynolds stress estimates for the 25th and 31st August, (top and bottom panel, respectively). Estimates based on sensors B and C are shown as dashed and dotted lines whilst estimations based on the new differencing technique are shown as solid lines.

Reynolds stress values were calculated using both individual sensors and the differencing technique. The results of this analysis are shown in Figure 5 for the 25th and 31st August, respectively. During the period without waves (Fig. 5, top panel) all three estimates of Reynolds stress (from the individual sensors and from their differences) are in agreement and show the same trend. The estimates from the differencing technique are almost the average of the two individual sensors.

During the 31st August, the surface waves contaminate the Reynolds stress estimates based on an individual sensor. The contamination is due to alignment errors of the sensors; the Reynolds Stress estimates are contaminated by the $\langle u_w w_w \rangle$ term of eq (2). When the contamination is severe the Reynolds stress estimations are even positive as happens during the period of weak mean current (e.g., 3:00 to 7:00 hours

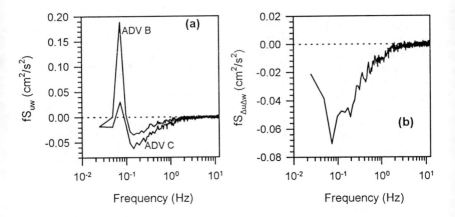

Figure 6. Area-preserving co-spectra of vertical (w) and downstream (u) velocities based on: (a) velocities recorded by the individual sensors; and (b) their difference.

and 17:00 to 22:00 hours). The difference in contamination between sensors B and C reflects the difference in misalignment. It is characteristic, though, that the use of the differencing technique gives much more realistic estimates of Reynolds stress even at times with weak mean current (e.g., 2:00 to 6:00 hours).

The effectiveness of the new technique is also shown in Figure 6. There, the 24-hour area-preserving co-spectra of u and w are shown for the 31st August data. Surface wave contamination of individual sensors' co-spectra is evident by the positive peaks at the frequency of 0.07Hz (Fig. 6a) which corresponds to the period of the surface wave field. The co-spectrum of the differenced signal (Fig. 6b) shows how successful this technique is in removing the surface wave signal. The co-spectrum remains negative even at the wave frequencies with a shape that is consistent with bottom generated turbulence.

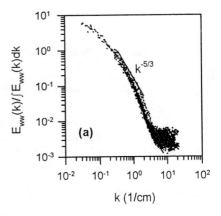

Figure 7. Normalised turbulent energy spectra of vertical velocity (all runs from 25th August). Notice the well developed -5/3 region.

Shear stress estimates from the new method were compared to shear stress estimates derived using the inertial dissipation technique on the spectra of the vertical velocity (Fig. 7). The comparison was undertaken on a selected number of bursts where the difference in mean velocity between sensors was less than 2% (Fig. 8a) and on the total number of bursts for the deployment period (Fig. 8b). The noteworthing characteristics are: (i) the shear velocity estimates from both techniques agree very well; (ii) the shear stresses from the inertial dissipation technique are higher at very low values (see best fit regression dashed line) indicating both the ability of the ADV to provide noise free Reynolds stress estimates (*see* Voulgaris and Trowbridge, in press) and the ability of the differencing technique to remove the wave contamination. Finally, the correlation between the two estimates is not altered when all of the bursts are included. This is because the effect of flow obstruction by the legs is diminished by averaging the estimates from all three pairs of sensors.

Conclusions

New technology and methods have been presented for obtaining accurate estimates of turbulent fluxes in the benthic boundary layer.

Sound speed measurements can be utilized for quantifying stratification and measuring turbulent temperature fluctuations in the benthic boundary layer. Data obtained with the acoustic thermometer can be used for estimating heat fluxes as described above, and additionally, temperature dissipation rates.

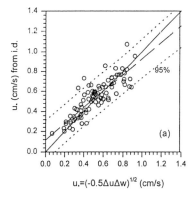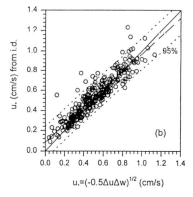

Figure 8. Comparison of shear stress estimates based on the inertial dissipation (i.d.) and the new differencing technique for (a) selected bursts (89) with differences in mean velocity between sensors less than 2%; and (b) for the total of the collected data (389 bursts). The 1:1 (solid) and the best fit line (dashed) with the 95% confidence level are also shown.

Inter-comparison of the records of the three ADV sensors revealed that measurements with benthic boundary layer tripods may suffer from contamination by turbulence induced by wakes from the structure.

The new technique for the removal of surface wave correlation from Reynolds stress estimates by differencing is a powerful tool for making direct estimates of momentum flux even under wave dominated conditions. Application of this technique with accurate sensors, such as the ADV, can enhance our understanding of bottom turbulence in the benthic boundary layer by acquiring high quality data even under extreme wave conditions.

Acknowledgments

The data presented and analyzed here were collected as part of the Coastal Mixing and Optics Program funded by ONR (Grant No. N900014-95-1-0373). Janet Fredericks is thanked for her invaluable assistance with the data retrieval and analysis. During the course of this study G.Voulgaris was recipient of a Woods Hole Oceanographic Institution Postdoctoral Award sponsored by the Andrew W Mellon Foundation. Traveling funds for G Voulgaris were obtained from the Dept. of Geology and Geophysics, Woods Hole Oceanographic Institution and the Andrew W Mellon Foundation, through the Rinehart Coastal Research Center.

Woods Hole Oceanographic Institution contribution 9529.

List of References

McKenzie, K.V., 1981. Nine-term equation for sound speed in the oceans. *J. Acoust. Soc. Am.*, 70(3): 807-812.

Shaw, W.,J., Williams,A.J.,III and Trowbridge,J.H., 1996. Measurement of Turbulent Sound Speed Fluctuations with an Acoustic Travel-Time Meter. Oceans'96 MTS/IEE, 105-110.

Trivett, 1991. Diffuse flow from hydrothermal vents. ScD. Thesis, MIT/WHOI, WHOI-91-23, 220pp.

Trowbridge, J.H., (in press). Note on a Technique for Measurement of Turbulent Shear Stress in the Presence of Surface Waves. *J. Atm. & Ocean. Tech.*

Voulgaris, G. and Trowbridge, J.H., (in press). Performance Evaluation of the Acoustic Doppler Velocimeter for Turbulence Measurements. *J. Atm. & Ocean. Tech.*

Williams, A.J., Tochko, J.S., Koehler, R.L., Grant, W.D., Gross,T.F. and Dunn, C.V.R., 1987. Measurements of Turbulence in the Oceanic Boundary Layer with an Acoustic Current Meter Array. *J. Atm. & Ocean. Tech.*, 4: 312-327.

A Comparison of Field Observations and Quasi-Steady Linear Shear Instabilities of the Wave Bottom Boundary Layer

D. L. Foster[1], A. J. Bowen[1], R. A. Beach[2] and R.A. Holman[2]

Abstract:

In this paper, we hypothesize that on flat seabeds, in the absence of depth penetrating wave breaking, the rapid generation of turbulence may be due to shear instabilities of the wave bottom boundary layer. Previous investigations have shown that the inflectional profile of the wave bottom boundary layer during and prior to flow reversal will support instabilities of appropriate length scales. In this paper, we hypothesize that beyond the effect on the instantaneous shear, the time-dependent nature of the wave bottom boundary layer has a direct and significant feedback to its instantaneous stability. A disturbance is defined as momentarily unstable when either the perturbation grows faster than the growth , or acceleration, of the background flow or when the perturbation is decaying less than the decaying, or decelerating, background flow.

The time-dependent model is evaluated with several synthetic wave cases and compared to field observations of near bed suspended sediment concentration and turbulent fluctuations. In all cases the time-dependent disturbance growth is significantly larger than the growth rate predicted by the quasi-steady model and was shown to increase with increasing wave skewness and asymmetry. In the nearshore region where the surface waves have strong skewness and asymmetry, we anticipate this effect to be particularly important. The model was also compared with Duck94 observations of suspended sediment concentration and turbulent variance. The spatial and temporal scales of the unstable perturbations are of correct order magnitude to be the mechanism for the generation of turbulence. Although, direct comparisons are difficult because much of the turbulent variance and suspended sediment may be advected past the sensors and not generated locally, qualitative comparisons are encouraging. Turbulent variance and suspended sediment events are generally coincident with peaks of model predicted growth rates.

[1] Department of Oceanography, Dalhousie University,Halifax, Nova Scotia, Canada, B3H 4J1.
internet: diane.foster@dal.ca

[2] Department of Oceanography, Oregon State University, Ocean Admin Bldg 104, Corvallis, Oregon, USA

1 Introduction

Nearshore suspended sediment clouds have sizes which vary from several centimeters to the water depth and are generally assumed to be rapidly suspended into the water column. We hypothesize that the rapid suspension is a result of rapid turbulence generation and is perhaps indicative of large vortical eddies. On flat beds, in the absence of wave breaking, a possible mechanism for the rapid generation of vortices is a shear instability of the oscillatory bottom boundary layer. With laboratory observations of an oscillatory boundary layer flow, Hino et al. (1983) concluded that increased levels of turbulence during the decelerating phase of the flow were triggered by a shear instability. The wave bottom boundary layer is the oscillatory boundary layer bordering the seabed which is forced by the surface gravity waves. Because the waves are periodic, the boundary layer reverses direction twice per wave period and consequently, is constricted to approximately the lowest 10 cm of the water column. Previous investigations have shown that the inflectional profile of the wave bottom boundary layer during and prior to flow reversal will support instabilities of appropriate length scales (Foster et al., 1994 and Foster et al., 1996). Both of these investigations used a quasi-steady approximation to examine each instantaneous profile for flow stability and neglected any interaction between the evolution of the background flow and the evolution of the disturbance.

According to Shen (1961), the background flow deceleration has a significant effect on flow stability beyond the effect of the shape of the instantaneous profile. He showed the quasi-steady approximation holds when the instantaneous time constant for the background flow divided by the time constant for the disturbance is significantly greater than instantaneous $Re^{1/3}$ or

$$\sigma_{\iota_c}/f >> Re^{1/3}. \tag{1}$$

where σ_{ι_c} is the critical quasi-steady perturbation growth rate, f is the wave frequency, $Re = u_m \delta/\nu$ is the instantaneous Reynolds number, u_m is the instantaneous maximum velocity, and δ is the boundary layer thickness. Even using averaged wave statistics, this relation would require that for a 100 cm/s free stream velocity, a 10 cm boundary layer thickness, and a 10 s wave period, the growth rate must satisfy, $\sigma_{\iota_c}(t) >> -4.6 \tan(\theta) |\cos(\theta)|^{1/3}$ s^{-1}, where θ is the wave phase. During flow reversal, when the accelerations are large and $\tan(\theta) \Rightarrow \infty$, the quasi-steady approximation is not valid.

In this paper, we examine the growth or decay of a disturbance relative to the growth or decay of the background flow field and compare with results from a quasi-steady model. This is accomplished following a method for considering the stability of time-dependent flows outline by Shen (1961). The time-dependent model is evaluated with several synthetic wave cases and compared to field observations of near bed suspended sediment concentration and turbulent fluctuations.

This paper is organized in the following manner. The theoretical formulation and solution method is given in section 2. Model evaluation and comparisons with observations are made in section 3. Finally, the conclusions are given in section 4.

2   Theoretical Formulation

Recall the linearized inviscid perturbation equation (Rayleigh and Strutt, 1880)

$$(\Delta^2 \Psi)_t + \tilde{u}(\Delta^2 \Psi)_x = \tilde{u}_{zz} \Psi_x \tag{2}$$

where Ψ is the perturbation stream function defined with perturbation velocities $u = \Psi_z$ and $w = -\Psi_x$, $\tilde{u} = f(z, t)$ is the oscillatory wave velocity, and the subscripts t, x, and z represent the partial derivative. The solution to this equation is

$$\Psi(x, z, t) = \phi(z) e^{i(kx - \sigma t)} \tag{3}$$

where k is the real perturbation wavenumber, $\sigma = \sigma_r + i\sigma_i$ is the complex perturbation frequency, and σ_i is the growth rate.

Previous investigations have made quasi-steady assumptions by examining the stability of the instantaneous profile. Here, we give the derivation for a background flow which evolves on time scales short enough to affect the growth or decay of given disturbances, following previous work by Shen (1961). The above equation may be simplified if background wave velocity may be decomposed with

$$\tilde{u}(z, t) \equiv \bar{u}(z) T(t) \tag{4}$$

A transformation which stretches or contracts the time variable is defined with

$$\tau \equiv \int T \, dt. \tag{5}$$

In the case of an oscillatory boundary layer, T is periodic and consequently τ is stretched and contract with the same periodicity. Substitution of (4) and (3) into (2) results in the simplified perturbation equation

$$(\Delta^2 \Psi)_\tau + \bar{u}(\Delta^2 \Psi)_x = \bar{u}_{zz} \Psi_x. \tag{6}$$

A perturbation solution of the classical form is assumed

$$\Psi(x, z, t) = \phi(z) e^{i(kx - \tilde{\sigma}\tau)}, \tag{7}$$

where k is the real perturbation wavenumber and $\tilde{\sigma} = \tilde{\sigma}_r + i\tilde{\sigma}_i$ is the complex perturbation frequency for the stretched temporal scales. In the classical quasi-steady problem, σ_i is the sole factor determining the growth or decay of a perturbation. In this situation the rate of growth of the perturbation is affected by the growth or decay of the background wave field. The energy ratio, E, is defined with the ratio of the volume integrated kinetic energy of the perturbation to the volume integrated kinetic energy of the wave is defined with

$$E(t) = \frac{\int_0^\delta \int_0^{\frac{2\pi}{k}} (u^2 + w^2) dx dz}{\frac{2\pi}{k} \int_0^\delta \tilde{u}^2 dz}. \tag{8}$$

where δ is the wave boundary layer thickness. The relative growth of the perturbation to the background wave field can be defined with (Shen, 1961 and Conrad and Criminale, 1965):

$$\Gamma \equiv \frac{1}{2E} \frac{dE}{dt}. \tag{9}$$

For separable background flow, as in (4), this becomes

$$\Gamma = \tilde{\sigma}_i T - T_t / 2T. \tag{10}$$

This definition of growth rate was chosen so that it coincides with the growth rate for steady flow, $\Gamma = \sigma_i$. Herein, the flow is considered stable if, under accelerating flow, the disturbance is not growing as fast as the background flow. Furthermore, under decelerating flow, the flow is considered stable if the disturbance is decaying faster than the background flow.

For the case considered here, the background flow is not simply separable in time and space. However, at any time, t_o, we may separate the velocity field into its spatial and temporal components. We assume that locally, in time, the evolution of the background flow may be characterized by the volume integrated kinetic energy of the background flow,

$$T(t_o) = \int_0^\delta \tilde{u} dz^2, \text{ and} \tag{11}$$

$$T'(t_o) = \frac{\partial}{\partial t} \int_0^\delta \tilde{u}^2 dz.$$

Unlike the separable case above where (6) must only be solved once, this require3 (6) be solved at each instant in time. The growth rate of the perturbation relative to the growth of the background wave field is

$$\Gamma = \sigma_i - \frac{\frac{d}{dt} \int_0^\delta \tilde{u}^2 dz}{\int_0^\delta \tilde{u}^2 dz}. \tag{12}$$

3   Results

First, we compare the predicted flow stability for the quasi-steady and time-dependent models for 4 synthetic wave cases. The 4 wave cases were chosen to adequately represent a range of skewed and asymmetric waves commonly found in the nearshore region. The boundary layer profiles were estimated with a time- and depth- dependent eddy viscosity model with

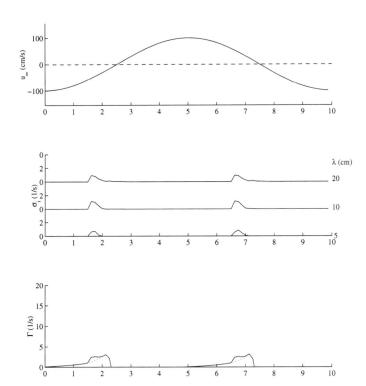

Figure 1: This figure shows the model results for a skewed wave: free stream velocity (top panel), quasi-steady perturbation at 3 separate wavelengths (middle panel); and the combined growth rate, Γ for a wavelength of 10 cm. The normalized wave skewness and asymmetry are S=0, and A=0.

arbitrary wave forcing, (Foster, 1996). Figures 1-4 show the flow stability for a monochromatic, skewed, asymmetric, and skewed and asymmetric free stream waves In the first three cases, the time-dependent growth rates are approximately 3 times higher than the quasi-steady growth rates. In the fourth case, Figure 4, the time-dependent model predicts an instantaneous growth rate which is nearly 30 times the quasi-steady model. The presence of skewed and asymmetry, synonymous with a wave bore, produces large instantaneous wave accelerations around the period of flow reversal which has a significant effect on flow stability and produces a large bias of the growth rates between the wave troughs and crests.

Secondly, we compare the model predictions with observations of near-bed turbulent velocity fluctuations and suspended sediment concentrations. The observations were made at the Army Corps of Engineers, Field Research Facility (FRF) in Duck, NC on August 17, 1994, as part of the Duck94 field experiment. The significant offshore wave height,

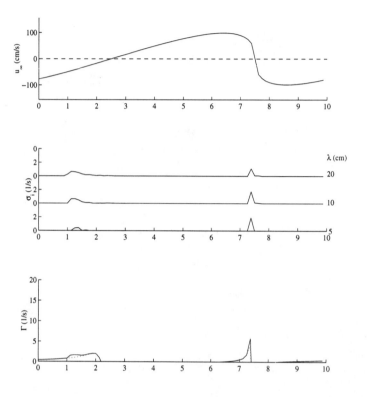

Figure 2: This figure shows the model results for a skewed wave: free stream velocity (top panel), quasi-steady perturbation at 3 separate wavelengths (middle panel); and the combined growth rate, Γ for a wavelength of 10 cm. The normalized wave skewness and asymmetry are S=0, and A=1.3.

angle, and period measured in 8 m water depth were 0.83 m, 50 from the southeast, and 4.54 s, respectively. The observations were made in 2 m water depth on the crest of the bar over flat bed conditions. Near bed velocity observations were made with an array of 5 hot-film anemometers, sampled at 256 Hz, separated with a 1 cm vertical spacing. The velocity outside the wave bottom boundary layer was measured with an electromagnetic current meter. The bed elevation and suspended sediment were measured with a vertical stack of 19 sensor fiber-optic backscatter probe, sampled at 16 Hz. The bed elevation is determined by examining each independent sensor for burial. A thorough presentation of field techniques and instrument calibration may be found in Foster et al. (1996).

Because the suspension events have a vertical extent of approximately 10 cm, we search for unstable perturbations which have a wavelength of 10 cm. At all times, the growth rates for the time-dependent model are at least an order of magnitude larger than those

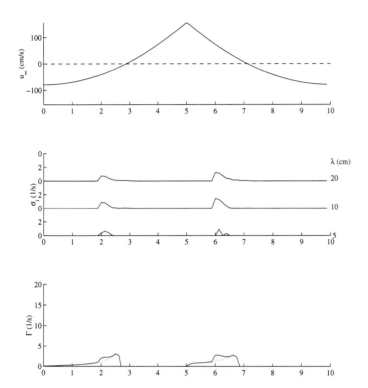

Figure 3: This figure shows the model results for a skewed wave: free stream velocity (top panel), quasi-steady perturbation at 3 separate wavelengths (middle panel); and the combined growth rate, Γ for a wavelength of 10 cm. The normalized wave skewness and asymmetry are S=.6, and A=0.

predicted by the quasi-steady model. A instantaneous growth rate of $10\ s^{-1}$ corresponds to an e-folding time scale of .1 s, and is of correct magnitude to be responsible for the rapid generation of turbulence and suspension of sediment for in the nearshore region.

The integrated turbulent variance shows a reasonable correlation to the model predictions. The four distinct concentration events are coincident with growing disturbances. Direct comparisons are difficult because the model only predicts the growth or decay of locally generated disturbances and much of the turbulent variance and suspended sediment may be advected past the sensors and not generated locally.

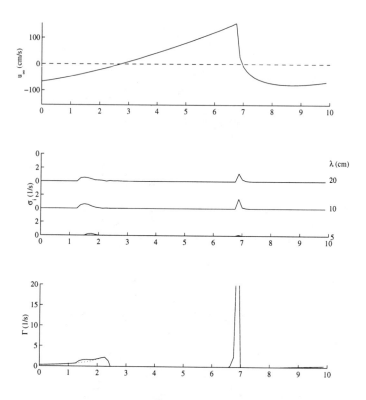

Figure 4: This figure shows the model results for a skewed wave: free stream velocity (top panel), quasi-steady perturbation at 3 separate wavelengths (middle panel); and the combined growth rate, Γ for a wavelength of 10 cm. The normalized wave skewness and asymmetry are S=.6, and A=1.3.

4   Conclusions

In this paper, we hypothesize that beyond the effect on the instantaneous shear, the time-dependent nature of the wave bottom boundary layer has a direct and significant feedback to its instantaneous stability. The time-dependent growth was defined as the growth or decay of a disturbance relative to the growth or decay of the background flow field (Shen ,1961 and Conrad and Criminale, 1965). A disturbance is defined as momentarily unstable when either the perturbation grows faster than the growth , or acceleration, of the background flow or when the perturbation is decaying less than the decaying, or decelerating, background flow.

The quasi-steady and time-dependent stability of four synthetic free stream wave profiles was examined. In all cases the time-dependent disturbance growth is significantly larger than the growth rate predicted by the quasi-steady model and was shown to increase

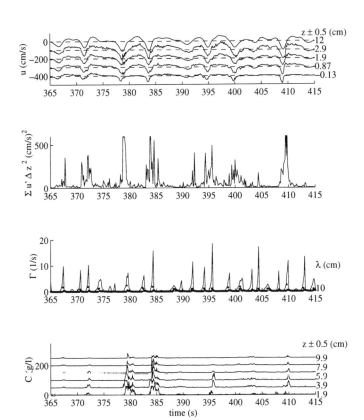

Figure 5: This example 50 sec time series shows near bed velocity as measured by four hot film anemometers and the free stream velocity as measured by an EMCM (top panel); depth integrated high frequency VITA velocity variance (upper middle panel); the growth rates predicted by the time-dependent model (---) and the quasi-steady line ($\cdots$) (lower middle panel); time dependent and concentration at 5 elevations above the bed (bottom panel). The average distance of each sensor as measured by the FOBS is listed in the right column. Each velocity, concentration, and predicted growth rate time series is offset by 100 cm/s, 100 g/l, and 2 Hz, respectively. Offshore directed flow is defined with positive velocities.

with increasing wave skewness and asymmetry. In the nearshore region where the surface waves have strong skewness and asymmetry, we anticipate this effect to be particularly important. Furthermore, we anticipated this effect will evolve as waves propagate across the beach.

The model was also compared with Duck94 observations of suspended sediment concentration and turbulent variance. Although, direct comparisons are difficult because the model only predicts the growth or decay of locally generated disturbances and much of the turbulent variance and suspended sediment may be advected past the sensors and not generated locally, qualitative comparisons are encouraging. Turbulent variance and suspended sediment events are generally coincident with peaks of model predicted growth rates.

Acknowledgments

This work was graciously funded by the Office of Naval Research, Coastal Sciences Program, and the Andrew Mellon Foundation. We wish to thank Keith Thompson for his insight into the numerical computations world. The first author also wishes to acknowledge the nursing staff at the Cardiff Royal Infirmary, without whose assistance this paper would not exist.

References

Conrad, P.W. and W. O. Criminale (1965),*The stability of time-dependent laminar flow: Parallel flows*, Z. Angew. Math. Phys., 16,233-254.

Foster, D.L., R.A. Beach, and R.A. Holman (1994), *Correlation sediment suspension events and shear instabilities in the bottom boundary layer of the surf zone*, Coastal Dynamics '94, ASCE, Barcelona, Spain.

Foster, D.L., R.A. Beach, and R.A. Holman (1996), *Field Observations of the Wave Bottom Boundary Layer*, submitted to Journal of Geophysical Research.

Foster, D.L. (1996), *Dynamics of the nearshore wave bottom boundary layer*, Ph.D. thesis, Oregon State University.

Hino,M., M. Kashiwayanagi, A. Nakayama, and T. Hara (1983), *Experiments on the turbulent statistics and the structure of a reciprocating oscillatory flow*, Journal of Fluid Mechanics, 131, 363-400, .

Rayleigh, L., and J.W. Strutt (1880), *On the stability, or instability of certain motions*, Proceedings of the London Mathematical Society, 11, 57-70.

Shen, S.F.(1961), *Some Considerations on the laminar stability of incompressible time-dependent basic flows*, Journal of Aerospace Science, 28, 397-404 and 417.

Mixing Processes due to Breaking Wave
Activity in the Coastal Zone

J. M. Pearson[1], I. Guymer[1], L. E. Coates[2], J. R. West[2]

Abstract

Data are presented from a series of laboratory based
experiments conducted on the UK Coastal Research Facility
(CRF) at HR Wallingford, Ltd. to characterise the solute
mixing processes due to breaking waves. The results show
that the transverse mixing is significantly increased,
almost 50 times, when compared to current only conditions.

Introduction

Industrial and sewage treatment works often discharge
their effluent into estuarine or coastal waters. With
increased awareness by both regulatory bodies and the
general public, it has become vital to assess the importance
of pollutants in these regions.

Frequently, the tools employed for the management of
the coastal environment are numerical simulations. However,
although considerable advances have been made in the
understanding of the hydrodynamic processes occurring within
this zone, few experimental data are available to validate
the mathematical predictions.

In numerical models of the coastal region, the
dispersal of plumes are simulated using predicted velocity
distributions and an estimated mixing coefficient. Even
fewer data are available, particularly with respect to
direct tracer measurements from which to estimate the mixing
coefficient. This results in an inability to accurately
predict the mixing occurring within the coastal zone.

This paper describes a series of fluorescent tracer
experiments and reports results for two known breaking wave
conditions undertaken on the CRF, to elucidate the magnitude
of mixing within the surf zone.

----

[1]Dept. of Civil & Structural Engineering, University of
Sheffield, Mappin street, Sheffield S1 4DT, UK.

[2]School of Civil Engineering, The University of Birmingham,
Edgbaston, Birmingham B15 2TT, UK.

Background

There have been comparatively few experimental studies to investigate aspects of mixing under waves in the coastal zone. Some site specific field studies have been undertaken, however, the contribution to mixing due to wave activity is difficult to interpret. An extensive review of nearshore mixing was undertaken by Bowen & Inman (1974). Despite being undertaken over two decades ago, little new information is available for studies within the coastal zone.

Bowen & Inman (1974) referred to previous investigations by Harris et ~l.(1963) who undertook a series of experiments in both field and laboratory to investigate the mixing of a solute when released into the surf zone. Both field and laboratory-based experiments produced results, which suggested that the mixing across the surf zone is proportional to H^2/T, where H is the crest to trough wave height and T is the wave period.

Bowen & Inman (1974) also refer to previous studies undertaken by Inman, Tait & Nordstrom (1971) in which they suggest that the mixing across the surf zone is much larger than in the longshore direction. They made observations in which a spot of dye released into the surf zone quickly dispersed across the zone, with most of the dye staying within the confines of the zone, and only returning seawards through rip currents. Based on the experimental results obtained, it was suggested that the mixing across the surf zone was a function of $H_b X_b/T$ where X_b is the width of the surf zone and H_b is the wave height at the breaker line.

Tanaka et al.(1980) constructed a physical model to calibrate the dispersion of cooling water from a power station and compared results to field tests. Although comparison of results with other experimental works are difficult, because of the site specific nature of the experiments, they found that the mixing across the surf zone produced similar results to those of Harris et al.(1963) and were one order of magnitude lower than those of Inman et al.(1971).

Bowen & Inman (1974) referred to previous work undertaken by Bowen (1969), Thornton (1970) & Longuet-Higgins (1970) in which the possible mechanisms for lateral mixing on gently sloping beaches were discussed. In these studies, they utilised measured longshore current distributions and applied theoretical diffusion coefficients which model the measured currents in mathematical predictions. Putrevu & Svendsen (1992) numerically demonstrated that the cross-shore current, generated by wave activity within the coastal zone, dominated the mixing. They suggest that this could exceed the contribution to mixing caused by turbulent activity by an order of magnitude, but report that few data are available to fully validate their findings.

This brief review suggests that although the likely mechanisms to describe the mixing within the nearshore zone are known, few experimental data are available. Hence, this paper describes detailed tracer measurements to characterise the solute mixing processes due to breaking wave activity.

Laboratory Study

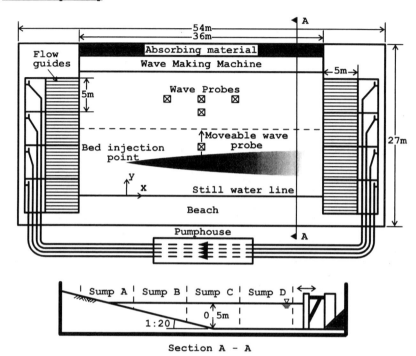

Figure 1. Layout of the Coastal Research Facility

The results presented in this paper relate to
measurements undertaken on the CRF (Figure 1). Tests were
performed on the transverse sloping section of the facility,
whereby a parabolic longshore current was created, and three
wave conditions of varying wave height, nominally
H{0,160,210}mm with a constant period of 1.2s were
superimposed. The current parallel to the shore was
generated by four individual computer controlled pumps.
Water from each pump was stilled in 5m wide sumps. Fine
tuning of the flow from each sump was undertaken by varying
the discharge through 10 individual adjustable undershot
sluice gates before entering flow straighteners. This
arrangement is repeated at the downstream end of the
facility. The waves on the facility are generated by 72
computer controlled paddles and investigations concentrated
on monochromatic waves, with the crests approaching parallel
to the shore. The bed of the facility consisted of 10mm
granite chippings embedded into cement mortar.

For the velocity measurements, a 3D Sontec Ultrasonic
Current Meter (UCM) was used to determine the mean flow
field for the longshore current condition. The instrument
was capable of measuring three orthogonal directions at a

maximum frequency of 25Hz, and has a quoted resolution of
±1mm/s. Measurements were taken at a constant depth of 80mm
below the still water level and a mean value was obtained
from 180s of velocity data. Measurements were undertaken at
three locations along the beach to describe the longshore
current. Unfortunately, due to the air entrainment of
breaking waves, the UCM was unsuitable for velocity
measurements with the waves superimposed. Hence, it was only
possible to obtain velocity measurements under the current
only condition.

Wave height measurements were undertaken at three
locations in the offshore section of the basin (15m from the
shore). A single wave probe was also stepped across the
centre line of the basin at 0.5 metre intervals to determine
the breaking point of the waves.

A constant head injection of Rhodamine WT dye was
introduced to the basin from a small tapping point flush
with the bed, 5m from the shoreline (centre of sump B) and
7.65m from the flow guides. The subsequent spreading of the
tracer was then recorded by pumping one litre samples into
sterilised containers for later analysis. The CRF has a recirculating flow system which causes the continual build-up
of background dye concentrations. Hence, to minimise the
temporal build-up during testing, ten discrete samples were
collected simultaneously with their sample tubes spaced at a
transverse distance of 100mm apart. The ten samples took
approximately two minutes to collect, and the array of tubes
were stepped across the plume at one-metre intervals. Up to
five transverse sections at different longitudinal locations
within the basin were investigated for each test condition
to describe the lateral mixing.

Results

Results for the velocity measurements under current
only conditions are shown in Figure 2. No longitudinal bed
slope exists on the facility, thus it was not possible to
get true uniform flow. The results appear to show that the
velocity increases by approximately 10% over the 20m control
section of the facility.

The offshore wave conditions were continuously logged
on three probes throughout the duration of testing and an
average height of 150 & 190mm for each wave condition was
recorded. Very small variations between the wave probes were
measured, less than 2% of the mean value. No attempt was
made to allow for the small amount of reflection from the
beach.

Figure 3 shows the ensemble averaged wave profiles of
approximately 100 waves, taken across the centreline of the
basin. The results indicate that the waves begin to break at
approximately 4.5m for the 150mm wave and 6m for the 190mm
wave.

The samples collected during the testing period were
later passed through a Series 10 Turner Designs Fluorometer.
By calibrating the instrument with known concentrations of
dye and fitting a least squares linear regression through
the points, the voltage output from the Fluorometer can be
converted to concentrations.

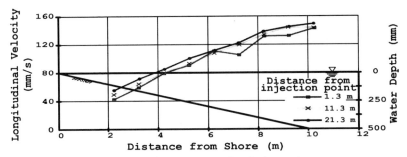

Figure 2.   Longshore velocities at
             three sections along beach

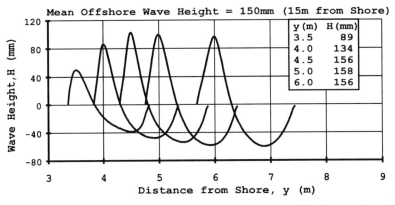

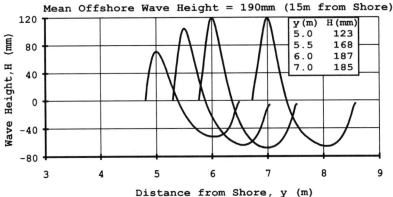

Figure 3.   Ensembled Wave Profiles
             Across Centreline of Basin

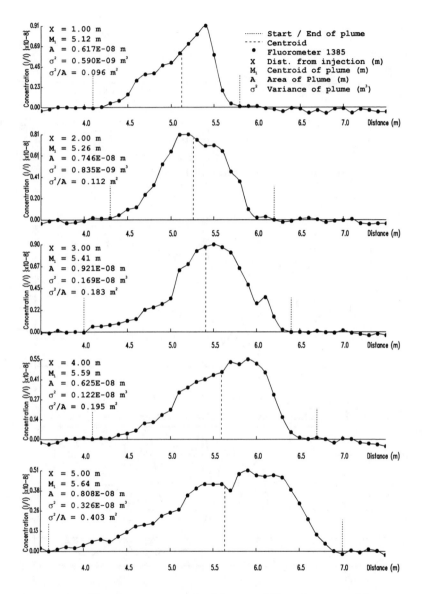

Figure 4.   Transverse concentration profiles for a continuous
             injection of tracer, 5m from the shore line under
             monochromatic waves with an offshore wave height
             of 150mm superimposed on a longitudinal current

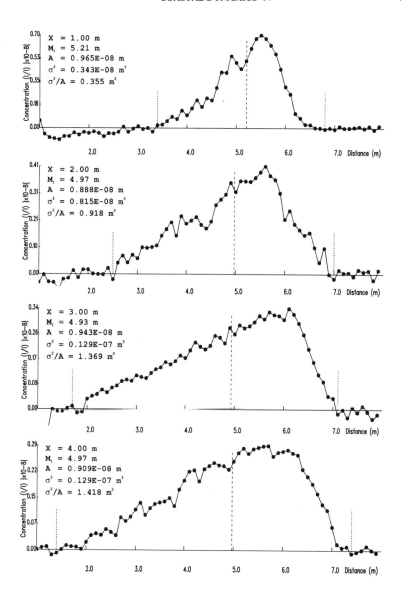

Figure 5.   Transverse concentration profiles for a continuous
             injection of tracer, 5m from the shore line under
             monochromatic waves with an offshore wave height
             of 190mm superimposed on a longitudinal current

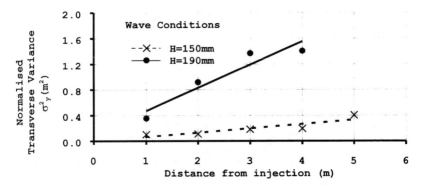

Figure 6.   Transverse mixing characteristics
under breaking waves

The problem with a re-circulating body of water, is the continual build-up of background dye concentrations. In the CRF, the problem is further complicated because there are four individual pumping systems to circulate the current, and thus, the background builds up at different rates in each sump. Therefore, during the data collection period, background samples were taken upstream of the injection point, inline with the array of extraction tubes which aided the subtraction of the background concentration for each individual profile. Mass recovery ratios of ±20% and ±5% of the mean values were obtained for the 150mm & 190mm wave conditions respectively, which due to the nature of the surf zone measurements seems reasonable.

Employing the concept of Taylor's (1953) turbulent diffusion analogy, which states:

$$\mathcal{E}_{sy} = \frac{U}{2} \frac{d\sigma^2}{dx} \qquad \qquad 1$$

an estimate of the transverse mixing coefficient $\mathcal{E}_{sy}$, combining both diffusion and dispersion or differential advection processes can be obtained.

Wave Condition	H=0mm (current-only)	H=150mm, T=1.2s Breaking at y=4.5m	H=190mm, T=1.2s Breaking at y=6.0m
Transverse Mixing Coefficient $\mathcal{E}_{sy}$ (cm²/s)	3.6 Potter et al. (1996)	32	170

Table 1.   Transverse mixing characteristics under varying wave conditions

Figure 2 shows that the velocity distribution varies over the width of the plume, but equation 1 does not account for the transverse variations in velocity, hence the validity of Taylor's theory in this case is questionable. It however, can be seen from Figure 6, that the increase in variance with longitudinal distance is approximately linear in the region observed. A velocity at the 5m injection point of 0.095m/s was adopted for the derivation of the transverse mixing coefficient and results are presented in Table 1.

Discussion

Results have been presented in Table 1 to illustrate the effect that increased wave breaking has on the transverse mixing characteristics of a longshore current. The proportion of surf zone mixing was artificially changed by increasing the wave height. On a sloping beach the associated shoaling process has the effect of moving the break point offshore. Nevertheless, in these tests it was not possible to maintain the dye wholly within the surf-zone. However, for the 150mm waves, the break point was shoreward of the injection point and for the 190mm case the waves were already breaking 1m further offshore than the injection point. Thus, the wave height should be viewed as an indirect index of the degree of breaking rather than a parameter explaining the increased mixing coefficient.

It is clear from the asymmetry of the concentration profiles in Figures 4 and 5, that the calculation of a global mixing coefficient is inappropriate. However, there does not seem to be an established technique for describing this situation. It may be seen, for example from the profile at x=3m in Figure 5, that considerably more mixing is occurring in the surf zone. The increase in global mixing brought about by shifting the break point 1.5m, together with the associated increase in wave height, is of the order of 5 times.

The analysis of Putrevu & Svendsen (1992) showed that interactions between oblique wave velocities and their inherent longshore induced currents provided a mechanism for much increased mixing. The results in this paper may show that the interaction between an imposed longshore current and orthogonal wave velocities produce similarly large mixing coefficients. However, comparison of the mixing coefficients in Table 1 with the field and laboratory data in Bowen & Inman (1974) show these new values to be low. A possible explanation is that the plume is only partially in the surf zone, but it may also suggest that the mechanism for orthogonal waves is weaker.

Further study of Figures 4 and 5 reveals that although the shoreward sides of the profiles exhibit obviously increased mixing the regions just outside the surf zone, say between y =5m and y = 7m, are still contributing to mixing which is considerably in excess of the current only coefficients. This conclusion also agrees with the predictions of Putrevu & Svendsen's (1992) model but is not predicted by the simpler models of wave-induced transverse variations of longshore currents.

What is now needed is a model that assumes, on some physical basis, a parameterised cross-shore variation of mixing coefficient and which can combine it with the introduction of a continuous point source of dye. The resulting concentration profiles could then be tuned against

the CRF measurements to reveal the optimum parameter set.
Putrevu & Svendsen's (1992) model may provide the basis for
such a procedure.

Conclusions

The addition of breaking and near-breaking shore-
orthogonal waves to a longshore current has been
demonstrated to lead to a dramatic increase in cross-shore
mixing. Surf zone mixing increases by at least 50 times over
the current only case. In addition, increasing the wave
height, and the subsequent increase in surf zone width,
produced a 5-fold increase in cross-shore mixing.

Acknowledgement

The authors gratefully acknowledge the UK Engineering and
Physical Sciences Research Council (Grant Ref. GR/J/12499)
for their financial support and the technical assistance of
HR Wallingford for the research undertaken on the CRF.

References

Bowen A. J., 1969. The generation of longshore currents on a
plane beach. J. Mar. Res., 27(2): 206-15.

Bowen A. J. & Inman D. L., 1974. Nearshore mixing due to
waves and wave-induced currents. Rapp. P.-v. Reun. Cons.
int. Explor. Mer, 167: 6-12.

Harris T. F. W., Jordan J. M., McMurry W. R., Verwey C. J. &
Anderson F. P., 1963. Mixing in the surf zone. Int. J. Air
Wat. Pollut., 7: 649-67.

Inman D. L., Tait F. J. & Nordstrom C. E., 1971. Mixing in
the surf zone. J. geophys. Res., 76: 3493-3514.

Longuet-Higgins M. S., 1970. Longshore currents generated by
obliquely incident sea waves, parts 1 and 2. J. geophys.
Res., 75: 6778-6801.

Potter R., Guymer I., Pearson J.M., West J.R. & Coates L.E.,
1996. Mixing processes within open channel flows containing
transverse depth variations. Proc. 2nd Int. Conf.
Hydrodynamics, Hong-Kong, 1013-18.

Putrevu U & Svendsen I. A., 1992. A mixing mechanism in the
nearshore region. Proc. 23rd Int. Conf. on Coastal
Engineering, Italy, 2758-71.

Tanaka H., Wada A., Komori S. & Tekeuchi I., 1980. Effects
of nearshore currents on diffusion in the surf zone. Coastal
Engineering in Japan, Vol 23, 231-50.

Taylor G. I., 1953. Dispersion of soluble matter in solvent
flowing slowly through a tube. Proc. R. Soc. London Ser. A,
219, 186-203.

Thornton E. B., 1970. Variation of longshore currents across
the surf zone. Proc. 12th Int. Conf. on Coastal Engineering,
Washington, 291-308.

DISPERSIVE MIXING IN THE NEARSHORE

Uday Putrevu [1] and Ib A. Svendsen [2]

Abstract: It is shown that the vertical nonuniformity of the short-wave-averaged horizontal velocities leads to mixing-like terms for the horizontal velocity in the depth-integrated equations of momentum. The mechanism is analogous to Taylor's (1953, 1954) shear-dispersion mechanism for solutes in a shear flow. The results presented here are an extension of the results found by Svendsen & Putrevu (1994) to the general case of unsteady flow over an arbitrary bottom topography.

1 Introduction

In a previous paper (Svendsen & Putrevu 1994, SP94 hereafter), we considered the case of steady longshore currents on a longshore-uniform coast and found that the vertical nonuniformity of the currents leads to a mixing-like term in the depth-integrated longshore momentum equation. The mechanism by which this happens is analogous to the shear-dispersion mechanism found by Taylor (1953, 1954) for the lateral spreading of solutes in a shear flow.

For the case considered in SP94, the lateral mixing caused by the shear-dispersion mechanism was an order of magnitude larger than the turbulent lateral mixing, even inside the surf zone. This result suggests that the dispersive mixing will be a major

---

[1] NorthWest Research Associates, PO Box 3027, Bellevue, WA 98009, USA, e-mail: putrevu@nwra.com.

[2] Center for Applied Coastal Research, Ocean Engineering Lab, University of Delaware, Newark, DE 19716, USA.

contributor to the total lateral mixing in the nearshore region. Therefore, it is desirable to extend the results of SP94 to the general case of unsteady flow over an arbitrary bottom topography. This paper presents such an extension.

The present paper is organized as follows. Section 2 discusses the depth-integrated, short-wave-averaged equations of continuity and momentum for the case in which the short-wave-averaged horizontal velocities are allowed to vary with the vertical coordinate. The evaluation of the extra terms that arise from the vertical nonuniformity of the short-wave-averaged horizontal velocities forms the subject of Sections 3 and 4. The final section is devoted to a few concluding remarks.

2 Depth-integrated, short-wave-averaged equations

We start with the depth-integrated, short-wave-averaged equations of continuity and horizontal momentum which allow for the short-wave-averaged velocities to vary with the vertical. These equations are derived following the steps given in Phillips (1977) or Mei (1989) and are minor extensions of the equations given therein [eqs. 3.6.4 and 3.6.11 of Phillips (1977, pp. 61–62) and eqs. 2.50 and 2.51 of Mei (1989, p. 463)]. (Phillips and Mei only considered situations where the short-wave-averaged velocities are uniform over the vertical.)

In the subsequent equations the instantaneous horizontal velocity u_α is divided into a wave component $(u_{w\alpha})$, a turbulence component (u'_α), and a short-wave-averaged velocity (V_α). The short-wave-averaged velocity is further split into two components. The first component is uniform over depth and is defined by

$$\tilde{V}_\alpha = \frac{1}{h}\overline{\int_{-h_0}^{\zeta} u_\alpha \ dz} \tag{1}$$

where an overbar denotes averaging over a short-wave period, h_0, and ζ, represent the still water depth, and instantaneous water surface elevation, respectively and $h = h_0 + \overline{\zeta}$ represents the total depth. The second component, $V_{1\alpha}$, accounts for the vertical variation of V_α and satisfies

$$\overline{\int_{-h_0}^{\zeta} V_{1\alpha} dz} = -\overline{\int_{\zeta_t}^{\zeta} u_{w\alpha} \ dz} = -Q_{w\alpha} \tag{2}$$

where ζ_t represents the elevation of the wave trough level, and $Q_{w\alpha}$ is short-wave-induced volume flux. Notice that, in addition to representing the vertical variation of

V_α, the depth-averaged value of $V_{1\alpha}$ is the part of the short-wave-averaged motion that compensates for the volume flux due to the short wave motion.

In terms of these variables, the depth-integrated, short-wave-averaged equations read

$$\frac{\partial \overline{\zeta}}{\partial t} + \frac{\partial}{\partial x_\alpha}\left(\tilde{V}_\alpha h\right) = 0 \tag{3}$$

and

$$\frac{\partial}{\partial t}\left(\tilde{V}_\alpha h\right) + \frac{\partial}{\partial x_\beta}\left[\tilde{V}_\alpha \tilde{V}_\beta h + \int_{-h_0}^{\overline{\zeta}} V_{1\alpha} V_{1\beta}\ dz + \overline{\int_{\zeta_t}^{\zeta}\left(u_{w\alpha}V_{1\beta} + V_{1\alpha}u_{w\beta}\right)\ dz}\right]$$
$$+\frac{1}{\rho}\frac{\partial S_{\alpha\beta}}{\partial x_\beta} + gh\frac{\partial \overline{\zeta}}{\partial x_\alpha} - \frac{\tau_\alpha^S}{\rho} + \frac{\tau_\alpha^B}{\rho} = 0 \tag{4}$$

where z is the vertical coordinate (measured from the still water level), and the indices α, β represent the x, y - directions. τ_α^S, τ_α^B, and $S_{\alpha\beta}$ are the surface shear stress, the bottom shear stress, and the short-wave-induced radiation stress, respectively. In (4) the radiation stress is defined by

$$S_{\alpha\beta} = \overline{\int_{-h_0}^{\zeta}\left(\rho u_{w\alpha}u_{w\beta} + p\delta_{\alpha\beta}\right)dz} - \frac{1}{2}\rho gh^2\delta_{\alpha\beta} \tag{5}$$

where p is the total pressure and $\delta_{\alpha\beta}$ is the Kronecker delta function. This definition of the radiation stress is slightly different from the one used by Phillips (1977, eq. 3.6.12, p. 62).

Note that since we have allowed for a time variation in (3) and (4), these equations model both mean currents and long waves. In the above, we have neglected the turbulent lateral stresses in the depth-integrated momentum equation in anticipation of the fact that the direct lateral mixing caused by the turbulence will be much smaller than the mixing caused by the shear-dispersion mechanism discussed here.

The terms involving V_1 in (4) represent the effects of the vertical nonuniformity of the short-wave-averaged velocities, and it is these terms that give rise to the dispersive mixing. The goal of the following is to express the dispersive terms in terms of $\tilde{V}_\alpha$ and the short-wave-related quantities ($S_{\alpha\beta}$, $Q_{w\alpha}$, $u_{w\alpha}$, etc.) so that (3) and (4) reduce to equations in which the only unknowns are $\overline{\zeta}$ and $\tilde{V}_\alpha$.

3 Vertical structure of $V_{1\alpha}$

The vertical structure of the short-wave-averaged horizontal velocity is governed by (Svendsen & Lorenz 1989, Putrevu & Svendsen 1997)

$$\frac{\partial V_{1\alpha}}{\partial t} - \frac{\partial}{\partial z}\left(\nu_t \frac{\partial V_{1\alpha}}{\partial z}\right) = -\left(\frac{\partial \tilde{V}_\alpha}{\partial t} + \tilde{V}_\beta \frac{\partial \tilde{V}_\alpha}{\partial x_\beta} + g\frac{\partial \overline{\zeta}}{\partial x_\alpha} + f_\alpha\right)$$
$$- \left(\tilde{V}_\beta \frac{\partial V_{1\alpha}}{\partial x_\beta} + V_{1\beta}\frac{\partial \tilde{V}_\alpha}{\partial x_\beta} + V_{1\beta}\frac{\partial V_{1\alpha}}{\partial x_\beta} + W\frac{\partial V_{1\alpha}}{\partial z}\right) \tag{6}$$

where f_α is the local contribution to the radiation stress and is given by

$$f_\alpha = \overline{u_{w\beta}\frac{\partial u_{w\alpha}}{\partial x_\beta}} + \overline{w_w \frac{\partial u_{w\alpha}}{\partial z}} \tag{7}$$

In (6) W is the short-wave-averaged vertical velocity, and ν_t is the turbulent eddy viscosity. [Equation 6 is derived by introducing an eddy viscosity closure to model the turbulent stresses in the local short-wave-averaged equation of horizontal momentum (see Svendsen & Lorenz 1989)].

The conditions associated with (6) are

$$\nu_t \frac{\partial V_{1\alpha}}{\partial z}\Bigg)_{z=-h_0} = \frac{\tau_\alpha^B}{\rho}, \qquad \int_{-h_0}^{\overline{\zeta}} V_{1\alpha}\ dz = -Q_{w\alpha} \tag{8}$$

and an appropriate initial condition.

If (6) subject to (8) can be solved in closed form, then the integrals required in (4) can be readily calculated. Thus, we concentrate on solving (6) subject to (8) below. It turns out that such a closed form solution is possible for situations in which

$$V_{1\alpha} \ll \tilde{V}_\alpha \tag{9}$$

Using a scaling analysis, Putrevu & Svendsen (1997) show that in cases in which (9) is violated, we typically have one of two situations

1. Either the dispersive terms are insignificant or

2. The variations in the alongshore direction are small. In such cases, the SP94 solution applies.

Therefore, considering the case in which (9) is satisfied is relevant. It also turns out that while the results below are derived assuming (9), they are robust and are also valid for situations in which the basic assumption is violated.

Putrevu & Svendsen (1997) show that (9) implies that we can solve (6) by expanding $V_{1\alpha}$ as follows

$$V_{1\alpha} = V_{1\alpha}^{(0)} + V_{1\alpha}^{(1)} \tag{10}$$

with $V_{1\alpha}^{(1)} \ll V_{1\alpha}^{(0)}$. Putrevu & Svendsen further show that the equations for $V_{1\alpha}^{(0)}$ and $V_{1\alpha}^{(1)}$ reduce to the following.

The $V_{1\alpha}^{(0)}$ problem

$$\frac{\partial V_{1\alpha}^{(0)}}{\partial t} - \frac{\partial}{\partial z}\left(\nu_t \frac{\partial V_{1\alpha}^{(0)}}{\partial z}\right) = F_\alpha \tag{11}$$

where F_α represents the difference between the depth-averaged and local values of the forcing on a fluid element and is given by

$$F_\alpha = \frac{1}{\rho h}\frac{\partial S_{\alpha\beta}}{\partial x_\beta} - f_\alpha - \frac{\tau_\alpha^S}{\rho h} + \frac{\tau_\alpha^B}{\rho h} \tag{12}$$

The conditions associated with (11) are

$$\nu_t \frac{\partial V_{1\alpha}^{(0)}}{\partial z}\Bigg)_{z=-h_0} = \frac{\tau_\alpha^B}{\rho}, \qquad \int_{-h_0}^{\bar{\zeta}} V_{1\alpha}^{(0)}\ dz = -Q_{w\alpha} \tag{13}$$

and an appropriate initial condition. The solution of the $V_{1\alpha}^{(0)}$ problem turns out to be completely analogous to the usual undertow problem and its solution is rather straightforward (see Putrevu & Svendsen 1997).

The $V_{1\alpha}^{(1)}$ problem

The $V_{1\alpha}^{(1)}$ problem reduces to the following

$$\frac{\partial V_{1\alpha}^{(1)}}{\partial t} - \frac{\partial}{\partial z}\left(\nu_t \frac{\partial V_{1\alpha}^{(1)}}{\partial z}\right) = -\left(\tilde{V}_\beta \frac{\partial V_{1\alpha}^{(0)}}{\partial x_\beta} + V_{1\beta}^{(0)}\frac{\partial \tilde{V}_\alpha}{\partial x_\beta} + W\frac{\partial V_{1\alpha}^{(0)}}{\partial z}\right) \tag{14}$$

subject to the conditions

$$\nu_t \frac{\partial V_{1\alpha}^{(1)}}{\partial z}\Bigg)_{z=-h_0} = 0, \qquad \int_{-h_0}^{\bar{\zeta}} V_{1\alpha}^{(1)}\ dz = 0 \tag{15}$$

and an initial condition.

To the order of accuracy of (14), it is consistent to approximate W in (14) by the following

$$W = -(h_0 + z)\frac{\partial \tilde{V}_\alpha}{\partial x_\alpha} - \tilde{V}_\alpha \frac{\partial h_0}{\partial x_\alpha} \tag{16}$$

(The above is derived by integrating the local short-wave-averaged continuity equation.)

Solution for $V_{1\alpha}^{(1)}$

A closed form solution of (14) subject to the appropriate conditions can be worked out following the procedure given in Carslaw & Jaeger (1959). This solution is discussed in Putrevu & Svendsen (1997) and is not discussed here due to space limitations. Instead, we consider the the simpler situation in which the time scale of $V_{1\alpha}^{(1)}$ is large enough so that the temporal acceleration term in (14) may be neglected. [It is shown in Putrevu & Svendsen (1997) that the solution given below represents the first approximation of the complete solution.] For this situation, it is straightforward to integrate (14) twice to get the following solution for $V_{1\alpha}^{(1)}$

$$V_{1\alpha}^{(1)}(z) - V_{1\alpha}^{(1)}(-h_0) = \tilde{V}_\beta \int_{-h_0}^{z} \frac{1}{\nu_t} \int_{-h_0}^{z'} \left(\frac{\partial V_{1\alpha}^{(0)}}{\partial x_\beta} - \frac{\partial h_0}{\partial x_\beta} \frac{\partial V_{1\alpha}^{(0)}}{\partial z''} \right) \ dz'' \ dz'$$

$$+ \frac{\partial \tilde{V}_\alpha}{\partial x_\beta} \int_{-h_0}^{z} \frac{1}{\nu_t} \int_{-h_0}^{z'} V_{1\beta}^{(0)} \ dz'' \ dz' - \frac{\partial \tilde{V}_\beta}{\partial x_\beta} \int_{-h_0}^{z} \frac{1}{\nu_t} \int_{-h_0}^{z'} \left[(h_0 + z'') \frac{\partial V_{1\alpha}^{(0)}}{\partial z''} \right] \ dz'' \ dz' \quad (17)$$

The constant of integration, $V_{1\alpha}^{(1)}(-h_0)$, can be determined from the integral constraint on $V_{1\alpha}^{(1)}$. The integrals required in (4) can now be calculated as follows

$$\int_{-h_0}^{\overline{\zeta}} V_{1\alpha} V_{1\beta} \ dz + \overline{\int_{\zeta_t}^{\zeta} (u_{w\alpha} V_{1\beta} + V_{1\alpha} u_{w\beta})} \ dz \approx \int_{-h_0}^{\overline{\zeta}} V_{1\alpha} V_{1\beta} \ dz + V_{1\beta}(\overline{\zeta}) Q_{w\alpha} + V_{1\alpha}(\overline{\zeta}) Q_{w\beta} \quad (18)$$

$$= \int_{-h_0}^{\overline{\zeta}} V_{1\alpha}^{(0)} V_{1\beta}^{(0)} \ dz + V_{1\beta}^{(0)}(\overline{\zeta}) Q_{w\alpha} + V_{1\alpha}^{(0)}(\overline{\zeta}) Q_{w\beta}$$

$$+ \int_{-h_0}^{\overline{\zeta}} \left(V_{1\alpha}^{(0)} V_{1\beta}^{(1)} + V_{1\alpha}^{(1)} V_{1\beta}^{(0)} \right) \ dz$$

$$+ V_{1\beta}^{(1)}(\overline{\zeta}) Q_{w\alpha} + V_{1\alpha}^{(1)}(\overline{\zeta}) Q_{w\beta} + O(V_1^{(1)})^2 \quad (19)$$

4 Results for the integrals

Substitution of (17) into (19) leads to the following, after some straightforward calculations

$$\int_{-h_0}^{\overline{\zeta}} V_{1\alpha} V_{1\beta} \ dz + V_{1\beta}(\overline{\zeta}) Q_{w\alpha} + V_{1\alpha}(\overline{\zeta}) Q_{w\beta} = M_{\alpha\beta} + A_{\alpha\beta\delta} \tilde{V}_\delta$$

$$- h \left(D_{\delta\beta} \frac{\partial \tilde{V}_\alpha}{\partial x_\delta} + D_{\delta\alpha} \frac{\partial \tilde{V}_\beta}{\partial x_\delta} + B_{\alpha\beta} \frac{\partial \tilde{V}_\delta}{\partial x_\delta} \right) \quad (20)$$

where the tensors A, B, D, and M are defined by the following

$$
A_{\alpha\beta\delta} = -\left\{ \int_{-h_0}^{\bar{\zeta}} \frac{1}{\nu_t} \left(\int_{-h_0}^{z} \frac{\partial V_{1\alpha}^{(0)}}{\partial x_\delta} - \frac{\partial h_0}{\partial x_\delta} \frac{\partial V_{1\alpha}^{(0)}}{\partial z'} \, dz' \right) \left(\int_{-h_0}^{z} V_{1\beta}^{(0)} dz'' \right) dz + \right.
$$
$$
\left. \int_{-h_0}^{\bar{\zeta}} \frac{1}{\nu_t} \left(\int_{-h_0}^{z} \frac{\partial V_{1\beta}^{(0)}}{\partial x_\delta} - \frac{\partial h_0}{\partial x_\delta} \frac{\partial V_{1\beta}^{(0)}}{\partial z'} \, dz' \right) \left(\int_{-h_0}^{z} V_{1\alpha}^{(0)} dz'' \right) dz \right\} \tag{21}
$$

$$
B_{\alpha\beta} = -\frac{1}{h} \left\{ \int_{-h_0}^{\bar{\zeta}} \frac{1}{\nu_t} \left(\int_{-h_0}^{z} (h_0 + z') \frac{\partial V_{1\alpha}^{(0)}}{\partial z'} \, dz' \right) \left(\int_{-h_0}^{z} V_{1\beta}^{(0)} dz'' \right) dz + \right.
$$
$$
\left. \int_{-h_0}^{\bar{\zeta}} \frac{1}{\nu_t} \left(\int_{-h_0}^{z} (h_0 + z') \frac{\partial V_{1\beta}^{(0)}}{\partial z'} \, dz' \right) \left(\int_{-h_0}^{z} V_{1\alpha}^{(0)} dz'' \right) dz \right\} \tag{22}
$$

$$
D_{\alpha\beta} = \frac{1}{h} \int_{-h_0}^{\bar{\zeta}} \frac{1}{\nu_t} \left(\int_{-h_0}^{z} V_{1\alpha}^{(0)} \, dz' \right) \left(\int_{-h_0}^{z} V_{1\beta}^{(0)} \, dz'' \right) \, dz \tag{23}
$$

$$
M_{\alpha\beta} = \int_{-h_0}^{\bar{\zeta}} V_{1\alpha}^{(0)} V_{1\beta}^{(0)} \, dz + V_{1\beta}^{(0)}(\bar{\zeta}) Q_{w\alpha} + V_{1\alpha}^{(0)}(\bar{\zeta}) Q_{w\beta} \tag{24}
$$

Substituting (20) into (4) we then get

$$
\frac{\partial}{\partial t} \left(\tilde{V}_\alpha h \right) + \frac{\partial}{\partial x_\beta} \left(\tilde{V}_\alpha \tilde{V}_\beta h \right) + \frac{1}{\rho} \frac{\partial S_{\alpha\beta}}{\partial x_\beta} + gh \frac{\partial \bar{\zeta}}{\partial x_\alpha} - \frac{\tau_\alpha^S}{\rho} + \frac{\tau_\alpha^B}{\rho}
$$
$$
+ \frac{\partial}{\partial x_\beta} \left[M_{\alpha\beta} + A_{\alpha\beta\delta} \tilde{V}_\delta - h \left(D_{\delta\beta} \frac{\partial \tilde{V}_\alpha}{\partial x_\delta} + D_{\delta\alpha} \frac{\partial \tilde{V}_\beta}{\partial x_\delta} + B_{\alpha\beta} \frac{\partial \tilde{V}_\delta}{\partial x_\delta} \right) \right] = 0 \tag{25}
$$

Thus, as in SP94, we find that the vertical variation of the short-wave-averaged horizontal velocities leads to mixing-like terms in the depth-integrated momentum equation and the mechanism by which this happens (the combination of vertical mixing and horizontal advection) is identical to Taylor's (1953, 1954) shear dispersion mechanism. An attractive feature of this result is that the resulting dispersion tensor can be readily calculated. This is important because, as in SP94, the lateral mixing caused by the shear dispersion mechanism is expected to dominate the lateral mixing in the nearshore.

Before proceeding further, we note that the expression for the dispersion tensor, $D_{\alpha\beta}$, can be rewritten in a form that is similar to the one given in Fischer et $al.$ (1979) for the dispersion of solutes. To do that, we define

$$
p = \int_{-h_0}^{z} V_{1\alpha}^{(0)} \, dz' \quad \text{and} \quad q = \int_{z}^{\bar{\zeta}} \frac{1}{\nu_t} \int_{-h_0}^{z'} V_{1\beta}^{(0)} \, dz'' \, dz' \tag{26}
$$

so that

$$\frac{1}{h} \int_{-h_0}^{\overline{\zeta}} \frac{1}{\nu_t} \left(\int_{-h_0}^z V_{1\alpha}^{(0)} \, dz' \right) \left(\int_{-h_0}^z V_{1\beta}^{(0)} \, dz'' \right) \, dz = -\int_{-h_0}^{\overline{\zeta}} p \, dq \tag{27}$$

$$= -pq]_{-h_0}^{\overline{\zeta}} + \int_{-h_0}^{\overline{\zeta}} q \, dp \tag{28}$$

$$= \frac{1}{h} \int_{-h_0}^{\overline{\zeta}} V_{1\alpha}^{(0)} \int_z^{\overline{\zeta}} \frac{1}{\nu_t} \int_{-h_0}^{z'} V_{1\beta}^{(0)} \, dz'' \, dz' \, dz \tag{29}$$

The boundary terms vanish since $p(-h_0) = q(\overline{\zeta}) = 0$. Therefore,

$$D_{\alpha\beta} = \frac{1}{h} \int_{-h_0}^{\overline{\zeta}} V_{1\alpha}^{(0)} \int_z^{\overline{\zeta}} \frac{1}{\nu_t} \int_{-h_0}^{z'} V_{1\beta}^{(0)} \, dz'' \, dz' \, dz \tag{30}$$

The structure of the expression (30) for the dispersion tensor is similar to that given in Fischer *et al.* (1979, eq. 4.64) for the dispersion of solutes in a two-dimensional shear flow, showing the close analogy between the the shear-dispersion of momentum considered here and the shear-dispersion of solutes initially considered by Taylor and expanded on by others [see Fischer *et al.* (1979, Chapter 4) for a discussion of the solute dispersion problem].

To estimate the importance of the mixing that results from the vertical nonuniformity of the short-wave-averaged horizontal velocities, consider using the approximation

$$V_{1\alpha}^{(0)} \approx \frac{-Q_{w\alpha}}{h} \tag{31}$$

[This corresponds to a uniformly distributed return flow to compensate for the volume flux Q_w. This is a relevant assumption because in the simple case of steady longshore currents we found that such an approximation leads to a result that is of the correct order of magnitude (Putrevu & Svendsen 1992)]. In this approximation, we have

$$A_{\alpha\beta\delta} = \frac{h^3}{3\nu_t} \frac{\partial}{\partial x_\delta} \left(\frac{Q_{w\alpha}Q_{w\beta}}{h^2} \right), \quad B_{\alpha\beta} = 0, \quad D_{\alpha\beta} = \frac{Q_{w\alpha}Q_{w\beta}}{3\nu_t}, \quad M_{\alpha\beta} = \frac{-Q_{w\alpha}Q_{w\beta}}{h} \tag{32}$$

For typical nearshore conditions ($\nu_t \sim 0.01 h \sqrt{gh}$, $Q_w \sim 0.1 H^2 \sqrt{gh}/h$, $H \sim 0.6 h$, $h \sim 1$ m), we have

$$D_{\alpha\beta} \sim \frac{Q_{w\alpha}Q_{w\beta}}{\nu_t} \sim \left(\frac{H}{h} \right)^4 h \sqrt{gh} \gg \nu_t \tag{33}$$

which shows that the lateral mixing provided by the $D_{\alpha\beta}$ terms is significantly larger than the direct lateral mixing due to turbulence.

5 Concluding Remarks

All the discussions in this paper have dealt with circulations induced by short-waves in the nearshore region. However, the equations derived here for shear dispersion of momentum, as well as the primary conclusion that the vertical nonuniformity of the horizontal velocities leads to enhanced horizontal mixing, should hold for a number of other flows modeled by depth-integrated equations. In that context, it is interesting to notice that Blumberg & Mellor (1987) observed a similar effect for mesoscale phenomena in the coastal ocean in their numerical experiments using a three-dimensional model. Describing their results, they write, "The relatively fine vertical resolution used in the applications resulted in a reduced need for horizontal diffusion because horizontal advection followed by vertical mixing effectively acts like horizontal diffusion in a real physical sense."

Acknowledgements

This work was funded by the U.S. Office of Naval Research, Coastal Sciences [Contracts N00014-95-C-0011 and N00014-96-C-0075 (UP), and N00014-95-C-0075 (IAS)].

References

Blumberg, A.F. & Mellor, G.L. 1987 A description of a three-dimensional coastal ocean circulation model. *Three Dimensional Coastal Ocean Models* (ed. N.S. Heaps), 1-16, AGU.

Carslaw, H.S. & Jaeger, J.C. 1959 *Conduction of heat in solids.* Oxford University Press.

Fischer, H.B., List, E.J., Koh, R.C.Y., Imberger, J. & and Brooks N.H. 1978 *Mixing in Inland and Coastal Waters.* Academic Press.

Mei, C.C. 1989 *The applied dynamics of ocean surface waves.* World Scientific.

Phillips, O.M. 1977 *The dynamics of the upper ocean.* Cambridge University Press.

Putrevu, U. & Svendsen, I.A. 1992 A mixing mechanism in the nearshore region. *Proc. 23rd International Conference on Coastal Engineering*, 2758-2771.

Putrevu, U. & Svendsen, I.A. 1997 Shear dispersion of momentum in the nearshore. Submitted for publication.

Svendsen, I.A. & Lorenz, R.S. 1989 Velocities in combined undertow and longshore currents. *Coastal Engineering*, **13**, 55-79.

Svendsen, I.A. & Putrevu, U. 1994 Nearshore mixing and dispersion. *Proc. Roy. Soc. Lond. A.*, **445**, 561-576.

Svendsen, I.A. & Putrevu, U. 1996 Surf zone hydrodynamics. *Advances in Coastal and Ocean Engineering*, **2**, 1-78.

Taylor, G.I. 1953 Dispersion of soluble matter in solvent flowing slowly through a tube. *Proc. Roy. Soc. Lond. A.*, **219**, 186-203.

Taylor, G.I. 1954 The dispersion of matter in a turbulent flow through a pipe. *Proc. Roy. Soc. Lond. A.*, **219**, 446-468.

COMPARISON BETWEEN A DETAILED SEDIMENT TRANSPORT MODEL AND AN ENERGETICS BASED FORMULA

Rolf Deigaard[1]

Abstract

Two models for sediment transport in combined waves and current have been compared: A detailed model (the Fredsøe model) based on a description of the physical processes involved and a model based on an 'energetics' approach (the Bailard formula). The comparison is made on the basis of results from model simulations. The two models are found to give rather similar results over a quite wide range of parameters. Cases of significant deviations are identified and explained.

Introduction

A rapid development has occurred in numerical modelling of coastal morphology. Typically a model for coastal morphology consists of modules for the simulation of waves, currents (tidal, wind- and wave-driven), sediment transport and the resulting bed evolution. The properties of different morphological modelling complexes may be assessed by comparing simulation results , cf. de.Vriend et al. (1993), and by considering the different modules in the complex. The hydrodynamic modelling has reached an advanced stage. The different modelling concepts are rather well defined, through the processes represented in the models and the simplifications introduced (such as the use of depth integrated flow equations), and their characteristics in many respects are known.

For the sediment transport, the modelling concepts are widely different, and the consequences of using one type of model instead of another are often not known. The sediment transport models can be largely empirical, or they may be based on a more or

---

[1]Dept. of Hydrodynamics and Water Resources (ISVA), Building 115, Technical University of Denmark, DK-2800 Lyngby, Denmark.

less detailed description of the physical processes involved. Some models simulate the variation in the hydrodynamics and sediment transport during a wave period, other describe only wave period averaged quantities. Some models calculate the sediment transport as composed of bed load and suspended load, while other try to determine the total load.

The purpose of this study has been to compare two sediment transport models that both are used intensively for coastal morphological modelling: the Fredsøe model, which is a detailed process based model, and the Bailard model, which is based on an 'energetics' approach. The comparison is made on basis of simulation results. It is an intercomparison between the models, and the model results are not compared with measurements.

The Models

The detailed sediment transport model: the Fredsøe model
The detailed model consists of two parts: a hydrodynamic module and a sediment transport module. The hydrodynamic module describes the turbulent boundary layer under combined waves and current, using the theory of Fredsøe (1984). This model includes the turbulent interaction between the wave- and the current boundary layer, and predicts the increased flow resistance for the mean current due to the wave boundary layer. The sediment transport module calculates the time-varying bed load and sediment concentration profile from the time varying bed shear stress and turbulent exchange coefficient obtained by the hydrodynamic module. The resulting bed load transport and suspended load transport is found as the time average of the instantaneous values. The sediment transport module is based on the work of Fredsøe et al. (1985) and Engelund and Fredsøe (1976). The numerical model applied is contained in the programme STP, which is part of the LITPACK software package of the Danish Hydraulic Institute.

For strong waves and for strong currents the bed is taken to be plane, but for weak waves and a weak current an option has been set to estimate the dimensions of wave ripples and their effect on the eddy diffusivity. The ripple predictions are based on empirical relations primarily obtained from the results of Nielsen (1979). For the present comparison the effect of the ripples is not very significant.

The energetics model: the Bailard formula
The energetics model (Bailard,1981; Bailard and Inman, 1980) that has been used for the comparison calculates the instantaneous sediment transport as function of the instantaneous flow velocity vector $\vec{u}$ taken 5 cm above the bed:

$$\vec{q}_b = \frac{e_b c_f}{(s-1) g \tan \phi} |\vec{u}|^2 \vec{u} \tag{1}$$

$$\vec{q_s} = \frac{e_s c_f}{(s - 1) g \; w_s} \; |\vec{u}|^3 \; \vec{u} \tag{2}$$

where $\vec{q_b}$ and $\vec{q_s}$ are the instantaneous bed load transport and suspended load transport, c_f is a friction coefficient, ϵ_b and ϵ_s are efficiency factors, ϕ is a friction angle, g is the acceleration of gravity, s is the relative density of the sediment, and w_s is the settling velocity of the sediment. The sediment transport is measured in m^3/ms, solid volume.

The following coefficients have been used in the simulations: $\epsilon_b = 0.1$, $\epsilon_s = 0.02$, $\tan \phi = 0.63$, s = 2.65, g = 9.81 m^2/s.

The friction coefficient is proportional to the wave friction factor proposed by Jonsson, cf. Jonsson and Carlsen (1976). When calculating c_f the bed roughness is taken to be 2.5 times the grain size d.

The Bailard model requires a hydrodynamic input. The mean flow velocity 5 cm above the bed is obtained from the detailed boundary layer model, Fredsøe (1984). For both models the near bed orbital motion is described by 1st order Stokes theory. The hydrodynamic conditions are thus consistent for the two models in the simulations.

The Comparison

The two models have been run for a wide range of input parameters.

Three grain sizes, d (and associated fall velocity, w_s):

d = 0.15 mm;	w_s = 0.013 m/s
d = 0.20 mm;	w_s = 0.022 m/s
d = 0.35 mm;	w_s = 0.042 m/s

Three wave periods, T (and associated water depth, D):

T = 3.5 s;	D = 4 m
T = 7.0 s;	D = 8 m
T = 14.0 s;	D = 8 m

Six wave heights: H = 0.3 m; 0.5 m; 1.0 m; 1.5 m; 2.0 m; 3.0 m.

Seven depth averaged mean current velocities: V = 0.1 m/s; 0.2 m/s; 0.4 m/s; 0.6 m/s; 1.0 m/s; 1.5 m/s; 2.0 m/s.

The wave conditions have only been chosen as the means to obtain a range of conditions for the near bed orbital motion in the sediment transport models. It is not considered if the conditions are realistic for example with respect to wave steepness or wave-current interaction. Apart from the turbulent interaction in the near bed boundary layer, no interaction between the current and the waves has been taken into account.

Co-directional waves and current

For the case of co-directional wave and current motion the total number (378) of simulations have been carried out.

Fig. 1 shows the comparison between the total load transport ($q_b + q_s$) calculated by the two models, with all data points. The agreement is quite good (within about a factor 5) over a wide range. For small transport rates the Bailard formula gives higher transport rates than the Fredsøe model, because it has no threshold for sediment motion. In fact a few data points are omitted because the detailed model predicted no transport at all. For high transport rates the detailed model gives a higher transport rate than the energetics model.

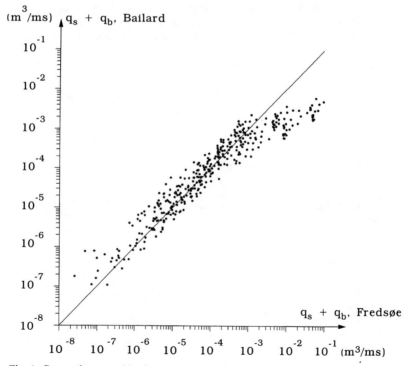

Fig. 1. Comparison, total load transport, co-directional waves and current.

Fig. 2 shows the comparison for the suspended sediment transport only. It shows the same pattern with an even larger difference at the low transport rates (where the suspended load is less important than the bed load).

The comparison for bed load transport only is shown in Fig. 3. This plot shows also a quite good agreement, with the energetics model giving more transport than the detailed model at high transport intensities.

The deviation between the total load transport rates predicted by the two models is analysed in Fig. 4. The largest deviations are found for high mean current velocities. For these conditions the detailed sediment transport model converges towards a pure current sediment transport model, with a logarithmic velocity profile determined by the grain roughness and a Rouse profile for the suspended sediment concentration. The distribution of the suspended sediment is thus determined by the Rouse parameter, Z:

$$Z = \frac{w_s}{\kappa\, U_f} \qquad (3)$$

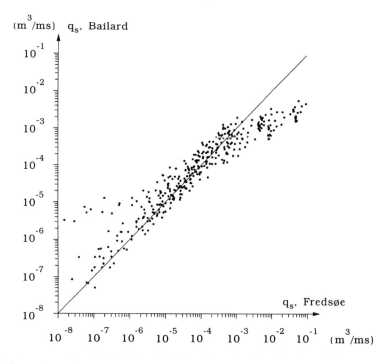

Fig. 2. Comparison, suspended load transport, co-directional waves and current.

where κ is v. Kármán's constant (κ = 0.4) and U_f is the friction velocity of the mean current, Rouse (1937).

In Fig. 4 the ratio between the transport rates determined by the two models is plotted against the parameter U_{fc}/w_s, where U_{fc} is the friction velocity based on the mean bed shear stress. It is seen that the ratio between the calculated transport rates decrease drastically for increasing values of U_{fc}/w_s, corresponding to decreasing values of the Rouse parameter. This reflects that the 'classical' pure current suspended load model at a certain stage predicts a stronger increase in the transport rate with increasing flow velocity than the power of 4 used in the Bailard model. Actually the situations with a dominating current are outside the range of applicability for the present version of the Bailard model, because the friction coefficient in this model is determined solely on basis of the wave orbital motion, which in these cases is negligible.

Waves and current at an angle
        Calculations with the waves propagating perpendicularly to the current direction, α equal to 90° showed no deviations from the pattern found for co-directional waves

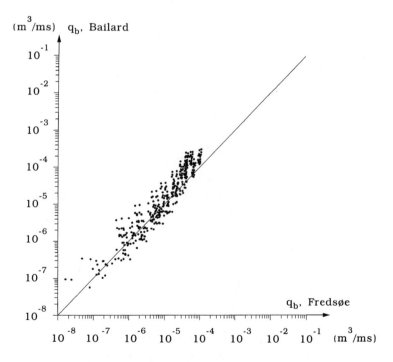

Fig. 3. Comparison, bed load transport, co-directional waves and current.

and current.

A series of simulations were made for the angle α equal to 45⁰. Figure 5 shows a comparison of the calculated total load transport rates in the current direction. This is very similar to the comparison for co-directional waves and current, Fig. 1.

For angles different from 0° or 90° there is an asymmetry which causes a transport component (drift) normal to the current direction. Figure 6 shows a comparison of the calculated total load transport rates normal to the current direction (the drift), by the two models. The agreement is rather good (within a factor of about 10), when considering that the transport normal to the current is a smaller secondary component. The energetics model tends to give a higher transport rate than the detailed model.

The drift of the bed load show larger discrepancies, with the Bailard model giving much larger transport rates at high wave intensities. This is found to be because the bed load transport model of Engelund and Fredsøe (1976), which is used in the detailed model, is less sensitive to variations in the bed shear stress at high values of the

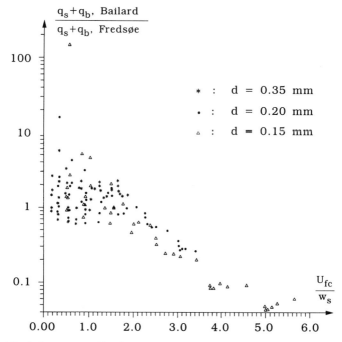

Fig. 4. Ratio between total load transport rates calculated by the two models, versus the parameter U_{fc}/w_s.

Shields' parameter. The large differences in the calculated bed load transport rates is only found in situations where the suspended load transport is totally dominant.

Conclusions

A large number of simulations have shown that the results obtained by the two sediment transport models for combined waves and current are in rather good agreement for the transport in the mean current direction. There is a tendency for the Bailard model to predict higher transport rates than the Fredsøe model at low transport intensity and smaller transport rates at high transport intensity. The largest discrepancy is found under conditions where the Bailard model is not expected to be valid: when the maximum bed shear stress is close to (or even smaller than) the critical shear stress or when the current is totally dominant over the wave motion. For the drift of sediment (cross current transport at an oblique angle between the waves and the current) the agreement is still rather good for the total load transport.

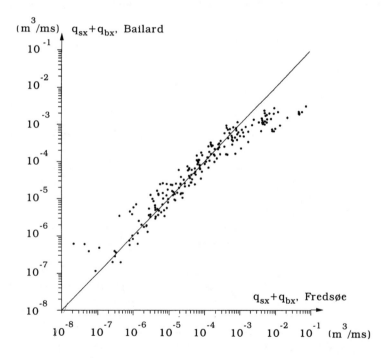

Fig. 5. Comparison, total load transport in the current direction, 45° between the waves and the current.

When comparing the models, it should be kept in mind that the relatively simple Bailard model requires a hydrodynamic input, which in the present applications has been obtained from the boundary layer model of Fredsøe (1984). The hydrodynamic module is a substantial part of the detailed process based model, which in this comparison is actually incorporated in the Bailard model.

Acknowledgement

This study has been supported by the Danish Technical Research Council (STVF) under the programme 'Marin Teknik'. The work was initiated in a task force led by Richard Soulsby, HR Wallingford, in the G8-Morphodynamics project under the MAST II programme.

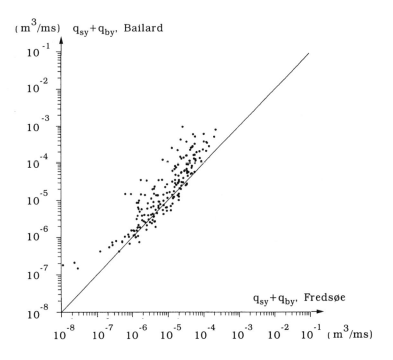

Fig. 6. Comparison, total load transport normal to the current direction, 45° between the waves and the current.

References

Bailard, J.A. (1981): An energetics total load sediment transport model for a plane sloping beach. J. of Geophys. Res., 86 (C 11) pp. 10938-10954.

Bailard, J.A. and Inman, D.L. (1980): An energetics bedload model for a plane sloping beach, local transport. J. of Geophys. Res., 86 (C 3): pp. 2035-2043.

DeVriend, H.J., Zyserman, J., Péchon, P., Roelvink, J.A., Southgate, H.N., and Nicholson, J. (1993): Medium term 2DH coastal area modelling. Coastal Engineering 21, pp. 193-224.

Engelund, F. and Fredsøe, J. (1976): A sediment transport model for straight alluvial channels. Nordic Hydrology 7: pp. 293-306.

Fredsøe, J. (1984): Turbulent boundary layers in wave-current motion. J. of Hydraulic Eng., ASCE, 110(HY8): pp. 1103-1120.

Fredsøe, J., Andersen, O.H. and Silberg, S. (1985): Distribution of suspended sediment in large waves. J. of Waterway, Port, Coastal and Ocean Eng. Div., ASCE, 111 (6): pp. 1041-1059.

Jonsson, I.G. and Carlsen, N.A. (1976): Experimental and theoretical investigations in an oscillatory rough turbulent boundary layer. J. of Hydr. Res., Vol. 14, No. 1, pp. 45-60.

Nielsen, P. (1979): Some basic concepts of wave sediment transport. Series Paper No. 20, Inst. of Hydrodynamics and Hydraulic Engineering, ISVA, Techn. Univ., Denmark.

Rouse, H. (1937): Modern conceptions of the mechanics of fluid Turbulence. Trans. ASCE, Vol. 102, pp. 463-543.

Spectral Test of Energetic Approach
for Suspended Sand Transport in Surf Zone

S.Yu. Kuznetsov[1] and N.V. Pykhov[1]

Abstract

Wide discussed problem of validity of the energetic approach for prediction of suspended sand transport in the surf zone is analyzed on base of the data of field experiment 'Norderney-94'. Follow the Bailard (1981) the theoretical concentration was calculated as proportional to the cube of the water velocity module. A theoretical suspended sand transport (SST) was calculated by multiplying the theoretical sand concentration and cross-shore velocity. The co-spectral analyze of the theoretical and the measured sand concentrations and the SST was performed. The comparison of the theory and experiment was performed and the possible reasons of disagreements and the possible ways of improving of energetic approach are discussed.

Introduction

Energetic approach was constructed by Bagnold and then adapted for wave flow by Inman, Bowen and Bailard. Energetic approach is very attractive due to simple formulae and almost obvious general idea - the more strong wave regime is the more sand is moved. Field investigations and energetic model estimations at its base show that we do not have a clear view on the validity of energetic approaches to the suspended sand transport (SST) prediction (for example - Foote et al., 1995, Russell et al., 1996) up to now. In some cases the "energetic" predictions are good, in other cases the SST values differ from the measurements or even are controversy. In some case the SST-part induced by gravity waves is well predicted, but the SST-part related to long waves is incorrect or even wrong. More wide tests of Bailard sediment transport formulae (Bailard, 1981) were performed during MAST-2 G8M project and show the disagreements between formulae and experiment for most type

---

[1]P.P.Shirshov Institute of oceanology, Russian Academy of Science, 36, Nakhimovski prosp., Moscow, 117851, Russia. E-mail: sergey@shelf.msk.ru; pykhov@coast.msk.ru

of waves and current regimes (Soulsby, 1995). The reason of this disagreement is not really known.

The aim of this paper is to check the energetic approach by methods of spectral analysis and show how the theory predictions of suspended sand flux fluctuations will follow to the measured velocity and concentration fluctuations at different time scales for irregular wave conditions. Also, we try to explain the divergence between predictions and measurements of SST on the base of sand suspending mechanisms.

Experiment

Field experiment was carried out on a beach section of Norderney Island, North Sea in October 1994. A joint German-Russian team of scientists carried out multipurpose survey at an especially equipped testing site, which is located on the north-western side of the island (Kunz et al., 1994). In this paper the results of measurements in points M1A and M1 (10 centimeters and more above the bottom) are presented and discussed (Fig. 1). The suspended sand concentration, the free surface elevation and the two horizontal components of water velocity were measured synchronously by optical turbidimeter, pressure gauge and two-component electromagnetic current meter at sampling rate of 4.5 Hz over 60 minutes. The suspended sand concentration was measured at level 8-10 cm above the bottom, cross-shore velocity was measured at level 18-20 cm above the bottom. Horizontal distance between the turbidimeter and current meter was 30 cm.

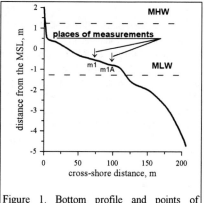

Figure 1. Bottom profile and points of measurements.

The turbidimeter measures the light beam attenuation at a distance of 6 centimeters that then is recalculated into the suspended sand concentration (Kos'yan et al., 1994). Two-component electromagnetic current meter and pressure sensors (Type NSW) had been used for measuring. Their linear calibration characteristics are in the frequency band 0 - 20 Hz. The diameter of the measuring head of the currentmeter is 6 centimeters.

The bed consists of sand with a mean diameter of 0.23 mm. The tidal range of about 2.4 m afforded to produce the measurements at a depth range from 0.5 to 2.5 m. Bottom slope is about one hundred parts between points M1A and M1. The measurements were carried out in the inner and middle parts of the surf zone under the wind wave and swell conditions. The 52 series of recording were obtained during the whole experiment.

Data processing

Experimental (measured, $Q_e(t)$) and theoretical (predicted, $Q_t(t)$) cross-shore suspended sand fluxes were calculated by multiplying of corresponding suspended sand concentrations ($c_e(t)$ and $c_t(t)$) and measured cross-shore velocity ($u(t)$). The net cross-shore suspended sediment flux (Q_n) was divided in a sum of fluxes determined by time mean current ($Q_m = <c(t)><u(t)>$), by long (Q_l), and short (Q_s), waves:, were < > denotes the time averaging. Experimental and theoretical values of Q_l were calculated by integration of co-spectrum of suspended sand concentration and cross-shore velocity ($Co_{cu}(f)$) in frequency range $0 < f < 0.05$ Hz, and Q_s - in frequency range $0.05 < f < 0.5$ Hz. The theoretical values of suspended sand concentration were calculated from formula of Bailard (1981) which was simplified to exclude the bottom slope, because it is negligible for our condition:

$$c_t(t) = \frac{\rho_s}{\rho_s - \rho_w} \cdot \frac{\rho \varepsilon_s f_w}{g h w_s} \cdot |U(t)|^3 ,$$

where $\rho_s = 2650$ kg/m^3, $\varepsilon_s = 0.02$, $f_w = 0.01$, $w_s = 0.028$ m/s.

It is necessary to note, that the quantitative comparison of the measured and predicted suspended sand fluxes is not fully correctly< because the measured fluxes are calculated only at one point at level ~ 10 cm above the bottom and fluxes predicted by energetic approach are depth averaged values. Because the main part of suspended sediment is transported in the near bottom layer, such comparison is possible to consider as a first approximation only.

The spectral analysis was performed with 0.01 Hz spectral window wide and 70 degrees of freedom.

The parameters of bottom ripples were calculated by formulae of Nielsen (1979) and type of ripples was established by diagram of Kaneco (1981).

Results

First of all we will look on the cases of different frequency structure of Q_e and Q_t. We will analyze three records produced in the inner and middle parts of surf zone with different values of outflow velocity <u>.

In the first case - run 25 - the <u> is small and long and short waves are responsible for the net flux. The experimental and theoretical values of Q_n, Q_s, Q_l, and Q_m are shown in Figure 2. The rms values of suspended sand flux fluctuations equal to 1.2 kg m^{-2}s^{-1} in this case. This is ten times more than net flux. The main input to

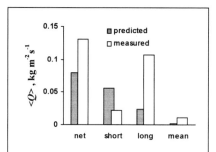

Figure 2. Cross-shore suspended sand flux components. Positive - directed onshore, negative - offshore. Run 25. Significant wave height $H_S = 0.80$ m, mean depth $<h> = 1.14$ m, outflow velocity $<u> = 0.01$ m/s.

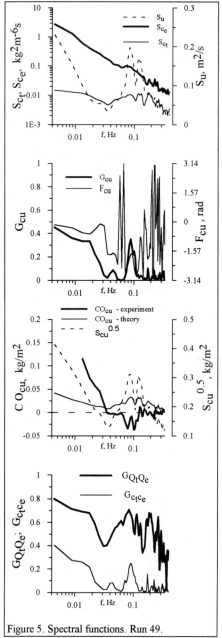

Figure 5. Spectral functions. Run 49.

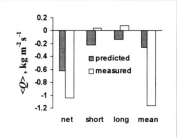

Figure 3. The same as Fig. 2. Run 37. H_S = 0.84 m, $<h>$ = 1.16 m, $<u>$ = -0.10 m/s.

Figure 4. The same as Fig. 2. Run 28. H_S = 1.42 m, $<h>$ = 2.30 m, $<u>$ = -0.40 m/s.

measured net suspended sand flux is by long waves. To another side, main input in the predicted net suspended sand flux give the short waves.

In the next case - run 37 - the measured fluxes by long and short waves almost compensate the flux by outflow and predicted fluxes by short and long waves have the opposite direction than measured (Fig. 3).

In the third case - run 28 - the measured flux by outflow is dominated (Fig. 4). The net sediment transport predicted and measured is almost the same, but it predicted frequency structure is wrong.

Let us consider the examples

of suspended sand transport for three different mechanisms of sand suspending (Pykhov, Kuznetsov, Kunz, 1997).

During the run 49 the two-dimensional sand ripples cover the bottom. At figure 5 the next spectral characteristics are shown for run 49: the spectra of cross-shore velocity (S_u), predicted (S_{ct}) and measured (S_{ce}) suspended sand concentration, coherence squared (G_{cu}) and phase spectrum (F_{cu}) of concentration and velocity, measured (CO_{ceu}) and predicted (CO_{ctu}) co-spectra of concentration and velocity, spectrum of measured flux fluctuation ($S_{cu}^{0.5}$) in degree of 0.5 to compare with co-spectra and coherence squared function for measured and predicted concentration and fluxes (G_{QtQe}, G_{ctce}). Coherence squared between concentration and cross-shore velocity is high enough (Fig. 5). Phase shift is about $-\pi/2$ at the main wave frequency. The mechanism of sand suspending from behind the ripple crest is this case. The prediction of direction of sand flux for long waves is correct, for short waves is wrong. The coherence between theoretical and measured concentration fluctuation is not so high, but coherence between fluctuations fluxes is excellent.

At the run 13 the three dimensional ripples cover the bottom. Coherence between measured and predicted concentration is high enough at short waves frequencies (Fig. 6). Phase shift is about $-\pi/4$. The mechanism of boundary layer

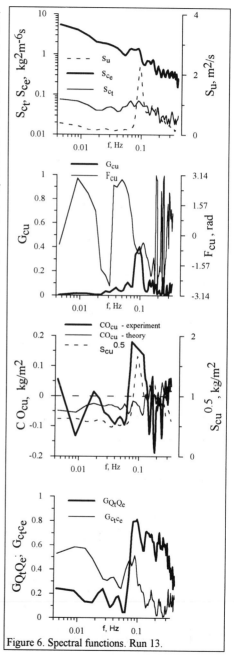

Figure 6. Spectral functions. Run 13.

instability at the phase of flow deceleration is dominated in this case. The predicted flux by short waves has wrong direction, but for long waves is good in spite the fact that coherence between predicted and measured fluxes is high for short waves and low for long waves.

Next case is the sheet flow condition under breaking waves (Run 28). Predicted and measured fluxes are opposite in direction but equal by modulus (Fig. 7). No coherence exists between predicted and measured concentration, but high coherence between measured and predicted fluxes exist due to high mean value of concentration.

Discussion

The possible reasons of divergence between the predicted and measured fluxes are:

1. Phase shift.

As it can be seen from figures 5 and 6, the main reason of divergence for first two mechanisms of sand suspending (the lee vortex outburst from behind the ripples crest and the boundary layer instability at the phase of flow deceleration) is the phase shift between concentration and cross-shore velocity. Phase shift control not only the absolute values of the flux, but its direction also. Phase shift is not taken into account in energetic approach formulae.

2. Measurement errors.

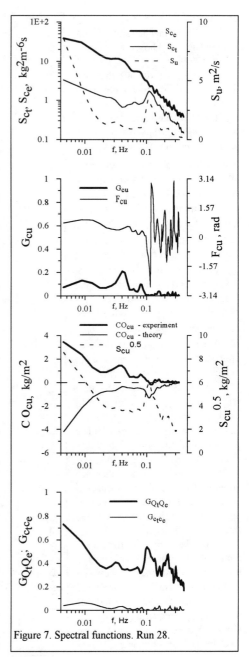

Figure 7. Spectral functions. Run 28.

It is difficult to correct measure the net sand flux in the case then the coherence between concentration and velocity is small (Fig. 7). In this case the fluctuations of suspended sand flux are one or two orders more than time means value and small errors in concentration and velocity measurements are dramatic for time mean flux.

3. Irregularity of sand suspending.

As it was calculated in experiment 'Shkorpilovtsy-85' (Kos'yan et al., 1993) the strong suspended events, which occupied 20% of time of measurements, give about 80% of net sediment flux. So, it is not important how the sand moves over 80% of time.

4. Advection of suspended sand.

Energetic approach does not take into account the sand suspended in another place and advected to the point of measurement.

Conclusions

1. Our results confirm the certain divergence between the prediction of suspended sand flux by formulae of energetic approach and results of measurements in the surf zone.

2. Probably it is possible to improve the energetic approach by taking in account the correspondent phase lag between concentration and velocity for first two mechanisms of sand suspending and by considering dependence the concentration from the energy dissipation by breaking instead by bottom friction inside the surf zone.

Acknowledgment

This research was supported by the grants 95-05-14343 and 97-05-74658 from the Russian Foundation for Basic Research.

References

Bailard J.A. (1981). An energetics total load sediment transport model for a plane slopping beach. *Journal of Geophysical Research*, v. 86, pp.10938-10954.

Foote Y.L.M., D.H. Huntley, T. O'Hare (1995). "Sand transport on macrotidal beaches". *Proceedings of Euromech 310 colloquium* (Le Havre), 360-374.

Kaneco A. (1981). Oscillation sand ripples in viscous fluids. *Proc. Jap. Soc. Civ. Eng.*, Vol. 307, p.113-124.

Kos'yan R., S. Kuznetsov, I. Podymov, N. Pykhov , O. Pushkarev, N. Grishin, D. Harisomenov (1995). "The optical device for suspended sand concentration measuring in the storm condition in the nearshore zone". *Oceanology*, Vol. 35, N 3, 463-469. (In Russian)

Kos'yan R.D., S.Yu. Kuznetsov, N.V. Pykhov (1994). Low-frequency fluctuation of suspended sand and wave groups in the surf zone. *Proc. of the Second Int.*

Symp. "Ocean Wave Measurements and Analysis", ed. by O.T.Magoon and
J.M.Hemsley. Pub. by ASCE, New York, 1994, 352-363.

Kunz H, R. Kos'yan, S. Kuznetsov, I. Podymov, O. Pushkarev, N. Pykhov (1994).
Field investigations of physical regularities and spatial-temporal scales of sand
suspending and transport in the coastal zone under the storm condition.
Norderney, Gelendgik, Moscow. 198 p.

Nielsen P. (1979). Some basic concepts of wave sediment transport. *Progr. Rept.
Inst. Hydrodyn. and Hydraul.* Eng. Techn. Univ. Denmark, 2, 160 p.

Pykhov N.V., S.Yu. Kuznetsov, H. Kunz (1997). Mechanisms of sand suspending
under non-breaking and under breaking irregular waves. *Proceedings of the
International Conference on Coastal Research in Terms of Large Scale
Experiments.* Coastal Dynamics' 97. ASCE, New York.

Russell P., Y. Foote, D. Huntley (1996). An energetics approach to sand transport
on beaches. *Proceedings of the International Conference on Coastal Research
in Terms of Large Scale Experiments. Coastal Dynamics' 95.* ASCE, New
York. 829-840.

Soulsby R.L. (1995). The 'Bailard' sediment transport formula: comparisions with
data and models. *Abstracts-in-depth of Final Overall Meeting the G8-Coastal
Morfodynamic Project* (MAST-II). 2-46 - 2-50.

EFFECT OF BED ROUGHNESS ON NET SEDIMENT TRANSPORT BY ASYMMETRICAL WAVES

C.Villaret[1] and A.G. Davies[2]

Abstract

The effects of wave asymmetry and of the wave-induced Eulerian drift on the net sediment transport rate are discussed, based on a review of existing data sets. Above rippled beds, these two effects oppose one another. According to some new hydrodynamical results, the wave-induced Eulerian drift above ripples consists of a strong onshore jet near the bed, with a reversal in the direction of drift occuring in the outer boundary layer flow. Here, the interaction between the positive mean flow vorticity and ejected vortices is proposed as an important mechanism influencing the net transport rate.

Introduction

Waves propagating in the nearshore region become non-sinusoidal. The resulting net wave-induced sediment transport is determined by various sources of asymmetry, which create differences in the flow intensity in the two halves of the wave cycle. In this paper, we discuss the effects of second-order wave asymmetry and of the wave-induced Eulerian drift on net sediment transport rate.

Despite numerous important applications like, for example, the prediction of beach profile evolution, sediment transport processes are still poorly understood, particularly when the bed is rippled. Small-scale sand ripples, which occur typically in low wave conditions, strongly affect both the flow hydrodynamics and the resulting sand transport. In the rippled bed regime, classical turbulent wave boundary layer models, and also sediment transport rate formulae, are no longer valid.

The relatively small wave-induced Eulerian drift, which is generated within the wave boundary layer, is expected to have a large effect on the net sediment transport rate. In the case of sinusoidal progressive waves, the drift above a flat bed is directed onshore, as described by Longuet-Higgins (1953).
. Above flat rough beds, where classical turbulent boundary layer concepts apply, the amount of sediment which is carried in suspension or as bed load, is governed by the

---

[1] E.D.F., Laboratoire National d'Hydraulique, 6 quai Watier, 78400 Chatou, France

[2] Univ. of Wales (Bangor), School of Ocean Sciences, Menai Bridge, Gwynedd LL595EY, U.K.

time-varying bed shear stress. Here wave asymmetry, and also the onshore directed drift, produce net transport in the onshore direction. Existing sand transport formulae, like for example the Bailard (1981) formula, can be used to predict the net transport rate.
. Above very rough and rippled beds, momentum transfer is governed by organised eddy shedding at flow reversal rather than by small-scale, random, turbulent processes. Here the net wave-induced sediment transport rate is governed by any asymmetry in the vortex shedding process which affects entrainment into the associated sand clouds. In the wave tunnel experiments of Sato and Horikawa (1986), the effect of wave asymmetry was to cause net transport in the offshore direction. This process was correctly reproduced by use of a detailed 2D discrete vortex model (see Perrier et al, 1994). However, this model did not include the effect of wave propagation and, therefore, failed to reproduce the large onshore sediment transport rate observed in the wave flume experiments of Villaret and Latteux (1992).

For practical applications, a 1-D vertical, spatially-averaged description of the flow was proposed by Davies and Villaret (1997a). In their formulation, the vortex shedding process was parameterized by use of a depth-invariant, time-varying, 'convective' eddy viscosity, with its peak values occuring near flow reversal. This model has recently been extended to include asymmetrical waves. In agreement with a large number of data sets, the new model solution for the drift produces a strong forward jet near the bed, with reversal in the direction of drift occuring within the wave boundary layer.

The objective of this paper is to examine the implications of these new hydrodynamical results for the net wave-induced sediment transport rate above rippled beds on the net wave-induced sediment transport rate. In Part I, we recall the theoretical background and previous model results, and present model comparisons with Eulerian drift profiles obtained in both wave flume and wave tunnel experiments. In Part II, we present a review of existing wave-induced sediment transport data sets above rippled beds. In Part III, the effect of the Eulerian drift on the net transport rate is explained qualitatively in terms of the interaction between the mean flow vorticity and ejected vortices.

1. Wave-induced Eulerian drift

In the case of <u>sinusoidal waves</u>, the effect of wave propagation is to give rise to vertical velocities within the wave boundary layer. The resulting mean stress drives a steady flow in the direction of wave propagation, as explained by Longuet-Higgins (1953) (hereafter 'LH'). His solution for the vertical profile of the drift $\overline{U_{LH}}$, tending at the edge of the boundary layer to the classical result $\overline{U_\infty} c/U_0^2 = 0.75$, where c is the wave celerity and U_0 the outer flow velocity amplitude, was obtained for a constant (laminar) viscosity. However, it can be shown that vertical variation in the eddy viscosity does not influence the drift profile, such that the LH result remains valid in the smooth turbulent regime.

The <u>effect of bed roughness</u> on the Eulerian drift was discussed by Davies and Villaret (1997b) (hereafter 'DV97') based on an extensive review of available experimental data:
. In the case of a flat rough bed, the effect of bed roughness is to reduce slightly the

outer flow drift value. Such an effect can be reproduced by including depth-variation in the eddy viscosity formulation, as well as a reference elevation ($z_0 = k_s/30$), where k_s is the equivalent bed roughness (see Trowbridge and Madsen, 1984).
. Above rippled beds, the effect of symmetrical time-variation in the eddy viscosity, with amplitude $|\varepsilon_2| \approx 1.3$ and phase $\phi_2 \approx \pi$, is to reduce the drift in comparison with the LH result, as shown by DV97. By extension of the LH argument, the drift component ($\overline{U^{(1)}}$) due to vertical velocities can be written:

$$\overline{U^{(1)}} = \overline{U_{LH}}(1 + \frac{1}{2}|\varepsilon_2|\cos(\phi_2))$$

This component alone, however, overestimates measurements of the drift at the outer edge of the boundary layer. This was attributed by DV97 to the fact that the waves in the experiments were not pure sinusoidal waves, but all displayed some degree of asymmetry.

The <u>effect of wave asymmetry</u> is to enhance the onshore velocities and, hence, create some asymmetry in the turbulence between the two halves of the wave cycle. This affects the time-averaged momentum balance and has a significant effect on the Eulerian drift.

The time-varying velocity at the outer edge of the wave boundary layer can be described by Stokes' second-order theory:

$$U_\infty(t) = \overline{U_\infty} + U_0(\cos(\omega t) + B\cos(2\omega t))$$

where U_0 is the horizontal, first-order, velocity amplitude near the bed, ω is the wave frequency and B is the asymmetry parameter.
. Above flat rough beds, the effect of asymmetrical time-variation in the eddy-viscosity was considered by Trowbridge and Madsen (1984), who found a significant reduction in the drift throughout the wave boundary layer, with a reversal in its direction for very long waves ($kh < 0.7$), where k is the wave number and h the water depth.
. Above rippled beds, the DV97 'convective' eddy viscosity model has recently been extended to second-order asymmetrical waves by including an extra asymmetry term ε_1 with phase angle ϕ_1 in the time-varying convective eddy viscosity:

$$K = \frac{K_0}{2}(1 + |\varepsilon_1|\cos(\omega t + \phi_1) + |\varepsilon_2|\cos(2\omega t + \phi_2))$$

For consistency with Stokes theory, the convective terms in the momentum equation are assumed to be small ($kA0 \ll 1$), where A_0 is the near-bed orbital amplitude ($A_0 = U_0/\omega$), allowing the problem to be decomposed into first and second-order problems in the wave slope. Moreover, in this perturbation approach, both $|\varepsilon_1|$ and $|\varepsilon_2|$ have been assumed to be O(1) or less.

In addition to the previous contribution due to vertical wave velocities ($\overline{U^{(1)}}$), the drift profile now includes two additional components. The second component ($\overline{U^{(2)}}$) is due to the effect of the asymmetry in the eddy viscosity:

$$\overline{U^{(2)}} = \frac{1}{2}U_0|\varepsilon_1|(e^{-\alpha z}\cos(\alpha z + \phi_1) - \cos(\phi_1))$$

where $\alpha = \sqrt{\omega/K_0}$. The third component ($\overline{U^{(3)}}$), which is due to the presence of the second-order velocity field, is a function of both the asymmetry parameter B and the propagation velocity c:

$$\overline{U^{(3)}} = \frac{1}{2}U_0|\varepsilon_2| \left[\frac{U_0}{2c} \left\{ \begin{array}{l} e^{-\alpha z}\cos(\alpha z + \phi_2) + \alpha z e^{-\alpha z}(\cos(\alpha z + \phi_2) + \sin(\alpha z + \phi_2)) \\ -e^{-\sqrt{2}\alpha z}\cos(\sqrt{2}\alpha z + \phi_2) \end{array} \right\} + B\left\{ e^{-\sqrt{2}\alpha z}\cos(\sqrt{2}\alpha z + \phi_2) - \cos(\phi_2) \right\} \right]$$

In order to reproduce the various experimental observations of near-bed drift, it was assumed that the phase of the maximum peak in K should occur 4° ahead of the flow reversal after passage of a wave crest. The intensities of both the ε_1 and ε_2 contributions were, as a first approximation, assumed to remain constant ($|\varepsilon_1|=|\varepsilon_2|=1.3$). The resulting time-varying eddy-viscosity is shown in Figure 1.

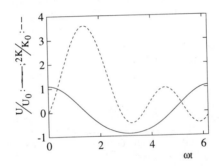

Figure 1: Time-varying eddy viscosity (B=0.1)

Figure 2a represents the model solution for the Eulerian drift profile, together with its three components. The second component (dashed-dotted line) is dominant and shows a strong onshore jet near the bed, with a reversal occuring within the wave boundary layer, beneath a negative drift in the outer flow. Such a vertical structure in the drift, which is consistent with observations, cannot be reproduced by, for example, the Trowbridge and Madsen (1984) model (hereafter 'TM'), which predicts a more uniform profile. A typical comparison between the model solutions and Test 'VP39' of Villaret and Perrier (1992) is shown in Figure 3. The present model can also be applied to calculate the drift in wave tunnel experiments by setting c= ∞. The first component of the drift due to vertical velocities is now zero. A typical model prediction is shown in Figure 2b. As expected, the drift profiles obtained in both wave flume and wave tunnel experiments are similar.

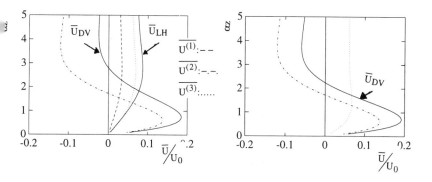

Figure 2: Model solution for the Eulerian drift profile
a) Wave flume (B=0.1; c/U_0=10) ; b) Wave tunnel (B=0.1)

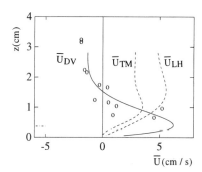

Figure 3: Comparison between the model prediction for the Eulerian drift profile and measurements (Test VP39: B= 0.135; c/U_0=5.8; k_s=2.31cm)

2. Review of existing sediment transport data sets

The net wave-induced sediment transport rate has been measured in a number of wave flume experiments: Inman and Bowen (1962) ('IB'), Russell and Dyke (1963)('RD'), Abou-Seida (1964) ('AS') and Villaret and Latteux (1992) (see also Villaret and Perrier, 1992, 'VP'). Test VP42 has been excluded from the last data set, due to the presence of secondary free waves. We consider here also the wave tunnel experiments of Sato and Horikawa (1986) ('SH'), and also the first series of experiments of Watanabe and Isobe (1990) ('WI'). We have excluded their second series of experiments with very coarse sand (D_{50}=0.87mm), where only bed-load sediment transport above rolling grain ripples was reported.

The net transport rate (Q) was generally estimated by collecting the sand deposited in sand traps located at both ends of a test section, except for RD who used a fluorescent tracer method. AS used only one sand trap, located downstream. The sand trap method was combined with detailed bed profile measurements along the flume by VP, SH and WI. In the IB data set, the net transport rates were corrected to account for the effect of wave attenuation along the flume test section. Detailed profile measurements of both time-varying concentration and velocity were also provided in VP data set, allowing sediment flux profiles to be determined. Net transport rates were then calculated by integrating these flux profiles over the depth. This method was found to be in reasonably good agreement with the one based on sand traps.

The ripples were generally 2D, becoming 3D when the wave size was increased, as in the WI and VP data sets. Ripple dimensions were measured in all cases, except for the RD and WI data sets. In general, fairly symmetrical ripples were observed, except in the SH experiments, despite the very asymmetrical wave motions. Note that in VP experiments, 'anorbital' ripples ($A_0/\lambda_R \approx 1.4\text{-}1.9$, where λ_R is the ripple length), were reported, whereas, in the less energetic RD data sets, 'orbital' equilibrium ripples would have been expected ($A_0/\lambda_R \approx 0.7\text{-}0.8$).

The experimental parameters and measured net transport rates are summarized in Table 1.
. The transport rate was directed onshore in most wave flume experiments, with a tendency to decrease when the wave asymmetry parameter (B) increased. This tendency is observed consistently in all individual data sets, except for that of AS where Q was very small. In the RD experiments, the net transport rate direction became offshore when B > 0.11.
. In all the wave tunnel experiments, the asymmetry parameter was very large (B > 0.2) and the net sediment transport was in all cases directed offshore. In the SH data set, the effect of increasing B was clearly to increase the intensity of the net transport rate in the offshore direction. In most of the WI experiments, although B was kept constant (B=0.2), the intensity of the net offshore sediment transport varied widely.

Such variability in the intensity of the transport rate can also be found by comparing, for example, the VP and RD data sets. For approximately the same value of B and also similar sand diameter, the onshore transport rate observed in the VP experiments was an order of magnitude larger than in the experiments of RD. This clearly indicates that, if B is an important parameter, it is not the only one; the net transport rate probably depends also on the detailed geometry of the ripples. However, this effect is presently difficult to evaluate, because of the lack of sufficiently precise experimental information.

Tests		D_{50} (mm)	T (s)	U_0 (cm/s)	B	Q x100 Kg/m/s
Wave	IB	0.2	1.4	23	0.04	+0.12
			2.0	30	0.15	+0.035
Flume	RD	0.095	1.0	10	0.025	+0.005
			-	-	-	-
			2.5	15	0.4	-0.2
	AS	0.145	0.77	8	0.04	+0.002
			-	-	-	-
			1.0	13	0.10	+0.016
	VP	0.09	1.5	30	0.04	+1.2
			2.	37	0.135	+1.1
Water	SH	0.18	3	28	0.26	-0.64
			-	-	-	-
			5	40	0.36	-2.52
Tunnel	WI	0.18	3	59		-0.07
			-	-	0.2	-
			6	86		-1.6
					0.4	-0.54

Table 1: Experimental parameters and net transport rates

3. Interpretation of the data

Importance of the asymmetry parameter:

Despite some uncertainty and variability in the results, the role of B on the net transport rate can be summarized as follows: in the case of very asymmetrical waves (B>0.15), the net transport rate is offshore whereas, in the case of fairly symmetrical waves (B<0.10), the net transport rate is directed onshore. Such results can be explained by the competing effects of the wave-induced Eulerian drift and wave asymmetry on the net transport rate.

The limiting value of B below which the effect of the wave-induced Eulerian drift becomes dominant is not known precisely, and probably depends upon other parameters (e.g. asymmetry in the ripples, relative roughness). For example, in both the Villaret and Latteux and also Inman and Bowen data sets, the net transport was still positive when B was about 0.15 whereas, in the Russell and Dyke data set, the net transport became offshore when B was greater than 0.11.

Effect of wave asymmetry:

The transport rate Q becomes increasingly negative (offshore), when B increases. The effect of second-order wave asymmetry can be explained in terms of asymmetry in the vortex shedding process between the two halves of the wave cycle. After the passage of a wave crest, the strong onshore velocity tends to enhance the negative vorticity, which is then ejected at the time of flow reversal and transported in the offshore direction. This mechanism which has already been explained by Sato and Horikawa (1986), is schematized in Figure 4a.

Effect of the wave-induced Eulerian drift:

When the asymmetry parameter B is relatively small (B<0.15), net tansport in the onshore direction is generally observed. As seen in the previous part, the vertical structure of the Eulerian drift corresponds to a positive mean vorticity $\Omega = -\dfrac{dU}{dz} > 0$, at least above the near-bed maximum in the drift velocity. This negative drift may be expected to interact with the ejected vortices and to enhace the positive vortex ejected after passage of a wave trough. Such interaction between the mean flow vorticity and the ejected vortices has been detected in some recent PIV measurements, for the case of waves superimposed on a current (Earnshaw, 1996). This mechanism, which is responsible for an onshore net transport rate, is schematized in Figure 4b (after Villaret and Perrier, 1992).

Dir. of wave propagation

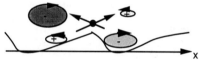

Figure 4a: Effect of wave asymmetry;

Figure 4b: Effect of the wave-induced
Eulerian drift

This effect was also detected in the phase-averaged time-varying concentration field measured at different elevations, by Villaret and Latteux (1992). As shown in Figure 5 for Test VP39, the two concentration peaks observed near the bed have similar intensity (upper curve). The signal then becomes clearly asymmetrical in the upper part of the wave boundary layer, the largest peak being transported in the onshore direction (U>0).

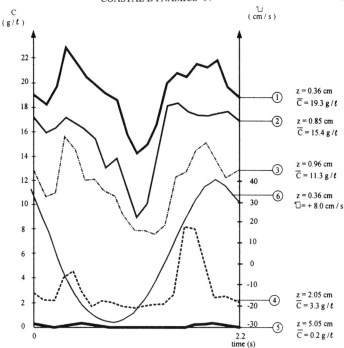

Figure 5: Phase-averaged concentration measured at different elevations above the bed (Test VP39: T=2s, k_s=2.3 cm, c/U_0=5.8, B=0.13, λ_R=6.1cm, δ_R=0.75cm). The full line shows also the near -bed time-varying velocity.

Conclusions

The effect of the wave asymmetry parameter B on the net wave-induced sediment transport rate has been observed consistently in all data sets. However, it is clearly not the only relevant parameter, and the picture remains at this stage rather qualitative. The effect of the wave-induced Eulerian drift can explain the onshore directed net sediment transport rate observed in some experiments involving fairly symmetrical waves (B<0.1-0.15), whereas, for very asymmetrical waves (B>0.2), the net transport rate is directed offshore.

According to the DV97 'convective eddy viscosity' model, the vertical profile of the Eulerian drift shows a strong jet near the bed directed onshore, with a reversal in the direction of the drift occuring within the wave boundary layer. The resulting positive mean vorticity interacts with the ejected vortices, enhancing the positive 'counter-clockwise' vortex, which is generated after passage of a wave trough and then transported onshore.

There is considerable variability in the intensity of the net transport rate such that, at this stage, it is not possible to quantify the transport at all precisely. In particular, the influence of ripple dimensions and of the parameter A_0/λ_r still need to be clarified.

References

M.Abou-Seida, 1964: Sediment transport by waves and currents, Univ. of California, *Report HEL-2-7.*

J.A.Bailard, 1981: An energetics total load sediment transport model for plane sloping beach, *Journal of Geophysical Research*, 86, C11,10938-10954.

A.G.Davies, C.Villaret, 1997a: Oscillatory flow over rippled beds : Boundary layer structure and wave-induced Eulerian drift, Chapter 6 in *Gravity Waves in Water of Finite Depth*, pp.215-254, ed. J.N.Hunt, Advances in Fluid Mechanics, Vol. 10.

A.Davies, C.Villaret, 1997b: Wave-induced currents above rippled beds and their effects on sediment transport, *Proc. 8th Int. Biennal Conf. on Phys. of Est. and Coast. Seas, The Hague, Sept. 1996.*

H. Earnshaw, 1996: *PHD Thesis*, Univ. of Edinburgh

D.I.Inman, A.J.Bowen.,1962: Flume experiments on sand transport by waves and currents, *Proc. I C.C.E.*, pp. 137-150.

M.S. Longuet-Higgins, 1953: Mass transport in water waves, *Phil. Trans. Roy. Soc. Lond., Series A*, 245 (903):535-581.

G.Perrier, E.A.Hansen, C.Villaret, R. Deigaard, J. Fredsoe, 1994:   Flow and sediment transport over a rippled bed in waves and current , *Proc. I.C.C.E.*, pp.2043-2057.

R.C.H.Russell, J.R.J.Dyke, 1963: The direction of net sediment transport caused by waves passing over a horizontal bed, *I.A.H.R. Congress,* pp.41-46.

S.Sato, K.Horikawa, 1986: Oscillatory boundary layer flow over rippled beds, *Proc. ICCE*, pp.2293-2309.

J.H.Trowbridge, O.S. Madsen, 1984: Turbulent wave bopundary layers. 2. Second-order theory and mass transport, *Jour. of Geophysical Res.*, 89, C5, pp.7999-8007.

C. Villaret, B. Latteux, 1992: Entrainment and transport of fine sand by combined waves and current: an experimental study, *Proc. I.C.C.E..*, pp. 2500-2512.

C. Villaret, G.Perrier, 1992: Transport of fine sand by combined waves and current: an experimental study , *LNH Report HE-42/92.68.*

A.Watanabe and M.Isobe, 1990: Sand transport rate under wave-current action, *Proc. ICCE*, pp.2495-2507.

Effect of Grain Size Gradation and Reference Concentration on Sediment Transport beneath Large Waves

Z. Li[1] and A.G. Davies[2]

Abstract

One-equation turbulent kinetic energy (k) numerical model is used to study the effect of graded sediment sizes on the distribution of suspended sediment concentration beneath large waves above a plane bed. 'Convective' suspended sediment concentration peaks at times of flow reversal observed in laboratory data are simulated by use of a newly-developed Two-equation ($k - \varepsilon$) model which includes a 'baseline' bed reference concentration in the bottom boundary condition for the suspension layer.

Introduction

If realistic medium- and long-term predictions are to be made by morphological models, accurate parameterization of sediment transport processes is required. However, our present understanding of the physical processes which affect net sediment transport rates beneath waves and combined wave-current flows remains incomplete due to the complexity of the physics of fluid-grain-bedform interaction. For example, bed sediment material is commonly represented by the median grain size, rather than by mixtures of grain size which are generally found in the natural environment. For given wave conditions above a plane bed, finer grains are more easily entrained into the suspension layer than coarser grains on account of the larger non-dimensional bed shear stress (i.e. Shields parameter). This, together with their smaller settling velocity, enables them to be diffused to greater heights in the water column.

In this paper, two numerical models are used to study sediment transport in plane bed (i.e. sheet flow) conditions beneath waves. The horizontally uniform

---

[1] Res. Ofcr., School of Oc. Sci., Univ. of Wales (Bangor), Menai Bridge, UK.
[2] Sr. Lect., School of Oc. Sci., Univ. of Wales (Bangor), Menai Bridge, UK.

1DV turbulent diffusion models, which solve the time-dependent momentum equation for horizontal velocity and the sediment continuity equation for suspended concentration, are formulated on the basis of (i) a One-equation turbulent kinetic energy (k) closure and (ii) a newly-developed Two-equation ($k - \varepsilon$) closure. In the k-closure, turbulence damping due to suspended sediment stratification has previously been incorporated by means of a turbulence buoyancy production term in the k equation (Li and Davies, 1996). Here this model is extended by the inclusion of graded sediment sizes. A reference-concentration boundary condition for the suspension layer is used in each model, whereby the instantaneous bottom concentration during the wave cycle is taken as the larger value of concentration obtained from (i) Engelund and Fredsoe's (1976) reference concentration formula and (ii) pure settling of sediment under gravity. These models have been developed during the EC MAST2 G8-M Coastal Morphodynamics project, and also the UK MAFF CAMELOT project, with a view to producing practical engineering predictors for sediment transport rates in coastal areas.

Graded Sediment Sizes

Instead of assuming that the sediment comprises a single uniform grain size, we consider the bed to comprise a mixture of three (or more) grain sizes. Given a knowledge of the grain size distribution, the model predicts the distribution of each fraction in suspension by solving the sediment continuity equation with a reformulated bottom boundary condition to account for the presence of different grain sizes. The total concentration is then obtained by summing the contributions from these three or more grain sizes throughout the suspension layer.

The formulation of the k-closure model for a single uniform grain size has been described by Li and Davies (1996), and Davies and Li (1997). For graded sediment grain mixtures, we first sub-divide the volumetric suspended sediment concentration (c) into N (≥ 3) volumetrically equal fractions. The horizontally uniform equation for the vertical distribution of suspended sediment concentration (c_i) for the ith fraction may be written as:

$$\frac{\partial c_i}{\partial t} = \frac{\partial}{\partial z}\left[w_{s,i} c_i + \varepsilon_s \frac{\partial c_i}{\partial z} \right] \qquad \text{for} \qquad i = 1,2,...,N \qquad (1)$$

where $w_{s,i}$ is the settling velocity of the ith grain size fraction, which is associated with the representative grain diameter D_i, ε_s is the sediment diffusivity, and $c = (1/N)\sum_{i=1}^{N} c_i$.

The first boundary condition accompanying (1) is the requirement of zero sediment flux at the surface, namely

$$\varepsilon_s \frac{\partial c_i}{\partial z} + w_{s,i} c_i = 0 \qquad\qquad (2)$$

The formulation of the second bottom boundary condition involves the use instantaneously (i.e. in a quasi-steady manner) of Engelund and Fredsoe's (1976) formula for the (volumetric) reference concentration at height $z = 2D_{50}$ (c.f. Davies and Li, 1997). Specifically, it is assumed that the instantaneous sediment concentration for each sediment fraction at the bottom of the suspension layer can be determined from the time-varying bed shear stress $\tau_b(t)$ at $z = z_0$ generated by the model. Here $z_0 = k_s / 30$ is the bottom roughness scale, where k_s is the equivalent bottom roughness here taken equal to $2.5D_{50}$ where D_{50} is the median grain diameter (50% by volume being finer) of the bed sediment. The reference concentration, $c_{b,i}$, has the form:

$$c_{b,i} = \frac{0.65}{(1+1/\lambda)^3} \qquad \text{at} \qquad z = 2D_{50} \qquad (3)$$

where λ, the so-called linear concentration, is given by

$$\lambda = \left(\frac{|\theta_i| - \theta_c - \pi p_* / 6}{0.027 |\theta_i| s} \right)^{1/2} \qquad \text{for} \qquad |\theta_i| > \theta_c + \pi p_* / 6 \qquad (4a)$$

and

$$\lambda = 0 \qquad \text{for} \qquad |\theta_i| \le \theta_c + \pi p_* / 6 \qquad (4b)$$

in which $s = \rho_s / \rho_w$ (ρ_s = sediment density, ρ_w = water density), and p_* is defined by

$$p_* = \left[1 + \left(\frac{\pi / 6}{|\theta_i| - \theta_c} \right)^4 \right]^{-1/4} \qquad\qquad (5)$$

where θ_i is the Shields parameter for the ith grain fraction. $|\theta_i|$ and θ_c in (4) and (5) are the magnitude of the time-varying Shields parameter and its critical value at the threshold of motion ($\theta_c = 0.045$), respectively. The Shields parameter θ_i is defined by

$$\theta_i = \frac{\tau_b}{\left[\left(\rho_s - \rho_w\right)g D_i\right]} \tag{6}$$

where D_i is the diameter of the ith grain size fraction. The response of a bed comprising a mixture of grain sizes to a given time-varying bed stress τ_b is, for example, to allow finer sediment to go to into suspension (as a result of its larger θ_i value), leaving coarser sediment immobile on the bed.

Further, the bottom boundary condition for suspended sediment concentration given by (3) at $z = 2D_{50}$ is accompanied by a correction procedure proposed by Hagatun and Eidsvik (1986), which accounts for the effect of the settling grains from above at times during the wave cycle when the Shields parameter, and hence the bottom concentration given by (3), is small. The implementation of this procedure has been described by Li and Davies (1996).

Before we can solve equation (1), the grain-size distribution at the bed needs to be either determined from measurements or prescribed. Here we follow the method described by Fredsoe and Deigaard (1992), in which the frequency or probability density function curve for the bed material is assumed to be represented in terms of grain diameter D by a log-normal distribution:

$$f(D) = \frac{1}{\sqrt{2}D\ln(\sigma_g)} \exp\left[-0.5\left(\frac{\ln(D/D_{50})}{\ln(\sigma_g)}\right)^2\right] \tag{7}$$

where σ_g is the geometric standard deviation, given by $\sigma_g = \sqrt{D_{84}/D_{16}}$ in which D_{84}, and D_{16} are the grain diameters for which 84%, and 16% of the grains are finer by volume (or by weight). The cumulative frequency curve, $F(D)$, is then given by

$$\begin{aligned}
F(D) = P(D' < D) &= \int_0^D f(D')dD' \\
&= 0.5\left[1 + erf\left(\frac{1}{\sqrt{2}}\frac{\ln(D/D_{50})}{\ln(\sigma_g)}\right)\right]
\end{aligned} \tag{8}$$

where erf is the error function, defined by $erf(y) = \dfrac{2}{\sqrt{\pi}} \displaystyle\int_0^y e^{-x^2} dx$. Thus the grain size distribution in (8) depends on D_{50} and σ_g, which may be determined from measurements and included as model input parameters. It may be noted that $\sigma_g = 1$ corresponds to a single uniform grain size. Thus the present formulation includes implicitly the formulation for a single uniform grain size based on D_{50} only (e.g. Li and Davies, 1996).

Finally, knowledge of the settling velocity $w_{s,i}$ in (1) is also required by the model. Here we use Hallermeier's (1981) empirical formula for sand grains. Defining the Archimedes buoyancy index as $A = \gamma' gD_i / \upsilon^2$, where υ is the kinematic viscosity, $\gamma' = s - 1$, the settling velocity has the form:

$$w_{s,i} = \frac{\upsilon}{D_i} \frac{A}{18} \qquad\qquad \text{for} \qquad A < 39 \qquad\qquad (9a)$$

$$w_{s,i} = \frac{\upsilon}{D_i} \frac{A^{0.7}}{6} \qquad\qquad \text{for} \qquad 39 < A < 10^4 \qquad\qquad (9b)$$

$$w_{s,i} = \frac{\upsilon}{D_i} \frac{A^{0.5}}{0.95} \qquad\qquad \text{for} \qquad 10^4 < A < 3 \times 10^6 \qquad\qquad (9c)$$

Figure 1 shows a comparison between the theoretical and actual grain size distributions for the sand used in several experiments carried out in sheet flow conditions in the Large Oscillating Water Tunnel at Delft Hydraulics, The Netherlands (Ribberink and Al-Salem, 1995). Here the theoretical curve is given by (8), with the log-normal curve (7) defined by $D_{50} = 0.21$ mm and $\sigma_g = 1.4$. It may be noted that, while there is an apparent discrepancy between the experimental data and the theoretical curve for the diameter range of 0.15-0.2 mm, the grain-size distribution curve is in close agreement with the data for the rest of the diameter range. In the case of Ribberink and Al-Salem's experimental data for a large symmetrical wave ['Condition 3': free-stream velocity amplitude = 1.7 m/s, wave period = 7.2 s], according to the criterion of Fredsoe and Deigaard (1992) that a particle is able to go into suspension from the bed if its settling velocity is less than $0.8u_*^m$, where u_*^m is the maximum bed friction velocity magnitude during the wave cycle, 99.79% of the bed material has the potential to be entrained into suspension. If we divide the sediment grain size range in suspension into three equal fractions, centred on $F(D) = 16.63\%$, 49.89% and 83.15%, these fractions correspond to three grain diameters 0.1516,

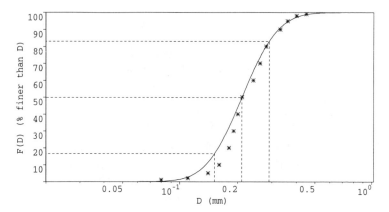

Fig. 1 Computed (full line) and measured (symbols) grain size distributions for
laboratory experiments carried out in sheet flow conditions at Delft Hydraulics.
For symmetrical wave 'Condition 3', the values of $F(D)$ and D for three
volumetrically equal grain sizes in suspension are indicated by the dashed lines.

0.2098, and 0.2901 mm (see Fig. 1), with settling velocities given by (9) of
18.5, 26.4 and 37.8 mm/s, respectively. For Condition 3, the median diameter
for the sediment grains in suspension is therefore almost the same as that of the
bed as a whole. Fig. 2 shows a comparison between the predicted time-series of
suspended sediment concentration at heights ($z = 0.42$, 2, 16, 31 mm) obtained
assuming (i) a single uniform grain size of $D_{50} = 0.21$ mm ($w_s = 26.4$ mm/s)
with turbulence damping included in the k equation only (original model of Li
and Davies, 1996), and (ii) the mixture of three grain size fractions referred to
above. Where available (Fig. 2c and 2d), the measured sediment concentration
data is also shown for comparison. It may be noted that reference level
$z = 2D_{50} = 0.42$ mm is the lowest computational level for the suspension layer.
Fig. 2a shows the sediment concentration at this reference level. Both here and
at level $z = 2$ mm (Fig. 2a and 2b), sediment size gradation has little effect on
the concentration time-series. However, since finer sediment grains are readily
diffused to greater heights above the bed due to their relatively low settling
velocity, the time-series in Fig. 2c and 2d show an enhanced concentration at
these higher levels. As shown in Fig. 3, the cycle-mean concentration resulting
from the mixture of grain size is larger than that for a single uniform grain size,
at least in the outer part of the suspension layer.

'Convective' Concentration Peaks

The accuracy of sediment transport rate predictions in unsteady flows depends
rather critically upon the accuracy of not only the magnitude but also the phase

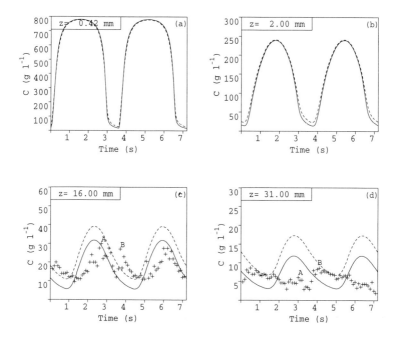

Fig. 2   Predicted (lines) and measured (symbols) time-series of suspended concentration at four levels ($z = 0.42, 2, 16, 31$ mm) for symmetrical wave 'Condition 3'. The model results are based on (i) a single uniform grain size, and (ii) three, volumetrically equal, grain size fractions.

of the time-varying concentration. Careful examination of the concentration time-series data at heights $z = 16$ and 31 mm (Fig. 2c and 2d) reveals that, in each half cycle, in addition to a 'diffusive' peak (labelled A), a second 'convective' peak (B) occurs near flow reversal (time $= 3.6$ s). This latter phenomenon, which appears to be associated with 'convective events' due to boundary layer instability, is not predicted here using the standard bottom condition for the suspension layer (see also Davies and Li, 1997). This is because the Shields parameter is smaller than its critical value near flow reversal, though the k-model does, in fact, predict relatively *strong* vertical mixing at around this time in the cycle (Li and Davies, 1997). Li and Davies (1997) showed that a $k - \varepsilon$ model provides an even more rapid vertical exchange of momentum in the very near-bed layer around flow reversal than that predicted by the k-model. However, in order to allow sediment to be entrained into suspension near flow reversal, an assumption such as the continuous availability of bed material during the wave cycle needs to be made (c.f. Savioli and Justesen, 1997), which may be physically plausible in vigorous

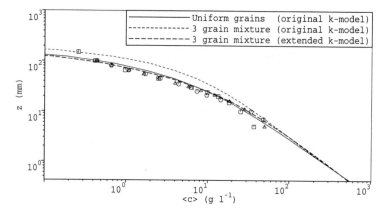

Fig. 3 Predicted (lines) and measured (symbols) vertical profiles of cycle-mean concentration for symmetrical wave 'Condition 3'. The model results have been obtained on the basis of (i) a single uniform grain size, (ii) three, volumetrically equal, grain size fractions, and (iii) the three grain size fractions together with mixing length and hindered settling corrections.

sheet flow conditions. Fig. 4 shows the concentration time-series obtained at $z = 16$ mm using a newly-developed $k - \varepsilon$ model based on (i) the original bottom boundary condition and (ii) a modified bottom boundary condition. In the latter case, the instantaneous bed concentration has been taken throughout the wave cycle as the *larger* value of concentration obtained from (i) Engelund and Fredsoe's (1976) reference concentration formula and (ii) a 'baseline' volumetric concentration of 0.20. As a result of the introduction of this 'baseline' bed concentration, as well as predicting the second concentration peak near flow reversal, the computed concentration is in better overall agreement with the data in terms of magnitude and phase. This is likely to lead to a significant improved prediction of the net sediment transport rate in cases involving asymmetrical waves, and combined wave and current flows. This will be discussed more fully elsewhere.

Conclusions

The effect of sediment gradation on the distribution of suspended sediment concentration beneath large waves above a plane bed has been considered on the basis of a turbulent kinetic energy (k) model. A cumulative grain-size distribution curve derived from the frequency density function for a log-linear distribution has been shown to agree well with the experimental data considered here. Compared with results for a single uniform grain size, the inclusion of a mixture of (three) grain sizes in the model leads to an increase in the

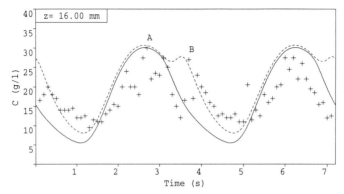

Fig. 4 Predicted (lines) and measured (symbols) time-series of concentration at height $z = 16$ mm in the suspension layer for symmetrical wave 'Condition 3'. The model results have been obtained by the $k - \varepsilon$ model using (i) the original bottom boundary condition (full line), and (ii) a new bottom boundary condition that includes a 'baseline' bed concentration of 0.20 (dashed line).

concentration of sediment in suspension in the outer suspension layer while, close to the bed, sediment gradation has a little effect on concentration.

To simulate 'convective peaks' observed in concentration time-series at flow reversal, a Two-equation ($k - \varepsilon$) model has been developed. In order to reproduce these convective peaks successfully, an assumption of availability of bed material at times when the bed shear stress is smaller than its critical value needs to be made. As pointed out by Li and Davies (1997), although this assumption may be plausible in vigorous wave conditions, its validity requires further investigation and justification. Moreover, while models based on turbulent diffusion arguments may provide a partial explanation for the concentration peaks observed at flow reversal, a combined convection-diffusion model is probably required to model fully these convective effects. The accurate simulation of convective peaks in concentration at flow reversal holds the key to improving predictions of the net sediment transport rate in unsteady flows above plane beds.

Finally, it should be emphasised that sediment gradation considered here is only one of the physical processes which are potentially important in determining the sediment concentration. As shown in Fig. 2 and 3, a model which includes some of the processes only can result in worse agreement with the data. In order to simulate sediment concentration and sediment transport successfully in unsteady flows, a fully consistent model incorporating the effect of graded sediment sizes, turbulence damping and hindered settling may be required. This is demonstrated in Fig. 3 which shows the cycle-mean concentration predicted by 'an extended k – model' including (i) a flux Richardson number correction term in the vertical mixing length, and (ii) a hindered settling correction. The inclusion of these additional processes leads to improved agreement between the predicted and measured vertical profiles of the cycle-mean sediment concentration.

References

Davies, A. G., and Li, Z. (1997). "Modelling sediment transport beneath regular symmetrical and asymmetrical waves above a plane bed." *Continental Shelf Research*, **17**(5), 555-582.

Engelund, F., and Fredsoe, J. (1976). "A sediment transport model for straight alluvial channels." *Nordic Hydrology*, **7**, 293-306.

Fredsoe, J., and Deigaard, R. (1992). *Mechanics of Coastal Sediment Transport*. World Scientific, Singapore, 369pp.

Hagatun, K., and Eidsvik, K. J. (1986). "Oscillating turbulent boundary layer with suspended sediments." *Journal of Geophysical Research*, **91**(C11), 13045-13055.

Hallermeier, R. (1981). "Terminal settling velocity of commonly occurring sand grains." *Sedimentology*, **28**, 859-865.

Li, Z., and Davies, A. G. (1996). "Towards predicting sediment transport in combined wave-current flow." *Journal of Waterway, Port, Coastal and Ocean Engineering*, ASCE, **122**(4), 157-164.

Li, Z., and Davies, A. G. (1997). Closure to the discussion of "Towards predicting sediment transport in combined wave-current flow". *Journal of Waterway, Port, Coastal and Ocean Engineering*, ASCE, in press.

Ribberink, J. S., and Al-Salem, A. A. (1995). "Sheet flow and suspension of sand in oscillatory boundary layers." *Coastal Engineering*, **25**, 205-225.

Savioli, J., and Justesen, P. (1997). "Sediment in oscillatory flows over a plane bed." *Journal of Hydraulic Research*, **35**(2), 177-190.

Near-Crest Pressure Gradient of Irregular Water Waves Approaching to Break

K. Nadaoka[1], O. Ono[2] and H. Kurihara[3]

ABSTRACT

The vertical pressure gradient near wave crest was examined through laboratory experiments by developing a new technique to estimate the pressure field of nonlinear irregular waves from the free surface fluctuations both in space and time. With this technique the vertical pressure gradient near the crest of nonlinear irregular waves was found to rapidly decrease and become nearly zero at the instant of the wave breaking, indicating that there arises a situation of free fall of water mass near the crest at the breaking and that the condition of the zero vertical pressure gradient may be used as an effective breaking criterion under general situations including nonlinear irregular waves. By incorporating this criterion into the Nadaoka et al.'s wave equation, a numerical simulation to predict the breaker depth was attempted, and the computational result was found to give close agreement with that by the empirical breaking criterion of Goda (1970).

INTRODUCTION

Recent progress in water wave modeling enables us to compute evolution of waves under general conditions; e.g., Nadaoka et al. (1994, 1997) developed fully dispersive nonlinear wave equations, which can describe nonlinear irregular waves with a broad-banded spectrum on an arbitrary depth. However accurate computation by a nonlinear wave model is still limited to non-breaking waves. Although several attempts have been made to incorporate breaking effects into a nonlinear wave equation such as Boussinesq equation (Schäffer et al.,1992; Karambas & Koutitas, 1992, etc.), the ambiguity in the breaking condition itself may provide appreciable inaccuracy in numerical simulation.

Although various breaking criteria exist, none of them can be employed in general situation including irregular wave field. As one of the famous breaking criteria, Rankine (1886) suggested that when a wave breaks the horizontal particle velocity at the wave crest becomes greater than the wave speed C. However the condition is not

[1]Professor and [2]Graduate Student, Graduate School of Information Science and Engineering, Tokyo Inst. of Technology, 2-12-1 O-okayama, Meguro-ku, Tokyo 152, Japan, [3]Tokyu Construction Company Limited.

suitable for the computation because it is difficult to define the wave speed C for irregular waves. Using the ratio of the particle velocity at the crest to the wave speed, Stokes (1847) indicated the limiting angle at the crest of the highest wave in case of a deep water wave train with a symmetric profile. But in general the actual waves are not symmetric nor periodic. Some empirical criteria, on the other hand, have been suggested but they can be applied only to periodic regular waves. Besides the empirical methods are not able to apply to waves over arbitrary bottom topography. Recently, Schäffer et al.(1992) proposed a breaking wave model in which waves are assumed to break when the wave front slope becomes steeper beyond a prescribed value. However according to our experimental result, as will be shown later, irregular waves do not break with a unique slope.

ZERO VERTICAL PRESSURE GRADIENT AS FREE FALL CONDITION

For deep water waves, "Stokes' 120 degrees angle" is one of the famous breaking criteria, as mentioned before. According to this theory, the crest angle becomes 120 degrees at the wave breaking. The corresponding vertical acceleration and hence the vertical pressure gradient divided by the fluid density ρ is $0.5g$ (Longuet-Higgins,1963). In these theories waves are assumed to be symmetric in their profiles around the crest even at the breaking. However this assumption is unrealistic because actual breaking waves show appreciable deformation with highly skewed and asymmetric profile.

Recently Yasuda et al. (1996) showed from the computational result by a BIM of deformation of a solitary wave on a rectangular reef that the maximum vertical acceleration of the water particle motion near the wave crest becomes g. This means that the minimum vertical pressure gradient around the crest becomes zero when the wave breaks. The flow field with the zero vertical pressure gradient is corresponding to the situation of free fall of water mass. For standing waves it is known that the free fall situation appears at the crest (Penny & Price, 19). In the present study, the free fall condition is assumed to apply as a breaking criterion to various wave field including progressive irregular waves. The validity of this assumption was confirmed through a laboratory experiment by analyzing the vertical pressure gradient of progressive irregular waves on a uniform slope.

It is difficult to directly measure the pressure gradient at the free surface in laboratory experiments. Instead one can use the vertical momentum equation to evaluate the pressure gradient. The vertical momentum equation at the free surface is

$$-\frac{1}{\rho}\frac{\partial p}{\partial z}\bigg|_{z=\eta} = g + \frac{\partial w}{\partial t}\bigg|_{z=\eta} + u_s\frac{\partial w}{\partial x}\bigg|_{z=\eta} + w_s\frac{\partial w}{\partial z}\bigg|_{z=\eta}, \tag{1}$$

where p is the pressure, t the time, x and z the horizontal and vertical coordinates, u and w the velocity components in the x and z direction, η the free surface displacement, and the suffix s indicates the variables at $z=\eta$. Because of the presence of the derivative of w with respect to z, Eq.(1) is complicated in its form to apply. However the derivatives of w with respect to x and t at the free surface are transformed as,

$$\frac{\partial w}{\partial t}\bigg|_{z=\eta} = \frac{\partial w_s}{\partial t} - \frac{\partial \eta}{\partial t} \cdot \frac{\partial w}{\partial z}\bigg|_{z=\eta}, \tag{2}$$

$$\frac{\partial w}{\partial x}\bigg|_{z=\eta} = \frac{\partial w_s}{\partial x} - \frac{\partial \eta}{\partial x} \cdot \frac{\partial w}{\partial z}\bigg|_{z=\eta}. \tag{3}$$

Substituting the above relations and the boundary condition at the free surface,

$$w_s = \frac{\partial \eta}{\partial t} + u_s \frac{\partial \eta}{\partial x} \tag{4}$$

into Eq. (1), we obtain

$$-\frac{1}{\rho}\frac{\partial p}{\partial z}\bigg|_{z=\eta} = g + \frac{\partial^2 \eta}{\partial t^2} + u_s^2 \frac{\partial^2 \eta}{\partial x^2} + 2u_s \frac{\partial^2 \eta}{\partial t \partial x} + \frac{\partial u_s}{\partial t}\frac{\partial \eta}{\partial x} + u_s \frac{\partial u_s}{\partial x}\frac{\partial \eta}{\partial x}. \tag{5}$$

To evaluate the vertical pressure gradient at the free surface from Eq.(5), the free surface fluctuation η and the horizontal velocity at the free surface u_s must be known. η can be easily measured in laboratory experiment, while the measurement of u_s needs a rather sophisticated method. Therefore u_s was estimated with a newly developed method described below.

A NEW TECHNIQUE TO ESTIMATE VELOCITY FIELD UNDER IRREGULAR WAVES

Dean's stream function method (1965) is a well-known technique to estimate a velocity field from the free surface displacement of nonlinear waves. However it can not be applied to irregular waves, because for irregular waves the water surface cannot be treated as a streamline in any translated coordinate system. Therefore Dean's method can not be employed for our purpose to evaluate the horizontal velocity at the free surface u_s in eq.(5). Hence in the present study an alternative new technique was developed to estimate the velocity field of irregular and nonlinear waves from the free surface fluctuations both in space and time.

If the variation in the water depth is so small that the depth may be assumed to be locally uniform, the general solution of the Laplace equation of the velocity potential ϕ can be given as

$$\phi(x,z,t) = \int_{-\infty}^{\infty} A(k,t)\frac{\cosh k(h+z)}{\cosh kh}\exp(ikx)dk, \tag{6}$$

where A is a time-varying wave number spectrum, k the wave-number, h the local depth, and i the imaginary unit. The kinematic boundary condition at the free surface is,

$$\frac{\partial \phi}{\partial z}\bigg|_{z=\eta} = \frac{\partial \eta}{\partial x} \cdot \frac{\partial \phi}{\partial x}\bigg|_{z=\eta} + \frac{\partial \eta}{\partial t}. \tag{7}$$

Substituting Eq.(6) into Eq.(7) and discretizing the resultant equation, we have

$$\left[1+\left(\frac{\partial \eta}{\partial x}\bigg|_m\right)^2\right]\sum_{n=1}^{N}A_n(t)k_n\,\frac{\sinh k_n(h_m+\eta_m)}{\cosh kh}\exp(ik_n x_m)\Delta k\big|_n$$

$$=\frac{\partial \eta}{\partial x}\bigg|_m\sum_{n=1}^{N}A_n(t)\left\{\left[\frac{\partial}{\partial x}\left(\frac{\cosh k_n(h+\eta)}{\cosh k_n h}\right)\right]_m\right. \tag{8}$$

$$\left.+\frac{\cosh k_n(h_m+\eta_m)}{\cosh k_n h_m}ik_n\right\}\exp(ik_n x_m)\Delta k\big|_n+\frac{\partial \eta}{\partial t}\bigg|_m .$$

The suffix m and n indicate respectively the discrete points on the space x and the wave-number k, and m runs from 1 to M. In the above equation, the unknown variable is only A_n.

Substituting $A_n=a_n+ib_n$ into Eq.(8) and taking the real parts, we finally obtain

$$\sum_{n=1}^{N}a_n\alpha_{mn}+b_n\beta_{mn}=\gamma_m,\quad (m=1,\cdots,M) \tag{9}$$

where,

$$\alpha_{mn}=\left\{-\left[1+\left(\frac{\partial \eta}{\partial x}\bigg|_m\right)^2\right]k_n\,\frac{\sinh k_n(h_m+\eta_m)}{\cosh k_n h_m}\sin k_n x_m\right.$$

$$\left.+\frac{\partial \eta}{\partial x}\bigg|_m k_n\,\frac{\cosh k_n(h_m+\eta_m)}{\cosh k_n h_m}\cos k_n x_m+\frac{\partial \eta}{\partial x}\bigg|_m\left[\frac{\partial}{\partial x}\left(\frac{\cosh k_n(h+\eta)}{\cosh k_n h}\right)\right]_m\sin k_n x_m\right\}\Delta k\big|_n ,$$

$$\beta_{mn}=\left\{\left[1+\left(\frac{\partial \eta}{\partial x}\bigg|_m\right)^2\right]k_n\,\frac{\sinh k_n(h_m+\eta_m)}{\cosh k_n h_m}\cos k_n x_m\right.$$

$$\left.+\frac{\partial \eta}{\partial x}\bigg|_m k_n\,\frac{\cosh k_n(h_m+\eta_m)}{\cosh k_n h_m}\sin k_n x_m-\frac{\partial \eta}{\partial x}\bigg|_m\left[\frac{\partial}{\partial x}\left(\frac{\cosh k_n(h+\eta)}{\cosh k_n h}\right)\right]_m\cos k_n x_m\right\}\Delta k\big|_n ,$$

$$\gamma_m=\frac{\partial \eta}{\partial t}\bigg|_m .$$

a_n and b_n in Eq.(9) may be determined from the following conditions

$$\frac{\partial S}{\partial a_n}=0,\quad \frac{\partial S}{\partial b_n}=0,\qquad (n=1,\cdots,N) \tag{10}$$

where S is defined as

$$S=\sum_{m=1}^{M}\left[\sum_{n=1}^{N}(a_n\alpha_{mn}+b_n\beta_{mn})-\gamma_m\right]^2 . \tag{11}$$

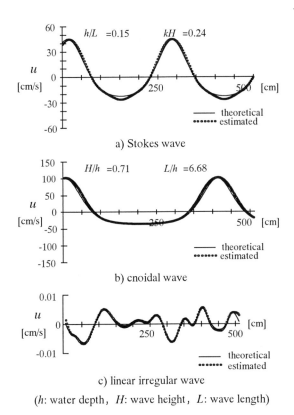

a) Stokes wave

b) cnoidal wave

c) linear irregular wave

(*h*: water depth, *H*: wave height, *L*: wave length)

Fig. 1 Theoretical and estimated velocities at the free surface.

The resultant equations to determine a_n and b_n are

$$\sum_{m=1}^{M} \alpha_{lm} \left[\sum_{n=1}^{N} \left(a_n \alpha_{mn} + b_n \beta_{mn} \right) \right] = \sum_{m=1}^{N} \alpha_{lm} \gamma_m , \qquad (12)$$

$$\sum_{m=1}^{M} \beta_{lm} \left[\sum_{n=1}^{N} \left(a_n \alpha_{mn} + b_n \beta_{mn} \right) \right] = \sum_{m=1}^{N} \beta_{lm} \gamma_m . \qquad (13)$$

Although the wave field to which we want to apply the present method is nonlinear irregular waves, we have examined the validity of this method by applying it to the third-order Stokes and the second-order cnoidal waves and linear irregular waves, because these waves have their theoretical expressions to be compared. Figure 1 shows the results of these comparison and indicates the good performance of the present method to estimate velocity fields under Stokes and cnoidal waves as typical nonlinear but regular waves and linear irregular waves. From these results one can

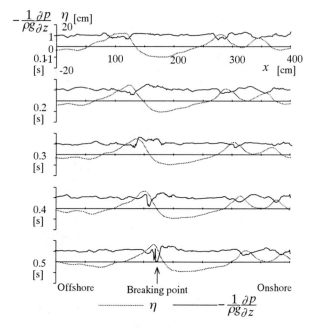

Fig.2 Time sequence of the free surface deformation and the corresponding vertical pressure gradient.

expect that this method is applicable also to nonlinear irregular waves.

LABORATORY EXPERIMENT

The wave channel used for the experiment is equipped with a flap-type wave generator at one end and a slope of 1:20 at another end. The width of the channel is 40 cm, and the still water depth was kept to be 40 cm. Irregular waves generated in the channel were recorded by video camera at the interval of 1/30 seconds from the side of the channel, and the free surface elevation was detected by an image analysis of the video data. The generated incident irregular waves have a Bretschneider type spectrum with the mean period of 1.0 second and the mean height of 5.3 cm. The instant of the wave breaking is defined in the experiment as that when a wave just starts to overturn.

Figure 2 shows a time sequence of the on-offshore distribution of the normalized pressure gradient (solid lines) and the water surface displacement (dotted lines) in which the breaking occurs at the location indicated. The waves propagate from the left to the right. As the wave approaches the break, the pressure gradient rapidly decreases and becomes nearly zero when the wave breaks. The similar behavior of the pressure gradient was observed in almost all other waves approaching the break.

Figure 3 represents the variations in the smallest vertical pressure gradient near the

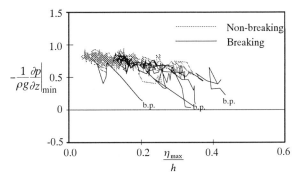

Fig.3 Variation in the normalized vertical pressure gradient and the corresponding value of η_{max}/h for 12 progressive waves; η_{max} denotes the crest height.

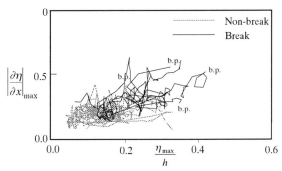

Fig.4 Variation in the steepest foreside surface slope and the corresponding value of η_{max}/h for 12 progressive waves.

wave crest and the corresponding change in η_{max}/h with the wave propagation, where η_{max} denotes the crest height. The waves broken within the observed section are classified as "breaking", and the others are classified as "non-breaking". The dotted lines represent the "non-breaking" waves, while the solid lines indicate the "breaking" waves. In the "non-breaking" waves, the vertical pressure gradient does not show appreciable decrease even when η_{max}/h becomes quite large. In contrast, the "breaking" waves indicate the abrupt decrease as the wave approaches the break and the vertical pressure gradient becomes nearly zero at the instant of the wave breaking. These results suggest that the vertical pressure gradient is more pertinent than the nonlinear parameter η_{max}/h as the breaking criterion and that the critical value of the pressure gradient is zero.

Figure 4 represents the variations in the steepest foreside slope and the corresponding change in η_{max}/h with the wave propagation. (It should be noted that the water surface slope was calculated by the approximation $\partial\eta/\partial x \cong \Delta\eta/\Delta x$ with $\Delta x = 4.5$cm, like the finite difference scheme for numerical simulation.) The dotted and

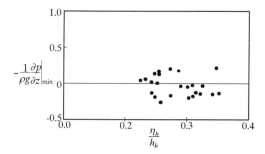

Fig.5 The minimum vertical pressure gradient at the instant of the wave breaking.

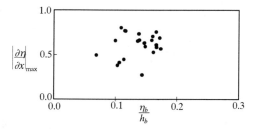

Fig.6 The steepest foreside surface slope at the instant of the wave breaking.

solid lines represent the "non-breaking" and "breaking" waves respectively. The steepest foreside values do not indicate appreciable distinction between "breaking" and "non-breaking" waves.

Figures 5 and 6 show respectively the minimum vertical pressure gradient within a wave and the steepest foreside surface slope at the instant of the wave breaking. The horizontal axis in each figure is represented with the ratio of the surface elevation to the water depth at the location where the pressure gradient is minimum for Fig.5 and the surface slope is steepest for Fig. 6. When the waves break the pressure gradients become nearly zero, whereas the foreside surface slopes show appreciable scatter. These results give another demonstration to show that the vertical pressure gradient can be used as an effective breaking criterion, which is applicable even to irregular waves.

A WAVE MODEL COMPUTATION OF BREAKER DEPTH

To examine the performance of the zero pressure gradient condition as a breaking criterion for numerical wave simulations, it was attempted to predict the breaker depth as the result of the numerical simulation of nonlinear shoaling of a wave train, using a nonlinear wave equation and incorporating the zero pressure gradient condition in the simulation. As the nonlinear wave equation, the single component version of the fully dispersive weakly nonlinear wave equations of Nadaoka et al.'s (1994,1997) was adopted. The single component version of the equations can describe the nonlinear evolution of regular and irregular waves with a narrow banded spectrum over an

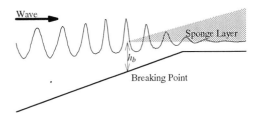

Fig.7 Sketch of the computational domain.

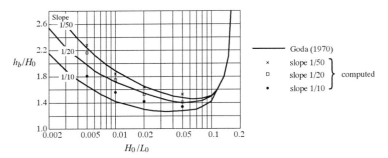

Fig.8 Computed breaker depth compared with the Goda's(1970) empirical criterion.

arbitrary depth. The continuity and momentum equations with a damping term to install a sponge layer, which will be described later, are written as

$$\frac{\partial \eta}{\partial t} + \frac{\partial}{\partial x}\left(\frac{C_p^{\,2}}{g} + \eta\right)u_0 = 0, \tag{14}$$

$$C_p C_g \frac{\partial u_0}{\partial t} + C_p^{\,2} \frac{\partial}{\partial x}\left[g\eta + \eta \frac{\partial w_0}{\partial t} + \frac{1}{2}\left(u_0^{\,2} + w_0^{\,2}\right)\right]$$

$$= \frac{C_p\left(C_p - C_g\right)}{k^2}\frac{\partial^3 u_0}{\partial^2 x \partial t} + \frac{\partial}{\partial x}\left[\frac{C_p\left(C_p - C_g\right)}{k^2}\right]\frac{\partial^2 u_0}{\partial x \partial t} - \mu C_p C_g u_0, \tag{15}$$

where C_p, C_g and k denote respectively the phase and group velocities and wave-number computed according to the linear dispersion relation for a prescribed dominant frequency ω and a local depth h. u_0 and w_0 are respectively horizontal and vertical velocity component at the still water level and μ is the damping coefficient.

The numerical simulation was performed for unidirectional waves on slopes of 1:50, 1:20 and 1:10. As shown in Fig.7, a sponge layer with a linearly increasing damping coefficient μ was installed at the region after the breaking to get rid of possible wave reflections otherwise the reflected waves may contaminate the computational wave field concerned. The computation was done for regular waves with the period of 1.0 second, because the existing breaking criteria are all for regular waves. The spatially variable grid, the size which is 1/30 of the local linear wavelength, was adopted for the computation.

The computed breaker depth was compared with the Goda's (1970) empirical breaking criterion. Figure 8 shows the breaker depth h_b relative to the offshore wave height H_0, taking the offshore wave steepness H_0/L_0 as the horizontal axis. The good agreements between the computed and empirical values of the breaker depth are found especially for the slopes of 1/50 and 1/20. This fact demonstrates that the zero vertical pressure gradient condition may be effectively used with a wave equation to predict the breaking point.

CONCLUSION

A new technique was developed to estimate the pressure gradient of nonlinear irregular waves from the free surface fluctuations both in space and time. With this technique the vertical pressure gradient of irregular waves approaching the breaks was analyzed in detail through laboratory experiments. The experimental results show the vertical pressure gradient near the crest rapidly decreases and becomes nearly zero at the instant of the wave breaking, indicating that there arises a situation of free fall of water mass near the crest at the breaking and that the condition of the zero vertical pressure gradient may be used as an effective breaking criterion under general situations including nonlinear irregular waves. By incorporating this criterion into the Nadaoka *et al.*'s wave equation, a numerical simulation to predict the breaker depth was attempted, and the computational result was found to give close agreement with that by the empirical breaking criterion of Goda (1970).

References

Dean, R. G. (1965) "Stream function representation of nonlinear ocean waves", *J. Geophys. Res.*, Vol.70, pp.4561-4572.

Goda, Y . (1970) "A synthesis of breaker indices", *Trans. Japanese Soc. Civil Eng.*, No.180, pp.39-49. (in Japanese)

Karambas, Th. V. & Koutitas, C.(1992) "A breaking wave propagation model based on the Boussinesq equations", *Coastal Eng.*, 18, pp.1-19.

Longuet-Higgins, M. S. (1963) "Generation of capillary waves by gravity waves", *J. Fluid Mech.*, Vol.16, pp.138-159.

Nadaoka, K., Beji, S. & Nakagawa, Y. (1994) "A fully-dispersive nonlinear wave model and its numerical solutions", *Proc. 24th Int. Conf. on Coastal Eng.*, ASCE, pp.427-441.

Nadaoka, K., Beji, S. & Nakagawa, Y. (1997) "A fully-dispersive weakly nonlinear model for water waves", *Proc. Roy. Soc. Lond. A*, Vol. 453, pp.303-318.

Penny, W.G. & Price, F.R.S. & A.T.(1952) "Finite periodic stationary gravity waves in a perfect liquid", *Phil. Trans. Roy. Soc. Lond. A*, Vol. 244, pp.254-284.

Rankine, W. J. (1864) "Summary of properties of certain stream line", *Phil. Mag., Ser.* 4, 29.

Schäffer, H. A., Deigaard, R. & Madsen, P. A. (1992) "A two-dimensional surf zone model based on the Boussinesq equations", *Proc. 24th Int. Conf. Coastal Eng.*, ASCE, pp.576-589.

Stokes, G. G. (1847) "On the theory of oscillatory waves", *Trans. Camb. Phil. Soc.*, Vol.8, pp.441-455.

Yasuda, T., Ohmiya,Y. & Mori, N. (1996) "A method of judging wave breaking applicable to field data in time domain", *Proc. Coastal Eng.*, JSCE, Vol.43(1), pp.61-65. (in Japanese)

Three Dimensional Flow Field Measurement of Breaking Waves

Junji HAMADA [1]   and   Akira MANO [2]

Abstract

A handmade system has been developed to measure the three dimensional (3D) flow fields of breaking waves. The measurement is based on the stereoscopic highspeed photographing. Correlation methods are applied to identify entrained certain air bubbles, chosen as the tracers, taken in the photographs different in time as well as view point. Several techniques are introduced to increase the accuracy and stability of the measurement. Simultaneous use of two sets of the systems from side and bottom for breaking waves gives transition from 2D to 3D behavior together with the fields of velocity, vorticity and 2D divergence.

1. Introduction

Recent flow visualizations have shown the existence of 3D coherent vortex motion in breaking waves, which could be linked with the mechanisms of energy dissipation and also of the sediment transport in the surf zone. Attentions have been paid on these vortex motions in the breaking waves by Nadaoka(1986), Yamashita et al.(1988), and Bonmarin(1989). The evaluation of the effects requires quantitative measurement of the motion and might change the dynamic models such as Svendsen et al.(1984), Basco(1985), and Nadaoka et al.(1989).

On 3D measurement of flow fields, the particle image velocimetry benefits as opposed to a time history at single point. Recently the studies of Mano et al.(1990), Quinn et al.(1995) and Watanabe et al.(1995) have been making these structure clear using particle image velocimeter.

The particle tracking method of that might be common, however it requires low density of tracers. It is formidably difficult to keep the density low and uniform in the turbulent breaking waves. Furthermore, huge amount of air bubbles

[1] Takenaka Corporation, 154 Ohtsuka, InzaiCity, Chiba Prefecture, 27043, Japan
[2] Tohoku University, Sendai, 98077, Japan

entrained in the breaking waves prevent the recognition of such traces. The bubbles also block the use of the laser doppler anemometer. This study utilizes, contrary, this cumbersome air bubbles as tracers of the correlation method.

The purposes of this study are to develop a 3-D measuring system and to apply the system to the breaking solitary wave to evaluate the kinematic behavior.

2. 3-D Flow Field Measuring System

2.1. Principle of the measurement

The 3-D flow field measuring system is based on stereoscopic highspeed photographing. The principle of the measurement is thus. Successive photographs of a tracer in time give necessary information on movement and stereoscopic photographs taken from different view points provide required information on the 3-D position. The combinations of these information enable us to compute 3-D movement of the tracer.

The key to realize the principle lies in identifying a certain tracer among many in two photographs. The same skill of identification would be applied to the two kinds of photographs, different in time and view. The skill relates closely with the density of tracers. When the density is low, the proper skill would be a kind of the particle tracking method. On the other hand, if the density is high, this is the case, the correlation method is superior. Here air bubbles entrained in breaking waves have been chosen as the tracers, because air bubbles cover breaking zone and have enough high density.

This choosing leads to two problems. The correlation method tends to grow errors during the computation. To prevent or decrease the errors, several techniques have been introduced, that will be shown later. The other problem is that we can see only the surface of the cloud of the air bubbles. As the common feature of the methods by using light, the measurement on this system is also confined near the surface of the cloud.

2.2. Hardwares

Hardwares used to measure 3-D flow field are thus. Two sets of cameras and a special wave flume were prepared to see the whole figure simultaneously (Fig.1). The flume is 0.8 m in width, 17.6 m in length, and 0.5 m in the maximum height with sloped bottom of 1/40. One side and the bottom were made of glass to take photographs. At one end of the flume is a wave generator of moving wall type driven by AC servo motor which is controlled by personal computer to make solitary waves.

The stereoscopic highspeed photographing are performed by the highspeed drum camera with the minimum switching time of frames, 2 ms, and a stereoscopic adapter. The framing is made by 25 flushes of the storoboscope during one revolution of the drum. The adapter is simple but has some aberration that was corrected numerically.

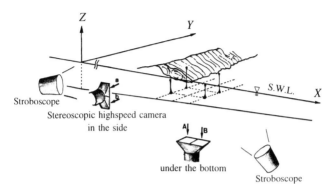

Fig.1   Experimental setup and coordinate system.

2.3. Softwares

High sensitive monochrome films used for the experiments were read by the scanner to give the data of 1024 by 1024 pixels and 256 tones of brightness. One pixel corresponded to 0.58 mm and 0.42 mm in real scale for the side and bottom views respectively. Figure 2 shows contour lines of the brightness inside the enclosed line in Photo 1. The correlation method computes the correlation for different subregions which we denote templates. Therefore this method requires the spatial distribution of brightness as shown in Photo 1. The relation between the correlation coefficient and dispersion of brightness in Fig.3 indicates that the coefficients increase with the dispersion.

To ensure accuracy and stability of the computation, we have introduced the following processes;

(1) To optimize the template size:

Figures 4 (a), (b), (c) show the effect of the template size for the computation of velocity vector fields which were obtained from the two photographs with the interval of 2 ms. Larger template stabilizes the velocity fields and increases the dispersion. This suggests much larger structure of clouds is existing. However, as the compensation it wastes more time for computation and loses the resolution, summarized in Table 1. To balance these effects, we decided the template size as 15 by 15 pixels that is 8.75 mm in the side view.

(2) To correct the aberration of the stereoscopic adapter:

The distortion by the adapter was tested by taking photograph of the right mesh and corrected by the fully third order polynomial function of spatial coordinate.

(3) To compare twin velocity fields:

Through two lenses of the stereoscopic adapter, we can get twin velocity fields. Although these fields are of different 3-D positions, they should be similar judging from the gradient of the velocity fields. Largely deviated velocities were eliminated as errors.

(4) To check by the correlation coefficient:
Velocities with the coefficient less than 0.6 were eliminated.

(5) To smooth the field by 4-D linear function:
We have employed 4-D linear function in terms of space(x,y,z) and time t in order to interpolate the data defected through the above procedures and to smooth the

$$u(x, y, z, t) = ax + by + cz + dt + e$$

where u is the velocity component, the coefficient a to e are determined by the best fit. The another benefit to introduce this function is that we can get stable partial derivatives of u by x to t, that give calculation of vorticity, divergence, and local acceleration. Principally we can get three components of velocity, however the component along the lens axis was not adopted because the signal is comparable with the error.

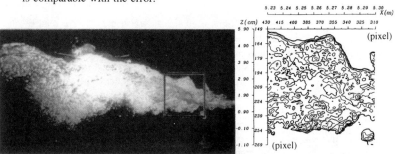

Photo 1   A sample photograph   Fig.2   Contour line of brightness.
taken from the side.

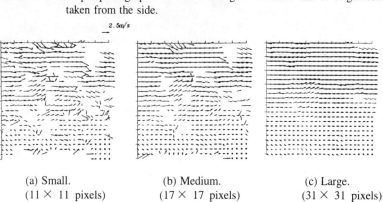

          (a) Small.   (b) Medium.   (c) Large.
    (11 × 11 pixels)   (17 × 17 pixels)   (31 × 31 pixels)
          Fig.4   Effect of the plate size on velocity fields.

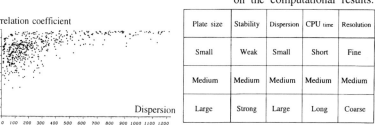

Fig.3 Relationship between correlation coefficient and dispersion in brightness.

Table 1 Effects of the template size on the computational results.

Plate size	Stability	Dispersion	CPU time	Resolution
Small	Weak	Small	Short	Fine
Medium	Medium	Medium	Medium	Medium
Large	Strong	Large	Long	Coarse

3. Experiments and Results

Experiments were carried out for the solitary wave with the incident depth of 0.2 m and height of 0.1 m. The origin of the x was set at toe of the slope and the wave broke at $x = 4.2$ m with the plunging type. Since measurement for one experimental run covers about 0.2 m in x direction, several runs with different positions of the equipment were conducted. Sensitive features of the breaking waves required us careful experiments with intervals more than 20 minutes to get the reproducibility.

3.1. Stereoscopic photographs

The wave transforms from the completely 2-D feature with long crest before breaking to 3-D feature near shore, that is our another interest. Before entering the precise analysis in the following subsections, here we show rather qualitative characteristics through photos. Photos 2(a), (b) show the stereographs taken simultaneously from the side and bottom at the initial stage. The plunging jet and the splash are displayed in the side view. This jet looks like a dark strip with uniform width at $x = 4.53$ m in the bottom view. The plunging line is slightly undulating in the transverse direction, but it is fundamentally two dimensional.

At the advanced stage of the breaker, Photos 3(a), (b) near $x = 5.0$ m show thick and long air bubble cloud in the side view and its complicated distribution in the bottom view.

3.2. Initial stage of breaking wave

Figure 5 shows the time series of the velocity and vorticity in the side view with a time span of 16 ms. The cross section is taken near the side wall. Behind the jet, the intense clockwise horizontal vortex created by the plunging jet can be seen, and the water mass which jumped into the front is also shown. Here clockwise vortex correspond to negative vorticity. These vortices are enlarging with time and second horizontal vortex is beginning to be created. And also, the splashed water seems to be pushed forward by the jet.

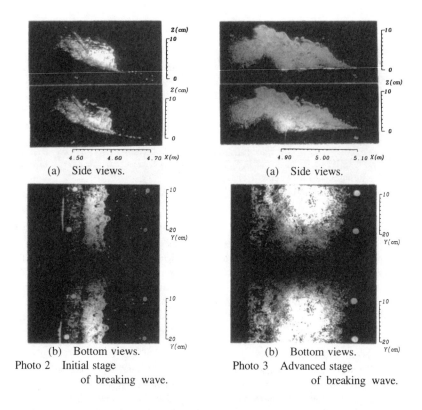

(a) Side views.   (a) Side views.

(b) Bottom views.   (b) Bottom views.
Photo 2   Initial stage   Photo 3   Advanced stage
            of breaking wave.   of breaking wave.

Figure 6 shows the bottom views at the same time as Fig.5. The upper figure
exhibits a positive area in 2-D divergence is transversely stretching behind the jet
at $x = 4.53$ m and a negative area is in front of the jet. Positive distribution, that is
source distribution, means that there is net inflow to the plane, and negative
distribution corresponds to net outflow. Therefore the positive and negative
distribution in this figure would show the source distribution by the entrainment
due to the jet and sink distribution by splash respectively. The distribution is
initially 2-D and gradually changes.

As for the vorticity, the distribution is rather 3-D in front of the jet from the
earlier stage. Positive vorticity indicates anti-clockwise rotation also in the bottom
view. The velocity of the frontal splash is nonuniform in the span-wise direction
leading undulating front line. The splash at the apex of the line would be pushed
out briskly by the jet. When the splash goes forward excessively, it would become
slow than the ambient water because of absence to push forward. The vortex pair
at the apex; at $x = 4.61$ m and $y = 17$ cm would show this state. However this
vortex pair diminishes in time, because the front is gradually caught up by ambient
water.

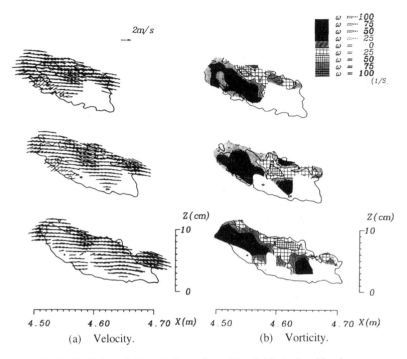

2m/s

ω ::::··· 100
ω :::···· 75
ω ::::··· 50
ω ::::··· 25
ω = 0
ω = 25
ω = 50
ω = 75
ω = 100
(1/s)

Z (cm)
10

0

Z (cm)
10

0

4.50   4.60   4.70 X(m)   4.50   4.60   4.70 X(m)

(a)   Velocity.   (b)   Vorticity.

Fig.5   Evolution of the velocity and vorticity fields in the side view.

3.3. Advanced stage of breaking wave

For the advance stage of the breaker, Figs.7, 8 show the velocity and vorticity distribution in the side view near the wall, at the range from $y = 0$ m, wall, to $y = 5.7$ cm. The distribution of the cloud surface for the y coordinate is plotted at the top of Fig.7. The surrounding line shows the projection of the whole cloud region. The area of the projection is much larger than that of the field. This indicates that the cloud undulates or changes its height in the spanwise direction. Later, we will discuss on the elevation distribution of the bottom surface of the cloud region. The plunging jet and its rebounds or splashes produce a wavy distribution in the velocity field, and not only three major clockwise vortexes, which correspond to $\omega < 0$, but also two anticlockwise vortexes among them. These two types of vortices function to conserve the angular momentum in the breaker zone.

For the bottom view at the same moment, the distribution of velocity, vorticity, and 2-D divergence in xy plane are shown in Figs.9, 10, 11. Though the first and second plunging lines are seen and indicated by the thick lines in Fig.9, the third does not clearly appear in these figures. And the first and second plunging lines can be seen clearly at $x \approx 4.9$, 5.0 m also in the side view. These

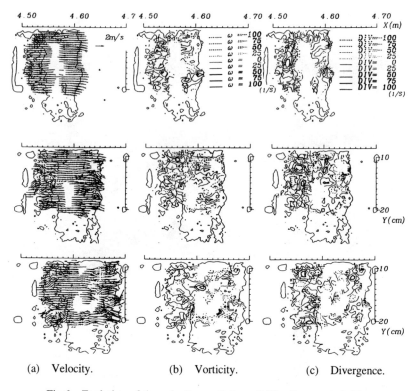

(a)   Velocity.   (b)   Vorticity.   (c)   Divergence.

Fig.6   Evolution of the velocity, vorticity and 2-D divergence fields
in the bottom view.

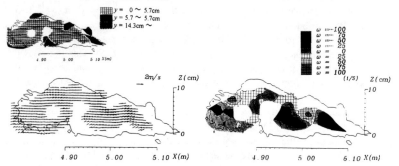

Fig.7   Velocity distribution
in the side view.
(y = 0 ~ 5.7 cm)

Fig.8   Vorticity distribution
in the side view.
(y = 0 ~ 5.7 cm)

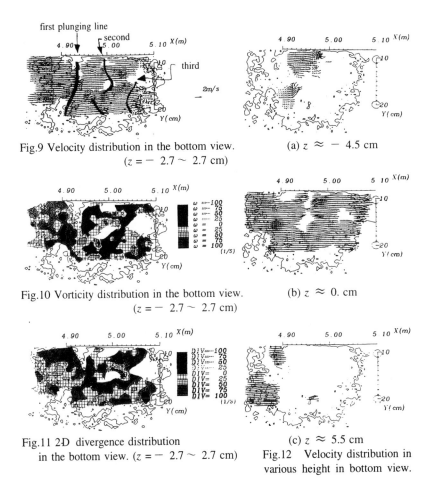

Fig.9 Velocity distribution in the bottom view.
($z = -\ 2.7 \sim 2.7$ cm)

(a) $z \approx -\ 4.5$ cm

Fig.10 Vorticity distribution in the bottom view.
($z = -\ 2.7 \sim 2.7$ cm)

(b) $z \approx 0.$ cm

Fig.11 2-D divergence distribution
in the bottom view. ($z = -\ 2.7 \sim 2.7$ cm)

(c) $z \approx 5.5$ cm

Fig.12   Velocity distribution in
various height in bottom view.

plunging lines incline from the span-wise direction, not perpendicular to the direction of wave propagation. The inclination or the 3-D feature increases as it goes forward.

At the apex of the second plunging line at x = 5.01 m and y = 14 cm, there exists positive intense 2-D divergence. Also a spotted pair of vortexes with opposite rotation distributes arround there. The configuration of this pair to wave propagation direction is inverse of one described in the previous section. The mechanism would be as follows; the splashed water caused by intense positive divergence is faster than the ambient water, or the plunging line is descending downward at that point.

Figures 12(a), (b), (c) show the velocity distribution in various height in bottom view. Although there are some duplicated areas over the different heights, this would be due to the effect of interporation. These show that the clouded surface undulates to span-wise. This phenomenon can be checked by the side view as well (the top of Fig.7).

4. Conclusions

A handmade system has been developed to measure the 3-D flow fields inside breaking waves. The main compositions of the system are the high speed camera with a stereoscopic adopter and supporting softwares which correct the aberration of the adapter, eliminate erroneouse velocities, balance the stability and resolution, and also enables stable calculation of deformation rate fields.

Two sets of the systems have been applied for the breaking solitary wave on a uniform slope. Simultaneous measurement from the side and bottom shows deformation of the plunging jet from 2-D to 3-D behavior, and distributions and their mutual relations of velocity, vorticity and 2-D divergence.

Acknowledgements

It is pleasure to acknowledge Professors Masaki Sawamoto, Tohoku University, and Masatomo Nagao, Ashikaga Institute of Technology, for valuable comments for this study.

References

[1] Basco, D.R. (1985): A qualitative description of wave breaking, ASCE. Jour. WW. Vol.111, No.2, pp.171-188.

[2] Bonmarin, P. (1989): Geometric properties of deep-water breaking waves, J.Fluid Mech., Vol.209, pp.405-433.

[3] Mano, A. and Kamio, S. (1990): Flow field analysis on air-entrained region of breaking wave by correlation method, Proc. Japan Soc. Civil Eng., No.423, pp.171-180 (In Japanese).

[4] Nadaoka, K. (1986): A Fundamental Study on Shoreling and Velocity Field Structure of Water Waves in the Near Shore Zone, Doctoral Dissertation, Tokyo Inst. Tech. (reproduced in Tech. Rep. of Dept. Civil Engng. Tokyo Inst. of Tech., No.36, pp.33-125)

[5] Nadaoka, K., Hino, M. and Koyano, Y. (1989): Structure of the turbulent flow field under breaking waves in the surf zone, J.Fluid Mech. Vol.204, pp.359-387.

[6] Quinn, P.A., Barnes, T.C.D., Lloyd, S.T., Greated, C.A. and Peregrine, D.H. (1995): Velocity measurements of post-breaking turbulence generated by plunging breakers, Int. Conf. Coastal Dynamics'95, pp293-304.

[7] Svendsen, I.A. and Madsen, P.A. (1984): A turbulent bore on a beach, J.Fluid Mech. Vol.148, pp.73-96.

[8] Watanabe, Y. and Saeki, H. (1995): Consideration on the kinematic mechanism in surf zone through image processing, Proc. Coastal Eng., JSCE, Vol.42, pp.116-pp.120(In Japanese).

[9] Yamashita, T., Tallent, J.R. and Tsuchiya, Y. (1988): Generation mechanism and movement charateistics of the horizontal vortexes in breaking waves, Proc. Coastal Eng., JSCE, Vol.35, pp.54-58(In Japanese).

Probability Distribution of Wave and Roller Energy in the Nearshore Zone

Marien Boers[1]

Abstract

The present paper describes a cross-shore model for the wave and roller energy in the nearshore zone. This model is based on the energy balance equations for wave and roller energy separately. The balance equation for wave energy consists of dissipation parameters for bottom friction, for the breaking process by wave instability and for the breaking process by the impact of the roller. Moreover there is a transfer of wave energy to roller energy by the instability process and by the impact of the roller. This transfer contributes as a source term to the balance equation for the roller energy. Dissipation of roller energy occurs by internal dissipation and loss of kinetic and potential energy to the wave below the roller. Some of these dissipation and transfer parameters are explicitly presented, others have still to be defined. These parameters have an important impact on the sediment transport, because of the generation of turbulence and the influence of the undertow by additional mass drift and a shear stress on top of the undertow profile. The energy balance equations are used to elaborate functions for the spatial gradients of wave and roller energy. These functions are used for the development of a cross-shore model for the probability distribution of wave and roller energy in the nearshore zone.

Introduction

For the morphological modelling of the nearshore zone of a sandy coast, an accurate description of the wave and roller energy is most essential. The wave energy has an important impact on the bed shear stress and on the currents in cross-shore and longshore direction (undertow, longshore currents and streaming). The roller energy largely contributes to the turbulence in the water column and importantly influences the undertow because of the additional mass transport [Svendsen (1984b)] and the shear stress performed on top of the undertow profile [Svendsen (1985)].

A number of well-known wave models for the surf zone don't explicitly describe the

[1] Delft University of Technology, Department of Civil Engineering,
PO Box 5048, 2600 GA Delft, The Netherlands

roller energy, like the wave averaged models of Battjes and Janssen (1978) and Thornton and Guza (1983), and the wave-by-wave model of Dally *et al.* (1985). Models which include the roller energy, like the model of Lippmann *et al.* (1996), are generally based on the overall energy balance equation of Svendsen (1984a):

$$\frac{\partial E c_g}{\partial x} + \frac{\partial E_r c}{\partial x} + D = 0 \qquad\qquad Eq.\ 1$$

(E is the wave energy, E_r is the roller energy, c_g is the wave group celerity, c is the wave celerity and D is the energy dissipation)

Because both the wave energy and the roller energy are unknown, an additional equation is required to solve this equation. Often, formulations are applied which assume a direct relationship between the roller characteristics and the local wave parameters. These formulations are either derived from measurements carried out in the inner surf zone [Svendsen (1984a)] or derived from a comparison with a stable bore [Deigaard *et al.* (1991)].

The problem with this type of formulations is that the generation of the roller and the cessation at a certain distance onshore of a breaker bar are not properly described. Therefore, the present paper explicitly analyses the dissipation of wave energy, the transfer of wave energy to roller energy and the dissipation of roller energy. This analysis results into equations for the wave energy and for the roller energy separately. Subsequently, formulations are given which describe the various dissipation and transfer of energy processes. Finally, these equations are applied in a cross-shore model which describes the probability distribution of the wave and roller energy at every position in the nearshore zone.

Description of the breaking processes

To quantify the transfer of wave energy to roller energy and the dissipation of wave energy and roller energy, it is necessary to study the process of wave breaking. The following processes, which play a role in the breaking process, are described in detail [Figure 1]:
- wave breaking because of an unstable wave front
- dissipation of wave energy due to the roller
- dissipation of roller energy

wave breaking because of an unstable wave front
When a wave front becomes too steep, the wave becomes unstable and starts to break. For plunging and spilling breakers, the structure of the orbital flow in the crest breaks down and is transferred into vortices in front of the wave crest. For spilling breakers, most of these vortices remain at the surface of the wave and form together the roller. For plunging breakers, a jet penetrates into the water column underneath the surface of the wave. Therefore, the ratio between the transfer of wave energy to roller energy and the

dissipation of wave energy into turbulence depends on the type of wave breaking. For spilling breakers, most dissipated wave energy is transferred to roller energy; for plunging breakers, a significant amount of the dissipated wave energy is transferred to turbulence energy below the roller.

dissipation of wave energy due to the roller
After the roller is generated, there is an interaction between the roller and the wave, which results into an additional loss of wave energy. Part of this wave energy is transferred into roller energy; the other part is dissipated to turbulence energy.

dissipation of roller energy
There are two processes responsible for the dissipation of roller energy. The first process is the dissipation of roller energy within the roller. The other process is the loss of kinetic and potential roller energy by the downward flow from the roller into the wave below.

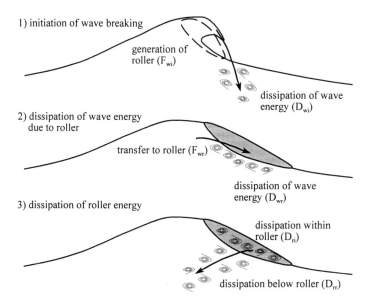

1) initiation of wave breaking

generation of roller (F_{wi})

dissipation of wave energy (D_{wi})

2) dissipation of wave energy due to roller

transfer to roller (F_{wr})

dissipation of wave energy (D_{wr})

3) dissipation of roller energy

dissipation within roller (D_{ri})

dissipation below roller (D_{rr})

Figure 1: Transfer and dissipation of wave and roller energy

The loss of wave energy consists of the dissipation by bottom friction D_f, the dissipation by wave breaking D_w and the transfer of energy to the roller F_w:

$$\frac{\partial E c_g}{\partial x} + D_f + D_w + F_w = 0 \qquad\qquad Eq. \ 2$$

The transfer of wave to roller energy F_w is a source term in the balance equation for the roller energy. Further, there is a sink term because of the dissipation of roller energy $\mathbf{D_r}$:

$$\frac{\partial E_r c}{\partial x} - F_w + D_r = 0 \qquad\qquad Eq.\ 3$$

So, the total dissipation of wave and roller energy yields:

$$D = D_f + D_w + D_r \qquad\qquad Eq.\ 4$$

It appears that two hydraulic processes are responsible for the dissipation and the transfer of wave energy. The first process is the wave energy dissipation because of an unstable wave front $\mathbf{D_{wi}}$; the second process is the wave energy dissipation caused by the interaction between the wave and the roller $\mathbf{D_{wr}}$. The dissipation of wave energy by wave breaking yields:

$$D_w = D_{wi} + D_{wr} \qquad\qquad Eq.\ 5$$

Also the transfer of wave energy to roller energy is composed of energy transfer due to wave instability $\mathbf{F_{wi}}$ and energy transfer due to interaction between wave and roller $\mathbf{F_{wr}}$:

$$F_w = F_{wi} + F_{wr} \qquad\qquad Eq.\ 6$$

It is important to know where the dissipation of wave and roller energy occurs, in order to understand the generation of turbulence energy in the water column below the roller. The dissipation of wave energy $\mathbf{D_w}$ occurs by definition below the roller. The dissipation of the roller energy $\mathbf{D_r}$ however, only partly contributes to the turbulence in the water column below the roller, because part of the roller energy dissipates within the roller itself. Therefore a distinction is made between the dissipation of roller energy within the roller $\mathbf{D_{rr}}$ and loss of roller energy to turbulence below the roller $\mathbf{D_{rw}}$:

$$D_r = D_{rr} + D_{rw} \qquad\qquad Eq.\ 7$$

So, in order to resolve the Equations 2 and 3 one requires one formulation for the dissipation by bottom friction, two formulations for the dissipation of wave energy by breaking and two formulations for the dissipation of roller energy. Furthermore, two formulations are required for the transfer of wave energy to roller energy. Some of these formulations are presently not properly described, but others are derived from mass, momentum and energy balances of the roller and the wave and from the measurements described by Boers (1996). These formulations are based on a roller model, which includes the following features:

– the energy transfer from wave to roller occurs by the upward flow of potential and kinetic energy, and by the hydraulic pressure of the wave on the roller;
– the roller celerity is not always equal to the wave celerity;
– the roller energy includes both the kinetic and the potential energy.

Figure 2 illustrates a number of the important parameters, which play a role in this roller model. Through the upward flow discharge $\mathbf{Q_{ui}}$ all water enters the roller. Subsequently, the flow in the roller moves downward along the surface of the wave η, where $\mathbf{u}_r$ is the horizontal component of the roller velocity. Because this roller velocity is larger than the orbital flow velocity $\tilde{u}_\eta$ below the roller, there is a shear stress between wave and roller τ_{rw}. Meanwhile, there is a loss of water because of the downward flow $\mathbf{q_d}$. This process occurs over the whole length of the roller $\mathbf{L_r}$.

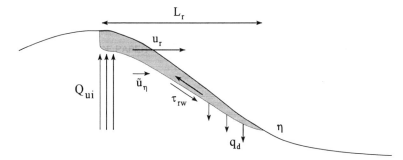

Figure 2: Interaction between wave and roller

The transfer of wave energy to roller energy by wave instability $\mathbf{F_{wi}}$ occurs by the discharge of upward flow $\mathbf{Q_{ui}}$ at the crest of the wave. There is a transport of kinetic and potential energy, where it is assumed that the orbital velocity below the wave crest $\tilde{u}_\eta$ equals the wave celerity $\mathbf{c}$:

$$F_{wi} = Q_{ui}(\rho\frac{c^2}{2} + \rho g \frac{H}{2})\qquad\qquad Eq.\ 8$$

Further this equation includes the mass density of water ρ, the gravitational acceleration $\mathbf{g}$ and the wave height $\mathbf{H}$. It is assumed that an upward flow in the breaking wave exists if the present wave front slope $\tan(\theta)$ exceeds the critical slope $\tan(\theta_{cr})$. The following empirical relation seems to work properly with $\beta_i \approx 0.40$, and with the critical wave front slope $\tan(\theta_{cr}) \approx 0.20$:

$$Q_{ui} = \beta_i c H [\tan(\theta) - \tan(\theta_{cr})]\qquad\qquad Eq.\ 9$$

The dissipation of wave energy by the wave instability $\mathbf{D_{wi}}$ appears almost zero for spilling breakers. The transfer of wave energy to roller energy by the roller impact $\mathbf{F_{wr}}$ and the dissipation of wave energy by the roller impact $\mathbf{D_{wr}}$ requires further investigation of the work performed by the shear stress and the hydraulic pressure between the roller and the wave. The dissipation of roller energy within the roller $\mathbf{D_{rr}}$ depends on the roller length $\mathbf{L_r}$, the shear stress between wave and roller τ_{rw} and the roller velocity $\mathbf{u_r}$:

$$D_{rr} = \int_{L_r} \tau_{rw} u_r dx\qquad\qquad Eq.\ 10$$

The loss of kinetic and potential roller energy by the downward flow $\mathbf{D_{wr}}$ is given by the following equation:

$$D_{wr} = \int_{L_r} q_d \rho (\frac{u_r^2}{2} + g\eta) dx \qquad\qquad Eq.\ 11$$

For the downward flow q_d an additional equation is required. It is assumed that this flow depends on the orbital velocity $\tilde{u}_\eta$ and the roller velocity u_r which is of the same order of magnitude as the wave celerity c. From the measurements described by Boers (1996), it appears that the following empirical relationship seems to work properly if the coefficient β_d has a value of about 0.03 - 0.05:

$$q_d = \beta_d(c - \tilde{u}_\eta) \qquad\qquad Eq.\ 12$$

With the formulations for the dissipation and transfer of energy parameters, the spatial gradient of wave and roller energy can be calculated. In the following of this paper a cross-shore model for the probability distribution of wave and roller energy is based on this approach.

Cross-shore model for the probability distribution of wave and roller energy

With the Equations 2 and 3, it is easy to elaborate a function **g** which describes the spatial gradient of the wave energy of a propagating wave. To symbolize that the parameters in this equation represent the characteristics of that particular wave, all wave parameters are represented in lowercase format:

$$g = \frac{de}{dx} = -\frac{e}{c_g}\frac{\partial c_g}{\partial x} - \frac{f_w + d_w + d_f}{c_g} \qquad\qquad Eq.\ 13$$

Further there is also a function **h** which describes the spatial gradient of the roller energy:

$$h = \frac{de_r}{dx} = -\frac{e_r}{c}\frac{\partial c}{\partial x} + \frac{f_w - d_r}{c} \qquad\qquad Eq.\ 14$$

With these equations, it is possible to calculate the wave and roller energy of individual waves, like the method of Dally et al. (1985). For the modelling of an irregular wave climate, such a calculation should be carried out many times with different waves until the wave climate is sufficiently reproduced. For an accurate computation of the probability distribution of wave and roller energy, the number of calculations will be very large. Therefore another method is introduced in order to save computational effort.

It is assumed that the waves in the surf zone are fully described if the wave frequency **f**, the wave energy **e** and the roller energy e_r are known. This means that the wave climate at a certain position can be described with a non-physical **FWR-space**, which is defined by these three parameters. When the probability density is known for every combination of **f**, **e** and e_r, this FWR-space can be used to describe the probability distribution of the wave frequency, the wave energy and the roller energy.

The Figures 3 and 4 illustrate the FWR-space in more detail. The left chart of Figure 3 shows the distribution of the probability density of the wave frequency. This probability density $\mathbf{p_f}$ contains all waves having a certain frequency. However, the waves with that particular frequency have different wave 'energies', as it is illustrated by the right chart of Figure 3.

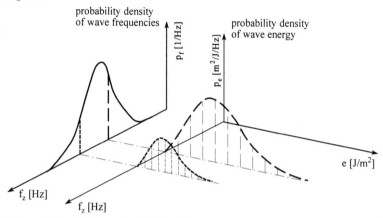

Figure 3: Distribution of probability density of wave frequency and wave energy

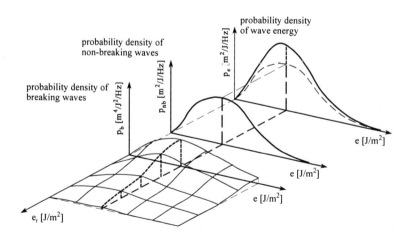

Figure 4: Distribution of probability density of non-breaking and breaking waves

It is easy to notice that the probability of the wave frequency is a cumulation of the probability density of wave energy $\mathbf{p_e}$ for that particular frequency:

$$p_f = \int_0^\infty p_e de \qquad\qquad Eq.\ 15$$

Waves having a certain frequency and a certain wave energy can further have a varying roller energy. If the roller energy is zero, the wave is non-breaking. For these non-breaking waves, there is a special probability density of non-breaking waves p_{nb}, which is illustrated with the middle chart of Figure 4. This probability density p_{nb} only depends on the wave frequency and wave energy.

If the roller energy is larger than zero, the wave is breaking. For these breaking waves a probability density of breaking waves p_b is introduced, which is illustrated by the front chart of Figure 4. This probability density p_b depends on the wave frequency, the wave energy and the roller energy.

So the probability density of wave energy p_e is a cumulation of the probability density of non-breaking waves p_{nb} and all breaking waves p_b:

$$p_e = p_{nb} + \int_{e_{r0}}^\infty p_b de_r \qquad\qquad Eq.\ 16$$

In this equation the parameter e_{r0} means a very small quantity of roller energy:

$$e_{r0} = \lim(e_r \downarrow 0) \qquad\qquad Eq.\ 17$$

Now the FWR-space is defined, it is possible to compute the probability distribution of wave and roller energy for every position in the nearshore zone if the following information is available:

–   the probability density of wave and roller energy at the seaward boundary. It is easiest to put this boundary at a position where no waves break.
–   the functions g and h [Equations 13 and 14]. These functions depend on the water depth, the wave frequency, the wave energy and the roller energy.
–   the probability that a non-breaking wave starts breaking.

Instead of the computation of the spatial gradient of the wave and roller energy of an individual wave, the computation is now performed on an area of the FWR-space. For non-breaking waves, this area has a size of $\Delta f * \Delta e$. Then the gradient of the fraction of waves present within this area is described as:

$$\frac{dp_{nb}\Delta f \Delta e}{dx} = (s_c - s_b)\Delta f \Delta e \qquad\qquad Eq.\ 18$$

The parameter s_c reflects the source term related to cessation of breaking waves and the parameter s_b the sink term related to the non-breaking waves which starts breaking. Subsequently, the spatial gradient of Equation 13 can be rewritten as:

$$\frac{d}{dx} = \frac{\partial}{\partial x} + g\frac{\partial}{\partial e}$$

Eq. 19

Where g is described by Equation 13 with the wave breaking parameters equal to zero. Further, it can be seen that:

$$\frac{\partial \Delta e}{\partial x} = \Delta e \frac{\partial g}{\partial e}$$

Eq. 20

With the Equations 19 and 20, Equation 18 is elaborated to:

$$\frac{\partial p_{nb}}{\partial x} + \frac{\partial g p_{nb}}{\partial e} = s_c - s_b$$

Eq. 21

Likewise, it is possible to compute the spatial gradient of an area with breaking waves. This area has the size of $\Delta f * \Delta e * \Delta e_r$. Sink and source terms are not present.

$$\frac{d p_b \Delta f \Delta e \Delta e_r}{dx} = 0$$

Eq. 22

Which is rewritten as:

$$\frac{\partial p_b}{\partial x} + \frac{\partial g p_b}{\partial e} + \frac{\partial h p_b}{\partial e_r} = 0$$

Eq. 23

The Equations 21 and 23 are based on the assumption that all waves have a constant wave frequency f during their propagation through the nearshore zone.

The sink term s_b in Equation 18 is dependent of the probability density of non-breaking waves p_{nb} and the probability that waves with the particular frequency and wave energy start breaking w_e, like it is described by Thornton and Guza (1983). This sink term also determines the boundary condition for the breaking waves with the roller energy e_{r0}. This boundary condition is required because initiation of wave breaking always means a positive value for the function h:

$$h p_b = s_b = w_e p_{nb}$$

Eq. 24

If breaking waves with a roller energy e_{r0} have a negative value for the function h, the wave breaking is ceasing. This means that the source term s_c is defined as:

$$s_c = -h p_b$$

Eq. 25

Further research

To make the cross-shore model applicable, it is necessary that the dissipation and transfer of energy parameters are elaborated to formulations which only depend on the wave frequency f, the wave energy e, the roller energy e_r and the water depth d. Therefore

more research is required to obtain the following formulations:
- the dissipation and transfer of energy by roller impact;
- the relations between hydraulic parameters like the wave front slope $\tan(\theta)$ and the roller length L_r, and the model parameters f, e, e_r and d.

Moreover, the formulations for the transfer and dissipation of energy are very applicable for an improved modelling of the following processes:
- mass drift by the roller with the upward flow discharge Q_{ui} and the downward flow velocity q_d;
- turbulence in the water column below the roller with the dissipation of wave energy to turbulence energy D_w and the loss of roller energy to the wave D_{rw};
- shear stress on top of the undertow profile by the exchange of momentum between wave and roller.

An improved modelling of the mass drift by the roller, the turbulence in the water column below the roller and the shear stress on top of the undertow profile results in a better modelling of the undertow and the mixing of the suspended sediment concentration. This leads to an improved morphological modelling of the nearshore zone.

References

Battjes, J.A. and J.P.F.M. Janssen. 1978. Energy loss and set-up due to breaking in random waves. *Proc. 16th Int. Conf. on Coastal Eng.*, ASCE, pp 569-587.

Boers, M. 1996. Simulation of a surf zone with a barred beach; Report 1: Wave Heights and Wave Breaking. *Communications on Hydraulic and Geotechnical Engineering*, No. 96-5, Delft University of Technology.

Dally, W.R, R.G. Dean and R.A. Dalrymple. 1984. A model for breaker decay on beaches. *Journal of Geophysical Research*, Vol. 90, No C6, pp 11917-11927.

Deigaard, R., P. Justesen and J. Fredsøe. 1991. Modelling of undertow by a one-equation turbulence model. *Coastal Engineering*, Vol. 15, pp. 431-458.

Lippmann, T.C., A.H. Brookins and E.B. Thornton. 1996. Wave energy transformation on natural profiles. *Coastal Engineering*, Vol. 27, pp. 1-20.

Svendsen, I.A. 1984a. Wave heights and set-up in the surf zone. *Coastal Engineering*, Vol. 8, pp. 303-329.

Svendsen, I.A. 1984b. Mass flux and undertow in the surf zone. *Coastal Engineering*, Vol. 8, pp. 347-365.

Svendsen, I.A. 1985. On the formulation of the cross-shore wave-current problem. *Proc. Workshop "European Coastal Zones"*, Athens, pp 1.1-1.9.

Thornton, E.B. and R.T. Guza. 1983. Transformation of wave height distribution. *Journal of Geophysical Research*, Vol. 88, No C10, pp 5925-5938.

Spectral Modelling of Nonlinear Wave Shoaling and Breaking over Arbitrary Depths

Serdar Beji[1] and Kazuo Nadaoka[2]

Abstract

A spectral model is formulated for the simulation of breaking and non-breaking unidirectional waves propagating over arbitrary depths. The evolution equations corresponding to the conservative wave model of Beji and Nadoka (1997a) are given first and then an empirical generalization of Battjes's (1986) energy dissipation model for breaking waves is introduced. The generalized dissipation terms are embedded into the conservative evolution equations so as to make the spectral model capable of simulating wave height decay due to breaking. The evolution equations with the breaking terms are then used to reproduce Horikawa and Kuo's (1966) and Beji and Battjes's (1993) laboratory measurements of breaking waves as well as the field measurements of Nakamura and Katoh (1992) in the surf zone. Despite the crude approximations involved in the formulation of the breaking dissipation, the comparisons show rather good agreements.

Introduction

With the rapid advance of computational facilities, recent years have seen an increasing interest towards the simulation of the nonlinear aspects of the surface waves, particularly in the nearshore region where these effects are observed to be most appreciable. Wave skewness related sediment transport, effects of harmonic generation on the characteristics of a wave field, and influence of breaking on the surf-zone processes are the most striking examples of such phenomena (Doering and Bowen, 1986; Freilich and Guza, 1984; Nadaoka et al., 1989). For practical applications the nonlinear wave models are not yet in common use, however, there are evidences that when augmented with appropriate generation and dissipation mechanisms, these models may well be the prototypes of the commercial models to come.

---

[1] Assoc. Prof., Department of Naval Architecture and Ocean Engineering, Istanbul Technical University, Maslak 80626, Istanbul, Turkey

[2] Professor, Graduate School of Information Science and Engineering, Tokyo Institute of Technology, 2-12-1 O-okayama, Meguro-ku, Tokyo 152, Japan

In this work a spectral model useable both for breaking and nonbreaking unidirectional waves is developed by employing the recently introduced wave equation of Beji and Nadaoka (1997a), which is a combined unidirectional form of the fully dispersive weakly nonlinear wave equations of Nadaoka $et\ al.$ (1994, 1997), in conjunction with Battjes's (1986) dissipation model. The resulting evolution equations are numerically solved for various test cases to demonstrate the capabilities of the model. Comparisons with the measurements reveal quite acceptable agreements even for considerably complicated cases.

Wave Model

Quite recently, Beji and Nadaoka (1997a) have introduced the following wave equation for weakly-nonlinear, narrow-banded, unidirectional waves travelling over arbitrary depths

$$
\begin{aligned}
&C_g\eta_t + \frac{1}{2}C_p(C_p + C_g)\eta_x - \frac{(C_p - C_g)}{k^2}\eta_{xxt} - \frac{C_p(C_p - C_g)}{2k^2}\eta_{xxx} \\
&+ \frac{1}{2}\left[C_p(C_g)_x + (C_p - C_g)(C_p)_x\right]\eta + \frac{3}{4}g\left(3 - 2\frac{C_g}{C_p} - \frac{k^2C_p^4}{g^2}\right)\left(\eta^2\right)_x = 0,
\end{aligned}
\tag{1}
$$

where C_p, C_g, and k are respectively the phase and group velocities and the wave-number computed according to the linear theory dispersion relation for a dominant wave frequency ω and a given local depth h, the subscripts x and t indicate partial differentiation with respect to space and time, respectively.

If the relative depth is small $C_p = C_g = (gh)^{1/2}$ and equation (1) reduces to the combined unidirectional form of Airy's nonlinear non-dispersive equations. Allowing the lowest-order dispersion by letting $C_p = (gh)^{1/2}(1 - k^2h^2/6)$ and $C_g = (gh)^{1/2}(1 - k^2h^2/2)$ leads to the KdV equation. For infinitely deep water waves the model equation admits the second-order Stokes waves as solution. Thus, the model equation may be viewed as a unified nonlinear wave model describing evolution of a narrow-banded wave field from infinitely deep to very shallow waters with smooth transition.

An important aspect in modelling wave transformations over variable sea bed is the linear shoaling characteristics of the wave model employed. If the incident wave frequency coincides with the prescribed wave frequency of equation (1) the linear shoaling is predicted exactly for any relative depth. This point may be easily demonstrated using the approach introduced by Madsen and Sørensen (1992). An incident wave of the from $\eta = a(x)\exp[\omega t - \int k(x)dx]$ is substituted into the wave equation and the higher derivatives of $a(x)$ and $k(x)$ are neglected so that an expression for the spatial variation of $a(x)$ is obtained. Carrying out this procedure for equation (1) results in $a_x/a = -(C_g)_x/2C_g$, which is exactly the expression obtained from the constancy of energy flux.

Evolution Equations

Using the wave equation (1) and expressing the surface elevation as a Fourier series with spatially varying amplitudes and phases, Beji and Nadaoka (1997b) derived a set of evolution equations describing the spatial changes of the component wave amplitudes of a prescribed incident wave field:

$$\frac{da_n}{dx} = -\frac{\alpha_s}{\alpha_1}a_n$$

$$+\beta \sum_{m=1}^{N-n} \alpha^-[(a_m b_{n+m} - a_{n+m}b_m)\cos\theta^- + (a_m a_{n+m} + b_m b_{n+m})\sin\theta^-]$$

$$+\frac{1}{2}\beta \sum_{m=1}^{n-1} \alpha^+[(a_m b_{n-m} + a_{n-m}b_m)\cos\theta^+ + (a_m a_{n-m} - b_m b_{n-m})\sin\theta^+]$$

$$(2a)$$

$$\frac{db_n}{dx} = -\frac{\alpha_s}{\alpha_1}b_n$$

$$-\beta \sum_{m=1}^{N-n} \alpha^-[(a_m a_{n+m} + b_m b_{n+m})\cos\theta^- - (a_m b_{n+m} - a_{n+m}b_m)\sin\theta^-]$$

$$-\frac{1}{2}\beta \sum_{m=1}^{n-1} \alpha^+[(a_m a_{n-m} - b_m b_{n-m})\cos\theta^+ - (a_m b_{n-m} + a_{n-m}b_m)\sin\theta^+]$$

$$(2b)$$

where

$$\alpha_1 = \frac{1}{2}C_p(C_p + C_g) - 2\frac{(C_p - C_g)}{k^2}\omega_n k_n + \frac{3C_p(C_p - C_g)}{2k^2}k_n^2,$$

$$\alpha_2 = \frac{(C_p - C_g)}{k^2}\left(\frac{3}{2}C_p k_n - \omega_n\right),$$

$$\alpha_s = \frac{1}{2}[C_p(C_g)_x + (C_p - C_g)(C_p)_x] + \frac{(C_p - C_g)}{k^2}\left(\frac{3}{2}C_p k_n - \omega_n\right)(k_n)_x,$$

$$\beta = \frac{3}{4}g\left(3 - 2\frac{C_g}{C_p} - \frac{k^2 C_p^4}{g^2}\right), \quad \delta^+ = k_{n-m} + k_m - k_n, \quad \delta^- = k_{n+m} - k_m - k_n,$$

$$\alpha^+ = \frac{k_{n-m} + k_m}{\alpha_1 + \alpha_2\delta^+}, \quad \alpha^- = \frac{k_{n+m} - k_m}{\alpha_1 + \alpha_2\delta^-}, \quad \theta^+ = \int_0^x \delta^+ dx, \quad \theta^- = \int_0^x \delta^- dx,$$

$$(3)$$

in which k_n for each component is computed from the linear dispersion relation of the wave model for the local depth $h(x)$ and the radian frequency $\omega_n = n\Delta\omega$, $\Delta\omega$ being the frequency of resolution:

$$C_g\omega_n - \frac{1}{2}C_p(C_p + C_g)k_n + \frac{(C_p - C_g)}{k^2}\omega_n k_n^2 - \frac{C_p(C_p - C_g)}{2k^2}k_n^3 = 0, \quad (4)$$

and $(k_n)_x$ appearing in α_s is obtained from (4) by differentiating it with respect to x.

The free index n runs from 1 to N, resulting in $2N$ number of nonlinearly coupled first-order differential equations for the unknown components $a_n(x)$ and $b_n(x)$. Once the $a_n(x)$s and $b_n(x)$s are obtained the free surface elevation may be constructed from $\eta = \Sigma a_n \cos \phi_n + b_n \sin \phi_n$, with $\phi_n = \omega_n t - \int k_n dx$. Various numerical integration techniques (e.g., Adams-Bashford-Moulton, Bulirsch-Stoer, Runge-Kutta) are available for the integration of equation (2). Here, the Runge-Kutta fourth-order formulation is preferred as it proved to be the fastest while being as reliable as the other more sophisticated integration methods.

Energy Dissipation Due to Breaking: A Generalization of Battjes's model

An empirical approach is now adopted to account for the dissipative role of wave breaking so as to render the spectral model operational even for breaking waves. To this end we first refer to Battjes's(1986) dissipation model for periodic waves:

$$\frac{\partial P}{\partial x} + D = 0 \qquad \text{with} \qquad D = \frac{B}{4\gamma^3} \frac{\rho g H^2}{T} \left(\frac{H}{h}\right)^4 , \qquad (5)$$

where ρ is the water density, g the gravitational acceleration, H the wave height, T the wave period, $P = EC_g$ the energy flux with $E = \frac{1}{8}\rho g H^2$ and $C_g = (gh)^{1/2}$ (shallow water). $B = O(1)$ is a calibration coefficient independent of any variable, and finally γ is the breaking coefficient which is *inversely* proportional to the height of the foam region in a breaker. The latter meaning of γ is especially emphasized for an *ad hoc* modification which is taken up later.

We have selected the above dissipation model of periodic waves rather than the stochastic description of Battjes and Janssen (1978) because our evolution equations are essentially based on a deterministic wave-by-wave analysis that traces the spatial changes of simple sinusoidal components with definite periods. From this simplified (the effects of phase-locking ignored) point of view it at once becomes evident that a dissipation model for periodic waves is the most appropriate choice.

Equation (5) is applicable only to shallow water waves and it is desirable to extend it to deep water so that it will be in accord with equation (1), which is valid for arbitrary depths. It is fully understood that such an extension is artificial and not in line with the physical considerations followed in derivation of (5); therefore, it is best to view this extension as purely empirical.

First of all the ratio H/h in (5) is viewed as the *wave steepness* for shallow water waves, the most important non-dimensional variable dictating the dissipation rate. A generalized wave steepness for waves over arbitrary depths has been introduced as gH/C_p^2 (Beji, 1995) with C_p denoting the wave phase speed according to linear theory. Thus, we propose the following modified form of (5) as the dissipation model

$$\frac{H_x}{H} = -\frac{B}{2\pi\gamma^3} \frac{\omega}{C_g} \left(\frac{gH}{C_p^2}\right)^4 , \qquad (6)$$

in which the linear shoaling term implicitly present in (5) has been removed since the wave model itself includes the linear shoaling properly, and C_g in its full form as defined by linear theory is adopted instead of the shallow water approximation. Note that for shallow waters $C_p^2 = gh$ and the ratio gH/C_p^2 reduces to H/h, which is identical with the original expression. On the other hand, for deep waters gH/C_p^2 tends to kH and, unlike H/h, still gives a non-zero dissipation due to breaking so long as the deep water wave steepness kH is appreciable.

If the wave is represented as $a(x)\cos\phi + b(x)\sin\phi$, ϕ being the phase angle, then, according to (6), the spatial rates of reduction of the amplitude components due to breaking are

$$\frac{a_x}{a} = -\frac{B}{2\pi\gamma^3}\frac{\omega}{C_g}\left(\frac{2g\sqrt{a^2+b^2}}{C_p^2}\right)^4, \qquad \frac{b_x}{b} = -\frac{B}{2\pi\gamma^3}\frac{\omega}{C_g}\left(\frac{2g\sqrt{a^2+b^2}}{C_p^2}\right)^4, \quad (7)$$

respectively for $a(x)$ and $b(x)$. The above formulation is valid only for simple sinusoidal wave forms; the evolution equations contain not only a single component but a number of components with different frequencies. It is therefore necessary to introduce some approximations to use the formulation (7) in (2). A plausible approach is to assume that each harmonic component is dissipated according to (7) while the main dissipation mechanism gH/C_p^2 comprises the total effect of all the components. In other words, *all* the wave components are dissipated in proportion to the *resultant* wave height, should it becomes large enough to induce breaking. Thus, for the wave breaking dissipation portions of (2a) and (2b) one can write

$$\left(\frac{da_n}{dx}\right)_d = -\frac{B}{2\pi\gamma^3}\frac{\omega_n}{(C_g)_n}\left(\frac{gH}{C_p^2}\right)^4 a_n, \qquad \left(\frac{db_n}{dx}\right)_d = -\frac{B}{2\pi\gamma^3}\frac{\omega_n}{(C_g)_n}\left(\frac{gH}{C_p^2}\right)^4 b_n,$$

$$(8)$$

where the resultant wave height is approximated simply as $H \simeq 2\sqrt{\Sigma(a_n^2+b_n^2)}$ while C_p is, as before, the local phase celerity computed for the dominant wave frequency ω and local depth h. $(C_g)_n$ denotes the group velocity corresponding to the frequency ω_n and local depth h according to linear theory.

Earlier it has been indicated that the parameter γ may be viewed as a measure of the extend of the foaming region in a breaker (*i.e.*, the smaller the γ the stronger the breaking). Battjes (1986) uses $\gamma = 0.6$ for constant depth and $\gamma = 0.7 + 5S$ (S is the bottom slope) for varying depth. In general there is no strict rule for the selection of γ (see Battjes and Stive, 1985). Here, we shall do a final modification and introduce the following definition of γ for use in our computations.

$$\gamma = 0.6\left[1 + \left(\frac{\mu}{1+\mu}\right)\gamma^*\right] + \left(\frac{1}{1+\mu}\right)5S, \qquad (9)$$

where $\mu = kh$ is the relative depth for the dominant wave, and γ^* a constant compensating for the relatively smaller dissipation rate of a deep water breaking wave by making γ larger. Typically, γ^* may be selected in the range of $0 \leq \gamma^* \leq 1$, and we use $\gamma^* = 0.3$ throughout.

Note that for very shallow water waves μ tends to zero and equation (9) becomes identical with Battjes's choice $\gamma = 0.7 + 5S$, except for the constant 0.7. The selection of 0.6 instead of 0.7 as a constant was done merely for including the constant depth value as a special case. For deep water waves (9) becomes

$\gamma = 0.6(1 + \gamma^*)$, which is independent of bottom slope and larger than 0.6, as deemed by the physical nature of white capping in contrast to depth-induced breaking.

After introducing such extensive modifications to Battjes's original formulation it becomes unavoidable to redefine the constant B. Battjes typically uses $B = 2$; after a few trials we have selected $B = 1$. In what follows we use $B = 1$, and γ as defined in (9) with $\gamma^* = 0.3$ for *all* the computational results presented. No other calibration is introduced despite the different nature of the cases considered. Furthermore, various tests with different γ^* values and different functional forms of γ have revealed no appreciable sensitivity and the overall dissipation model has been found to be quite robust.

Thus, after incorporating the dissipative effect of wave breaking by simply adding the expressions in (8) to the right-hand sides of (2a) and (2b) the final evolution equations may simulate breaking waves. It is worthwhile to indicate that the evolution equations with the breaking terms may be used for nonbreaking waves as well since the waves steepness term gH/C_p^2, which acts as a breaking criteria, becomes small for small amplitudes thus rendering the dissipation terms inactive.

Sample Simulations

The evolution equations formulated above are now used for sample simulations of breaking waves. Nonbreaking waves are covered in Beji and Nadaoka (1997b) and will not be considered here.

(a) Periodic waves breaking on constant depth

For a simple but fundamental demonstration we compare the results of the numerical solution of the present wave model with the exact analytical solution of Battjes's original model for an experimental case of Horikawa and Kuo (1966) for breaking waves over a horizontal bottom. Battjes shows that his analytical solution agrees well with the measurements of Horikawa and Kuo (1986).

Figure 1 shows the comparison for $h = 0.1$ m depth and $H_b/h = 0.8$ for the linearized equations, that is, when $N = 1$ in the evolution equations. The agreement with Battjes's original model (hence with Horikawa and Kuo's experimental data) is good and supports the validity of the approximations made in deriving (8).

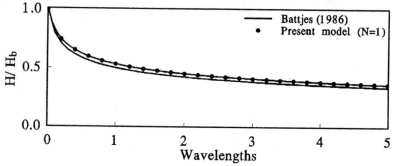

Figure 1: Wave height decay due to breaking over a horizontal bottom (Horikawa and Kuo, 1966). $h = 0.1$ m and $H_b/h = 0.8$, H_b is the initial breaking waveheight.

(b) Random waves breaking over a submerged bar

Beji and Battjes (1993) performed laboratory experiments for investigating the nonlinear transformations of random breaking waves travelling over a submerged bar. Their experimental measurements for random waves initially having a JONSWAP type wave spectrum with peak frequencies of $f_p = 0.4$ Hz (long waves) and $f_p = 1.0$ Hz (short waves) are simulated here.

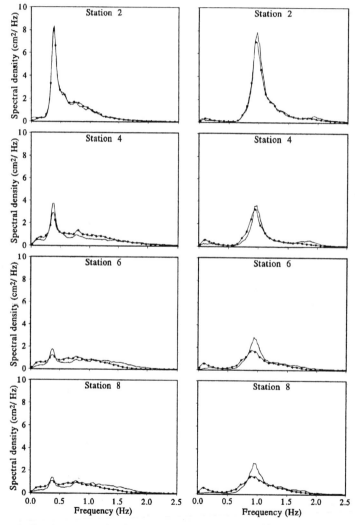

Figure 2: Random waves breaking over a submerged bar, (-) measurement and ($\bullet$) computation. Left column: long waves, right column: short waves.

The initial wave amplitudes were such that most of the breaking waves were *plunging* type. (*Spilling* type breakers were also simulated successfully but the comparisons could not be included here due to space limitations.) For long waves the dominant frequency for the model equation was set to the mean frequency of the incident spectrum while for short waves the peak frequency was used since the frequency width of the spectrum was comparatively narrower. The records of the surface elevation at Station 1 were divided into 80 segments of 512 data points and then each segment was Fourier transformed. Out of the 256 unique pairs the first 150 Fourier components, which covered a frequency range of 0.0195-3.0 Hz, were found to be quite sufficient to represent the incident wave spectrum hence the spectral model was run for 80 different realizations with $\Delta\omega = 2\pi \times 0.0195$ rd/s and $N = 150$, using the measured Fourier components as the incoming boundary condition at Station 1.

Figure 2 compares the measured and computed spectra at four selected stations for the long (left column) and short (right column) wave cases, respectively. The spectra were obtained after ensemble averaging all the realizations and frequency smoothing three neighbouring components. Each spectrum then has 480 degrees of freedom and 6.5 % normalized standard error. The agreement of the computations with the both sets of measurements is remarkably good, especially if the complicated nature of the wave transformations is considered.

(c) Breaking waves in surf zone

Nakamura and Katoh (1992) carried out field measurements from 25 February 1989 to 1 March 1989 at the Hazaki Oceanographical Research Facility (HORF) near Kashima, Japan. The site of the field observations is a natural sandy beach facing the Pacific Ocean. Ten ultrasonic wave gauges were used, of which seven were installed on the 427 m-long observatory pier while the remaining three were deployed at distances of 1.3, 2.1, and 3.2 km from the shoreline. Data was continuously sampled at a rate of 2 Hz for two-hour durations, at six-hour time intervals. A quantitative assessment made by Nwogu *et al.* (1992) using the maximum entropy method provides convincing evidence that for practical purposes the wave field in this particular region may be considered unidirectional.

Since the first seven measurement stations were located in the surf zone the majority of the waves were breaking and these stations were most suitable for testing the wave model. Thus, only the Stations 7, 6, 5, 4, and 3 were considered; the measured data at Station 7 served as the incoming boundary condition. Comparisons were made for the data of February 28, which was recorded in the aftermath of a severe storm and represented a rather rough sea state.

The collected data at Station 7 was segmented into 24 groups of 512 data points and Fourier transformed. Of the 256 unique transformed pairs, the first 128 components which covered a frequency range between 0.0039 to 0.5 Hz were considered sufficient to represent the incident spectrum. The dominant frequency of the wave model was set to the peak frequency of the incident wave spectrum and the computations were performed for 24 different realizations with $\Delta\omega = 2\pi \times 0.0039$ rd/s and $N = 128$. Figure 3 shows the measured and computed spectra at Stations 7, 6, 5, 4, and 3. Each spectrum has 192 degrees of freedom and 10% normalized standard error. The agreement is quite reasonable, especially if allowances are made for the uncertainties involved in the unidirectionalty of waves and the exact form of the bottom topography.

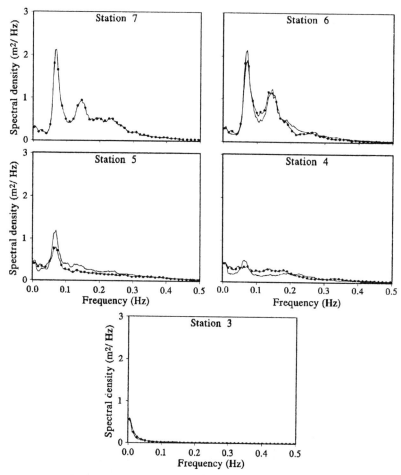

Figure 3: Breaking waves in the surf zone (Nakamura and Katoh, 1992), (-) measurement and (●) computation.

Concluding Remarks

A spectral model which is employable for both breaking and nonbreaking waves over arbitrary depths has been developed by modifying the spectral domain formulation of a weakly-nonlinear unidirectional wave equation through a generalized version of Battjes's (1986) dissipation model. The performance of the spectral model for breaking waves has been tested for several different cases and found to be acceptable. Quite importantly, the generalized dissipation mechanism has proved to be insensitive to the calibration coefficients; a single initial calibration was enough for diversely different simulations.

Acknowledgment

The first author would like to acknowledge the grant received from the Japan Society for Promotion of Science (JSPS) during the completion of this work at Tokyo Institute of Technology.

References

Battjes, J.A., 1986. Energy dissipation in breaking solitary and periodic waves, *Comm. on Hydraulic and Geotech. Eng.*, Internal Report, Delft Univ. of Tech., Report num. 86-5.

Battjes, J.A. and J.P.F.M. Janssen, 1978. Energy loss and set-up due to breaking of random waves. *Proc. 16th Coastal Eng. Conf.*, 569-587.

Battjes, J.A. and M.J.F. Stive, 1985. Calibration and verification of a dissipation model for random breaking waves. *J. Geophys. Res.*, **90**, C5, 9159-9167.

Beji, S., 1995. Note on a nonlinearity parameter of surface waves. *Coastal Eng.*, **25**, 81-85.

Beji, S. and J.A. Battjes, 1993. Experimental investigation of wave propagation over a bar. *Coastal Eng.*, **19**, 151-162.

Beji, S. and K. Nadaoka, 1997a. A time-dependent nonlinear mild-slope equation for water waves. *Proc. Roy. Soc. Lond.* A, **453**, 319-332.

Beji, S. and K. Nadaoka, 1997b. A spectral model for unidirectional nonlinear wave propagation over arbitrary depths. *J. Waterway, Port, Coastal, and Ocean Eng.*, submitted.

Doering, J.C. and A.J. Bowen, 1986. Shoaling surface gravity waves: A bispectral analysis. *Proc. 20th Coastal Eng. Conf.*, 150-162.

Freilich, M.H. and R.T. Guza, 1984. Nonlinear effects on shoaling surface gravity waves. *Phil. Trans. Roy. Soc. Lond.* A, **3**, 1-41.

Horikawa, K. and C.-T. Kuo, 1966. A study on wave transformation inside surf zone, *Proc. 10th Coastal Eng. Conf.*, **1**, 217-233.

Madsen, P.A. and O.R. Sørensen, 1992. A new form of the Boussinesq equations with improved linear dispersion characteristics. Part 2: A slowly-varying bathymetry. *Coastal Eng.*, **8**, 183-204.

Nadaoka, K, M. Hino and Y. Koyano, 1989. Structure of the turbulent flow field under breaking waves in the surf zone. *J. Fluid Mech.*, **204**, 359-387.

Nadaoka, K., S. Beji, and Y. Nakagawa, 1994. A fully-dispersive weakly-nonlinear wave model and its numerical solutions. *Proc. 24th Coastal Eng. Conf.*, 427-441.

Nadaoka, K., S. Beji and Y. Nakagawa, 1997. A fully dispersive weakly nonlinear model for water waves. *Proc. R. Soc. Lond.* A, **453**, 303-318.

Nakamura, S. and K. Katoh, 1992. Generation of infragravity waves in breaking process of wave groups. *Proc. 23th Coastal Eng. Conf.*, 990-1003.

Nwogu, O., T. Takayama, and N. Ikeda, 1992. Numerical simulation of the shoaling of irregular waves using a new Boussinesq model. *Rep. of Port and Harbour Res. Inst.*, **31-2**, 3-19.

Wave transformation in the nearshore zone: Comparison between a non-linear model and large scale laboratory data

F. Ozanne[1] , A. Chadwick[1], D. Huntley[2] and J. McKee Smith[3]

ABSTRACT: A previously developed one dimensional time domain Boussinesq-type model is applied to large scale laboratory data. Computed free surface and depth-averaged velocity oscillations are compared with shoaling and surf zone measurements from the SUPERTANK experiment. Particular emphasis is put on assessing the capability of the model to predict high order surface elevation and velocity moments, and on assessing its sensitivity to the empirical parameters involved in the simulation of wave breaking, in particular the critical local slope of the surface elevation ϕ_B. Comparisons of measured and computed results for regular waves show that the numerical model can predict surface elevation and velocities reasonably well for both shoaling and breaking. The results for the irregular waves show discrepancies with the measurements, especially for the skewness and kurtosis of the surface elevation. These are mainly due to the limitation of the model to weakly non-linear waves. Not surprisingly, the results for the regular waves show a significant dependence on ϕ_B, whereas results for the irregular waves show very little sensitivity to ϕ_B.

INTRODUCTION

It is well known that asymmetries in the oscillatory wave field control cross-shore sediment transport. In particular, nearshore sediment transport models based on the energetics approach rely on the importance of various moments of the velocity field. Some of the high order velocity moments of importance in a Bailard formulation are the third (skewness) and fourth (kurtosis) moments. The prediction of sediment transport therefore relies on a description of hydrodynamics that predicts these terms. For a Gaussian distribution of the surface elevation (or velocity), the skewness is 0 and the kurtosis is 3; a deviation from these figures indicates a deviation from a Gaussian distribution. The skewness and kurtosis are therefore measures of the deviation of a wave field from the linear random sea assumption. The objective is to assess a non-linear shoaling and breaking wave model in a way that may be useful for sediment transport models and for practical application. Models based on the

---

[1] SCSE, University of Plymouth, Palace St, Plymouth PL1 2DE. Email: fozanne@plymouth.ac.uk
[2] IMS, University of Plymouth, Drake Circus, Plymouth PL4 8AA.
[3] Res. Hyd. Engr., USAE Waterways Experiment Station, CERC, Vicksburg, MS 39180 USA.

Boussinesq equations are very promising wave shape, hence velocity moments, predictors. The time domain Boussinesq models are based on depth-integrated equations that (typically) describe the propagation of weakly non-linear shallow water waves. Besides other improvements, they were recently extended to include the description of wave breaking using various concepts (see Hamm et al.,1993 for a review & more recent work by Wei & Kirby,1995).

The model used in this study is based on time domain Boussinesq equations for which the application was extended to shorter waves by Madsen & Sørensen (1992) and to spilling breaking by Schäffer et al. (1993). Schäffer et al. extended their Boussinesq model to the surf zone by using a surface roller concept. However, the model still retains the same restrictions on the degree of non-linearity. They evaluated their model for the prediction of the evolution of the wave height during shoaling and breaking, and low order wave statistics (mean water level and standard deviation of the water surface) with small scale laboratory data. The results are generally good but do produce an underestimation of the wave height in the region of the breaking point. No evaluation of the model's capabilities to predict the skewness and kurtosis of the surface elevation and the depth-averaged velocity has been published.

Using two runs from the SUPERTANK experiment with a complex bathymetry and with regular and random waves condition, some high order moments are compared to results from the Boussinesq model. The model-data comparison involves the evaluation of the performance of the model for the prediction of free surface elevation and depth-averaged velocity, in a one-dimensional situation. Furthermore, particular attention is given to assess the sensitivity of the solution to the free parameters involved in the description of wave breaking on the solution.

THE NUMERICAL MODEL

The numerical model is based on the Boussinesq equations as derived by Madsen et al. (1991) and Madsen and Sørensen (1992), with free surface elevation and depth-integrated velocity as variables. It allows slowly varying bathymetries and contains correction terms to improve the dispersion for shorter waves periods, and thus improve the shoaling properties of the model. This modification was shown by Madsen and Sørensen (1993) to be important for the correct simulation of wave-wave interactions. The model does not include bottom friction. It also is limited to the description of weakly non-linear waves, that is waves for which $\varepsilon \ll 1$, where ε is the ratio of the wave amplitude to the water depth.

Wave breaking is modelled using the concept of surface roller as formulated by Schäffer et al. (1993). This concept assumes spilling breakers. The surface roller is regarded as a passive bulk of water being carried by the wave, at the wave celerity. The roller generation is assumed to introduce a non-uniform velocity profile. The effect of wave breaking is incorporated in the depth-integrated momentum equation through an additional convective term resulting from the new non-uniform velocity profile. The breaking criterion, which is related to the local slope of the water surface, determinates the thickness of the roller δ from simple geometry: breaking is initiated when the local slope of the free surface ϕ reaches a threshold ϕ_B. The water above the tangent of the slope $\tan\phi$ is assumed to belong to the roller, thus defining δ. ϕ then exponentially decays to ϕ_o as described by equation 1. This time variation of ϕ is chosen by virtue of the fact that breakers often transform rather quickly into the bore-like stage:

$$\tan \phi = \tan \phi_o + (\tan \phi_B - \tan \phi_o) \exp\left(-\ln 2 \frac{t - t_B}{t^*} \right) \tag{1}$$

t^* controls the rate of decay of ϕ. t_B is the time at which breaking is initiated. The fourth parameter involved in the simulation of wave breaking is the roller shape parameter f_δ. After the determination of the surface roller, δ is multiplied by f_δ, purely to compensate for the simple way of separating the roller from the rest of the flow. Breaking ceases if the local slope becomes less than $\tan\phi$ again. The waves are absorbed by a numerical sponge layer at the shoreline boundary. The model therefore assumes that there is no reflection from the inner surf zone and swash zone. Full details of the model and the numerical scheme used to solve the equations are in Madsen & Sørensen (1992) and Schäffer et al. (1993) and references therein.

Although this is a simple wave breaking model that relies on several empirical parameters and that still assumes weakly non-linear waves, Schäffer et al. showed its capabilities to simulate various quantities and surf zone processes. Initiation and cessation of wave breaking, cross-shore evolution of the wave height and mean water level, horizontal shift between the break point and the point at which the set up in the mean water level is initiated are all reasonably well predicted. Computed and measured time series of surface elevation were also visually compared and showed satisfactory wave shape and harmonics generation predictions. These validations were carried out for both regular and irregular waves measured in small scale laboratory experiments. In the current study, the capabilities of this numerical model to predict the skewness and kurtosis of both the surface elevation and the velocity field are assessed.

Presumably, the main drawback of this model is that the wave breaking simulation relies on four empirical parameters. The figures recommended by Schäffer et al. for ϕ_B, ϕ_o, t^* and f_δ are $20°$, $10°$, $T/10$ and 1.5 respectively, where T is a typical wave period. Another objective of this study is to test the sensitivity of the model to these four free parameters. It was found, for the two cases presented here, that the results show very little sensitivity to ϕ_o, t^* and f_δ. ϕ_B however had some significant effects on the solution for regular waves. It is important to note that since this model is able to predict weakly non-linear waves fields only, it tends to underestimate the wave height in the region of breaking. Consequently, the local slope of the water surface will tend to be underestimated in this region. The value recommended by Schäffer et al. for ϕ_B is therefore chosen smaller than the actual wave steepness at the breaking point.

THE DATA: SUPERTANK EXPERIMENT

To achieve the objectives described in the previous section, comparisons are made with the data from the SUPERTANK Laboratory Data Collection Project run in 1991 in the Wave Research Laboratory, Oregon State University (Kraus & McKee Smith, 1994). This experiment was chosen for several reasons: first it provides data for conditions comparable to those in the field. The total length of the channel is 104 m, its width is 3.7 m and depth 4.6 m. It contained a 76 m long sandy beach. This on the other hand meant that in the numerical tests, the bed profile was assumed to be stable for the duration of the experiment (typically 3 minutes). This data set was also chosen because it provided a comprehensive set of cross-shore wave and current measurements, in the shoaling zone and throughout the surf zone (figure 1). The free surface oscillations were measured using 16 resistance gauges (gauges 1 to 16) and 10 capacitance gauges that measured the swash oscillations mainly (gauges 17 to 26). The numerical experiment stopped after the second capacitance gauge (i.e. gauge 18), where the sandy bed was never exposed. The most offshore wave gauge signal was used as input in the numerical model at the seaward boundary. The velocities were measured using 18 electromagnetic current meters set in vertical arrays of 1 to 4 meters (figure 1).

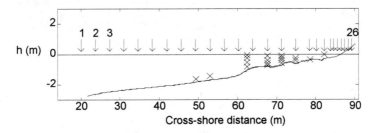

Figure 1 . Experimental set-up. Wave gauges are numbered 1 to 26. Gauges 1 to 16 are resistance gauges, gauges 17 to 26 are capacitance gauges. The 18 electromagnetic current meters are represented by a cross. The profiles at the beginning of both experiment are drawn as plain lines.

Data selection was made considering the model's assumptions of weak non-linearity, spilling breakers, no reflection from the region beyond the landward boundary of the model, and slowly varying bathymetry. Obviously, the aim here being to assess the model for hydrodynamics prediction in the surf zone, the regime is expected to be highly non-linear, and the weak non-linearity assumption to be violated.

		Hrms (m)	Tp (s)	S_o	S_{br}	Breaker Type	ε_o	ε_{br}	μ_o
A1215A	irregular	0.6	8			spill/plunge	0.1	0.28	0.01
A1216A	regular	0.6	8	0.6	6	spill/plunge	0.1	0.47	0.01

Table 1 . Wave characteristics. the subscripts 'o' and 'br' refer to the wave generation and breaking regions respectively. The breaker type is estimated with the surf similarity parameter computed using the outer surf zone slope.

The results of comparison with two runs is presented here: one of regular waves, and one of irregular waves (broad spectrum). The wave characteristics of both runs are presented in table 1. The regular run was executed immediately after the irregular one. As the latter hardly resulted in any net sediment transport, the bed profiles for both runs are very similar (figure 1). The beach profile is complex, the main features being one main bar and a smaller one inshore. The beach is fairly shallow, with a slope of 1 in 30 and 1 in 25 in front and beyond the main bar respectively. However, the typical wave period for both runs being as high as 8 sec., the beach slope parameter S ($=h_xL/h$) is very large, and reflection from the slope is anticipated. The dispersion parameter μ ($\approx h/gT^2$) is typical of very long waves. The non-linearity parameter ε (= Hrms/2h) is as high as 0.47 at the breaking point for the regular wave case. It reaches 0.28 for the irregular waves case. However, the computation of ε is based on Hrms; ε computed with Hmax gives 0.6 in the breaking region for this case.

Reflections from the region from gauge 18 (landward boundary of the model) up to the shoreline are a matter of concern. The typical wave period for both runs being large, and the foreshore being steep, it is probable that significant reflection occurred in this region. This presumption is confirmed by the fact that the cross-shore variation of the measured wave height show clear modulations, especially in the inner surf zone and swash zone. These modulations, when compared to a Bessel function solution (quick analytical solution), correspond to the predicted locations of nodes and antinodes.

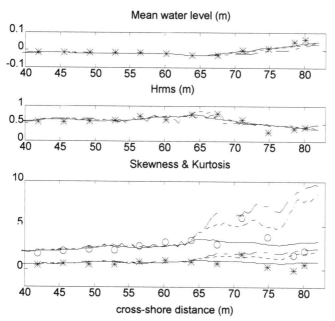

Figure 2 . Regular run: comparison for <u>free surface elevation</u> predictions. Gauges 7 to 18 only are plotted here. The stars and circles are the measured data. The solid, dotted and dashed curves are computed results for $\phi_B = 20°$, $26°$ and $30°$ respectively. The bottom plot shows the kurtosis (top three curves and circles) and skewness (bottom three curves and stars) of the surface elevation.

THE RESULTS

The comparison of the computed and measured data consisted of the study of the cross-shore variation of several quantities.

Regular waves test
Figure 2 shows the cross-shore variation of the mean water level, the root-mean-square wave height Hrms and the skewness and kurtosis of the surface elevation, as predicted by the model, using three different ϕ_B (20° being the figure recommended by Schäffer *et al.*), and compared with measurements. The breaking point is located above the first bar, at about 65m. The results for shoaling show very good agreement with the measurements for the mean water level and Hrms, but also for third and fourth order statistics. Therefore, this analysis concentrates on wave breaking. The set up is slightly underestimated in this case. The effect of ϕ_B on Hrms is as expected: increasing ϕ_B from 20° to 26° has resulted in a shift shoreward of the breaking point, and a better fit with measurements. However, it has also resulted in a further underestimation of the mean water level. Increasing ϕ_B further to 30° results in fact in a decrease and modulations in Hrms. This is explained by the fact that allowing the waves to shoal further increases non-linearity and the generation of higher harmonics.

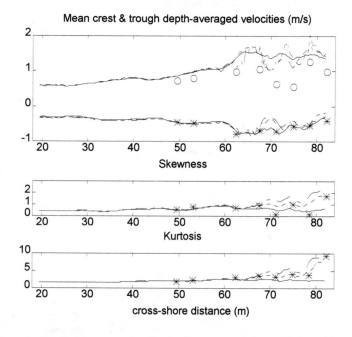

Figure 3 . Regular run: comparison for <u>depth-averaged velocity</u> predictions. The stars and circles are the measured data. The solid, dotted and dashed curves are computed results for $\phi_B = 20°$, 26° and 30° respectively. The top plot shows the maximum onshore (positive, i.e. top three curves and circles) and minimum offshore (negative, i.e. bottom three curves and stars) velocity predictions.

The results for Hrms show some sensitivity to ϕ_B, but it is on the skewness and kurtosis of the free surface elevation (S_η and K_η respectively) that ϕ_B has the most significant effect.

Both these parameters are best predicted for $\phi_B = 20°$. As ϕ_B increases, so do S_η and K_η. Again the effect of increasing ϕ_B is that the waves are allowed to shoal further, and therefore become more non-linear. Thus it is expected that S_η and K_η should increase with increasing ϕ_B. However, an increase of ϕ_B by 6° only results in a very significant increase in S_η and K_η, both being overpredicted in this case. It is also interesting to notice the strong correlation between S_η and K_η and the water depth.

Figure 3 shows the cross-shore variation of the mean maximum (shoreward) and mean minimum (offshore) depth-averaged velocities (Umax and Umin respectively), and the skewness and kurtosis of the depth-averaged velocity. The results are fairly similar to those for the surface elevation in that they show the same sensitivity to ϕ_B. The prediction of Umin is fairly good and generally better than that of Umax, which is overestimated in the surf zone. This is probably due to the underestimation of the crest elevation in the breaking region, which the model cannot properly predict due to the weak non-linearity assumption. This discrepancy can also be explained by the way the computed and measured velocities were compared. Indeed, the velocity variable output from the model is the depth-integrated velocity, from which the depth-averaged velocity is calculated. The measurements of cross-shore velocity consisted of vertical arrays of 1 to 4 current meters as shown on figure 1. Here, in order to obtain a depth-averaged velocity from the measurements, it was chosen to simply take the average value of all current meters in each array for comparison with model output. Hence some discrepancies are expected. For instance, the current meters located at the same locations as the wave gauges 16 and 17 were set in the bottom half of the water column, close to the bed. Therefore, it was expected, and found, that the calculation of the measured velocity at this location leads to an underestimation.

It is interesting to observe again the prediction by the model of strong interactions of the waves with the bathymetry: the modulations in Umax and Umin perfectly follow the bathymetry, Umax and |Umin| increasing over the bars, and decreasing over the troughs.

The skewness and kurtosis of the depth-averaged velocity (S_U and K_U respectively) show the same strong sensitivity to ϕ_B as S_η and K_η, but in this case, the best fit for K_U seems to be for a ϕ_B of 26° (which also happens to be the best fit for Hrms), as opposed to 20° for S_η, K_η and S_U.

Irregular waves

Figure 4 shows the cross-shore variation of the mean water level, Hrms, the maximum wave height Hmax, S_η and K_η for the test of irregular waves. The breaking point here is slightly further offshore, at about 63 m. The results for this test show no important dependence on ϕ_B. The first observation to make here is the under-estimation of the set-down and set-up. Hrms is fairly well simulated, but the results show a clear underestimation of Hmax in the breaking area, which is as expected for this weakly non-linear model.

The results for S_η and K_η are not as good as for the regular waves case, both S_η and K_η being underestimated in the shoaling region. Table 1 gives $\varepsilon = 0.1$ at 20m and based on Hrms. ε based on Hmax is already as high as 0.2 at 20m and reaches 0.6 at the average breaking point. Clearly for the highest waves in the wave train the weak non-linearity assumption is violated, which probably explains the underestimation of S_η and K_η. The prediction of S_η and K_η in the breaking region is more startling. Both S_η and K_η are slightly underestimated from the average breaking point at 63 m up to about 76 m, after which they are largely overestimated by the model. The reason for this sudden jump has not been established yet.

The sudden increase in S_η corresponds to an underestimation of the MWL and a correct prediction of Hmax and Hrms. These discrepancies might be due to boundary effects, though preliminary tests suggest that this may not be the problem. Another possible cause for this could be the model's limitation in this region where non-linear effects dominate over dispersive effects.

Figure 5 shows the cross-shore variation of Umin, Umax, S_U and K_U. The same comments as for the regular test can be made here, with the exception that no dependence on ϕ_B appears. The overall fit for S_U and K_U is reasonable, S_U being overestimated at gauge 17, and K_U underestimated at gauge 18.

In terms of percentage difference between measurements and computed results, S_U and K_U were generally better predicted than S_η and K_η. However, since the discrepancies found for S_η and K_η are greater for predictions in the surf zone, the relatively good fit for S_U and K_U, compared to S_η and K_η, suggests that a possible cause for these disagreements could be the difficulty to measure free surface elevations in the surf zone due to turbulence and aeration.

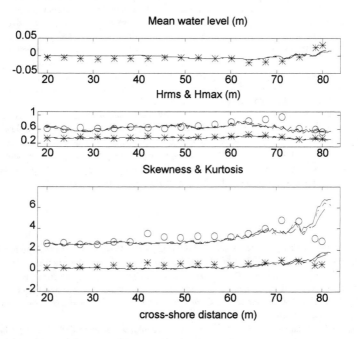

Figure 4 . Irregular run: comparison for <u>free surface elevation</u> predictions. The stars and circles are the measured data. The solid, dotted and dashed curves are computed results for $\phi_B = 20°$, $23°$ and $17°$ respectively. The middle plot shows the comparison of results for Hmax (top three curves and circles) and Hrms (bottom three curves and stars). The bottom plot shows the kurtosis (top three curves and circles) and skewness (bottom three curves and stars) of the surface elevation.

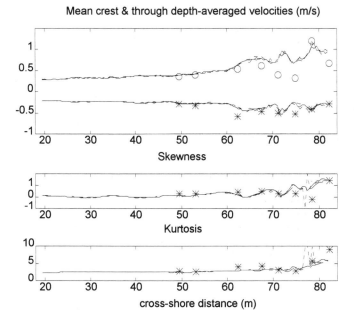

Figure 5. Irregular run: comparison for <u>depth-averaged velocity</u> predictions. The stars and circles are the measured data. The solid, dotted and dashed curves are computed results for $\phi_B = 20°$, $23°$ and $17°$ respectively. The top plot shows the maximum onshore (positive, i.e. top three curves and circles) and minimum offshore (negative, i.e. bottom three curves and stars) velocity predictions.

CONCLUSION

This Boussinesq model successfully simulates third and fourth order moments of surface elevation during the shoaling of regular waves. The results for the irregular waves show an underestimation of those moments probably due to the limitation of the model to weakly non-linear waves.

Wave decay modelling is reasonably good for both cases. However, this study highlights two points:

- the predictions of the regular waves test are very sensitive to ϕ_B, especially the terms of importance to sediment transport predictions (S_η, K_η, S_U and K_U). Schäffer *et al.* recommended a ϕ_B of 20°. For the case tested here, this corresponds to an optimum prediction of S_η, K_η, S_U and the mean water level. A ϕ_B of 26°, on the other hand, gave optimum results for Hrms and K_U. The fact that the irregular waves test showed no sensitivity to ϕ_B is encouraging for practical application.

- the solution of S_η and K_η close to the seaward boundary for the irregular wave test highlighted the need to investigate boundary effects, and the limitations of this model where the assumption $\varepsilon/\mu^2 = O(1)$ does not hold.

The velocity predictions are reasonable, particularly in view of the simple way of calculating the depth-averaged velocity from measurements. However, the model clearly tends to overestimate Umax. The probable cause for that is, again, the limitation of the model to weakly non-linear waves. It should however be remembered that the calculation of the depth-averaged velocity from the measurements led to underestimations at some locations. The predictions of S_U and K_U for both runs seem reasonable. More measurement in the surf zone would help assess these predictions, but the qualitative cross-shore variation of the predicted velocity is realistic.

Analysis of a wider range of wave conditions is now required in order to assess the generality of these preliminary results. Additionally, a closer look at the effects of reflection in the tank and numerical model is necessary. It is not clear to what extent the assumptions of weak non-linearity, weak dispersion and the neglect of reflection from the inner surf and swash zones can account for the discrepancies between laboratory data and the model. A comparison with the latest Boussinesq models that allow for highly non-linear waves and include swash oscillations (Wei et al.,1995; Madsen et al.,1996; Madsen et al.,1994) is now required.

REFERENCES

Hamm L., Madsen P.A. and Peregrine D.H., 1993. Wave transformation in the nearshore zone: a review. Coastal Eng., 21, pp 5-39

Kraus N.C. and McKee Smith J.,1994. SUPERTANK Laboratory Data Collection Project. US Army Corps of Engineers, Waterways Experiment Station, TR CERC-94-3.

Madsen P.A., Murray R. and Sørensen O.R.,1991. A new form of the Boussinesq equations with improved linear dispersion characteristics (part 1). Coastal Eng.,18, pp 183-204

Madsen P.A. and Sørensen O.R.,1992. A new form of the Boussinesq equations with improved linear dispersion characteristics.Part 2: a slowly varying bathymetry. Coastal Eng.,18, pp 183-204

Madsen P.A. and Sørensen O.R., 1993. Bound waves and triad interactions in shallow water. Ocean Eng., 20, pp 183-204

Madsen P.A., Sørensen O.R. and Schäffer H.A., 1994. Time domain modelling of wave breaking, run-up and surf beats. ICCE '94, pp 399-411.

Madsen P.A., Banijamili B., Schäffer H.A. and Sørensen O.R.,1996. Higher order Boussinesq type equations with improved dispersion and non-linearity. Abstract for ICCE '96.

Schäffer H.A.,Madsen P.A. and Deigaard R.,1993. a Boussinesq model for wave breaking in shallow water. Coastal eng.,20,pp 185-202

Wei G. and Kirby J.T.,1995. A coastal processes model based on time-domain equations, Coastal Dynamics '95, Gdansk.

Wei G., Kirby J.T., Grilli S.T. and Subramanya R., 1995. A fully non-linear Boussinesq model for surface waves. Part 1. Highly non-linear unsteady waves. J. Fluid Mech., vol 294, pp 71-92.

ANALYSIS OF THE ROLLER IN HYDRAULIC JUMPS

J. Veeramony and I. A. Svendsen [1]

Abstract: The paper describes results of accurate LDV measurements in a hydraulic jump. In particular, the velocities and shear stresses in the roller region are considered. The results for the roller flow are believed to be relevant for breaking waves in the surf-zone.

1 Introduction.

Wave breaking is a natural phenomena that is most commonly observed in the nearshore region. Peregrine and Svendsen (1978) pointed out the similarities between hydraulic jumps and breaking waves in the surf-zone and Svendsen et al. (1978) analyzed the dynamics of this similarity further. From the perspective of getting information for modelling surf-zone waves, it is therefore useful to study hydraulic jumps that have their Froude number in the same range as can be expected in a breaking wave (Froude number 1.2 ~ 2.0).

2 Measurements in hydraulic jump.

Velocity measurements obtained at a number of different locations in three different hydraulic jumps were reported by Bakunin (1995). The horizontal and vertical velocities at each location were measured using a two-component LDA. Along with the velocity measurements, the water surface location was also measured using a capacitance type wave gage. Figure 1 shows the schematic representation of the locations of measurement and table 1 gives the parameters of each hydraulic jump. The toe of the hydraulic jumps is not at a fixed location, but moves slightly back and forth in time. The measurement location in front of the jump is therefore not at the toe of the roller, but $10\,mm$ upstream of the toe, so that the inflow conditions for the jump are measured without interference from the fluctuation of the toe. $x/h_1 = 0$ is placed at the mean position of the toe.

The instantaneous velocities (u, v) are obtained as a time series at each (x, z) location. The time series of the velocity fluctuations (u', v') are then obtained by subtracting the mean velocities $(\overline{u}, \overline{v})$ from (u, v). Since the (u, v) velocities are obtained simultaneously, the turbulent shear stress $\overline{u'v'}$ is also obtained.

[1] Center for Applied Coastal Research, Ocean Engineering Lab, University of Delaware, Newark, DE 19716, USA. Correspondence e-mail: jay@coastal.udel.edu

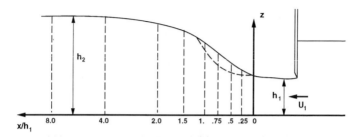

Figure 1: Schematic representation of the hydraulic jumps and the locations for the measurement of velocities. The dashed vertical lines show the measuring stations.

Froude number	h_1	U_1
	(m)	(m/s)
1.28	0.072	0.9483
1.44	0.064	1.0597
1.60	0.059	1.1621

Table 1: Parameters for the three hydraulic jumps.

The air bubbles entrapped in the roller region interferes with the data acquisition via the LDA. Hence, no velocity measurements could be obtained in the upper part of the roller and at the free surface. To perform detailed computations we therefore have to rely on extrapolations into the roller region. To facilitate this, a best fit to the vertical profile of the measured velocities was created. In the case of the horizontal velocities, which are much larger than the vertical velocities, we can also utilize the constraint that the volume flux at each cross section is the same.

3 Velocity profiles and roller thickness in the jump.

To obtain the best fit, each vertical section is heuristically divided into four regions:

1. The wall region (from the bottom ($z = 0$) to the first measurement location in the vertical ($z = z_1$))

2. The boundary layer (from $z = z_1$ to $z = \delta$, where δ is the boundary layer thickness)

3. The middle region (from $z = \delta$ to the elevaton at which the shear stress is non zero ($z = z_2$))

4. The upper region (from $z = z_2$ to the free surface ($z = \overline{\eta}$))

At the locations where there is no roller, the middle region extends all the way up to the free surface. In the wall region, the 1/7 power law is used to fit the data. In the boundary layer, the log-law is used to fit the velocity distribution. In the middle region, a fourth-order polynomial fit is used. It was found (Madsen and Svendsen

1983) that the velocity profile in the upper region can be described by a third order polynomial, which we use in the upper region. Continuity in velocity and velocity gradient are used to connect the regions.

It is found that this approach gave an excellent fit to the measured velocities. Figure 2 shows the results of the approximation. The total volume flux is defined as

$$Q = \int_0^\zeta \tilde{u}\, dz, \tag{1}$$

where $\tilde{u}$ is the turbulent averaged local velocity defined by dividing the total velocity u into a mean component $\tilde{u}$ and a turbulent component u' and ζ is the height of the mean free surface above the bottom. As a measure of the accuracy of the approximations, figure 3(a) shows that the total volume flux for each cross section is constant to within 0.1%. The mean volume flux in the jump is then determined as the average of the fluxes at all the sections in figure 3(a).

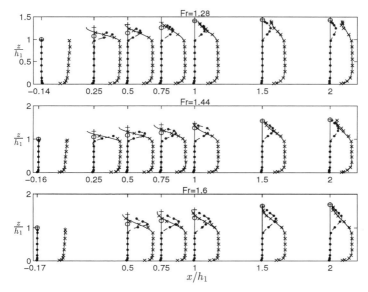

Figure 2: The vertical lines show the measurement locations, the free surface (+) and the lower limit of the roller. The measured velocity ($\times$) and the approximate velocity profiles (solid curve) are nondimensionalized by the mean upstream velocity U_1 and scaled down by a factor of 5. The measured shear stress ($\bullet$) and the approximate shear stress (dashed curve) are nondimensionalized by ρU_1^2 and scaled up by a factor of 5.

The velocities can be integrated from the bottom upwards. Since the net volume flux in the roller is by definition zero, the lower edge of the roller can be defined as the level below which $\int_0^z \tilde{u}\, dz$ equals Q. This can be used to obtain the thickness of the roller at each cross section. In figure 3(b) this thickness, non-dimensionalized by the downstream depth, has been plotted as a function of the the distance from the toe. The length l_r of the roller has been estimated as the distance from the toe to the point where the roller intersects the free surface. It is seen that the thickness non-dimensionalized this way is the same for all Froude numbers, which suggests that within the range of Froude number investigated, the roller thickness for a breaking wave can be expressed as a function of the wave height and the roller length.

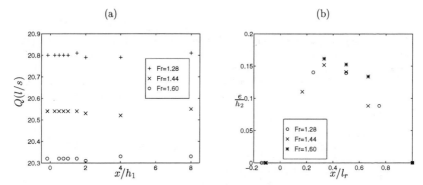

Figure 3: (a) The volume flux $Q(l/s)$ calculated from the approximate velocity profiles for the three Froude numbers at each location. (b) The nondimensional roller thickness at the measurement locations.

4 Momentum balance and the influence of the roller.

Having obtained a reliable velocity profile, the momentum balance in the hydraulic jumps can now be checked. The momentum flux through each vertical section is

$$M(x) = \int_0^{h(x)} \left(\rho u^2 + p \right) \, dz, \qquad (2)$$

which, apart from the contribution from the bottom shear stress τ_b should be constant at all sections. We define the mean velocity $U(x)$ and the momentum correction factor $\alpha(x)$ at each section by

$$U(x) \equiv \frac{1}{h(x)} \int_0^{h(x)} \tilde{u}(x,z) \, dz; \qquad \alpha(x) \equiv \frac{1}{h(x)U^2} \int_0^{h(x)} \tilde{u}^2 \, dz = \frac{h(x)}{Q^2} \int_0^{h(x)} \tilde{u}^2 \, dz. \qquad (3)$$

In the present context, pressure may be assumed hydrostatic (see discussion in Hornung et al. 1995). (2) then becomes

$$M(x) = \alpha(x)\rho \frac{Q^2}{h(x)} + \rho \int_0^{h(x)} \widetilde{u'^2} \, dz + \frac{1}{2}\rho g h^2(x). \qquad (4)$$

At the inflow section, $\alpha(0)$ is virtually unity and the turbulence weak so that $\widetilde{u'^2} \sim 0$. The momentum flux for the inflow section then is

$$M_1 = M(0) = \rho\frac{Q^2}{h_1} + \frac{1}{2}\rho g h_1^2 \tag{5}$$

The momentum equation for a control volume including the inflow section, the bottom and an arbitrary section at x then gives the momentum flux $M(x)$ as

$$M(x) = \rho\left[\int_0^h \widetilde{u}^2\,dz + \int_0^h \widetilde{u'^2}\,dz + \frac{1}{2}gh^2\right] = M_1 - \int_0^x \tau_b(x)\,dx \tag{6}$$

Figure 4(a) shows the contribution of the different terms to this momentum flux. It is seen that the balance is primarily between the pressure and the convective acceleration whereas the contribution from the turbulent normal stress is negligible. The contributions from the bottom and side wall friction were found to be the same order as the turbulent intensity and are not shown in the figure. At the upstream section however, the curvature of the streamlines was found to increase the total pressure considerably and this is included in the calculation of momentum flux at the inflow section.

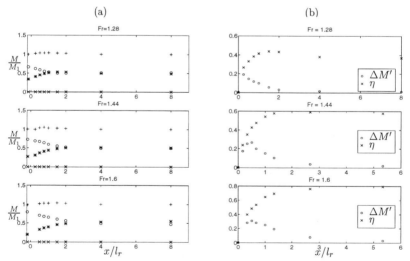

Figure 4: (a) The nondimensional momentum flux at each location ($+$), the contribution from the pressure ($*$), the volume flux contribution (o) and the contribution from the turbulent shear stress ($\times$). (b) The nondimensional breaker induced momentum flux $\Delta M'(= \Delta M/M - 1)$ (o) and the nondimensional water surface elevation $\eta(= \zeta/h_1 - 1)$ ($\times$).

As mentioned earlier, the breaking in the hydraulic jump enhances the momentum flux. The excess momentum flux ΔM is defined as the difference between the total flux and the flux we would have had with depth uniform velocities and hydrostatic pressure everywhere as in nonlinear, non-dispersive shallow water waves. Hence, ΔM is defined as

$$\begin{aligned}
\Delta M &\equiv M - \left[\rho\frac{Q^2}{h(x)} + \frac{1}{2}\rho g h^2(x)\right] \\
&= (\alpha - 1)\rho\frac{Q^2}{h(x)} + \rho \int_0^{h(x)} \widetilde{u'^2}\,dz
\end{aligned}$$

Figure 4(b) shows the variation of ΔM through the jump calculated for the three Froude numbers. The abscissa is x/l_r where l_r is the length of the roller. First of all we see that the value of ΔM quickly reaches a maximum value halfway or less through the jump. The effect of the roller is however felt much farther downstream than the roller region, as is evident from the non-zero value of ΔM. It appears that the overwhelming part of ΔM comes from the velocity part of the flux whereas the pressure appears to be virtually hydrostatic. As pointed out in Svendsen and Madsen (1984), this means that the breaking is creating the additional momentum flux required to prevent the jump from steepening furthere, as it would, had the nonlinear shallow water conditions of uniform velocity and hydrostatic pressure existed.

5 Stresses in the roller.

Based on the information collected, the stresses in the roller region can be analyzed by considering the momentum equation for the flow.

In integral form, the equation for the conservation of momentum is

$$\frac{\partial}{\partial t}\int_{V(t)} \rho\vec{u}\,dV + \int_{S(t)} \rho\vec{u}(\vec{u}.\vec{n})\,dS = \int_{V(t)} \rho\vec{g}\,dV + \int_{S(t)} \sigma\vec{n}\,dS \tag{7}$$

Here, $\vec{u}$ is the velocity vector of the flow, $V(t)$ is the control volume with the boundary $S(t)$, $\vec{g}$ is the acceleration due to gravity, σ, τ are the normal and tangential stresses at the boundary, and $\vec{n}$ is the local normal along the control volume.

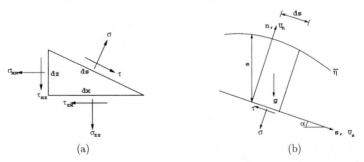

Figure 5: Definition sketch showing the control volume and the force balance. $\overline{\eta}$ is the free surface measured along the n-axis from the dividing streamline.

First, we express the stresses (σ, τ) along a mean streamline in terms of the stresses in the fixed (x, y) coordinate system. At any level below the surface, the relation between τ, σ and the horizontal and vertical stresses can be determined by considering the triangular control volume shown in figure 5(a). With $ds = dx/\cos\alpha = dy/\sin\alpha$, $\sigma_{xx} = -p + \tau_{xx}$, $\sigma_{yy} = -p + \tau_{yy}$ and $\tau_{xy} = \tau_{yx}$, the balance of forces in the horizontal and vertical direction gives

$$\sigma_{nn} = -p + \tau_{xx}\sin^2\alpha + \tau_{yy}\cos^2\alpha + \tau_{xy}\sin 2\alpha \qquad (8)$$

$$\tau_{ns} = \frac{1}{2}(\tau_{xx} - \tau_{yy})\sin 2\alpha + \tau_{xy}\cos 2\alpha \qquad (9)$$

Next we consider the control volume shown in figure 5(b), where the lower boundary is a mean streamline inside or at the bottom of the roller, the sides are perpendicular to this direction and the upper boundary is the mean free surface. Since the lower boundary by choice is a mean streamline, the n-component of the momentum balance in the control volume gives

$$p = \rho g e - \rho\widetilde{v'^2_n} \qquad (10)$$

For the s-component, we get

$$\tau_{ns} = \rho g e \tan\alpha - \frac{\partial}{\partial s}\left(\frac{1}{2\cos\alpha}\rho g e^2\right) + \frac{\partial}{\partial s}\left(\int_0^{\eta_s}\rho\widetilde{v'^2_n}\,dn\right)$$
$$- \frac{\partial}{\partial s}\left(\int_0^{\eta_s}\rho\left(v_s^2 + \widetilde{v'^2_s}\right)\,dn\right) \qquad (11)$$

The terms in this equation represent contributions from gravity, pressure, the turbulent normal stress and the mean momentum flux. From the measurements, it was evident that the turbulence is almost isotropic so that $\widetilde{v'^2_n} \sim \widetilde{v'^2_s}$, and therefore, the shear stress along the streamline given in (11) reduces to

$$\tau_{ns} = \rho g e \tan\alpha - \frac{\partial}{\partial s}\left(\frac{1}{2\cos\alpha}\rho g e^2\right) - \frac{\partial}{\partial s}\left(\int_0^{\eta_s}\rho v_s^2\,dn\right) \qquad (12)$$

Figure 6 shows the variation of the shear stress in the roller region (solid curve) for all three jumps together with the measured shear stresses (∗). Each diagram corresponds to one of the x-locations where measurements have been taken in the roller region. The vertical coordinate is $z = (z - e)/(\zeta - e)$ where e is the thickness of the roller and the shear stress values have been nondimensionalized by ρU_1^2.

The roller is considered a volume of water that is held in position at the front slope of the jump or bore by the stresses along its lower boundary, which according to the definition used here is also a mean streamline. Figure 7(a) shows the measured shear stress at this lower "dividing" streamline and the equivalent shear stresses predicted

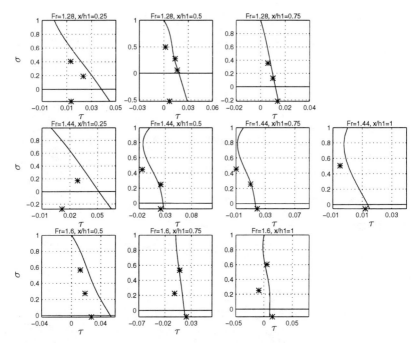

Figure 6: Nondimensionalized shear stress in the roller $(\tau/\rho U_1^2)$ for Froude numbers 1.28, 1.44 and 1.60 with the measured values $(*)$, the prediction according to (12) (solid line). The y-axis is $\sigma = (z - e)/(\zeta - e)$ where e is the vertical thickness of the roller. Hence, the zero on the y-axis corresponds to the lower edge of the roller and 1 is the free surface.

by (12). We see that the prediction of the shear stress is quite good, except in the region of the roller where the surface curvature is quite large.

The computations show that near the toe of the roller all three terms in (12) are important, whereas farther away from the toe the contribution from the curvature of the surface is negligible. Unfortunately, data for ζ as a function of the distance from the toe of the jump is limited to the measurement locations, and the curvature of the free surface is quite large in this region compared to the rest of the jump. Small errors in the estimation of the free surface curvature can therefore lead to large errors in the shear stress predicted using (12), which may be the reason for relatively poor agreement at the cross section closest to the toe. Figure 7(b) shows the comparison between the nondimensionalized measured shear stress for all three Froude numbers. The shear stress is normalized as $\tau'_{ns} = \tau_{ns}/(U_1^2 \xi)$ where $\xi = h_2/h_1$ and we see this scaling causes the values for all three Froude numbers fall on the same curve.

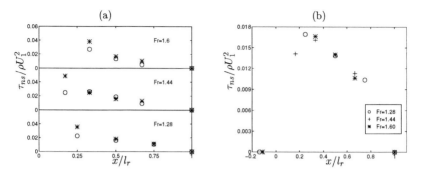

Figure 7: (a) Measured (o) and computed (∗) τ_{ns} at the lower edge of the roller. (b) Non-dimensionalized shear stress for all three Froude numbers.

6 Vorticity and eddy viscosity.

Having obtained the velocity profiles in the hydraulic jump, the vorticity distribution can also be calculated. The vorticity is given by

$$\omega = \frac{\partial u}{\partial z} - \frac{\partial w}{\partial x} \qquad (13)$$

From the data, we found that the vertical velocities are small and the variation along the jump of the vertical velocities were negligible in comparison to $\partial u/\partial z$ in the region covered by the measurements. Thus, the main contribution to the vorticity comes from the vertical gradient of the horizontal velocity. Figure 8(a) shows the vorticity thus calculated, at the bottom of the roller. The vorticity is non-dimensionalized as $\omega' = \omega U_1/(h_2\xi)$, which causes the non-dimensional values for all three Froude numbers to fall on the same curve.

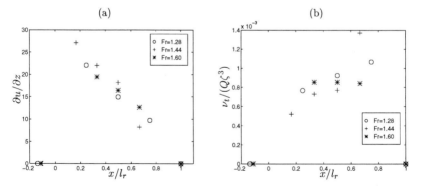

Figure 8: (a) The nondimensional vorticity at the lower edge of the roller. (b) The nondimensional turbulent eddy viscosity (ν_t) at the lower edge of the roller.

If we assume the shear stress and the velocity gradient in a turbulent flow are related as

$$\tau = \rho \nu_t \frac{\partial u}{\partial z}$$

where ν_t is the turbulent eddy viscosity, we expect from the scaling of τ and $\partial u/\partial z$ that the non-dimensional eddy viscosity would scale as $\nu_t' = \nu_t/(Q\xi^3)$, and furthermore, that the eddy viscosity obtained thus for all Froude numbers would fall on the same curve. This is illustrated in figure 8(b).

7 Conclusions.

Hydraulic jumps and bores have a lot in common with breaking waves. Measurements were made in hydraulic jumps with Froude numbers 1.28, 1.44 and 1.60. A careful analysis of the measurements has made it possible to assess flow details in the roller region, although few direct measurements were made in that region. From this analysis, several characteristic properties of the roller such as the roller thickness and the shear stress, vorticity and eddy viscosity distributions along the lower limit of the roller are obtained. It is found that all of these properties can be consistently scaled to uniform distributions for all the Froude numbers investigated. For the momentum balance, it was also found that the shear stress at the lower limit of the roller is not just carrying the weight of the roller. The momentum flux in the roller and the change in pressure due to the curvature of the free surface gave important contributions to the momentum balance.

Acknowledgements.

This work was sponsored by the National Science Foundation under the grant OCE-9203277 and by the US Army Reasearch Office, University Research Initiative under contract No. DAAL 03-92-G-0016. The United States government is authorized to produce and distribute reprints for government purposes notwithstanding any copyright notation that may appear herein.

References

Bakunin, J. (1995). Experimental study of hydraulic jumps in low froude number range. Master's thesis, Center for Applied Coastal Research, University of Delaware, Newark, DE19711.

Hornung, H. G., C. Willert, and S. Turner (1995). The flow field downstream of a hydraulic jump. *Journal of Fluid Mechanics 287*, pp. 299–316.

Madsen, P. A. and I. A. Svendsen (1983). Turbulent bores and hydraulic jump. *Journal of Fluid Mechanics 129*, pp. 1–25.

Peregrine, D. H. and I. A. Svendsen (1978). Spilling breakers, bores and hydraulic jumps. In *Proceedings of the 16th International Conference of Coastal Engineering, Hamburg*, pp. 540–550.

Svendsen, I. A. and P. A. Madsen (1984). A turbulent bore on a beach. *Journal of Fluid Mechanics 148*, pp. 73–96.

Svendsen, I. A., P. A. Madsen, and J. B. Hansen (1978). Wave characteristics in the surf zone. In *Proceedings of the 16th International Conference of Coastal Engineering, Hamburg*, pp. 520–539.

Sand Transport over a Breaker Bar in the Surf Zone

Bart T. Grasmeijer[1] and Leo C. van Rijn[2]

Abstract

Physical processes related to sand transport under breaking waves and breaker bar behaviour are described on the basis of results of a series of flume experiments. A moderate storm is simulated over a breaker bar profile in the flume. Near the bar crest, the net transport rate between the lowest measurement point and the water surface (measured zone) is dominated by the time-averaged (current-related) transport rate due to the undertow. Outside the bar area, the net transport rate in the measured zone is dominated by the high-frequency (wave-related) transport rate, which is offshore directed due to phase differences between peak orbital velocities and concentrations. Comparison of the measured transport rates and those derived from bed level changes, indicates that the transport rate in a rather thin layer near the sand bed (unmeasured zone), strongly affects the morphological behaviour of the breaker bar. It is expected that in particular the high-frequency transport rate in this unmeasured zone near the bed will dramatically change both the magnitude and direction of the net transport rate.

Introduction

The nearshore topography of a more or less dissipative beach is often characterized by the presence of one or more breaker-bars parallel to the shoreline. Waves break on these nearshore bars reducing the level of wave energy reaching the shore, in this way acting as a natural breakwater. The morphological behaviour of breaker-bars and the associated net sediment transport rates under the influence of wave action and wave- and tide-induced currents are hardly understood. In this paper the physical processes related to sand transport under breaking waves and breaker bar behaviour are analyzed on the basis of results of a series of flume experiments.
The generation and maintenance of longshore bars is commonly associated with the shoaling and breaking of high-frequency waves and the generation of low-frequency wave effects in the surf zone. Discussions of bar behaviour processes have been given

[1] PhD student, Institute for Marine and Atmospheric Research Utrecht, Utrecht University,
  P.O.Box 80.115, 35508 TC Utrecht, the Netherlands, e-mail: b.grasmeijer@frw.ruu.nl
[2] Sr.Engr., Delft Hydraulics, P.O.Box 177, 2600 MH Delft, the Netherlands

by Greenwood and Davidson-Arnott (1979), by Dean (1973), by Dally (1987), by Holman and Sallenger (1993) and O'Hare and Huntley (1993, 1994).

To demonstrate the relative importance of the various transport processes, in this paper the suspended sediment transport is analyzed in terms of instantaneous and time-averaged quantities. The net transport at height z above the bed can be obtained by time-averaging of the instantaneous values (indicated by an overbar), yielding:

$$\overline{uc} = \overline{u}\,\overline{c} + \overline{\widetilde{u}_{high}\widetilde{c}_{high}} + \overline{\widetilde{u}_{low}\widetilde{c}_{low}} \tag{1}$$

with:

Time-averaged transport (current-related): $\overline{u}\,\overline{c}$ (2)

High-frequency transport (wave-related): $\overline{\widetilde{u}_{high} \cdot \widetilde{c}_{high}}$ (3)

Low-frequency transport (wave-related): $\overline{\widetilde{u}_{low} \cdot \widetilde{c}_{low}}$ (4)

The depth-integrated transport rates (S_{net}, $S_{average}$, S_{high}, S_{low}) can be obtained by integration in vertical direction of the components. The physical interpretation of the various transport components is discussed by Van Rijn and Havinga (1995).

Experimental set-up

        The experiments were conducted in the Large Research Flume of the Laboratory of Fluid Mechanics of the Faculty of Civil Engineering (Delft University of Technology). The flume has a length of 45 m, a width of 0.8 m and a depth of 1.0 m. Irregular waves (Jonswap spectrum) were generated with a spectrum peak period of $T_p = 2.3$ s (± 0.2 s).

Two different series of experiments were performed. During the first series of experiments (series A) the sand transport in case of a plane sloping bed (1:100) was studied. In particular, the influence of wave breaking on the suspended sediment transport was studied. Measurements were performed in a measurement section at 18 m from the beginning of the sand bed. The water depth in this section was 0.30 m. The wave height, H_s, varied between 0.10 and 0.16 m; the current strength, U_m, between 0.0 and 0.30 m/s. Currents were generated in the same direction as the wave propagation.

During the second series of experiments (series B) sediment transport processes related to breaking waves over a breaker bar were examined. The breaker bar is shown in Fig. 3f. The bed profile varies in depth from 0.60 m seawards of the bar to 0.30 m at the bar crest (Fig. 3f). The water depth in the trough landward of the bar crest is 0.50 m. The breaker bar has a steep seaward slope of 1 to 20 and a steep landward slope of 1 to 25. The bed slope landward of the bar trough is 1 to 63. Measurements were performed in 10 different cross-sections along the breaker bar.

For all experiments sand was used with the following grain size characteristics: $D_{10} = 0.076$ mm, $D_{50} = 0.095$ mm, and $D_{90} = 0.131$ mm. The representative (50%) settling velocity of the sediment is $W_s = 0.008$ m/s. Detailed velocity and concentration data are presented by Grasmeijer and Sies (1995).

A measurement carriage was placed above the flume on which a pump sampler, a wave height meter (WHM), an electro-magnetic fluid velocity meter (EMF), a profile follower (PROFO) and an acoustic sediment transport meter (ASTM) were mounted. The velocities and concentrations were measured at 10 elevations above the bed, the lowest position being about 0.01 m above the crest level of the bedforms. To eliminate small-scale ripple-related variations in velocity and concentration, the measurement carriage was moved forward and backward over a few ripple lengths during the sampling period (5 min.). The velocity of the moving carriage was small (approximately 0.01 m/s) compared to the fluid velocity and large compared to the bedform migration velocity. A second WHM was placed more landward of the measurement section to check the uniformity of the wave height at the measurement section.

Results of plane sloping bed tests

Shown in Figure 1a are sand concentration profiles for three different wave heights in case of waves superimposed on a current (U_m = 0.20 m/s). It can be observed that an increase of the wave height results in an increase of the concentrations and a more uniform concentration profile. Generally, the sediment concentration distribution over vortex ripples shows that the profiles are straight lines in the usual semi-logarithmic plots in which c decays exponentially away from the bed, see Eq.(5).

$$c(z) = c_o \exp\left(\frac{-z}{l_s}\right) \tag{5}$$

In the present experiments the log-linear region of the concentration profile is found between 0.02 and 0.1 m above mean bed. The concentrations at higher elevations above the bed are larger than may be expected for a log-linear profile. Fig. 1b shows the mixing length (l_s) derived from Eq.(6):

$$l_s(z) = \left[\frac{d \ln c}{dz}\right]^{-1} \tag{6}$$

According to Nielsen (1990), the vertical length scale is closely related to the ripple height r for sharp crested ripples. Osborne and Vincent (1996) found the length scales to be proportional to the ripple height by a factor 2. In Fig. 1b, the vertical length scales in the log-linear region near the bed are about 3 times the bedform height ($l_s \approx 0.035$ m; $r \approx 0.009$ to 0.012 m). At higher elevations above the bed ($z >$ 0.1 m), the vertical length scale (l_s) increases with the height above the bed. In this region the influence of the bedform on the vertical length scale becomes less pronounced while wave-, current- and breaker-induced mixing become more important.

When comparing concentration profiles for different current velocities it appeared that the concentrations in the near bed layer slightly decreased with increasing current velocities while at higher elevations above the bed the concentrations were found to be increasing with increasing current strength. Current-induced mixing causes according to Van Rijn and Havinga (1995) the increasing

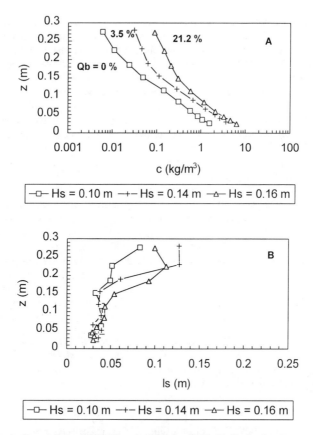

Figure 1. Concentration profile (A) and mixing length (B) distribution for waves superimposed on a current (Um = 0.20 m/s)

concentrations outside the near-bed layer for increasing velocities. They found that the concentrations in the near-bed layer were not noticeably affected by the current velocity increase. Similar results are obtained in the present study.

It is noted that an increasing wave height (H_s) from 0 to 0.16 m, which is related to an increase of the fraction of breaking waves (Q_b) from 0 to 21%, hardly affects the vertical length scales in the log-linear region of the concentration profile near the bed (Fig.1b), which suggests that breaker-induced mixing is of minor importance in this region.

Nielsen (1984) describes two extremes with respect to breaker-induced mixing. Waves that plunge heavily on shallow bars or on the step of steep beaches can form very strong jets that penetrate right through to the bed and thus introduce very strong external turbulence into the boundary layer itself. Mixing due to bores and spilling breakers has a very different character. Nielsen shows that, in case of monochromatic waves, breaker turbulence changes the upper part of the concentration profile drastically while the lower part is unchanged, so the concentration magnitude and the near-bed vertical length scale (l_s) are almost the same under spilling breakers as under non-breaking waves.

In the present experiments the influence of breaking of waves on the concentration profile was observed also, although less pronounced compared to Nielsen. This difference is probably caused by the fact that Nielsen used monochromatic waves for which the fraction of breaking waves is 100%. The fraction of breaking waves in the present experiments is much smaller.

The present data suggest that the mixing effect of spilling breakers might be of importance in case of relatively small breaking waves ($H_s/h < 0.5$) in the presence of no or weak current ($U_m < 0.1$ m/s), especially in the upper part of the concentration profile ($z/h > 0.33$). In case of a larger wave height and/or a stronger current the suspended sediment distribution is predominantly determined by wave- and current-induced mixing, despite the sometimes-larger fraction of breaking waves. The effect of breaker-induced mixing is relatively small.

To get a better insight in the relationship between the transport rate and the fraction of breaking waves an empirical formula was fitted to the present data, expressing the depth-integrated time-averaged transport rate as a function of the mean velocity, the relative wave height, the peak near-bed velocities and the fraction of breaking waves (Q_b). From the empirical formula (with a regression coefficient of 0.99) it appeared that the influence of spilling breakers on the depth-integrated transport rate is moderate. The depth-integrated transport rate was found to be roughly proportional to $(1-Q_b)^{-1}$, see Fig. 2. For example, an increase of Q_b from 0 to 30% leads to an increase of the depth-integrated transport rate of approximately 40%. In comparison, the increase of the relative wave height related to these values of Q_b leads to an increase of the transport rate of more than 170%.

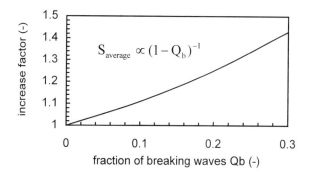

Figure 2. Influence of fraction of breaking waves on time-averaged transport rate.

Results of breaker bar tests

Wave height, fraction of breaking waves and transport rates are presented in Fig. 3 as a function of the location in the flume. It can be observed that waves are shoaling towards the bar crest (Fig.3a). Maximum wave height is found near the bar crest.

Moving further landwards, the wave height decreases both due to the increasing water depth and due to energy dissipation caused by breaking of waves. Nearly all breaking waves were of the spilling breaker type. Occasionally near the bar crest, a plunging breaker occurred. It is interesting to note that the fraction of breaking waves is largest just landward of the bar crest (Fig.3b). Breaking waves of the spilling breaker type start breaking near the bar crest but continue breaking over several wave lengths while propagating further landwards. Thus, waves initially breaking near the bar crest are also included in the fraction of breaking waves more landward of the crest.

The depth-integrated transport rates calculated from velocity and concentration measurements are compared with transport rates measured from bed profile changes over short time intervals (Fig.3c-e). In contrast with Van Rijn and Havinga (1995), the transport rates calculated from velocity and concentration measurements were not extrapolated to the unmeasured zone between the lowest measuring point and the mean bed level. Thus, the calculated depth-integrated values represent the suspended transport rather than the total transport rates (bed-load transport excluded).

From Fig. 3c it can be observed that the net suspended transport rate, S_{net}, is offshore directed in all measurement sections and increasing when moving towards the bar crest. Decreasing transport rates are found when moving further landwards. Although the same trends can be observed for the time-averaged suspended transport rate ($S_{average}$, Fig. 3c), occasionally small transport rates were found to be onshore directed as a result from the time-averaged velocity near the bed being onshore directed (Longuet-Higgins streaming) and the sediment not being stirred up to high elevations above the bed where the time-averaged velocity is offshore directed (return flow or undertow).

The high-frequency suspended transport rate (S_{high}) is found to be offshore directed in all measurement sections (Fig. 3d) and increasing when moving toward the bar crest where the largest high-frequency suspended transport rates are found. Moving further landwards a pronounced decrease of S_{high} can be observed. The high-frequency suspended transport rate is found to be negligible in the measurement sections landward of the bar, where the fraction of breaking waves is largest. The offshore direction of the high-frequency suspended transport rate is likely to be related to the mechanism of sand suspension over rippled bed forms. From the analysis of the concentration signals and from visual observations it appeared that, on the time scale

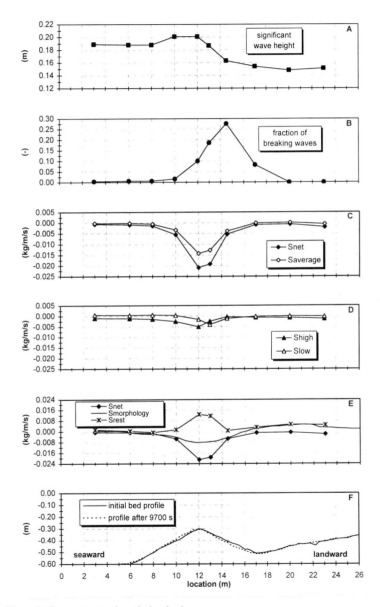

Figure 3. Transport rates along the breaker bar

of a single wave cycle, sediment was eroded from the bed at the onshore stroke of the wave and mobilized in vortices between the ripples. At flow reversal from on- to offshore direction the vortex cloud was lifted and advected leading to high concentrations at the offshore stroke of the wave. This variation in timing of concentration maxima and minima results in a vertical distribution in which the high-frequency transport rate tends to be onshore directed near the bed and offshore directed at higher elevations above the bed. Measurements close to the bed could not be made. Hence, the transport rate in the near bed layer (bed-load transport) could not be estimated from the data, but this transport rate is believed to be onshore directed.

From Fig. 3d it can be observed that in the measurement locations just landward of the bar crest, the low-frequency suspended transport rate is found to be relatively large and offshore directed. In these measurement sections the low-frequency suspended transport rates may not only be caused by the bound-long wave phenomenon. The undertow generated by irregular breaking waves may also show long-period velocity oscillations. Large concentration magnitudes under high-amplitude waves combined with the undertow oscillations induced by breaking waves results in a relatively large offshore-directed low-frequency suspended transport rate at locations just landward of the bar crest.

The wave-related suspended transport is dominated by the high-frequency component in nearly all measurement locations seaward of the breaker bar. At locations just landward of the bar crest, however, the low-frequency suspended transport rate exceeds the high-frequency component. It is interesting to note that the dominance of S_{low} over S_{high} occurs at locations were the fraction of breaking waves is relatively large, indicating that long period oscillations of the undertow are important.

Transport rates based on morphological changes were determined by measuring the bed level at the flume window every 0.25 m along the profile. In Fig. 3f the initial bed level and the bed level after 9700 s of wave action are plotted for series B2. It can be observed that morphological changes take place especially at the seaward slope of the bar (accretion), and at the landward slope of the bar (erosion). The bar crest is moving seawards. Transport rates were calculated from the bed level changes assuming that the transport rates at deep water seaward of the bar are zero. Transport rates based on these morphological changes ($S_{morphology}$) are plotted in Fig. 3e, as well as the time-averaged net suspended transport rates based on instantaneous velocity and concentrations measurements (S_{net}) and the difference between both curves (S_{rest}). The S_{net}-value is the same curve as shown in Fig. 3c. It can be observed from Fig. 3e that $S_{morphology}$ is onshore directed landward of the bar and offshore directed near the bar crest. At locations near the bar crest S_{net} and $S_{morphology}$ are both found to be offshore directed. The rest term, S_{rest}, is onshore directed in nearly all measurement locations. Largest values are found at locations near the bar crest, while S_{rest} is negligible in the measurement section where the fraction of breaking waves is largest.

It is noted that the transport rates derived from the bed-level changes ($S_{morphology}$) represents both the suspended and the bed load transport rate, whereas the transport rates derived from the velocity and concentration measurements (S_{net}) only represents the suspended transport.

The discrepancy between the transport rate based on morphology and that based on velocity and concentration measurements probably represents a combination of wave-related transport and current-related transport in the unmeasured zone near the bed. From analysis of the vertical distribution of transport rates it appeared that in particular the high-frequency wave-related transport rate near the bed tends to be onshore directed. The relatively large high-frequency velocities with onshore near-bed peak velocities larger than offshore near-bed peak velocities, together with the large concentrations magnitudes (and variation in concentration) may lead to large high-frequency transport rates in onshore direction. It is expected that the high-frequency transport rate in the unmeasured zone near the bed will dramatically change both the magnitude and the direction of the total net transport rate.

Conclusions

The main findings of the study are summarized in the following conclusions:
- The effect of the fraction of breaking waves (Q_b) on the depth-integrated and time-averaged (mean) transport rate was found to be relatively small. An increase of Q_b from 0 to 30% gave an increase of the transport rate by about 40%.
- Near the bar crest, the net transport in the measured zone is dominated by the time-averaged (current-related) transport rate due to the undertow.
- Outside the bar area, the net transport in the measured zone is dominated by the high-frequency (wave-related) transport rate, which is offshore directed due to phase differences between peak orbital velocities and concentrations.
- With the exception of the region just landward of the bar crest, where the fraction of breaking waves is relatively large, the low-frequency transport rate is negligible.
- The morphological behaviour of a breaker bar and the associated net transport rates are strongly related to the wave-related transport rates in a rather thin layer near the sediment bed.
- Measurement of instantaneous velocities and concentrations in this layer are of crucial importance to determine the net transport rates and hence the net migration rate of the bar. It is, however, hardly possible to measure these transport rates with the available measurement techniques, not even in flume conditions.

Acknowledgments

The present study was a joint project of Delft Hydraulics, Delft University of Technology, Rijkswaterstaat of the Dutch Ministry of Public Works and the MAST-program of the EEC.

Appendix

Dally, W.R. (1987). "Longshore bar formation - surf beat or undertow ?". Proceedings of Coastal Sediments '87, pp. 71-85.

Dean, R.G. (1973). "Heuristic models of sand transport in the surf zone". Conf. on Eng. Dynamics in Coastal Zone". Sydney, Australia.

Grasmeijer, B.T. and Sies, E.M. (1995). "Sediment concentrations and sediment transport in case of irregular breaking waves over a plane sloping bed and a barred profile". Part H and I. Delft Hydr., Delft Univ. of Technol., Coastal Eng. Dep., Delft, The Netherlands.

Grasmeijer, B.T. (1996). "Sediment concentrations and transport in case of irregular breaking waves and currents over plane and barred profiles". Part J and K. . Delft Hydr., Delft Univ. of Technol., Coastal Eng. Dep., Delft, The Netherlands

Greenwood, B. and Davidson-Arnott, R.G.D. (1979). "Sedimentation and equilibrium in wave-formed bars: a review and case study". Canadian Journal of Earth Sc., Vol. 16, pp. 312-332

Holman, R.A. and Sallenger, A.H. (1993). "Sand bar generation: a discussion of the Duck experiment series". Journal of Coastal Research, SI 15, pp. 76-92.

Longuet-Higgens, M.S., 1953. "Mass transport in water waves". Royal Society Phil. Trans., London, Vol. 245, A903, pp. 535-581.

O'Hare, T.J. and Huntley, D.A. (1994). "Bar formation due to wave groups and associated long waves". Marine Geology, Vol. 116, pp. 313-325.

O'Hare, T.J. and Huntley, D.A. (1993). "Sand bar evolution beneath partially-standing waves: laboratory experiments and model simulations". Continental Shelf Research. Vol.13, No.11, pp. 1149-1181.

Osborne, P.D. and Vincent, C.E. (1996). "Vertical and horizontal structure in suspended sand concentrations and wave-induced fluxes over bedforms". Marine Geology, Vol. 131, pp. 195-208

Van Rijn, L.C. and Havinga, F.J. (1995). "Transport of fine sands by currents and waves". Journal of Waterway, Port, Coastal and Ocean Engineering, Vol. 121, No. 2, pp. 123-133.

Dynamics of One Bar and Multibar Beach Profiles

Grzegorz Różyński,[1] Zbigniew Pruszak,[1] Pierluigi Aminti[2]

Abstract

Two coasts with different bar systems are compared in the study. The former, situated on the Baltic Sea coast at Lubiatowo, Poland, is featured by a system of 3-5 bars. The latter lies on the North Thyrrenian Sea at Cecina Mare, Italy and is basically a one bar entity. Random sine functions and correlation analysis were applied to both sites. The aim of the study was to find similarities and differences of bar behaviour.

Introduction

Surf zone dynamics is controlled by longshore bars, which highly affect wave transformation and nearshore bed changes. Two major groups of theories try to explain bar formation and evolution, one associating them with wave breaking, cf. Dally (1987) or Dean et al. (1992), and the other relating bar phenomena to long waves or links them to components of the wave spectrum and non-linear wave effects, together with energy exchange between those components, cf. Howd et. al. (1991) or Doczai-Karakiewicz et al. (1995). An original hypothesis of bar formation and dissipation was presented by Wijnberg (1995), who describes continuous offshore bar migration, from its formation in the vicinity of shoreline, to dissipation on greater depths. Multiple bars can be expected when bottom inclination is gentle and the shore is composed of sandy, fine sediment. In case of steeply sloping bottom, with coarse sediments, one bar is likely to appear. For example, Larson and Kraus (1992) found that the shore at Duck Point, North Carolina, USA, with a mean slope $m = \tan\alpha \approx 1.5$ % and average grain diameter $D_{50} = 0.2 \div 0.4$ mm, usually exhibits two distinct bars. For much milder bottom tilt at Chesapeake Bay, Maryland, USA, explored by Dolan and Dean (1985) with $m = \tan\alpha \approx 0.52$ %, $D_{50} = 0.12 \div 0.24$ mm, a system of up to 17 small, yet distinct bars was detected.

The sites chosen for the current study match this tendency. One site, at the Coastal Research Facility (CRF) Lubiatowo, Poland is featured by a system of $3 \div 5$ bars for $m = \tan\alpha \approx 1$ % and $D_{50} = 0.22$ mm. The other site lies on the Thyrrenian Sea coast at Cecina Mare, Italy. Aminti et al. (1996) usually observe only one fairly large

[1] Snr Res. Assoc., Assoc. Prof., Polish Academy of Sciences'Institute of Hydro-Engineering IBW PAN,
7 Kościerska St., 80-953 Gdańsk, Poland
[2] Professor, University of Florence, Dept. of Civil Engng., v. Santa Marta 3, 50139 Florence Italy

bar at this steep and coarse shore m = tanα ≈ 2 %, D_{50}= 1 mm. The shore at
Lubiatowo has been thoroughly explored cf. Pruszak *et al.* (1997) and Pruszak &
Różyński (1997). In the former, correlation analysis (CA) was employed for the
investigation of behaviour of basic bar parameters. The latter dealt with the random
sine function technique (RSF), in which residuals about the equilibrium profile were
treated as sine functions with random envelope, bar wave number and phase. In the
current study the same tools were applied to the shore at Cecina Mare to produce the
results, which could be compared with studies on Lubiatowo beach.

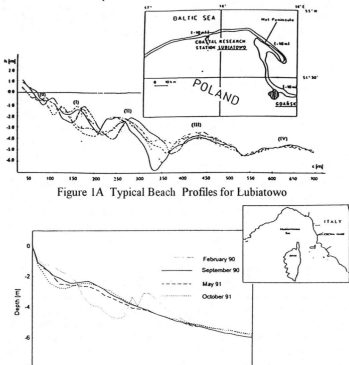

Figure 1A Typical Beach Profiles for Lubiatowo

Figure 1B Typical Beach Profiles for Cecina Mare

Sites and Field Data

The coast at CRF Lubiatowo is a natural, unprotected, multibar, dissipative
shore. Beach profiles exhibit 3÷5 bars. The bars are approximately parallel to the
shoreline and to each other (Fig.1A). The Baltic Sea is a non-tidal water body.
Average storm surges occur a couple of times a year and hardly exceed 0.5 m.
Significant wave heights reach H_s= 3.5÷4 m and rarely more during heavy storms.

The corresponding periods are between T = 5÷7 s. Significant waves break over the outermost bar and then over inner bars. Smaller waves, H = 0.5÷1.5 m, break over inner bars. The site has been sampled since 1964, usually once a year. The sampling area covers the shore stretch between km 163.2 and km 164.1 of the national coast chainage. Three neighbouring, profiles in its middle (km 163.6, 163.7 and 163.8), yielded the bulk of data. Additionally, an extra profile, referred to as 0-th profile, was also included but only into the correlation analysis. Profiles 163.6 and 163.7 were surveyed 24 times and profile 163.8 had 23 surveys between 1964 and 1994. The 0-th profile was surveyed 10 times, so 81 surveys were used in the correlation analysis vs. 71 in the RSF study. Earlier surveys reached about 600m offshore, which later was extended to 850 m. The sampling interval Δx was 10 m.

Cecina Mare beach lies on the North Thyrrenian Sea (Fig.1B). It occupies a 2.6 km wide unit. The beach profile normally has one bar, but a second bar is also present at times. Beyond the depth of 4 m isobaths run parallel to the coastline. Deep water waves have been measured some 2 km offhore since 1983. The records provide the wave climate pattern, where 63% of time account for calm periods (H < 0.5 m), 25 % correspond to mild conditions H = 0.5÷1.25 m, 10 % are in the interval of growing sea: H = 1.25÷2.5 m. The remaining 2 % represent heavy storms with wave heights in the range H = 2.5÷4 m and periods rarely exceeding 6s. More extreme conditions are very rare and last about 18 hours a year. Eighteen beach profiles, evenly distributed over the sampling area, were surveyed in Feb.'90, Sep.'90, May'91 and Oct.'91. In all, 73 surveys were available, because one extra survey, from Sept.'89 was added to the dataset. The cross-shore range of surveys was 400 m. Sampling intervals varied in order to thoroughly map the bed configuration, but they were interpolated in order to obtain datasets sharing the same sampling interval $\Delta x = 10$ m.

The amount of data for both sites is similar. However, one should be remembered that the time span of data collection for Cecina Mare is only a fraction of that for Lubiatowo. Therefore, the Cecina Mare data features rather short and medium-term phenomena vs. long-term effects reflected by the Lubiatowo dataset.

Random Sine Functions (RSF)

Let us introduce a Dean equilibrium profile:
$$y = A \cdot x^{2/3} \tag{1}$$
where A is a site-specific coefficient and x stands for the distance offshore. The equilibrium profile has residuals, which can be written as:
$$r(x) = A(x) \cdot \sin(k(x) \cdot x + \phi(x)) \tag{2}$$
where: $r(x)$ is the residual, $A(x)$ an amplitude (envelope), $k(x)$ - a bar wave number and $\varphi(x)$ a phase of the sine function. The residuals are random, so the sine function in Eq.2 can be referred to as the random sine function. This equation can be differentiated with respect to x:
$$r'(x) = A'(x) \cdot \sin(k(x) \cdot x + \phi(x)) + A(x) \cdot [k(x) + x \cdot k'(x) + \phi'(x)] \cdot \cos(k(x) \cdot x + \phi(x)) \tag{3}$$

When consecutive residuals $r(x)$ do not differ much, the variability of envelope, wave number and phase is low, cf. Pruszak & Różyński 1997. Hence, $A'(x)$, $k'(x)$ and $\phi'(x)$ can be assumed to be equal to zero. The formula (3) then simplifies:

$$r'(x) = A(x) \cdot k(x) \cdot \cos(k(x) \cdot x + \phi(x)) \tag{4}$$

The combination of formulas (2) and (4) allows to calculate the envelope:

$$A(x) = \sqrt{r^2(x) + \left(\frac{r'(x)}{k(x)}\right)^2} \tag{5}$$

The derivative $r'(x)$ can be approximated by the quotient:

$$r'(x) \approx \frac{r(x + \Delta x) - r(x - \Delta x)}{2\Delta x} \tag{6}$$

Since $A'(x), k'(x)$ and $\phi'(x)$ are zero within an interval $x - \Delta x \leq x \leq x + \Delta x$, the equation (2) must be true for $x - \Delta x$, x and $x + \Delta x$, leading to three equations from which $k(x)$ and $A(x)$ can be determined directly, while the phase $\varphi(x)$ can be computed from a non-linear relationship:

$$k(x) \cdot \Delta x = \arccos\left(\frac{r(x+\Delta x)+r(x-\Delta x)}{2r(x)}\right) \tag{7}$$

$$A(x) = \sqrt{r^2(x) + \left(\frac{r(x+\Delta x)-r(x-\Delta x)}{2k(x)\cdot\Delta x}\right)^2} \tag{8}$$

$$\frac{r(x)}{A(x)} = \sin(k(x) \cdot x) \cdot \cos(\varphi(x)) + \cos(k(x) \cdot x) \cdot \sin(\varphi(x)) \tag{9}$$

RSF can be especially useful for description of local, geometrical bar parameters. The crucial advantage is that two basic parameters i.e. bar length and height can be described at each point x along a beach profile. Hence, the variability of large, rhythmic bed forms can be investigated depending on their location on a profile. It should be noted that empirical orthogonal functions (EOF) represent a rather static description of multibar profiles, whereas the RSF approach permits for a quasi-dynamic analysis. Another merit, is that RSF can be applied to datasets, in which the number of realisations in time is limited.

Bar wave number $k(x)$ permits to evaluate the lengths of bars and troughs, related to a location x. However, such reasoning may be somewhat abstract, because $L(x)$ is significant with respect to the whole profile length and it represents the properties of point x 'stretched' over $L(x)$. In reality bar parameters are controlled by a set of forcing factors, representative for a given profile section. The envelope describes mean profile deviation from the curve, treated as the reference level. Thus, qualitative and quantitative bottom deformations can be examined over the whole profile. Moreover, the spots of the most intensive evolution (envelope peaks) can be distinguished. $A(x)$ represents the whole set of bars, describing their gradient of height variability across the whole surf zone. Phases are of secondary importance; they have no physical interpretation they only show whether a function is closer to sine or cosine.

Correlation Analysis (CA)

The correlations were computed to assess mutual relationships of bar parameters, so they were computed for pairs of different parameters of the same bar and for the same parameters of different bars. The latter could only be done for cases with sufficient number of occurrences, so no correlations between parameters of different bars were computed for Cecina Mare. The correlations were obtained by the formula:

$$r(p_1, p_2) = \frac{\text{cov}(p_1 p_2)}{\sigma_{p1} \sigma_{p2}} \quad (10)$$

where p_1 and p_2 are bar parameters and σ_1, σ_2 are their standard deviations. The covariance was defined as:

$$\text{cov}(p_1 p_2) = \frac{1}{n} \sum_{i=1}^{n} (p_{1i} - \bar{p}_1)(p_{2i} - \bar{p}_2) \quad (11)$$

where n stands for the number of occurrences of a given bar parameter and the overbarred symbols denote their mean values.

Reference Profiles and Bar Parameters

Up to 81 or 73 values of a given parameter could be found for each bar. The shape of a bar and its location on a profile can be described by the set of basic bar parameters: bar length (L_r), bar height (z), water depth to bar crest (h_r), distance from shoreline to shoreward bar extremity (bar origin x_p), distance from shoreline to seaward extremity (bar end x_k), distance from shoreline to bar crest (x_g), distance from shoreline to bar centre of gravity (x_c) and bar volume (V). For Lubiatowo they were found from intersections between the reference profile and a given beach profile (Fig.2). For every profile, i.e. 163.6, 163.7, 163.8, and 0-th, the reference equilibrium profile was computed by a modified Dean-type equation in order to avoid the infinite slope at the origin. The modified equation (1) reads:

$$y = m \cdot x \qquad \text{for} \quad x < x_o$$
$$y = A \cdot x^{2/3} \qquad \text{for} \quad x \ge x_o \quad (12)$$

where $m = 0.04$ is the slope at shoreline and $x_0 = (A/m)^3$. The coefficient A was least square fitted into the corresponding data subsets. In all cases the value of x_0 was in the range of 10m, so practically the results were identical to those obtained by Eq.1. The equilibrium profiles were used in the CA analysis. For RSF one average profile was computed by fitting the A coefficient into 71 surveys. The values used in the CA analysis proved to be very similar, i.e. 0.084 for 163.6, 0.090 for 163.7, 0.084 for 163.8. and 0.080 for 0-th, vs. 0.084 employed in the RSF analysis.

Determination of bar parameters from intersections of equilibrium and beach profiles failed for Cecina Mare, because part of equilibrium profile was above and the rest below a given beach profile. Such a behaviour appears to be typical of one bar profiles, because it usually minimises the sum of squares. Therefore, the bar parameters had to be identified independently of an equilibrium profile, which was achieved upon the analysis of differenced beach profiles:

$$z_{dif}(x) = z(x + \Delta x) - z(x - \Delta x) \quad (13)$$

For constant Δx Eq.13 behaves similarly to a profile's derivative. It is shown in Fig.3, together with an exemplary beach profile. Bar origin is identified at a point, where Eq.13 reaches maximum, corresponding to maximum shoaling rate on a profile. Likewise, minimum value of Eq.13 stands for bar end at a point of maximum deepening rate. Gradually passing from positive to negative values, the Eq.13 is equal to zero at a point, which designates a crest. The RSF analysis of Cecina Mare was identical to that of Lubiatowo: one average Dean equilibrium curve, with $A = 0.114$, was least square fitted into all surveys.

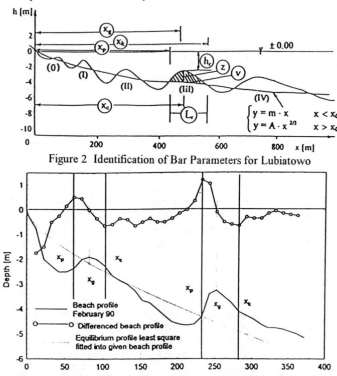

Figure 2 Identification of Bar Parameters for Lubiatowo

Figure 3 Bar Identification from Differenced Profile for Cecina Mare

Bar Features and Correlations

For Lubiatowo five bars (0, I, II, III, IV) emerged in the surveys. The minimum, average and maximum values of bar parameters are put together in Tab.1, upper section. Mean bar length depends on bar location on cross-shore profiles. It is short (25-30 m, max. 60 m) for the ephemeral, innermost bar 0 and the first stable bar I (30-35 m on the average, max. 75 m). The longest outermost bar IV is 6-8 times longer (200-210 m on the average, max. 370 m). The maximum standard deviation was also observed for the outermost bar, indicating high variability of its length. Taking into account its high length and small height, the bar IV is a large, flat and irregular bed form. Such a shape can be attributed to spatially distributed wave

breaking over this segment of the shore profile. The spacing between the bar crest and
the centre of gravity is usually small. The centre of gravity is situated several meters
offshore of the bar crest for bars (0, I, II, III), with the maximum of 40 m for bar IV,
showing the degree of asymmetry. The maximum standard deviations were observed
for the outermost bar, and it again can be assigned to spatial distribution of wave
breaking, especially during heavy storms. The spacing between bars grows in the
offshore direction. Thus, the inner bars lie closer to each other and their spacing is
equal to 75 m between bars 0 and I, 120 m between I and II and 200 m between III
and IV. Bar height is defined as an elevation of crest above the equilibrium profile.
Maximum bar heights may reach 1.5 m above that line but, if counted from trough to
crest, they may even exceed 2.5 -3 m. The height is greatest for the central bar II - the
mean height of that bar is 40% greater than that of the inner bar I, 10% more than for
the outer bar III and 20% more than the outermost bar IV. The standard deviation of
bar height is also maximum for bar II. Thus, bar II is the most conspicuous bed form
of a multibar profile and the one undergoing the strongest morphodynamic changes.
The mean water depth over consecutive bars grows with offshore distance. At the
ephemeral bar, the depth over its crest varies from 0.1 m (direct shoreline proximity)
to 1.7 m. On the other hand, depth oscillations over the crest of the outermost bar
vary from 3.8 m to 7.3 m. Mean bar volume, determined by elevation of the bed
profile above the equilibrium line, grows with the distance from shoreline. The volume
of the outermost bar IV is ten times greater than that of bar 0, six times than for bar I
and twice as much as the neighbouring bar III. The volume strongly depends on
geometrical parameters of bars, particularly their length. Bar IV also exhibits the
strongest standard deviation of volume despite its remoteness from shoreline and the
highest depth.

Table 1 Average Bar Parameters and their Standard Deviations

Bar #	Bar length L_r [m]	Bar height z [m]	Depth over bar crest h_r [m]	Location of crest x_k [m]	Loc. centre of gravity x_c [m]	Bar volume V [m3/m]
LUBIATOWO						
0	26±14	0.35±0.20	0.6±0.33	39±13	45±13	6±5
I	34±15	0.55±0.29	1.4±0.42	113±35	116±35	13±11
II	57±24	0.77±0.35	2.4±0.65	235±53	240±52	30±19
III	99±44	0.71±0.34	3.8±0.55	394±49	407±46	47±28
IV	204±84	0.64±0.31	5.0±0.54	586±80	628±64	76±34
CECINA MARE						
inner	39.5±14.7	0.41±0.18	1.8±0.4	70.7±25.9	76.6±25.8	9.8±5.3
outer	54.5±16.1	0.51±0.25	3.1±0.46	173.8±38.6	176.6±36.5	17±10.1

The correlations calculated for different couples of bar parameters are generally
low, with a few exceptions. High correlations of 0.7-0.95 were found for $V \Leftrightarrow z$ and $V
\Leftrightarrow L_r$, which is obvious. Notable are also the correlations $V \Leftrightarrow x_g$, and $z \Leftrightarrow x_g$
belonging to bar I. The first correlation indicates that the more offshore bar I, the
greater its volume. This can be caused by the exchange of sediment between bars I
and 0 (and the beach as well). During storms the sediment moves seawards, and bar I
can accumulate sand from bar 0, which is ephemeral and disappears at times. On the

other hand, during calm periods the sediment moves shorewards and bar I may convey sand to bar 0, which emerges occasionally. The latter then stores temporarily the sediment migrating from beach towards the system of more stable bars. Correlations for other pairs of parameters are much lower, being about 0.3-0.5 for bar I and 0.0-0.3 for other bars. The correlations are all put together in Tab.2, upper section. In order to see whether bars act independently or as a system, correlations of basic bar parameters were computed for different bar couples. The symbols R_0, R_I, R_{II}, R_{III} and R_{IV} denote the respective bars that were set in different combinations for which correlations were computed (see Tab.3). The analysis shows that the correlations are not high and decrease with the distance offshore. The highest values (although below 0.6) have been obtained for $R_0 \Leftrightarrow R_I$ (L_r, x_g, x_c, V) and for $R_I \Leftrightarrow R_{II}$ (x_g, x_c, z, V). The correlations of x_c and x_g computed for all consecutive pairs of bars imply that the cross-shore migration of the inner bars 0 to II is somewhat correlated while the outer bars III and IV are not statistically coupled with the movement of the inner bar system. Thus, it is only the inner bar subsystem that may be claimed to move as an entity either offshore or onshore. Smaller values of correlations may partly stem from the high instability and mobility of bar 0 and a relatively high stability of bar IV. The latter is particularly vulnerable to substantial evolution at extreme events. Bar lengths are coupled only within $R_0 \Leftrightarrow R_I$, whilst bar heights and volumes are correlated significantly within the subsystem $R_I \Leftrightarrow R_{II}$. This may indicate once again that the inner bars 0, I and II are much more conjugated than the outer bars III and IV. It can also be concluded that bars III and IV change their size and volume independently of other bars. Crucial causes of their evolution may be either rare, extremely heavy storms or long-term shoreline changes and the respective disturbances of sediment transport. Therefore, the evolution of outer bars is associated with greater spatial and temporal scales. It is worth noting that, on the contrary, the inner bars 0, I and II are subject to the most rapid and dynamic bed changes due to coastal phenomena of smaller scales. Hence, a system of multiple bars can be split up into two subsystems, controlled by coastal processes of different scales.

Table 2 Correlations between Parameters of the same Bar

Bar #	L_r vs. x_g	z vs. x_g	V vs. x_g	V vs. z	V vs. h_I	V vs. L_I
LUBIATOWO						
0	-0.27	0.06	-0.18	0.59	-0.50	0.89
I	0.31	0.44	0.49	0.92	-0.26	0.84
II	0.28	0.00	0.19	0.79	-0.21	0.82
III	-0.22	-0.09	-0.07	0.82	-0.16	0.75
IV	-0.34	0.46	0.11	0.53	-0.26	0.67
CECINA MARE						
inner	0.27	0.16	0.33	0.54	0.08	0.69
outer	-0.10	0.07	-0.11	0.83	-0.21	0.58

Table 3 Correlations of the same Parameters of Different Bars for Lubiatowo

Parameters	bar 0	bar I	bar II	bar III	bar I	bar II
x_g	0.37	0.49	0.42	0.36	0.36	0.06
L_r	0.46	0.12	0.17	-0.18	-0.15	0.13
V	0.55	0.54	0.05	-0.20	-0.14	-0.16
z	0.10	0.39	0.02	-0.14	-0.13	-0.24
x_c	0.44	0.52	0.39	-0.53	0.40	0.06

The same parameters were investigated Cecina Mare. Since one bar with widely varying location was predominantly observed, the investigation of so much varying bar as one entity did not look promising at the beginning and two bars were defined depending on their cross-shore location. One of them covered a steeper part of the profile stretching up to 100m offshore. The second one was extended further offshore, where the bottom tilt is milder. For such a division 32 instances of the inner bar and 58 of the outer bar were detected, of which there were 19 instances where both were present. As shown later, most of bar parameters were similar for both sub-zones, so the bar actually could be analysed as one entity. Bar length for the inner bar equals 39.5 m on average, with standard deviation 14.2 m, vs. 54 m of mean length and with 16.2 m standard deviation for the outer bar. Although the lengths match a well known pattern that more offshore bar locations correspond to greater lengths, the mean length of outer bar is only slightly greater than the length of the inner bar. Consequently, rather small standard deviations indicate that bar length is a stable feature. Bar crests and centres of gravity lie very close to each other for both bars; the gravity centres are situated a few meters offshore of crests (70.7 m and 76.6 m for inner bar vs. 173.8 and 176.6 m for outer bar). Slightly greater scatter of both parameters for inner bar suggests that when approaching the shore, the bar becomes only a bit more asymmetrical. Standard deviations are nearly identical for both parameters and equal 26 m for inner bar and 37 m for the outer bar. It may be claimed that greater standard deviations for outer bar can be attributed to more intensive wave action for more offshore locations. Bar height is defined as distance between the crest and a line crossing the bar extremities. For inner bar the mean height equals 0.41 m vs. 0.51 m for the outer bar. Standard deviations equal 0.18 m and 0.25 m respectively. Thus, both bars are fairly conspicuous bed forms with similar heights having similar standard deviations. Depths to crest grow with offshore distance and on the average are equal to 1.8 m for the inner bar and 3.1 m for outer bar. Standard deviations are practically the same for both instances and equal 0.42 m and 0.46 m respectively. This suggests that the variation of depth to crest is more or less the same along the profile. Mean volume of the inner bar is equal to 9.8 m³/m vs 17 m³/m for the outer bar. This can be explained by greater L_r and z for outer bar and nearly the same asymmetry of both bars; the ratio $V_I/V_{II} = 0.58$ and the corresponding ratio $L_{rI}·z_I/L_{rII}·z_{II} = 0.57$. Bearing in mind greater standard deviations of these quantities for the outer bar, it is not surprising that also the volume of the outer bar is characterised by greater standard deviation (5.3 m³/m for inner bar and 10.2 m³/m for outer bar).

The parameters of inner and outer bar are put together in Tab.1, lower section. The analysis proved that Cecina Mare is basically a one bar profile. Mean values of linear bar parameters (L_r and z) are similar, the ones for the outer bar being slightly greater and having greater standard deviations. The resultant bar volume exhibits the same behaviour. Standard deviations of distance to bar crest and centre of gravity are also greater for outer bar. The same pattern is valid for water depths to bar crest, but here the standard deviation for outer bar is only a bit greater. All that can be attributed to more intensive wave action over offshore segment of beach profile, but this conclusion can not be supported by long-term data. The investigation has shown that the moving bar practically does not change its shape, i.e. its asymmetry. In addition to small changes in bar length and height, it can be inferred that at least part of it moves like a 'rigid' body. Correlations were computed for the same basic bar features. Generally, they are low with some obvious exceptions. Thus, significantly correlated are bar length and height with bar volume. High correlation of distance to crest and depth over crest is also not surprising. The correlations are put together in Tab.2, lower section.

Residuals of Beach Profiles in Terms of RSF

The RSF analysis for Lubiatowo aimed at finding the distribution of bar wave numbers and envelope over the beach profile. The wave number at a point x was calculated with Eq.7 as an average of all events for that location. The numbers form a series with a clear exponentially decaying trend. Upon least square fit, the bar wave number pattern was adopted in form of equation:

$$k(x) = 0.086exp(-0.0025x) \tag{14}$$

where 0.086 m^{-1} may be defined as a scale parameter and 0.0025 m^{-1} as a decay rate. For such a pattern the envelope was calculated from Eq.8. It is plotted in Fig.5 together with mean empirical deviations about the Dean curve. Since both the envelope and mean deviations describe profile variability, they should be similar over the whole profile. The exponential wave number pattern guarantees this requirement be satisfied. The envelope is characterised by two peaks some 250 and 650m offshore, which matches the observed breaker locations. The outer peak is less pronounced and spatially distributed. They are located over the outermost bar IV and over bar II, because the outer part of the beach profile is only modelled during rare, heavy storms and the inner one experiences constant changes caused by small and medium waves.

The analysis for Cecina Mare produced similar bar wave number pattern with the scale parameter 0.057 m^{-1} and decay rate 0.0031m^{-1}. The envelope is similar to mean deviations about average Dean profile, which confirms that the exponential bar wave number pattern is the pertinent assumption. The envelope has only one peak, some 100 m offshore. The peak is not that much conspicuous, due to the varying location of bar crest. Like before, it defines breaker locations. Graphical representation of RSF for both sites is presented in Figs.4 and 5.

Comparison of Results & Conclusions

The purpose of the comparison of two sites was to reveal similarities and dissimilarities of the behaviour of one and multibar beach profiles. Regular sampling at Cecina Mare started only a couple of years ago, so the comparison may be distorted by different time scales. It should also be added that the surveyed cross-shore area for Lubiatowo is about twice as wide as the one for Cecina Mare which, in light of the RSF analysis may not be sufficient, despite the steep sloping of the Cecina Mare beach and narrower surf zone.

Bar length exhibits similar behaviour at both sites; it grows with the distance offshore, and so grow the corresponding standard deviations. On the other hand, the bar height of remote bars is small. The comparison of envelopes of RSF for remote locations is equally interesting: the second peak of such envelope for Lubiatowo is spatially distributed and its values are not very high, and the offshore values of envelope for Cecina Mare are not high either. Since the envelopes depict mean deviations about the reference levels, it can be seen that offshore locations rather do not undergo dramatic changes of depth. It can all be related to spatial distribution of breakers during heavy storms over outer bars.

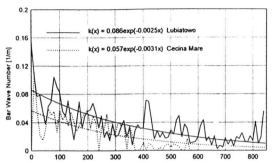

Figure 4 Exponential Bar Wave Number Patterns for Lubiatowo and Cecina Mare

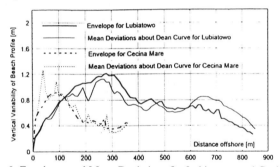

Figure 5 Envelopes and Mean Deviations for Lubiatowo and Cecina Mare

The analysis of bar height for other bars supplements the previous conclusion. The height of central bar II at Lubiatowo is the greatest. The height of outer bar at Cecina Mare is also greater than the height of inner bar. Both bars are the most conspicuous bed forms and it is striking how similar they are; $L_r^{Lub} = 57m$ vs. $L_r^{Cec}=54.5m$, $z^{Lub}=0.77m$ vs. $z^{Cec}=0.51m$, $x_g^{Lub}=234m$ vs. $x_g^{Cec}=174m$. It looks that bar shape in general is governed by breakers in the whole surf zone. Spatially distributed breakers of deep water waves over deeper parts of beach profile produce long, flat bars. During heavy storms the waves transform and break again over inner bars, so energy is dissipated. Breakers over inner bars also occur for milder waves, which pass unhindered over outer bars. However, those inner breakers are much less spatially distributed, so shorter yet higher inner bars are produced. Energy dissipation over inner bars is thus much more concentrated, so they should undergo the most intensive evolution and constant remodelling. Peaks on the envelopes of RSF indicate spots of intensive evolution, it can be seen that energy dissipation for Lubiatowo takes place in two sub-zones vs. only one for Cecina Mare (Fig.5).

The behaviour of bar volume is controlled by bar length, height and to some extent bar asymmetry. For both sites a general rule holds true that offshore bars have greater volumes with greater standard deviations. The asymmetry of inner bars at Lubiatowo and all bars at Cecina Mare is nearly the same. Only the outermost bar at Lubiatowo has much more vivid asymmetry.

Bar wave numbers in the RSF analysis follow the exponentially decaying pattern, so offshore bars are always longer than the inner ones. It would therefore be worth studying the behaviour of wave length along both beach profiles to see if the wave lengths obey the same trend.

The study on Cecina Mare beach also materialised in finding a novel way of bar identification from differenced beach profile surveys. It may be applied to any beach profile type, but it appears to be especially useful for steeply sloping beaches.

References

Aminti P., Pruszak Z., Zeidler R.B. 1996 *Multiscale Shore Variability at Two Coasts*, Proc. Coastal Dynamics'95, ASCE, New York, 617-628.

Boczar-Karakiewicz B., Forbes D.L. and Drapeau G., 1995. *Nearshore Bar Development in Southern Gulf of St. Lawrence*, Journal of Waterway, Port, Coastal and Ocean Engineering, Vol 121, No.1, p. 49-60.

Dally W., 1987, *Longshore Bar Formation - Surf Beat or Undertow ?*, Proc. Coastal Sediments'87, ASCE, 1987, p. 71-86.

Dean R., Srinivas R., Parchure T., 1992, *Longshore Bar Generation Mechanisms*, Proc. of 23rd Coastal Engineering Conference, Venice 1992, p.2001-2014.

Dolan T.J. and Dean R.G. 1985, *Multiple Longshore Sand Bars in the Upper Chesapeake Bay*, Estuarine, Coastal and Shelf Science 21, 727-743.

Howd P., Bowen T., Holman R. and Oltman-Shay J. 1991. *Infragravity Waves, Longshore Currents and Linear Sand Bar Formation*, Proc. Coastal Sediments'91, ASCE, p. 72-84.

Larson M., Kraus N., 1992, *Dynamics of Longshore Bars*, Proc. of the 23rd Coastal Engineering Conference, Venice, 1992, p. 2219-2232.

Pruszak Z., Różyński G., 1997, *Variability of Multibar Profiles in Terms of Random Sine Functions*, Journal of Waterway, Port, Coastal and Ocean Engineering, in press.

Pruszak Z., Różyński G., Zeidler R.B., 1997, *Statistical Propeties of Multiple Bars*, Coastal Engineering, International Journal for Coastal, Harbour and Offshore Engineers, in press.

Wijnberg K. (1995) *Morphologic Behaviour of a Barred Coast over a Period of Decades*, Ph.D. Thesis, University of Utrecht 1-245.

Remote Sensing and Coastal Wave Research

Paul A. Hwang[1] and Edward J. Walsh[2]

Abstract

Airborne sensors provide effective coverage of a broad region and are suitable for large-scale experiments. In this paper, a scanning sensor using direct ranging technique to measure the surface wave displacement is described. On a aircraft flying at 100 m/s, the sensor can complete one run across a 100-km continental shelf in 17 minutes. A case study is presented using the resulting two-dimensional spatial measurements of the surface topography to derive wave damping due to bottom friction and wave breaking. The results are in very good agreement with analytical prediction. This study demonstrates that remote sensing measurements can be used for rapid characterization of waves on continental shelves and coastal regions.

Introduction

Water wave research in the past has placed emphasis either in the deep ocean to develop global wave modeling or in the coastal water to study impact on the nearshore circulation and beach erosion. The continental shelf between the deep water and the surf zone had been largely left out and a critical gap obviously exists. For the study of waves in the littoral environment, the area of coverage spans from the edge of the continental shelf to the coastline. Because of the large expanse of the continental shelves (the shelf width in the East Coast of U.S. is on the order of 100 km), the application of remote sensing for studying wave propagation on the continental shelf is an attractive option. With airspeed on the order of 100 m/s, airborne sensors provide very efficient two-dimensional surface wave measurements over a large area in the ocean. For example, the NASA P-3 aircraft can cross the

[1] Oceanography Division, Naval Research Laboratory, Stennis Space Center, MS 39529.

[2] NASA/GSFC/WFF, Wallops Island, VA 23337, Presently on assignment at NOAA/ETL, R/E/ET6, 325 Broadway, Boulder, CO 80303.

continental shelf of 100-km width near Duck, NC in 1000 s. Such broad range coverage is difficult to achieve with in-situ sensors. In this article, applications of the airborne remote sensing using a scanning radar are discussed. Section 2 describes the measurement principles of the remote sensing technique and the relevant resolution issues. Examples of applications to the study of wave dynamics are also given in this section. Section 3 presents a case study of deriving the dissipation of shoaling swells using the spatial data. Section 4 is summary and conclusion.

Remote sensing by radio and optical ranging

The data to be analyzed in this paper were collected during DUCK94 using a scanning radar altimeter (SRA). The SRA and its predecessor, the surface contour radar (SCR) use the direct ranging technique. By measuring the time-of-flight of the return signal from the air-water interface, the elevation topography of the ocean surface can be derived (Walsh, 1977; Walsh et al., 1978, Kenney et al., 1979).

The SRA operates at 36 GHz (Ka-band) and scans its 1° (two-way) beam from left to right across the ground track at 10 Hz, generating 64 points evenly distributed across the swath. The scanning range is ±22 degrees; therefore, the SRA provides a cross-track coverage approximately 0.8 times of the aircraft elevation. At an aircraft elevation of 520 m and speed of 100 m/s, the cross-track spacing and the along-track spacing are about 6.7 and 10 m, respectively. Another factor determining the horizontal resolution is the footprint size of the radar waves impinging on the ocean surface. For the 1° beam width at a range of 520 m, the footprint diameter is approximately 9 m. The vertical resolution of the system is determined by many factors, including the accuracy of monitoring the aircraft motion, electronic design, and system noise. For typical operating conditions, it is estimated that the accuracy of the elevation output is approximately 0.2 m. The system is an equivalence of a linear array of 64 non-intrusive elevation sensors moving along the flight track. The an electronic circuit provides fast readout of the sensor suite every 0.1 s. Earlier comparisons of SCR and an airborne optical lidar data with nearby buoy measurements show very positive agreement (e.g., Walsh et al. 1985, 1987). SCR also provided very unique spatial wave measurements to test the hypotheses of wave directional distribution (Walsh et al., 1985) and wave growth with fetch (Walsh et al., 1989).

The high-resolution mapping of the surface may be used to identify breaking patches and other nonlinear properties of the wave field such as skewness and asymmetry of the waveforms. These nonlinear properties have been identified to be important parameters of wave transformation and breaking dissipation in the surf zone (e.g., Lippman et al., 1996). The spatial images of the wave topography may also be useful for quantifying and ranking of various dissipation mechanisms. An example of applying the spatial data to calculate the dissipation of shoaling swells will be illustrated in the next section. The calculated results are found to be in very good agreement with analytical expressions given in Thornton and Guza (1983).

A case study – dissipation of shoaling swells

To further explore the application of airborne spatial measurements to study the propagation of ocean waves in the continental shelf with non-trivial bathymetric variations, an example is presented here. The data were taken on October 18, 1994, two days after a Northeaster hitting the area. Wind speed was low (3 m/s) and long swells propagating from the East (peak period ~ 12 s, significant wave height ~ 2 m) dominated the wave field. One of the flight tracks was along the direction of swell propagation and covers the broad region from approximately 70-km offshore to the coastline. The US Army Field Research Facility at Duck, NC is a couple miles to the north of the track. The bottom slope of the region is rather mild with the continental shelf approximately 100 km wide. The aircraft was flying into the shore. The water depth of the offshore location was less than 40 m. The elevation of the aircraft was at 608 m, the flight speed was 100 m/s and the data acquisition time for the 70-km coverage was less than 12 minutes, from 08:09:27 to 08:20:48 of October 18, 1994. Examples of the two-dimensional images, one from offshore and one near the coast, are shown in Figure 1.

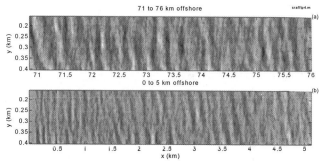

Figure 1. (a) An example of the two-dimensional wave topography obtained by the SRA. The image covers a region 71 to 76 km from the coastline. (b) Same as (a) but the image covers the nearshore region, from 0 to 5 km to the coastline.

The effect of shoaling on wave propagation can be clearly seen from comparing these two images. In the 5 km region closest to the coastline (Figure 1b) the reduction of wavelength from offshore toward the beach is quite evident. The variation of the wave topography in the offshore region (Figure 1a) is considerably less than the nearshore waves. Indeed, as will be seen later, the major transformation of the wave conditions occurs within 30 km from the coastline. The long-crested swell feature is very obvious from these images. Since the flight track is coincident with the wave propagation direction, the distortion of the wave direction due to finite aircraft speed is minimized.

This case is ideal for the investigation of swell dissipation during the shoaling process. The energy balance governing the dynamics of wave propagation can be expressed by the action density conservation equation

$$\frac{\partial A(k)}{\partial t} + \nabla\left(A(k)C_g(k)\right) = \sum Q_i \tag{1}$$

where A is the action density, k wavenumber, t time, C_g the wave group velocity vector and Q_i represent various source and sink terms. The most important source and sink functions are wind generation, breaking and frictional dissipation, and nonlinear wave-wave interaction. Assuming a quasi-steady condition and one-dimensional propagation, and ignoring the wind generation and nonlinear interaction, the action density conservation can be expressed as

$$\frac{\partial\left(AC_g\right)}{\partial x} = -\varepsilon_A = -\beta_{fA} A \tag{2}$$

where C_g is the group velocity in the direction of wave propagation (x), ε_A the action density dissipation and β_{fA} the dissipation rate, assuming an exponential decay. For the swell system in this case study, breaking occurs only in the nearshore region. As seen in the photographs taken simultaneously with the radar measurement, the breaking activity is limited to within 3 to 4 wavelengths, or less than 500 m, from the coastline. The dissipation mechanism outside the surf zone can be safely assumed to be due to bottom effects, either by bottom friction or bathymetric scattering. The former represents a net loss of the wave energy, the latter represents loss of wave energy in the flight direction due to waves scattering into other propagation directions. Further assuming that the swell frequency remains constant during the observation period and the observation range, (2) becomes an energy conservation equation

$$\frac{\partial\left(EC_g\right)}{\partial x} = -\varepsilon = -\beta_f E \tag{3}$$

where EC_g is the energy flux, ε the wave energy dissipation and β_f the dissipation rate.

From the two-dimensional spatial measurements of the surface topography, it is straightforward to calculate the wavenumber spectra and the flux terms on the LHS of (3) for computing wave dissipation. Figures 2a to 2b show the results from spectral analysis. The spectra were computed using strips of images of approximately $500 \times 5120 \text{ m}^2$. Figure 2a shows the peak wavenumber, k_p, and Figure 2b the spectral density, $S(k_p)$, along the 60-km flight track. There is a general trend of increasing wavenumber and decreasing spectral density shoreward (the West End of the track), as expected from the wave shoaling process. Although these trends are quite distinctive, the scatters of data points shown in Figures 2a and 2b are rather large. This is considered to be due to the fact that each $500 \times 5120 \text{ m}^2$ topographic image used for the spectral computation represents a merely 51.2 seconds of data. To reduce data scatter from the random nature of ocean waves, multiple passes are needed to provide a suitable number for ensemble average. Unfortunately, the only other flight along the same track started at almost 30 km from the coastline and flew seaward, missing the region of most active changes of the wave properties.

The group velocity on the LHS in Eq. (3) can be computed from k_p with the knowledge of the swell period and the local water depth. The bathymetry (Figure 2c) needed for this computation is from the survey by R/V Hatteras performed during the experimental period. The calculated group velocity is shown in Figures 2d. In the following, computations will be carried out with the raw data of k_p, C_g and $S(k_p)$ shown as circles in Figures 2a and 2b, as well as a "smoothed" version of k_p, C_g and $S(k_p)$ derived from best-fitting of the raw data. The smoothed curves are shown with dashed curves in Figures 2a, 2b and 2d. These smoothed "data" represent the hypothetical situation that there were sufficient number of repeated flights along the same track and that the data scatter due to the random nature of the wave field is reduced through ensemble average of these repeated measurements.

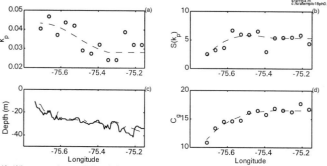

Figure 2. The variation of the (a) peak wavenumber, and (b) spectral density at the peak wavenumber, in the direction of wave propagation. (c) The water depth along the flight track: solid line; survey by R/V Hatteras during the period of experiment: dashed line; depth obtained from the bathymetry database. (d) Group velocity of the swell calculated from peak wavenumber shown in (a). Dashed curves represent the best fits of the data used as "smoothed" data in later computations.

In the subsequent presentation, the wave energy, E, of the swell will be represented by the spectral density at the peak wavenumber, $S(k_p)$. The energy flux is computed as the product of C_g and $S(k_p)$, shown in Figure 3a. It is noted that the calculated energy flux using the smoothed version of C_g and $S(k_p)$ follows reasonably well the calculated values based on the raw data.

In a conserved system, the flux of energy will remain constant. The large decrease of the energy flux in the coastal region indicates strong damping taking place. Figure 3b shows the gradient of the energy flux along the flight direction, which is also the direction of wave propagation. The general trend of increasing wave dissipation shoreward is clearly shown in the results processed with the "smoothed" energy flux distribution (dashed curve). Deviations from this general trend as revealed from the calculation using the raw data appear to be associated with major features of the bottom topography. For example, the dip near 75.2°W is in the lee side of a bottom "mound" and the spike near 75.6°W seems to be related to the approaching steep slope of the bottom profile. For reference, the bathymetry along the flight track is reproduced in Figure 3d. Clarification of such fine features will require additional measurements from repeated passes along the same flight tracks, as discussed earlier. In any event, the energy dissipation, ε, of shoaling swells with 12 s period is found to be approximately 5×10^{-3} m2/s near the coastal region and gradually decreases offshore. The magnitude reduces to about 1×10^{-3} at 25 km from shore with a local water depths of 27 m. The dissipation rate, β_f, is calculated and shows in Figure 3c. The magnitudes of 2×10^{-3} s^{-1} in the coastal region and 2×10^{-4} s^{-1} at 27-m depth are many orders of magnitude higher than the growth rate due to the wind speed present at the time of the experiment.

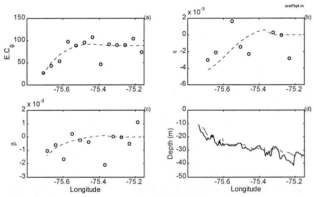

Figure 3. The (a) energy flux, (b) energy dissipation, and (c) dissipation rate calculated from raw data (circles) and smoothed data (dashed curve). (d) Water depth along the flight track.

Both energy dissipation ε, and dissipation rate and β_f, are found to be closely related to the local water depth. Figure 4a presents β_f plotted against water depth. There is an apparent trend of decreasing dissipation with increasing depth. Using the data points in the shallow region (water depth less than 30 m), the following formula for β_f is derived from least square fitting

$$-\beta_f = 1.18 \times 10^{-6} h^2 + 3.23 \times 10^{-5} h - 1.89 \times 10^{-3} \tag{4}$$

These results can be compared with analytical expressions of dissipation due bottom friction and wave breaking. Thornton and Guza (1983) use the bore model for simulating breaking waves. They apply conservation of mass and momentum at the regions of flow upstream and down stream of a fully developed hydraulic jump and find that the energy dissipation due to wave breaking is

$$\varepsilon_{bTG} = -\frac{\omega}{8\pi} \rho g \frac{(BH)^3}{h} \tag{5}$$

where ω is the angular frequency, ρ the density of water, g the gravitational acceleration, H the wave height, h the local water depth and B an empirical constant of $O(1)$. For the bottom friction, using the quadratic formulation of bottom shear stress, the energy dissipation is found to be

$$\varepsilon_{fTG} = -\frac{\rho c_f}{6\pi} \left(\frac{\omega}{\sinh kh}\right)^3 H^3 \tag{6}$$

where c_f is the friction coefficient of order 0.01.

Substitute $\rho g H^2/8$ for the wave energy, the dissipation rates can be expressed as

$$\beta_{bTG} = -\frac{\omega B^3 H}{\pi h} \tag{7}$$

and

$$\beta_{fTG} = -\frac{4c_f}{3\pi g} \left(\frac{\omega}{\sinh kh}\right)^3 H \tag{8}$$

Further expressed in dimensionless form, the dissipation coefficients become

$$\frac{\beta_{bTG}}{\omega} = -\frac{B^3 H}{\pi h} \tag{9}$$

and

$$\frac{\beta_{fTG}}{\omega} = -\frac{4c_f}{3\pi} \frac{kH}{\cosh kh \sinh^2 kh} \tag{10}$$

where the dispersion relation was used to simplify the expression of (10). For a quick check, take $H/h = 0.4$, $\omega = 2\pi/12$, then $\beta_{bTG} = 6.7 \times 10^{-2}$, which is about one to two orders of magnitude greater than the wave dissipation due to bottom friction. As noted earlier, breaking of swells in this experiment occurs only in the surf zone with a width of less than 500 m. Beyond the surf zone, dissipation due to bottom friction is probably the dominant mechanism for shoaling swells, except in regions of complicated bottom bathymetry. The dimensionless dissipation coefficients based on the airborne radar data and the analytic expressions are shown in Figure 4b (where the dimensionless parameter is presented as $\beta/\omega/kH$. The analytical calculations indicate that for the given experimental conditions, dissipation rate due to wave breaking (solid curve) is one to two orders of magnitude greater that that due the bottom friction (dashed-and-dotted curve). The curve of the predicted bottom friction represents an excellent lower bound of the measured dissipation rates, and it is considered that the agreement between the measured and the analytical results is quite good. This is especially gratifying given the fact that these calculations were based on a single flight pass and less than 12 minutes data. This illustrates the incredible capability of airborne remote sensing for rapid acquisition of high-resolution field data to study the ocean wave dynamics.

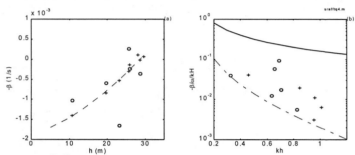

Figure 4. (a) The calculated energy dissipation plotted as a function of local water depth. (b) The dimensionless dissipation rate vs. dimensionless water depth. Circles: measurements (raw data); crosses: measurements (smoothed data); Solid curve: analytical solution of breaking dissipation, Eq. (9); Dashed-and-Dotted curve: analytical solution of bottom dissipation, Eq. (10). The analytical model is based on Thornton and Guza (1983).

Summary and Conclusions

In this paper, an airborne remote sensing devices (SRA) using direct ranging technique to measure the two-dimensional ocean surface topography was described. The SRA applies linear scanning method and obtains two-dimensional coverage through translating the linear scan by the aircraft motion – a push-broom approach. The measurements are equivalent to those from 64 non-intrusive wave elevation sensors traveling at 100 m/s. The resulting topographic images are in rectangular grids and can be easily processed by conventional two-dimensional FFT transform to obtain wavenumber spectra.

A case study is presented using the spatial measurement of the surface elevation from SRA to derive the energy dissipation of shoaling swells. The procedure to calculate the energy flux and the flux gradient in the direction of swell propagation is relatively straightforward using the two-dimensional surface topographic data. For the Duck, NC area, the dissipate rate, β_f, of the 12-s swells is found to be 2×10^{-3} s^{-1} in the coastal region, decreasing to 2×10^{-4} s^{-1} at 30 km offshore (27 m local water depth) based on the formulation of exponential decay (Eq. 3). An empirical formula relating the dissipation rate to the local depths is given in Eq. (5).

Very good agreement is found in the comparison of the measured results and the analytical model of quadratic bottom shear stress formulation (Thornton and Guza,

1983). This is an incredible achievement as the analysis is based on less than 12 minutes data. Within that duration, the airborne remote sensor covers an area of 0.5 × 68 km², demonstrating the capability for the rapid characterization of the ocean environment.

Finally, it is noticed that the analysis from single-pass data tends to have large data scatter. This is attributed to the random nature of ocean waves. The data scatter can be reduced through ensemble average, which suggests conducting experiments with multiple passes along the same tracks.

Acknowledgement

This work is supported by the Office of Naval Research (NRL JO 6800-07). The bathymetry data were provided by Bill O'Reiley, Bob Guza, and Tom Herbers. The authors would like to acknowledge the stimulating discussions with Jim Kaihatu. This is NRL Contribution number PP/7332-96-0021.

References

Kenney, J. E., E. A. Uliana, and E. J. Walsh, The surface contour radar, a unique remote sensing instrument, IEEE *Trans. Microwave Theory and Tech.*, **MTT-27**, 1080-1092, 1979.

Lippmann, T. C., A. H. Brookins, and E. B. Thornton, Wave energy transformation on natural profiles, *Coastal Eng.*, **27**, 1-20, 1996.

Thornton, E. B. and R. T. Guza, Transformation of wave height distribution, *J. Geophys. Res.*, 88 (C10): 5925-5938, 1983.

Walsh, E. J., Limitation on oceanographic use of beam-limited target-referenced radars, IEEE *Trans. Antennas Prop.*, **25**, 312 318, 1977.

Walsh, E. J., E. A. Uliana, and B. S. Yaplee, Ocean wave heights measured by a high resolution pulse-limited radar altimeter, *Bound.-Layer Meteorol.*, **13**, 263-276, 1978.

Walsh, E. J., D. W. Hancock, D. E. Hines, R. N. Swift, and J. F. Scott, Directional wave spectra measured with the surface contour radar, *J. Phys. Oceanogr.*, **15**, 566-592, 1985.

Walsh, E. J., D. W. Hancock, D. E. Hines, R. N. Swift, and J. F. Scott, Wave-measurement capabilities of the surface contour radar and the airborne oceanographic lidar, *Johns Hopkins APL Tech. Dig.*, **8**, 74-81, 1987.

Walsh, E. J., D. W. Hancock, D. E. Hines, R. N. Swift, and J. F. Scott, An observation of the directional wave spectrum evolution from shoreline to fully developed, *J. Phys. Oceanogr.*, **19**, 670-690, 1989.

AN AERIAL VIDEO SYSTEM FOR RAPIDLY MEASURING SAND BAR MORPHOLOGY

C. R. Worley[1], T. C. Lippmann[1], J. W. Haines[2], A. H. Sallenger[2]

Abstract

A low-cost, accurate aerial video system to rapidly measure the spatial scales of nearshore sand bars within several hundred meters of the coastline and extending 100's of kilometers along the coast is described. The system allows us to sample the spatial and temporal scales of nearshore sand bar and shoreline evolution over a vastly increased range, and with higher resolution, than previous proven video or remote sensing techniques.

Aerial video of the nearshore wave breaking patterns can be used to examine the behavior of the very large scale, O(10-100 km), nearshore morphology in response to episodic storms and seasonal changes in the wave climate. The measured patterns in sand bar morphology are inferred from average wave breaking patterns computed digitally using the measured image-to-ground coordinate transformation parameters.

Introduction

Substantial variability of the shoreline position and nearshore sedimentation occurs over several orders of magnitude in space and time, ranging from a few meters to hundreds of kilometers alongshore and evolving over time periods of a few wave periods (at small scale) to decades (at largest scale). In the dynamic region of the nearshore, sand bars constitute a large portion of the variability, particularly at time scales ranging a few hours to several years. Over the past 3 decades researchers have been conducting nearshore field work aimed at a better understanding of the details of the fluid-sediment interaction which produces sand bars, as well as the long term evolution of nearshore sand bar behavior.

[1]Center for Coastal Studies, Scripps Institution of Oceanography, University of California San Diego, 9500 Gilman Dr., La Jolla, CA, 92093-0209, USA.
[2]Center for Coastal Geology, U. S. Geological Survey, 600 4th St. South, St. Petersburg, FL, 33701, USA

Often short-term intensive field experiments focus on sand bar response to storm cycles, with typical periods of 1-2 weeks. The acquisition of longer time series spanning a decade or more have provided insight into time scales associated with inter annual or seasonal variability, as well as longer trends in large scale morphologic evolution (which could be cyclic at longer time scales). For example, recent data from Duck, North Carolina, (Birkemeier, 1984; Lippmann, et al., 1993) have shown that the timing of very large storm events (which occur every 1-2 years) with the configuration of the nearshore bar system can have a profound effect on long term morphology behavior. The phase of the observed morphology patterns within any larger alongshore variability, $O(10^1$-10^2 km), which are known to exist along the North Carolina coast (Lester, 1980), was not considered and is not well understood.

Very long term data sets spanning several years that also encompass very large alongshore distances are rare. One notable exception is the JARKUS profile data obtained along the full length of The Netherlands coastline, about 350 km (e.g., Wijnberg, 1995). These data have shown apparent cyclic behavior in the nearshore sand bars, with length scales of up to 10-20 km, that dominates the morphologic variability over the 30 years of JARKUS measurements. Interestingly, and perhaps surprisingly, the influence of man-made structures on the variability of the large scale morphology patterns at adjacent sites many kilometers up and down coast of the structure was found to be quite pronounced (Wijnberg, 1995). The generalization of the observations to other geographical areas is difficult in the absence of other data sets. The principal limiting factor is that the acquisition of long time series of large scale morphology is costly and logistically difficult, often requiring several months to complete one survey.

We have developed a quantitative aerial video system which can be used to rapidly sample the large scale morphology patterns over distances much greater than a few kilometers. The system can be mounted in (for example) easily accessible Cessna 172 or 182 aircraft, and is capable of sampling the morphology over a 100 km stretch of beach in 2-4 hours, the time to fly the coast (round-trip) in the aircraft. The nearshore morphology is inferred from digitally constructed time-exposure images which reveal the time-averaged wave breaking patterns. The technique is analogous to time-exposure imagery generated from fixed cameras overlooking a 1-2 km length of beach (Lippmann and Holman, 1989). The ground coordinates (determined using measured transformation parameters) of the edge of the dune, the landward bound of the wet/dry line on the beach face, the approximate shoreline (inferred from the shore break image intensity patterns), and any offshore sand bars (inferred from the offshore image intensity patterns) are determined using pattern recognition algorithms, image correlation techniques, or manually.

Aerial Video System

The aerial system rapidly (on the order of a few hours) and accurately (+/- 5-10 m) samples the spatial variation of the wave breaking patterns over distances of 100 km or more. Nearshore morphology patterns are inferred from the time-averaged wave breaking patterns recorded in the video (Lippmann and Holman, 1989; 1990), and are transformed from image coordinates to a real world coordinate system using measured transformation parameters (Holland, *et al.*, 1997).

The video data is post-processed in the laboratory equipped with the imaging hardware capable of analyzing the data. The raw video frames are individually rectified into true ground coordinates using the measured geometrical orientation of the aerial camera relative to the mean sea surface. Successive, overlapping images can then be averaged over 1-4 minutes (depending on aircraft ground speed and elevation, and the field-of-view of the lens) to reduce statistical uncertainty associated with random wave heights and breaking distributions. The precise occurrences of the over flights is optimal when the weather is mild enough to allow good visibility of wave breaking and the wave field is large enough (approximately 2-3 m significant wave height) to break over any submerged sand bars in 4-5 m depth.

The system is mounted in easily accessible Cessna 182 aircraft (with an installed luggage door which allows downward looking video to be made; Figures 1 and 2), and is capable of sampling the morphology over a 100 km stretch of beach in 2-4 hours, the time to fly the coast (round-trip) in the aircraft.

Figure 1. Photograph of the Cessna 182Q aircraft used in the pilot California surveys. The FAA approved modification to the rear luggage door is installed, and visible in the approximate center of the aircraft.

Figure 2. Aft view of the aircraft showing the approximately 6 inch relief of the modified luggage door. The mounted positioning device is located in the interior of the aircraft and orients the camera downward with unobstructed view through a hole in the luggage door.

The system includes a video camera mounted on an articulated pedestal (Figure 3) which positions and measures the orientation of the camera look angle relative to gravity with a gyro-stabilized digital inclinometer and flux-gate compass, video cassette recorder, small viewing monitor, video (VITC) time-stamping devices, digital altimeter, hand-held differential GPS positioning and timing, and a laptop PC computer to record aircraft position and camera orientation.

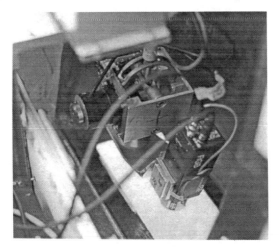

Figure 3. Interior view (from above) of the camera mounted in the aircraft.

The precision and accuracy of the differential GPS, digital altimeter, and gyro-stabilized compass and inclinometer are summarized in Table 1. Assuming precise positioning, and for a typical deployment using a 2/3 inch format video camera with 4.8 mm lens, the expected image resolution for a range of flying altitudes ranges between 1-10 meters depending on flight altitudes.

Table 1. Instrument resolution and accuracy.

Instrument	Resolution	Accuracy
Altimeter (z)	< 0.5 m	< 1 m
DGPS (x, y)	< 2 m	+/- 1-5 m
Inclinometer & Compass	+/- 0.1 deg	+/- 1 deg

With a typical aircraft ground speed of 100 knots (~50 m/sec) we expect averaging times in one flyby to be about 17-186 seconds for flight altitudes ranging 457-2438 m (1500-8000 ft). Actual values are summarized in Table 2. The averaging time can be increased linearly by making several flyby's of the same ground area.

Table 2. Image resolution and averaging time as a function of aircraft altitude. Resolutions are computed for a 2/3 format camera with a 4.8 mm lens. The averaging time is assuming an aircraft ground speed of 100 knots (~50 m/sec).

Altitude z (m)	Image Resolution $\overline{\Delta x}$,	$\overline{\Delta y}$ (m)	Averaging Time ΔT (sec)
457	1.6	1.1	17
914	3.3	2.2	34
1524	5.5	3.7	56
2438	8.8	5.9	89

The prototype system was initially tested in a series of over flights over the southern California coast near La Jolla (Scripps beach). The flight track of the over-flights is shown in Figure 4.

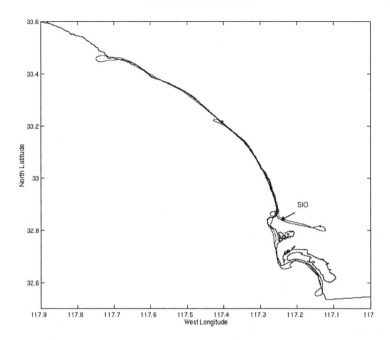

Figure 4. Over-flight track along the Southern California Bight between Dana Pt. California and the California-Mexican border on 2 June 1997. The example rectified aerial time exposure (Figure 5) is taken from the location of Scripps Beach, shown by the arrow.

An example rectified time exposure of 34 second duration from an over flight near Scripps beach (flight altitude of 947 m) is shown in Figure 5. The transformed location of the image coordinates is expected to be accurate to within +/- 5-10 m (Tables 1 and 2), although in practice random errors in positioning will result in an averaged image with highly accurate transformations, but slightly blurred imagery. Surveyed coordinates on the Scripps research pier, clearly visible in the images, were used to verify the accuracy of the aircraft positioning systems, and the image processing and averaging algorithms.

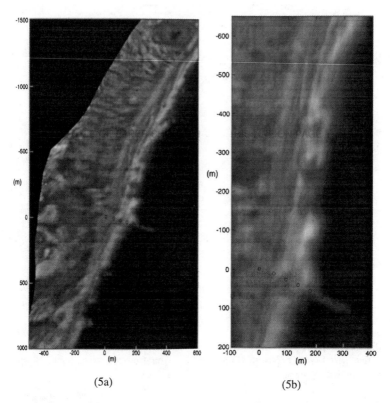

(5a)   (5b)

Figure 5. Rectified time-exposure image of 34 second duration obtained from an aerial over flight of Scripps Beach in Southern California on 2 June 1997 using the prototype aerial video system. (a) Image spanning 2.5 km approximately alongshore and 1.1 km across-shore. (b) Enlarged view near the center of (5a). The location of the Scripps research pier is clearly visible and the location of surveyed ground points along the pier and on the Scripps campus (the CCS building) is shown with the circles. The axes are in UTM coordinates relative to the CCS building with Northings along the vertical axis and Eastings along the horizontal axis. The pixel resolution is approximately 2.5 m in each direction.

<u>Summary</u>

A quantitative airborne video-based measurement system to rapidly sample the very large scale nearshore morphology over distances much greater than a few kilometers is described. The video system is composed of a downward looking high-resolution color CCD video camera with auto-iris fixed focal length lens, video cassette recorder, and dual viewing monitors. The camera is mounted on a central pedestal assembly comprised of a digital gyro compass, digital inclinometer, and a motor control unit with angular accuracy of better than +/- 1.0 degree. The platform keeps the camera vertical and with fixed orientation relative to the ground as the pilot makes turns to maintain optimal field of view over the nearshore region of interest. The aircraft location in space and time is determined with GPS positioning to within +/- 5-10 m. GPS timing is written into the vertical interval of each video frame and simultaneously digitally recorded on an onboard lap-top computer along with the position, tilt, and compass readings.

The camera geometry relative to the fixed ground coordinate is computed for each frame of the video. The video frames are then digitized with an image processing system and rectified into the ground-based ortho-normal coordinate system using the measured transformation parameters. Successive overlapping rectified video frames are then digitally added. The summation of all overlapping points is retained in its ortho-normal reference grid and stored on disk as a succession of rectified time exposure images. The return flight of the aircraft produces a second series of images that is again combined with the first pass to increase the averaging time. For typical ground speeds (50 m/s), camera field of view (85 degrees), and ground resolution within +/- 5-10 m, we expect to obtain averaging times of 2-4 minutes, long enough to smooth out possible breaking variations due to the passage of wave groups. The pixel resolution is less than about 2.5 m for these flight parameters and instrument precision.

Acknowledgments

This work was funded by the U. S. Geological Survey and the California Department of Boating and Waterways. We would like to thank Chuck Rosenfeld of Oregon State University for valuable assistance in the initial stages of the aerial deployments, Kimball Millikan for expert piloting of the over-flights and general assistance, and Fred Wright who wrote the data acquisition software package.

References

Birkemeier, W. A., 1984, Time scales of nearshore profile change, in *Proceedings of the 19th International Conference Coastal Engineering*, American Society of Civil Engineers, New York, 1507-1521.

Holland, K. T., R. A. Holman, J. Stanley, and N. Plant, 1997, Practical use of video imagery in nearshore oceanographic field studies, *IEEE J. of Ocean. Eng.*, 22(1), 81-92.

Lester, M. E., 1980, Aerial investigation of longshore bars along the Outer Banks of North Carolina, U. S. Army Corps of Engineers, CERC, Fort Belvoir, VA, unpublished.

Lippmann, T. C., and R. A. Holman, 1989, Quantification of sand bar morphology: A video technique based on wave dissipation, *J. Geophys. Res.*, 94(C1), 995-1,011.

Lippmann, T. C., and R. A. Holman, 1990, The spatial and temporal variability of sand bar morphology, *J. Geophys. Res.*, 95(C7), 11,575-11,590.

Lippmann, T. C., R. A. Holman, and K. K. Hathaway, 1993, Episodic, nonstationary behavior of a double bar system at Duck, North Carolina, U.S.A., 1986-1991, *J. Coastal Res.,* Special Issue 15, 49-75.

Peterson, C., and C. Rosenfeld, 1994, Littoral cell development in the Cascadia Margin of the Pacific Northwest, *SEPM Special Publication #46*, in Shepard Commemorative Volume, "From the Shoreline to the Abyss", Society for Sed. Geol., 17-35.

Wijnberg, K. M., 1995, Morphologic behavior of a barred coast over a period of decades, *Ph.D. Thesis*, Universiteit Utrecht, pp. 245.

Strange Kinematics of Sandbars

Nathaniel Plant and Rob Holman

ABSTRACT
Surfzone sandbar behavior is paradoxical. Bar response to wave height variations often seems unpredictable, yet at the same time sandbar response often steps sequentially through well described morphodynamic states. Feedback may be responsible. We have defined feedback based on the correlation between beach profiles and the overlying, time-averaged sediment transport patterns. The sense of this correlation (e.g. positive or negative) can be estimated from observed bar response. We found that sediment transport patterns were typically either in phase or out of phase with the topographic patterns of sandbars. The choice of phase (0 or π) appeared to be a strong function of wave breaking intensity, and switched when the wave height became saturated (depth limited). Strong feedback that is sensitive (nonlinearly related) to system state (breaking intensity, in our case) is a characteristic of dynamical systems that exhibit strange behavior, including deterministic chaos. Understanding sandbar behavior likely depends on understanding more fully the role that feedback plays.

INTRODUCTION
Surfzone sandbars are common features along many wave-dominated coastlines and play a significant role in modifying the incoming wave field, which, in turn, affects beach change. The migration of bars is associated with large fluxes of sediment, affecting a large portion of the nearshore profile. And, the horizontal structure (alongshore variability) of bars affects the alongshore variability of beach response (Komar 1993). It is the ultimate goal of many nearshore researchers to predict beach response, including sandbar behavior. A general assumption is that sandbars represent morphologic equilibrium with the overlying fluid processes. When the fluid processes change, so must the sandbar in order to maintain equilibrium (Holman and Bowen 1982). However, attempts to predict bar response, based on both theoretical and empirical relationships have met with limited success, even when models have been forced with detailed descriptions (observed or modeled) of the relevant parameters (Thornton 1995). Paradoxically, the short term (e.g. less than an annual cycle) behavior of bars appears to be constrained to relatively few, ordered states. Observations show that bar response tends to progress sequentially through these states (Lippmann and Holman 1990; Wright et al. 1985), implying that morphologic feedback is important.

College of Oceanography, Oregon State University, Corvallis, OR 97331, USA

There appears to be a similar paradox that applies to long term (interannual) sandbar variability. Over the long term and over larger spatial scales comprising multiple sandbars, a simple cyclic response has been observed (Ruessink and Kroon 1994). A cycle consists of bar birth from the shore (see Figure 1, Egmond deviations, bar 3, 1976), offshore migration (bar 2), and then decay (bar 1). An inner bar to outer bar transition tends to take place following the decay of an existing outer bar. The duration of a cycle ranges from several years (Wijnberg and Terwindt 1995) to a decade (Birkemeier 1985; Lippmann and Holman 1993; Ruessink and Kroon 1994). An investigation by Lippmann et al. (1993) of the wave climate variations and the beginning of a new cycle (inner to outer bar transition) found little correlation between long term variations in the forcing (such as extreme wave events) and bar behavior. Although the range of behavior appeared constrained (cyclic), the details of bar response appeared unpredictable. Again, morphologic feedback may be the key to understanding sandbar behavior on a variety of time scales. Can we describe the role of feedback by utilizing observations of sandbar response?

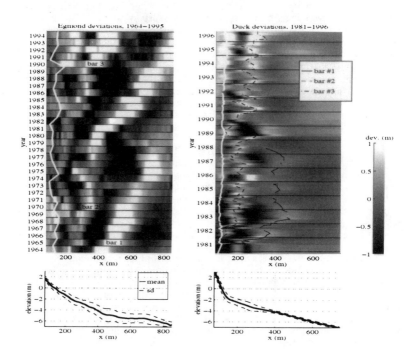

Figure 1. Time stack of beach profile deviations. Shoreline is marked by heavy white line.

In this paper, we describe two data sets that can be used to investigate morphologic feedback. One data set consists of daily beach surveys sampled over one month and provides the highest short term resolution. The other data set comprises monthly beach surveys sampled over a 16 year period, and resolves long term behavior. In order to investigate feedback, we define it with a simple, parameterized model, and we describe how the observations may be used to estimate the parameters. Next, we apply the estimation technique to both the short term and long term data sets. A comparison of the results acts as a consistency test and helps in the interpretation of the results. Finally, we discuss the implications of the results.

METHODS

Data Description

Long term behavior in morphologic systems is derived simply from the integration of the short term behavior. However, in the presence of strong morphologic feedback, even relatively simple and deterministic short term behavior does not necessarily lead to simple long term behavior. Simple behavior could be measured by the correlation between a forcing variable (wave height in the case of sandbars) and bar response (bar position, for example). Therefore, to gain insight into the role of feedback in the long term behavior, we should look for feedback in the short term behavior. We will do this using two data sets. Both data sets consist of bathymetric surveys (Figure 2) that were collected by researchers at the Army Corps of Engineers' Field Research Facility (FRF), located in Duck, NC, USA. All surveys used the FRF's CRAB (Coastal Research Amphibious Buggy (Birkemeier and Mason 1984)), which has a 10 m footprint, 10 cm accuracy, and can survey to depths of about 8 m if wave heights are less than about 2 m. Corresponding wave data were collected from a pressure sensor array located in 8 m water depth.

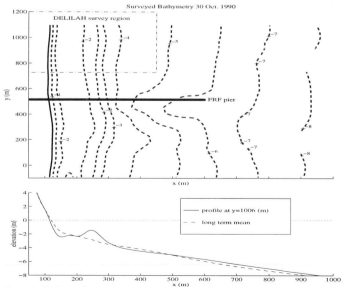

Figure 2. Bathymetric contour map (contour interval is 1 m) and example profile

The first data set was collected in October, 1990, during the DELILAH experiment (Birkemeier 1991). It consisted of 19 daily surveys along cross-shore transects that extended to depths of about 4 m. The transects were separated by about 20 m in the alongshore direction and spanned about 400 m of shoreline (Figure 2).

We narrowed our investigation to cross-shore migration of the single sandbar on the beach during this experiment (Figure 2). To do this, the survey data were interpolated (using statistical interpolation) to a regularly spaced grid and averaged in the alongshore direction (defined as perpendicular to mean beach gradient). Thus, effects of alongshore variability associated with length scales less than about 400 m were removed from analysis. Our analysis of morphologic feedback focuses on observed sandbar response to variations in wave height. As we will show, the relevant measures of sandbar response include migration and growth rates. We delay descriptions of how we estimated these variables until after describing how we analyze them.

The second data set consisted of monthly beach surveys at the FRF. These have been sampled since 1981 along a series of 26 profiles, spanning 1 km of thr shoreline, centered about the FRF pier (Figure 2). The profiles extend from the dune line (>4 m elevation NGVD) to 8 m depth. As before, we have interpolated these data and alongshore averaged them, utilizing the north-most (positive y-direction) 400 m of the study site.

Figure 1 (Duck example) shows the long term mean profile and the elevation variance about the mean ($\pm$ 1 SD) estimated from these data. Note that the elevation variance was highest in the region between the mean shoreline (h=0) and a position corresponding to a mean depth of about 3 m. Much of this elevation variance was due to the fluctuations associated with sand bars. Figure 1 also shows a time-stack (space-time map) of deviations from the mean elevation profile, in which it is possible to track the response of several sandbars. We have located the bar crests. A crest was defined as the point of maximum deviation from the mean profile. Two inner to outer bar transitions are visible in this figure, one in early 1982, and one in 1989. During the 1989 transition, an outer bar disappeared (bar #1), an inner bar jumped seaward (bar #2), and a new bar is formed near the shoreline (bar #3).

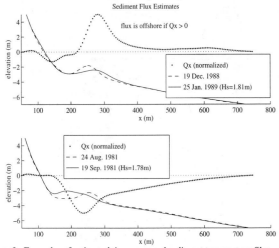

Figure 3. Examples of estimated time-averaged sediment transport profiles.

Morphologic Feedback

We propose that morphologic feedback manifests itself in the spatial correlation between bathymetric patterns (i.e. sandbars) and time-averaged sediment transport patterns. Conveniently, the observed beach profiles can be used to estimate sediment transport patterns. Then, these patterns can be compared to the bathymetry itself. The strength and sense of correlation provide excellent descriptions of morphologic feedback.

The sediment continuity equation relates changes in bathymetry (Z_b) to gradients in the sediment transport (Q):

$$\frac{\partial Q}{\partial x} = -\frac{\partial Z_b}{\partial t} \qquad (1)$$

Given observations of the bathymetric change from surveyed beach profiles, this equation can be integrated in the cross-shore (x) direction and in time to estimate the sediment transport profile (Q(x)). Figure 3 shows two examples of sediment transport estimates. In one case, sediment transport reached a maximum near the bar crest, and the spatial scale of the sediment transport pattern was similar to the scale of the bars. This situation corresponded to offshore migration of the bar. In the other case, the sediment transport magnitude was again maximum at the bar crest, but the sign of the transport was negative. This corresponded to onshore migration of the bar crest. These examples suggest that morphologic feedback manifests itself in the correlation between bathymetric and sediment transport profiles. As we will show, this correlation, including a possible phase shift, controls the behavior of sandbars.

A simple example of morphologic feedback consists of sinusoidal variations of bathymetry and a corresponding sinusoidal sediment transport pattern:

$$Z_b(x,t) = A(t) \, Sin[k(x - C\,t)] \qquad \text{(bathymetry)} \qquad (2)$$

and

$$Q(x,t) = B(t) \, Sin[k(x - C\,t - \lambda)], \qquad \text{(volume flux)} \qquad (3)$$

where k is a wavenumber, C is a celerity, and λ is a spatial lag. Inserting these equations into the continuity equation (equation 1), two terms, ϕ (= k λ) and $\frac{B}{A}$, describe the response of the system, measured by the celerity and growth rate of the bathymetry:

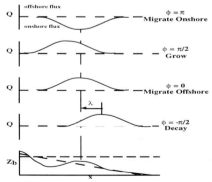

Figure 4. Schematic diagram of sediment transport phase shifts

$$C = \frac{B}{A} \, \mathrm{Cos}[\phi] \tag{4}$$

$$\frac{\partial A}{\partial t} = -(B \, k) \, \mathrm{Sin}[\phi] \tag{5}$$

The sediment transport phase shift, ϕ (Figure 4), controls the ratio of bar migration to growth, while the sediment transport magnitude controls the migration and growth rates. Observed morphology and morphologic change can be projected onto this system, and used to estimate the sediment transport magnitude and phase shift. Four observed variables are required: amplitude, wavelength, celerity, and growth rate. Then,

$$\phi_j = \mathrm{Tan}^{-1} [\frac{-\Delta A}{k \, \Delta t} / (C \, A)]_j \tag{6}$$

and

$$B_j = [(\frac{\Delta A}{k \, \Delta t})^2 + (C \, A)^2]_j^{1/2}, \tag{7}$$

where the index, j, refers to data from a pair of surveys, separated in time by Δt.

Bar amplitude and wavelength were defined using a so-called z-transform:

$$\widetilde{\zeta}(x) = \zeta(x) + i \, \zeta_H(x). \tag{8}$$

$\widetilde{\zeta}(x)$ is a complex variable whose real part is defined as the deviation of a surveyed profile from the long term mean. The imaginary part is the Hilbert transform of the profile (Bendat and Piersol 1986). Thus, at the bar crest location, an amplitude is defined by the magnitude of the complex variable ($A_j = (\zeta_j^2 + \zeta_{H_j}^2)^{1/2}$) and wavenumber by the cross-shore gradient in the phase of this variable ($k_j = \frac{\partial \Phi_j}{\partial x}$, $\Phi_j = \mathrm{Tan}^{-1}[\zeta/\zeta_H]_j$). To apply equations 6 and 7, migration and growth rates were obtained by backwards differences of the bar position and amplitudes, and the amplitude and wavelengths from each observation pair were averaged.

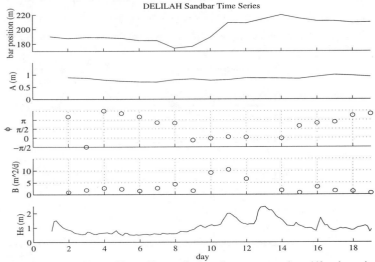

Figure 5. Time series of bar position and bar amplitude, sediment transport phase shift and magnitude, and significant wave height during the DELILAH experiment.

RESULTS

Figure 5 shows a time series of bar response variables and the corresponding feedback parameters estimated using the DELILAH data. Associated with the onset of a storm (day 9), the sandbar moved offshore. After the passage of the last peak in the wave height (day 15), the bar began to migrate onshore. This behavior was reflected in the estimates of the sediment transport phase shift. At the onset of the storm, the sediment transport phase shift jumped from π, corresponding to onshore migration, to 0 (offshore migration). As the storm passed, the phase shift increased first to $\pi/2$ (bar growth), then π (onshore migration), then nearly $3\pi/2$ (bar decay). Interestingly, bar growth occurred during onshore migration. The behavior of the sediment transport magnitude was also interesting; it reached a maximum on day 11, prior to the wave height maximum (day 13). Perhaps this is an indication of equilibrium behavior. If wave height adequately describes the hydrodynamic forcing of sediment transport over sandbars, how does it control the strength and sense of hydrodynamic feedback?

Figure 6 shows the sediment transport phase shift and magnitude, plotted as a function of significant wave height (H_s). The phase shifts were clustered about 0 and π. As expected, when wave heights were low, the bar migrated onshore. When wave heights were high, the bar migrated offshore. The transition between these regions occurred for $H_s \sim 1$ m. This corresponds to saturated breaking over sandbar crests, which can be shown by plotting the sediment transport phase shift vs the ratio of rms wave height to water depth (γ) at the bar crest. To estimate the wave height at the bar crest, we used the model described by Thornton and Guza (1983) to predict the evolution of the rms wave height through the shoaling and breaking regions. Figure 7 indicates that when the intensity of breaking was expected to be low (low values of γ), the sediment transport phase shifts were scattered about π. For γ between 0.35 and 0.4, the phase shift dropped from π to 0. In this γ range, we would expect intense breaking and that wave height would be nearly saturated (Sallenger 1985; Thornton and Guza 1982). At large values of γ, the phase shift remained close to 0. In general, bar response magnitude increased as either H_s or γ increased.

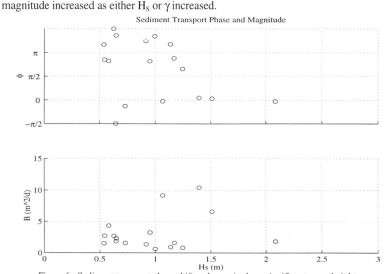

Figure 6. Sediment transport phase shift and magnitude vs significant wave height.

We repeated this analysis on the monthly surveys. Figure 8 shows the response (ϕ) of three sandbars. Bar #1, an outer bar, existed from 1981 to 1989 (see Figure 1, Duck). Bar #2 was the corresponding inner bar. Bar #3 was an inner bar that existed from 1989 to 1996. The behavior of ϕ as a function of γ was similar to the behavior observed in the DELILAH data (perhaps surprising, since the sampling strategies differed markedly). At low values of γ, bars migrated onshore. At high γ values, bars migrated offshore. There were systematic differences in the location of the onshore-offshore migration transition, which was indicated by the phase shift jumping from π to 0. The transition occurred at a lower value for the outer bar (bar#1), than for either of the inner bars (bars #2 and #3), or the bar during the DELILAH experiment (which was actually bar #3). This difference in the transition value of γ was consistent with expected variations in the saturated breaking criteria (γ_c), which increases with increasing beach slope.(Sallenger 1985). Bar #1 existed at a mean depth of about 4 m, where the slopes were milder than those of the inner bars (located in about 2 m mean depth). Again, the magnitude of bar response (not shown) increased as γ increased. These results are consistent with the hypothesis that morphologic feedback is important to sandbar behavior. It appears that the sense of the feedback (phase shift of 0 or π) is strongly controlled by the intensity of wave breaking over a bar crest.

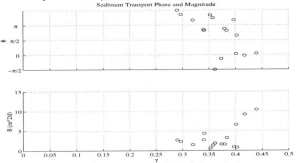

Figure 7. Sediment transport phase shift and magnitude vs γ.

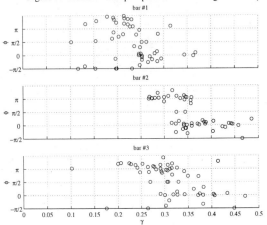

Figure 8. Sediment transport phase shift estimated from monthly surveys

DISCUSSION

The observed sensitivity of the sediment transport phase shift to wave breaking is likely related to the balance of two competing sediment transport processes. Offshore transport is most effectively driven by strong, mean return flow (undertow), while onshore transport is likely driven by skewed wave orbital velocities, in which the maximum onshore velocity is greater than the maximum offshore velocity (Roelvink and Stive 1989; Thornton 1995). We might expect that these processes would vary continuously with variations in the wave height, since each should be linearly related to the local wave height. However, it appears that at the onset of wave breaking, the resulting sediment transport processes were strong, nonlinear functions of wave height.

Strong feedback in a nonlinear system can lead to interesting behavior (May 1976). Often, the complex behavior that arises from such systems can be captured with very simple descriptions of the governing physical processes. We have constructed a very simple picture of nearshore sandbars, which were described as simple sinusoidal perturbations on a mean profile. Perhaps this description can be used to search for a deterministic source for the types of behavior that have been observed in both short term and long term data sets.

We have ignored an important aspect of sandbar evolution: alongshore variability. Changes in alongshore bar structure are associated with cross-shore response of sandbars. However, it is reasonable, as a first step, to omit alongshore variability. For example, Lippmann and Holman (1990) quantified the relative contributions of cross-shore and alongshore variability to variations of sandbar crest position over a 3 year period. The alongshore variability accounted for only 25% of the total variance. In our analysis of the monthly surveys, only 10% of the variability (over the entire profile) was due to an alongshore component. The presence of strong alongshore variability implies that both alongshore and cross-shore components of sediment transport have contributed to morphologic evolution. This may result from horizontal circulation, such as that associated with rip currents. Clearly, alongshore variability is worth including in future analyses.

CONCLUSION

We have described a form of feedback that affects cross-shore behavior of sandbars. We have taken a very simple approach, utilizing short term observations (profile changes) to estimate clearly defined feedback parameters. These parameters described the sense of correlation between spatial variations of a beach profile and the overlying sediment transport profile, in terms of a spatial phase shift. The behavior of the phase shift parameter (ϕ) with respect to the ratio of wave height to water depth (γ) indicated that the onset of saturated breaking divided regimes of onshore and offshore bar migration. The transition zone (in γ space) was narrow (nearly a step function), clearly indicating a nonlinear response to the wave height (normalized by depth). The combination of strong feedback and nonlinearity provides the ingredients for interesting dynamical behavior, including deterministic chaos. Thus, it is possible that strange sandbar behavior at a variety of scales may stem from the type of feedback that we have resolved in short term (profile change) observations.

ACKNOWLEDGMENTS

We are indebted to all of the researchers at the FRF. In particular, we thank Bill Birkemeier, Clif Barron, and Mike Leffler for collecting and distributing the CRAB data. Kathelijne Wijnberg generously provided the data from Egmond, NL (which looked really great in color). We thank the CRAB too. This work was funded by The Office of Naval Research, Coastal Dynamics Program (grant # N00014-960237).

REFERENCES

Bendat, J. S., and Piersol, A. G. (1986). *Random Data: Analysis and Measurement Techniques*, Wiley-Interscience, New York.

Birkemeier, W. A. (1985). "Time scales of nearshore profile change." 19th International Conference on Coastal Engineering, ASCE, New York, 1507-1521.

Birkemeier, W. A. (1991). "Samson and Delilah at the FRF." *The CERCular*, CERC-91-1, 1-6.

Birkemeier, W. A., and Mason, C. (1984). "The CRAB: A unique nearshore surveying vehicle." *Journal of Survey Engineering*, 110, 1-7.

Holman, R. A., and Bowen, A. J. (1982). "Bars, bumps and holes: Models for the generation of complex beach topography." *Journal of Geophysical Research*, 87(C1), 457-468.

Komar, P. D., and S. Shih. (1993). "Cliff erosion along the Oregon Coast: a tectonic-sea level imprint plus local controls by beach processes." *Journal of Coastal Research*, 9, 747-765.

Lippmann, T. C., and Holman, R. A. (1990). "The spatial and temporal variability of sand bar morphology." *Journal of Geophysical Research*, 95(C7), 11,575-11,590.

Lippmann, T. C., and Holman, R. A. (1993). "Episodic, non-stationary behavior of a two sand bar system at Duck, NC, USA." *Journal of Coastal Research*, SI(15), 49-75.

May, R. M. (1976). "Simple mathematical models with very complicated dynamics." *Nature*, 261, 459-467.

Roelvink, J. A., and Stive, M. J. F. (1989). "Bar-generating cross-shore flow mechanisms on a beach." *Journal of Geophysical Research*, 94(C4), 4785-4800.

Ruessink, B. G., and Kroon, A. (1994). "The behavior of a multiple bar system in the nearshore zone of Terschelling, the Netherlands, 1965-1993." *Marine Geology*, 121, 187-197.

Sallenger, A. H. J., and R.A. Holman. (1985). "Wave energy saturation on a natural beach of variable slope." *Journal of Geophysical Research*, 90, 11,939-11,944.

Thornton, E. B., R.T. Humiston, and W. Birkemeier. (1995). "Bar/trough generation on a natural beach." *Journal of Geophysical Research*, 101, 12,097-12,110.

Thornton, E. B., and Guza, R. T. (1982). "Energy saturation and phase speeds measured on a natural beach." *Journal of Geophysical Research*, 87(C12), 9499-9508.

Thornton, E. B., and Guza, R. T. (1983). "Transformation of wave height distribution." *Journal of Geophysical Research*, 88(C10), 5925-5938.

Wijnberg, K. M., and Terwindt, J. H. J. (1995). "Extracting decadal morphological behavior from high-resolution, long-term bathymetric surveys along the Holland coast using eigenfunction analysis." *Marine Geology*, 126.

Wright, L. D., Short, A. D., and Green, M. O. (1985). "Short-term changes in the morphodynamic states of beaches and surf zones: An empirical predictive model." *Marine Geology*, 62, 339-364.

Quantitative Estimations of Bar Dynamics from Video Images

Stefan G.J. Aarninkhof[1], Pieter C. Janssen[1] and Nathaniel G. Plant[2]

Abstract

This paper addresses a new monitoring technique, based on hourly video observations of the nearshore zone. The work focusses on the quantitative interpretation of observed patterns of wave breaking in order to estimate the bathymetry of sand bars. To that end, methods to identify the waterline from video images are needed. Moreover, the latter can be applied to directly monitor intertidal beach morphodynamics. Typical estimation errors were of order 20 to 50 cm for bar bathymetry and 10 cm for the the shoreline.

Introduction

Since 1992 hydrodynamics and beach morphodynamics have been studied from ARGUS video observations, a technique developed at the Coastal Imaging Lab, Oregon State University. ARGUS video cameras are present at eight beach locations world wide. Every daylight hour a snapshot, a 12 minute time-exposure and a variance image are collected automatically and fed into a central database, covering up to 10 years of observations. In this way nearshore coastal processes can be studied over a wide range of space and time scales, varying from a storm-event (hours-days) to longer term beach development (years-decades).

---

[1] Netherlands Centre for Coastal Research, Delft University of Technology c/o Delft Hydraulics, P.O. box 177, 2600 MH Delft, The Netherlands
[2] College of Oceanic and Atmospheric Sciences, Oregon State University, Corvallis OR 97331, USA

Fig. 1. *Time exposure image of Duck (NC)*

Time-averaged video images (Fig. 1) show bright, longshore bands, clearly indicating the locations where waves preferably break. In the nearshore zone the breaking of waves is generally caused by depth-limitation, so bright image intensity patterns can be assumed to reflect the underlying bar bathymetry. This relationship was already indicated qualitatively by Lippmann and Holman [1989].

This work focusses on a technique to extract quantitative bathymetric data from video observations of breaking waves. To that end, methods to identify the waterline from ARGUS images are needed. Both tools will be addressed in this paper.

A model to quantify bar bathymetry from video observations

We wish to relate observed intensity patterns of breaking waves quantitatively to actual bar bathymetry. For this reason image intensities are sampled along a cross-shore array of pixels. An exemplary realisation of the resulting intensity profile is shown in Fig. 2 (left-hand side), where x represents the cross-shore coordinate, positive in seaward direction. Two regions of wave breaking can be distinguished, each corresponding to a peak in the intensity profile and hence a breaker bar.

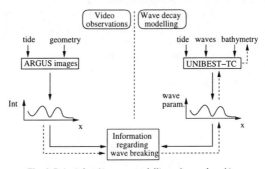

Fig. 2 *Principle of inverse modelling of wave breaking*

Knowing off-shore water levels and wave conditions, the wave transformation through the surf zone can be simulated with the help of a wave decay model. When the cross-shore distribution of wave breaking related parameters (like dissipation of wave energy) is investigated, again cross-shore profiles are found with maximum values around the sand bars (Fig. 2, right-hand side). Both the intensity profile and the wave parameter profile contain information regarding the process of wave breaking.

Given a tidal level and wave conditions measured just outside the surfzone, the cross-shore bottom profile can be determined via inverse modelling of a known cross-shore distribution of the wave breaking parameter. We aim at using image intensities as a proxy for the wave parameter (dotted path in Fig. 2), so the central question reads 'How do image intensities quantitatively match the wave parameter'. As image intensities are indexed between zero and one, this implies a scaling operation. Following this approach, the model to quantify bathymetry from video images comprises three steps (Aarninkhof [1996]).

1. Identification of an intensity-matching wave parameter

Intensity-matching wave parameters need to be found from wave decay modelling trough the surf zone. The wave transformation model applied is the ENDEC model (Battjes and Janssen [1978]), upgraded with the roller formulation according to Svendsen [1984] of which the dissipation formulation was sligthly adapted according to the appendix of Stive and De Vriend [1994]. Furthermore a breaker delay function according to Roelvink et al. [1995] was incorporated.

If the process of 'self-aeration' (Longuet-Higgins and Turner [1974]) is adopted as the dominant mechanism for the entrainment of air, then it is reasonable to assume that the thickness of the foamy layer formed at the water surface does not depend on any wave height or depth scale, but is only related to capillary effects which do not vary in cross-shore direction; hence the roller height H_r is assumed to be constant. Assuming furthermore a binary intensity model - intensity is one in presence of a roller and zero in absence of a roller - then it follows that time averaged pixel intensities should be related to the wave parameter $I_d = E_r/c^2$, in which E_r is the roller energy and c the phase speed (Aarninkhof [1996]). Apart from these physical considerations, statistical investigation regarding the correlation between normalized cross-shore intensity profiles and the variation of I_d confirms the selection of I_d as the intensity matching wave parameter.

2. Scaling of the indexed intensity values

In order to obtain a quantitative match between modified intensity values and I_d, the raw intensity data, having indexed values between zero and one, need to be scaled. This operation was performed by means of a three parameter model (I_{base}, r, SF). The background intensity level I_{base} and the linear trend removal parameter r were derived from raw image intensities, while the upscaling factor SF was related to a dimensionless waveheight at the seaward boundary of the model, just outside the surfzone. Finally, the upscaled intensity profile was lifted such that its seaward boundary value matched the corresponding value of I_d.

3. Inverse modelling of the wave parameter

The wave decay model mentioned above essentially comprises three balance equations, viz. the balance of organized wave energy $E = \frac{1}{8}\rho g H_{rms}^2$, the momentum balance in terms of radiation stress S_{xx} and the balance of roller energy $E_r = \frac{\rho A c^2}{2L}$, where A is the roller area, c the phase speed and L the wave length. Shoreward integration of the roller energy balance in terms of E_r/c^2 yields

$$\left(\frac{E_r}{c^2}c^3\cos\theta\right)_{i+1} = \left(\frac{E_r}{c^2}c^3\cos\theta\right)_i + \frac{1}{2}\left(D_w - D_r\right)_i \Delta x \qquad (1)$$

where subscript i increases in shoreward direction, D_w is the dissipation of organized wave energy according to Battjes and Janssen [1978] and D_r the dissipation of roller energy according to the appendix of Stive and De Vriend [1994]. Snell's Law is applied to account for obliquely incident waves with angle θ. Knowing the wave conditions $(H_{rms}, T_p, \theta)_i$, as well as a waterdepth h_i and a value for $(E_r/c^2)_i$ at a location just ouside the surfzone, the right hand side of equation (1) is fully known at the seaward end of the model. As furthermore the cross-shore distribution of the E_r/c^2 parameter is known from intensity data, we can solve the phase speed c_{i+1} and the angle of incidence θ_{i+1} iteratively from Eq. (1) and Snell's Law. Next the waterdepth h_{i+1} is determined from the dispersion equation, the set-up η_{i+1} is solved from the momentum balance and a new estimate of local bottom elevation can be obtained from

$$z_{b,i+1} = z_{tide} - h_{i+1} + \eta_{i+1} \qquad (2)$$

In this way the bottom elevation z_b along a cross-shore transect is modelled inversely, using a forward stepping integration method and based on intensity data obtained from one single ARGUS image. However, as the morphological time scale is of the order days-weeks while we have one image per hour at our disposal, successive estimates of bar bathymetry can be combined in order to improve the final result. The latter is done by means of a statistical filter technique, while making use of 12 images covering a whole tidal cycle.

Performance of the inverse model to quantify bar bathymetry from video images

The inverse model has first been applied to the single-barred beach system at Duck, NC, because of the availability of a high-quality data set. It has been calibrated against data obtained from the Duck'94 field campaign, yielding an exponential relation for the upscaling factor SF as a function of a non-dimensionless off-shore waveheight. Based on this relation the model has been tested for 27 different cases. For situations within the range of calibrated wave conditions, reliable estimates of bathymetry were produced (Fig. 3): deviations at the top of the bar amounted 10 to 20 cm, while the mean difference across the bar was 30 to 40 cm.

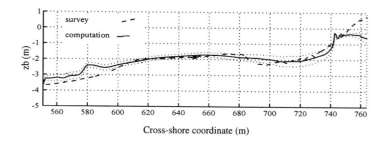

Fig. 3 *Estimate of bar bathymetry Duck NC, Oct. 11, 1994*

For situations outside the range of calibrated wave conditions, model performance was worse which can be explained immediately from a poorly acting scaling operation. Around the waterline the depth was systematically overestimated, probably due to an incorrect modelling of the overall amount of dissipation. Knowledge of a shoreward bottom level might solve for this problem.

Why do we need to know a shoreward bottom elevation?

Investigation of model results learns that inaccurate estimates of bar bathymetry generally are caused by a poorly acting scaling operation, in particular an incorrect value of the upscaling factor SF: underestimation of SF yields an underestimation of the total amount of dissipation and hence, computed bottom elevations which are too low. As the model equations are solved in shoreward direction, errors due to an incorrect estimate of SF propagate through the model and show up at the waterline.

Imagine a bottom elevation to be known at the shoreward boundary of the model. In that case the model can be 'forced' to compute this elevation correctly via modification of the upscaling factor SF, a higher value of SF giving a higher bottom elevation at the shoreward boundary. In this way the bar bathymetry is estimated iteratively, see Fig. 4.

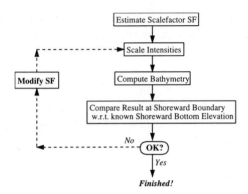

Fig. 4 *Iteratively scaling of image intensities*

Principally, the waterline can be used as a proxy for the shoreward boundary condition. The time averaged location (x_{WL}, y_{WL}) of the waterline in the horizontal plane can be observed from ARGUS video images, while the adjoining vertical coordinate z_{WL} can be estimated from tide and wave conditions measured off-shore. Two methods to do this are treated below.

Methods to detect the waterline from ARGUS video images

Recently two methods have been developed to identify the waterline from ARGUS video images. The Shore Line Intensity Maximum method (Plant and Holman, in press) was based on data from the Duck'94 field experiments; due to the relatively steep beach profile a clear shoreline break can often be seen at the time exposure images. Unfortunately such a shoreline break is often absent at ARGUS images from Noordwijk, because of the milder beach slope and emerging inner bar. For this reason a second method was developed, the Pixel Variance Method (Janssen [1997]).

The Shore Line Intensity Maximum (SLIM) method

The Shore Line Intensity Maximum Method or SLIM method (Plant and Holman, in press) estimates the location of a shoreward bottom level from the so-called Shore Line Intensity Maximum (SLIM). The SLIM is a peak in the cross-shore intensity profile, related to the ultimate dissipation of wave energy or shoreline break (Fig. 5).

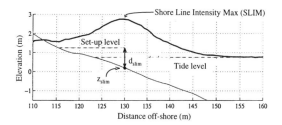

Fig. 5. *Defenition sketch of SLIM*

The horizontal location (x_{slim}, y_{slim}) of the SLIM can be determined from ARGUS time exposure images. The adjoining vertical coordinate z_{slim} is obtained by estimating the waterdepth d_{slim} at the location of the SLIM. In this way the shoreward bottom elevation can be computed from

$$z_{slim} = z_{tide} + \eta - d_{slim} \tag{3}$$

where z_{tide} is the tidal level and η the wave set-up. Key element of the technique is the estimate of d_{slim}. From accurately surveyed beach elevations during the Duck'94 field campaign d_{slim} was found to depend on the hydrodynamic regime at the shoreline. In case of reflective hydrodynamic conditions set-up turned out to be negligible and d_{slim} could be approximated by

$$d_{slim} \approx 0.5 \cdot H_{shore} \tag{4}$$

where H_{shore} is the wave height at the shore. In case of dissipative conditions the empirical relationship for d_{slim} reads

$$d_{slim} + \eta \approx 0.5 \cdot H_{shore} + 0.2 \cdot H_b \tag{5}$$

where H_b is the wave height at the break point. The latter term accounts for wave set-up. Based on these empirical expressions estimates of beach elevation at the shoreline have been made over a wide variety of conditions (using 366 ARGUS images in total), resulting in the measurement error being about 0.10 m in vertical sense.

Pixel Variance Method

The Pixel Variance Method (Janssen [1997]) makes use of both ARGUS time exposure images and variance images. Variance images are obtained by computing the variance of the time-varying pixel intensity during the twelve minutes of time exposure (Fig. 6). Consequently they indicate locations of highly varying conditions, like the region of initial wave breaking and the swash zone.

Fig. 6. *Timex exposure and variance image Noordwijk, Sep. 21, GMT 08:43*

Again, horizontal coordinates (x_{wl}, y_{wl}) of the waterline are determined from ARGUS video images. Both the cross-shore intensity profile and the variance profile are taken into account. Generally a phase shift between the profiles is present at the wet part of the beach, most obviously seen from the seaward shift of the variance peaks relatively to the mean intensity peaks. This observation is investigated by determining the optimal shift of the variance profile (called 'lag'), as necessary to get the best, local linear correlation between the profiles. Results of this operation, performed at any cross-shore location, are presented in Fig. 7.

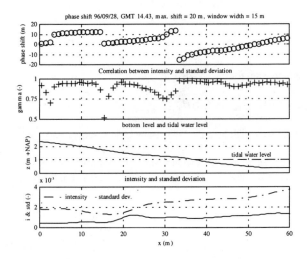

Fig. 7. *Lag and correlation between intensity and variance*

Around the waterline a characteristic feature shows up: moving in shoreward direction, the lag suddenly jumps from a negative value to a positive one (around x = 33 m). Furthermore a small decrease in the computed correlations is observed, slightly shoreward of the 'jump'. The negative value can be explained from the presence of a variance peak in the region of initial breaking, slightly seaward of the shoreline break, while the positive lag is caused by a small variance peak in the swash zone. Hence the location of the jump indicates a position just shoreward of the ultimate wave overtopping, which matches the seaward boundary of the swash zone.

The vertical coordinate z_{wl} is determined by estimating the wave and wind set-up (with the wave decay model described above) and a representative swash height, comprising surfbeat and individual wave run-up. For the wave run-up empirical relationships according to Battjes [1974] have been applied, while the empirical relation for surfbeat is given by Goda [1975]. Assuming a zero waterdepth at this location, the local bottom elevation is found.

Application of this technique at different longshore locations yielded a close approximation of the time-averaged position of the waterline (Fig. 8): mean differences in horizontal sense amounted 4-8 m, while the vertical deviation was 0.05-0.20 m. From Fig. 8 it can furthermore be seen that if the location of the jump was not uniquely defined (like at x = -700 m), it is still possible to identify the location of the waterline taking into account longshore consistency.

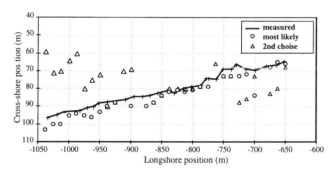

Fig. 8. *Waterline Noordwijk 97/03/13 GMT 07:43*

Conclusion

Recent work has focussed on the quantification of information obtained from ARGUS video imaging of the nearshore zone. An inverse model to quantify bar bathymetry from time exposure images has been developed, which produces accurate results in case of a single barred beach. In addition, two methods have been developed to identify the location of the waterline from video images. Incorperation of these techniques into the inverse model enables iterative determination of bar bathymetry, which is expected to improve model results considerably.

Bathymetric surveys from ARGUS images, available with high resolution in time and space, will be of vital importance to progress in the understanding of nearshore morphodynamics and in furthering the capabilities of process-based profile and planform morphodynamic models.

Acknowledgements

The ARGUS video technique has been developed with funds generated by the Coastal Imaging Lab, Oregon State University. Furthermore, work is undertaken in the PACE and SAFE projects, in the framework of the EU-sponsored Marine Science and Technology Programme (MAST-III), under contractno. MAS3-CT95-0002 and MAS3-CT95-0004 respectively. It is cosponsored by the Dutch Ministry of Transport and Public Works (Rijkswaterstaat) and DELFT HYDRAULICS.

References

Aarninkhof, S.G.J. (1996). Quantification of bar bathymetry from video observations. *Master Thesis, Delft University of Technology (Delft Hydraulics*, Report H2443).
Battjes, J.A. (1974). Surf Similarity. *Proc. of 14th Conf. on Coastal Eng.*, ASCE, pp. 466-480
Battjes, J.A. and Janssen, J.P.F.M. (1978). Energy loss and set-up due to breaking in random waves. *Proc. of 16th Conf. on Coastal Eng.*, ASCE, pp. 569-587
Goda, Y. (1975). Irregular wave deformation in the surf zone. *Coastal Eng. in Japan,* Vol. 18, pp. 13-26
Holman, R.A., Sallenger Jr, A.H., Lippmann, T.C. and Haines, J.W. (1993). The application of video image processing to the study of nearshore processes. *Oceanography*, Vol. 6, No 3.
Janssen, P.C. (1997). Intertidal beach level estimations from video images. *Master Thesis, Delft University of Technology (Delft Hydraulics*, Report Z2079).
Lippmann, T.C. and Holman, R.A. (1989). Quantification of sand bar morphology: A video technique based on wave dissipation. *J. Geophys. Res.*, 94(C1):995-1011
Longuet-Higgins, M.S. and Turner, J.S. (1974). An 'entraining plume' model of a spilling breaker. *J. Fluid. Mech.*, 63(1):1-20
Plant, N.G. and Holman, R.A. (in press). Intertidal beach profile estimation using video images. *Accepted with minor revision*, Marine Geology
Roelvink, J.A., Meijer, Th.J.G.P., Houwman, K., Bakker, R. and Spanhoff, R. (1995). Field validation and application of a coastal profile model. *Proc. Conf. Coastal Dynamics*, Gdansk.
Stive, M.J.F. and De Vriend, D.J. (1994). Shear stresses and mean flow in shoaling and breaking waves. *Proc. Int. Conf. Coastal Eng.*, Kobe, ASCE, pp. 594-608
Svendsen, I.A. (1984). Wave heights and set-up in a surf zone. *Coastal Eng.*, 8:303-329

CYCLIC BAR BEHAVIOR VIEWED BY VIDEO IMAGERY

Kathelijne M. Wijnberg[1,2] and Rob A. Holman[1]

ABSTRACT. Several multiple bar systems along the Dutch coast exhibit cyclic behavior. On a time scale of years the bar system moves in a net offshore direction, but at some distance offshore the outer bar decays while a new bar is generated near the shoreline. This study deals with the bar system near Noordwijk (The Netherlands) which completes one cycle in about 4 years. In this paper we present the preliminary results of an analysis of about 2 year of hourly sampled video time-exposures of the Noordwijk surf zone, focusing on the decay of the outer bar.
It appeared that the decay of the outer bar as recorded by the annual surveys of the nearshore bathymetry (JARKUS data base) is the net result of a series of morphological changes. Unfortunately, the analysis could not yet be conclusive about the hydrodynamic conditions that forced the decay of the outer bar. Further, support was found for the hypothesis that the outer bar controls the long term net behavior of the inner bar. Finally, the large scale, three-dimensional structure observed in the long term development of the bar system emphasizes the need for large scale monitoring of bar systems to understand their long term behavior.

INTRODUCTION

Previous studies on the long term behavior of multiple bar systems along the Dutch coast have shown that the bars behave in a very systematic way (Bakker and De Vroeg, 1988; Ruessink and Kroon, 1994; Wijnberg and Terwindt, 1995). On a time scale of years, all bars propagate in a net offshore direction, the outer bar decays offshore and a new bar is generated near the shoreline. This behavior results in a

[1] College of Oceanic and Atmospheric Sciences, Oregon State University, 104 Ocean Admin Building, Corvallis OR 97331-5503, USA.
[2] Department of Physical Geography, Institute for Marine and Atmospheric Research Utrecht (IMAU), Utrecht University, P.O.Box 80115, 3508 TC Utrecht, The Netherlands.

cyclic return of cross-shore bar system configurations, because after some time the net offshore propagating bars will move into the position of their predecessors. The time required to complete one such cycle varies between bar systems. For three different bar systems time spans of 4, 12, and 15 years have been observed. This cyclic behavior of the bar system appears to be essentially a cross-shore redistribution of sediment in the nearshore zone, implying that the sediment of the offshore decaying bar is transported onshore (Wijnberg, 1995).

The above findings are based on annual, ship-based sounding of the nearshore bathymetry along cross-shore transects with a 250 m longshore spacing (JARKUS data base). A few transects were surveyed more frequently, up to 8 times per year. The latter surveys sampled the bar system sufficiently to confirm the overall long-term behavior of the bars. The sampling was too sparse, however, to pinpoint conditions during which important changes in the bar system occurred. Therefore, explanations for the observed cyclic behavior in terms of processes involved, such as put forward by Wijnberg (1995), could not be verified against field observations. For the latter we need more frequent observations of the bar morphology, over long time spans, and under all types of wave conditions. An observation technique that seems to handle these requirements to a large extent is video time-exposure imagery collected by a so-called 'Argus' station (Lippmann and Holman, 1989).

In March 1995, an Argus station with two video cameras was installed at Noordwijk (The Netherlands) to monitor the multiple bar system that has been observed, based on 30 years of annual depth surveys, to complete one cycle in 4 years. In this paper we will present the first results of the analysis of these time series of video observations. Since one of the main points Wijnberg (1995) put forward in explaining the cyclic bar behavior was the key role of the outer bar, we focused our analysis on the behavior of the outer bar.

Wijnberg (1995) hypothesized that the systematic offshore decay of the outer bar controls the systematic changes in the inner nearshore: i.e. the net offshore migration of the bars and the subsequent generation of a new bar near the shoreline. She further hypothesized that the outer bar is only active during conditions with some minimum degree of wave breaking; a somewhat arbitrary lower limit was chosen of 0.1% of the waves breaking over the bar. Wave conditions with highly asymmetric waves over the bar (0.1-5% of the waves are breaking) were thought to favor decay of the bar, while conditions with more than 5% wave breaking over the bar should maintain the topography (i.e. growth, migration, or no change). These ideas were tentatively supported by exploratory model calculations (Wijnberg, 1997).

In this study we will document the decay of the outer bar in a more detailed way than was previously possible with the annual bathymetric surveys in the JARKUS data base. In addition, we will try to pinpoint the conditions that affect the outer bar topography, based on the Argus video observations and data on waves and tides, to evaluate the ideas regarding bar-decaying and bar-maintaining conditions.

ARGUS VIDEO TIME-EXPOSURES

An Argus station consists of one or more video cameras hooked up to a PC. Each day-light hour a snapshot and a 12 minute time-exposure are taken from the surf zone and stored in a data base. The spatial patterns of breaking waves reflect the positions of nearshore bars (Lippmann and Holman, 1989). The oblique images can be rectified to obtain plan view images (Fig. 1) using standard photogrammetric relationships (Holland et al., 1997).

Near Noordwijk, two cameras are looking in opposite directions, each monitoring up to about 3 km of beach and surf zone. In this paper we will focus on the image time series of the north-looking camera. We analyzed the video data from 15 March 1995 until 31 May 1997, with emphasis on the period between 5 July 1995 and 19 August 1996. On these dates bathymetric surveys of the area were available (JARKUS data base, see Wijnberg and Terwindt (1995) for survey properties). These surveys provided a bathymetric reference for the state of the bar system as well as a larger scale framework for the video observations of the bar morphology.

Fig.1: Example of oblique and rectified time-exposure image, Noordwijk.

PARAMETERISATION OF NEARSHORE FORCING CONDITIONS

The tide and wave data (significant wave height and period, and mean direction) available for this study are measured offshore at about 18 m water depth. However, the hypothesis to be evaluated requires the hydrodynamic conditions to be expressed in terms of percentage of waves breaking over the outer bar. Therefore, we have to determine whether during given offshore measured wave and tide conditions more than 5% of the waves break over the outer bar (maintaining the topography) or between 0.1 and 5% (reducing the topography).

Using the offshore wave and tide data and a simple wave transformation model (monochromatic, linear shoaling, refraction, no bottom friction, no breaking), a relative wave height H/h over the outer bar crest can be determined, where h = 'mean water depth over the bar crest + tidal elevation' and H is the computed significant wave height in h meter water depth. The relative wave height parameter can be used as a proxy for the amount of wave breaking, and is thus an indication for the presence of bar-decaying or bar-maintaining conditions. Because a monochromatic wave model calculates no useful percentage of breaking waves, a relationship between H/h and the percentage of breaking waves was approximated by using results from the exploratory model calculations presented in Wijnberg (1995).

Using a more advanced probabilistic wave transformation model (WAVIS, Van Rijn and Wijnberg, 1996), Wijnberg derived a set of wave and tide conditions that produced between 0.1 and 5% of breaking waves over an outer bar at 4 m mean water depth. In the bar-decay case analyzed in this paper, the crest of the outer bar before its decay is located at a mean water depth of 4 m as well. For that same set of offshore input conditions we calculated H/h with the simple shoaling-refraction model. The relation between H/h derived by the simple model and the percentage of waves breaking over the outer bar derived by WAVIS is shown in Fig. 2. It was than estimated that, using the simple wave model, bar-decaying conditions should occur for values of H/h between about 0.27 and 0.47 (Fig. 2). Note that these values are used as proxy measures only, and are not intended to be values that we should observe in nature for conditions with only 0.1-5% breaking waves

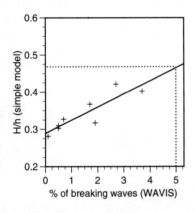

Fig. 2: Empirical relationship (solid line) between H/h derived by the simple model and the percentage of breaking waves derived by WAVIS.

RESULTS

Annual bathymetric surveys in the period 1994-1996 (JARKUS data base) show that the net bar behavior observed by the video cameras over the considered 2 year period involves the part of the bar behavior cycle that deals with the decay of the outer bar (Fig. 3). It also shows there is longshore variability involved. This is not in contradiction with the earlier statements about long term cyclic behavior, because those concerned large scale trends and did not consider the local, short term details of the behavior. The bar behavior as viewed by the cameras will obviously include a considerable component of this longshore variability, consisting of an oblique attachment of the decaying outer bar with the inner bar.

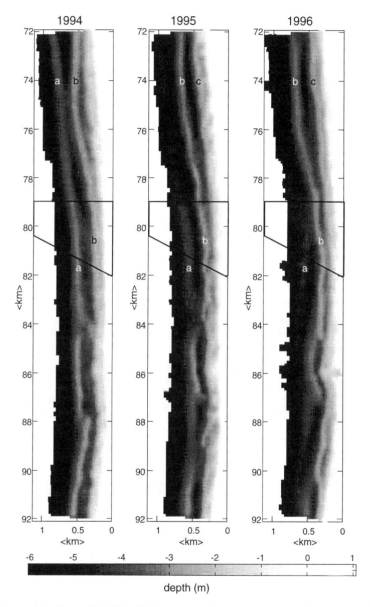

Fig. 3: Nearshore bathymetry (JARKUS) near Noordwijk. Boxes indicate the field of view of the Argus camera. Symbols (a, b, c) indentify bars in different years.

To analyze the offshore decay of the outer bar from the video images we first focused on the area near km 79.75 because it most resembled the idealized decay phase of an outer bar (Fig. 3 and Fig. 4). Fig. 5 shows a time stack of pixel intensity cross-sections sampled from the rectified images at longshore position km 79.75 (Fig. 6). Fig. 5 only contains the intensity profiles on which wave breaking in the outer bar region (about 600-700m offshore) was visible on the rectified time-exposure image covering 3 km of beach. Because of natural variation in overall brightness

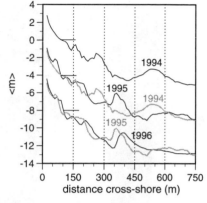

Fig.4: JARKUS surveys 1994-1996, line 79.75 (Profiles offset by 4 m)

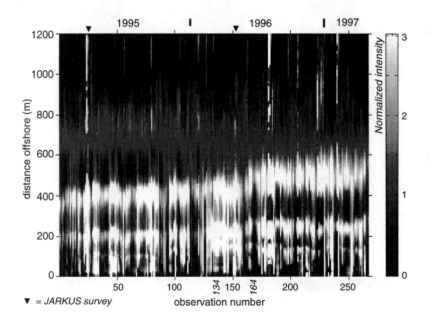

Fig. 5: Time stack of video image cross-sections at km 79.75 (see Fig. 6), normalized on pixel intensity at 670 m offshore.

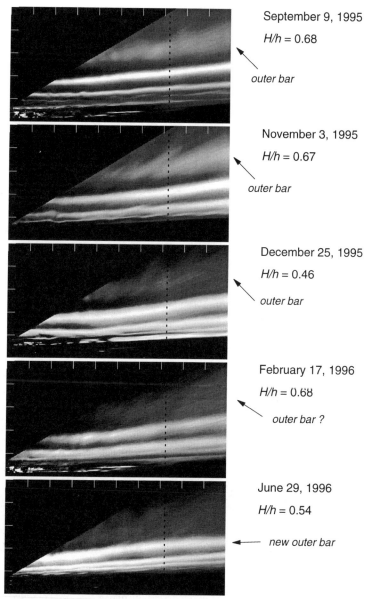

September 9, 1995
$H/h = 0.68$

outer bar

November 3, 1995
$H/h = 0.67$

outer bar

December 25, 1995
$H/h = 0.46$

outer bar

February 17, 1996
$H/h = 0.68$

outer bar ?

June 29, 1996
$H/h = 0.54$

new outer bar

*Fig. 6: Rectified video time-exposure sequence illustrating decay of the outer bar.
Distance between tick marks is 300 m; dotted lines indicate the position of
pixel intensity cross-sections shown in the time stack (km 79.75).*

of the image (e.g. time of day, cloudiness) each intensity profile was normalized with the intensity observed 670 m offshore, being the time-averaged position of the most seaward located pixel intensity maximum.

Fig. 5 reveals that the outer bar feature, clearly visible in 1995, has disappeared near observation 134 (19 June 1996). Subsequently, the former inner bar, which now became the new outer bar, jumped offshore near observation 164 (29 September 1996). It further appears that the annual decay of the outer bar feature is the net result of a series of morphological changes. This is indicated by the reduced visibility over time of the intensity minimum landward of the outer bar, which is to some extent a measure for the difference in depth between the bar and the trough. Analysis of the time series of the full images confirms this observation (Fig. 6).

It appeared to be difficult to pinpoint the conditions that induced the observed decay of the outer bar. The decay of the outer bar-trough topography is most obvious from the images of the high energy events in February 1996 and June 1996 (Fig. 6). However, it could not be determined whether the morphological changes occurred during the February storm event or in the period between the February and June observations, during which a series of hypothesized bar-decaying conditions occurred (Fig. 7). The changes in breaking patterns observed during the February storm event could either be related to changes in the tide and wave conditions or to bathymetric change. For comparison, the maximum depth change related to the decay

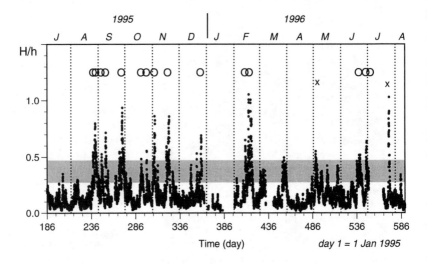

Fig. 7: Nearshore forcing conditions, period 5 July 1995 - 19 August 1996. The gray band indicates hypothesized bar-decaying conditions. The circles indicate the events during which the outer bar was observed on the time-exposures, the crosses indicate higher energy events during which image data was missing.

of the outer bar is about 0.5 m (Fig. 4) whereas the tidal range is 1.7 m. To distinguish between these two possibilities more advanced analysis tools are needed, such as the inverse modeling technique proposed by Aarninkhof (1996), which estimates bathymetry from time exposure images. However, that technique is still in an experimental stage. Further, it was not possible to analyze changes in bar topography during the hypothesized bar-decaying conditions, because the amount of wave breaking over the outer bar during those conditions was too small to be visible on the time-exposure images.

DISCUSSION AND CONCLUSIONS

Two years of hourly video sampling of the nearshore wave breaking patterns, as a proxy for the nearshore bar morphology, showed that the breaker bars exhibited the expected long term behavior of net offshore progression and offshore decay of the outer bar. These expectations were based on the extrapolation of 30 years of annual bathymetric surveys. The video data allowed a more detailed analysis of the process of bar decay than was possible with the bathymetry surveys, because of the high temporal and spatial resolution obtained by the Argus video monitoring system.

It appeared that the decay of the outer bar recorded in the annual bathymetric surveys (JARKUS) was the cumulative result of a series of morphological changes, rather than the effect of a single storm event. Unfortunately, we could not yet be conclusive about the conditions that induced the decay of the outer bar. This was partly because the outer bar appeared to be invisible on the time-exposure images during the hypothesized bar-decaying conditions. Also, more advanced analysis tools are needed for the interpretation of small changes in the time-averaged wave breaking patterns. Currently, it is hard to determine whether changes in these patterns are due to changes in the wave and tide conditions or due to bathymetric change. The inverse modeling technique proposed by Aarninkhof (1996) might be useful here. In addition, different video sampling techniques might be applied which are more suited to scan the underlying bar topography when only few waves are breaking (Stockdon, 1997).

Further, more support was found for the leading role of the outer bar in the cyclic behavior. It was observed that only after the outer bar decayed the inner bar exhibited net offshore migration.

Finally, the changes in the plan view bar morphology, emerging from both the bathymetric surveys and the video time-exposures, emphasize that longer term dynamics of bar systems can only be analyzed appropriately with a sufficiently large scale of monitoring. Therefore, the further development of monitoring techniques that can deal in a cost-effective way with regular surveys of such long stretches of coast (e.g. Worley et al., 1997) is of great importance.

ACKNOWLEDGEMENTS. K.M.W. was funded by the EU-sponsored Marine Science and Technology Programme (MAST-III), as part of the PACE-project under contract number MAS3-CT95-0002. The work was co-sponsored by the Andrew Mellon Foundation. Support for R.A.H. and for the Argus data collection has also been provided by the Coastal Geology Program of the U.S. Geological Survey and by the Coastal Dynamics program of the Office of Naval Research (grant number N00014-9610237). We also want to acknowledge the Dutch Ministry of Transport and Public Works for kindly providing the bathymetric data, special thanks go to Lia Walburg, and the wave and tide data.

REFERENCES

Bakker, W.T. and H.J. De Vroeg, 1988. Is de kust veilig? Analyse van het gedrag van de Hollandse kust in de laatste 20 jaar. Nota GWAO 88.017, Rijkswaterstaat, The Netherlands 42 pp.

Aarninkhof, S.G.J. 1996. Quantification of bar bathymetry from video observations. MSc-thesis, Delft Technical University, The Netherlands, 91 pp.

Holland K.T., R.A. Holman, T.C. Lippmann, J. Stanley, N. Plant, 1997. Practical use of video imagery in nearshore oceanographic field studies. *IEEE J. of Oceanic Engineering 22(1)*, pp. 81-92.

Lippmann, T.C. and R.A. Holman, 1989.Quantification of sand bar morphology: a video-technique based on wave dissipation. J. of Geophysical Res. 94 (C1), p. 995-1011.

Ruessink, B.G. and A. Kroon, 1994. The behaviour of a multiple bar system in the nearshore zone of Terschelling, The Netherlands: 1965-1993. Marine Geology 121, p. 187-197.

Stockdon, H.F., 1997. Estimation of wave phase speed and bottom bathymetry using video techniques. MSc-Thesis, Oregon State University, USA, 155pp.

Van Rijn, L.C. and Wijnberg, K.M., 1996. One-dimensional modelling of individual waves and wave-induced longshore currents in the surf zone. Coastal Engineering 28: 121-146.

Wijnberg, K.M., 1995. Morphologic behaviour of a barred coast over a period of decades. PhD-thesis Utrecht University, The Netherlands, 245 pp.

Wijnberg, K.M., 1997. On the systematic offshore decay of breaker bars. Proceedings Int. Conf. Coastal Engineering 1996, Orlando, ASCE, p. 3600-3613.

Wijnberg, K.M. and J.H.J. Terwindt, 1995. Extracting decadal morphological behaviour from high-resolution, long-term bathymetric surveys along the Holland coast using eigenfunction analysis. Marine Geology 126, p. 301-330.

Worley, C.R., Lippmann, T.C., Haines, J.W., and Sallenger, A.H., 1997. An aerial video system for rapidly measuring the spatial variability of very large scale (10^1-10^3 km) sand bar morphology. Book of Extended Abstracts of Coastal Dynamics '97. University of Plymouth, p. 228-229.

The Evaluation of Large Scale (km) Intertidal Beach Morphology on a Macrotidal Beach Using Video Images

[1]M. Davidson, [1]D. Huntley, [2]R. Holman & [1]K. George

Abstract: A method is described for the evaluation of large scale (km) changes in the intertidal beach morphology using video images. Data is presented from an Argus station (Holman, 1994) located at a strongly macrotidal beach (maximum tidal range ≈ 7.5m) at Perranporth, Cornwall, UK. Shoreline positions are digitised from spatial differentials of the rectified time-exposure images recorded at equal increments in the predicted tidal height. The horizontal (x-y) coordinates of the shoreline are obtained directly from these images and the vertical coordinate (z) is taken from an accurate tidal model for the area. Corrections for the vertical offset of the video survey relative to a measured beach profile which arise due wave setup, swash and atmospheric pressure variations are removed using a small amount of ground truth survey data. Comparisons between the video and a total station survey show an excellent agreement in the estimated 3-dimensional intertidal morphology with standard errors of less than 11cm. The method shows great promise for quantifying large scale (in this case 1000m by 500m), long term (weeks, months, years) changes in the intertidal beach morphology.

Introduction

In the last 3 decades field measurements using fast response sensors (e.g. pressure transducers, EMCMs, OBSs etc.) have significantly advanced our knowledge of coastal processes which has in turn led to more sophisticated bottom-up (process based) models. However, in spite of several large-scale multi-agency studies and significant modelling advances we are still unable to predict accurately large scale 3-dimensional coastal change over intermediate time-scales (weeks, months, years). Accurate measurement of the 3-dimensional beach topography over intermediate time-scales is costly and time consuming to obtain. However, if we are to be successful in our quest to predict morphological change at these scales, data of this type is essential. Video imaging of the nearshore potentially provides a means of effectively evaluating long term, large scale changes in beach morphology negating

[1] Institute of Marine Studies, Univ. Of Plymouth, Plymouth, Devon, PL4 8AA, UK.
[2] College of Oceanography, Oregon State University, Corvallis, OR 97331.

the need for expensive overflights (photogrametric surveys) or time-consuming manual surveys of the beach.

The video surveying method described in this contribution is appropriate to the intertidal beach area and therefore is potentially of most use in macrotidal environments. The intertidal profile forms the ultimate buffer zone between the sea and land. Therefore changes in this area of the beach are of great importance in terms of coastal protection.

Firstly, a description is given of the latest site (at the time of writing) in the Argus network, the video system used here, and the mode of monitoring. This is followed by a description of the video surveying method which is then tested against data obtained using a conventional total station.

Site Description

The field site at Perranporth, Cornwall is a macrotidal (maximum range ≈ 7.5m), high energy (mean $H_s=1.4$m) dissipative beach (gradient<0.02) situated on the north coast of the south-west peninsula of the United Kingdom (Figure 1). It is exposed to both locally generated sea and swell from the North Atlantic. The intertidal zone typically spans a

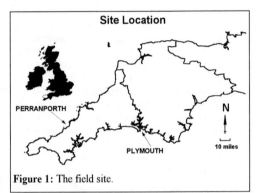

Figure 1: The field site.

cross-shore distance of 500 metres during spring tides. The beach at Perranporth, orientated with its longshore (y) axis approximately north-south, is comprised of some 3.5km of uninterrupted medium sand ($D_{50}=0.48$mm) bounded by headlands to the north (Ligger Point) and south (Droskyn Point). There is a plentiful supply of beach sediment from the extensive sand dune system at the head of the beach.

[2]Two video cameras are situated at a site originally designed for wave measurements some 47m above ordinance datum on Droskyn Point. The cameras are aimed northwards avoiding the direct sunlight. Both cameras are needed (even with a wide angle lens) in order to cover the broad intertidal profile. Example images from the high-tide (camera 1) and low-tide (camera 2) cameras are shown in Figure 2 illustrating the extremely broad intertidal range at this site. Each camera is triggered to sample every time the tide changes by a fixed height increment (of say 0.5m) as predicted by an accurate (+/- 0.2m) tidal model for the area. Every time the camera is

---

[2] This site at Perranporth was where some of the first observations of surf-beat were made by Tucker in 1950.

Figure 2: Views from camera 1 (top) at high-tide and camera 2 (bottom) at low-tide.

triggered 3 types of images are recorded from both cameras. These include, a snap shot, a time-exposure and a standard deviation image. The tidal scheduling is required in order to facilitate the even collection of data as the tide moves across the intertidal profile. If samples are taken at equal time increments the result is numerous measurements around both high and low water and very little data around mid-tide.

Figure 3 shows a typical 5 minute time-exposure image recorded at low water on the 28/6/96. The Figure shows a well developed offshore bar system punctuated by rip-currents. This low-tide bar system has been omnipresent since the installation of the camera system in August 1996 (1 year ago), although the precise form and cross-shore location of the bar has changed during this period. During storm conditions a second outer bar is sometimes visible. The rip current system labeled in Figure 3 is normally to be found at this longshore position and is likely to result from a rip-current draining out of the embayment at the top of the beach (see Figure 2, camera 1) during the high-water phase of the tide.

The intertidal beach flat is relatively featureless compared to the well formed bar system seawards of the low-water stand. Bar formation here is suppressed (smoothed out) by the rapid propagation of the tide over the profile, although some subtle ridge and runnel features of relatively low relief (<2m) are sometimes found here. Also common on the intertidal portion of the profile is a strong 'hummocky' structures which is particularly common around the mid-tide portion of the beach and during spring tides. These features are typically up to 15cm in amplitude and 1-2m in wavelength and therefore introduce an aliasing problem to the measurement of beach surveys which were collected on a 25m grid.

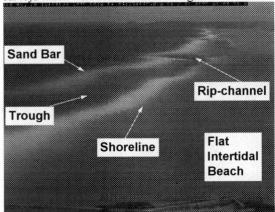

Figure 3: A time-exposure image from the low-tide camera.

The Video Surveying Method

The video surveying technique relies on the detection of the shoreline location (which is in effect a contour line) at a number of instances during a tidal cycle. However, what do we define as the shoreline, and what does it look like on a video image? There are several possible answers to these questions. For example the shoreline might be defined as the mean sea level or the position of maximum run-up. Whatever the definition the feature must be easily identifiable on images from different sites and under different wave and weather conditions. It must also be possible to evaluate the vertical coordinate associated with the chosen shoreline. Plant and Holman (1996) mapped the ShoreLine Intensity Maximum (or SLIM) located on time exposure images and related this feature to the mean sea level. Janssen (1997) examined the changes in phase resulting from a spatial cross-correlation between shore-normal profiles of intensity in standard deviation and time-exposure images to locate the position of maximum run-up.

Analysis of the Perranporth data showed that the shoreline features identified by Plant and Holman (1996) and Janssen (1996) were not always recognisable. One of the most consistently observed shoreline features in both the standard deviation and time-exposure images is the sharp increase in intensity with offshore distance which is theoretically located close to the point of maximum run-up. The rapid increase in intensity in the time-exposure images relates to the bright breaking foam in the shorebreak / swash zone contrasting strongly with the relatively dark beach. With the standard deviation images the spatial variation of intensity at the shoreline relates to the temporal change in intensity (bright areas in standard deviation images) associated with the wetting and drying of the beach in the swash zone relative to the fairly constant intensity (dark areas in standard deviation images) of the dry beach. The approach in this contribution is to map the maximum shoreward extent of the swash as indicated by the aforementioned rapid spatial intensity change. This feature is best observed in the spatial differential of the intensity image computed by:

$$\nabla I = \frac{\partial I}{\partial x} + \frac{\partial I}{\partial y} \tag{1}$$

Either time-exposure or standard deviation images can be used in this way to locate the shoreline. The relative merits of time-exposure and standard deviation images is still under analysis (Davidson *et al.*, 1997). Time-exposure images have been used for the day under consideration here. The standard deviation images proved difficult to analyse due to the cloudy weather conditions which led (unusually) to a bright beach area relative to the breaking waves in the swash zone and hence a reversal of the intensity gradient.

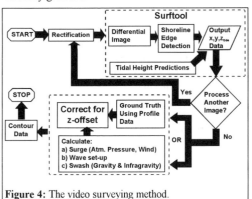

Figure 4: The video surveying method.

Figure 4 shows a flow chart illustrating the data analysis procedure used in the video surveying process and the results of this process are shown in Figure 5a & b. First the shoreline portion of an oblique time-exposure image is rectified using the method outlined by Holland, *et al.*, (1997) (Fgure 5a).

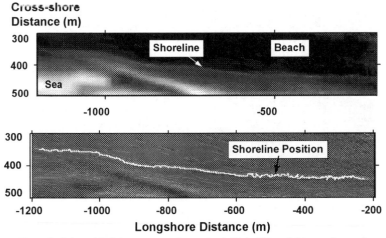

Figure 5: a) A rectified time-exposure image (top). b) A differential image (bottom).

Next a differential image computed using Equation 1. An example is shown in Figure 5b with the detected shoreline marked in white. Here the differential image has been normalised in order to aid visualisation such that the mean intensity over the data area is 0.5 (grey) and the maximum value is 1.0 (white). The differential image is then scanned in a longshore direction (column by column) to find the maximum change in intensity with distance associated with the shoreline position.

Each of the x-y coordinates from the edge detection process are assigned a z-value equivalent to the tidal height (z_{tide}) at the time of the observations. In the examples cited here the tidal level comes from a model although it is planned that a tide gauge will be deployed for future work. The above process is repeated until all the images have been processed and whole series of (x, y, z_{tide}) shoreline coordinates are saved to a file.

Next it is necessary to correct the z_{tide}-coordinates for the vertical offset arising from set-up (wave and [3]wind), swash (gravity and infragravity) and [3]atmospheric pressure variations. This can be done in two ways: Firstly, via theoretical calculation of the above (*c.f.* Janssen, 1997), although more experimentation is needed in order to validate the theoretical offset values for the mode of shoreline detection used here. This is a promising area of research in progress (Davidson *et al.*, 1997). Alternatively, a small amount of ground truth survey data may be used to determine the vertical offset. This is the approach used here. The offset correction is then applied to all the vertical shoreline coordinates and the data is contoured to provide a 3-dimensional map of the intertidal beach. Correcting the vertical coordinate in this way also removes any offset errors present in the tidal model.

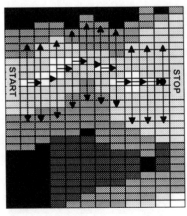

The simple edge detection procedure used here is shown schematically in Figure 6. Here each cell represents a pixel. In this example the longshore and cross-shore axes run approximately parallel to rows and columns respectively. The start position of the edge detection is user defined by a single mouse operation. It is hoped that in future the start position may be automated by using a combination of a known survey (say the previous days) and the tidal elevation. The routine then searches for the intensity maximum within a fixed window width. The window width can be automated to define an area where the intensity is above a threshold value, where the threshold is a

Figure 6: The edge detection method.

---

[3] The wind set-up and atmospheric pressure variation is only necessary when using predicted rather than measured tidal elevation.

fixed percentage (say 90%) of the maximum intensity value in the previous column. Alternatively, as illustrated in Figure 6 a fixed window width may be used. Experimentation with both methods yielded equivalent shorelines. However, if an inappropriate threshold value or window width is used it is possible for the detection routine to pick up on and trace other areas of rapidly changing intensity such as regions where bar systems occur in close proximity to the shoreline. Typical window widths used here were of the order of $\pm 20m$ ($\pm 10pixels$). The shoreline position is computed via a centre of mass calculation for that column, then the next column is scanned with the centroid of the window in the row corresponding to the shoreline position in the previous column. The edge detection stops when there is deemed to be no significant intensity maximum within the scan-window.

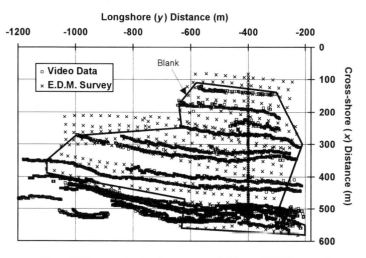

Figure 7: Diagram showing the coverage of video and EDM survey data.

More sophisticated edge detection procedures (Lim, 1990) were experimented with but this simple method was found to provide the best means of tracing a continuous shoreline. Complex 2-dimensional edge detection methods were found to be too sensitive to threshold conditions. However, unlike some of the more sophisticated edge detection methods this simple method is not capable of reversing direction and therefore may not work well for very complex topography. Additionally, the detection method used here will not perform optimally if the shoreline is at a large oblique angle to the defined y-axis. A truly 2-dimensional search routine for a continuous edge will be an area of future development.

Field Measurements
Data is presented here for spring tides (giving the broadest intertidal area) on the 7 May 1997. A total station (Electronic Distance Meter, *EDM*) survey of the intertidal

beach was conducted on a 25m grid covering an area 1000m by 500m in the
longshore and cross-shore direction respectively. A detailed cross-shore transect with
a resolution of 5m was also made at y=-400m to facilitate accurate 'ground truthing'
of the video survey. Additional virtual 'ground-control' points were also surveyed to
aid the accurate rectification of the images. The technique described above was then
applied to give a series of shorelines. The results of both the total station survey and
the video survey are shown in Figure 7. The solid polygon in Figure 7 represents a
blanking area where there is good coverage of both video and EDM survey data.
This blank is used later to compare contour maps resulting from both surveys.

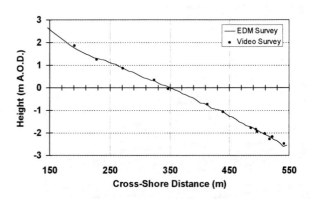

A regression was analysis conducted between the survey and video data (uncorrected for the vertical offset) collected along the detailed cross-shore profile line at y=-400m. Encouragingly, there is excellent agreement between the sets of survey data with a correlation coefficient (R^2)

Figure 8: A comparison of a cross-shore profile measured using an E.D.M. and corrected video data at y=-400m.

and standard error of 0.998 and 0.07m respectively. The line of 'best fit' between the
total station (*EDM*) and video survey is given by; *Video*=0.9914*EDM*-0.5417. The
gradient of the best fit line is approximately one suggesting a constant vertical
correction is appropriate across the profile. The intercept value from the regression
analysis of 54cm is the magnitude of vertical offset required to correct the *z*-
coordinate of the video survey for wave and wind set-up, swash and atmospheric
pressure variations.

The offset value (of 54cm) from the regression analysis was applied to the whole of
the video survey. A comparison of the *EDM* and corrected video survey data along
the profile line at y=-400m is shown in Figure 8. Contour maps of both the corrected
video and the *EDM* survey data were plotted (Figure 9) for the blanked area shown
in Figure 7.

Here a Krigging interpolation method has been used to interpolate both the video and
survey data onto equivalent regular 25m square grids. Inspection of Figure 9 shows
that qualitatively there is an excellent agreement between the video and total station
surveys. A more quantitative analysis of the agreement of the two surveys is shown in

Figure 10. Here a regression analysis between the survey and video data has been carried out for each of the data points in the regular grid in Figure 9. The correlation coefficient (R^2) of 0.993 and standard error of 0.11m indicates that there is an excellent agreement between the two surveys. Given the potential aliasing problems in the *EDM* survey associated with the small-scale bed forms this error is surprisingly good. Notice also that the gradient of the graph is 1.0 and the intercept is near zero. This indicates that the ground truth measurements made at y=-400m are appropriate across the whole study area. A close inspection of Figure 10 shows that in some areas of the beach (particularly the lower beach) there is greater scatter of the data around the best-fit line than in others. These areas may be largely explained due to interpolation errors corresponding to regions where data from either the video or *EDM* survey is sparse.

Conclusions

Time-exposure video images recorded over a one year period at a strongly macrotidal beach show a persistent, well developed low-tide bar system punctuated by rip-currents located approximately 50m seaward of the mean low-water springs mark. The broad (up to 600m) intertidal beach area by comparison remains relatively flat due to the lateral smearing of bed forms by the rapid advection of the tide across the intertidal profile.

A method of remotely sensing the intertidal beach topography using video images has been outlined here. The technique provides inexpensive, long term (weeks, months, years), large scale (kms) surveys with a standard error of approximately 11cm. The method currently relies on a small amount of survey data to ground truth the video measurements, correcting for the vertical offset due to wave set-up, swash and atmospheric pressure variations. Work is in progress to develop a semi-empirical method for evaluating the vertical offset correction, thus eliminating the need for any ground truth measurements.

References

Davidson, M.A., Kilborn, D., Huntley, D.A., and Holman, R.A., 1997. Field observations of large scale (km) changes in the intertidal beach morphology on a high energy macrotidal beach. Paper in preparation for Marine Geology.

Holland, K.T., Holman, R.A., Lippmann, T.C., Stanley, J. and Plant, N., 1997. Practical use of video imagery in nearshore oceanographic field studies. Submitted: IEEE J. Oceanic Eng. (special issue on image processing).

Holman, R.A. (1994). The Argus Program. Bulletin of the Coastal Imaging Lab., College of Oceanography, Oregon State University, Corvallis, OR 97331.

Janssen, P.C., 1997. Intertidal beach level estimation using video images. M.Sc. Thesis, Delft Univ. of Tech., Faculty of Eng., Hydraulic and Geotech. Eng. Div.,

Lim, J.S., 1990. Two-dimensional signal and image processing. Englewood Cliffs, NJ: Prentice Hall. p478-480.

Plant, N.G. and Holman, R.A., 1996. Intertidal beach profile estimation using video images. Accepted with revision: Marine Geology.

Tucker, M.J., 1950. Surf beats: sea waves of 1 to 5 min. period, 1952. Proc. Royal Soc. London A, 202, p565-573.

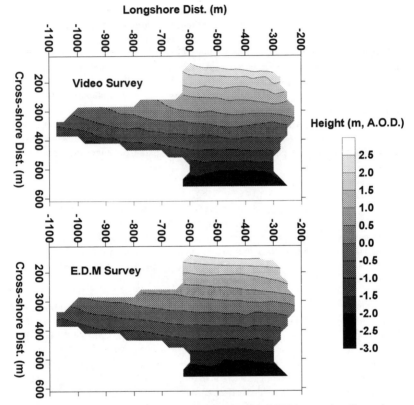

Figure 9: Contour maps produced using video data (top) and *EDM* survey data (bottom).

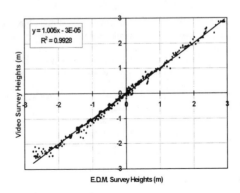

Figure 10: Regression analysis between video and EDM survey data.

Experimental Study Of Mixing Processes Using Images

Andrés Rodriguez [1], Eduardo Bahia [1], Margarita Díez [1],
Agustín Sánchez-Arcilla [1], José Manuel Redondo [2], and Marc Mestres [1]

Abstract

A new technique and its application to the processing of video images for the study of tracer dispersion in the nearshore region is presented. The images are transformed using a combination of two computer programs, and relevant parameters are then quantified using numerical and statistical methods. The procedure has been applied to images recorded during field experiments at different locations at the Ebro Delta, and the longitudinal and tranversal (relative to the shoreline) dispersion coefficients, and their evolution, have been obtained.

Introduction

This paper presents the methodology developed for the processing of digitised video images, together with some experimental results of the horizontal mixing of tracers in coastal waters, based on measurements obtained during the December'93, November'96 and April'97 field campaigns at the Ebro delta (Spain). The general methodology and some preliminary analyses have been presented in Redondo et al. (1994) and Rodriguez et al. (1995a); however, the complex physical mechanisms involved in the dispersion process continue to be analysed in detail using numerical and experimental methods. The digitised video images technique has been applied to hydrodynamic studies and, in particular, to the estimation of anisotropic horizontal mixing coefficients and the influence of wave breaking on dispersion processes.

[1] LIM, Polytechnical University of Catalonia, c/ Gran Capità s/n, Barcelona, Spain.
[2] Lab. Dinàmica de Fluids, DFA, Polytechnical University of Catalonia, c/ Gran Capità s/n, Barcelona, Spain.
e-mail: rodrigueza@etseccpb.upc.es

Moreover, a comparison of the calculated results of the horizontal diffusion and the values obtained from analysing macroturbulence series measured with six electromagnetic currentmeters (Rodriguez et al., 1995a), has been done.

Many researchers have based their studies of different hydro-morphodynamic characteristics on the analysis of video recordings. Lippman and Holman (1991), for instance, identified spatial and time variations of the bathymetry by averaging images and calibrating the intensity associated to the breaking-wave foam with the corresponding water depth. Other authors, such as Walton (1993), also apply the processing and the spectral analysis of intensity time series to estimate the characteristics of the run-up and the swash zone. On the other hand, field experiments to study surf zone (SZ) mixing with the aid of infrared cameras were carried out by Horikawa et al. (1978).

The developed technique permits to analyse series of images of tracer clouds to study their internal structure and quantify the local dispersion coefficients. A specific processing of the images, presented in this paper, is required to calculate these coefficients. Other hydrodynamic parameters, such as the advection, the broken-waves fraction, the width of the surf zone or the variation of the wave energy spectrum can also be estimated using video images.

Data acquisition campaigns

The hydrodynamic data were acquired during three field campaigns in the coastal zone of the Ebro delta: Delta'93, Delta'96 and Delta'97. In these experiments different parameters - including bathymetry, wave field, currents and tracer dispersion- were measured, allowing a complete study, analysis and numerical simulation of the relevant hydrodynamic processes (see Rodriguez et al., 1994).

During the Delta'93 experiments, hydro-morphodynamic measurements were taken at the Trabucador Bar, on the southern hemidelta. The location is of particular interest due to its multi-barred profile and uniform longitudinal conditions. This campaign was one of the first in Spain in which video images and hydrodynamic data were simultaneously taken, and it was used to develop and test the methodology required for the analysis of tracer dispersion. The video images were recorded using a remote camera atop a 20m high crane. Five sea states (12 tests) ranging from weak to medium energy wave fields coexisting with strong currents (≈ 1 m/s) were measured, and two particular cases have been selected and analysed: case II (Dec. 15th), corresponding to calm conditions without wave breaking, and case IV (Dec. 16th), with spilling-plunging wave breaking.

Delta'96 was undertaken by the LIM in collaboration with the Shirshov Institute, from the Russian Academy of Sciences, and the CEPYC-CEDEX (Sánchez-Arcilla et al., 1997), in the same working area as Delta'93. The environmental conditions measured in 45 tests included strong mean winds (10-15 m/s) and weak to medium wave fields ($H_s < 0.75$ m). The morphology of the beach and the bathymetry, together with wind, turbidity, currents and wave-heights across the SZ were measured, and mobile and fixed sediment traps were set to quantify the sediment transport. In addition, tracer clouds released in the SZ were recorded using cameras atop a 40m high crane and on a balloon suspended over the area.

The last campaign, Delta'97, involved the LIM, the UP at Valencia, Barcelona University and the French oceanographic ship Thetys. During this campaign the estuary and the region near the Ebro river mouth were studied, analysing the behaviour of the salt wedge in the river and of the river plume in the sea. Hydrodynamic and bio-chemical measurements were taken simultaneously to the video recording of tracer clouds in the open sea. In this case, the video images were recorded from atop a ship's mast, about 10 meters above the sea surface.

Figure 1 schematises the procedure followed for the acquisition (left) and processing (right) of video images, whereas an example of a digitised video image, showing a pollutant patch, is given in figure 2.

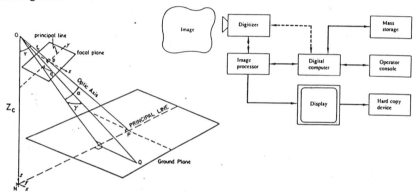

Figure 1: Schemes for the acquisition (left) and digitalisation (right) of video images.

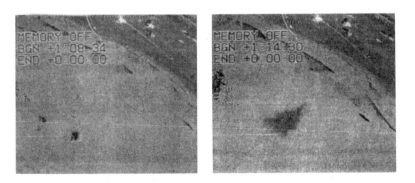

Figure 2: Digitised video images of a tracer cloud, showing its spreading with time.

The equipment used to record and process the video images included two remote SVHS video cameras with a fisheye lens, two Panasonic 7350 SVHS video recorders, two television screens, a DT2861 digitiser card, a controller interface and a PC. Images with a spatial resolution of 512x512 pixels, and a time resolution of 25 Hz, with an intensity variation range from 0 to 255, could be obtained with this configuration.

Image processing

The analysis of video images has been applied to the study of several hydrodynamic parameters, such as a) the SZ geometry, b) the broken wave fraction, c) Lagrangian tracer (buoys) advection, d) PIV -particle image velocimetry-, e) boundary waves and swash zone, and f) horizontal dispersion coefficients. The present work focuses on point f) above; several of the remaining applications have been described in Redondo et al. (1994) and Rodriguez (1997), whereas a more detailed description of the general use of video images can be found in Bahia (1997) and Díez (1997).

A large variety of software has been developed for image processing; *DigImage* (CERC, UK), in particular, is a system designed for the investigation of flow dynamics, and is especially useful to study turbulent dispersion, jets and plumes, or stratified flows. Even though this program allows a great number of applications, it lacks the ability to process local information at a scale of a few pixels. To overcome this limitation a program called *TICE* was designed and developed as a tool to automate the processing of data files corresponding to digitised images, allowing an interactive filtering of the images and simplifying the processing operation by dividing them into smaller windows.

Once the images have been obtained , the most interesting tests are selected according to the parameters to be studied (e.g., dispersion, wave-field, etc.). The corresponding images and the windows containing the tracer cloud are then chosen. An example of the combined processing of an image window using *DigImage* and *TICE* is shown in figure 3.

The real magnitude of the distances between pixels on the image is distorted due to both the camera's angle and lens; the former distorts mainly the transversal magnitude, while the lens distorts the image contour. Therefore, to work with real coordinates (X,Y), a function which maps the real coordinates of a point onto the the corresponding image pixels must be found. On the basis of the transformation-induced error (smaller than one and 0.5 meters in the cross- and longshore directions, respectively) a bi-quadratic type expression (eq. 1) has proved adequate. The coefficients A,..., H are calculated by a mean-squares algorithm using fiducial points topographically measured on the field, during the experiments.

$$X_{wc} = A + BX + CY + DXY + EX^2 + FXY^2 + GX^2Y + HY^2 \qquad (1)$$

The described method allows to digitise instantaneous or time-averaged images, as 512x512-pixel intensity arrays, or smaller windows in the particular case of the SZ where the effects due to the foam must be eliminated. These arrays are filtered and smoothed out using the program *TICE* to wipe out any existing interferences, and then the required numerical and statistical analyses are performed on the filtered and corrected images.

Estimation of the dispersion coefficients

Using Einstein's analogy of turbulent diffusion with Brownian motion, the dispersion coefficient K in a given direction can be estimated as a function of the time derivative of the intensity variance of a characteristic patch length scale in that direction:

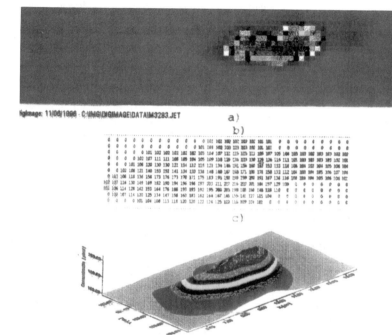

Figure 3: Patch analysing procedure. (a) digitised selected window, (b) filtered intensity array, and (c) three-dimensional plot of the patch.

$$K = \frac{1}{2}\frac{dV}{dt} \qquad (2)$$

With the video images, their digitalisation and the necessary corrections for the above mentioned distortions, the real time evolution of the tracer cloud can be tracked and the corresponding dispersion quantified. To do this, however, the following numerical-statistical procedure must be observed:

i) The relevant windows on the recorded images are selected. A linear relationship is established between the intensity of each pixel and the tracer concentration at that point. A series of images, taken from the recorded film at small time steps, are used to observe the tracer decay as the cloud is dispersed, and the window including the tracer is extracted. In what follows, variables marked with a sub-index "m" refer to image windows.

ii) The distortion of the tracer cloud, mainly in the transverse direction, due to the motion of the free surface is corrected. For the lateral coefficient, K_{trans} , a "ficticious plume" is generated by time-averaging the images (over 30 sec) in the cases which include wave breaking and a strong longitudinal current ($\approx$ 1 m/s in tests 8 and 9). The longitudinal (or time) variation of the plume width allows to estimate the time variation of the respective variance and, thus, to estimate K_{trans}. For K_{long} the variation in the longitudinal direction is obtained by averaging the tracer clouds over a wave period and filtering the intensity decay using a non-linear filter.

iii) The interference due to the background nonuniformity is corrected. The intensity of each spot is distorted by the intensity, brightness, and roughness of the background, which vary in time and space. To correct this effect, for each window containing the tracer another window about the same size but without tracer is selected from the same image. In the case of a plume, the background must be averaged over the same time interval as the tracer image. The variables referring to the background have a sub-index "b".

Once the data files corresponding to the image digitalisation and filtering have been obtained using Dig*Image*, they are processed by *TICE*.

iv) Tracer cloud filtering: The background is erased by means of an automatic filtering process which eliminates points whose intensity is smaller than a predefined cut value, estimated as the value of the intensity which better defines the cloud's boundary. If the waves break over the patch, the resulting foam may temporally hide all or part of the cloud, and this interference must be filtered manually, pixel by pixel, on each image. It is therefore convenient to select those windows for which this effect is minimum.

After performing the aforementioned corrections, the tracer dispersion is quantified by processing the intensity arrays.

v) Computation of the weighted variances: The centre (ξ_x ,ξ_y) of the tracer patch in the filtered window is calculated as the first-order moment of the intensity field. The variances V_x , V_y ,and V_{xy} are calculated as the second-order moment of the intensity field, weighted by the intensity.

When wave breaking existed, the cloud was dragged from right to left on the image due to a longitudinal current. The plume centerline was then defined as the locus of the points of maximum intensity within each array column. If the plume axis can be related to the current speed and the time elapsed since the tracer release, this intensity I_p^l , where the superindex denotes position l along the axis, gives an estimation of the concentration decay. All variables referred to the plume are denoted by a sub-index "p".

The time t_p^k associated to a position k along the plume axis is

$$t_p^k = t_m^l + F_{IT}\left(t_m^{l+1} - t_m^l\right) \qquad F_{IT} = \frac{X_m^l - X_p^k}{X_m^l - X_m^{k+1}} \qquad (3)$$

where t_m^l is the time at which the image number l was taken, and F_{IT} is an interpolation factor. The intensities at position k are calculated from the smoothed-out instantaneous images (I_{suavp}) and from the background (I_{bp}) as:

$$I_{suavp}{}^k = I_{suavm}{}^l + F_{IT}\left(I_{suavm}{}^{l+1} - I_{suavm}{}^l\right)$$

$$I_{bp}{}^k = I_{bm}{}^l + F_{IT}\left(I_{bm}{}^{l+1} - I_{bm}{}^l\right) \tag{4}$$

with I_{suavm} and I_{bm} the smoothed-out intensity of the image and the background intensity, respectively.

The corrected value of the plume intensity $I_{plumacor}$ must be multiplied by two smoothing factors, one associated to the intensity maxima and one to the background:

$$I_{plumacor}(i,j) = I_{pluma}(i,j)\frac{I_{suavp}{}^k}{I_p{}^k}\frac{I_{bmed}}{I_{bp}{}^k} \qquad I_{bmed} = \frac{1}{M}\sum_{k=1}^{M}I_{bm}{}^k \tag{5}$$

where I_{bmed} is the mean background intensity, and M is the number of images which have been averaged to obtain the plume.

The lateral dispersion coefficient is calculated using equation (2), where V is now the variance in the transverse direction V_y and the time derivative is represented by a first-order finite difference approach

$$K_{trans} \approx \frac{1}{2}\frac{\Delta V_y}{\Delta t} = \frac{1}{2}\frac{V_y{}^{k+1} - V_y{}^k}{t^{k+1} - t^k} \tag{6}$$

The time increments are found dividing the coordinate X of the plume by the horizontal component of the advective velocity $VELX$:

$$t^{k+1} - t^k = \frac{X^{k+1} - X^k}{VELX^k} \qquad VELX^k = \frac{X_m{}^{l+1} - X_m{}^l}{t_m{}^{l+1} - t_m{}^k} \tag{7}$$

For the longitudinal coefficient, the evolution of longitudinal variances is found as was previously mentioned. In this case, the estimation is easier since the tracer clouds are averaged over a wave period ($T\approx5$-6 sec).

Results analysis and discussion

The dispersion coefficients estimated from the analysed cases (tests 8 and 9) show that, in the presence of breaking waves, the time behaviour of the longitudinal and the transversal spreading differs considerably (figure 4). On the one hand, the transverse dispersion coefficient is nearly constant with time at the beginning of the mixing process, but increases shortly after the release. The estimated values for this coefficient lie within the range $0.035 <K_{trans}(m^2/s)<0.376$. On the other hand, the dispersion in the longitudinal direction is almost time-independent throughout the whole measurement period, with values ranging from 0.115 m^2/s to 0.125 m^2/s.

As can be seen in figure 4, the overall horizontal dispersion is approximately constant during the initial mixing phase. This could be so because the tracer, released at the surface, mixed initially in the vertical direction, and started spreading only after it had occupied most of the water column. The similarity between the Eulerian EMC and the Lagrangian video measurements of the longitudinal currents, confirming the existence of an intense vertical mixing in the SZ, further support this hypothesis. It should also be stressed that the released clouds were small or medium (with a mean diameter of about 3 m) with respect to the

402   COASTAL DYNAMICS '97

dominant physical length scales: the water depth and the wave height were of the order of 1 m, whilst the wavelength was of the order of 20 m, and the width of the SZ of about 50 m.

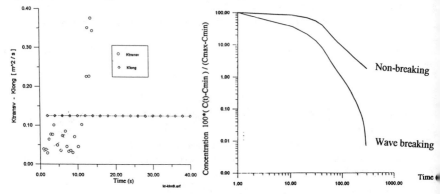

Figure 4: Time evolution of the dispersion coefficients, in a case with breaking waves.
Figure 5: Comparison of the mixing process with and without wave breaking.

Figure 5 plots the relative concentration decay for the cases of breaking and non-breaking waves. It can be seen that the estimated mixing coefficients, and their anisotropy, are larger in the former than in the latter.
For the series of experiments performed in similar wind and wave conditions but at different distances from the shoreline the transverse dispersion coefficient was found to be $0.005 < K_{trans} < 0.035$ m²/s. Similar values, with a similar data scatter, were found by Horikawa (1978) in his SZ experiments. In the SZ, the longitudinal mixing coefficients can vary up to one order of magnitude, for similar conditions; nevertheless, a dependence with the distance to the shoreline was observed, due to the longitudinal current and the wave conditions.
The time and spatial variations of the dispersion coefficients can be quantified similarly to an eddy viscosity coefficient, and can be used in the hydrodynamic modelling of coastal phenomena such as circulation, sediment transport and dispersal of pollutants. Figure 6 shows the comparison between the results obtained from a numerical simulation using the 2D dispersion model DISPER (Bahia, 1997) and those found from the digitalisation and processing of video images.

Conclusions

Quantitative information on dispersion coefficients is easily obtainable from video recordings through the described method of analysis; however, even though its theoretical basis and its formulation are fairly simple, its usage is very laborious and time-consuming. This method is able to quantify accurately the spatial and time variations of the intensity fields generated by natural (foam) and artificial (dyes and buoys) tracers, and can be used in a great number of hydro-morphodynamic applications. In the particular case of mixing processes in the SZ, the methodology described herein has led to the acquisition of data which are very sparse in the literature due to the difficulty in their measuring. This technique also allows to study the space and time variations of the turbulence effects in the SZ.

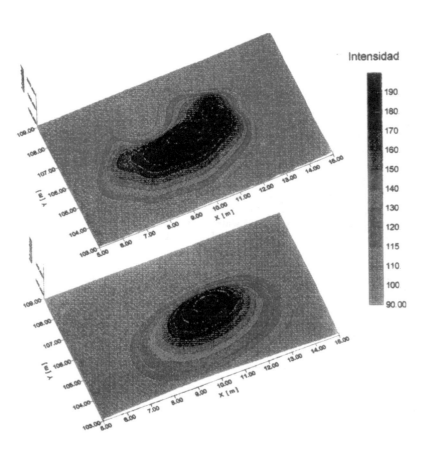

Figure 6: Measured (above) and modelled (below) dispersion of a dye spot.

Another aspect of practical importance is the selection of the tracer to be used. It should be carefully chosen taking into account the climatic conditions in order to guarantee a good contrast with the water. During the experiments potassium permanganate, rhodamine, fluoresceine and milk were compared, and in most of the tests a mixture of fluoresceine and milk was chosen because of its high persistence and contrast.

The time-averaging of images, and the time series of intensity in transects (horizontal or vertical lines on an image) have proved to be a convenient technique to characterise tracer dispersion in the presence of waves and currents, even in such complex regions as the SZ. As an example, the observed values of the mixing coefficients, and their trend, show that inside the SZ a "4/3 law"-type relationship cannot be applied.

Finally, it must be mentioned that the presentation of general results is awaiting the analyses of the numerous tests available, which continue to be processed (Díez, 1997).

<u>References</u>

- Bahia E. (1997), *Estudio numérico experimental de la dispersión de contaminantes en aguas costeras*, PhD Tesis, Polytechnical University of Catalunya, Barcelona.
- Díez M. (1997), *Estudio de hidodinámica en la zona costera mediante el análisis digital de imágenes*, Master Thesis, Polytechnical University of Catalunya, Barcelona.
- Horikawa K., Lin M. and Sasaki T. (1978), *Mixing of heated water discharged in the surf zone*, ICCE, ASCE, 2563-2583.
- Lippmann T and Holman R. (1991), *Phase speed and angle of breaking waves measured with video techniques*, Coastal sediments, ASCE, 542-556.
- Redondo J., Rodriguez A., Bahia E., Falqués A., Gracia V., Sánchez-Arcilla A. and Stive M. (1994), *Image analysis of surf-zone hydrodynamics*, Coastal Dynamics '94, ASCE, 350-365.
- Rodriguez A., Sánchez-Arcilla A., Collado F., Gracia V., Coussirat M. and Prieto J. (1994), *Waves and currents at the Ebro Delta surf zone: measured and modelled*, Proc. ICCE, ASCE, 2542-2556.
- Rodriguez A., Sánchez-Arcilla A., Redondo J., Bahia E. and Siera J. (1995a), *Measurements and modelling of pollutant dispersion in the nearshore region*, Water Science and Technology, IAWQ, **32** (9/10), 169-178.
- Rodriguez A., Sánchez-Arcilla A., Gómez J., Bahia E. (1995b), *Experimental study of surf zone macroturbulence and mixing using Delta'93 field data*, Coastal Dynamics '95, ASCE, 305-316.
- Rodriguez A. (1997), *Estudio experimental de la hidrodinámica en zona de rompientes*, PhD Thesis, Politechnical University of Catalunya, Barcelona.
- Sánchez-Arcilla A., Rodriguez A., Santás J., Redondo J., Gracia V., Kuznetsov S., K'osyan R. and Mösso C. (1997), *Ebro Delta nearshore hydromorphodynamics*, Coastal Dynamics '97, ASCE, (in press).
- Walton T. (1993), *Ocean City, Maryland, wave runup study*, J. Coastal Research, **9** (1), 1-10.

On the development of complete flow models for wave-current interactions

N. Madden[1], M. Stynes[1] & G.P. Thomas[2]

Abstract

The present paper considers the interaction between regular waves and a current within the linear wave regime, with particular attention directed towards the near-bed region. A model of the complete flow is developed and solved using an innovative numerical approach not previously applied to wave problems. Accurate solutions are obtained for models that describe both molecular viscosity alone and a combination of molecular viscosity and turbulence. These predictions are compared with available experimental data and found to give reasonable agreement.

Introduction

Consider a regular wave train of frequency ω progressing over water of constant mean depth h, in two dimensions. The z-axis is vertically upwards, with the origin in the mean water level, and the x-axis coincides with the direction of wave propagation. The configuration is shown schematically in Figure 1.

For all points below the wave trough, defined by $z = z_t$, the horizontal component of velocity u can be written as the superposition of a time-independent current term and a wave-like term

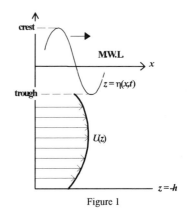

Figure 1

$$u(x,z,t) = U(z) + u_w(x,z,t) \qquad -h \leq z \leq z_t \quad (1)$$

[1] Dept. of Mathematics, University College Cork, Cork, Ireland
[2] Dept. of Applied Mathematics, University College Cork, Cork, Ireland

with the average value of u_w over a wave period being zero. At a fixed value of x the wave-like velocity component can also be written as a harmonic series

$$u_w(x,z,t) = \sum_{n=1}^{\infty} u_n(z) \cos n(kx - \omega t) \qquad (2)$$

in terms of a phase function $kx - \omega t$, valid for $-h \leq z \leq z_t$, with the frequency ω regarded as a positive real constant. The wavenumber k will be real for inviscid flows but complex for viscous flows, with the imaginary part of k being associated with dissipative effects. For comparative purposes it is the magnitude of the first harmonic velocity components, as indicated by $|u_1|$ and $|w_1|$, that are usually of general interest; linear models only predict these quantities and they can be readily extracted from the output of nonlinear models.

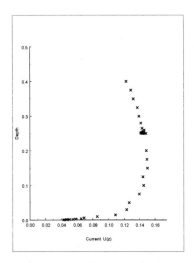

Figure 2. The current distribution $U(z)$ from Klopman (1994) appropriate for waves following a current and the wave data specified by (3).

For comparative purposes it is convenient to employ the experimental data set obtained by Klopman (1994) against which to test the model predictions. The current profile $U(z)$, plotted in figure 2, corresponds to the case when regular waves follow the current with data values

$$\begin{aligned} h &= 0.500 \text{ m,} \\ \omega &= 0.6944 \text{ Hz,} \qquad (3) \\ a &= 0.05987 \text{ m,} \end{aligned}$$

where a is the amplitude of the first harmonic of the surface elevation. This data set will be employed extensively to provide a reference for the models described later in the paper and the wave regime is weakly nonlinear.

Inviscid models, such as those of Thomas (1981, 1990), exist for arbitrary $U(z)$ for both linear or nonlinear wave regimes in two dimensions and these can be used to predict the wavelike velocity components and the wavelength. The prediction from Thomas (1981) for the first harmonic of the wavelike horizontal velocity component $|u_1(z)|$, as defined in (1) and (2), is shown in figure 3(a) for the data prescribed by figure 2 and equation (3). Note, that as the waves are weakly nonlinear, very little difference is expected between the predictions of the linear and nonlinear models. Also shown in figure 3(a) is the prediction from an irrotational constant current model, based upon the depth averaged current, and this is seen to provide surprisingly good estimates over the water depth. This confirms an assertion made by Ismail (1984) that reasonably good engineering predictions for waves following currents can be obtained by approximating the physical current by one that is uniform with depth. However, this assertion cannot apply to the complete flow field and will also depend upon how the current profile was generated.

Although figure 3(a) shows that the inviscid models provide good estimates over the vast majority of the flow domain, it is also clear that this is not true near to the bed. Figure 3(b) shows the bottom 5% of figure 3(a) on an enhanced scale so that the shape of the data and the failure of the inviscid models can be readily identified. From a comparison of figures 2 and 3, it can be seen that the inviscid model works reasonably well to within the layer defined by the bottom 2% of the water column; corresponding to a physical distance of 0.01m.

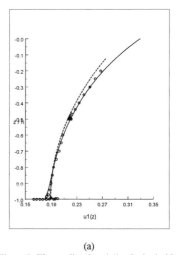
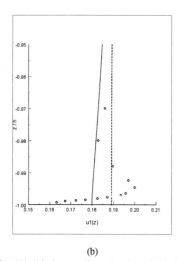

(a) (b)

Figure 3. The predicted variation by inviscid models of $|u_1(z)|$ with the non-dimensional depth z/h for the current profile in figure 2 and the specifications of equation (3), —— Thomas (1981); - - - constant depth-averaged current; o Klopman (1994). (a) complete profile, (b) bottom 5%.

The standard practice in modelling wave-current interactions is to divide the fluid domain into two distinct regions: the main body of the fluid, which is effectively inviscid, and the shear layers at the free surface and near the bed where viscosity and turbulence are considered important. The applicability of this approach has been confirmed by the figures 3(a) and 3(b) presented above. Not surprisingly the region near to the bed has received considerably more attention than the shear layer at the free surface. A detailed description of previous inviscid work is provided by Thomas & Klopman (1997). In addition it has already been established by previous authors that an inviscid model can provide a very good description of the flow kinematics over the vast majority of the flow domain even when the current is considered to be weakly turbulent, with current and wave data obtained by averaging over many wave periods. However such studies also identify the inherent weaknesses in inviscid models that are associated with the kinematics near to the bed and the inability to predict bed shear stresses.

The bottom boundary layer models usually assume that the motion in the oscillatory bottom boundary layer can be modelled by allowing the presence of a uniform current at the upper extremity of the boundary layer, so that the vertical shear in the current distribution is neglected. A considerable number of turbulence and viscosity models have been developed to predict the bed shear stresses but this is readily acknowledged to be a very difficult task for flow over roughened beds, even with the constant current approximation. An excellent state-of-the-art assessment has been provided by Soulsby *et al* (1993).

The purpose of the present study is to develop complete flow models capable of describing the kinematics of the whole flow field and the bed shear stresses simultaneously. In this initial work attention is restricted to the case of regular waves interacting with a current in two dimensions, over a smooth horizontal bed, so that the waves either follow or are adverse to the current and the current can be represented as a mean flow term $U(z)$. This approach assumes that $U(z)$ is deemed known beforehand and that knowledge of the generative processes that produce $U(z)$ is not required. In the absence of a better terminology, models of this type can be labelled *reactive* and most wave-current interaction models fall into this category. The second kind, designated *generative*, attempt to determine the function $U(z)$ which exists under certain specified conditions; an example of a generative model valid under specific approximations is given by Klopman (1992, 1997).

The present work assumes that the waves are sufficiently small so that linear wave theory remains valid. However, the definitive experimental studies, such as those of Kemp & Simons (1982, 1983) and Klopman (1994), are predominantly targeted at roughened beds and turbulent boundary layers as this is the case of greatest physical interest. The inclusion of a roughened bed is inappropriate in an initial study of the kind described here because it introduces a further order of complexity into an already difficult area. Kemp & Simons provide some excellent smooth bed data but their wave regime is more nonlinear than that represented by the Klopman data and the Kemp & Simons data is not yet available in harmonic form. For this reason the Klopman data is employed in this paper although the difficulty associated with the roughened bed is acknowledged; furthermore Klopman does not present bed-stresses and we will concentrate instead on the near-bed behaviour.

Formulation

In a linear wave approach we assume that the wave slope $a|k|$ is a small parameter and that the surface boundary conditions can be evaluated at the mean water level $z = 0$ by means of Taylor series expansions. It is convenient to represent the wavelike quantities u_1 and w_1 in terms of a real-valued streamfunction $\Psi(x,z,t)$ and then introduce a complex-valued function $\psi(z)$ following a harmonic decomposition, i.e.

$$u_1 = \frac{\partial \Psi}{\partial z}, \qquad w_1 = -\frac{\partial \Psi}{\partial x}, \qquad \Psi(x,z,t) = \mathrm{Re}\left\{\psi(z)\exp(i\theta)\right\}, \qquad (4)$$

where $\theta = kx - \omega t$ is the phase function.

Following the approach adopted by Klopman (1992) for weakly turbulent flow, the streamfunction can be shown to satisfy the equation

$$
\begin{aligned}
\frac{d^2\psi}{dz^2} &- \left(k^2 - \frac{k}{\omega - kU}\frac{d^2U}{dz^2}\right)\psi - \frac{i\nu}{\omega - kU}\left(\frac{d^2}{dz^2} - k^2\right)^2\psi \\
&= \frac{i}{(\omega - kU)}\left\{\left(\frac{d^2}{dz^2} + k^2\right)\left\{\nu_t\left(\frac{d^2\psi}{dz^2} + k^2\psi\right)\right\} - 4k^2\frac{d}{dz}\left(\nu_t\frac{d\psi}{dz}\right)\right\}
\end{aligned}
\qquad (5)
$$

where ν is the coefficient of molecular viscosity and $\nu_t(z)$ is a variable eddy viscosity coefficient associated with the standard Boussinesq turbulence model. The equation can be written in a number of ways and the form presented in (5) is adopted so that only turbulence terms appear on the right-hand-side. In the absence of turbulence, corresponding to $\nu_t = 0$, the equation reduces to the Orr-Sommerfeld equation of hydrodynamic instability.

At the free surface the kinematic condition corresponds to
$$\psi = a(c - U) \qquad \qquad \text{on } z = 0 \qquad (6)$$
where c is the local complex-valued phase velocity.

The dynamic free surface conditions are more complicated as they correspond to zero tangential stress and continuity of normal stress. Once the basic equations have been obtained they can be written in a number of different representations as (6) can be used for simplification; a suitable set is

$$\left(\nu + \nu_t\right)\frac{d^2\psi}{dz^2} = -a\left(\left(\nu + \nu_t\right)\frac{d^2U}{dz^2} + \frac{d\nu_t}{dz}\frac{dU}{dz} + k\left(\nu + \nu_t\right)(\omega - kU)\right)$$

$$i\left(\nu + \nu_t\right)\frac{d^3\psi}{dz^3} + i\frac{d\nu_t}{dz}\frac{d^2\psi}{dz^2} - \left\{\omega - kU + 3ik^2\left(\nu + \nu_t\right)\right\}\frac{d\psi}{dz} \qquad \text{on } z = 0 \quad (7)$$

$$= a\left((\omega - kU)\left\{\frac{dU}{dz} - ik\frac{d\nu_t}{dz}\right\} - gk\right).$$

The bottom boundary condition is a no-slip condition and applying this to the horizontal and vertical velocity components, via (4), gives

$$\psi(z) = \frac{d\psi}{dz} = 0 \qquad\qquad \text{on } z = -h \qquad (8)$$

The problem is defined so that we need to obtain solutions to (5), subject to the boundary conditions (6) - (8). The wave amplitude a, the frequency ω, current $U(z)$ and molecular viscosity v are all considered known either by specification or measurement. A turbulence model has not been prescribed and as this is an initial attempt at generic modelling then it is permissible to employ the function $v_t(z)$ derived by Klopman (1997), for the current-alone case, to provide an initial approximation for $v_t(z)$. Thus the unknowns are the streamfunction ψ and the wavenumber k. The model may be considered as a very non-trivial extension of the inviscid model developed by Thomas (1981), to include weak turbulence and viscosity.

Numerical Considerations

The principal difficulty with the numerical solution of equation (5) is the presence of a small parameter, in this case the viscosity, which is typically associated with the highest derivative in the equation. The solution of such an equation, with layers at the boundaries and also possibly internally, is difficult to obtain accurately and it is quite possible for numerical schemes to have appeared to have converged yet to produce solutions that are extremely inaccurate, with relative errors of two or more orders of magnitude within the layers. Recent work by numerical analysts, such as that described by Roos, Stynes & Tobiska (1996), has addressed these problems and provided usable grids that avoid these potentially catastrophic errors. This builds upon the pioneering work of Shishkin, who first proposed the mesh structure for problems of this type; an introduction to these methods is given in the recent book by Miller, O'Riordan & Shishkin (1996). These methods are not yet available within the general numerical solution domain.

The present study is innovative in two respects. It firstly represents the first application of these new numerical schemes to water wave problems and secondly, the nontrivial extension of methods originally developed for second order equations to fourth order.

A brief description of the new approach is now provided and more details can be obtained from Roos *et al* (1996) and Miller *et al* (1996). In terms of the non-dimensional water depth z/h, the solution domain becomes transformed to the interval [-1, 0]. A Shishkin mesh is now constructed by dividing the interval [-1,0] into the three subintervals [-1, -1+σ], [-1+σ, -σ] and [-σ, 0], corresponding to the shear layer at the bed, the interior domain and the surface shear layer respectively. Each sub-interval is now divided uniformly and the grid in each chosen so that, if a total of N intervals are employed, then N/4 occur in each of the end layers and N/2 in the interior layer. The value of σ is important as it defines the width of the shear layers and hence the width of the interior domain; it must clearly satisfy the inequality $0 < \sigma < 0.5$ for three distinct regions. However, σ is not arbitrary and is defined by

$$\sigma = \min\left\{\frac{1}{4}, \frac{3\varepsilon}{\alpha} \log N\right\}$$

where ε is the coefficient of the fourth order term in the equation and α is the minimum value of the absolute value of the coefficient of the second-order term. In practice we expect $\sigma \ll 1$ and the shear layers to be narrow at the surface and the bed. The aim is therefore to use a uniform mesh in each region, suitable for finite difference or finite element approaches, and match the solutions at the layer interfaces.

The solution strategy is similar to that of Thomas (1981). For given k the equation (5) is solved subject to (8) and two of the surface conditions in (6) and (7). The process is then repeated for differing k as part of an iterative process until the final surface condition has been satisfied and the correct value of k obtained; clearly the key element in this approach is to be able to solve (5) accurately. A finite element method is used to solve (5) for given k, implementing a procedure due to Sun & Stynes (1995) in which cubic Hermite basis functions are applied on a fitted mesh of Shishkin type. A Fortran program with a specification of quadruple precision was employed and this was required to avoid floating point errors that can occur when a high-order problem is solved on a very fine mesh.

On a typical mesh, with 1024 intervals, the width of the fine and course meshes was 2.015d-5 and 9.654d-4 respectively. The coefficient of the 4th-order term, being of $O(10d-6)$ for molecular viscosity alone, is still small relative to the mesh diameter and a uniformly convergent numerical method is desired. For this to occur the error in a given norm must be bounded independently of ε. For sufficiently small N, independent of ε, we have the uniform convergence estimate provided by Roos *et al*

$$\| \psi - \psi_N \| \leq C (N^{-1} \log N)^2,$$

where ψ_N is the solution for a mesh with N internal points and C is a real positive constant. This result is supported by the numerical experiments.

N	Fine Mesh	Course Mesh	Computed k
64	1.934d-4	1.543d-2	2.1499605826805 + 4.4390804818124d-4 i
128	1.128d-4	7.700d-3	2.1499648941008 + 4.4398338288498d-4 i
256	6.448d-5	3.842d-3	2.1499651753468 + 4.4399569269635d-4 i
512	3.267d-5	1.917d-3	2.1499651915667 + 4.4399773147026d-4 i
1024	2.015d-5	9.564d-4	2.1499651924098 + 4.4399781658032d-4 i
2048	1.108d-5	4.772d-4	2.1499651924579 + 4.4399781337206d-4 i
4096	6.045d-6	2.381d-4	2.1499651924607 + 4.4399781291831d-4 i
8192	3.275d-6	1.188d-4	2.1499651924609 + 4.4399781288216d-4 i

Table 1. The computed values of k against the mesh size N for a generic wave-current problem

The method was tested extensively, initially against the inviscid linear model of Thomas (1981) and subsequently by comparison with analytic solutions of the Orr-Sommerfeld equation, which can be obtained from (5) when $v_t \equiv 0$ and $U(z) =$ constant.   Excellent agreement was obtained in all cases.   Table 1 shows the convergence rates for the wavenumber k, with grid intervals N, for a current profile similar to that in figure 2 with molecular viscosity alone and $v = 1.138d\text{-}6$. Based upon this and similar comparisons, the value $N = 1024$ was used for most calculations, with the computed value of the wavenumber considered correct to ten significant figures.

Results
The numerical results depend upon the choices made for the molecular viscosity v and the turbulent eddy viscosity coefficient $v_t(z)$. The Klopman (1994) experiments quote temperature ranges of 11.5°C - 14°C and values of the molecular viscosity appropriate to this range were employed.

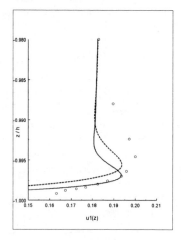
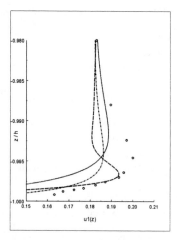

Figure 4. The profile of $|u_1(z)|$ against z/h in the absence of turbulence, —— $v = 1.0d\text{-}6$; – – – $v = 2.0d\text{-}6$; o Klopman (1994).

Figure 5. The profile of $|u_1(z)|$ against z/h with turbulence, – – – $v = 1.2d\text{-}6$, $v_t = 0$; —— $v =$ 1.2d-6, v_t = Klopman (1997); –·–·– $v = 0$, v_t = Klopman (1997); o Klopman (1994).

Figure 4 shows the variation of $|u_1(z)|$ with the non-dimensional depth z/h in the bottom 2% of the flow field in the absence of turbulent viscosity. Note that the agreement over the upper 98% of the flow domain is essentially the same as that shown in figure 3(a). Predictions are presented for v values of 1.0d-6 and 2.0d-6, corresponding to solid and broken lines respectively. It is clear from these two curves that the viscosity-alone model correctly predicts the general shape of the curve near to the bed but that layer is not of the correct width, which is only to be expected in the absence of turbulence. Another important feature is that the smaller value of two values of v gives not only a narrower layer but also one which occurs nearer to the bed. It should be noted that the variation in the value of v is usually associated with temperature changes.

Three curves of $|u_1(z)|$ against z/h are shown in figure 5 to illustrate the influence of turbulence. The benchmark curve shown to represent viscosity alone appears as a broken line and corresponds to $v = 1.2d-6$, being a good approximation to the experimental conditions. The second and third curves indicate turbulence alone (dot-dash line) and turbulence plus viscosity (solid line) respectively. No one curve can be considered universally good. While the viscosity-alone curve has considerable merit near to the bed, it does not work very well away from the bed. The turbulence curves do not approximate the near-bed behaviour very well, even when viscosity is included, but do appear to predict the width of the turbulent layer quite well while underestimating its strength. However, it is readily acknowledged at this stage that the eddy-viscosity coefficient $v_t(z)$ requires refinement although this is a major task.

The wavenumber predictions for figures 4 and 5 are shown in table 2, which also includes the inviscid prediction corresponding to figure 3(a) for comparison; the wavelengths can easily be determined from this if required. There are two striking features illustrated by the values shown. Firstly, the real part of the wavenumber predictions do not vary very much for the values shown. Secondly, the imaginary parts show considerable variation, with those values associated with turbulence being an order of magnitude greater than those with viscosity alone.

Figure	v	v_t	Real k	Imaginary k
3	0	0	2.1526	0
4	1.0d-6	0	2.1530	4.2394d-4
	2.0d-6	0	2.1532	6.0471d-4
5	1.2d-6	0	2.1530	4.6534d-4
	1.2d-6	Klopman (1997)	2.1531	2.4437d-3
	0	Klopman (1997)	2.1528	2.0797d-3

Table 2. The wavenumber values corresponding to figures 3 - 5.

Conclusions

There are two primary conclusions to be drawn. Firstly, the numerical method has been shown to provide highly accurate solutions throughout the flow. Secondly, it is clear that the viscous sublayer of the near-bed region can be reasonably well modelled provided that the current $U(z)$ is accurately described as input data. However, the modelling of the turbulent characteristics is not yet of acceptable standard and better models of the eddy-viscosity coefficient $v_t(z)$ are required. Steps to improve the existing models are being undertaken although the possibility must be acknowledged that the eddy viscosity formulation may not be the most appropriate one for the near-bed behaviour associated with wave problems.

Acknowledgements

This work was conducted as part of the project *The Kinematics and Dynamics of Wave-Current Interactions* and financial support was provided by the Commission of the European Communities, under contract MAS3-CT95-0011 of the MAST Programme. The authors also wish to express their appreciation to Gert Klopman, formerly of Delft Hydraulics and now of Albatross Flow Research, for much helpful advice.

References

Ismail N.M. 1984 *Wave-current models for design of marine structures.* J. Waterway, Port, Coastal & Ocean Eng., ASCE, 110, 432-447

Kemp P.H. & Simons R.R. 1982 *The interaction between waves and a turbulent current: waves propagating with the current.* J. Fluid Mech. 116, 227-250.

Kemp P.H. & Simons R.R. 1983 *The interaction between waves and a turbulent current: waves propagating against the current.* J. Fluid Mech. 130, 73-89.

Klopman G. 1992 *Vertical structure of the flow due to waves and currents.* Progress Report H840.30, Part 1, Delft Hydraulics.

Klopman G. 1994 *Vertical structure of the flow due to waves and currents.* Progress Report H840.30, Part 2, Delft Hydraulics.

Klopman G. 1997 *Oscillatory flow kinematics for waves on a current.* Paper presented in Coastal Dynamics 97.

Miller J.J.H., O'Riordan E. & Shishkin G.I. 1996 *Fitted numerical methods for singular perturbation problems.* World Scientific Press.

Roos H.-G., Stynes M. & Tobiska L. 1996 *Numerical methods for singularly perturbed differential equations.* Springer-Verlag, Berlin.

Soulsby R.L., Hamm L., Klopman G., Myrhaug D., Simons R.R. & Thomas G.P. 1993 *Wave-current interaction within and outside the bottom boundary layer.* Coastal Eng. 21, 41-69.

Sun G. & Stynes M. 1995 *Finite-element methods for singularly perturbed high-order elliptic two-point boundary value problems. I: reaction--diffusion-type problems.* IMA J. Numer. Anal. 15, 117-139.

Thomas G.P. 1981 *Wave-current interactions: an experimental and numerical study. Part 1. Linear waves.* J. Fluid Mech. 110, 457-474.

Thomas G.P. 1990 *Wave-current interactions: an experimental and numerical study. Part 2. Nonlinear waves.* J. Fluid Mech. 216, 505-536.

Thomas G.P & Klopman G. 1997 *Wave-current interactions in the near-shore region.* In 'Gravity waves in water of finite depth', ed. by J.N. Hunt, Advances in Fluid Mechanics, Computational Mechanics Publications, 255-319.

MODELLING OF WAVE-CURRENT INTERACTION
AT A RIVER MOUTH

B. A. O'Connor[1], S. Pan[2] and N. J. MacDonald[3]

ABSTRACT
In order to provide a group of reliable data sets for the study of wave-current interaction a series of laboratory experiments involving waves, tidal flows and river discharge in the vicinity of a river mouth was carried out. Details of the experiments and the methodologies used in the data collection, analysis and validation, and the results of a test case with obliquely-incident monochromatic waves and a river discharge are presented in this paper. Simulations using numerical models were also performed and comparisons between the physical and numerical model results are made.

INTRODUCTION
Most engineering design studies for coastal works rely on computer models in which the full interaction between waves and current fields is ignored despite the importance of such process, see Yoo et al (1986). The importance of these interactions in morphodynamic modelling has been shown by De Vriend et al (1993) who compared the results of a number of numerical models for the case of a river discharging into a plane beach. Another related area is the modelling of waves interacting with tidal flows. If the wave-induced longshore current is sufficiently strong it can govern the direction of the currents in the nearshore zone. Although operating both waves and tidal current dynamically is possible (see Yoo et al, 1988), more research is warranted. However, at present little comprehensive experimental data exists to test and validate the existing models in the above situations.

In order to achieve a better understanding of the hydrodynamics of the nearshore zone, a collaborative research project, called LUCIO, was undertaken by a consortium of four UK universities, Liverpool, Imperial College, University College London and Oxford. The project involved several series of experiments in the large-scale Coastal Research Facility (UKCRF) basin at HR Wallingford (Simons et al,

[1]Professor, [2]Research Associate, [3]Lecturer, Department of Civil Engineering, The University of Liverpool, PO Box 147, Liverpool, L69 3BX, UK

1995). The work carried out by the University of Liverpool focused on the investigation of wave-current interaction effects around a river mouth with and without the presence of tidal currents.

This paper describes details of the experimental facilities and conditions. Results of one of the tests (RUN3), the case with obliquely-incident monochromatic waves without tides are presented and discussed. Comparisons between experimental and numerical results are also made.

FACILITIES

The experiments were conducted in the UKCRF at HR Wallingford in Jan/Feb 1996. The UKCRF consists of a 22m x36m basin equipped with a 72 paddle random wave generator which is capable of generating regular waves (MON) and long-crested random waves (LCR) up to 30° to shore-normal, as well as short-crested random seas (SCR). The UKCRF also has facilities to circulate a current in the longshore direction using four independently recirculating pumps. The distribution of flow from each sump is controlled by 10 undershot gates, see Fig 1. The slope of the beach is 1:20 and, for the present experiments, was covered with a single layer of 10 mm (nominal) green granite chipping to encourage rough turbulent conditions.

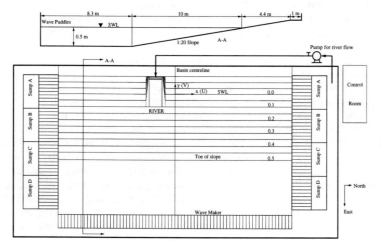

Fig 1 UKCRF and river layout

The river channel was constructed 3.0m south of the basin centreline, normal to the beach as shown in Fig 1. The channel was 2.0m wide at the base and 3.0m at the top with a variable slope in both side banks. A recirculating pump was used to provide river discharge with a maximum flowrate of 60 l/s by taking water from north end of the basin. A block of porous material was inserted at the shoreward end of the river outlet to ensure that the flow was uniformly distributed. The river discharge was monitored continuously using a mini-propeller current meter (MPCM) throughout the tests.

Velocity and wave measurements were focused on the area around the river mouth, where the strongest wave-current interactions were taking place, using a programmable three-dimensional positioner. Mounted on the positioner were a wave gauge and three Sontek acoustic doppler velocimeters (ADV). Each ADV has the capacity to measure three velocity components simultaneously. Velocity and free surface elevation measurements were taken at 32 principle measuring points in plan and the flow measurements at an additional 64 secondary points. Velocity measurements were taken at levels of 0.4D and 0.6D (D = water depth) from the bed where possible, although some local adjustments were necessary in shallow water. Additionally, wave measurements were taken at eleven fixed locations around the basin, see Pan et al (1997) for details.

Data acquisition was performed using PC-based DatsPlus V2 software at a sampling rate of 50 Hz. The sampling length was taken as 150 waves for monochromatic seas and 500 waves for the random seas.

EXPERIMENTAL CONDITIONS
The experiments were conducted using combinations of four sea states and two tidal conditions. In total, seven tests were performed in the study (see Table 1). A fixed river discharge of 60 l/s was used for all tests.

Table 1 Experimental conditions

Test	Wave				Longshore Current	Tides	River Discharge (l/s)
	Type	H (m)	T (s)	Dir (°)			
RUN1	LCR	0.125	1.2	70	Yes	No	60
RUN2	LCR	0.100	1.4	70	Yes	No	60
RUN3	MON	0.088	1.2	70	Yes	No	60
RUN4	SCR	0.125	1.2	90	No	No	60
RUN5	LCR	0.125	1.2	90	No	No	60
RUN6	LCR	0.125	1.2	70	Yes	F	60
RUN7	LCR	0.125	1.2	70	Yes	O	60

Notes: F = Following tide; O = Opposing tide; Wave direction was measured anti-clockwise from north (see Fig 1).

The results of RUN3 are presented below. The incident wave height of 0.088 m was selected to achieve equivalence in wave energy with the long-crested random waves of RUN1.

DATA ANALYSIS

<u>Time-mean velocity</u> The time-mean velocities were obtained by simple averaging of the de-spiked measured velocity time series. De-spiking was particularly important when the measurements were taken near the water surface where unrealistic signals were collected when the probes were out of water for a short period time.

<u>Wave characteristics</u> Wave heights and periods were obtained using a zero up-crossing method both offshore and in the surf zone. This method uses the real elevation time series and, since it does not rely on the wave height distribution, is believed to give accurate results, particularly in the surf zone. For the monochromatic waves, an ensemble averaging method was used to obtain the intra-period wave characteristics.

<u>Low-pass filter problem</u> After the tests were performed, it was discovered that a low-pass filter had been fitted into the ADV system by the instrument supplier to smooth the digital signal outputs. This resulted in the measured signals being reduced in magnitude and shifted in phase from very low frequencies. Examination of the filtered and unfiltered velocity spectra at selected points has shown significant differences between the measured and original signal spectra from as low as 2 Hz, which means that all data needed correction.

<u>Turbulent kinetic energy</u> The measured instantaneous velocity can be considered to consist of the components of the time-mean velocity, the wave orbital velocity and the turbulent velocity fluctuations. To separate the turbulent kinetic energy (TKE), defined as $k = 0.5\left(\overline{u'u'} + \overline{v'v'} + \overline{w'w'}\right)$, where u, v and w = velocity components in longshore, cross-shore and vertical direction respectively (see Fig 1), from the total energy the spectrum splitting technique of Soulsby & Humphery (1990) was used. This technique is based on the separation of the variance of the first three harmonics of the wave orbital velocity from the variance of the instantaneous velocity in the frequency domain.

RESULTS

<u>Currents</u> Fig 2 shows the horizontal velocity vectors at levels of 0.4D and 0.6D in the area defined by the position of the still wave level (SWL). This figure clearly shows that the river flow was deflected to the right due to the effect of the obliquely-incident waves. Although due to depth limitations,

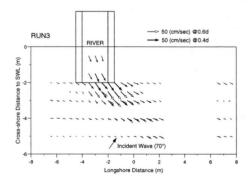

Fig 2 Current vectors

the ADV could not be deployed closer than 2 m to SWL, a gyre (not shown) was also observed to form downstream of the river. It can also be seen that the flow near the surface was deflected to a greater extent than the flow near the bed. This is believed to be due to differences in wave-induced mass flux between the two levels in the water column.

Waves Fig 3 shows the measured wave height contours. The position of the break point is approximately 3 m offshore of the SWL. It appears that the river discharge has little effect on the wave height prior to breaking and on the position of the break point. However, the wave height decay in the river flow path immediately after the breaker is greater due to the

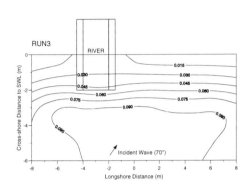

Fig 3 Wave height contours

interaction with the outflowing current during the course of wave breaking. A shadow area can also be seen on the upstream side due to the obliquely-incident waves and the upstream boundary of the basin.

Turbulent Kinetic Energy The distribution of TKE obtained by using the spectrum splitting technique for RUN3 is shown in Fig 4. This figure indicates that for monochromatic waves, the highest TKE values were concentrated well within the surf zone and the area of the highest current velocity. The two peaks in TKE along a line 2m offshore correspond to the areas of

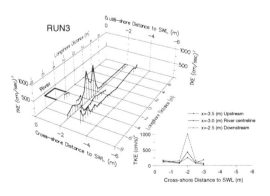

Fig 4 TKE distributions

high shear at the lateral boundaries of the river outflow.

Bed Shear Stress Although two shear plates were deployed in the tests, no useful information of the near-bed shear stress was obtained due to technical difficulties.

However, there are a variety of methods (Soulsby & Humphery, 1982) to estimate the bed shear stress (τ) using measured velocity data. Amongst these, the relationship between shear stress and turbulent kinetic energy:

$$\tau/\rho = 0.19k \tag{1}$$

is believed to be the most reliable. Relating the shear stress in Eq. 1 to the mean current velocity gives the drag coefficient C_d:

$$C_d = \frac{\tau}{\rho R^2} \tag{2}$$

where R = horizontal current speed and ρ = fluid density.

Fig 5 shows a plot of C_d versus current velocity. For a rough bed, a constant C_d is expected in the higher current velocity range as shown in the figure. However, similar to the results reported by Williams et al. (1996) with field measurements, the apparent enhancement of the drag coefficient in the small current velocity range is also shown in

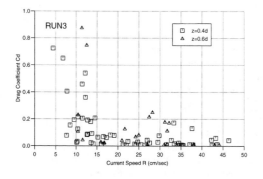

Fig 5 Drag coefficient vs current speed

this plot. While no clear explanation of this phenomenon has yet been proposed, it may suggest that Eq. 1 is not applicable at such a low velocity range.

Reynolds Shear Stress The Reynolds stress ($\overline{u'v'}$) represents the lateral shear in the combined wave-current flow. The relationship of the Reynolds stress to the wave-current ratio is shown in Fig 6. This figure shows that the Reynolds stress and the wave-current ratio are well correlated and that the lateral shear increases significantly with the

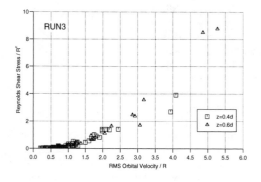

Fig 6 Reynolds stress vs wave-current ratio

increase of wave-current ratio. It is also found that this behaviour is independent of water depth, in particular for the fully developed flow, since the results from the two depth levels show little difference.

Intra-period Characteristics These results of the intra-period free surface elevation and wave orbital velocity were compared with the theoretical results computed by using both linear wave theory and first-order cnoidal wave theory.

Cnoidal wave theory includes the effects of non-linearity and dispersion, which are essential for the description of waves in shallow water. The formulations and methodology given by Isobe (1985) were used in this study.

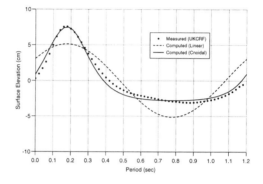

Fig 7 Surface elevation profiles

Comparisons of measured and theoretical intra-period wave characteristics were made at a nearshore point (3 m from SWL and 1.5 m north to the river centreline). Fig 7 shows the comparison of the measured and theoretical profiles of the surface elevation. The computed surface elevation using the cnoidal wave theory agrees with the measurements much better than that computed using linear wave theory due to the non-linearity of

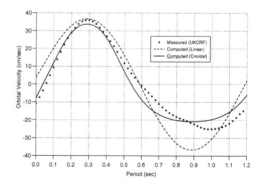

Fig 8 Wave orbital velocity profiles

the shallow water waves. Fig 8 shows the measured and computed wave orbital velocities at level of 0.4D. The results show that both wave theories produce close predictions for the onshore peak orbital velocity, while for the orbital velocity in the offshore direction, the results of the cnoidal wave theory are better. This nevertheless justifies the use of linear wave theory in many numerical models in shallow water.

NUMERICAL MODELLING

WC2D, a 2D wave-current model, which includes the effects of full wave-current interaction, shoaling, combined diffraction-refraction was applied to the test situations. The model also employs a two-equation (k-ε) turbulence closure (Yoo & O'Connor, 1988).

Fig 9 shows comparisons of the computed depth-averaged velocities with the measured current vectors at level of 0.4D. The results of the velocity are in good agreement in general, but the magnitude is slightly over-predicted in the shallow area and under-predicted in the deep water area. This may be due to a poor calculation

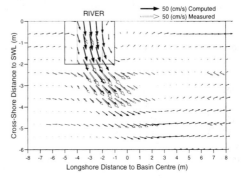

Fig 9 Comparison of current vectors

of the surface slope at the river upstream boundary 2m beyond SWL, therefore improvements regarding this aspect are needed. A gyre downstream of the river outflow also agrees with the experimental observation mentioned previously.

Comparisons of wave height along the river centreline (x=-3m) and basin centreline (x=0m) are given in Fig 10. Both show good agreement, although the wave height around the break point at the basin centreline was less satisfactory since the waves are very non-linear close to breaking. The model performs better along the river centreline, where the interaction with the current dominates the

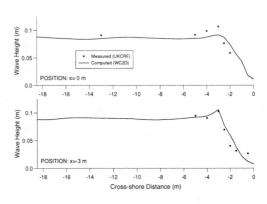

Fig 10 Comparison of wave height

wave height distribution. In addition, the inclusion of the area breaking was found to be necessary in the model. It involves the dissipation of wave energy over a distance of 2 wave lengths shoreward of the break point until a stable wave height is reached.

The computed TKE values are compared with the measured values at a level of 0.4D in Fig 11, along the same sections as those used in Fig 10. Good agreements are obtained in the TKE distributions. However, since the measured TKE values at 0.4D level were used in the comparisons, the computed depth-averaged results appear higher in magnitude and this can be explained by examining the vertical TKE distributions.

The vertical TKE distribution was calculated using a wave-period averaged version of the 1d k-equation model given by Fredsøe and Deigarrd (1992). Modifications were made to include bed frictional effects and take account of turbulence produced by rollers. The suggested maximum mixing length of 0.07D was used in this study. Computations performed at two points along the river centreline (x= -3 m) are shown in Fig 12. Point 1 is inside the surf zone, 2.5m offshore of the SWL. Points 2 is outside the surf zone, 3m offshore of the SWL. The vertical profiles of the computed results from the 1d model, together with the measured data and the depth-averaged TKE values from the 2d model (WC2D) are shown in Fig

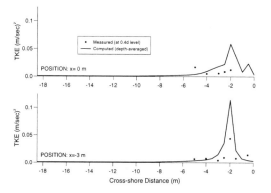

Fig 11 Comparison of TKE profiles

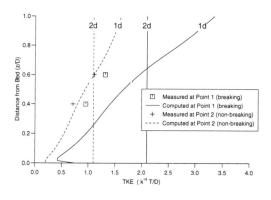

Fig 12 Vertical TKE profiles

12. For the non-breaking case the 2d model provides an excellent estimate of the depth-averaged TKE. For the breaking case, the 2d model also provides a reasonable representation for the depth-averaged value when the vertical distribution, as shown by the 1d model, is taken into consideration. This illustrates the adequacy of the depth-averaged k-ε approach for plan area modelling.

CONCLUSIONS

Laboratory tests investigating wave-current interaction in the presence of a river flow and tides were carried out. The data collected in tests were analyzed, verified and compared with 1d and 2d numerical models. The results showed that the wave height was more affected by the river outflow in the surf zone that outside it and that the river flow was deflected by the waves to varying degrees in the water column. Bed shear stresses and drag coefficients were derived from the measured TKE data. Comparison of the intra-period wave characteristics showed that linear wave theory provides a reasonable estimate of the maximum onshore wave orbital velocity for the shallow water waves. The results of the 1d k-equation model and 2d depth-averaged k-ε model agreed well with the measurements, which indicates the quality of the experimental data sets for the detailed aspects of wave-current interaction around river mouths.

ACKNOWLEDGEMENTS

The authors would like to acknowledge the UK EPSRC for their financial support of this project (GR/J76361), and HR Wallingford for their co-sponsorship of the UKCRF.

REFERENCES

De Vriend, H. J., Zyserman, J., Pechon, P., Roelvink, J. A., Southgate, H. N. and Nicholson, J. (1993), Medium-term 2DH coastal area modelling, *Coastal Engineering*, **21**, pp 193-224

Fredsøe, J. and Deigarrd, R. (1992), *Mechanics of Coastal Sediment Transport*, World Scientific Publishing Co., Singapore

Isobe, M. (1985), Calculation and application of first-order cnoidal wave theory, *Coastal Engineering*, **9**, pp 309-325

O'Connor, B. A., Sayers, P. and MacDonald, N. J. (1995), Wave-current interaction in random waves and currents. Rep No CE/10/95, Dept. of Civil Eng., University of Liverpool.

Pan, S., MacDonald, N. J. and O'Connor, B. A. (1997), A physical model study of wave-current interaction in a coastal river outflow. Rep No CE/4/97, Dept. of Civil Eng., University of Liverpool

Simons, R., Whitehouse, R., MacIver, R., Pearson, J., Sayers. P., Zhao, Y. and Channel, A. (1995), Evaluation of the UK Coastal Research Facility. *Proc. Coastal Dynamics '95*, Gdansk Poland, pp 161-172

Soulsby, R. L. and Humphery, J. D. (1990), Field observations of wave-current interaction at the sea bed. *Water Wave Kinematics*, (Eds) Torum, A. and Gudmestad, O. T., Klumer Academic Publishers, London, pp 413-428

Yoo, D. and O'Connor, B. A. (1986), Mathematical modelling of wave-induced nearshore circulations. *Proc. 20th Int. Conf. on Coastal Engrg.*, pp 1667-1682

Yoo, D. and O'Connor, B. A. (1988), Turbulence transport modelling of wave-induced currents, *Computer Modelling in Ocean Engineering*, (Eds) Schrefler, B. A. & Zienkiewicz, O. C., Balkema, Rotterdam, pp 151-158

Yoo, D., O'Connor, B. A. and Hedges, T. S. (1988), Numerical modelling of waves in an estuary. *6th Congress A & P Div., IAHR*, pp 65-72

Williams, J. J., Humphery, J. D., Moores, S. P., and Clipson, D. (1996), Analysis of STABLE data from deployment 1, Holderness, UK, Oct. 1994. Proundman Oceanographic Laboratory, Report No. 42.

WAVE REFRACTION ON A HORIZONTALLY SHEARED CURRENT IN THE UKCRF

R. D. MacIver[1] & R. R. Simons[1]

ABSTRACT

There is a lack of experimental data available for the verification of wave-current interaction models. This is particularly so for the refraction of waves by currents. The present paper addresses this problem. Measurements of the refraction of near linear regular waves by a narrow jet-like current characterised by large transverse velocity gradients are discussed.

INTRODUCTION

As a wave propagates through a region of horizontally sheared current the characteristic properties of the wave, namely length λ, height H, speed C, and direction θ, are modified. The situation can be complicated further by vertical shear in the current and, when the waves propagate in intermediate depth or shallow water where $h/\lambda < 0.5$, by variation in the water depth h.

Wave-current interaction is a complicated problem involving wave propagation in an inhomogeneous, non-isotropic, dispersive, dissipative, and moving medium that interacts with the wave (Jonsson (1990)). Despite this, significant progress has been made in the development of theoretical models describing the phenomenon of wave transformation by variations of current and depth: Smith (1983), Kirkby (1986), McKee (1977; 1987) and the review by Jonsson (1990). As the full linear problem is intractable, simplifying assumptions are necessary. Smith (1983) and Kirkby (1986) have used discrete vortex sheets to represent the current shear layer. An alternative approach is to restrict attention to situations where the current properties vary slowly over a lateral scale of many wavelengths which leads to WKB-type solutions (McKee (1977; 1987); Jonsson (1990)).

---

[1] Coastal Research Group. Department of Civil & Environmental Engineering, University College London, Gower Street, London. WC1E 6BT.

Ocean currents vary slowly over the length scale of a wavelength and satisfy the assumptions of the models mentioned. Jonsson (1990) refers to the observations of several field studies. In contrast, in the near-shore environment, river outlets and tidal races can produce currents characterised by rapid variation in strength occurring over length scale of a wavelength. In this situation the models are no longer strictly valid. However, there are insufficient studies of wave transformation by currents with a strong horizontal shear to permit a comparison. This paper describes measurements made in the UK Coastal Research Facility (UKCRF) of refraction of regular near linear waves in constant depth by jet-like currents characterised by strong horizontal shearing.

EXPERIMENTAL FACILITY

The sparsity of laboratory observations is due, in part, to the difficulty of constructing suitable experimental facilities. One is the UK Coastal Research Facility, of which Simons et al. (1995) provide a detailed description. The layout of the UKCRF is shown in schematic form in figure 1. In cross-section it consists of a flat bed region between a multi-element wave generator and a beach region with a 1-in-20 slope. Waves can be generated at angles between ±30° to the shore-normal. Shore-parallel currents are produced using four independent pumps, each of which draws from and discharges into a dedicated sump. For each sump the discharge into, and out of, the basin is further controlled through ten undershot sluice gates and guide flumes allowing fine control of the cross-shore distribution of the current (see figure 1). For the present experiment the water depth was 500mm.

Near linear regular waves with periods of 0.8s and 0.9s propagating at angles of incidence of ±30° to the shore normal have been used here. The incident wave, propagating from the still water region into the jet current, can experience partial reflection while the transmitted wave will be refracted as it propagates through the jet current. On passing through the current, the wave continues up the beach, where it shoals and breaks. Littoral currents generated in the surf-zone are recirculated independently of the jet current.

Wave refraction can be determined directly from measurements of the wave propagation direction as the wave moves through the current. One technique for determining wave direction is the directional wave spectrum. This can be established from temporally coincident measurements of: the surface elevation at several known locations; or co-located measurements of surface elevation and velocity. Establishing the refraction within the rapidly varying flow of the jet region would require a much finer resolution than is achievable with a spatial array. In this instance the co-located method is more suitable. However, a simpler and more immediate method is to establish the wave direction from the horizontal velocity vector. Measurements were made to

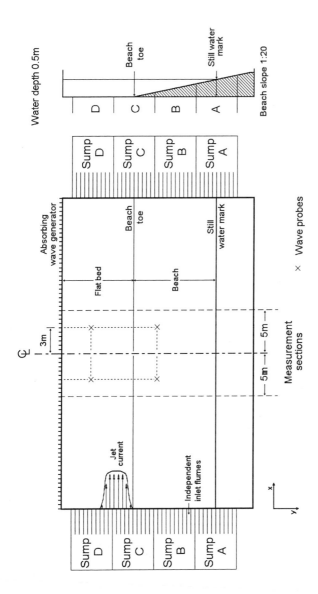

Figure 1: Schematic of the experimental arrangement in the UKCRF.

allow any of the above methods to be used. However this paper will restrict attention to refraction inferred from the horizontal velocity vector.

Velocity measurements were made using acoustic Doppler velocimeters (ADVs). These instruments produce coincident measurements of the three orthogonal velocity components. Surface elevation was recorded with an array of eight resistance-type wave probes arranged in a fixed pattern. The velocity measurements were made directly below a wave probe in the array to provide spatially coincident velocity and surface elevation information. Measurements were made at up to eleven locations on a shore normal section through the jet current. Surface elevation was also measured at four fixed locations, two offshore and two onshore of the current (see figure 1) to provide additional information of the incident and transmitted wave field respectively.

DEVELOPMENT OF THE JET CURRENT

A jet-like shore-parallel current was established in the flat bed part of the basin. The location of the current was determined by several factors. The offshore extent of the jet current should leave a region of still water immediately in front of the wave generator to allow incident wave conditions to be determined before the influence of the current. The inshore extent of the jet current should be within the limits of the flat bed region to avoid the influence of depth variation. Generating the current across the boundary between the two offshore sumps enabled a greater current centreline flow rate to be produced. Larger flow rates were desirable as this would enhance the wave refraction, although the jet current became unstable at large flow rates.

The location of the jet current inlet is shown in figure 1. The distribution of flow across the jet current at the inlet to the basin consisted of a central region of constant velocity, 2m wide, with the flow diminishing in a Gaussian form over a distance of one metre on either side. A free turbulent jet flow will nearly always spread into the surrounding fluid and is necessarily inhomogeneous in the stream direction as well as in the transverse direction (Townsend, 1976). The depth averaged mean horizontal velocity components of the jet current across the three measurement sections follow this description (figure 2). The centreline streamwise velocity U_{cl} decreases while the velocity half-width b increases with increasing distance from the inlet. The transverse distribution of streamwise velocity is self-similar at each section and is approximated by Reichardt's solution (Dracos $et\ al.$ (1992))

$$U(y) = U_{cl}\exp\left(-A\left(\frac{y}{b}\right)^2\right)$$

[1]

where y is the transverse distance from the flow centreline and A is a constant taken as -$Ln(0.5)$. The spanwise velocity distribution is not symmetric about the centreline. A small off-shore flow is present in that part of the jet off-shore of the centreline while in-shore of the centreline the flow is negligible, indicating shore parallel flow. This is probably due to the restraining influence of the beach which may also have been responsible for the under-prediction of the streamwise velocity nearest the beach by equation [1].

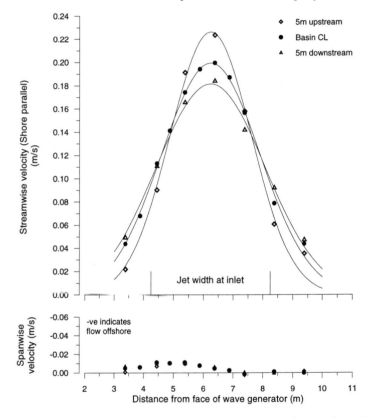

Figure 2: Development of the jet current velocity distribution through the measurement section. Values are depth averaged. Solid lines are equation [1].

WAVE REFRACTION IN THE JET CURRENT

The work presented here concentrates on one of the wave conditions studied. The incident wave had a period T of 0.8s and height H of 40mm and propagated at two angles of incidence to the jet current, 30° (following current) and -30° (opposing current).

The horizontal velocity vector of the wave in the absence of the jet current is shown in figure 3 at three locations across the centreline measurement section. The vectors are an ensemble average of the measured time series. Two distinct forms of the vector can be observed, one linear and one ellipsoidal in form.　Long-crested waves, which are essentially two-dimensional with the fluid motion remaining within a vertical plane parallel to the wave propagation direction, would produce a linear velocity vector. Departure from linear form suggests the presence of other wave components. Two sources for these secondary waves are known: reflection of the incident wave from the beach, and generation of spurious waves at the side walls of the basin as the incident wave passes the mouth of the current inlet/outlet flumes. The form of the velocity vector at a particular position is dependent on the phase relationship between the contributing wave components.

The wave direction was established with a linear least squares best fit to the velocity vector. The recovered directions agree well with the nominal propagation direction of $-30°$ (see table 1).

The horizontal velocity vector and mean flow components across the centreline measurement section for the case of the waves propagating into an opposing current are shown in figure 4 and for the case of waves propagating into a following current in figure 5.　Both forms of vector identified in the "without current" case are present in the "with current" cases. With an opposing current the propagation direction becomes more shore normal as the wave moves into the jet current. A minimum is reached near the centreline of the current before the direction becomes more shore parallel as the wave moves out of the jet current. With waves following the current the opposite behaviour is seen: the propagation direction initially becomes more shore parallel, reaching a maximum before becoming more shore normal as the wave leaves the jet current. Recovered wave directions are presented in table 1.

Test Condition	Measurement Location										
	1	2	3	4	5	6	7	8	9	10	11
No current	-31.7	--	--	--	-32.0	--	--	--	-28.3	--	--
Opposing current	-28.6	-25.5	-24.9	-25.5	-24.7	-27.9	-27.4	-29.9	-31.3	-30.7	-32.6
Following current	29.4	32.9	35.5	39.0	33.9	39.9	38.8	39.7	38.3	37.0	35.0

Table 1: Wave propagation direction in degrees established from velocity vectors. Measurement location 1 is nearest the wave generator.

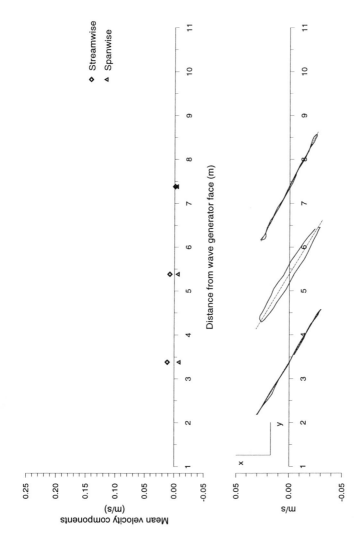

Figure 3: Distribution of horizontal velocity components on a transverse section. Waves alone. Wave conditions: T=0.8s, incidence=-30°. Upper figure: mean components, Lower figure: oscillatory components.

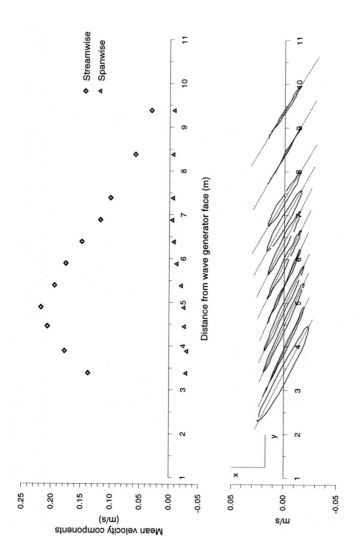

Figure 4: Distribution of horizontal velocity components on a transverse section. Waves opposing the jet current. Wave conditions: T=0.8s, incidence=-30°. Upper figure: mean components, Lower figure: oscillatory components.

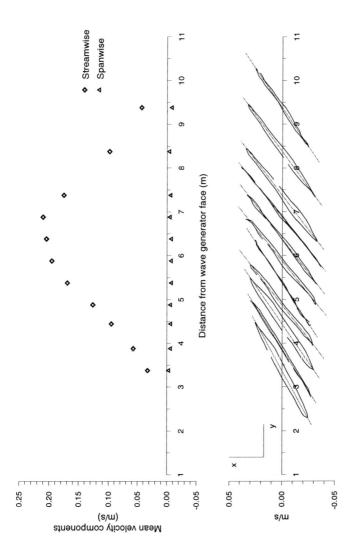

Figure 5: Distribution of horizontal velocity components on a transverse section. Waves following the jet current. Wave conditions: T=0.8s, incidence=30°. Upper figure: mean components, Lower figure: oscillatory components.

It is clear that the jet current affects the wave motion. What is also evident is the influence the waves have on the jet current. While the strength remains essentially the same in the presence of a following and an opposing wave, the location of the jet is affected. With a following wave the jet centreline has been shifted slightly on-shore. However, with an opposing wave the jet centreline has been moved approximately 1.5m off-shore and exhibits significant mean spanwise velocity components indicating the flow is no longer shore parallel through the test section.

A SIMPLE WAVE REFRACTION MODEL

Wave refraction on a straight shear layer - two regions of homogeneous flow separated by the shear layer - on constant depth is determined by Snell's law

$$\frac{L_1}{\sin \theta_1} = \frac{L_2}{\sin \theta_2} \qquad [2]$$

where L_i and θ_i, $i = 1,2$, are the wavelength and angle of incidence in the two regions (Jonsson, 1990). Given the quantities wave period T, water depth h, current velocities U_i, and angle of incidence to the current orthogonal θ_1 are known, the wavelengths L_i, $i = 1,2$, can be determined from (Jonsson, 1990)

$$\sqrt{\frac{h}{L_i} \tanh k_i h} = \sqrt{\frac{h}{L_0}} \left(\frac{1 - U_i \sin(\theta_1) T}{L_1} \right) \qquad [3]$$

where L_0 is the deep-water wavelength. The angle of incidence θ_2 is then given directly by Snell's law.

Equations [2] and [3] show that a wave propagating into a region of stronger and following current is refracted to a more current parallel direction, whereas a region of stronger and opposing current refracts the wave to a more current orthogonal direction. Qualitatively this is consistent with the trends observed in the jet current. Taking U_1 as zero, U_2 as the jet current centreline velocity and the angle of incidence θ_1 as 30° (-30°) for following (opposing) waves yields an estimate of 36° (-25.4°) for θ_2, the angle of incidence at the jet current centreline. This underestimates the case of a following wave but gives reasonable agreement for the case of an opposing wave. As incident wave direction without a current present may only be accurate to ±2° using the velocity vector method, the differences are not considered significant.

SUMMARY

A stable jet-like current has been established. The refraction of a regular wave by this jet-like current has been determined through measurements of the horizontal velocity vector. Qualitatively, the change of wave direction through the jet behaves as expected and is in satisfactory agreement, within the accuracy of the measurement technique, with a simple wave refraction model. The influence of waves on the jet current has also been observed where the position of the jet centreline is shifted from the "without waves" location.

Work is now proceeding with a directional wave analysis of the wave probe array and coincident velocity/surface elevation measurements.

REFERENCES
Dracos, T., Giger, M., Jirka, G.H. (1992). *Plane turbulent jets in a bounded fluid layer*. J. Fluid Mech., 241, 587-614.

Jonsson, I.G. (1990). *Wave current interactions*. In B. Le Mehaute & D.M. Hanes (Editors), The Sea, Vol 9A. Wiley-Interscience, New York, pp.65-120.

Kirkby, J.T. (1986). *Comments on "The effects of a jetlike current on gravity waves in shallow water"*. J. Phys. Oceanogr., 16, 395-397.

McKee, W.D. (1977). *The reflection of water waves by shear currents*. Pure Appl. Geophys., 115, 937-949.

McKee, W.D. (1987). *Water wave propagation across a shearing current*. Wave Motion, 9, 209-215.

Simons, R.R., Whitehouse, R.J.S., MacIver, R.D., Pearson, J., Sayers, P.B., Zhao, Y., & Channel, A.R. (1995). *Evaluation of the UK Coastal Research Facility*. Proc. Coastal Dynamics '95, Gdansk, Poland, pp.161-172.

Smith, J. (1983). *On surface gravity waves crossing weak current jets*. J. Fluid Mech., 134, 277-299.

Townsend, A. A. (1976). *The structure of turbulent shear flow*. Cambridge University Press. Cambridge.

ACKNOWLEDGEMENTS
This work was funded by the Commission of the European Communities Directorate General for Science, Research and Development under contract number MAS3-CT-95-0011. The authors would like to thank HR Wallingford for the use of the UKCRF for these experiments.

Non-linear modelling of shear instabilities

Ad Reniers[1] and Jurjen A. Battjes[2]

abstract

Numerical modelling is used in combination with laboratory basin measurements
to examine the generation, growth and equilibrium conditions of shear instabilities.
The wave transformation is modelled with a 2D-wave roller model to enable a correct prediction of the wave forcing. The velocity field, driven by the wave forcing,
is modelled with non-linear shallow water equations. Using the bottom friction as a
fit-parameter, good correspondence between measured and computed longshore current velocities is obtained. Thus calibrated, the model exhibits the generation and
growth of shear instabilites in the velocity field, comparable to the measurements.
Next the numerical model is used to virtually extend the basin in order to examine
the further development of the shear instabilities and longshore current profile as they
reach equilibrium conditions.

Introduction

Shear instabilities are low-frequency motions in the surfzone velocity field, first observed by Oltman-Shay et al. (1989). A first theoretical explanation of the observed
phenomena was put forward by Bowen and Holman (1989). Based on their analysis,
they stated that the presence of shear instabilities is expected to have a significant
influence on the resulting mean longshore current velocity profiles. The idea is that
the presence of shear instabilities leads to a cross-shore transfer of momentum, resulting in a smoothing of the velocity profile. This potential for momentum transfer,
often referred to as mixing, has been assesed by a number of other authors (Dodd and
Thornton, 1990, Putrevu and Svendsen, 1992, Church and Thornton, 1993), suggesting that the shear instabilities could result in a different longshore current velocity
profile from the initial profile. More detailed numerical analyses were made using
weakly non-linear (Dodd and Thornton, 1992) and fully non-linear models (Nadaoka

---

[1]Dutch Center for Coastal Reseach (NCK), Department of Civil Engineering, Delft University of Technology c/o DELFT HYDRAULICS, P.O. Box 177, 2600 MH, Delft, The Netherlands, ad.reniers@wldelft.nl
[2] Department of Civil Engineering, Delft University of Technology, Delft, The Netherlands

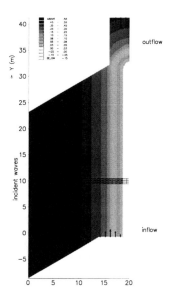

Figure 1: Model definition and bathymetry

and Yagi, 1995, Deigaard et al., 1994, Allen et al., 1995, Ozkan and Kirby, 1997) to describe the development of shear instabilities in wave-driven longshore currents. These studies confirmed the expected momentum transfer, resulting in a smoothing of the initial longshore current velocity profile. This potential momentum transfer was also examined in detail by Reniers and Battjes (1996), using data obtained from a laboratory experiment. The instabilities being infinitesimally small near the inflow opening, grew exponentially while propagating along the beach, reaching horizontal velocity amplitudes near the outflow of the same order as the incident wave orbital velocities. The instabilities at this point were stable, i.e. not breaking up into separate eddies (as followed from a visual check using dye). The analysis showed the presence of significant momentum transfer and corresponding forcing at the downstream end, however no significant changes in the longshore current velocity profile were observed, possibly due to the limited length of the basin. The aim of the present work is to make a quantitative analysis of the effect of shear instabilities on the cross-shore distribution of the longshore current. To this end these measurements have been used to calibrate a numerical model representing the laboratory basin, with the idea to virtually extend the basin to examine the further development of the shear instabilities and longshore current profile as they reach equilibrium conditions. The wave transformation is modelled with a 2D-wave roller model to enable a correct prediction of the wave forcing (Reniers and Battjes, 1997). The calibration is performed with the measured cross-shore distribution of wave heigth and set-up. The velocity field, driven by the wave forcing, is modelled with non-linear shallow water equations, where the bottom friction is used as a fit-parameter to obtain a good correspondence between measured and computed longshore current velocities and resulting generation and growth of shear instabilities in the velocity field.

Wave transformation

Model equations

The model domain is shown in Figure 1. The cross-shore grid spacing varies from 0.5 m offshore to .15 m in the surfzone. In the alongshore direction the grid spacing is equidistant with a mesh size of .35 m. Model computation time 45 minutes. The bathymetry is also shown in Figure 1, where the bar crest and trough can be recognised by the sequence of dark and light shades. At the downstream end the bottom profile curves around to obtain a smooth transition toward deeper water, which is slightly different from the physical model where a steeper slope at the downstream end was used. The oblique side boundaries of the numerical model correspond with the wave guides in the physical model. The up-wave boundary corresponds to $x=0$ in Figure 1. The wave transformation is modelled with a roller model which causes a spatial lag between the location of wave breaking and the actual dissipation (Nairn et al., 1990). Considering a stationary wave field, the balance for the wave energy on a variable bathymetry is given by:

$$\frac{\partial E_w c_g \cos(\theta)}{\partial x} + \frac{\partial E_w c_g \sin(\theta)}{\partial y} = S \tag{1}$$

where E_w represents the wave energy, S the wave energy dissipation and x and y the cross-shore and alongshore coordinates respectively. The incident wave energy is introduced at the up-wave boundary. Inside the computational domain the propagation is governed by the precalculated wave incidence angle, θ, and group velocity, c_g, based on a bottom refraction model. Current refraction is not taken into account. To model the wave energy dissipation due to wave breaking, we have used the following expression (Roelvink, 1993):

$$S = 2\alpha f_p (1 - e^{-\left(\frac{\sqrt{\frac{8E_w}{\rho g}}}{\gamma h}\right)^n}) E_w \tag{2}$$

where α, n and γ are the breaker parameters, f_p the peak frequency, ρ the water density, g the gravitational acceleration and h the total water depth. The wave energy dissipation serves as input in the balance for the roller energy, E_r:

$$S + \frac{\partial 2E_r c \cos(\theta)}{\partial x} + \frac{\partial 2E_r c \sin(\theta)}{\partial y} = D \tag{3}$$

where c is the wave phase speed and D the roller energy dissipation is expressed analogous to Deigaard (1993), leaving a single unknown in eq. 3, E_r, which is taken to be zero at the upwave boundaries. The wave forcing is obtained from the radiation stress tensor, including the roller contribution:

$$S_{xx} = \left(n \left(1 + \cos^2(\theta) \right) + \frac{1}{2} \right) E_w + 2 \cos^2(\theta) E_r \qquad (4)$$

$$S_{xy} = \sin(\theta) \cos(\theta) (E_w + 2E_r) \qquad (5)$$

$$S_{yy} = \left(n \left(1 + \sin^2(\theta) \right) + \frac{1}{2} \right) E_w + 2 \sin^2(\theta) E_r \qquad (6)$$

where the forcing of set-up and currents is now obtained from:

$$F^x = \frac{1}{\rho d} \left(\frac{\partial S_{xx}}{\partial x} + \frac{\partial S_{yx}}{\partial y} \right) \qquad (7)$$

$$F^y = \frac{1}{\rho d} \left(\frac{\partial S_{yy}}{\partial y} + \frac{\partial S_{xy}}{\partial x} \right) \qquad (8)$$

Calibration of wave model

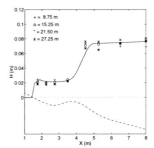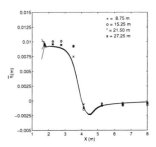

Figure 2: Left panel: Comparison of computed wave transformation with measurements. Computational results obtained for $\alpha = 1$, $\gamma = .55$ and $n = 10$. Right panel: Comparison of computed wave set-up for $\beta = .1$ and measurements

The breaker coefficients associated with the wave model are calibrated using the measured cross-shore distribution of the wave height. The results (see Figure 2) show that the computed wave height corresponds well with the measurements. There is no alongshore variation in the computed wave height distribution. This is in contrast to the measurements where small differences in wave height are bound to occur due to partial reflections, wave current interaction, small differences in bathymetry, etc. Given the fact that the roller adds to the cross-shore momentum balance, we can calibrate the roller model with the measured set-up. The computed set-up, obtained for $\beta = 0.1$, is shown in the right panel of Figure 2. The agreement between computations and measurements is good, though the measured maximum set-up is underestimated by $O(10\%)$. The final wet point shows some small wiggles associated with numerical accuracy. These are not believed to have an effect on the flow characteristics.

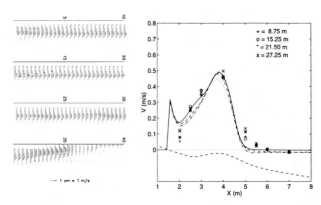

Figure 3: Left panel: Synoptic view of surfzone current velocities at $T = 40$ minutes. Inflow opening located at the upper left corner. Right panel: comparison of mean longshore current profiles. Computations at 8.75 m (solid), 15.05 (dashed), 21.35 m (dash-dotted) and 26.95 m (dotted line) obtained with $k_N = 0.0035$ m. Bottom profile shown as reference (lower dashed line)

Current velocity

The current velocity field is computed with non-linear shallow water equations. The continuity equation is given by:

$$\zeta_t + (hu)_x + (hv)_y = 0 \tag{9}$$

where ζ represent the short-wave averaged water level, u and v the depth averaged velocity components in the x and y direction respectively, and subscripts denote partial differentiation. The cross-shore momentum balance is given by:

$$u_t + uu_x + vu_y = -F^x - g\zeta_x + \nu_t(u_{xx} + u_{yy}) - \tau^x \tag{10}$$

where τ^x represents the combined wave and current bottom shear stress in the cross-shore direction. The bottom shear stress is computed using the parametrisation of Soulsby et al. (1993) of the friction model of Bijker (1967). The alongshore momemtum equation is given by:

$$v_t + uv_x + vv_y = -F^y - g\zeta_y + \nu_t(v_{xx} + v_{yy}) - \tau^y \tag{11}$$

where τ^y represents the combined wave and current bottom shear stress in the alongshore direction. The eddy viscosity, ν_{tb}, associated with lateral turbulent mixing is obtained from (Battjes, 1975):

$$\nu_t = h\left(\frac{D}{\rho}\right)^{\frac{1}{3}} \tag{12}$$

where the roller energy dissipation, D, is obtained from the wave transformation model. The measured alongshore flow velocities are prescribed at the inflow opening (see Figure 1), whereas at the outflow opening a constant water level has been used. The normal velocity component is set to zero for the remaining boundaries. The coupled differential equations for the wave energy, roller energy and cross-shore and alongshore momentum are solved numerically (Petit, 1997).

Calibration of flow model

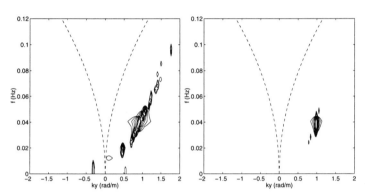

Figure 4: $f - k_y$-spectral estimates for the measured (left panel) and computed (right panel) alongshore velocities at the downstream end. Contour lines for spectral density at .25 .5 1. 2.5 5. 10 25 50 $(m/s)^2/(m^-1$ Hz). Zero-mode edge wave dispersion lines indicated by dashed lines.

To calibrate the flow model the Nikuradse roughness length scale, k_N, is varied, with the objective to obtain a good match with:

* the measured cross-shore distribution of the longshore current
* the alongshore development of the shear instability intensity
* the frequency-wavenumber signature of the shear instabilities

where the last two are strongly related to the back shear of the longshore current velocity. A number of roughness values have been applied, ranging from 0.002 m to 0.005 m, with the best match obtained obtained for a k_N value of 0.0035 m. A synoptic view of the computed flow field is given in Figure 3. This shows some small adjustments occurring near the inflow, where the flow velocities prescribed at the boundary adjust to the wave-driven longshore current. Up to the middle of the basin ($y = 25$ m) the flow is approximately alongshore uniform, after which the first indication of a slow oscillation becomes apparent. This disturbance increases in amplitude as the flow reaches the downstream end of the basin until the deeper part at the outflow has been reached. The estimated wave length is $O(7)$ m. The comparison between computed and measured mean longshore current is shown in Figure 3. The overall correspondence is good, showing strong velocities near the bar crest. Looking in detail shows that the back shear is overestimated with $O(10\%)$ compared to the measurements. A second peak at the water line is present in the computations in a region where there are no measurements (because of depth limitations). A visual check using dye confirmed the existence of a second peak during the experiment, though no quantitative information on the velocities was obtained. The observed changes in the measured mean longshore current velocity are small, without a clear indication of momentum transfer associated with the shear instabilities. This is also mostly true for the computed velocity profiles. Only the velocity profile nearest to the outflow opening has become slightly smoother.

Next we examine whether the computed flow exhibits similar behaviour regarding the generation and growth of shear instabilities as observed in the measurements. Using a spectral analysis technique based on maximum entropy the frequency- alongshore wave number signature of the shear instabilities can be analysed (Reniers et al., 1997). This analysis is performed on the alongshore velocities at the downstream end of the basin at an offshore distance $x = 4.25$ m, analogous to the measurements. The spectral density distribution corresponding to the shear instabilities (see Figure 4) is clearly outside the gravity domain indicated by the zero-mode edge wave dispersion curves. The measurements show a broader, approximately linear, distribution of the energy density with an estimated propagation speed, c, in the order of 30 cm/s. The peak of the spectrum is located at $k_y = .91$ rad/m with a frequency of .0407 Hz. In contrast, the computations display a narrow distribution. Still, the spectral peak is loacted at $k_y = .98$ rad/m, with a corresponding wave length of 6.4 m, and a frequency of 0.0366 Hz which compares well with the observed spectral peak from the measurements. Similar narrow distributions of the computed spectral density were present for all friction factors used.

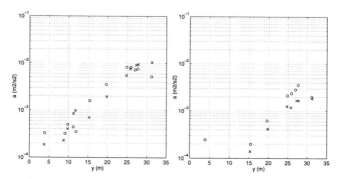

Figure 5: Alongshore development of integrated spectral density based on the along-shore (left panel) and cross-shore (right panel) shear instability velocities. Computed results, obtained with $k_N = 0.0035$ m, indicated by circles vs. measurements indicated by crosses.

To compare the alongshore development of the shear instabilities the contribution to the variance in the low-frequency range between 0.0 and 0.1 Hz denoted by a, is plotted as function of the alongshore position of the current velocity meters (see

Figure 5). A clear exponential growth along the beach can be observed. The growth rate and the total spectral density reached at the downstream end obtained for both the cross-shore and alongshore velocities are in fair agreement with the measurements. There is no clear indication that an equilibrium is reached at the downstream end of the basin, though both measurements and computations suggest a levelling off at the downstream end of the basin.

Equilibrium conditions

In the following the calibrated numerical model is extended to approximately four times its original length to examine the alongshore development of the modelled shear instabilities and corresponding mean longshore current toward equilibrium conditions. Only the relevant part of the basin is shown in the following figures. The alongshore

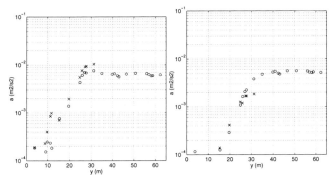

Figure 6: Alongshore development of integrated spectral density based on the alongshore (left panel) and cross-shore (right panel) shear instability velocities for the extended basin. Computed results, obtained with $k_N = 0.0035$ m, indicated by circles vs. measurements indicated by crosses.

development of the integrated spectral density for the extended basin is shown in Figure 6. This shows that an equilibrium for the alongshore shear instability velocity is indeed reached at approximately 27 metres from the inflow opening. This is not the case for the cross-shore component, which reaches an equilibrium at approximately 35 m. The expected asymptotic behaviour near the equilibrium is not present for the alongshore velocity component, showing a sudden levelling off at 27 m.

A synoptic view of part of the computed velocity field in the extended basin after approximately 40 minutes is shown in Figure 7. No significant changes occurred further downstream. The overall view shows an oscillation increasing in intensity up to approximately $y = 32$ m which is comparable to the length of the basin during the laboratory experiments, after which the oscillations are stable resulting in a more or less constant meandering of the longshore current. The alongshore development of the mean longshore current for the present case, shown in Figure 7, shows a rapid adjustment of the mean velocity profile between the transects located at 26.95 m

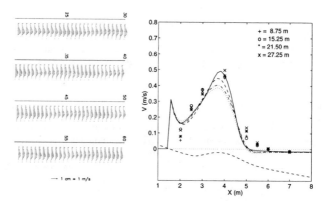

Figure 7: Left panel: Velocity field in part of extended basin. Right panel:Alongshore development of mean longshore current velocity profile obtained for $k_N = 0.0035$ m. Transects located at 15.05 m (solid), 26.95 m (dashed), 33.25 m (dash-dotted) and 51.75 m (dotted) from the inflow. Bottom profile as a reference (lower dashed line)

and 33.25 m from the inflow opening (compare Figures 3 and 7), indicating that equilibrium is not yet reached. After 33.25 metres only minor changes occur and an equilibrium is established, analogous to the observed behaviour for the cross-shore velocity component of the shear instabilities. Given the limited length of the basin during the laboratory experiment, this could not be observed at the time. It is therefore expected that if the basin had been longer, the anticipated smoothing of the longshore current velocity profile would have occurred.

Conclusions

The measurements obtained from a physical model experiment have been used to calibrate a 2D numerical model. Once calibrated the model exhibits the correct behavior for wave transformation, mean flow characteristics and shear instabilities. The computed growth rate corresponds well with the measurements, though the computed spectral density distribution is too narrow. Still, the peak quantities correspond to the measurements, with length scales in the order of 7 metres and a frequency of $O(25)$ s.

The computations with the extended numerical basin indicate that equilibrium for the shear instabilities was not yet reached during the experiment. The cross-shore component of the shear instability velocity reaches an equilibrium at approximately 35 m from the inflow opening (where the available length of the basin was 32 m), as does the mean longshore current profile. The latter exhibits the anticipated smoothing of the velocity profile.

Acknowledgements.
The experiments, performed in the large multidirectional wave-basin at Delft Hydraulics, were funded by the European Community in the framework of the European Large Installations Plan (LIP), project 19 M. Part of the work done by Ad Reniers was funded by the Burgers Foundation.

References

Allen, J.S., P.A. Newberger and R.A. Holman, 1996: Nonlinear shear instabilities of alongshore currents on plane beaches. J.Fluid Mech., vol 310, pp. 181-213.

Battjes, J.A., 1975: Modelling of turbulence in the surfzone. Proc. Symp. Model. Techniques, San Francisco. pp. 1050-1061.

Bowen A.J. and R.A. Holman, 1989. Shear instabilities of the mean longshore current, 1. Theory, *J.Geophys.Res.* 94, C12, 18023-18030.

Bijker, E.W., 1967: Some considerations about scales for coastal models with movable bed. Dissertation, Delft University of Technology, Delft, The Netherlands.

Church, J.C., E.B. Thornton and J. Oltman-Shay, 1992: Proc. 23rd Int. Conf. Coastal Eng., Venice, pp. 2999-3011.

Deigaard, R., 1993: A note on the three dimensional shear stress distribution in a surf zone. Coastal Eng., 20, pp. 157-171.

Deigaard, R., E.D. Christensen, J.S. Damgaard and J. Fredsoe, 1994: Numerical simulation of finite amplitude shear waves and sediment transport. Proc. 24th Int. Conf. Coastal Eng., Kobe, pp. 1919-1933.

Dodd, N. and E.B. Thornton, 1990. Growth and energetics of shear waves in the nearshore. *J.Geophys.Res.*, 95, C9, 16075-16083.

Dodd, N. and E.B. Thornton, 1992. Longshore current instabilities: growth to finite amplitude.*Proc. 23rd Int. Conf. Coastal Eng.*, Venice, 2655-2668.

Nadaoka, K. and H. Yagi, 1995: Modeling shallow water turbulence with horizontal large-eddy motion. Proc. Int. Symp. on Mathematical modelling of turbulent flows, Tokyo, Japan, pp. 245-250.

Nairn, R.B., J.A. Roelvink and H.N. Southgate, 1990: Transition zone width and implications for modelling surfzone hydrodynamics. Proc. 22nd Int. Conf. Coastal Eng., New York, pp. 68-81.

Oltman-Shay J., P.A. Howd and W.A. Birkemeier, 1989. Shear instabilities of the mean longshore current, 2. Field observations, *J.Geophys.Res.* 94, C12, 18031-18042.

Ozkan-Haller, H.T. and J.T. Kirby, 1996: Numerical study of low frequency surf zone motions. Proc. 25th Int. Conf. Coastal Eng., Orlando, pp. 1361-1374.

Petit, H., 1997: Wave-averaged enrgy transport, transformation to roller energy and geneartion of wave forces. Delft Hydraulics report in preparation, M3007.30.

Putrevu U. and I.A. Svendsen,1992. Shear instability of longshore currents: A numerical study, *J.Geophys.Res.* 97, C5, 7283-7303.

Reniers, A.J.H.M. and J.A. Battjes, 1996: Cross-shore momentum flux due to shear instabilities. Proc. 25th Int. Conf. Coastal Eng., Orlando, pp. 175-185.

Reniers, A.J.H.M, J.A. Battjes, 1997: A laboratory study of longshore currents over barred and non-barred beaches. J. of Coastal Eng., vol 30, pp. 1-22.

Reniers A.J.H.M, J.A. Battjes, A. Falqués and D.A. Huntley, 1997: A laboratory study on the shear instability of longshore currents. J. Geophys. Res., vol 102, no. C4, 00. 8597-8609.

Roelvink, J.A., 1993: Dissipation in random wave groups incident on a beach. J. Coastal Eng, vol 19, pp. 127-150.

Soulsby, R.L., A.G Davies, J. Fredsoe, D.A. Huntley, I.G. Jonsson, D. Myrhaug, R.R. Simons, A. Temperville and T. Zitman: 1993: bed shear stresss due to combined waves and currents. G8M (Mast), abstracts in depth, Grenoble, 2.1.

Shear instability of the longshore current as a function of incoming wave parameters

Miquel Caballería [1], Albert Falqués and Vicente Iranzo [2]

Abstract

The purpose of this work is a theoretical investigation of simple and general rules on the beach parameters and the incoming wave parameters for shear wave occurrence on plane beaches. The main parameters governing shear instability are found to be the incidence angle at the breaking line, α_b, the frictional parameter, $r = c_d/\beta$, (where c_d is the drag coefficient for bottom friction and β is the bottom slope) and the dimensionless eddy-viscosity coefficient, N (Longuet-Higgins, 1970). For any set of β, wave heigh H_b, breaking index $\gamma = H_b/h_b$ (where h_b is the total depth), α_b and r values there exists a threshold value of N such that instability requires lower values. As expected, increasing eddy viscosity or bottom friction tends to reduce the instability while increasing α_b favours instability. The heigh of the waves, H_b, and the beach slope, β, give the lengthscale and the timescale for shear waves, which are proportional to H_b/β and to $\sqrt{H_b}/\beta$ respectively, but do not affect the instability threshold. Finally, it is found that shear instability requires values of N so small ($N \sim 0.001$) that shear waves can hardly be generated on plane beaches.

1. Introduction

Waves approaching a straight coastline at an oblique angle generate a longshore current (Fig. 1). As it is well known, this longshore current has typically a sheared profile and may thus be unstable giving rise to travelling eddies called shear waves (Bowen and Holman, 1989). The aim of the present investigation is to find, given any particular beach, which parameters related to the beach and to the wave conditions determine the presence/absence of

---

[1] Universitat de Vic, Sagrada Familia 7, Vic 08500, Spain

[2] Departament de Física Aplicada, Universitat Politècnica de Catalunya (UPC), 08034 Barcelona, Spain

446

shear waves. It is known that the instability is favoured by a strong backshear, i.e., the gradient at the see-face of the current profile, while it is damped by bottom friction and lateral momentum diffusion due to breaking wave turbulence (Falqués and Iranzo, 1994a). The bottom topography can have an important influence too. Field observations (Oltman-Shay et al., 1989) and laboratory experiments (Reniers et al., 1997) suggest that barred beaches are more conducive to instability than monotone beaches.

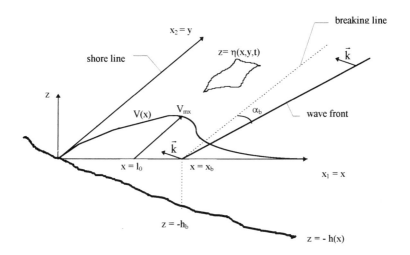

Figure 1: sketch of the geometry, coordinate system and longshore current.

Dodd and co-workers, 1992, 1994, recognized the competition between the maximum backshear and the friction coefficient regarding the onset of the instability. However, their analysis was restricted to some particular field experiments and they did not give general threshold values of the parameters. Falqués and co-workers, 1994b, found general transition criterions in terms of the frictional parameter, $r = c_d/\beta$, and the Reynolds number associated to the maximum backshear, the turbulent lateral momentum diffusion and the width of the current, $Re = f_s l_m^2/\nu_{tm}$. Here, c_d is the drag coefficient for bottom friction, β is the (mean) beach slope, f_s is the maximum backshear, $x = l_m$ is the cross-shore location of the peak longshore current (a measure of the width of the current, $\sim 2l_m$) and ν_{tm} is the maximum lateral mixing coefficient (see below). The transition lines in the $Re - r$ plane show that high values of Re and small values of r favour shear instability. The stability analysis was done by assuming plane sloping beach and analytical profiles

$$V(x) = axe^{-(x/l_m)^3} \qquad (01)$$

for the current and

$$\nu_t = \nu_{tm}(h(x)/h_b)^{3/2} \quad if \; x \leq x_b \qquad , \qquad \nu_t = \nu_{tm}e^{-\sigma(x-x_b)} \quad if \; x \geq x_b \tag{02}$$

for the eddy viscosity, where a, x_b, h_b, σ are constants. However, since these profiles do not differ qualitatively from what is predicted and/or observed on monotone and barred beaches the stability properties can be used in principle as estimate for any beach by tuning the parameters $a, x_l, \nu_{tm}, h_b, \sigma$ for a best fit to the physical profiles. Comparison of field observations at several sites (Duck, Leadbetter) and the prediction based on the transition lines in the $Re - r$ diagram revealed a reasonable agreement.

In the previous analysis explained above, the bottom topography, the lateral mixing, the current and the bottom shear stress were dealt with as if they were independent. However, given the bathymetry, the bottom roughness and the incident wave field, the current and the eddy viscosity are determined. Therefore, given a particular beach and some incoming wave conditions it will be possible to determine whether shear waves will be generated or not and how the different beach and wave parameters influence the instability. This approach, which seems the most natural one to shear wave generation, is the aim of the present work.

2. Longshore current modelling

Since a general transition criterion for shear wave occurrence was lacking we feel the simplest situation, i.e., regular waves approaching a plane sloping beach, is worth investigating. Thus, the Longuet-Higgins model (1970), which allows to deal with simple analytical expressions for a wide range of parameter values, has been considered.

The shallow water equations for momentum and mass conservation:

$$\frac{\partial v_i}{\partial t} + \sum_{j=1}^{2} v_j \frac{\partial v_i}{\partial x_j} + g\frac{\partial \eta}{\partial x_i} = \frac{1}{\rho\zeta}(\tau_{bi} + \tau_{ri} + \tau_{vi}) \qquad i = 1,2 \tag{1}$$

$$\frac{\partial \eta}{\partial t} + \sum_{j=1}^{2} \frac{\partial}{\partial x_j}(\zeta v_j) = 0 \tag{2}$$

where the coordinate system is described in Fig. 1 are considered as governing equations. The bottom profile is given by $z = -h(x)$ and the free surface elevation by $z = \eta(x, y, t)$. The total depth is $\zeta = h + \eta$, ρ stands for the water density, t for time, $\vec{v}$ for the depth averaged horizontal velocity. The forcing from bottom shear stress, from wave radiation stress and from turbulent lateral momentum diffusion is $\vec{\tau}_b, \vec{\tau}_r, \vec{\tau}_v$ respectively. The surf zone is assumed to be saturated, that is, $H = \gamma\zeta$, for $\zeta \leq H_b/\gamma$.

The basic undisturbed state consists of a longshore current, $v_1 = 0$, $v_2 = V(x)$ and a set-up/set-down of the free surface, $\eta = \eta_0(x)$, which are a steady solution of eqs. (1)-(2) with:

$$\tau_{b\,1} = 0 \quad , \quad \tau_{b\,2} = -\rho c_d \frac{\gamma}{\pi} \sqrt{gh}\, V \qquad (3)$$

$$\tau_{v\,1} = 0 \quad , \quad \tau_{v\,2} = \frac{\partial}{\partial x}(\rho \nu_t h \frac{dV}{dx}) \qquad (4)$$

$$\tau_{r\,1} = -\frac{dS_{11}}{dx} \quad , \quad \tau_{r\,2} = -\frac{dS_{21}}{dx} \qquad (5)$$

where the radiation stress tensor is given by:

$$S_{11} = (\frac{1}{2} + cos^2\alpha)\,E \quad S_{21} = -E\,sin\,\alpha\,cos\,\alpha \quad S_{22} = (\frac{1}{2} + sin^2\alpha)\,E \qquad (6)$$

α is the angle of wave propagation with respect to the cross-shore direction and the wave energy density is $E = \rho g H^2/8$. According to Longuet-Higgins, the lateral momentum diffusion or eddy viscosity coefficient is assumed to be:

$$\nu_t = N x \sqrt{gh} \qquad (7)$$

where N is a non-dimensional parameter. A horizontal length scale L_h and a velocity scale v_0 are choosen as

$$L_h = x_b = \frac{H_b}{\gamma\beta} \quad , \quad v_0 = \frac{5\pi\gamma}{16r}\sqrt{gh_b}\,sin\,\alpha_b \qquad (8)$$

where $r = c_d/\beta$ is the frictional parameter and then, a time scale

$$T = \frac{x_b}{v_0} = \frac{16}{5\pi}\frac{r}{\beta\gamma^{1.5}}\sqrt{\frac{H_b}{g}} \qquad (9)$$

results.

The non-dimensional velocity profile has a well known analytical expression (Horikawa, 1988) and depends only on the parameter $P = \pi N/\gamma r$ (see Fig. 2). The set-up, $\eta_0(x)$, is linear in the cross-shore coordinate and can be included through an effective beach slope, $\beta_e = \beta/(1 + 0.375\gamma^2)$, in the total depth: $\zeta_0(x) = h + \eta_0 = \beta_e x$, where the coordinate x has been shifted in order the origin coincides with the effective shoreline. The set-down will be neglected.

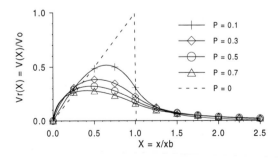

Figure 2: current profiles.

3. Linear stability analysis

The present research deals with the tendency of the beach and wave conditions to develop or not shear instabilities. This concerns the initial growth so that the standard linear stability analysis will be sufficient to our purpose. Thus a small perturbation

$$v_1 = u(x)e^{ik(y-ct)} \quad , \quad v_2 = V(x) + v(x)e^{ik(y-ct)} \quad , \quad \eta = \eta_0(x) + \xi(x)e^{ik(y-ct)} \tag{10}$$

is superimposed to the basic steady flow and free surface elevation, with an alongshore periodicity $2\pi/k$ and a complex phase speed c. Its real part gives the alongshore celerity of the shear wave and its complex part the growth rate. Substitution of (10) into the governing equations (1)-(2) and linearization result in a set of ordinary differential equations which is an eigenproblem where c is the eigenvalue and $(u(x), v(x), \xi(x))$ are the eigenfunctions.

Writting the eigenproblem requires to find the expressions for the bottom shear stress and the lateral momentum diffusion. Following Dodd (1994), the weak-current approximation and the small angle of incidence assumption are used and upon linearization

$$\frac{\tau_{b1}}{\rho\zeta} = -2\frac{f}{\zeta_0}u \quad , \quad \frac{\tau_{b2}}{\rho\zeta} = c_d\frac{\gamma}{\pi}\sqrt{\frac{g}{\zeta_0}}V - \frac{f}{\zeta_0}v + \frac{fV}{2\zeta_0^2}\xi \tag{11}$$

is obtained, where $f(x)$ is a function which depends on the drag coefficient and the total depth:

$$f(x) = c_d\frac{\gamma}{\pi}\sqrt{g\zeta_0(x)} \quad if \ x \leq x_b \quad , \quad f(x) = c_d\frac{\gamma}{\pi}\sqrt{g\zeta_b} \quad if \ x \geq x_b$$

Beyond the breaking line $f(x) = ct.$ is assumed that means a uniform wave orbital velocity.

Following the Longuet-Higgins modelling of the basic undisturbed state, equation (7) will be used to describe lateral mixing distribution. However, since an increase of lateral mixing seaward of the surf zone is obviously unrealistic, an exponential decay will be assumed beyond the breaking line:

$$\nu_t = Nx\sqrt{g\zeta_0} \quad if \ x \leq x_b \quad , \quad \nu_t = \nu_{tm}e^{-\sigma(x-x_b)} \quad if \ x \geq x_b \tag{12}$$

where $\nu_{tm} = Nx_b\sqrt{g\zeta_b}$ is the maximum eddy viscosity and $\sigma = 4/x_b$ has been used for the computations. Battjes (1975) gave a more realistic model of the lateral mixing where the eddy viscosity reads

$$\nu_t = M_B\zeta(\frac{D_b}{\rho})^{1/3} \tag{13}$$

where D_b is the breaking dissipation rate and M_b is a dimensionless constant. In fact, taking into account the depency of D_b on the water depth shows that equation (13) coincides with (7) where both viscosity parameters are related by:

$$N = (\frac{5}{16}\gamma^2\beta^4)^{1/3} M_B \tag{14}$$

Battjes formulation seems more firmly based on physical grounds and the parameter M_B will be therefore used for the final discussion. However, as it will be seen, the parameter N is more convenient for the stability analysis.

Finally, taking into account (10), (11), (12) and upon linearization, (1)-(2) result in a system of three ordinary differential equations that are not written here for the sake of brevity (see (13),(14),(15) in Reniers et al., 1997). These equations are made non-dimensional by means of the the time scale given in (9), the horizontal length scale in (8) and a vertical length scale

$$L_v = \beta x_b = \zeta_b = \frac{H_b}{\gamma} \tag{15}$$

so that the dimensionless parameters are $r = c_d/\beta, \gamma r$, a characteristic Froude number,

$$F = \frac{v_0}{\sqrt{g\zeta_b}} = \frac{5\pi}{16}\frac{\gamma}{r} sin\, \alpha_b \tag{16}$$

and a Reynolds number related to the eddy viscosity,

$$Re = \frac{x_b v_0}{\nu_{tm}} = \frac{5\pi}{16}\frac{\gamma}{Nr} sin\, \alpha_b = \frac{F}{N} \tag{17}$$

After discretization by spectral methods, the stability eigenproblem is solved by standard rutines to obtain the dispersion and instability curves (see Falqués and Iranzo, 1994a).

4. Results

The input parameters in our model are the slope, β, and the drag coefficient, c_d, related to the beach, the wave height, H_b, the incident angle, α_b, and the breaking index, γ, related to the incident waves, and the viscosity coefficient N and the decay parameter, σ, related to turbulent lateral mixing. However, only F, Re and γr appear in the nondimensional stability equations and the nondimensional current profile depends only on P. Therefore, shear instability will depend on α_b, N, r and γ. Notice that using the Longuet-Higgins viscosity coefficient N instead of Battjes parameter, M_B, the beach slope, β, drops out from the instability equations. The influence of β will thus be discussed more easily by looking later on at its influence on r and N.

Due to the relatively narrow range of variability of the breaking index, $0.6 \leq \gamma \leq 1$, and because γ is not expected to have a significant influence, $\gamma = 0.8$ has been fixed, and the instability has been investigated only as a function of α_b, N and r. The viscosity decay parameter has also been kept fixed to $\sigma = 4/x_b$.

The numerical tests have been carried out with $\beta = 0.02$ or $\beta = 0.04$, with $H_b = 0.5m$ or $H_b = 1m$ and values of the angle between $\alpha_b = 2^o$ and $\alpha_b = 25^o$. According to experimental data from Deguchi et al., 1992, the realistic range of N would roughly be $0.001 \leq N \leq 0.008$. Thus, $N = 0.001, 0.002, 0.005$ has been chosen.

For any couple of values of α_b and N, high values of r lead to stability and low values to instability, so that a critical value $r_c(\alpha_b, N)$ is found. The main results are summarized by means of the transition lines between stability and instability plotted on the $\alpha_b - r$ diagram as a function of the other parameters (Fig. 3).

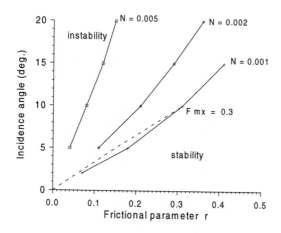

Figure 3: Transition lines for shear instability.

Each transition line divides the $\alpha_b - r$ plane in two regions: a stability region (lower part) and an instability region (upper part). Following these lines, α_b increases with r, and the slope of these curves increases with N. Therefore, a large incidence angle and a small frictional parameter favours instability. Since decreasing N widens the instability region, small N enhances instability too.

The influence of the beach slope, β, on the critical conditions has to be found through the parameters N and r in the stability diagram, Fig. 3. Experimental results by Deguchi and co-workers., 1992, suggest that N increases with the slope, β. Assuming a more "universal" character of the Battjes parameter, M_B, i.e., that it does not depend on β, Eq. 14 shows that N should be proportional to $\beta^{4/3}$. As a result, gentle slopes would favour shear instability. The influence of β through r is not so clear. The definition, $r = c_d/\beta$ would suggest a decrease of r for increasing β. However, c_d and β are, in fact, not independent. Large values of c_d correspond to high roughness, i.e., coarse sand, and are therefore found on steep beaches. Small values correspond to fine sand, and are thus found on gentle beaches. Therefore, c_d is an increasing function of β. The precise relationship between both parameters is not necessarily linear, so that r may still depend somewhat on β. This relationship, yet unknown, is probably very mild, not so strong than $\beta^{4/3}$. Thus, the dependence of N on β would dominate the possible dependence of r on β, and we could conclude, therefore, that gentle beaches would be more conducive to shear instabilities than steep ones.

From the diagram shown in Fig. 3, the possibility that waves incoming obliquely on a plane beach develop longshore currents with shear instabilities can be discussed. Let us first note that the Froude number of longshore currents on natural beaches hardly exceeds $F \sim 0.3 \sim 1ms^{-1}/\sqrt{10ms^{-2} \times 1m}$. In addition, eq. 16 shows that the constant F lines in the $\alpha_b - r$ diagram are roughly straight lines for moderate α_b. Therefore, beach and wave conditions corresponding approximately to the region above the dotted line in Fig. 3, which is the $F = 0.3$ line, are not usually observed. This can be due to the fact that: i) wave angle at breaking can not be very large because of topographic wave refraction and ii) as argued above, only a narrow range of $r = c_d/\beta$ values are possible. But then, we realize that in the "realistic" region, $F \leq 0.3$, instability can only exists for N of the order of 0.001 or less. This value of N is too small, in the lower limit of what can be realistic. Therefore, we conclude that shear instabilities can hardly be developped on plane beaches. Field data from Leadbetter Beach (Oltman-Shay et al., 1989; Dodd et al., 1992) and laboratory experiments (Reniers et al., 1997) seem to corroborate that plane beaches are not conducive to the generation of shear waves.

At first sight, the critical conditions for shear instability would seem to depend on the wave height, H_b. But recalling that shear waves have been observed with waves $1m$. high (Duck, Oltman-Shay et al., 1989) and with waves $0.08m$. high (LIP experiment, Reniers et al., 1997) shows that the onset of instability can not depend very much on H_b. In the analysis we present, H_b appears only in the length and time scaling, eqs. (8), (9), (15), and does not therefore influence the transition conditions. According to (8) and (9), the wavelength of shear waves will be proportional to $H_b/\gamma\beta$ and its period will be proportional to $(r/\beta\gamma^{1.5})\sqrt{H_b/g}$. One has to have in mind, however, that our study is restricted to plane beaches, where there is a linear relationship between the vertical scale given by H_b and the horizontal scale given by the width of the surf zone, x_b. Extrapolation to barred beaches, where this relationship does not hold anymore, has to be handled with care.

Table 1 shows a comparison of shear waves observed at Duck with shear waves observed in the LIP experiment. The ratio between wave heights is $H_b^*/H_b^{**} \sim 13$. The ratio between wavelengths of the shear waves is $\lambda^*/\lambda^{**} \sim 14$, very close to H_b^*/H_b^{**}. The ratio between periods is $T^*/T^{**} \sim 4$

	$H_b(m.)$	$\lambda(m.)$	$T(s.)$
Duck (*)	1	100	100
LIP (**)	0.08	7	25

Table 1: shear wave scaling; natural beach versus laboratory beach.

close to $\sqrt{H_b^*/H_b^{**}} \sim 3.5$. So, even though the beach topography was barred while the present analysis assumes planar topography, a reasonable accordance is found.

5. Conclusions

A stability diagram where the occurrence of shear waves on plane beaches can be predicted has been obtained. The main parameters that determine the onset of shear instability have been found to be the wave incidence angle α_b, the frictional coefficient, $r = c_d/\beta$, and the Longuet-Higgins viscosity coefficient, N. Shear instability requires high α_b and low r and N. The influence of the beach slope, β, is not so clear but it seems that gentle beaches would favour instability. Under realistic conditions for natural beaches, it has been found that instability requires values of N so small that shear wave generation would be almost impossible on plane beaches. The scaling of the model shows that the wave height does not affect the critical conditions for instability. Instead, the wave height determines the horizontal length scale of shear waves, which is proportional to H_b/β, and the time scale, which is proportional to $\sqrt{H_b}/\beta$. It turns out that this prediction, obtained assuming plane beach, is quite robust since experimental data suggest that it is approximately valid for barred beaches too. In addition to the assumption of planar topography, the present analysis has several limitations: regular waves has been assumed for the basic current modelling and small angle of incidence and weak current has been assumed for bottom friction. We however believe that the predictions we have obtained give the basic trends of the onset of shear instability on plane beaches that could be extrapolated to other situations. Quantitative predictions, in the case of irregular waves, for instance, would need however a similar modelling but with the adequated longshore generation.

Acknowlodgements

This work has been funded by the Spanish government (DGICYT) under contract PB93-0948. Interesting remarks by A.Reniers, J.Battjes and C.Vidal are gratefully acknowledged.

References

Battjes, J.A., 1975. A note on modeling of turbulence in the surf zone. Proc. Symp. Modeling Tech., ASCE, 1050-1061.

Bowen, A. and Holman, R., 1989. Shear instabilities of the mean longshore current, 1, Theory. J.Geophys.Res., 94, 23-30.

Deguchi, I., Sawaragi, T. and Ono, M., 1992. Longshore current and lateral mixing in the surf zone. Proc. 23th Int.Conf.Coastal Eng., ASCE. Ch.202, 2642-2654.

Dodd, N., Oltman-Shay, J. and Thornton, E., 1992. Shear instabilities in the longshore current: A comparison of observation and theory. J.Phys.Oceanography, 22, 62-82.

N.Dodd, 1994. On the destabilization of a longshore current on a plane beach: bottom shear stress, critical conditions, and onset of instability. J.Geophys.Res., 99(C1), 811-824.

Falqués, A. and Iranzo, V., 1994a. Numerical simulation of vorticity waves in the nearshore. J.Geophys. Res., 99 (C1), 825-841.

Falqués, A., Iranzo, V. and Caballería, M., 1994b. Shear instability of longshore currents: effects of dissipation and non-linearity. Proc. 24th Int.Conf.Coastal Eng., ASCE. Ch.143, 1983-1997.

K.Horikawa(editor), 1988. Nearshore Dynamics and Coastal Processes. University of Tokio Press.

Longuet-Higgins, M.S., 1970. Longshore currents generated by obliquely incident sea waves. J.Geophys.Res., 75, 6778-6801.

Oltman-Shay, J., Howd and P., Birkemeier, W., 1989. Shear instabilities of the mean longshore current, 2, Field observations. J. Geophys.Res., 94, 31-42.

A.J.H.Reniers, J.A.Battjes, A.Falqués and D.A.Huntley, 1997. A laboratory study on the shear instability of longshore currents. J.Geophys.Res., 102(C4), 8597-8609.

HORIZONTAL LARGE-EDDY COMPUTATION OF LONGSHORE CURRENTS

Hiroshi Yagi[1] and Kazuo Nadaoka[2]

Abstract

The developments of horizontal large-scale (HLS) eddies in the longshore currents under uni- and bi-directional wave conditions were simulated by using the SDS-2DH model, which has been recently developed by the authors to simulate shallow-water turbulent flows. The HLS eddies were demonstrated to have dominant contribution to momentum transport, although their turbulence intensity is small as compared with the local 3-D turbulence generated by wave breaking. In contrast to uni-directional waves, which induce only the *internal* shear instability mechanism in the longshore currents, bi-directional waves were found to provide also the *external* forcing and coupling mechanism by the longshore torque pattern migrating with the wave group.

INTRODUCTION

Resent studies on the nearshore currents have found out the existence of the shear instability mechanism in longshore currents (Oltman-shay et al., 1989, Reniers et al., 1994) and shown that the shear waves may result in appreciable cross-shore momentum transport and hence play an important role in the nearshore current dynamics (Bowen & Holman, 1989). Although nonlinear development of the shear waves has been studied by several researchers (e.g., Falques, et al., 1994), further developments of the shear waves to horizontal large-scale (HLS) eddies and their

---

[1] Department of Civil Engineering, Tokyo Institute of Technology,
2-12-1 O-okayama, Meguro-ku, Tokyo 152, Japan.
[2] Department of Mechanical and Environmental Informatics, Graduate School of
Information Science and Engineering, Tokyo Institute of Technology,
2-12-1 O-okayama, Meguro-ku, Tokyo 152, Japan.

subsequent merging as a strongly nonlinear process should be also investigated to find entire history of the eddy evolution.

The possibility of the development of the HLS eddies suggests the necessity to observe the surf zone "turbulence" from more integrated aspects. Namely the surf zone turbulence should be regarded to be characterized by the coexistence of the local turbulence caused mostly by wave breaking and the HLS eddies produced by the shear instability mechanism. The former is essentially 3-D turbulence with length scale less than the water depth, while the latter is horizontal 2-D eddies with much lager length scale. Such a *double-structural* and *strongly non-isotropic* character is common to various shallow water turbulent flows like river flow and tidal currents.

Recently the authors have developed a computational model named "SDS&2DH model" to simulate shallow-water turbulent flows and thereby to present a new framework to properly evaluate horizontal mixing process of mass and momentum (Nadaoka & Yagi, 1993a,b, 1995, 1997). Using this model, the authors succeeded in simulating the development of HLS eddies and their subsequent merging in the longshore currents under the condition of *uni*-directional wave train (Nadaoka & Yagi, 1993a,b).

In the present paper, some additional descriptions will be given first on the computational results mentioned above, such as relative contribution of the HLS eddies and the local turbulence due to wave breaking to cross-shore momentum transport in the longshore current. Then the results in case of *bi*-directional wave train will be shown and discussed from the viewpoint of the dependence of the HLS eddy development on the characteristics of the wave group, which migrates also in the longshore direction, with a special emphasis on the *external* forcing and coupling mechanism by the longshore torque pattern migrating with the wave group.

COMPUTATIONAL MODEL FOR LONGSHORE CURRENTS

Previous models for the horizontal mixing process

The horizontal mixing in the longshore currents is usually expressed with the horizontal eddy viscosity v_t. The turbulent velocity and length scales constituting v_t are expressed in various ways as shown in the following three typical models of v_t.

Thornton(1970), Jonsson et al.(1974) $v_t = M_T \, U_w \, d,$ (1)

Longuet-Higgins(1970) $v_t = M_L \sqrt{gh} \; x,$ (2)

Battjes(1975) $v_t = M_B \, h \left(D/\rho \right)^{1/3},$ (3)

where U_w and d are the velocity amplitude and excursion length of the water particle by wave motion, x the distance from the shoreline, g the gravitational acceleration, h the water depth, ρ the fluid density, D the mean rate of the wave energy dissipation by wave breaking, and M_T, M_L, M_B dimensionless coefficients. Thornton (1970) and Jonsson (1974) employ U_w and d as the turbulent velocity and length scales of v_t. However this idea is inconsistent with the physical process of mixing, because U_w

and d are the parameters of wave motion, and hence not directly related to the horizontal mixing process of longshore currents. The well-known model eq.(2) by Longuet-Higgins (1970) is similar to the mixing length model for the usual wall turbulent flow if the shoreline is regarded as the rigid wall. However, the role of the shore boundary is much different from that of the rigid wall boundary, because the shore boundary is not the source of turbulence like the usual rigid wall. Then, it is not reasonable to introduce the usual mixing length model for the wall turbulence to the evaluation of the horizontal mixing in the longshore current.

On the other hand, in the model eq.(3), Battjes(1975) regards the turbulence generated by wave breaking as the governing factor of the horizontal mixing in the surf zone. The numerical and laboratory experiments by Visser (1984) show that the dimensionless coefficient M_B should be 3 to fit the computational results of the cross-shore distribution of longshore current to the experimental data. However, as the turbulence length scale of wave breaking may not be larger than the water depth, 3 is unreasonably large value for M_B. This suggests that some other mechanisms in addition to the local turbulence due to wave breaking may be involved in the horizontal mixing process in the surf zone.

Outline of SDS&2DH model

The essential feature of shallow-water turbulence, which has a double-structural and strongly non-isotropic character as stated before, may be more clearly illustrated in a spectral domain as shown in Fig. 1. As compared with the spectral structure of turbulence to which usual LES may be applied, the shallow-water turbulence in a low wavenumber range has 2-D nature and hence may show cascade-up of spectral energy. It should be emphasized that these 2-D eddies have a direct attenuation mechanism due to energy dissipation by bottom friction. Some part of the dissipated energy may be supplied to 3-D turbulence in a higher wavenumber range, to which other possible sources like wave breaking and wind stress may directly provide turbulent energy.

For reasonable modeling of shallow-water turbulence described above, the local 3-D turbulence, whose length scale is smaller than the water depth, must be treated separately in an explicit manner. In the SDS&2DH model, a pertinent concept, subdepth-scale (SDS) turbulence", is introduced and shallow-water turbulence is assumed to consist of HLS eddies and SDS turbulence as shown in Fig.2. Hence the total kinetic energy of turbulence k_T is to be expressed as eq.(4) in Fig.2, where k_L and k_D denote the kinetic energy of HLS eddies and SDS turbulence, respectively.

Introducing a turbulence model *only* for the SDS turbulence with a depth-averaged k_D-equation and the turbulence length-scale expression of eq. (5) in Fig.2, the SDS&2DH model can directly simulate the evolution of the HLS eddies with the depth-averaged continuity and momentum equations, in which the eddy viscosity v_D is evaluated *only* with the SDS turbulence.

Specification of bi-directional wave filed and computational conditions

The wave height or wave amplitude of the bi-directional wave train, which is needed to estimate the radiation stresses in the momentum equations, is not easy to compute because no reliable methods exist to evaluate the wave breaking of bi-directional wave trains and their deformation in the surf zone. Therefore in the present study a rather simple method is employed; i.e., the wave field is expressed by simply superimposing the two wave trains, each of them is calculated with the linear theory, and in the surf zone, the wave amplitude of each wave train is assumed to attenuate according to the following equation.

$$a_i = a_i' \frac{H_b}{H'}, (i = 1,2),$$ (6)

in which a_i is the attenuated wave amplitude of each wave train, a_i' and H' are respectively the computed non-breaking wave amplitude and height of each and superimposed wave train at the location concerned, and H_b is the breaking wave height of the superimposed waves and is evaluated here by using the frequently used simple assumption,

$$H_b = 0.8h.$$ (7)

The computation was executed for the bottom slope of 1/50, the computational grid size of 2.5m, and the time step of 0.1s . To specify the initial flow condition for the numerical simulation, preliminary 1-D computations were made, in which the eddy viscosity was evaluated only with the SDS turbulence. To provide numerical "seeds" of flow instability, small disturbances with a magnitude less than 1% of the maximum mean velocity, which were generated with a set of pseudo-random numbers, were superimposed on the initial flow field around the breaker line. At the longshore boundaries, a periodic boundary condition was applied.

RESULTS AND DISCUSSTION

Uni-directional wave train
As previously mentioned, the authors succeeded, by using the SDS&2DH model, in simulating the development of HLS eddies and their subsequent merging in the longshore currents under the condition of uni-directional wave train (Nadaoka & Yagi, 1993a,b). Figure 3 shows the mean profile of the longshore current velocity, which becomes milder with time around the breaking point. A close examination of the mean profile and the corresponding eddy structure reveals that the merging of eddies causes further smoothing of the mean longshore-current velocity. This is a direct manifestation to show that the HLS eddies induce the appreciable momentum exchange and hence have an important role to establish the longshore current profile. Figures 4 and 5 show respectively the cross-shore distributions of the turbulence energy and Reynolds stress, indicating that the contribution of the SDS turbulence is dominant to the turbulence energy but negligibly small to the Reynolds stress. This is due to the fact that the length scale of the SDS turbulence is limited to be less than the water depth.

Bi-directional wave train
The wave condition of the bi-directional wave train is listed in Table 1, in

which Wave-1 represents the dominant wave train and has the same condition with that of the unidirectional wave train described above. By keeping the dominant wave condition constant, three different conditions of Wave-2 were selected. In the bi-directional wave field obtained by linearly adding Wave-1 and Wave-2, a wave group migrating in the longshore direction may appear. In the present condition, the alongshore lengths of the wave group for the three cases are the same to be 100m, while the longshore migrating speeds of the wave group are different each other.

Figure 6 shows the computational results of the velocity field for Case1, in which the longshore speed of the wave group is nearly equal to the advection velocity of the HLS eddies. The upper and lower figures show the velocity fields at t=50s and 300s, respectively. In this case, the HLS eddies with the length corresponding to the wave group develop from the initial stage of the computation and reach an equilibrium state relatively quickly without merging of the HLS eddies. The intensity of the HLS eddies is stronger than that in case of the unidirectional wave train. These facts suggest that the nonlinear coupling of the HLS eddies with the wave group may have a significant influence on the development process of the HLS eddies.

The left column of Fig. 7 represents the evolution of the velocity field for Case 2, in which the wave group speed is smaller than the advection velocity of the HLS eddies. In this case, the HLS eddy intensity varies periodically in time, while the longshore eddy size keeps to be nearly the same with the wave group length.

Figure 8 shows the results for Case3, in which the wave group propagates in the direction opposite from the HLS eddies with much different speed as compared from the eddy advection velocity. In this case, the longshore eddy size is not locked to be the wave group length. The development of the HLS eddies exhibits no straightforward coupling with the wave group and even at t=800s the flow field do not reach the equilibrium state.

The dependence of the eddy evolution on the wave group characteristics may be more clearly found by examining the relation between the eddy location and the spatial distribution of the torque due to the radiation stress. In Case 1, the HLS eddies are located nearly at the peaks of the torque in the longshore direction and both of them move with the same speed. In Case 2, on the other hand, the relative position between the HLS eddies and the peaks of the torque varies with time. As shown in the right column of Fig. 7, when the centers of the HLS eddies are close to the peaks of the torque the eddies are strengthened, while when the HLS eddies depart from the peaks of the torque the eddies are weakened.

These results for bi-directional waves indicate that the longshore torque pattern migrating with the wave group provides the *external* forcing and coupling mechanism. This is a highly contrasted aspect as compared with the HLS eddy development in case of uni-directional waves, in which *only* the *internal* shear instability mechanism exists.

CONCLUSIONS

1. The SDS&2DH model, which has been recently developed by the authors to

simulate shallow-water turbulent flows, was applied to the simulation of the development of horizontal large-scale (HLS) eddies in the longshore currents under the conditions of uni- and bi-directional waves. The HLS eddies were demonstrated to have dominant contribution to cross-shore momentum transport in the longshore currents, although their turbulence intensity is small as compared with the local 3-D turbulence generated by wave breaking.

2. In contrast to uni-directional waves, which induce only the *internal* shear instability mechanism in the longshore currents, bi-directional waves provide the *external* forcing and coupling mechanism as well by the longshore torque pattern migrating with the wave group.

References

Battjes. (1975): Modeling of turbulence in the surf zone, *Modelling 75, Proc. 2. Symp. Modelling Techniques, ASCE*, San Francisco, 1050-1061.

Bowen, A.J. and Holman, R.A. (1989): Shear instability of the mean longshore current, 1. Theory, *J.Geophys. Res.*, 94, 18023-18030.

Falques, A., Iranzo, V. and Caballeria, M. (1994): Shear instability of longshore currents: effects of dissipation and non-linearity, *Proc. of 24th Int. Conf. on Coastal Eng.*, *ASCE*, 1983-1997.

Jonsson, I.G., Skovgaad, O. and Jacobsen ,T.S. (1974): Computation of longshore currents, *Proc. of 14th Int. Conf. on Coastal Eng., ASCE*, 699-714.

Longuet-Higgins, M. S. (1970): Longshore currents generated by obliquely incident sea waves, 1, 2, *J. Geophys. Res.*, 75, 6778-6801.

Nadaoka, K and Yagi, H. (1993a): A turbulent-flow modeling to simulate horizontal large eddies in shallow water, *Advances in Hydro-Science & Eng.*, 1[B], 356-365.

Nadaoka, K. and Yagi, H. (1993b): A turbulence model for shallow water and its application to large-eddy computation of longshore currents., *J. of Hydraulic, Coastal and Environmental Engineering, JSCE*, No.473, 25-34.(*in Japanese*)

Nadaoka, K and Yagi, H. (1995): Modeling shallow-water turbulence with horizontal large-eddy motion, *Proc. Int. Sympo. of Turbulent Flows, JSCFD*, 245-250.

Nadaoka, K and Yagi, H. (1997): Shallow-water turbulence modeling and horizontal large-eddy computation of river flow, *J. of Hydraulic Eng., ASCE*. (*in press*).

Oltman-Shay, J., Howd, P.A. and Berkemeier, W.A. (1989): Shear instability of the mean longshore current, 2, Field observation., *J. Geophys. Res.*, 94, 18031-18042.

Reniers, Ad, Battjes, J.A., Falques, A. and Huntley, D.A. (1994): Shear-wave Laboratory experiment, *Proc. of Int. Symp. Waves- Physical and Numerical Modelling*, 356-365.

Thornton, E. B. (1970): Variation of longshore current across the surf zone, *Proc. of 12th Int. Conf. on Coastal Eng., ASCE*, 291-308.

Visser, P. J. (1984): Uniform longshore current measurements and calculations, *Proc. of 19th Int. Conf. on Coastal Eng., ASCE*, 2192-2207.

Table 1 Computational conditions

	Wave-1			Wave-2		
	Height (m)	Period (s)	Incident angle (°)	Height (m)	Period (s)	Incident angle (°)
Case 1					8.55	-19.4
2	2.0	8.0	45	0.15	8.13	-17.5
3					6.90	-12.5

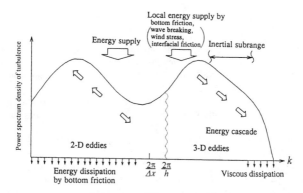

Fig.1 Conceptual spectral structure of shallow-water turbulent flow.

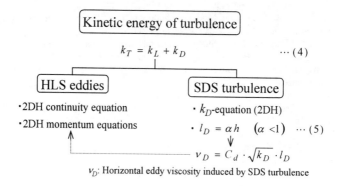

v_D: Horizontal eddy viscosity induced by SDS turbulence

Fig.2 Conceptual illustration of SDS&2DH model.

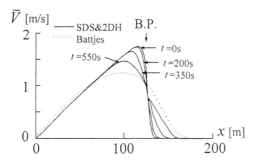

Fig.3 Cross-shore profile of longshore current.

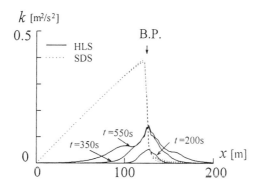

Fig.4 Cross-shore distribution of turbulent kinetic energy.

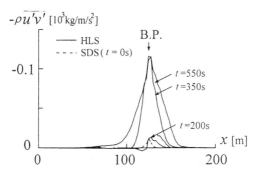

Fig.5 Cross-shore distribution of Reynolds stress.

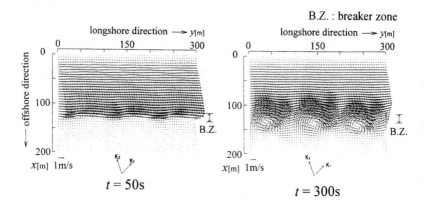

Fig.6 Velocity field for Case1.

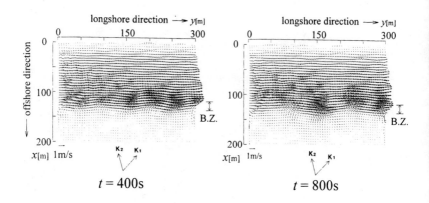

Fig.8 Velocity field for Case3.

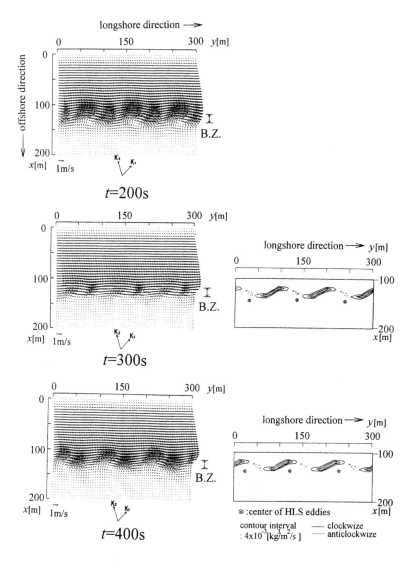

Fig.7 Velocity field, spatial distribution of torque and
location of HLS eddies (Case2).

Shear Instabilities of Longshore Currents: Flow Characteristics and Momentum Mixing During Superduck

H. Tuba Özkan-Haller and James T. Kirby[1]

Abstract

The shear instability climate during the SUPERDUCK field experiment is simulated. Due to uncertainties in the friction and lateral mixing coefficients, numerical simulations are carried out for a realistic range of values for these coefficients. The resulting flow structures can be characterized as unsteady vortices propagating in the longshore direction. These vortices interact, occasionally merge and are shed offshore. During the shedding process, locally strong offshore directed currents are generated. Lateral momentum mixing induced by the finite amplitude shear instabilities is analyzed and found to be an important mixing mechanism in the surf zone.

Introduction

Oltman-Shay *et al.* (1989) observed a meandering of the longshore current during the SUPERDUCK experiment over time scales up to $O(1000$ sec). Using data from a longshore array of current meters located in the trough region of the shore-parallel bar at SUPERDUCK, Oltman-Shay *et al.* (1989) constructed frequency-longshore wavenumber spectra of the velocities. Such a spectrum for the longshore velocities on October 16 is reproduced in Figure 1. The dispersion line for a 0-mode edge wave is also shown for reference. The motions in question, termed shear waves, can be readily distinguished from edge waves due to their nondispersive character and can be seen to dominate the frequencies less than 0.01 Hz. For the purposes of defining a dispersion line of the motions at low frequencies, cut-off lines of shear wave energy are defined and are shown in Figure 1. The equation for the shear wave dispersion line is noted above the plot. For more information about the estimation of the propagation speeds of the observed motions, the reader is referred to Özkan-Haller (1997).

Bowen and Holman (1989) performed an analytic study and showed that a shear instability of the longshore current can reproduce the nondispersive character and meandering nature of the motions observed by Oltman-Shay *et al.* (1989). Although several other mechanisms have been proposed to explain the experimental observations, the instability theory has, so far, been the most studied alternative. Linear instability computations as well as nonlinear computations have been carried out by various investigators for realistic current and bottom profiles. Much has been learned about the instability properties of longshore currents as well as the nature of the fully developed fluctuations. However, most studies analyzing finite amplitude behavior have been carried out using analytic bottom and longshore current profiles and show

---

[1]Center for Applied Coastal Research, University of Delaware, Newark, DE 19716, USA. E-mail: ozkan@coastal.udel.edu

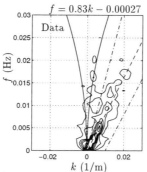

Figure 1: Frequency-cyclic longshore wavenumber spectra $S(f, k)$ (m^3/s) for longshore velocities from measurements on October 16. The upper and lower cut-off lines of the shear wave energy ($-\cdot-\cdot-$) and the 0-mode edge wave dispersion lines for a plane beach slope of 0.05 (——) are also shown. The equation for the best fit dispersion line ($-\,-\,-$) is noted above the figure.

that the fully developed fluctuations can behave in a variety of ways ranging from equilibrated small amplitude fluctuations to energetic random fluctuations involving strong vortices depending on the values of the frictional coefficients (e.g. Slinn *et al.*, 1997). In this study we seek to find out which type of behavior is most likely to explain the observations at SUPERDUCK.

We simulate four days, but discuss one day, from the SUPERDUCK field experiment for a range of realistic friction and lateral mixing coefficients and seek to identify friction and mixing coefficients that will reproduce the observed motions at SUPERDUCK. We had previously established (Özkan-Haller and Kirby, 1996) that shear instabilities cause lateral momentum mixing. In this paper, we also seek to assess the importance of this lateral mixing in comparison to the contribution by more traditional mixing mechanisms.

Approach

The short wave and depth-averaged horizontal momentum balance equations and the continuity equation form a two-dimensional model for the time varying behavior of surf zone currents and are given by

$$\eta_t + \boldsymbol{\nabla} \cdot d\mathbf{u} = 0$$
$$\mathbf{u}_t + \mathbf{u} \cdot \boldsymbol{\nabla}\mathbf{u} = -g\boldsymbol{\nabla}\eta + \tilde{\tau} + \tau' - \tau_b. \qquad (1)$$

Here, η is the short wave-averaged water surface elevation, h is the still water depth, $d = h + \eta$ is the total water depth, $\mathbf{u}$ is the current velocity vector with offshore (x) component u and longshore (y) component v.

The parameters $\tilde{\tau}$, τ' and τ_b represent the effects of short wave forcing, lateral momentum mixing and frictional damping, respectively. We purposefully choose to include these effects in a rudimentary fashion so that we can identify the resulting motions in the mathematically simplest setting.

The short wave forcing is modeled utilizing the radiation stress formulation and linear wave theory. The bathymetry at SUPERDUCK is characterized by a steep fore-shore and a shore-parallel bar formation approximately 60 m offshore. Assuming straight-and-parallel bottom contours, the measured SUPERDUCK bathymetry at a transect is used to compute the wave height decay due to breaking utilizing the wave height transformation model by Whitford (1988).

Lateral momentum mixing due to turbulence as well as the Taylor dispersion process identified by Svendsen and Putrevu (1994) are modeled utilizing an eddy viscosity formulation. The eddy viscosity ν represents a combined viscosity $\nu = \nu_t + \nu_D$, where ν_t is the turbulent eddy viscosity and ν_D is the viscosity induced by the Taylor mixing process. The parameter ν_D is a function of the depth variations of the current velocities and its specification requires the computation of the depth profiles using a profile model. In this study, we choose to include an order of magnitude estimate of the effect of ν_D by parameterizing the combined eddy viscosity ν following Battjes (1975) as

$$\nu = Md\left(\epsilon_b/\rho\right)^{1/3}, \tag{2}$$

where ϵ_b is the ensemble-averaged energy dissipation due to wave breaking and M is a mixing coefficient. This relationship was originally formulated to account for the cross-shore variation of the turbulent eddy viscosity ν_t. Therefore, the order of magnitude of M needs to be reevaluated when parameterizing the combined viscosity ν. The order of magnitude of M for field applications is estimated to be

$$0.06 < M < 0.48. \tag{3}$$

A more detailed discussion of the treatment of the mixing terms as well as the estimation of the order of magnitude of M can be found in Özkan-Haller (1997).

The bottom friction is modeled using linear damping terms in the momentum equation with an empirical friction coefficient c_f such that $\boldsymbol{\tau_b} = (2/\pi d)u_0 c_f \mathbf{u}$, where u_0 is the amplitude of the wave orbital velocity and can be related to the local wave height. Thornton and Whitford (1996) used current, wind and wave measurements obtained during SUPERDUCK to determine an appropriate friction coefficient by examining the longshore momentum balance for the period of October 15 through October 18. They obtained values for the friction coefficient c_f in the range

$$0.001 < c_f < 0.004. \tag{4}$$

We simulate the shear instability climate by initiating the computations with the fluid at rest except for small perturbations in the velocities for the range of friction and mixing coefficients given by (3) and (4), respectively.

Flow Properties

We first simulate the shear instability climate for October 15 through 18 at SU-PERDUCK for a range of friction coefficients. The mixing coefficient is kept constant for these simulations at $M=0.25$. The resulting time series in the bar trough region for October 16 are shown in Figure 2. Also shown are low-pass filtered velocity time series from the current meter located closest to the cross-shore transect. The time series show that the instabilities reach finite amplitude about 30 minutes into the simulation. For $c_f = 0.003$ the instabilities reach finite amplitude slightly sooner than for the other cases. The mean (longshore-averaged) current in each case is generated at the same time as the spin-up of the instabilities and displays some time variability. For all three cases, short time scale oscillations are observed during the first 30 minutes of the simulation after the instabilities are initiated. Further into the simulations, oscillations with longer time scales are observed. The longshore current oscillations are very energetic and are negative at times. This behavior is not observed in the time

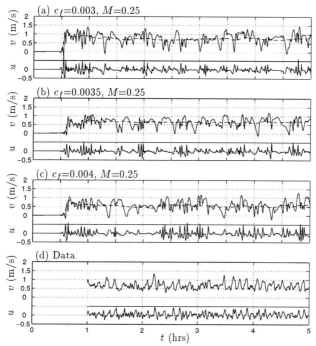

Figure 2: Time series of velocities u, v (———) and $\langle v \rangle$ (— — —) for October 16 at $(x, y) = (35$ m, $L_y/2)$ and time series of velocities u and v of data.

series of the data where longshore velocities are always observed to be larger than 0.25 m/s.

The frequency-longshore wavenumber spectra for the longshore velocities are shown in Figure 3. As the friction factor is decreased, the range of frequencies affected by the instabilities increases. The propagation speed of the motions also increases. The propagation speed seen in the data (see Figure 1) is reproduced to within 5% with $c_f = 0.0035$. Analysis of the mean longshore current profile for the cases (not shown) reveals that, as expected, a stronger current results for lower friction. Furthermore, the cross-shore distributions of the kinetic energy associated with the cases (not shown) shows that the resulting fluctuations are more energetic for lower friction.

Figure 4 shows a sequence of snapshots of the vorticity field, $q = v_x - u_y$, at several time levels for the simulation with $c_f = 0.0035$ and $M = 0.25$. The first two layers of positive and negative vorticity are observed to be closely confined to the shoreline. The outer two layers of vorticity display complicated behavior. Regions of concentrated positive and negative vorticity are observed. They are irregularly spaced in the longshore direction and display the character of vortex pairs and propagate in the longshore direction at a rate of about 250 m in 5 minutes (0.85 m/s). Vortices appear to be shed propagating multiple surf zone widths offshore. During the shedding process locally strong offshore directed currents are generated. The offshore vortices

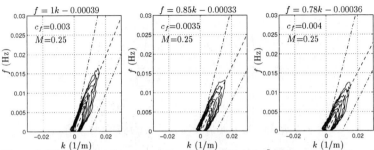

Figure 3: Frequency-cyclic wavenumber spectra $S(f, k)$ (m³/s) for computed longshore velocity on October 16 at $x =35$ m. See Figure 1 for more explanations.

appear in the form of pairs (see $t + 5$ min, $(x, y) \approx (400, 500)$ m) or single positive vortices surrounded by a region of negative vorticity (see $t + 10$ min, $(x, y) \approx (250, 1750)$ m). The behavior seen here corresponds to a category of flow termed turbulent shear flow by Slinn *et al.* (1997).

Next, simulations are carried out by fixing the friction coefficient at the value that reproduces the propagation speed seen in data (c_f=0.0035) and varying the mixing coefficient M. Frequency-longshore wavenumber spectra of the longshore velocities are shown for three cases (M=0, 0.25, 0.5) in Figure 5. It is observed that the propagation speeds vary by less than 10% for higher values of M. The size of this variation is equivalent to the uncertainty in the estimates of the propagation speed from the spectra (see Özkan-Haller, 1997). The propagation speed inferred from the data is well reproduced. It is noted that the spectra do not exhibit significant differences for the different M values. However, analysis of the kinetic energy for the cases (not shown) reveals that the resulting fluctuations are more energetic for lower M, especially in the region seaward of the bar trough.

For $M = 0$, a snapshot of the vorticity displayed in Figure 6 shows the presence of strong vortices both in the nearshore as well as offshore whereas vorticity fields for $M = 0.25$ and $M = 0.5$ show that the length scales associated with the disturbances increase since the fast oscillations are damped out by the diffusional effect of the eddy viscosity mixing term. The vortices associated with the higher M-values can also be observed to be weaker. In general, a more energetic instability climate results for lower M. However, it should be noted that all three simulations show evidence of highly transient, nonlinear vorticity waves that strengthen, weaken and interact, and vortices are frequently shed offshore. During this process the flow structures exhibit properties of transient rip currents.

Momentum Mixing due to Instabilities

In the absence of any fluctuating motions a steady mean longshore current $V(x)$ results from the momentum balance in the y direction. In the presence of fluctuating motions, the mean momentum balance in the longshore direction that leads to the generation of a longshore current can be obtained by longshore- and time-averaging the y-momentum equation. The resulting mean balance given by

$$\overline{\langle u\frac{\partial v}{\partial x}\rangle} + \langle\frac{\overline{\mu}}{d}v\rangle - \langle\overline{\tau_y'}\rangle = \langle\overline{\bar{\tau}_y}\rangle \tag{5}$$

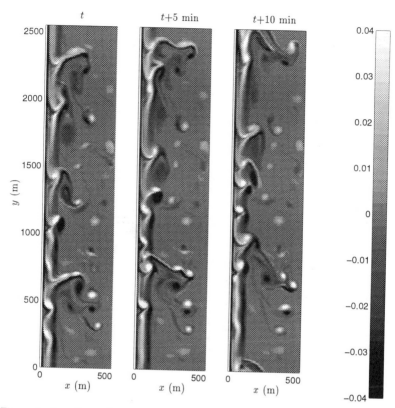

Figure 4: Snapshots of contour plots of vorticity q (1/s) at 5 minute intervals for $c_f = 0.0035$ and M=0.25.

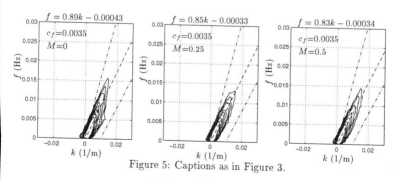

Figure 5: Captions as in Figure 3.

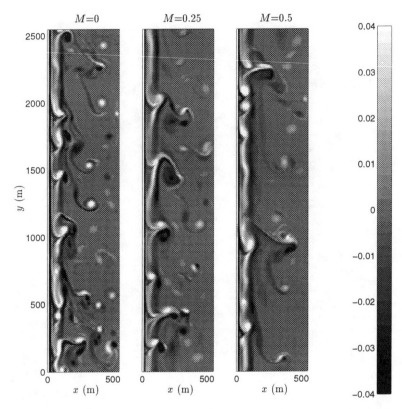

Figure 6: Contour plots of vorticity q (1/s) at $t = 5$ hrs for $c_f = 0.0035$ and $M = 0, 0.25, 0.5$.

states that the short wave forcing $\langle \overline{\tau}_y \rangle$ is balanced by bottom friction $\langle \overline{(\mu/d)v} \rangle$ and lateral mixing caused by the shear instabilities $\langle \overline{u \, \partial v / \partial x} \rangle$ as well as lateral mixing caused by other processes discussed earlier $\langle \overline{\tau_y'} \rangle$.

The time and longshore-averaged longshore currents for the three cases involving $M = 0, 0.25, 0.5$ are shown in Figure 7. Also shown is the current profile V that would result in the absence of any fluctuating motions for $M=0.5$. Examining the mean current velocities in the presence of the instabilities $\langle \overline{v} \rangle$, we notice that the strength of the current peak as well as the seaward shear of the current profiles are very similar although the instability climates that generated them are very different. Current measurements from the sled in the bar and trough regions are reproduced. The predicted maximum current is also of the measured size.

We examine the mean momentum balance in the longshore direction for the three cases utilizing Figure 8. The contribution of the bottom friction is very similar for all three cases confirming that the resulting mean longshore current profile for all the cases should also be similar. The short wave forcing term is the same for all

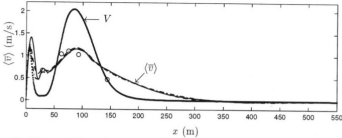

Figure 7: Time and longshore-averaged longshore currents $\langle \overline{v} \rangle$ for October 16 for $c_f = 0.0035$ and $M = 0$ (———), $M = 0.25$ (— — —), $M = 0.5$ (—·—·—) and sled data (○). Also shown is the mean current in the absence of shear instabilities V for $c_f = 0.0035$ and $M = 0.5$ (● ● ●).

three cases. The mixing caused by the instabilities in the trough region is almost identical in all three cases, confirming that differences between the cases should be minor in the bar trough region. However, differences in the induced mixing are more pronounced offshore of the bar trough. The eddy viscosity mixing term $\langle \tau'_y \rangle$ makes no contribution for $M = 0$. Its contribution increases as M increases. The term is especially active around the shoreline jet, where the shear instabilities do not induce mixing, and around the longshore current peak. We note that the mixing caused by the instabilities is much larger than mixing induced due to the $\langle \tau'_y \rangle$ term. However, the contribution due to the shear instabilities decreases as the contribution of the $\langle \tau'_y \rangle$ term increases.

Summary and Discussion

In the scope of this study, the effects of short wave forcing, bottom friction, turbulent momentum mixing as well as the effects of the depth variations of the current velocities were included in a rudimentary fashion. We incorporated simplifying assumptions that reduced the effects of these processes to the mathematically simplest formulations. We found that a stronger mean longshore current, more energetic fluctuation velocities and faster propagation speeds result if the friction factor is decreased. For three days from the SUPERDUCK data set, observed propagation speeds were reproduced. The general shape of the frequency-longshore wavenumber spectra of the velocities was reproduced for frequencies less than 0.01 Hz.

When linear instability theory was first proposed investigators such as Dodd *et al.* (1992) and Reniers *et al.* (1997) proceeded by examining the linear instability of current profiles that were measured in the field or in the laboratory. They obtained good predictions for the range of unstable wavenumbers leading to the concept that the observed fluctuations may be due to weakly nonlinear disturbances. However, we show that for a friction coefficient that reproduces the propagation speed inferred from the data and for a realistic range of mixing coefficients, the resulting motions all have the characteristics of highly transient, nonlinear vorticity waves. These results suggest that the shear wave climate observed during SUPERDUCK may be of a highly nonlinear nature.

The predicted motions cause significant momentum mixing in the surf zone and alter the resulting longshore current profile significantly (see Figure 7). The momen-

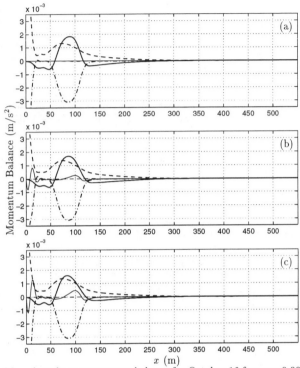

Figure 8: Mean longshore momentum balance for October 16 for $c_f = 0.0035$ and (a) $M = 0$, (b) $M = 0.25$ and (c) $M = 0.5$. $\langle u(\partial v/\partial x) \rangle$ (thick ———), $-\langle \tilde{\tau}_y \rangle$ (—·—·—), $-\langle \overline{\tau_y'} \rangle$ (thin ———), $\langle \frac{\mu}{D} v \rangle$ (thick – – –), residual (thin – – –).

tum mixing due to the instabilities is especially pronounced around the longshore current peak. Mixing is also induced in the bar trough region causing the generation of significant mean longshore current velocities even in the absence of other more traditional mixing mechanisms. In the presence of other mixing mechanisms, that we parameterize using an eddy viscosity formulation, the mixing due to the instabilities decreases. We find that the total amount of mixing in the surf zone is a constant for a fixed friction coefficient, so that virtually the same mean longshore current profile is generated regardless of the value of the mixing coefficient. If the mixing coefficient is increased, the amount of mixing due to the instabilities decreases proportionally. An increase in M causes relatively small variations in the propagation speeds, but an overall decrease in the total energy of the fluctuations. However, differences in the cases are less pronounced in the nearshore region, especially in the bar trough. However, the cases associated with the different mixing coefficients are very different since an increase in the mixing coefficient results in longer, less energetic motions with weaker vortices.

Compared to the fluctuation-free current profile V, the resulting current profile including the effect of the finite amplitude instabilities $\langle \bar{v} \rangle$ displays a weaker peak along with a milder back shear (see Figure 7). It is important to note that the measured mean longshore current profile includes the effects of momentum mixing due to the fluctuations. The linear stability characteristic of the measured current is, therefore, not necessarily descriptive of the shear instability climate that exists along with it. Rather, the stability characteristics of the often unknown, but relevant, fluctuation-free initial state should be considered. This observation is also noted by Slinn *et al.* (1997), who show that linear instability computations of the final state may lead to good predictions of the range of unstable wavenumbers; however, the growth rates will be underpredicted.

Examining the mean longshore momentum balance, we found that the mixing induced by the instabilities is larger than mixing due to the eddy viscosity terms for reasonable mixing coefficients. We note that the cross-shore distribution of the momentum mixing due to the instabilities is such that the location of the longshore current peak is not altered. For all simulations carried out for the SUPERDUCK experiment, the longshore current peak was located over the bar crest. Sled measurements during SUPERDUCK also indicate that the current maximum occurred on the bar crest. It is noted that a useful engineering tool would be obtained if an appropriate parameterization of the mixing induced by the shear instabilities can be constructed. The mean longshore current profile in the presence of the instabilities can then be determined without carrying out lengthy computations of the time-dependent nature of the flow.

Acknowledgments. Thanks are due to Drs. J. Oltman-Shay and N. Dodd for providing the SUPERDUCK data as well as the software to obtain estimates for two-dimensional spectra of the data. This research has been sponsored by the Office of Naval Research, Coastal Sciences Program.

References

Battjes, J. (1975). "Modeling of turbulence in the surf zone." *Proc. Symposium on Modeling Techniques*, San Francisco, 1050–1061.

Bowen, A.J. and R.A. Holman (1989) "Shear instabilities of the mean longshore current. 1. Theory." *J. Geophys. Res.*, **94**, 18023–18030.

Dodd, N., Oltman-Shay, J. and Thornton, E. B. (1992) "Shear instabilities in the longshore current: A comparison of observation and theory." *J. Phys. Oceanography*, **22**, 62–82.

Oltman-Shay, J., P.A. Howd and W.A. Birkemeier (1989). "Shear instabilities of the mean longshore current. 2. Field Observation." *J. Geophys. Res.*, **94**, 18031–18042.

Özkan-Haller, H.T. (1997). "Nonlinear evolution of shear instabilities of the longshore current." Ph.D. dissertation, University of Delaware.

Özkan-Haller, H.T. and Kirby, J.T. (1996). "Numerical study of low frequency surf zone motions." *Proc. 25th Intl. Conf. Coastal Eng.*, 1361–1374.

Reniers, A.J.H.M., J.A. Battjes, A. Falqués, D.A. Huntley (1997). "A laboratory study on the shear instability of longshore currents." *J. Geophys. Res.*, **102**, 8597–8609.

Slinn, D.N., J.S. Allen, P.A. Newberger and R.A. Holman (1997). "Nonlinear shear instabilities of alongshore currents over barred beaches." *J. Geophys. Res.*, in review.

Svendsen, I.A. and U. Putrevu (1994). "Nearshore mixing and dispersion." *Proc. Roy. Soc. Lond. A*, **445**, 561–576.

Whitford, D.J. (1988). "Wind and wave forcing of longshore currents across a barred beach." Ph.D. Dissertation, Naval Postgraduate School.

Morphological Effect of Spring-Neap Tidal Variations on Macrotidal Beaches

P.R.Fisher, T.J.O'Hare & D.A.Huntley

Abstract

A simple morphodynamic model is presented which adopts an energetics based approach for the determination of cross-shore structure of sediment transport for the investigation of beach profile evolution.

This approach is not new, but the translation of a transport function across the beachface, emulating the dynamic regime of a macrotidal environment, represents a departure from previous contributions. The development of this stable, high resolution model facilitates the investigation of the morphodynamic effect of a large tidal range and the interaction of various harmonic components, which become significant outside of a microtidal regime.

The model predicts the formation of bars at the approximate break-point, although these formations only occur at tidal standing positions. Once formed, high and low water bars converge, with offshore movement of the high water bar. Interestingly, with the introduction of a second tidal component at S_2 frequency, formations occur at the break-point positions relating to high and low water of *full springs* and *full neaps*, suggesting that a period of several tidal cycles is required for the establishment of these formations. Temporary formations occur in the interim tidal standing locations but are quickly broken down by the translation of the surf and shoaling zones.

The model offers many possibilities for investigating the morphological effect of wave/tide interaction, as well as surf/shoaling dominance and tidal harmonic interactions.

Institute of Marine Studies, University of Plymouth, Drake Circus, Plymouth, UK, PL4 8AA

Introduction

Researchers have recently focused on the effect of large tidal ranges on beach sediment dynamics (e.g. Short, 1991; Masselink, 1993). It has been suggested that rapid immersion and emergence of the beach-face in the mid-tidal zone suppresses net transport rates in a macro-tidal regime, and that the characteristic profiles of macrotidal beaches are a result of tidal translation of the swash, surf and shoaling wave zones.

Fisher & O'Hare (1996) used a simple profile evolution model to investigate the relative importance of tidal range and wave height on the profile shape. The model was based on field measurements from the British Beach and Nearshore Dynamics experiment (Davidson *et al*, 1992). Using the energetics approach to sediment transport modelling (Bailard, 1987), Russell and Huntley (1996) evaluated bed load and suspended load transport significant velocity moments. They proposed the existence of a quasi-universal cross-shore structure for sediment fluxes on high energy beaches. Combining data from reflective, intermediate and dissipative sites they produced a 'shape function' predicting net onshore transport seaward of the breakpoint and net offshore transport shoreward of it. In Fisher & O'Hare (1996) the shape function is used to evolve beach profiles by translating it at a tidal celerity across the nearshore region and calculating the resulting bed elevation changes at each time-step. The effects of weighted residence times of the swash, surf and shoaling zones across the beach-face mean that temporal variations in cross-shore beach profiles are a function of the tide signal, as well as variations in the incident wave field.

This contribution presents a development of the shape-function approach, so that sediment fluxes are quantitatively correct assuming the application of Airy wave theory across the nearshore zone. The introduction of a wave energy dissipation model, such as that of Battjes and Janssen (1978) is necessary to stabilise the profile evolution over run periods of more than a few hours. It is suggested that tidal variation is particularly significant in a macro-tidal regime, where surf and swash zone dominance can switch within a spring-neap cycle. With no tidal range the shape-function model predicts generation and offshore migration of sand waves through the surf-zone, and developments in time-lapse video surveillance have shown this same phenomena in a micro-tidal environment (Lippman *et al*, 1993). However, the 'masking' of this process which occurs within environments with large tidal excursions is difficult to assimilate.

A useful overview of the B-BAND experiment is given by Davidson *et al* (1992). This three-year experiment investigated surf zone processes at three high energy macrotidal locations around the UK. The sites were chosen to represent dissipative, intermediate and reflective beaches following the morphodynamic classifications of Wright and Short (1984). The locations are illustrated in Figure 1 along with typical beach profiles and are described briefly in turn.

Figure 1. Location of B-BAND field sites around the UK.

Llangennith: A dissipative high energy beach with a shallow beach gradient (0.014-0.020) with fine to medium quartz sands (D_{50}=0.21mm). This beach is west facing, has a tide range up to 9 metres, and broad surf (up to 350m) and intertidal ($\sim$ 500m) zones.

Spurn Head: An intermediate beach located near the end of a 5km long spit facing the North Sea. The beach profile comprises a steep high tide beach ($\tan\beta \sim$ 0.0975) and a low tide terrace ($\tan\beta \sim$ 0.023, and D_{50}=0.35mm). This beach has a tide range up to 6m and experiences strong longshore tidal currents which have a significant effect on surf zone dynamics.

Teignmouth: A reflective beach site facing south-east into the English Channel sheltered from Atlantic swell. The beach is backed by a sea wall, and has a gradient of 0.067-0.142. Cross -shore variation of sediment size at this location is significant, with a D_{50} value of 0.24mm. Tidal range is 6 metres.

Shape Function Approach To Sediment Transport

Foote (1994), Foote *et al* (1994), and Russell and Huntley (1996) examined field measurements of cross-shore sediment transport from these three locations. They calculated velocity moment contributions to sand transport, by assuming the instantaneous velocity field to be composed of the following contributions:

$$u = \bar{u} + u_S + u_L \tag{1}$$

where $\bar{u}$ is the mean flow, u_S is the incident wave component of velocity, and u_L is the long wave component. Using Bagnold's (1963, 1966) unidirectional stream flow total load sediment transport model, modified by Bailard (1981, 1987) for cross shore sediment transport under bi-directional flow, Guza and Thornton (1985) used field measurements to examine the relative importance of the velocity moment terms in Bailard's model, and found the normalised transport significant terms to be:

$$\overline{|u|^2 u}\Big/\left(\overline{u^2}\right)^{3/2} \qquad \text{for bedload transport, and} \qquad (2)$$

$$\overline{|u|^3 u}\Big/\left(\overline{u^2}\right)^{2} \qquad \text{for suspended load transport} \qquad (3)$$

Foote *et al* (1994) examined the cross-shore distribution of the significant moments for bedload and suspended load and found in both cases that the sum predicted a net onshore movement shoreward of the breakpoint and a net offshore movement seaward of the breakpoint. To clarify this pattern for each case they normalised the water depth by the depth of wave breaking. The position of wave breaking was determined from the position of maximum wave height from pressure transducer records and this was checked with visual observations in the field. Combining data from all three field sites they noted that not only was there a pattern of net onshore transport shoreward of wave breaking and a net offshore transport seaward of it, but that a curve could be reasonably fitted through the velocity moments calculated for the interim positions.

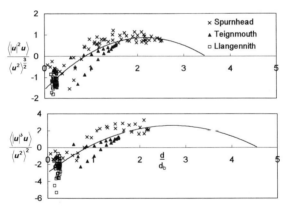

Figure 2. Sediment transport shape functions obtained from
the B-BAND data, bedload (top) and suspended load
(bottom).

Foote *et al* subsequently proposed the existence of quasi-universal, cross-shore sediment transport spatial distribution curves for bed load and suspended load on high energy beaches. These sediment 'shape functions' are illustrated in Figure 2 with second order polynomials fitted through each data set. O'Hare (1994) proposed that this 'shape function' approach be adopted to develop a tentative cross-shore sediment transport model, suggesting that the patterns observed in the B-BAND data are in fact required by the break-point bar hypothesis, and that characteristic macrotidal profiles may be constructed by the repeated tidal excursion of the shape

function. This contribution presents the preliminary results from the development of
such a model.

The model

The model simply translates the flux profiles shown in Figure 2 across an initially
linear beach face according to a tidal signal with M_2 and S_2 components. At each
time-step (determined using a Courant type stability criterion) a new tide level is
evaluated using a simple harmonic model, from which a cross-shore depth profile is
calculated. From this, the wavenumber k is calculated using the general Airy

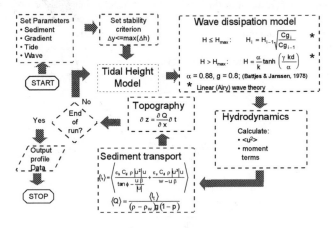

Figure 3. Model Flowchart.

expression and a Newton-Raphson iteration which is then used to calculate shoaling
and breaking wave heights. Wave breaking occurs at criteria suggested by Battjes &
Janssen (1978), (*dissipation coefficient $\alpha = 0.88$, breaking coefficient $\gamma = 0.8$*). The
wave height profile H is used to calculate time-averaged values for u^2 (denoted in
Figure 2 as $\langle u^2 \rangle$) again assuming Airy wave theory (linear waves). These values are
used to un-normalise the shape functions in Figure 2, to give expressions for $\langle |u|^2 u \rangle$
and $\langle |u|^3 u \rangle$ which are then used to determine the immersed weight sediment
transport rate (from Bagnolds energetics-based formula) It, volumetric sediment
transport rate, Q, and bed level change δz. The beach profile is modified accordingly,
and the model time is incremented. Tide level is re-evaluated and the new profile is
used in the calculation of water depth for the next iteration. By using a small spatial
resolution ($\delta x \sim 1m$) and relative temporal resolution ($\delta t \sim < 5$ *incident wave periods*),
it is possible to keep the profile evolution stable, despite the very high number of
iterations required to evolve the profile over several spring-neap cycles. The
mechanics of the model are summarised in Figure 3.

Results

Figure 4 is a 3-D illustration of a model evolution of a linear beach over 60 tidal cycles with a monochromatic (at M_2 frequency) tidal signal. The white line near the top of the beach illustrates the high tide position, beach profiles that make up this illustration were sampled at every high tide, so there are negligible 'between tide' changes which are not illustrated here. The illustration shows the rapid formation of two bars, one at the high tide break-point, and one at the low tide break-point, (the break-points are the points of transport 'convergence') there is substantial erosion at the shore line and around depth of closure. Through time there is a convergence of the two bars, as the high water bar gradually migrates offshore, and the low water bar steepens. Figures 5 and 6 are the same evolution, but from a 2-D perspective;

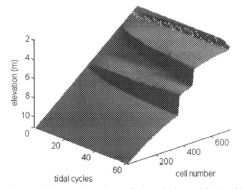

Figure 4. Linear beach profile evolution with M_2 tide only and 1m waves.

dark areas represent erosion, and light areas represent accretion. Figure 6 is the same as Figure 5, but the pattern is overlain with the break-point excursion, represented by a narrow white line. These figures reveal the same bar formation patterns, but Figure 6 shows that the positions of the bars follow closely the positions of high tide and low tide standing of the break-point.

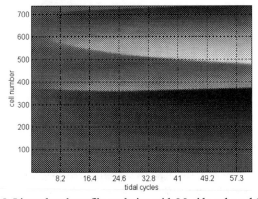

Figure 5. Linear beach profile evolution with M_2 tide only and 1m waves.

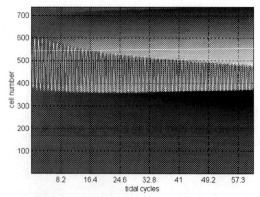

Figure 6. Linear beach profile evolution with M_2 tide only and 1m waves.

Figure 7 is a 3-D illustration of a model evolution of a linear beach over 60 tidal cycles with a *bi-chromatic* tidal signal (with components at M_2 and S_2 frequency). The white line again illustrates the high tide position, so that during this evolution the tide moves from spring to neaps, and then back to springs again. With the introduction of the second tidal component the evolution is 'subdued' by the continual variation of the level of tidal standing, and the result after 60 tidal cycles is not as extreme as for a monochromatic tide. Similar bar formations are observed at high and low water break-points, but this time, as positions of high and low water converge from spring tides through to neaps, new bars are formed at neap high and low tide. There are temporary

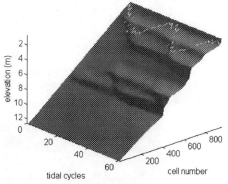

Figure 7. Linear beach profile evolution with M_2 and S_2 tide and 1m waves.

formations during the interim period between full springs and full neaps, but these are not substantial due to the limited residence of tidal standing of these levels, and they are quickly broken down by tidal translation.

As before, the subsequent Figures 8 and 9 are a 2-D view of the same evolution from a bi-chromatic tide. Figure 8 is overlain with the oscillating break-point excursion, represented by a narrow white line. These figures highlight the accretional nature of the high water bar (shown as light areas) and the erosion at the

bottom of the low water bar (shown as dark areas). Examination of Figure 8 shows
that again the accretional patterns follow closely the tidal standing position of the

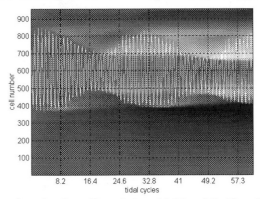

Figure 8. Linear beach profile evolution with M_2 and S_2 tide and 1m waves

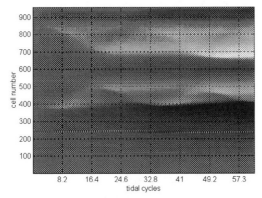

Figure 9. Linear beach profile evolution with M_2 and S_2 tide and 1m waves.

break-point. Because of the 'smoothing' effect of a continual variation of tidal
standing levels, the offshore migration of the high water bar is not as rapid as the
previous monochromatic case, but it is significant. As before all bar formations tend
to converge toward each other, with the focus of convergence being at around mean
water level.

Figure 10 is a 3-D illustration of a model run again with an M_2-S_2 tidal signal,
but on this occassion the amplitude of the two signals was double that of the model
illustrated in Figure 7, giving a greater separation between the spring and neap tidal
standing levels. The sediment flux efficiency factors ε_b and ε_s were also increased, to
try and facilitate the formation of structures between spring and neap tidal standing

levels. The result is that, despite the modification to tidal dynamics and sediment transport rates, the profile in Figure 10 is very similar to that in Figure 7, with significant formations still only at high and low water break-points of full spring and full neap tides, with intermediate formations negligible (not really visible in Figure 10). The result indicates that in the model, tidal dynamics are very much the dominant mechanism for the formation of morphological features.

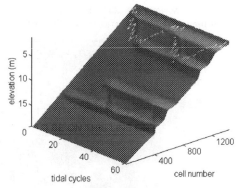

Figure 10. Linear beach profile evolution with large M_2 and S_2 tide and 1m waves.

Discussion

This contribution discusses some of the early results of a simple morphodynamic model that will allow us to examine some of the short-term (tidal cycle) and long term (spring-neap cycle) morphological effects of tidal translation of a cross-shore structure of sediment transport. These early results show the development of some morphological features that occur in the natural environment, as well as some that do not; particularly striking is the formation of substantial bar structures on the high tide portion of the beach, these are mostly absent within the environments described by the B-BAND experiment.

Several questions immediately bear some consideration; for example, what are the effects of introducing a non-linear profile (of say, a Dean form, or an actual beach profile) at the start? It is possible that a shoreward steepening profile will suppress transport on the high water part of the beach and lead to more realistic formations in this region. In addition, the results presented are all formed from a rather unrealistic continuous 1m wave field. It will be interesting to examine the beach response to a significant change in this field. It is anticipated that the beach will erode during a transition to larger waves, but the model's ability to recover and move sediment onshore during quiescent periods is as yet untested. The other question which arises is the validity of a generic approach to determining the sediment transport shape function. On one of the beaches examined as part of the B-BAND investigation (Llangennith), it is the stirring action of long period waves which is the most significant source for transport by the mean flow. The model results presented here are for a linear beach of similar gradient and sediment grain size as the one at Llangennith. Examination of the cross-shore structure of velocity moments for long wave stirring and transport by the mean flow reveal that offshore transport in the surf zone is much more significant than for the other two beaches, where stirring by incident waves is the dominant mechanism. It may be the case that a specific description of the velocity moment structure at each site is required for a reasonable prediction of the beach form, and only when a large data-set of a wide variety of 2-D beaches is obtained, that we may start to generalise with regard to the likely appearance of certain morphodynamic features under certain tidal and hydrodynamic regimes. This model offers ample scope for experimenting with such data sets.

References

Bagnold, R.A. (1963). Mechanics of marine sedimentation, In: *The Sea, Ideas and Observations*, Wiley-Interscience Publishers, New York, **3**, 507-528.

Bagnold, R.A. (1966). An approach to the sediment transport problem from general physics, *U.S. Geol. Surv. Prof. Paper* 422-1.

Bailard, J.A. (1981). An energetics total load sediment transport model for a plane sloping beach, *Journal of Geophysical Research* **86**, 10938-10954.

Bailard, J.A. (1987). Surfzone wave velocity moments. *Proceedings of Coastal Hydrodynamics, ASCE.* 328-342.

Battjes, J.A. & Janssen, J.P.F.M. (1978). Energy loss and set-up due to breaking of random waves. *Coastal Engineering 1978: Proceedings of the 16th international conference ASCE*, 569-587.

Davidson, M.A., Russell, P.E., Huntley, D.A., Hardisty, J. & Cramp, A. (1992). An overview of the British Beach and Nearshore Dynamics (B-Band) programme. *Coastal Engineering 1992: Proc. of the 23rd international conference ASCE 2*, 1987-2000.

Fisher, P.R., & O'Hare, T.J. (1996). Modelling sand transport and profile evolution on macrotidal beaches. *Coastal Engineering 1996: Proceedings of the 25th international conference ASCE 3*, 2994-3005.

Foote, Y.L.M. (1994). Waves, Currents and Sand Transport Predictors on a Macro-Tidal Beach, *Ph.D. Thesis.* Institute of Marine Studies, University of Plymouth.

Foote, Y.L.M., Huntley, D.A. and O'Hare, T.J. (1994). Sand transport on macro-tidal beaches, *Proceedings of the Euromech 310 colloquium*, Le Havre, 360-374.

Guza, R.B. and Thornton, E.B. (1985). Velocity moments in the nearshore zone, *Journal of Waterways, Port, Coastal and Ocean Engineering*, **111**, 235-256.

Lippman, T.C., Holman, R.A. & Hathaway, K.K. (1993). Episodic, nonstationary behaviour of a double bar system at Duck, North Carolina, U.S.A., 1986-1991. *Journal of Coastal Research*, **15**, 49-75.

Masselink, G. (1993). Simulating the effects of tides on beach morphodynamics. *Journal of Coastal Research*, **15**, 180-197.

O'Hare, T.J. (1994). What is the origin of the macro-tidal beach profile? *Unpublished Report,* Institute of Marine Studies, University of Plymouth.

Russell, P.R., & Huntley, D.A. (1996). The spatial distribution of cross-shore transport on high energy beaches. *(in prep.)*

Short, A.D. (1991). Macro-meso tidal beach morphodynamics - an overview. *Journal of Coastal Research*, **72**, 417-436.

Wright, L.D. and Short, A.D. (1984). Morphodynamic variability of surf zones and beaches: a synthesis, *Marine Geology* **56**, 93-11.

Simultaneous Observations on Irregular waves, Currents,
Suspended Sediment Concentration and Beach Profile Changes
in Large Wave Flume

Takao Shimizu[1] and Yoshiaki Kawata[2]

Abstract

Large scale experiments on beach profile change due to
random waves were performed in the CRIEPI FLUME, and the 48
channels data including wave gauges, current meters, turbidity
meters and runup gauge were observed and recorded
simultaneously. Beach profile changes, random waves
deformations in the surf zone and the swash zone were
discussed.

Introduction

In order to clarify the mechanism of beach change, it
is essential to investigate the phenomena that occurs not
only in the surf zone but also in the swash zone. For this
purpose, accurate and detailed observation in the field is
ideal, but it's too expensive and dangerous especially in
the surf zone. Additionally, in the surf zone and swash
zone, wave motion including undertow seems very complex,
because the low frequency component of the wind waves grows
up to significant component while the high frequency component
is breaking. So, in the Large Wave Flume of CRIEPI, large
scale experiments on cross-shore beach profile change due
to random wave incidence were performed and simultaneous
observations of irregular waves, currents, suspended sediment
concentration and beach profile changes were executed by
means of a lot of sensors and human powers

[1]Research Fellow, Central Research Institute of Electric
Power Industry, 1646 Abiko, Abiko-city, Chiba 270-11,
Japan. Tel.+81-471-82-1181. Fax.+81-471-84-7142. E-mail
s-takao@criepi.denken.or.jp.
[2]Professor, Research Center for Disaster Reduction Systems,
DPRI, Kyoto University, Gokasho, Uji, Kyoto 611, Japan.
Tel.+81-774-32-3111. Fax.+81-774-31-8294.

gathering from many researchers. This is the joint experiments by the authors and Dr. Masaaki Ikeno, Dr. Akio Okayasu, Mr. Yoshiaki Kuriyama, Dr. Shinji Satoh, Dr. Hiroaki Shimada, Mr. Takuzo Shimizu, Mr. Satoshi Takewaka, Dr. Ryuichiro Nishi. In these experiments, beach profile changes due to accretional and erosional random wave trains were investigated. Similar projects have been performed in the DELTA FLUME or the Supertank after Arcilla et al.(1993), Kraus et al.(1992). The experience of these projects was a great help to the author's project. The sloshing of water in the flume was prevented by controlling of free long waves due to paddling of the wave generator, and the data sampling from all sensors was completely synchronized.

Experimental method and conditions

As shown in Fig.1, 1/30 uniform slope of sand was piled up in the flume and 4 m depth of the flume was filled up with water. Median diameter of sand particle was 1 mm. 22 capacitance type wave gauges, a capacitance type runup gauge, 11 electromagnetic current meters of dual components type and 3 optical turbidity meters were set up. Completely simultaneous 48 channels of data obtained by these sensors were sampled by a couple of synchronized digital data recorders on sample hold mode. Wave gauge W0 was set up on the initial shoreline. Wave gauges W1 to W5 were placed at intervals of 5 m from the initial shoreline toward the offshore. Wave gauge W6 was 8 m offshore from the position of W5, and W7 to W14 were placed at the intervals of 10 m from the position of W6. Wave gauges W15 to W17 were set up on the flat floor. Wave gauges WA,WB and WC were set up in the swash zone, and W18 was installed on the measuring vehicle. Current meter CA was set up 2 cm above the sand surface on the shoreline making a line with W0. In addition, a sand trap system of plankton net, S, after Shimizu et al.(1996) was set up on the shoreline too. Current meters C1 to C6 were set up 20 cm over the sand surface of the initial slope making lines with the wave gauges W4, W6, W7, W8, W9 and W10, respectively.

Current meters C7, C8 and C9 were set up at the intervals of 40 cm together with turbidity meters of back scattering type T1 and transmission type T2 and T3 on the installing arm of measuring vehicle going up and down. These turbidity meters were calibrated in a mixing tank using the sand and water in the large wave flume. But near the water surface in the surf zone, suspended sediment and bubbles can not be measured separately by the turbidity meters. So, additionally, two sets of pumps and tubes for sampling water with suspended sediment were set up side by side with T2 and T3, after Shimizu et al.(1996). Current meter C10 was set up with T7. The current meters CA and C10 are cross-shore horizontal and alongshore horizontal, and others are cross-shore horizontal and vertical. 4 video

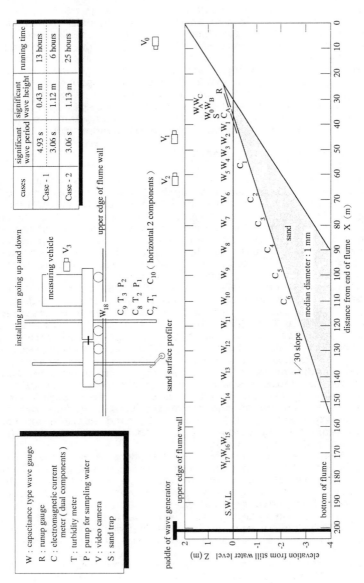

cases	significant wave period	significant wave height	running time
Case - 1	4.93 s	0.43 m	13 hours
	3.06 s	1.12 m	6 hours
Case - 2	3.06 s	1.13 m	25 hours

W : capacitance type wave gauge
R : runup gauge
C : electromagnetic current meter (dual components)
T : turbidity meter
P : pump for sampling water
V : video camera
S : sand trap

Fig.1   arrangement of experiments

cameras were set up along the flume.   Beach profile of
some stages of cases were measured by a couple of wheel
type sand surface profilers.
     Two cases of experiment were performed with the conditions
as shown in Fig.1 using random waves of JONSWAP spectrum.
In the Case-1, 13 hours of running with accretional waves
were followed by 6 hours of running with erosional waves.
In the Case-2, reformed uniform slope of sand was affected
by 25 hours of running with erosional waves.   In both
cases, free long waves due to paddling of piston type wave
generator were controlled by bounded long wave reproduction
system "BLOWRES" after Ikeno et al.(1996).   Fig.2 shows the
record of W1 before and after the stopping of wave generation.
There's a random long wave before the stopping and a
periodic oscillation due to wave set-up after the stopping.
But, no periodic oscillation is recognized before the
stopping.   A random waves train repeats with the interval
of 820 seconds.   So, the each measurements of 48 channels
executed during 820 seconds.   Sampling interval is 0.05
seconds.
     Above mentioned experimental method and conditions are
characterized by coarse sand, uniform initial slope, BLOWRES
and completely synchronized data sampling, comparing to the
other projects in the DELTA FLUME or the Supertank.

Experimental results

     Fig.3 and Fig.4 show beach profile changes, distributions
of significant wave height H1/3 and numbers of waves during
820 seconds of measurement in the Case-1 and Case-2,
respectively.   Ripple marks on the profiles were smoothed
for convenience to compare the profiles to the initial
slope.   In the Case-1, the initial uniform slope changed to
a step profile for the first 13 hours.   For the following 6
hours, the step profile was changing to a bar profile.   In
the initial stage of the Case-1, the wave height took the
maximum value at X=62m, and the number of waves took the
maximum value at X=42m that is about 4 m offshore from the

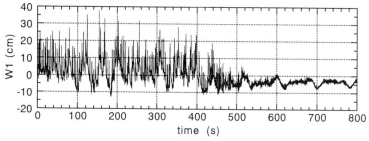

Fig.2   record of W1 before and after stopping of wave
        generation

initial shoreline at X=38m.   The position of the step at
X=50m is almost center of the surf zone.   In the Case-2,
the initial uniform slope changed to a bar profile for 25
hours.   The wave height distribution didn't show a peak,
and the wave height began to reduce at X=120m.   The wave
number took the maximum value at X=47m that is 8 m offshore
from the initial shoreline X=39m.   The position of the bar
at X=56m is about 1/5 of the width of the surf zone,
offshore from the initial shoreline.   In both cases, the
position of the step or the bar looks like just onshore of

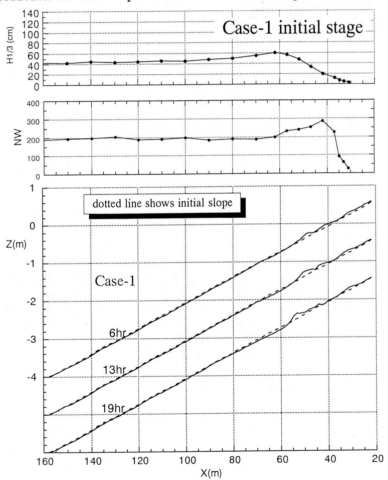

Fig.3   beach profile change and distribution of wave
        hight and number of waves in the Case-1

the position where the number of waves increases.
      Fig.5 shows waves propagating from the offshore zone
to the surf zone in the Case-2. Identifications of sensors
are indicated in the left side of the figures. The wave
gauge W11 is almost on the border of the offshore zone and
the surf zone. The interval of horizontal dotted grids is
1 m. From Fig.5, we can find the breaking process of each
waves in the random waves train. Large waves broke at

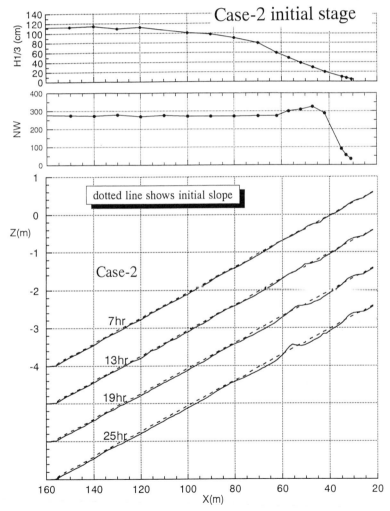

Fig.4   beach profile change and distribution of wave
      hight and number of waves in the Case-2

various places in the surf zone, so the difference of each wave heights becomes small. And at the wave gauge W4, a lot of waves broke primarily or secondarily. This position almost agrees with the bar formation. From the wave gauge W4 to W0, almost all of the wave heights decrease.

Fig.6 shows horizontal components of current velocities and suspended sediment concentrations measured by turbidity meters in the surf zone in the Case-2. Cross-shore horizontal component of velocity is identified by the subscript "u". Alongshore component is identified by 'v'. The current meter C10 and the turbidity meter T1 are placed at X=100m, Z=-1.9m. This position is same to the current meter C5. C8 and T2 are placed at X=100m, Z=-1.5m, and C9

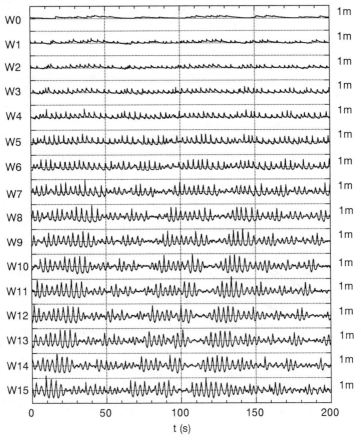

Fig.5   waves propagating from the offshore zone to the surf zone in the Case-2

and T3 are placed at X=100m, Z=-1.1m. Scales of measured values are shown in the right side of the figures as the interval of the horizontal dotted grids. The sediment concentration T1 increases when the amplitude of the current velocity C5-u and the turbulence C10-v increase. But, the sediment concentration T2 only 40 cm above T1, or T3 80 cm above T1 doesn't correspond to T1. The current velocities C1-u to C4-u remarkably include low frequency fluctuations. According these facts, it's suggested that the cloud of suspended sediment lifted up from the sand bed ascend drifting onshore-offshore.

Fig.7 shows waves propagating from the surf zone to the swash zone, W5 to WC, wave runup, R, swash and backwash velocity, CA-u, and sediment transport due to swash, S, in the Case-2. The base line of records of the wave gauges including W0 to WC is the still water level. The baseline of the record of R is the initial shoreline. The record of the current velocity CA-u was masked during the head of the sensor is not submerged. The heights of triangles on the record of S indicate the sand volumes trapped during each swash. The left edges of the triangles are starts of trapping, and the right edges of the triangles are finishes of trapping. In the surf zone, the low frequency components

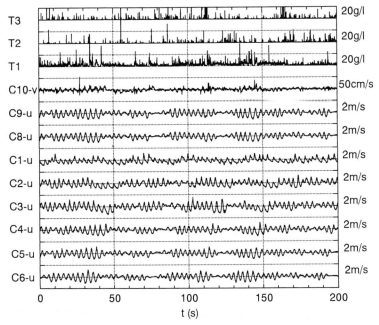

Fig.6 current velocity and suspended sediment concentration
      in the surf zone in the Case-2

of waves increase while the high frequency elements decrease.
The wave profiles change drastically between W1 and W0.
From the record of runup, it can be read that the waves
which do not reach to the shoreline propagate individually,
but the waves which pass across the shoreline are enveloped.
It's clear that the shoreline is the border between the
surf zone and the swash zone. The waves in the swash zone
are grouped, and the groups well correspond to the crest
phases of the low frequency components of the waves in the
surf zone. Comparing the records CA-u to W0, it's recognized
that the swash velocity of the first wave and the backwash
velocity of the last wave in each group are quite large.

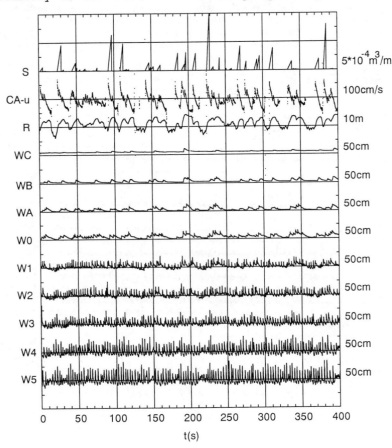

Fig.7   waves propagating from the surf zone to the swash
        zone, wave runup, swash and backwash velocity and
        sediment transport due to swash in the Case-2

So, the sediment transport due to the first swash in a group is quite large. The sediment transport due to backwash is the same.

Conclusions

❶Long wave motion plays great role in suspended sediment transport in the surf zone together with stationary component.
❷Short waves run up into the swash zone in the crest phases of long waves in the surf zone and the bed load sediment transport in the swash zone is motivated by swash and backwash motion of these waves.
❸Significant career effect was not recognized in erosional change of beach profile after accretional change.
❹Beach profile changes due to random waves were occasionally characterized by the bar formation in the middle of the surf zone far from the breaker zone.

Dataset obtained in these experiments will be open to the public for the various advances by all of the researchers in the world. Please send the E-mail with the subject "[request #03]" to the address "s-takao@criepi.denken.or.jp", if you want to access the dataset.

References

Ikeno, M. and H. Tanaka(1996): A new experimental system for reproduction of infragravity waves bounded by irregular wave groups and their wave deformation, Report of Central Research Institute of Electric Power Industry, No.U95042.(in Japanese)

Shimizu, T. and M. Ikeno(1996):Experimental study on sediment transport in surf and swash zones using large wave flume, Proc. 25th Int. Conf. on Coastal Eng., Vol.3, pp.3076-3089.

Arcilla, A.S., J.A.Roelvink, B.A.O'Connor, A.Reniers and J.A.Jimenez (1993): The DELTA FLUME'93 EXPERIMENT, Proc. Coastal Dynamics '94, ASCE, pp.488-502.

Kraus, N.C., J.M.Smith and C.K.Sollitt (1992): Supertank laboratory data collection project, Proc. 23th Coastal Eng. Conf., ASCE, pp.2191-2204.

MORPHOLOGICAL AND SEDIMENT BUDGET CONTROLS
ON DEPTH OF CLOSURE AT DUCK, NC

Robert J. Nicholls[1] and William A. Birkemeier[2]

ABSTRACT

Field observations at Duck show that d_ℓ (Hallermeier, 1981) defines an upper limit to observations of depth of closure due to erosional events and over annual time scales. In addition to extreme wave height, other factors influence closure. For erosional events associated with large waves ($d_\ell > 7$ m), pre-event outer bar volume partly control closure. Based on analysis of annual closures, the sediment budget has some control on closure, an accreting profile produces relatively deeper closures and *vice versa*. Further, the results suggest that the observed closure might exceed d_ℓ . under accreting conditions.

INTRODUCTION

Repetitive beach-nearshore profiles (henceforth beach profiles) define an envelope of variation which declines with depth. The seaward limit of this envelope is generally termed depth of closure. Its precise location is a matter of definition, and is usually based on the accuracy of the measurements. This depth has often been used by coastal engineers as an empirical measure of the seaward limit of significant cross-shore sediment transport. Using an argument based on the critical value of a sediment entrainment parameter, Hallermeier (1981) developed a simple analytical approach to estimate an annual depth of closure (d_ℓ). It is a function of extreme wave conditions and in a generalized time-dependent form is:

$$d_{\ell,t} = 2.28 H_{e,t} - 68.5 (H_{e,t}^2 / g T_{e,t}^2) \qquad (1)$$

1. Middlesex University, Enfield, London EN3 4SF, UK.
2. US Army Engineer Waterways Experiment Station, Field Research Facility, 1261 Duck Road, Kitty Hawk, NC 27949, USA.

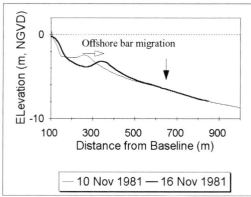

Figure 1. An example of 6-cm closure (black arrow) due to offshore bar migration during an erosional event on PL188.

where $d_{\ell,t}$ is the predicted depth of closure over t years, referenced to Mean Low Water (MLW); $H_{e,t}$ is the non-breaking significant wave height that is exceeded 12 hours per t years; $T_{e,t}$ is the associated wave period; and g is the acceleration due to gravity. This method explicitly recognizes that some sediment movement will occur seaward of d_{ℓ}.

Duck, N.C. is a wave-dominated sandy oceanic beach where high precision profiles have been collected to 8-m depth every 2 to 4 weeks and after storms since 1981 (Lee and Birkemeier, 1993). Profile data from profile lines 62 and 188 are considered here (termed PL62 and PL188, respectively). One or two bars are usually present (Birkemeier, 1984; Lippmann et al., 1993).

Nicholls et al. (1997a; 1997b) used 12 years (1981 to 1993) of Duck data to critically examine the characteristics and interpretation of depth of closure (or D_c). A 6-cm change criteria appears to be a better definition of closure than larger criteria, based on comparison with surf zone widths. Using this criteria, D_c could be defined for 97% of short term erosional events (defined by consistent offshore bar migration as illustrated in Figure 1) and 65% of annual time periods. Therefore, it is concluded that D_c is a meaningful and useful concept. d_{ℓ} provides a conservative <u>limit</u> to the resolved erosional event and annual observations (Figures 2 and 3). Further, for the non-closing cases, $d_{\ell} > 8$ m, which is consistent with the 8-m depth limit of the measurements.

The scatter of observations beneath the limit defined by d_{ℓ} is large and using a best-fit approach, the observed D_c is 69%

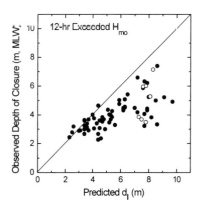

Figure 2. Observed (averaged alongshore) versus predicted closure for erosional events at Duck, NC . The cases considered in Figure 5 are shown as open circles.

and 76% of d_ℓ for erosional and annual cases, respectively. The scatter suggests that in addition to H_e, other factors are controlling the observed D_c. This paper explores possible controls of the scatter shown in Figures 2 and 3 to gain insight into improved predictions of D_c (cf. Capobianco et al., 1997). The possible controls which are considered are (1) storm characteristics, (2) morphological factors, including slope, and (3) the sediment budget.

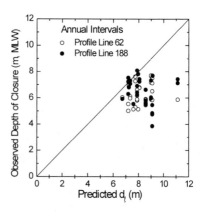

Figure 3. Observed versus predicted annual depth of closure.

The unexplained scatter is described by the residual (R_u):

$$R_u = d_\ell - \text{observed } D_c \qquad (2)$$

In all cases, data is presented for the individual profiles (PL62 and PL188) referenced to MLW (which is 0.42 m below NGVD). Cross-shore distance is referenced to the FRF baseline, rather than the actual shoreline.

EROSIONAL EVENTS

Nicholls et al. (1997a) analysed the scatter of seven erosional events which

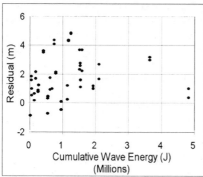

have similar predictions of d_ℓ (7 to 8 m), but a large scatter for D_c (indicated in Figure 2) . The first hypothesis was storm duration or wave energy. However, this does not appear to be a controlling factor of R_u. This is illustrated in Figure 4, which shows R_u versus cumulative wave energy for the more energetic events from 1981 to 1991 (taken from Table 2 in Lee et al., 1997). No relationship is apparent. This suggests that H_e is an effective parameter to describe wave forcing.

Figure 4 Residuals (R_u) for selected erosional events (using wave height) at PL62 and PL188 versus cumulative wave power for each event. Power is integrated when $H_s > 2$ m (after Lee et al., 1997).

Examination of profile shape for the same seven erosional events suggests that the pre-event bar configuration, particularly bar position and volume, controls observed D_c (Figure 5). The largest observed D_c is associated with a large single bar at about 400 m, while the smallest D_c is associated with a smaller single bar at about 200 m. Intermediate D_c are associated with two-bar configurations: the outer bar is intermediate in size and position.

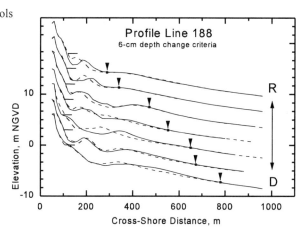

Figure 5. Pre- and post-storm morphology (closure is arrowed) for the seven erosional events indicated in Figure 2. R -- Reflective Extreme. D -- Dissipative Extreme.

This relationship was further explored using estimates of bar location and volume developed for Duck from 1981 to 1991 (Capobianco and Larson, 1997). These estimates were automatically extracted by determining a linear base to the bars derived directly from the profile. The bar crest is defined as the largest deviation from this base and is not necessarily the highest point on the bar. The bar size and location refers to the more seaward bar in cases where two bars were present. The control suggested in Figure 5 could be explained by the outer bar controlling the upper shoreface slope (cf. Lee et al., 1995; 1997), and hence rates of cross-shore sediment transport. Therefore, a representative pre-event slope on the upper shoreface was measured from 600 to 700 m. (Volume change in each event was also considered, but the changes are relatively minor compared with the annual changes: see below).

In this analysis, two depth bins defined with d_ℓ were considered: (1) Bin 1: d_ℓ = 5 to 7 m; and (2) Bin 2: $d_\ell > 7$ m (in round terms equivalent to H_e =2.5 to 3.5 m and $H_e > 3.5$ m, respectively). This gives 14 and 17 erosional events, respectively. (Events where $d_\ell > 5$ m are not considered). Figure 6 shows R_u versus the pre-event position of the bar crest position, the pre-event volume of the bar and pre-event upper shoreface slope for both depth bins. There are no significant relationships in Bin 1. In Bin 2, R_u is negatively correlated with the bar volume R_u (sig > 98%), but there is considerable scatter and only 16% of the variance is explained. Bar crest position and upper shoreface slope show no significant relationship.

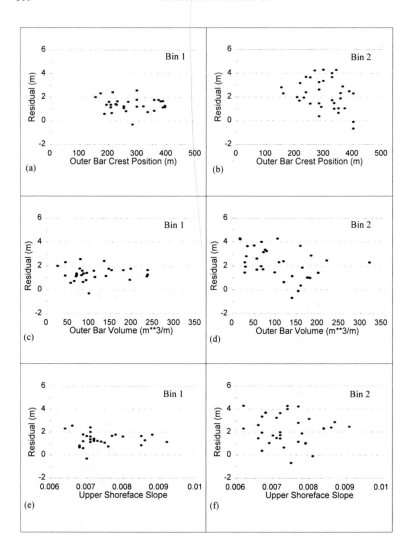

Figure 6. Selected erosional events from 1981 to 1991 for PL62 and PL188 for two depth bins (Bin 1: d_ℓ = 5 to 7 m, and Bin 2: d_ℓ > 7 m). Residuals (R_u) are shown versus (a) and (b) pre-event outermost bar crest cross-shore position; (c) and (d) pre-event outermost bar volume; and (e) and (f) pre-event upper shoreface slope.

In conclusion, these results provide some support for the observation that the pre-event profile influences D_c for the deepest predictions ($d_\ell > 7$ m).

ANNUAL PERIODS

Annual closure is an integrated response to both erosional and accretional processes. At Duck, several large and abrupt seaward translations of the profile have had a profound influence on the evolution of the profile envelope (Lee et al., 1997; Nicholls et al., 1997a). In the annual periods when these translations occurred, closure using a 6-cm change criteria could not be defined.

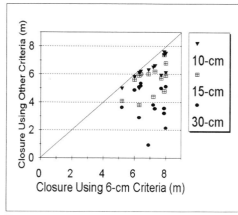

In this analysis, annual closures are measured from July to July for 12 annual periods: 1981 to 1993. Closures are defined with 6-cm, 10-cm, 15-cm and 30-cm change criteria. The larger change criteria produce estimates of closure that are landward of the smaller change criteria (Nicholls et al., 1997b) and hence they have a larger R_u. The relationship between the closure data due to the different change criteria are illustrated in Figure 7. The 10-cm and 15-cm data have similar properties to the 6-cm data, but average about 90% and 80% of the 6-cm values, respectively. The 30-cm data is more scattered and only averages 53% of the 6-cm values, but it does provide a complete data set with no missing cases. To maximize data availability, the 30-cm results are used. Six cases (25%) are excluded when $D_c < 3$ m -- in these cases the data appears to be decoupled from the closure values defined with smaller criteria.

Figure 7. Scatterplot comparing closures derived using a 6-cm change criteria with those determined with a 10-cm, 15-cm and 30-cm change criteria.

Figure 8 shows R_u for 6-cm and 30-cm annual closure estimates versus the corresponding net annual volume change on the profile (calculated from 100 m to the seaward limit of the profiles). For the 30-cm change data, there is a statistically significant negative relationship (>95%) between R_u and volume change explaining 33% of the variance: the deepest closure is associated with accreting profiles. However, as already discussed, a 6-cm criterion appears to be a better definition of closure. If we offset the relationship for a 30-cm change criteria to pass through the mean for the 6-cm data, we can develop an empirical estimate of the relationship between R_u for 6-cm closure and annual volume change. Both relationships are

indicated in Figure 8. Given
the scatter shown in Figure
7, the 6-cm relationship
should only be considered as
a qualitative indication of
behaviour. It suggests that on
(I) on a volumetrically
constant profile, $d_{\ell,I} \approx$
observed D_c, (ii) on an
eroding profile, $d_{\ell,I}$ will
overestimate D_c, while (iii)
on an accreting profile, $d_{\ell,I}$
will underestimate D_c! (A
correlation using the 6-cm
data suggests similar
behaviour at 90%
significance).

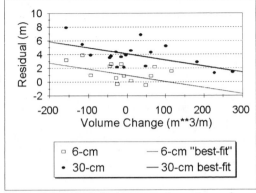

Figure 8. Residuals (R_u) versus volume change for annual closures on PL62 and PL188.

 The shoreface slope was also considered as a controlling factor (e.g., Figure 9). As for the erosional events, there was no statistically significant relationship.

DISCUSSION

 These results give further insight into the factors which control closure in addition to extreme wave height. It is well-known that pre-event profile conditions are an important control on
profile response close to the
shoreline (Wright and Short,
1984; Lippmann and Holman,
1990). This work shows that
profile response can be
sensitive to pre-event
morphology at the seaward
limit of the profile envelope.
Our conceptual model is that
d_ℓ provides an estimate of an
"equilibrium" or "endpoint"
value for D_c for both
erosional events and over
annual time scales (cf.
Nicholls et al., 1997a). As
single storms at Duck are
generally too short to achieve
an equilibrium, d_ℓ acts as a
limit. During erosional

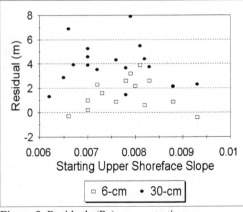

Figure 9. Residuals (R_u) versus starting upper shoreface slope for 6-cm and 30-cm annual closures on PL62 and PL188. The slope was measured from 700 to 800 m.

events, D_c is usually produced by offshore bar migration (Figures 1 and 5) and as shown here, the pre-event volume of the seaward bar (rather than position) exerts some control on D_c if $d_\ell > 7$ m (implying in round terms that $H_e > 3.5$ m).

Table 1. Observed and predicted closure for Depth Bins 1 and 2.

Depth Bin (d_ℓ)	Observed D_c, relative to MLW (m)		
	Minimum	Median	Maximum
1 (5 to 7 m)	3.3	4.1	6.2
2 (>7 m)	3.3	5.1	7.5

The absence of a similar bar volume control in the shallower depth bin is interesting and probably related to the relative location of D_c and bar morphology. As shown in Table 1, when d_ℓ is 5 to 7 m, D_c ranges from 3.2 to 6.2 m below MLW. Much of this zone is greatly influenced by back and forth bar migration. Therefore, D_c often occurs in complex morphologic situations such as on an existing bar crest or flank and morphological control, if present, is difficult to describe. When $d_\ell > 7$ m, D_c ranges from 3.2 to 7.5 m below MLW and many closures occur where the morphology is simple and near linear. Therefore, any morphological control is more simple to identify. The authors hypothesize that morphological control also exists in shallower water, but we are uncertain what are the key parameters (cf. Nicholls et al., 1997a).

Other factors such as profile shape are more difficult to describe and were not considered here. However, they may also be significant and should be considered in future work. The morphological control on closure might be further investigated using the reflective/dissipative morphodynamic model of beach profiles (Lippmann and Holman, 1990; Lippmann et al., 1993). Given the same wave conditions, dissipative profiles with large outer bars would tend to yield deeper closures than reflective profiles with a smaller single bar, as suggested in Figure 5.

Given that cross-shore slope is generally considered to be a major control on the rates of cross-shore transport it is surprising that no slope control is apparent at Duck. This result agrees with the earlier conclusion of Hallermeier (1978) about the negligible influence of slope on d_ℓ. However, the magnitude of the upper shoreface slopes at Duck are relatively low (range 0.006 to 0.01). Measurement of closure on profiles with steeper slopes would be useful to further explore this factor.

On natural beaches, significant gains and losses of sand due to gradients of longshore transport is an important process. The suggestion that such gains and losses are an additional control on closure is significant. Good agreement between $d_{\ell,1}$ and D_c for constant volume conditions provides further support for Hallermeier (1981), as Equation 1 does not address gains or losses of sediment. However, if volume change does occur, D_c would be expected to deviate from d_ℓ. Analysis of profile observations on accreting spits demonstrates that D_c is independent of d_ℓ and significantly deeper under strongly accretive conditions (Garcia et al., 1997). The Duck data suggests that this effect is always present when volume change occurs, resulting in relatively shallower closures in areas subject to erosion and relatively deeper closures in areas subject to accretion. This effect is unlikely to be apparent at the event scale, as volume changes are usually relatively small. However, as the time interval increases (e.g. one year as considered here), so this effect becomes apparent. The quantitative influence of this effect requires further evaluation.

Continued high quality profile observations are required from a range of sites to help improve our understanding of closure (and related morphodynamic properties). These can usefully be combined with the efforts described by Capobianco et al. (1997).

CONCLUSIONS

While closure is primarily controled by extreme wave conditions at the erosional event and annual scales considered, other factors also play a role. During large erosional events, the pre-storm outer bar volume, and possibly other morphological factors, are one control. The gain and/or loss of sediment (i.e. the sediment budget) also appears to exert a significant control on the behaviour of closure. Further measurements which embrace a wider range of conditions than those described here are required, including further evaluation of the factors considered here.

Acknowledgments: This study represents a collaboration between the PACE-project in the framework of the EU-sponsored Marine Science and Technology (MAST-III) under contract no. MAS3-CT95-002 and the Field Research Facility Measurement and Analysis Work Units under the Coastal Research and Development Program of the US Army Corps of Engineers. Michele Capobianco is thanked providing the bar statistics for Duck. Permission was granted by the Chief of Engineers to publish this information.

References

Birkemeier, W.A., 1984. Time scales of nearshore profile change. *Proceedings 19th ICCE*, Houston, ASCE, New York, pp. 1507-1521

Capobianco, M. and Larson, M. 1997. Longshore bar morphology at Duck, North Carolina. First Annual SAFE Workshop, Abstracts-in-Depth, Abstract B14, 6p.

Capobianco, M., Larson, M., Nicholls, R.J. and Kraus, N.C., 1997. Depth of closure: A contribution to the reconciliation of theory, practice and evidence. These Proceedings.

Garcia, V., Jimenez, J.A. and Sanchez-Arcilla, A., 1997. Temporal and spatial variability of the depth of closure in the Ebro delta coast. First Annual PACE Workshop, Abstracts-in-Depth, Abstract #1.4, 4p.

Hallermeier, R.J., 1978. Uses for a calculated limit depth to beach erosion. *Proceedings 16th ICCE*, Hamburg, ASCE, New York, pp. 1493-1512.

Hallermeier, R.J., 1981. A profile zonation for seasonal sand beaches from wave climate, *Coastal Engineering*, 4, 253-277.

Lee, G. and Birkemeier, W.A., 1993. Beach and nearshore survey data: 1985-1991, CERC Field Research Facility. *Technical Report-93-3*, US Army Corps of Engineers, Waterways Experiment Station, Vicksburg, MS., 26 pp.

Lee, G., Nicholls, R. J., Birkemeier, W.A. and Leatherman, S.P., 1995. A conceptual fairweather-storm model of beach-nearshore profile evolution at Duck, North Carolina, U.S.A. *Journal of Coastal Research*, 11, 1157-1166.

Lee, G., Nicholls, R. J. and Birkemeier, W.A., 1997. Storm-driven variability of the beach-nearshore profile at Duck, North Carolina., 1981-1991. *Marine Geology. in press.*

Lippmann, T. C. and Holman, R.A., 1990. The spatial and temporal variability of sand bar morphology. *Journal of Geophysical Research*, 95(C7), 11575-11590.

Lippmann, T.C., Holman, R.A. and Hathaway, K.K., 1993. Episodic, nonstationary behaviour of a double bar system at Duck, N.C., U.S.A., 1986-1991. *Journal of Coastal Research*, Special Issue No. 15, 49-75.

Nicholls, R. J., Birkemeier, W. A. and Lee, G.. 1997a Evaluation of depth of closure using data from Duck, NC, USA. *Marine Geology. in press*

Nicholls, R.J., Birkemeier, W.A. and Hallermeier, R.J., 1997b. Application of the depth of closure concept. *Proceedings 25th ICCE*, Orlando, FL, ASCE, New York, pp. 3874-3887.

Wright, L.D. Short, A.D., 1984. Morphodynamic variability of surf zones and beaches: A synthesis. *Marine Geology*, 56, 93-118.

DEPTH OF CLOSURE: A CONTRIBUTION TO THE RECONCILIATION OF THEORY, PRACTICE, AND EVIDENCE.

Michele Capobianco[1], Magnus Larson[2], Robert J. Nicholls[3], Nicholas C. Kraus[4]

Abstract

Our long-term research objective is to reconcile morphodynamic theory, engineering practice, and field observations concerning the concept of depth of closure. As a first step, we have analysed cross-shore profile variation at Duck, North Carolina. The result has been used to derive the return period of different magnitudes of vertical depth change across the profile, and hence explore return-period-dependent estimates of closure. From this analysis, some modelling concepts are presented.

Introduction

A central concept in coastal engineering is known as "depth of closure", or more practically, the seaward limit of the profile envelope. It divides the submerged beach profile into two discrete zones: (1) a nearshore zone which is considered to be morphological active, and (2) an offshore zone where morphological activity is much less and hence is considered insignificant. The concept is based on observations: change in bottom elevation tends to decrease offshore and beneath a certain nearshore depth, bottom elevation shows "negligible" change, generally defined by the profile measurement resolution (Figure 1). If bottom elevation change is limited, it is reasonable to infer that there is limited cross-shore transport and the depth of closure indicates the seaward limit of "significant cross-shore sediment transport". The analytical formulation of Hallermeier (1981), which is based on extreme wave height, is a widely applied formulation of the concept. While it has been validated at one site (Duck), the interpretation and application of results remain subjective and unstandardised.

---

[1] R&D Division, Environment, Tecnomare SpA, San Marco 3584, 30124, Venezia, Italy. Tel: +39 41 796711, Fax: +39 41 796800, E-Mail: capobianco.m@tecnomare.it
[2] Dept. of Water Resources Eng., University of Lund, Box 118, S-22100, Lund, Sweden.
[3] Flood Hazard Research Centre, Middlesex University, London EN3 4SF, UK.
[4] U.S.A.E. Waterways Exp. Sta., Coastal Engrg. Research Center, 3909 Halls Ferry Road, Vicksburg, MS 39180-6199, USA.

In contrast, coastal geologists find evidence of sand exchange across the entire shoreface (see Wright et al., 1991). Therefore, they dispute the existence of an inactive sedimentary zone as assumed in the engineering models (Pilkey et al., 1993; Riggs et al., 1995), but the time scale of some of the relevant processes is still uncertain.

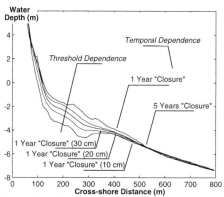

Figure 1. Definition Sketch for Depth of Closure (1 Year and 5 Years evolution "mean envelope") Dependence on Time and Threshold.

Engineers often motivate the concept of closure by asserting that what happens over geologic time scales is of little consequence to them. However, the expanding time scale of coastal engineering and management activities, with time frames of 50 to 100 years becoming more common (cf. Stive et al., 1992), calls this view into question. Therefore, coastal engineers require improved tools to predict seaward boundaries for sandy coastal shorelines which utilise and reconcile the two perspectives presented above considering the lack of applicable detailed process knowledge. As a first step, this paper examines cross-shore profile variability at Duck, North Carolina using the unique, high-quality data set (Lee and Birkemeier, 1993). Closure is a relative rather than an absolute concept and depends on definition. Instead of looking explicitly at closure, we examine the variation with time scale of the entire profile envelope with emphasis on the return period of depth changes (e.g. Figure 1). Depth of closure is simply one of these depth changes, selected for a specific application. This stresses that closure can be considered in a time-dependent manner.

The Field Evidence at Duck

We consider here profile lines 62 and 188 at Duck from 1981 to 1991. In Figure 2 the standard deviation of depth change (SDDC) over the whole 11-years period is plotted with reference to the local depth of the mean profile. In previous work, analysis of the Duck data set has shown that robust closures (using a 6-cm change criteria) can be defined for short-term erosional and accretional events, and over annual and longer periods up to at least 8 years (Nicholls et al., 1997a). However, as time scale increases, so a smaller proportion of closures are resolved by this data which probably reflects the depth limit of the measurements at Duck (8-m depth). Hallermeier's (1981) formulation was explicitly designed for the annual time scale. It is found to define a robust *limit* to the erosional events and over annual and longer periods: agreement appears best at the annual scale (Nicholls et al., 1997a; 1997b). As time scale increases from one year to 30 months, the predictions grow more rapidly than the observations. Thus, the limit nature of the relationship is not violated, but the practical value of the formulation declines. Therefore, closure as a function of time scale remains poorly understood.

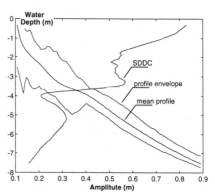

Figure 2. Standard Deviation of Depth Change (line 188)

The characteristic profile behaviour can be seen in Figure 3 where the variation in depth at constant (cross-shore) distances are plotted for the lower part of the measured profile (line 188). Two regimes of morphological evolution can be distinguished: (1) a background evolution when depths slowly increase and (2) short-lived extreme episodic changes when the depth decreases suddenly. These changes are due to slow onshore sediment transport under fair-weather conditions, and abrupt offshore sediment transport associated with a few groups of storms, respectively (Lee et al., 1997). The abrupt depth decreases in early 1987 and early 1989 contribute to the variation of the profile envelope below 6-m depth in Figure 2. In the following analysis, we focus on the profile behaviour over "time intervals", also denoted as "time windows" or as a function of "return period".

The mean maximum displacement around the mean values (MMD) across the profile as a function of variable time window from one month to 10 years is shown in Figure 4 (with enlarged results below 5-m depth in Figure 4c-4d). As the time window increases, the shape of the displacement curve changes and more closely resembles the standard deviation of change in Figure 2, but there is also less data and the statistical significance decreases. The minimum at 4-m depth deserves some investigation; it corresponds to the seaward limit of offshore movement of the outer bar crest, and has been noted by previous authors (Larson and Kraus, 1994; Nicholls et al., 1997b).

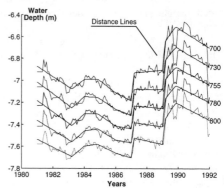

Figure 3. Variation in Depth with Time at Fixed Cross-shore Distances (line 188)

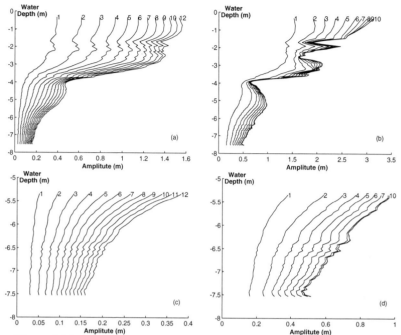

Figure 4. Mean maximum displacement as a function of time scale: (a) and (c) monthly time windows, (b) and (d) yearly time windows.

Figure 5a shows the MMD as a function of the return period for various (mean) profile depths, whereas Figure 5b shows the ratio between MMD and probable maximum wave height in the same return period. Figure 5a indicates that for a given depth, the behaviour resembles the "probable significant wave height" as a function of return period. This type of data can be empirically "fitted" with a theoretical curve in order to derive some rule of behaviour. As an exercise, the lognormal cumulative distribution function was used, and by assuming that they are composed of single month events, the Longuet-Higgins (1952) formula for expected value of the maximum wave height in N events may be used. The results still need to be fully analysed; but the approach of finding probability distribution functions (pdf) for displacements at various depths in relation to the pdf of the significant wave height looks promising. Assuming (reasonably) that there is a cause-effect relationship between wave data and morphological evolution, Figure 5b highlights that the upper part of the profile tends to respond to the wave forcing independent of time period, while the lower part, where closure is defined, tends to show a greater response over longer time periods than shorter time periods. The prediction of such behaviour is linked to the estimation of extreme events, which remains an important and active research subject (e.g., Ochi, 1992). The Hallermeier formulation of 1981 computes closure depth by considering the extreme wave conditions exceeded 12 hours over one year, or more generally the period of interest. It is unclear if the extreme wave conditions should *exceed 12 hours continuously or not* over the period of interest.

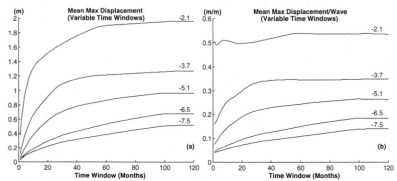

Figure 5. Vertical Displacement as a Function of Return Period for Selected Depths (line 188 - Absolute Values and Relative to Wave Forcing)

Figure 6 summarises the change with depth of different vertical changes (noted as "threshold" values) as a function of return period. This shows that at 7.0-m depth, a 20 cm change has a return period of 2 years, while a 100-cm change has a return period >9 years at 4.0-m. These results support the notion of a time-dependent closure. These calculations considered the MMDs over the variable time period. In the context of time-dependent closure this might not be the best parameter, and using less conservative parameters to describe profile change would displace all the threshold values.

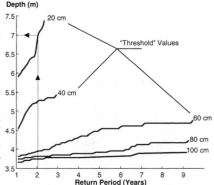

Figure 6. Depth Change as a function of Return Period for line 188 (various Threshold values)

Modeling Depth of Closure

The previous analysis showed characteristic patterns in time and space of the profile variability that are intimately linked to the definition and properties of the depth of closure. These patterns are functions of the local forcing, where short-period waves are one important component that often dominates the sediment transport in the nearshore. A purely probabilistic approach could define closure for any site-specific data set of beach profiles, irrespective of forcing, for the time and space scales inherent in the measurements (cf. Nicholls et al., 1997b). However, the possibility of generalizing the observations to other sites would be limited, and considering the limited availability of profile measurements, such an approach would only be feasible for very few locations. Thus, in order to develop a general predictive model for the depth of closure the forcing and the profile variability must be connected. It is suggested that such connections can be modelled at three levels:

- relating statistical properties of the depth variation to some overall parameter that characterizes the forcing;
- computing local forcing parameter values and through simple, physically-based models derive the local depth variation;
- employing process-based numerical models of beach evolution to derive a time series of depth variation.

Presently, state-of-the-art long-term process-based numerical models cannot produce sufficiently robust results to develop reliable information on the depth variation. Instead, a simple approach to relate the statistics of depth variation to the statistics of the offshore wave height will be presented in the following. Also, a method will be outlined for deriving the depth of closure based on calculated local wave parameters and equilibrium profile theory. Following the success of the Hallermeier formula as a limit at Duck, it is assumed that the wave height is the most important measure of the forcing. The mean significant wave height over 12 hr is employed as a suitable parameter, although other quantities may be chosen. The depth of closure is defined based on a maximum absolute depth change between consecutive surveys. By selecting specific values of this change, (5, 10, or 20 cm), time series of depth change were produced using the Duck data (in this case profile line 62, which has the largest number of surveys). Assuming a general relationship between the waves and the depth variation (i.e., higher waves produce deeper changes and a larger depth of closure), it is possible to predict exceedance probabilities for a certain depth change at a certain depth, if the wave climate is known. Thus, empirical distribution functions were derived for the waves and the depths corresponding to specific depth changes, and exceedance probabilities were matched between these functions (the basic idea is that, for example, the 10% highest waves would produce the 10% largest depth changes).

Figure 7 illustrates the results using the Duck data where the depth of closure is given as a function of the mean 12-hr significant wave height for selected depth change criteria (depth of closure is referenced to MLW in accordance with Hallermeier). Empirical expressions were also fitted to the data to allow for easy application and extrapolation to larger waves. A power function with a power value of 2/3 gave a good overall fit, with the multiplier being 3.4, 2.8, and, 2.1 for a depth change of 5, 10, and 20 cm, respectively. For small depth changes a lot of scatter around the mean trend is seen in deeper water attributed to measurement accuracy. These points were disregarded in the present simplified fits but will certainly be worth some deeper investigation in the future. The statistical relationship between waves and depths seems to change in shallow water, and since this region is of less interest in predictive models it was also neglected in the fitting. Once a design wave height has been selected (related to a certain exceedance probability), Figure 7 may be used to predict the corresponding depth of closure depending on the depth change of interest. The curves were developed specifically for Duck; however, assuming that the statistical relationship between waves and depth change is reasonably representative for wave-dominated coasts, such curves might be employed at other sites where the wave statistics are known. This assumption needs to be tested -- analysis at Duck suggests that closure on swell-dominated coasts might have different characteristics to closure on storm-dominated coasts (Nicholls et al., 1997a).

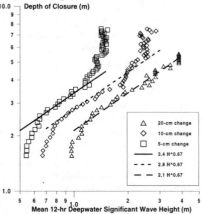

A more general model to calculate the depth of closure should involve local forcing as well as relationships based on physical considerations to derive the depth variation from this forcing (e.g. Hughes and Borgman, 1987). Again, assuming that short-period waves are the main control of depth change, local wave characteristics have to be computed

Figure 7. Depth of Closure as a Function of Mean 12-hr Significant Wave Height (line 62)

using a wave decay model. In the present study the random wave transformation model developed by Larson (1995) was employed to compute nearshore wave properties. The 11-year wave time series measured in the offshore at Duck (waves typically recorded every 6 hr) was used as input and statistical analysis was performed for various local wave parameters at points across the profile. The mean beach profile based on 11 years of measurements was used in the wave calculations, without taking into account temporal perturbations in the profile shape, in order to capture the overall variation in the forcing. As an example, Figure 8 displays the mean monthly wave energy dissipation as a function of the water depth. On the average, the largest wave energy input at Duck occurs in March and the smallest in July (see thick lines in Figure 8). The two subdued nearshore bars in the mean profile at Duck are clearly seen as two local dissipation peaks. The calculations also show significant dissipation close to the shoreline, where the slope is steep and smaller waves dissipate all their energy. The curves in Figure 8 may be employed to determine seasonal values for the depth of closure once a model is employed that relates local depth change to wave energy dissipation (or a similar forcing parameter). In this respect the equilibrium theory by Dean (1977) may be employed assuming quasi-steady conditions and a profile that is in balance with these conditions (Larson and Kraus 1996).

A probabilistic approach to model the depth of closure should rely on the statistics of the local forcing. Thus, the return period of specific events is established and the associated depth variation is determined based on a physical model. In order to explore this approach, empirical distributions for the local forcing conditions were derived that provided information on the return period for specific events. As an example, Figure 9 shows curves of exceedance probability for the 6-hr mean energy dissipation at different water depths, ignoring events of 'zero' dissipation. The probability of a certain

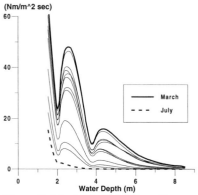

Figure 8. Mean Monthly Wave Energy Dissipation (line 62)

dissipation being exceeded is less in deeper water implying a longer return period for a specific depth change (under the assumption that depth change is related to the dissipation). Figure 9 may be used to directly compare the conditions at different depths. For example, if a certain dissipation is needed for a certain depth change, the curves provide information on the probability of this depth change taking place. However, to allow for absolute predictions the relationship between depth change and energy dissipation (or some other forcing parameter) has to be established. Again, using equilibrium profile theory would be one possible approach. Profile shapes for specific forcing conditions could be obtained, and changes in these shapes would yield temporal and spatial information on depth change from which the depth of closure could be derived.

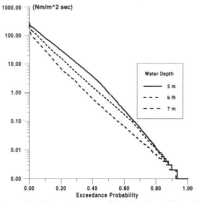

Figure 9. Mean Energy Dissipation (line 62)

Discussion and Further Work

As engineers working on long-term morphodynamics, we recognise that variability and uncertainty are intrinsic in the idea of "significant sediment transport" which is embedded in the closure concept, particularly as there is no a priori specification of "significant for what". Within long-term morphodynamics, the need to assess the relative importance of (1) slow tendencies of evolution, and (2) extreme events, is increasingly recognised as a general problem, particularly to define predictability.

Different processes act across the profile and, over any time period, certain processes can be considered more or less important than others.

As both this analysis and earlier work has shown, depth of closure is both a time- and scale-scale dependent concept (Kraus and Harikai, 1983; Stive et al., 1992; Larson and Kraus, 1994; Nicholls et al., 1997a; 1997b). Hallermeier (1981) assumes annual closure on sandy beaches is primarily controlled by extreme wave conditions. Based on the previous analysis at Duck (Nicholls et al., 1997a; 1997b) this appears to be correct. The control on closure of other coastal characteristics such as slope and grain size is unclear, but in non-equilibrium artificial situations such as beach nourishment we expect them to be important factors. This is being examined using profile data from the Ocean City, Maryland beach nourishment project

In summary, further work will proceed in a number of directions. Theoretical analysis will develop the original Hallermeier formula, particularly the issue of the appropriate wave height in shallow water (cf. Nicholls et al., 1997b). Field evidence will be used to help us interpret the response of profiles to extreme events and the different time scales of bottom elevation change as a function of depth. Modelling will be used to examine the role of slow tendencies of evolution (cf. Figure 3). It is envisaged that the end result will be a generalised modification of the Hallermeier formula with potential application over decadal (≤ 50 year) time scales. Our goal is an engineering tool which helps specify the site specific conditions governing the depth of closure, including their time and space scale dependencies, and hence, probabilistic output of maximum value to the user community.

Conclusions

While these results and thoughts are preliminary, the idea of attaching a return period to depth of closure appears to be a sound concept. In the present paper we presented some results from the first evaluations that will help us to sustain the field evidence and to better address our theoretical efforts. We are seeking relationships of the computed time-dependent changes with mean characteristics of the profile or characteristics of the equilibrium profile and for relationships with statistics of wave forcing.

Acknowledgements

MC, ML and RJN were supported by the PACE-project in the framework of the EU-sponsored MAST-III Program under contract no. MAS3-CT95-002. The contribution of NCK was performed under the Coastal Inlets Research Program of the US Army Corps of Engineers (USACE). He appreciates permission to publish this information given by Headquarters, USACE. Mr. William Birkemeier and his coworkers at the US Army Field Research Facility in Duck, North Carolina, are gratefully acknowledged for supplying the high-quality data used for this analysis.

References

Dean, R.G. (1977) "Equilibrium Beach Profiles: U.S. Atlantic and Gulf Coasts," *Ocean Engineering Report No. 12*, Department of Civil Engineering, University of Delaware, Newark, DE.

Hallermeier, R.J. (1981). "A profile zonation for seasonal sand beaches from wave climate", *Coastal Engineering*, 4, 253-277.

Hughes S.A., Borgman L.E. (1987). "Beta-Rayleigh Distribution for Shallow Water Wave Heights", *Proc. Conference on Coastal Hydrodynamics*, ASCE, University of Delaware, USA, June 28-July 1, pp. 17-31.

Kraus N.C., Harikai S. (1983). "Numerical Model of shoreline change at Oarai Beach". *Coastal Eng.*, 7(1), 1-28.

Larson, M. (1995). "Model for Decay of Random Waves in the Surf Zone," *Journal of Waterway, Port, Coastal and Ocean Engineering*, Vol 121, No. 1, pp 1-12.

Larson M., Kraus N.C. (1994). "Temporal and spatial scales of beach profile change, Duck, North Carolina", *Marine Geology*, 117, 75-94.

Larson M., Kraus N.C. (1996). "Advancement of Equilibrium Beach Profile Concepts and Numerical Mod. Tech.". *Interim Report*, Vol. I&II, Texas A&M Res. Found.

Lee G., Birkemeier, W.A. (1993). "Beach and Nearshore Survey Data: 1985-1991. CERC Field Research Facility." *Technical Report CERC-93-3*, CERC, US Army Engineer Waterways Experiment Station, Vicksburg, MS.

Lee G., Nicholls, R.J. and Birkemeier, W.A. (1997). "Storm-driven variability of the beach-nearshore profile at Duck, NC, 1981-1991". *Marine Geology*, in press.

Longuet-Higgins M.S., (1952). "On the Statistical Distribution of the Height of Sea Waves". *Journal of Marine Research*, v. 11, pp. 245-266.

Nicholls, R.J., Birkemeier, W.A. and Lee, G. (1997a). "Evaluation of depth of closure using data from Duck, NC, USA". *Marine Geology*, in press.

Nicholls, R.J., Birkemeier, W.A. and Hallermeier, R.J. (1997b). "Application of the depth of closure concept". *Proc. 25th Int. Conf. on Coastal Eng.*, ASCE, New York, 3874-3887.

Ochi M.K., (1992). "New Approach for Estimating the Severest Sea State from Statistical Data". *Proc. 23rd Int. Conf. on Coastal Eng.*, pp. 512-525.

Pilkey, O.H., Young, R.S., Riggs, S.R., Smith, A.W.S., Wu, H. and Pilkey, W.A., (1993). "The concept of shoreface profile of equilibrium: A critical review". *JCR.*, 9, 255-278.

Riggs, S.R., Cleary, W.J. and Snyder, S.W. (1995). "Influence of inherited geological framework on barrier shoreface morphology and dynamics". *Marine Geology*, 126, 213-234.

Stive M.J.F., De Vriend H.J., Nicholls R.J., Capobianco M., (1992). "Shore Nourishment and the Active Zone: a Time Scale Dependent View", *Proc. 23rd Int. Conf. on Coastal Eng.*, ASCE, New York, 2464-2473.

Wright L.D., Boon J.D., Kim S.C., List J.H., (1991). "Modes of Cross-Shore Sediment Transport on the Shoreface of the Middle Atlantic Bight", *Marine Geology*, n.96, pp. 19-51.

Hydraulic Design of a
Large-Scale Longshore Current Recirculation System

David G. Hamilton[1],
Peter J. Neilans[2], Julie D. Rosati[2], Jimmy E. Fowler[2], and Jane M. Smith[2]

ABSTRACT
This paper documents the hydraulic design of a large-scale longshore current recirculation system required to maintain a uniform longshore current within the testing region of a new longshore sediment transport laboratory facility. Twenty vertical turbine pumps, with a total discharge capacity of 1,250 litres/sec, are being used to recirculate the longshore current, while twenty independent piping systems provide the capability to operate any combination of the twenty pump-and-piping systems simultaneously. Variable speed pumps combined with a low and high flow measurement mechanism in each piping system provides the capability to accurately control and recirculate between 10% and 100% of the maximum design capacity at all twenty cross-shore locations. This design allows a wide range of wave-induced uniform longshore current magnitudes and distributions to be generated within the new facility.

1. INTRODUCTION
The present work is part of a research program to improve the U.S. Army Corps of Engineers capability to predict local and total longshore sediment transport rates and to evaluate errors associated with these predictions. The first goal of this research is to develop a world-class Longshore Sediment Transport Facility (LSTF) at the U.S. Army Engineer Waterways Experiment Station, Coastal and Hydraulics Laboratory (formally Coastal Engineering Research Center) for conducting three-dimensional moveable-bed experiments. The LSTF will simulate nearshore hydrodynamic and sediment transport processes at a relatively large geometric scale. Further information on the objectives of this research program can be found in Rosati et al. (1995).

This paper describes the hydraulic design of a large-scale longshore current recirculation system required to minimize the laboratory effects resulting from the lateral boundaries of the new facility. The objective of this system is to precisely control and recirculate the longshore current with the goal of maintaining a uniform longshore current magnitude and cross-shore distribution within the testing region of the new facility.

---

[1] Coastal Engineer, C5-310 Cain Ridge Road, Vicksburg, Mississippi, USA, 39180,
   Tel: 601 634-3029, Fax: 601 634-4314, E-mail: D.Hamilton@CERC.WES.ARMY.MIL
[2] Research Hydraulic Engineer, USAE Waterways Experiment Station, Coastal and Hydraulic Laboratory,
   3909 Halls Ferry Road, Vicksburg, Mississippi, USA, 39180-6199

Section 2 gives a brief description of the longshore sediment transport facility. Section 3 gives a brief overview of seven types of longshore current recirculation systems previously or presently being used for laboratory investigations of longshore currents. Sections 4 and 5 discuss the system requirements and the required pumping capacity of the longshore current recirculation system, respectively. Section 6 provides a detailed description of each of the five sub-systems which make up the longshore current recirculation system. Section 7 describes how the system was designed with the capability to recirculate between 10% and 100% of the maximum design capacity at all twenty cross-shore locations. Section 8 provides conclusions.

2. DESCRIPTION OF THE NEW LABORATORY FACILITY

The LSTF has dimensions of 30-m cross-shore, 50-m longshore, and is 1.4 m deep (Figure 1). The wave generators are synchronized to generate unidirectional long-crested waves up to 0.5 m in height for wave periods up to 3.0 sec, and offshore angles of wave incidence up to 20 degrees. A fixed-bed beach having a longshore dimension of 30 m, a cross-shore dimension of 21 m, with a constant 1:30 slope, was accurately constructed (tolerance of ± 2.0 mm) with parallel contours. During future moveable-bed experiments a 0.3 m thick layer of quartz sand will be placed on top of the fixed-bed beach and will extend approximately 18 m offshore. Figure 1 also shows the primary components of the longshore current recirculation system including: 20 independent upstream flow channels and 20 downstream flow channels and sumps; 20 vertical turbine pumps; 20 independent piping systems; and 40 flow measurement sensors, two in each pump-and-piping system. Wave, current, and bathymetric data will be collected using a suite of sensors located on a 21 m long fully automated instrumentation bridge which can traverse the entire length of the facility, as described by Rosati et al. (1995).

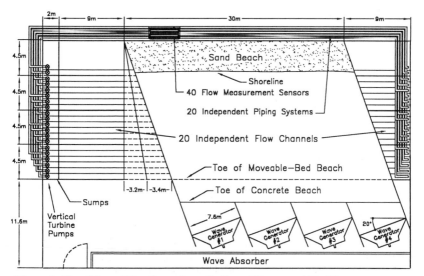

Figure 1: Plan View of New Longshore Sediment Transport Facility

3. OTHER LONGSHORE CURRENT RECIRCULATION SYSTEMS

During the past five decades, seven types of longshore current recirculation systems have been used to conduct laboratory investigations of longshore currents (Figure 2). The earliest systems were fairly simplistic whereas the most recent systems are more complex and are, in general, physically larger which makes them substantially more expensive to design and construct. Each of these seven types of systems will be discussed briefly in this section. Visser (1991) and Svendsen (1991) also provide a description of several of these systems.

Type 1: Putnam et al. (1949) used a completely enclosed wave basin and allowed the longshore current to recirculate in the offshore section of the wave basin.

Type 2: Galvin and Eagleson (1965) and Mizuguchi and Horikawa (1978) terminated the downstream wave guide at about the location of wave breaking, allowing the longshore flux to exit the testing region. The longshore current recirculated outside of the downdrift wave guide and re-entered into the testing region beneath the wave generators.

Type 3: Brebner and Kamphuis (1963) terminated both the upstream and downstream wave guide near the breaker point. The recirculation is driven by a water level difference between the upstream and downstream end of the facility.

Type 4: Kamphuis (1977) used a procedure almost identical to Type 3 except that the wave generators were raised slightly above the basin floor allowing the longshore current to also re-enter the testing region beneath the wave generators.

Type 5: Dalrymple and Dean (1972) used a circular wave basin with spiral wave generator and a circumferential beach. This is a unique method of avoiding the traditional boundary effects resulting from the upstream and downstream lateral boundary of a wave basin.

Type 6: Visser (1982 and 1991) was the first to use an active recirculation system driven by a pump. As with the Brebner and Kamphuis (1963) experiments, the wave guides are open at both ends of the beach. However, in this case 12 independent weirs are used to input the required cross-shore distribution of longshore current at the updrift boundary.

Type 7: H.R.Wallingford (1994) and Simons et al. (1995) provide a description of the Coastal Research Facility at H.R.Wallingford. This facility is a further improvement upon Type 6 in that adjustable weirs are used at both ends of the facility. Therefore, the cross-shore distribution of longshore current can be controlled at both lateral boundaries.

4. SYSTEM REQUIREMENTS

Six primary requirements were identified for the longshore current recirculation system: 1) The system must have a pumping capacity capable of recirculating the longshore flux associated with a wide range of wave, water level, and beach conditions; 2) A Type-7 recirculation system should be used since it has the capability to control the cross-shore distribution of longshore current at both lateral boundaries, and will therefore maximize the length over which the longshore current is uniform; 3) The system must have the flexibility to recirculate relatively low longshore current magnitudes associated with, for example, the offshore tail of the longshore current distribution. Therefore, the system should have the capability to recirculate between 10% and 100% of the maximum design pumping capacity at each cross-shore location; 4) The system should be capable of maintaining constant discharge rates for several hours of continuous operation to accommodate the time scales associated with conducting moveable-bed experiments at this relatively large geometric scale; 5) The system should require minimal time and labor to adjust the magnitude and cross-shore distribution of the longshore current being recirculated. This is of paramount importance for moveable-bed experiments since the cross-shore distribution of recirculated longshore current may need to be adjusted in response to the changing beach profile; 6) The system should have a design life of 15-20 years given the long-term requirements of the facility.

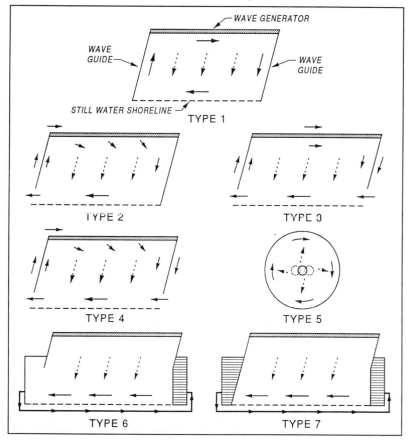

Figure 2: Seven Types of Longshore Current Recirculation Systems
Illustration of Type 1 through Type 5 reproduced from Figure 1 of Visser (1991)
(dashed arrows denote wave direction and solid arrows denote mean current direction)

5. DESIGN PUMPING CAPACITY

The methodology used to estimate the required pumping capacity of each pump in the longshore recirculation system is documented in Hamilton et al. (1996). Figure 3, shown below, indicates the required and design pumping capacity of the system. To reduce the number of pumps near the shoreline, where the longshore flux was estimated to be relatively small, one larger pump was used in place of four smaller pumps, as shown in Figure 3. This can also be seen in Figure 1. The required and design pump capacities are slightly different primarily due to the need to minimize cost by selecting pumps with capacities which were commercially available, and by a desire to have additional pumping capacity at the offshore end due to uncertainties in the estimating the longshore current magnitude at that location, as discussed in Hamilton et al. (1996).

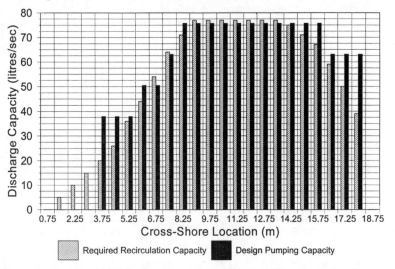

Figure 3: Required and Design Pumping Capacity of Recirculation System

6. DESIGN OF SYSTEM COMPONENTS

This section provides a detailed technical description of each of the five sub-systems that make up the longshore current recirculation system.

6.1 VERTICAL TURBINE PUMPS AND MOTORS

Vertical turbine pumps were chosen for this application based on a desire to alleviate problems with priming and air entrainment. These problems are eliminated since, by definition, the impeller on a vertical turbine pump is located below the mean water level.

Figure 4 provides an oblique view of the twenty vertical turbine pumps and motors prior to installing the piping systems. Each vertical turbine pump consists of a single-stage impeller and bowl assembly, a vertical column assembly, a discharge head, and an electric drive motor. Closed-type Francis-Vane impellers were used to eliminate the requirement of having to adjust the clearance beneath the impellers to maintain optimum hydraulic efficiency. The impellers are constructed of high strength silicon bronze and the suction bowls have vitra glass-lined waterways for maximize abrasion resistance. Minimizing long-term abrasion in the pumps was an important design criteria since we anticipate small quantities of sand passing through the system during moveable-bed experiments.

Figure 4: Oblique View of Twenty Vertical Turbine Pumps and Motors

Table 1 lists the primary design specifications of each pump including: discharge capacity at the best efficiency point; total dynamic discharge head; discharge pipe diameter; impeller diameter, optimized to the best efficiency point of each pump; hydraulic efficiency; specific speed; and the electric power required to drive each pump at maximum discharge. The specific speed of all four models of pumps range from 2,880 to 4,340. Hence, each of the pumps can be classified as mixed-flow pumps which generate discharge head through a combination of centrifugal and axial forces. The electric motors are inverter duty motors manufactured using special inverter grade insulation so that the motors do not overheat during low speed operation. Pump No. 1 is located at the onshore end and pump No. 20 is at the offshore end of the system.

Pump Location Number	Discharge Capacity (litres/sec)	Total Dynamic Head (m)	Discharge Pipe Diameter (mm)	Optimum Impeller Diameter (mm)	Hydraulic Efficiency (1800 rpm)	Specific Speed (-)	Required Power (KW)
1 - 3	37.8	9.8	150	149	79%	2,880	4.6
4, 5	50.5	12.2	150	176	81%	2,580	7.4
6,18 - 20	63.1	8.8	200	149	79%	4,340	6.9
7 - 17	75.7	11.9	200	178	80%	3,930	11.0

Table 1: Vertical Turbine Pump and Motor Specifications

6.2 VARIABLE SPEED MOTOR CONTROLLERS

To meet system requirement No. 3 (see Section 4) each pump-and-piping system required the capability to recirculate between 10% and 100% of the maximum design discharge. Variable speed motors, as opposed to throttling valves or weirs, were determined to be the most desirable option for controlling the discharge rate of each pump-and-piping system, for two primary reasons: 1) there is a linear relationship between flow rate and pump speed giving accurate control of discharge over a wide range of flow rates, and 2) pump discharge rates could be remotely controlled and automated using a personal computer. As a result, twenty individual pulse width modulated variable frequency alternating current motor controllers were used to control the speed of the pumps. Eleven of the motor controllers are 11.25 KW systems and nine are 7.5KW systems. The output frequency of the motor controllers, and hence the speed of the pumps, can be manually adjusted using a human interface key pad, with digital display, mounted on the front panel of each controller, or remotely adjusted using a personal computer and RS-485 protocol. Electrical power is provided by a 400 ampere, 480-volt, 3-phase, 60-Hz, electrical service.

6.3 PIPING SYSTEM

The piping system consists of fifteen 200-mm and five 150-mm diameter pipelines, with a total pipe length of approximately 1,500 m. Selection of the discharge pipe diameter was based primarily on economics. The cost of various pipe sizes was compared with the associated pump and motor size required to overcome the resulting head loss through the piping system at maximum discharge. All of the pipe, fittings, and valves are constructed of rigid polyvinyl chloride. The advantages of polyvinyl chloride is that it has good pressure bearing capability, exceptional long-term corrosion resistance, and smooth interior walls to minimize head loss. Transparent polyvinyl chloride pipe was used at the pump discharge heads and immediately upstream of the flow sensors as observation windows to determine if air or sediment is being pumped through the piping system. This needs to be monitored since the pumps are designed to withstand only small quantities of sand passing though the system and because the flow sensors, discussed in Section 6.4, are somewhat sensitive to air and sediment entrainment.

6.4 FLOW MEASUREMENT SYSTEM

Midway along each pump-and-piping system, the main pipeline branches out into two smaller pipelines (total of 40) running parallel for approximately 5.0 m before converging back into the main pipeline. These smaller pipelines are the flow measurement sections and each contains a true union ball valve, a transparent section of polyvinyl pipe used for monitoring of air and sediment entrainment, and an impeller-type flow sensor. System No's. 1, 2 and 3 have 75-mm and 38-mm diameter flow measurement sections, whereas system No's. 4 through 20 have 100-mm and 50-mm diameter flow measurement sections. For high flow conditions in a given system, the valve in the smaller pipe is closed and the flow rate is measured using the flow sensor in the larger of the two parallel pipes. In contrast, for low flow conditions, the ball valve in the larger pipe is closed and the flow rate is measured using the flow sensor in the smaller pipe. It should be mentioned that none of the valves are used to throttle the flow rate; they are either fully closed or open. The reason for using two parallel pipes with individual flow sensors was two fold: 1) to increase the water velocity at the flow sensor during low flow conditions to minimize measurement error, and 2) to increase the range of discharge rates over which the vertical turbine pumps will operate (see Section 7).

Each of the flow sensors has a six-blade impeller with a non-magnetic sensing mechanism. The forward-swept design coupled with the absence of magnetic drag provides improved accuracy and repeatability at lower flow rates. The manufacturer claims the following specifications: calibration range 0.3 m/s to 9.1 m/s; accuracy $\pm 1\%$ of full scale; repeatability $\pm 0.5\%$; and linearity $\pm 0.5\%$. The frequency of the output signal, a low impedance 8-volt DC square wave, is proportional to the magnitude of the flow rate through the pipe, and is transformed into a digital signal and transmitted to a personal computer using RS-485 protocol. The desired flow rate is compared to the measured flow rate and can be adjusted by changing the pump speed.

6.5 FLOW CHANNELS AND INTAKE SYSTEM
As shown in Figure 1, flow channels are used at the upstream end of the facility to guide water from the discharge pipes to the upstream boundary of the beach. Likewise, on the downstream end of the facility, flow channels guide the longshore current from the downstream boundary of the beach to the vertical turbine pumps. Each of the flow channels is 0.75 m in width. The two side walls of the channels consist of fabricated aluminum flow guides. A continuous neoprene gasket is used to seal the interface between the flow guides and the reinforced concrete floor and sidewalls of the facility, so that each flow channel is independent and water tight. To ensure maximum flexibility of the facility, the recirculation system has been designed with the capability to conduct experiments with water levels ranging from 0.3 to 1.0 m. As a result, twenty sumps were constructed to ensure vortex free operation of the vertical turbine pumps at maximum discharge, and hence maximum drawdown, with a minimum operating water level of 0.3 m. Each sump is constructed of reinforced concrete and is 0.75 m wide, 1.5 m long and 1.2 m deep. Hence, design of the intake system to each pump consists of a straight open flow channel with a sump at the downstream end of the channel. Turns and obstructions in the intake system were minimized to avoid the possibility of eddy currents causing submerged vortices under high flow conditions.

7. HYDRAULIC PERFORMANCE OF PUMP AND PIPING SYSTEMS
By the use of variable speed motors the discharge of each pump can be varied to suit the recirculation requirements for a given testing condition. Figure 5 shows the performance curve for pump No. 10 operating at speeds of 540, 1080, 1440, and 1800 RPM, which correspond to 30, 60, 80 and 100% of the nominal operating speed of 1800 RPM. The performance curve for a pump speed of 1800 RPM was supplied by the manufacturer. The performance curves representing slower pump speeds were calculated using the standard homologous pump equations

Figure 4 also shows two characteristic curves for the No. 10 piping system. These two curves represent head loss through the 70 meter long, 200-mm diameter main pipeline combined with the head loss through the 5 meter long flow sensor sections of 100 mm diameter and 50 mm diameter (FS = Flow Sensor in legend of Figure 5). The intersections of the pump performance curves with the two characteristic pipe curves define the range of operating conditions of the combined pump-and-pipe system.

The following technique is used to meet the design requirement of being able to pump between 10% and 100% of the maximum discharge of each pump (see Section 4). Point No. 1, in Figure 5, shows that the maximum discharge through system No. 10 is approximately 75 litres/sec. The pump speed can then be reduced to 30% (540 RPM) of the nominal pump speed of 1800 RPM depicted by Point No. 2. However, if the pump speed is reduced to less than approximately 30% of the nominal speed, the pump discharge will suddenly reduce to zero due to inadequate centrifugal and axial forces at the impeller. Therefore, to obtain lower discharge rates, flow through the piping system is diverted through the smaller flow sensor section which increases the head loss in the piping system. To overcome this additional head loss the pump speed is increased to approximately 1680 RPM (Point No. 3) on the second characteristic pipe curve. To obtain lower flow conditions, the pump discharge can again be reduced until Point No. 4 is reached, corresponding to 30% of the nominal speed. At Point No. 4 the discharge through the piping system is approximately 7.5 litres/sec, which is 10% of the maximum discharge of 75 litres/sec, at Point No. 1. All twenty pump-and-piping systems were designed to have this same functional capability.

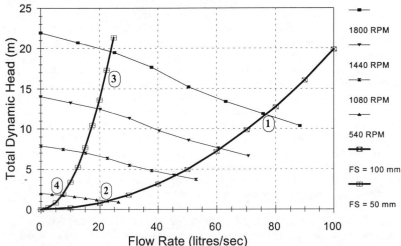

Figure 5: Combined Pump Performance and Characteristic Pipe Curves

8. SUMMARY

This paper documents the hydraulic design of a large-scale longshore current recirculation system required to maintain a uniform longshore current within the testing region of a new moveable-bed longshore sediment transport facility. Twenty independent pump-and-piping systems, with a total discharge capacity of 1,250 litres/sec, provide the unique capability to operate any combination of the twenty systems simultaneously. Using a combination of variable speed vertical turbine pumps along with a low and high flow measurement mechanism in each piping system, the recirculation system has the capability to accurately recirculate between 10% and 100% of the maximum design capacity at all twenty cross-shore locations. This unique design allows a wide range of longshore current magnitudes and cross-shore distributions to be accurately controlled and recirculated within the facility. This paper should be beneficial to those involved in designing similar recirculation systems in the future.

9. ACKNOWLEDGMENTS

The authors would like to thank Richard Whitehouse for generously providing technical information about the Coastal Research Facility at H.R. Wallingford. In addition, we would like to thank Paul Visser at Delft University of Technology for sharing his experience with laboratory measurements of uniform longshore currents. This research is funded by the "Large-Scale Laboratory Investigation of Longshore Sediment Transport" work unit of the Coastal Sediment and Dredging Program, under the Coastal Navigation and Flood Damage Reduction Research Area. Permission to publish this paper was granted by the Chief of Engineers, U.S. Army Corps of Engineers.

10. REFERENCES

Brebner, A., and Kamphuis, J.W., 1963. Model Tests on the Relationship Between Deep-Water Wave Characteristics and Longshore Currents. Report No. 31, Department of Civil Engineering, Queen's University at Kingston, Canada.

Dalrymple, R.A., and Dean, R.G., 1972. The Spiral Wave Maker for Littoral Drift Studies. Proc. of 13th Int. Conf. Coastal Eng., pp. 689-705, Vancouver, Canada.

Galvin, C.J., and Eagleson, P.S., 1965. Experimental Study of Longshore Currents on a Plane Beach. Technical Memorandum No. 10, US Army Waterways Experiment Station, Coastal Engineering Research Center, Vicksburg, Mississippi, USA.

Hamilton, D.G., Rosati, J.D., Smith, J.M., and Fowler, J.E., 1996. Design Capacity of a Longshore Current Recirculation System for a Longshore Sediment Transport Laboratory Facility. Proc. of 25th Int. Conf. Coastal Eng., pp. 3628- 3641, Orlando.

H.R. Wallingford, 1994. Understanding the Nearshore Environment. A publication describing the Coastal Research Facility at HR Wallingford, 4 pg.

Kamphuis, J.W., 1977. Discussion on Wave-Induced Circulation in Shallow Basins. Journal of Waterway, Port, Coastal and Ocean Engineering, ASCE, 103: pp. 570-571.

Mizuguchi, M., and Horikawa, K., 1978. Experimental Study on Longshore Current Velocity Distribution. Bull. Fac. Sci. Eng., Chuo Univ., Tokyo, Japan, 21: pp. 123-150.

Putnam, J.A., Munk, W.H., and Traylor, M.A., 1949. The Prediction of Longshore Currents. Transactions of the American Geophysical Union, Vol. 30 (3), pp. 337-345.

Rosati, J.D., Hamilton, D.G., Fowler, J.E., and Smith, J.M., 1995. Design of a Laboratory Facility for Longshore Sediment Transport Research. Proc. of Coastal Dynamics '95, pp. 771-782, Gdansk, Poland.

Simons, R.R., Whitehouse, R.J.S., MacIver, R.D., Pearson, J., Sayers, P.B., Zhao, Y., and Channell, A.R., 1995. Evaluation of the UK Coastal Research Facility. Proc. of Coastal Dynamics '95, pp. 161-172, Gdansk, Poland.

Svendsen, I.A., 1991. Development of a Comprehensive Plan for Modeling Longshore Current Generation in the Laboratory. Unpublished Contract Report for the Waterways Experiment Station, Coastal Engineering Research Center, Vicksburg, Mississippi.

Visser, P.J., 1982. The Proper Longshore Current in a Wave Basin. Report No. 82-1, Department of Civil Engineering, Delft University of Technology, The Netherlands.

Visser, P.J., 1991. Laboratory Measurements of Uniform Longshore Currents. Coastal Engineering 15, pp. 563-593.

Flow modelling using an inverse method with direct minimisation

Graham Copeland [1]

Abstract
An inverse flow model is described which is able to assimilate large quantities of flow data into solutions of the depth averaged equations of motion. Applications are given of the model to both laboratory-scale flows and to full-scale tidal flows in coastal areas. The benefits of the inverse approach are shown to be the lack of dependence on boundary data in the flow variables and solutions which are in good agreement with measurements.

Introduction
The paper describes a flow model based on an inverse method which uses direct minimisation as a solution procedure. Applications are shown including flow past a model headland and tidal flow in a coastal area.

The term 'inverse' refers to a model's construction, inverse models use both measured data and the governing equations. Formally, an initial or boundary value problem becomes inverse if some of the boundary or initial data is missing (i.e. the problem is ill-posed) and is replaced by information (data) about the desired solution located within the model domain. This situation is typical of coastal modelling problems where boundary data is often poorly defined yet there exists good data within the model area.

The paper describes an operational model, called WOLF/FLOW, based on an inverse method which can accommodate large quantities of measured data within the model domain and does not require initial or boundary data in the state variables. These are benefits that overcome the weaknesses in conventional direct models which can be ill-posed because of inadequate boundary data and which cannot make use of extensive survey data. The method uses a direct minimisation technique and the governing equations are included as a weak constraint on the solution. This allows the relative influences of data and equations on the final solution to be defined.

----

[1] Department of Civil Engineering, University of Strathclyde, Glasgow, G4 0NG
    Tel. + 44 141 548 3252, Fax + 44 141 553 2066, email g.m.copeland@strath.ac.uk

The Model (WOLF$_{/FLOW}$)

The model solves the depth averaged shallow water equations of motion for tidal flows in coastal waters, and by the addition of radiation stress gradient terms to the momentum equations it   will be developed to solve the governing equations for wave-induced flows near the shoreline.

The important aspect of these solutions is that they are constrained to be close to the flow field described by the data. That is, the solution found is that which lies closest to a 'first guess' provided by interpolation of the data and in which flow variables remain close to data values at points in the solution domain where they are provided.

The main objective of the WOLF$_{/FLOW}$ operational model is that it should be able to assimilate a large quantity of full-scale survey data into numerical computations of tidal flow and so produce good simulations of the flow fields and thereby produce reliable calculations of transport phenomena. In this way details of a flow regime revealed in measurements can be reliably reproduced in a numerical simulation.

A second objective is that it should be able to assimilate and assist with the interpretation of laboratory scale data. Steady and periodic flows past model headlands are being studied and the model will be extended to allow mean wave-induced currents measured around model breakwaters in the UK Coastal Research Facility at HR Wallingford to be studied.

A further application will be its use to assist with the interpretation of airborne survey data (CASI images) of coastal waters at selected locations on the coast of England provided by the UK Environment Agency

Inverse Modelling

Conventional or direct models need initial and/or boundary data in the flow variables but need no data within the domain. The solution proceeds by some form of numerical integration of the equations, see for example Davies, Dakin and Falconer 1995.

A modelling problem is inverse when some of the   boundary or initial values are unknown, then the problem is ill-posed. In such cases the missing data are replaced by data within the domain.

The principle of the inverse method used in this work is outlined in Figure 1. The principle of the solution method is that a function is minimised according to constraints that the solution is a good fit to the governing equations and is close to the data values, see Bennett 1992.

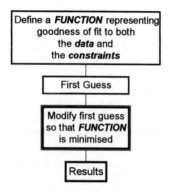

Figure 1 Outline of the inverse method

The Function

A quantity **F** is defined which represents the goodness of fit in a least-squares sense of the solution to the governing equations everywhere in the domain and its proximity to the data where provided. The solution and data are defined in terms of flow variables: surface elevation η (m) and discharges p, q (m^{2}s^{-1}). The constrained cost function, **F**, is :

$$\mathbf{F} = \sum_k \sum_i \sum_j \begin{array}{l} w1_{i,j}^k s1(continutity)^2 \\ +w1_{i,j}^k s2(x\_momentum)^2 + w1_{i,j}^k s2(y\_momentum)^2 \\ +w2_{i,j}^k s3(p_{i,j}^k - p_{oi,j}^{\ k})^2 + w3_{i,j}^k s3(q_{i,j}^k - q_{oi,j}^{\ k})^2 \\ +w3_{i,j}^k s4(\eta_{i,j}^k - \eta_{oi,j}^{\ k})^2 \end{array} \qquad \text{eq. 1}$$

where $w1_{i,j}^k$, $w2_{i,j}^k$, $w3_{i,j}^k$ represent the weights, $s1$, $s2$, $s3$, $s4$ represent the scaling factors and '*continuity*', '*x_ momentum*' and '*y_ momentum*' are defined as the (residuals of the) discretized equations of depth-integrated unsteady flow. The weights are real numbers which can be varied independently as required in order to correctly specify the relationship between the data and the equations. The scaling factors are used to make the terms in **F** the same order of magnitude and dimensionally consistent. The units of **F** are m^{2}s^{-2} and so $s1$ is unity and dimensionless.

The Solution Procedure

The procedure is described schematically in Figure 2 below, see Bayne 1997.

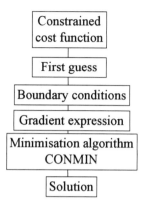

Figure 2.   The solution procedure.

The solution procedure starts with a first guess to the solution and needs to use some method of modifying the first guess so that **F** is minimised such that the solution satisfies the governing equations (i.e. minimises their residuals) whilst remaining close to the data. The following discussion describes the elements of the procedure: the constraints, the first guess, boundary conditions and the minimisation technique.

Constraints
Inverse methods are constrained minimisation problems; that is, a conditional minimum is sought. The conditions can either be applied as strong or weak constraints. Strong constraints require that that the constraint equations (the governing equations of shallow water flow) are satisfied exactly, this can be seen as a special case. Weak constraints, on the other hand, require that the constraint equations are satisfied only approximately. A system of weights, $w1$, $w2$ and $w3$ described above, allow for a degree of flexibility in the solution. This can be seen as the general case. Methods which impose strong constraints are provided by the calculus of variations. Weak constraint methods as used here are provided by direct minimisation using gradient techniques, Bennett 1992.

The First Guess
The inverse method seeks the best solution closest to the first guess. In practice, the closer the first guess is to the final solution the quicker the convergence occurs. A good first guess should set its values equal to the data where available in the solution domain and interpolate or estimate any missing neighbouring values. Discontinuities, inconsistent or incompatible data should be avoided in the first guess. These will either lead to a much longer convergence time or to a failure of the minimisation process. Interpolation either by a power law weighted average or by a tiling method, see Lloyd, Ball and Stansby 1995, is suitable.

Boundary Conditions
The solution requires that the residuals of the constraint equations are minimised. The problem becomes a boundary value problem in these residuals within the domain described in x, y space and time t for unsteady flows. The boundary conditions which apply are therefore values or gradients in the residuals of the three governing equations. Note that no boundary values in the actual flow variables p, q and η are required. This is a very great advantage of this method.

The minimisation technique described below modifies the flow variables (towards a solution of the governing equations) by increments defined by functions of the gradients in the residuals. The meaning of the boundary values given for these residuals can be understood in these terms. Either a residual is given a value (of zero) which is the Dirichlet condition; this leaves the gradient free to take a non-zero value determined by the calculations and so provide a modification to the first guess. This boundary value is suited to an open boundary in the model either in space or time. Or, the gradient in a residual is set to zero which is the Neumann condition; this ensures that the first guess cannot be modified by the minimisation process. This boundary condition is suited to a closed model boundary at which the first guess value of zero normal flow is preserved. Or finally, a periodic boundary condition can be imposed in time when a complete tidal cycle is being computed.

Minimisation Technique
The calculus of variations is a well established method of finding a conditional minimum but leads in this case, via arduous algebra, to a system of extremely large elliptic equations,   Monteiro and Copeland 1995. Direct minimisation is another method and provides a more practicable procedure, Bennett 1992.   This uses the gradient of **F** with respect to each of the flow variables to modify those variables successively in the direction of minimum **F**. This requires a numerical algorithm such as the conjugate gradient method; CONMIN was used in this model. Because the solution procedure is a substitution method and the equations of motion are not integrated, it has the advantage of not being subject to the Courant stability condition.

Validation
The model has been validated using idealised topographies in terms of data assimilation, flow diversion by topography and tidal flooding and drying over a shoal, Bayne 1997,. Copeland and Bayne 1996. The validation tests checked continuity, friction slope, Coriolis effect at a coastal boundary and convective acceleration terms for flow over a submerged weir. The validation process continues through the application of the model to laboratory scale data.

Applications
A series of half elliptical model headlands, with geometry after Signell & Geyer 1991, were installed in a laboratory flume. The geometry and scales were as described in Table 1 and Figure 3. A steady flow of approximately 0.06 ms^{-1} was used and the

velocity field measured using an impellor type flow meter and a series of floats tracked by video. The floats revealed the structure of the flow separation zone in the lee of the headland and indicated the reattachment point. A series of headland aspect ratios were used (b/a = 0.4, 1.0, 4.0, 14, the value of 14 represents a thin sheet) , flow separation occurred in all cases. The data shown in Figure 4 a) was used by interpolation to produce the first guess flow field shown in Figure 4 b). The first guess and the data were used in the inverse flow model to produce the solution shown in Figure 4 c). Visual inspection of first guess and minimised solution shows only minor differences. However, the substantive differences between them are revealed by simulated flow tracks shown in Figure 5 a) and 5 b). The tracks in the first guess show singularities, whereas tracks in the solution show closed eddies.

It can be seen that the solution contains a good representation of the structure of the flow separation zone. This is because the solution was guided by good data. If the model is run with no data and a poor first guess describing a uniform flow along the channel even (but unphysically discontinuous) in the lee of the headland then the solution produced is continuous and similar to the potential flow solution but shows no separation. This aspect of the validation process shows that for steady flow the model cannot predict flow separation without *a priori* knowledge. This may be because the solution procedure does not integrate the governing equations forward in time and hence the separation features cannot evolve through transport of vorticity, see for comparison Davies, Dakin and Falconer 1995. It has been noted, however, that for periodic unsteady tidal flows over complex topography including headlands the inverse model can produce eddies and flow separation features. These features may occur in such conditions because adverse pressure gradients form more readily as the tide reverses.

SCALING (Froude):	Prototype	Model	MODEL RANGES:	
Main-stream velocity (ms^{-1})	0.5	0.042	Main stream velocity (ms^{-1})	0.038 - 0.075
Depth - D (m)	10	0.071	Depth of water (m)	0.060 - 0.080
Headland length(m)	200	0.14	Reynolds Number	2400 - 6000
Froude No.	0.00253	0.00253	Froude Number	0.0082 - 0.00184
Reynolds No.	6.9×10^7	4183	Manning's Number	0.011 (Perspex)

Table 1 Scaling data and ranges of flow variables studied

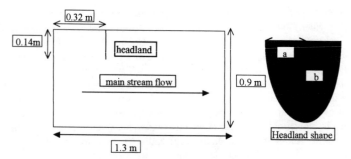

Figure 3. Flume and headland geometry.

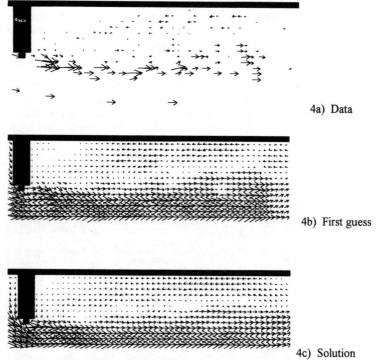

4a) Data

4b) First guess

4c) Solution

Figure 4. Assimilation of laboratory scale data into an inverse model solution on a 0.02m grid for the headland with b/a = 4.0.

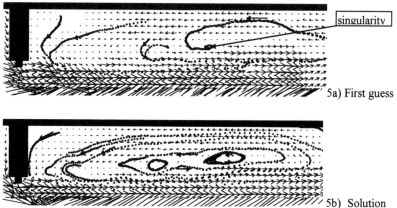

Figure 5. Simulated float tracks in first guess and solution flow fields.

A further example of assimilation of laboratory scale data is shown in Figure 6. The data was taken at the University of Manchester, England, Lloyd, Ball and Stansby 1995, and shows flow separation around a model conical island of base diameter 0.53 m and side slope angle 22.2° . The flume width was 1.5 m, the main stream velocity 0.088 ms⁻¹ and the water depth away from the island was 0.08 m. The data was taken using a video based particle tracking velocimetry (PTV) technique. The data shown is part of a sequence from the unsteady development of eddies. For clarity, Figure 6 shows only a small part of the solution domain of the inverse model.

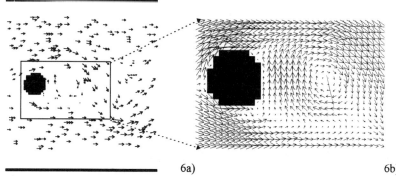

Figure 6   a) Particle Tracking velocimetry data for flow around a conical island, Lloyd et al 1995.   b) a detail from the inverse model solution based on the PTV data.

An application of the model to a full-scale coastal area was made in which a large quantity of survey data and published information was available from drogue tracks and from an Admiralty Tidal Streams Atlas. The model area was a central section of the Firth of Forth, Scotland near the road and rail bridges, see also Copeland 1997. One of the 25 half-hourly flow fields is shown in the Figure 7 where the data and the final solution are given. This application demonstrated the benefits that data assimilation and the removal of the need for flow boundary data bring to engineering practice.

 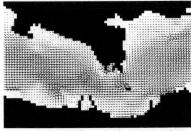

7a) Data on the ebb tide at HW+3 hrs   7b) Computed flow field at HW+3 hrs
                                                using the data in 6a)

Figure 7.Data assimilation by the inverse model in part of the Firth of Forth, Scotland.

Future Applications

The inverse model will be applied to a coastal study at a site on the east coast of UK. A hydrographic survey has produced a large quantity of good quality flow data using free floating drogues, an acoustic Doppler current profiler (adcp) and recording current meters (rcm). The vectorised drogue track data for all stages of the tides during the survey are shown in Figure 8. These data will be scaled to a common tidal range using rcm data, split into time intervals through a tidal cycle and used to create first guess flow fields for the inverse model and to control the solution. The drogue and adcp data revealed complex flow features such as eddies and strong horizontal shears especially off the headland. The inverse model is able to preserve these features in the solution. Again, the inverse model has benefits for applications in engineering practice where good use of survey data supports the financial investment in the survey and lends confidence to the design process.

Conclusions

An inverse model of shallow water flow has been developed and applied to both laboratory-scale and full-scale data. The model is able to assimilate data into the solution of the depth averaged flow equations such that the final solution is in good agreement with the data and retains important features of the measured flow. This aspect of the model ensures that best use is made of survey data. The other benefits of the method are that the solution procedure does not require boundary data in the flow variables nor is it bound by the Courant stability criterion. A limitation of the model at present is that it cannot predict flow separation in the lee of a headland in steady flow, a good solution can be found in such circumstances by including data about the separation.

Acknowledgements
Funding was provided by the Engineering and Physical Sciences Research Council.
The assistance of graduate student Scott Couch is gratefully acknowledged, as is
Peter Lloyd for his provision of the PTV data. The co-operation of R.H. Cuthbertson
and Partners, Consulting Engineers, Edinburgh is also gratefully acknowledged.

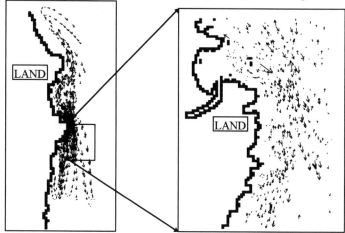

Figure 7. Vectorised drogue track data taken at a location on the east coast of UK

References
Bayne, G.L.S., 1997. 'A direct minimization model to determine tidal flows on
coastal waters', PhD Thesis, Dept. of Civil Engineering, University of Strathclyde
Bennett, A.F., 1992 . 'Inverse models in physical oceanography', CUP .
Copeland, G.J.M. & Bayne, G. L. S., 1996. 'Data Rich Models of Tidal Flows using
Inverse Methods', Hydroinformatics 96, ed. Muller, pub. Balkema 1996, ISBN 90 5410
8525, pp. 239-244.
Copeland, G.J.M., 1997. 'Computer modelling of tidal flows and effluent dispersion in the
Firth of Forth, Scotland using an inverse method and particle tracking techniques', (in
press) J. Mar. Env. Eng'g.
Davies, P.A., Dakin, J.M. and Falconer, R.A., 1995. 'Eddy formation behind a coastal
headland', J. of Coastal Research, 11, 1 pp154-167.
Lloyd, P.M., Ball, D.J.and Stansby, P.K., 1995. 'Unsteady surface-velocity field
measurements using particle tracking velocimetry', J. of Hyd. Res. Vol. 33, Nr.4,
pp519-534.
Monteiro, T.C.N. & Copeland, G.J.M., 1995. 'A Hydrodynamic model of unsteady tidal
flow in coastal waters based on an inverse method', Arch. Mech., 47, 6, pp 1057-1071
Signell, R.P. & Geyer. W.R., 1991. 'Transient eddy formation around headlands'. J.
Geo. Res. Vol. 96 No. C2 pp 2561-2575

CHANGES OF MEAN VELOCITY PROFILES
IN A GLM FORMULATION

J. Groeneweg[1], M.W. Dingemans[2] and G. Klopman[3, 2]

ABSTRACT
A model is presented to describe the mechanism of wave-current interaction. The model is based on the Generalized-Lagrangian-Mean (GLM) formulation introduced by Andrews and McIntyre (1978), which enables splitting of the mean and oscillating motion over the whole depth in an unambiguous and unique way. In the present formulation an explanation is given from a mathematical point of view for typical, measured wave-induced current velocities profiles in wave-current flumes. The wave-induced driving force for the mean motion is shown to be dominated by two contributions, the wave Reynolds stress and the Stokes correction of the shear stress, i.e. the total transfer of horizontal momentum in vertical direction.

INTRODUCTION
The importance of the interaction between waves and currents has long been recognized and much has already been done in this field. Two important review articles by Peregrine (1976) and Jonsson (1990) cover large parts, and recently Thomas and Klopman (1997) summarized some analytical and numerical models which are suitable to predict the combined motion in the nearshore region. They considered both regular and random gravity waves on currents with an arbitrary vertical velocity profile. Some attention was directed at experimental studies as well.

The importance of experiments in wave-current interactions is well known. Apart from validating theoretical models they also improve the understanding of phenomena which are acknowledged to exist but are difficult to model mathematically. In this light several authors, e.g. Van Doorn (1981), Kemp and Simons (1982, 1983), Klopman (1994) have found in their laboratory flume experiments that the Eulerian-mean velocities under progressive, non-breaking waves have completely different distributions than those which are the result of linear addition of the separately obtained current and wave velocities. For currents following a monochromatic wave field the Eulerian-mean

[1] Delft University of Technology, Department of Civil Engineering, P.O. Box 5048, 2600 GA Delft, The Netherlands, E-mail: J.Groeneweg@ct.tudelft.nl
[2] Delft Hydraulics, P.O. Box 177, 2600 MH Delft, The Netherlands.
[3] Netherlands Centre for Coastal Research, P.O. Box 5048, 2600 GA Delft, The Netherlands.

horizontal velocity reaches a maximum value at a level between bottom boundary layer and wave trough and decreases towards the free surface (see Figure 1). In contrast, for waves opposing the current the Eulerian-mean horizontal velocity increases more rapidly towards the free surface than in absence of waves.

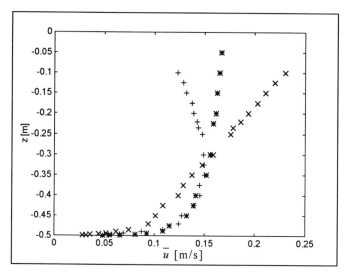

Figure 1: Eulerian-mean velocity profiles for the situation of no waves (*), following waves (+) and opposing waves (×) of the same size (depth 0.5m, wave period 1.44s, wave amplitude 0.06m). After Klopman (1994).

To the authors' knowledge only Nielsen and You (1996) and Dingemans et al. (1996) presented theoretical models to explain the wave-induced changes in the Eulerian-mean horizontal velocity profiles. The model of Nielsen and You (1996) is based on a local force balance. In a steady two-dimensional flow the vertical variation of the total shearing force per unit area of a cross section was balanced by the horizontal variation of the total normal stress. Assuming linear wave theory, expressions were derived for $\langle \tilde{u} \tilde{w} \rangle$ and for the radiation stress. Although their model gives a qualitative explanation of the physical mechanisms involved, quantitative agreement with Klopman's results was obtained only after a significant ad hoc enhancement of $\langle \tilde{u} \tilde{w} \rangle$ by a linearly depth-dependent empirical factor. The empirical adjustment is based on the fact that the interaction with a current induces extra vorticity of the wave motion. Dingemans et al. (1996) developed a quasi-3D model, the results of which were compared with the wave flume experiments by Klopman (1994). The effect of waves has been incorporated by adding the so-called Craik-Leibovich (CL) vortex force, consisting of $\overline{u}^S \times \overline{\omega}$ with $\overline{u}^S$ the Stokes drift and $\overline{\omega} = \nabla \times \overline{u}$ the mean vorticity. Under certain assumptions the vortex force is the main term in the ensemble-averaged momentum equations in a GLM formulation (see Leibovich, 1980). Dingemans et al. (1996)

observed in their simulations that secondary circulations in lateral direction induced by the CL vortex force caused changes in the vertical structure of the mean flow. However, due to poor estimates of the Stokes drift and the CL vortex force in the boundary layers, quantitative agreement with Klopman's experimental results was not obtained.

The aim of this work is to enlarge the insight in the mechanism of wave-current interaction and to give an explanation from a mathematical point of view for the changes in the Eulerian-mean velocity profiles due to the presence of waves as observed in several experimental studies. Groeneweg and Klopman (1997) have developed a two-dimensional model which is based on the Generalized-Lagrangian-Mean (GLM) theory. The momentum equations in GLM formulation derived in that paper are used here. An analysis of these equations shows some similarities with the one given by Nielsen and You (1996). It even may give a theoretical explanation for the underprediction of vorticity of the wave motion and the resulting incorrect velocity profiles, which resulted from their model before the empirical adjustment.

GENERALIZED LAGRANGIAN MEAN FORMULATION

In this paper the following notation is used. The velocity vector u is given by $u = (u_1, u_2, u_3)^T$ as a function of time t and space vector $x = (x_1, x_2, x_3)^T$. The velocity is denoted as $u = \bar{u} + \tilde{u} + u'$. Here $\bar{u} = \langle u \rangle$ is the Eulerian-mean velocity, defined in any of the usual senses (time, space, ensemble averaging); $\tilde{u}$ and u' represent the wave motion and turbulent motion. Two-dimensional flow refers to motion in the vertical plane. The velocity and space vectors are denoted as $u = (u, w)^T$ and $x = (x, z)^T$. Furthermore, Einstein's summation convention is used, where is summed over repeated indices. Latin indices take the value 1,2 or 3. For convenience Greek indices are employed for horizontal variables only and thus take the value 1 or 2.

For a good understanding of processes that are responsible for the modifications in a mean current profile due to the interaction with waves, a consistent way of splitting the mean from the oscillating motion is a necessity. In the conventional Eulerian formulation problems arise in defining the mean motion in an otherwise oscillating field in a unique, unambiguous, consistent and meaningful way, since there is no water at points in the region between wave trough and wave crest part of the time. This can be circumvented by using a purely Lagrangian framework; however, this formulation cannot be applied in any exact sense, if the Lagrangian-mean velocity is required at a specific point in space. This is due to the fact that the particle to be followed will generally wander away from this point. In order to be able to consider the Lagrangian-mean velocity as a "field", Andrews and McIntyre (1978) introduced the Generalized Lagrangian Mean description, which enables splitting of the mean and oscillating motion over the whole depth in a unique and unambiguous way, also in the region between wave trough and crest. An essential part in the GLM theory is the definition of the particle displacement ξ associated with the waves, as a function of the position x and time t. The GLM formulation combines the advantages of a purely Lagrangian and purely Eulerian approach. The essential idea is to average over positions displaced by the disturbance motion, i.e. the wave motion. After that the resulting mean quantity is assigned to a fixed position.

The following properties are discussed extensively in Andrews and McIntyre (1978).

A simpler discussion is contained in McIntyre (1980) and Dingemans (1997). If φ is a quantity to be averaged in the GLM formulation and $\langle (\ ) \rangle$ denotes an Eulerian average operator at fixed position, then the corresponding GLM quantity $\overline{\varphi}^{L}$, is defined as

$$\overline{\varphi (x, t)}^{L} = \langle \varphi (x + \xi(x, t), t) \rangle , \tag{1}$$

which entails averaging over the disturbed positions $\Xi(x,t) = x + \xi(x,t)$. Andrews and McIntyre (1978) required ξ to be a true disturbance-associated quantity, i.e.

$$\langle \xi(x, t) \rangle = 0 . \tag{2}$$

Together with relations (1) and (2) the following equations form the basic equations of the generalized Lagrangian mean theory,

$$u^{\ell}(x, t) \equiv u(x + \xi(x, t), t) - \overline{u}^{L}(x, t) , \tag{3}$$

$$\overline{D}^{L}\xi \equiv \left(\frac{\partial}{\partial t} + \overline{u}_{j}^{L} \frac{\partial}{\partial x_{j}} \right) \xi = u^{\ell} . \tag{4}$$

The material derivative denotes the rate of change following the GLM flow, instead of the total flow. Relation (4) between the disturbance-associated field ξ and u^{ℓ} not only defines ξ but also validates the claim to regard ξ as a disturbance-associated particle displacement. Finally, Eulerian-mean and generalized-Lagrangian-mean quantities are related to each other by the generalized Stokes correction $\overline{\varphi}^{S}$, which is defined as

$$\overline{\varphi(x, t)}^{S} = \overline{\varphi(x, t)}^{L} - \overline{\varphi(x, t)} , \tag{5}$$

and can be written in terms of disturbance-associated quantities.

FLOW EQUATIONS IN GLM FORM
In Groeneweg and Klopman (1997) the three-dimensional Reynolds-averaged Navier-Stokes equations for an incompressible fluid are transformed in a GLM framework. After some manipulations the GLM flow equations have been shown to read as,

$$\frac{\partial \overline{u}_{j}^{L}}{\partial x_{j}} = -\overline{D}^{L}(\log J) , \tag{6}$$

$$\overline{D}^{L}\overline{u}_{i}^{L} + \frac{1}{\rho} \frac{\partial \overline{p}^{L}}{\partial x_{i}} - \frac{1}{\rho} \frac{\partial \overline{\tau}_{ij}^{L}}{\partial x_{j}} - \overline{F}_{i}^{L} = \overline{S}_{i}^{L} , \tag{7}$$

where the right-hand sides of (6)-(7) depend on quantities which correspond to the wave motion. The Jacobian J of the mapping $x \rightarrow x + \xi$ is given by $J = \det(\partial \Xi_{j} / \partial x_{i})$ and

$$\overline{S}_{i}^{L} = \frac{1}{\rho} \frac{\partial \overline{p}^{S}}{\partial x_{i}} - \frac{1}{\rho} \frac{\partial \overline{\tau}_{ij}^{S}}{\partial x_{j}} - \frac{\partial}{\partial x_{j}} \left(\overline{u_{i}^{\ell} u_{j}^{\ell}} \right) + \frac{\partial}{\partial x_{j}} \left(\overline{D}^{L} \left(\overline{u_{i}^{\ell} \xi_{j}} \right) \right) + O(|\xi|^{3}) . \tag{8}$$

The shear-stress components $\overline{\tau}_{ij}^{L}$ are coupled directly to the deformation tensor by using Boussinesq's hypothesis. The eddy-viscosity has been modeled by using a one-equation q-l turbulence model, describing the transport of turbulence kinetic energy q. The turbulence length scale $l(z)$ is prescribed. The turbulence model for the total motion has been simplified, so that the total eddy viscosity is obtained from the problem that represents a situation of only a current without waves.

This form of the GLM equations differs from the one derived by Andrews and McIntyre (1978) in the following way. Before averaging they multiplied the equations, which are evaluated in the disturbed positions, by the tensor $\partial \Xi_j / \partial x_i$. In this paper the main interest concerns interaction of waves and a turbulent current. The divergence of the shear-stress tensor plays an essential role in modeling the turbulent current. If the approach of Andrews and McIntyre (1978) had been followed, lengthy expressions for the shear stress tensor in GLM form would have been obtained. This would certainly have led to difficulties in interpretation of the meaning of the GLM shear stress tensor. The derivation of GLM equations (6)-(8) can be found in Groeneweg and Klopman (1997). In the latter paper conservation equations for the wave motion in terms of GLM quantities are also given, together with proper boundary conditions at the bottom and the free surface.

In Groeneweg and Klopman (1997) equations (6)-(8) together with the equations for the disturbed motion have been solved by applying a combination of a perturbation series approach and the WKBJ method. The latter method is based on the assumption that the amplitude function varies much more slowly in time and horizontal direction than the phase function. After Chu and Mei (1970) the following form for each quantity is assumed,

$$\varphi(x, t) = \sum_{n=0}^{\infty} \varepsilon^n \sum_{m=-n}^{n} \hat{\varphi}^{(n, m)}(X_\alpha, z, t) E^m , \qquad (9)$$

where

$$E = \exp(i(k_\beta X_\beta - \omega T)/\delta) . \qquad (10)$$

The slowly varying coordinates are defined as $X_\alpha = \delta x_\alpha$, $T = \delta t$ and the modulation parameter δ is assumed to be of the same order of magnitude as the wave slope $\varepsilon = k a$, i.e. $\delta \sim \varepsilon$. By expressing each quantity in the form given by (9) and substituting them both into equations (6)-(8) and into the conservation equations for the wave motion, a cascade of systems of ordinary differential equations has been obtained, the vertical coordinate z being the only independent variable. These systems have been solved numerically by using the trapezoidal rule.

Since the main interest concerns the explanation of the changes of the mean velocity profiles, the second-order zero-th harmonic solution ($n = 2$, $m = 0$) is considered. Furthermore, since we are only interested in the local solution of the wave-current motion in a flume, the assumption of slow variation in time and horizontal direction is justified.

Many theoretical models solve the problem by first considering the problem in the bottom and free surface boundary layers. The solutions in these regions are used as boundary conditions for the equations in the core region. Instead of splitting up into three layers, the present model predicts the combined wave-current motion of the complete flow field.

MODEL RESULTS

Model results are compared with laboratory flume measurements, obtained by Klopman (1994). In this case Laser Doppler velocimetry (LDV) flow meters were used to measure horizontal and vertical velocities of the total turbulent flow. In the present model a turbulent current with a cross-sectionally averaged horizontal mass transport velocity $Q = 0.16$ m/s was generated in a flume with a still-water depth $d = 0.50$ m. A monochromatic wave field with a wave period $T = 1.44$ s and wave amplitude $a = 0.060$ m is imposed on the current.

The modifications of the mean horizontal velocity profile due to waves are shown in Figure 2. Here the Eulerian-mean velocity profiles in case of waves following and opposing the current are compared to the current profile in the situation without waves. Comparing the model results with the experimental data of Klopman (1994) not only a qualitative agreement can be observed, but the computed velocity profiles show quantitative correspondence as well. The changes of the mean velocity profiles due to the presence of following or opposing waves are significant. The waves propagating in the current direction cause a reduction of the mean velocity shear, or vertical gradient of the mean horizontal velocity, whereas waves opposing the current increase the velocity shear.

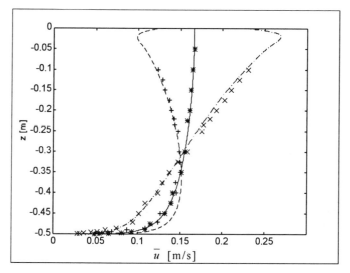

Figure 2: GLM results (present model) and experimental results (Klopman, 1994) for Eulerian-mean horizontal velocity profile.

MATHEMATICAL EXPLANATION

In order to be able to explain the model results given in Figure 2, the momentum equations (6)-(7) are simplified by neglecting horizontal variations ($\partial/\partial x_\beta = 0$), except for the horizontal GLM pressure gradient, which is assumed to depend on the horizontal gradient of the GLM free-surface elevation. Up to second order each GLM quantity can be written as the sum of its basic solution, corresponding to the situation of a current without waves, and a second-order correction,

$$\overline{\varphi}^{L} = \overline{\varphi}^{(0)} + \overline{\varphi}^{(2)} . \tag{11}$$

The horizontal momentum equation at second order reduces to

$$\frac{\partial \overline{\tau}_{xz}^{(2)}}{\partial z} = \rho g \frac{\partial \overline{\zeta}^{(1)}}{\partial X} + \frac{\partial \overline{\tau}_{xz}^{S}}{\partial z} + \rho \frac{\partial}{\partial z} \left(\overline{u^{\ell} w^{\ell}} \right),$$

$$\overline{\tau}_{xz}^{(2)} = \rho \nu_{T} \frac{\partial \overline{u}^{(2)}}{\partial z} . \tag{12}$$

The z-independent variable $\partial \overline{\zeta}^{(1)}/\partial X$ (index 1 due to the assumption of slow variation in x-direction) is used to fulfil the restriction that the horizontal mass transport velocity

$$Q = \frac{1}{d + \overline{\zeta}^{(2)}} \int\limits_{-d}^{\overline{\zeta}^{(2)}} \left(\overline{u}^{(0)} + \overline{u}^{(2)} \right) dz \tag{13}$$

is fixed. Integration over depth, using the boundary condition stating vanishing shear stress at the free surface or $\tau_{xz}^{(2)} = 0$ at $z = \overline{\zeta}^{(2)}$, leads to

$$\nu_{T} \frac{\partial \overline{u}^{(2)}}{\partial z} = g \frac{\partial \overline{\zeta}^{(1)}}{\partial X} (z - \overline{\zeta}^{(2)}) + \frac{\overline{\tau}_{xz}^{S}}{\rho} + \left\langle u^{\ell} w^{\ell} \right\rangle - \left[\frac{\overline{\tau}_{xz}^{S}}{\rho} + \left\langle u^{\ell} w^{\ell} \right\rangle \right]_{z = \overline{\zeta}^{(2)}} . \tag{14}$$

All terms in the right-hand side of (14) can be evaluated, since the model first solves the basic problem and the first order wave problem. After correction in order to obtain the Eulerian quantities, the latter solution is almost equal to the potential solution, except in the regions near the bottom and the free surface. In Figure 3 the vertical distribution of $\left\langle u^{\ell} w^{\ell} \right\rangle$, which will be denoted as the "wave Reynolds stress", and the Stokes correction of the shear stress $\overline{\tau}_{xz}^{S}/\rho$ are given, for the situation that waves are propagating in the flow direction of the current.

Nielsen and You (1996) decomposed $\left\langle u^{\ell} w^{\ell} \right\rangle$ into two components, the first due to phase shifting of velocities in the bottom boundary layer, resulting in the same expression as derived by Longuet-Higgins (1953) for boundary layers with constant eddy viscosity, and the second due to wave height decay. For Klopman's (1994) tests they show both contributions (Figure 4 in their paper). The distribution of the total wave Reynolds stress, i.e. the sum of the two contributions, looks quite similar to the one given in Figure 3, despite the fact that it is given in GLM formulation. However, Nielsen and You (1996) needed an empirical adjustment of the contribution of the wave Reynolds stress which is due to the phase shifting of velocities in the bottom boundary layer. However, in the present model the Stokes correction of the shear stress tensor prevents the total wave-induced driving force from increasing towards the free surface, which is the case in Nielsen and You's (1996) model before adjustment.

Figure 4 depicts the horizontal pressure gradient term, in which the horizontal gradient of the free surface elevation is chosen such that relation (13) is fulfilled with Q

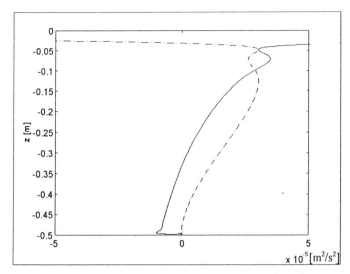

Figure 3: Vertical distributions of wave Reynolds stress $\langle u^{\prime} w^{\prime}\rangle$ (——) and Stokes correction of shear stress $\overline{\tau}_{xz}^{S}/\rho$ (–·–·) for situation of waves propagating in current direction

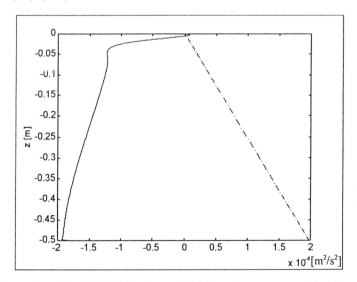

Figure 4: Vertical distributions of horizontal pressure gradient term $\frac{g}{}(z - \overline{\zeta}^{(2)}) \partial \overline{\zeta}^{(1)}/\partial X$ (–·–·) and wave-induced driving force $\overline{\tau}_{xz}^{S}/\rho + \langle u^{\prime} w^{\prime}\rangle - \left[\overline{\tau}_{xz}^{S}/\rho + \langle u^{\prime} w^{\prime}\rangle\right]_{z = \zeta^{(2)}}$ (——).

prescribed (in Klopman's experiments $Q = 0.16$ m/s), and the remaining terms in the right-hand side of equation (14). It shows that the right-hand side of (14) will be negative except in the small region near the bottom. Therefore $\overline{u}^{(2)}$ will decrease over almost the whole depth. Since no analytical expressions for $\langle u^\ell w^\ell \rangle$ and $\overline{\tau}_{xz}^S / \rho$ exist, equation (14) with $\overline{u}^{(2)} = 0$ at $z = -d$ has to be solved numerically. Subtraction of the Stokes drift $\overline{u}^S$ provides the second order correction of the Eulerian-mean horizontal velocity, see Figure 5. By adding the basic solution (current alone) the velocity profile as given in Figure 2 is almost obtained for waves following the current. A slight difference is observed due to the fact that horizontal variations are not neglected in the present model.

A similar analysis can be given for waves opposing a current. However, the correction is in opposite direction, so the gradient of the mean horizontal velocity increases.

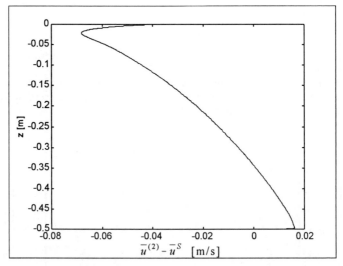

Figure 5: Second order correction of Eulerian-mean horizontal velocity $\overline{u}^{(2)} - \overline{u}^S$

CONCLUSIONS
In this paper a brief outline is given of the GLM theory as introduced by Andrews and McIntyre (1978), as well as an application of this theory in the field of wave-current interaction. A two-dimensional flow model for the combined wave-current motion has been developed by Groeneweg and Klopman (1997) in a GLM framework. The model results agree with wave-current flume measurements and are also used as a tool to give an explanation for the wave-induced changes in the mean velocity profiles. With a brief analysis it is shown that the wave Reynolds stress and the Stokes correction of the shear stress are the dominant terms in the wave-induced driving force for the mean horizontal velocity.

However, the given explanation is purely mathematical. It has to be investigated in the near future, how both contributions should be interpreted from a physical point of view. The difficulty is that they are given in terms of GLM quantities, while in our reasoning we are used to the Eulerian framework. Taylor series expansions are needed to evaluate the expressions in terms of Eulerian quantities. Apart from this, the meaning of the Stokes correction of a shear stress tensor, which appears as a kind of disbalance in the GLM momentum equations, has to investigated carefully as well.

REFERENCES

Andrews, D.G. and M.E. McIntyre, 1978: An exact theory of nonlinear waves on a Lagrangian-mean flow. *J. Fluid Mech.* **89**, 609-646.

Chu, V.H. and C.C. Mei, 1970: On slowly-varying Stokes waves. *J. Fluid Mech.* **41**, 873-887.

Dingemans, M.W., J.A.Th.M. van Kester, A.C. Radder and R.E. Uittenbogaard, 1996: The effect of the CL-vortex force in 3D wave-current interaction. In *Proc. 25th ICCE*, Orlando, 4821-4832.

Dingemans (1997), M.W.: *Water wave propagation over uneven bottoms.* World Scientific, Singapore.

Groeneweg, J. and G. Klopman, 1997: Changes of the mean velocity profiles in the combined wave-current motion in a GLM formulation. Submitted for publication.

Jonsson, I.G., 1990: Wave-current interactions. In *The Sea, Ocean Engineering Science*, **9A**. Eds. B. LeMehaute and D.M. Hanes, 65-120. J. Wiley and Sons, New York.

Kemp, P.H. and R.R. Simons, 1982: The interaction of waves and a turbulent current; waves propagating with the current. *J. Fluid Mech.*, **116**, 227-250.

Kemp, P.H. and R.R. Simons, 1983: The interaction of waves and a turbulent current; waves propagating against the current. *J. Fluid Mech.*, **130**, 73-89.

Klopman, G., 1994: Vertical structure of the flow due to waves and currents. Delft Hydraulics, Progress Report, H 840.30, Part 2.

Leibovich, S., 1980: On wave-current interaction theories of Langmuir circulations. *J. Fluid Mech.*, **99** (4), 715-724.

Longuet-Higgins, M.S., 1953. Mass transport in water waves. *Phil. Trans. Roy. Soc. London*, **A245**, 535-581.

McIntyre, M.E., 1980: Towards a Lagrangian-mean description of stratospheric circulation and chemical transports. *Phil. Trans. R. Soc. London*, **A296**, 129-148.

Nielsen, P. and Z-J You, 1996: Eulerian mean velocities under non-breaking waves on horizontal bottoms. In *Proc. 25th ICCE*, Orlando, 4066-4078.

Peregrine, D.H., 1976: The interaction of water waves and currents. *Adv. Appl. Mech.*, **16**, 9-117.

Thomas, G.P. and G. Klopman, 1997: Wave-current interactions in the nearshore regions. In *Gravity waves in water of finite depth*, chapter 7, 255-319. Ed. J.N. Hunt, series Adv. in Fluid Mech., Computational Mech. Publ.

Van Doorn, T., 1981: Experimental investigation of near bottom velocities in water waves without and with a current. Delft Hydraulics, Report M1423.

Coastal groundwater dynamics

Peter Nielsen
Associate Professor, Ph D, D Eng, Department of Civil Engineering, The University of Queensland, Brisbane, Australia 4072, fax: +61 7 3365 4599, email: P.Nielsen@mailbox.uq.edu.au

ABSTRACT
Wave runup on beaches have the capacity to raise the watertable near the beach up to several metres depending on the wave height and period. Similarly the asymmetric nature of the tidal in/ex-filtration on a sloping beach tends to raise the coastal watertable above mean sea level. Due to this combined effect of waves and tides, the shape of the watertable and the salinity structure in coastal barriers of width less than one kilometre are quite different from the classical scenarios. On this scale the extra watertable height caused by wave runup on the ocean side can drive a significant landward ground water velocity. A primary consequence of this is that any wastewater released into the aquifer, including oil spills on the beach, will travel towards the land rather than towards the ocean. Secondly, this landward flow of salty groundwater makes the freshwater lens much thinner than the "Ghyben-Herzberg thickness". The possible implications for sediment transport and beach erosion of the in/ex-filtration through the beach face are also reviewed. A final answer is several years away, mainly because the structure of swash zone flow in general, and its boundary layer in particular, are not well understood. We do however, by now, know the general magnitude of the in/ex-filtration velocities. Individual waves running up onto an unsaturated beach may generate short bursts of infiltration equal to the hydraulic conductivity K, and the average over 10 to 20 minutes during rising tides may reach $0.6K$. If the beach is saturated, the in/ex-filtration velocities are always much smaller. When porewater pressures in and landward of saturated beaches are observed to oscillate with large amplitude (tens of centimetres) at 100 to 300 second periods, this is usually not associated with large amounts of flow through the beach face. These oscillations are driven by the alternating appearance and disappearance of miniscusses just below the runup limit, which generate a pressure (head) signal of the order of the capillary height of the sand. Observations of the effect of seepage on sediment transport are discussed in terms of an adapted Shields parameter.

1.   WATERTABLE HEIGHTS AND SALINITY
The action of waves and tides on sandy beaches tend to raise the coastal groundwater levels. With large waves and/or large tides on a flat slope, the overheight may be several metres. This means that the ground water levels and the salinity structure often as shown in Figure 1.

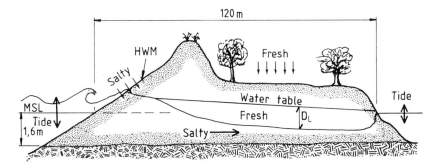

Figure 1: Watertable heights and depth of the freshwater lens under the northern end of Bribie Island, north of Brisbane, Australia. Measured October 6 1995. The tidal range at the site varies between 1m and 2m and wave heights are usually of the order 0.8m to 1.5m with periods 4s to 6s.

The asymmetry will be greater the more the landward side is protected from the waves and the tides. Also, if the landward side is steeper than the ocean side, the tidal superelevation is further reduced, cf Nielsen (1990). For a given difference in levels from the exposed to the protected side it is also clear that the watertable slope and hence the landward groundwater flow will be greater the narrower the barrier is.

The dynamic range of the freshwater lens is very large. Between 1994 and 1997 observations of its volume have varied by almost a factor 3.

The general picture shown in Figure 1 has a number of significant consequences for the environmental management of coastal barriers. Firstly, it is noted that any wastewater released into the aquifer will travel towards the continent rather than towards the ocean. Secondly, the fact that there is a net inflow of water through the beach face means that pollutants landing on the beach face will have a strong tendency to enter the aquifer under the barrier. Thus, the use of detergents to disperse oily pollutants from barrier beaches must be discouraged since it will, more than anything, make the pollutants enter the barrier groundwater system with increased speed. Thirdly, the freshwater lens under such a barrier is asymmetrical and much thinner than in the classical Ghyben-Herzberg scenario and the vegetation may be subject to salt poisoning under extreme conditions of large waves after a period with low rainfall. Thus, there was some evidence of several trees suffering from the very salty conditions measured in 1994, Figure 1.

2. THE OCEAN BOUNDARY CONDITION FOR THE WATERTABLE

The watertable a few tens of metres inland from the high water mark on a beach will be considerably higher than the Mean Sea level (MSL) even if there is no outflow due to rainfall on the land. This overheight is partly due to waves and partly due to tides. The situation is illustrated in Figure 2 and a few useful definitions will be given below.

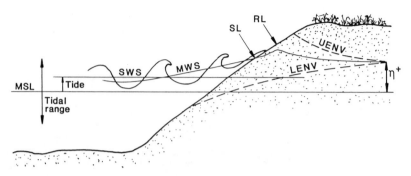

Figure 2: Definition sketch for terms used in formulating the coastal boundary
condition for groundwater modelling.

    The long term average level of the ocean surface outside the surf zone is
the mean sea level, MSL. The still water surface, SWS is the flat (on the scale
considered) surface which would exist in the absence of winds and waves. It moves
up and down due to astronomical tides and changes in barometric pressure. Local
short time (15 to 20 minutes) averaging of the water level defines the mean water
surface, MWS, which intersects the beach at the shoreline (SL) and becomes the
watertable. The MWS/watertable is not a flat surface, it rises towards the beach due
to wave setup and wind setup. The watertable rises further landward of the
shoreline. On a low tide, this rise may be caused partly by stored water from the
previous high tide. On a rising tide, however, it is entirely due to the infiltration and
this gives rise to the humped shape of the watertable with the hump being near the
runup limit RL, cf Kang et al (1994a,b). Landward of the high water mark (= the
high tide runup limit) the watertable oscillates inside an envelope (UENV, LENV)
which tapers off to define an average superelevation η^+ above the mean sea level.
The landward range of the different oscillations depends on their period. Wind
waves, surf beats and even tides are not felt more than a few tens of metres
landward of the high water mark but oscillations due to wave height changes over
several days reach further. See the data of Nielsen et al (1988), Nielsen (1990) and
Kang et al (1994a). While the oscillations have a limited range, the time averaged
consequences of waves and tides may influence the groundwater on a regional
scale.
    The fact that salty sea water is poured in on top throughout the swash
zone means that the classical large scale scenario shown, e g, by Cooper (1959) -
his Figure 5, must be supplemented with an area of sea water salinity on top of
fresher water in the swash zone. For a tide free situation, the general flow pattern
which is driven by the waves on the scale of the surf zone was illustrated and
modelled by Longuet-Higgins (1983).
    The nature of the tidal watertable fluctuations on the two sides of a
coastal barrier were shown in Figure 3. Both gauges were of the order 10m inland
of the respective high water marks. The watertable variation on the ocean side is
largest when the waves are big because the wave setup and runup bring "the action"

closer to the well. The shape of the tidal signal is skewed towards a saw tooth shape, and the watertable difference between the wells is seen to correlate with the offshore wave height. See also the data of Turner et al (1996).

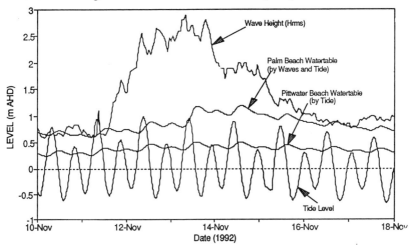

Figure 3: Watertable time series from the exposed (Palm Beach) and the protected (Pittwater) sides of the Palm Beach Isthmus, Sydney, Australia. Data courtesy of the New South Wales Department of Soil and Water Conservation. The apparently delayed response of the watertable to the large wave event is probably due to the large waves initially travelling almost parallel to the coast and gradually becoming almost shore normal. The larger effect later in the storm is also partly due to the wave period becoming longer.

3. WAVE FORCING OF THE WATERTABLE

As indicated by Figure 3, the contribution to the watertable height by wind waves varies on the time scale of hours as the height, H, period, T and direction of the off shore waves changes. The changing beach topography will also play a role. The data set in Figure 3 suggests the estimate

$$\eta_w^+ = 0.44\sqrt{H'_{o,rms}L_o}\tan\beta_F \qquad (1)$$

for steep beaches ($\tan\beta_F > 0.1$) like Palm Beach. However, the result (1) which is derived from a set of steep-beach-data should not be uncritically adopted to dissipative conditions where the inner surf zone hydrodynamics are qualitatively different. Hence, it is unlikely that η_w^+ should ever be less than the shoreline setup which according to Hanslow & Nielsen (1993) is of the order $0.4H_{o,rms}$ or $0.05\sqrt{H_{o,rms}L_o}$ irrespective of $\tan\beta_F$. In fact, the effect of surf beat swash is likely to lift the watertable above the shoreline level on flat beaches so that

$$\eta_w^+ \geq 0.05\sqrt{H'_{o,rms}L_o} \qquad (2)$$

for all beaches.

4. TIDAL FORCING OF THE WATERTABLE

Contrary to the wave effects, the tidal effects on the coastal watertable have been measured directly in the field, e g, the "Pittwater data" in Figure 3 and the dataset of Nielsen (1990). Also, some analytical modelling has been acheived using the Dupuit-Forchheimer assumptions and ignoring the capillary fringe, i e, by considering solutions to the Boussinesq Equation

$$n\frac{\partial h}{\partial t}=K\frac{\partial}{\partial x}\left(h\frac{\partial h}{\partial x}\right)$$ (3)

where h is the local height of the watertable above a horizontal, impermeable base, K is the hydraulic conductivity, n is the porosity, x is a shore-normal, horizontal coordinate and t is time.

Phillip (1973) showed that a tide of amplitude A_{tide} acting on a vertical beach with undisturbed aquifer depth D, will create an asymptotic overheight compared to the MSL of

$$\eta_v^+ = \sqrt{D^2+\frac{1}{2}A_{tide}^2}-D \approx \frac{A_{tide}^2}{4D}$$ (4)

and Nielsen (1990) showed by different means that this height is approached exponentially as

$$\overline{\eta}(x) = \eta_v^+(1-e^{-2k_Bx})$$ (5)

where k_B is the Boussinesq wave number

$$k_B =\sqrt{\frac{n\omega}{2KD}}$$ (6)

This overheight is due to non-linearity of the Boussinesq equation in the interior. Nielsen (1990) also found that an extra overheight is generated if the beach, instead of being vertical, forms the angle β with the horizontal. This overheight is approximately

$$\eta_\beta^+=0.5\varepsilon A_{tide}$$ (7)

a result which is accurate for $\varepsilon = k_B A_{tide}/\tan\beta < 0.5$ provided no seepage face is formed. A seepage face formed during the falling tide will cause a further increase of η_β^+. See the measurements of Nielsen (1990), Turner (1993) and Baird et al (1996). For very flat beaches of low permeability with no waves η_β^+ would approach the tidal amplitude, but not exceed it.

$$\eta_\beta^+ \to A_{tide} \quad for \quad \tan\beta \to 0$$ (8)

5. HIGH FREQUENCY PORE PRESSURE FLUCTUATIONS

Several authors, e g, Waddell (1973), Lewandowski & Zeidler (1978) and Hegge & Masselink (1991) have measured pressure fluctuations in beaches with frequencies in the range 0.003Hz to 0.01Hz. These watertable waves had two unexpected features which were unresolved for a long time: They decayed much more slowly than predicted by shallow aquifer theory (exp(-k_Bx) ref Eq 6) and they were practically standing waves, i e, the phase shifts between measuring stations on a shore normal transect were very small. An explanation in terms of the effect of a capillary fringe on top of a shallow aquifer was given by Aseervatham (1994) and by Li et al (1997). A different explanation was suggested by Nielsen et al (1996) who showed that in a deep aquifer, without a capillary fringe, the pressure waves in the aquifer become standing waves and decay as $\exp(-\pi x/2d)$, i e, much slower than predicted by shallow aquifer theory. Waddell (1973), Lewandowski & Zeidler (1978) and Hegge & Masselink (1991) did not measure the actual positions of the watertable but rather the pressure at some depth. Hence it is not known at present how the watertable, as opposed to the non-hydrostatic pressure at some depth, actually behaves under unsaturated conditions landward of the runup limit.

Pressure measurements in the saturated sand just seaward of the runup limit on a falling tide show other surprising features, see the data of Turner & Nielsen (1997). Firstly, the pressure head fluctuations are disproportionally large (30cm to 40cm) compared with the swash depths (2cm to 5cm) on top of the sand which might have been expected to drive them. Secondly, the spectral peak frequency for the pressure fluctuations did not correspond to any of the peaks (wind waves or surf beat) in the surface elevation spectrum in the inner surf zone. The dominant peak in the pressure spectrum was at 170s while the surf beat was at 40s to 50s. An explanation for the first feature was given by Nielsen et al (1988), see Figure 4.

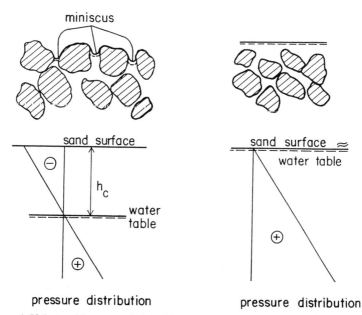

Figure 4: If the sand is saturated, the addition of a thin film of water to eliminate the surface miniscusses will lead to a jump in the pressure head below of the order of the capillary height h_c of the sand which may be more than half a metre in fine beach sand. After Nielsen et al (1988).

The pressure increases due to the elimination of the miniscusses are transmitted downwards as pressure waves with very little flow of water. The overall picture of the upper swash zone generating watertable waves by this mechanism which subsequently travel inland with or without significant assistance from a capillary fringe remain to be explored.

6. IN/EX-FILTRATION EFFECTS ON SEDIMENT TRANSPORT.

It has been observed that beach areas from which groundwater is pumped will experience extra accretion during periods with low or average wave activity. This indicates that lowering the watertable with the resulting increased infiltration has a stabilizing effect on beaches.

It has however also been observed that, at the arrival of storms where the waves start to run up higher onto the beach and when infiltration is therefore strong, the beach will usually erode rapidly. Thus, although enhanced infiltration has some stabilizing effect, it seems that other factors usually dominate under extreme erosion conditions.

Some data from controlled experiments are now available, e g, Dean (1989) and Curtis et al (1996) for pumped systems and, Davis et al (1992) and Katoh et al (1994) for systems relying on gravity drainage. The extent to which drains provide erosion protection during storms is however still not clear.

From a basic sediment transport point of view the in/ex-filtration has two opposing effects: Infiltration will make the boundary layer thinner and thereby increase the skin friction shear stress τ'. At same time however, the infiltration velocity will generate a stabilizing force on the sediment.

Martin (1970) reviewed the literature on boundary layer thinning and seepage forces and carried out experiments on initiation of motion in pipe flow with sand and nickel pellets of the same diameter: $d_{50}=0.58$mm. He found that the overall effect was destabilizing for the heavy (specific gravity $s = 8.75$) nickel particles, i e, movement started at lower flow rates with greater infiltration velocities. On the other hand, the overall effect on the quartz sand was stabilizing. The experiments covered seepage velocities in the range $-20<w/K<0$.

Martin (1970) suggests that the seepage drag on surface particles is reduced by 50% compared to buried particles and that the effect of the seepage velocity on the effective drag is linear. On that basis one is lead to consider a revised Shields parameter of the form

$$\theta = \frac{u_{*_o}^2 (1-\alpha_s \frac{w}{u_{*_o}})}{gd_{50}(s-1-0.5w/K)} \qquad (9)$$

where w is the seepage velocity, negative as infiltration, u_{*_o} is the shear velocity without seepage. That is, the infiltration is assumed to increase the shear stress linearly and the stabilizing effect, cf Nielsen 1992 p 102, is simply added to the specific gravity and buoyancy terms in the denominator.

This revised Shields parameter attempts to quantify the two opposing effects of an infiltration velocity (negative w): The numerator is increased corresponding to the thinning of the boundary layer and the denominator is increased due to the downward seepage drag on the particles.

The opposite net effects for Martin's quartz and nickel particles can be explained by the different relative effect a certain seepage velocity, say $w=-K$, has on the denominator. For quartz with $s=2.65$ the change is 33%. For nickel with $s=8.75$ the relative change is only 6.4%.

The effect on the numerator would be the same for the two like-sized sediments (same K), and since denominator changes of respectively 33% and 6.4% gave opposite net results, the numerator's relative change

$$\alpha_s \frac{w}{u_{*_o}} = \alpha_s \frac{K}{u_{*_o}} \frac{w}{K} \tag{10}$$

must have been in the range [6.4%; 33%] or since we consider $w/K=1$:

$$0.064 < \alpha_s \frac{K}{u_{*_o}} < 0.33 \tag{11}$$

Equation (9) also indicates that for fixed sediment density, decreasing grain size, and correspondingly decreasing K, will mean increased stability dominance. Hence, finer quartz sands ($d_{50} < 0.58mm$) are likely to be stabilized by infiltration. For shingle or pebble beaches with much larger K, the effect on the numerator may be dominant so that the net effect of infiltration may be destabilizing for such beaches.

With respect to swash zone sediment transport, the conclusions above should be taken as indications only since Martin's experiments are from flow in a circular pipe which may be significantly dissimilar from swash zones in several respects.

7 CONCLUSIONS

Coastal groundwater dynamics are of interest with respect to both groundwater modelling and beach erosion modelling.

Waves and tides provide groundwater forcing on all time scales longer than wind wave periods.

The oscillating components with periods up to several days influence sediment transport processes and exchanges of salt and oxygen through the beach face but they do not directly influence groundwater hydrology on the regional scale.

The time averaged effects of these oscillations, i e, the steady overheights (η^+) due to wave runup, beach slope and non-linearity are however significant for regional groundwater modelling. That is, the groundwater level at the coastal boundary may be metres above mean sea level and the groundwater may be salty at the top due to wave runup.

In coastal barriers, less than 1km wide, the coastal overheight can drive a significant net groundwater flow towards the land side.

The understanding of the nature of short groundwater waves has been greatly advanced in recent years. The forcing mechanism in the upper swash zone (Figure 4) is qualitatively understood. Also the observed deviations from "Dupuit-Forchheimer"–behaviour with respect to damping rate and phase speed has been accounted for, either as an effect of vertical flow or as a result of a capillary fringe.

The beach face infiltration velocities may in short bursts equal the hydraulic conductivity K, while 10-20minute averages can reach $0.6K$ during rising

tides. Under saturated, i e falling tide conditions, infiltration velocities are much smaller although the watertable may be oscillating tens of centimetres (Figure 4).

Beach drainage with ensuing enhanced infiltration is generally agreed to enhance accretion of sandy beaches during fair weather conditions but the beneficial effects of drains during storm erosion events is not yet clear.

Infiltration affects sediment transport in two ways: Increased shear stress through boundary layer thinning and stabilization via seepage drag. Pipe flow laboratory measurements, which showed both net stabilization and net destabilization, explain that the net effect is stabilization for sand, at least for $d_{50} < 0.58$mm. They do however, also indicate that the net effect is most likely destabilizing for shingle and pebble beaches.

8 REFERENCES

Aseervatham, A M, (1994): *Tidal dynamics of coastal watertables*. Ph D Thesis, The University of Queensland, 253pp.

Baird, A J, & D P Horne (1996): Monitoring and modelling groundwater behaviour in sandy beaches. *J Coastal Res, Vol 12, No 3*, pp 630-640.

Cooper, H H (1959): A hypothesis concerning the balance of fresh water and salt water in a coastal aquifer. *J Geophys Res, Vol 64,No 4*, pp 461-467.

Curtis, W R, J E Davis & I L Turner (1996): Independent evaluation of a beach dewateringsystem: Nantuckett Island, Massachusetts. *Proc 25th Int Conf Coastal Eng, Orlando*, A S C E.

Davis, G A, D J Hanslow, K Hibbert & P Nielsen (1992): Gravity Drainage: A new method of beach stabilisation through drainage of the watertable. *Proc 23rd Int Conf Coastal Eng, Venice*, A S C E, pp11229-1141.

Dean, R G (1989): Independent analysis of beach changes in the vicinity of the Stabeach System at sailfish point Florida. Consultants report, 15pp

Fetter, C W (1988): *Applied Hydrogeology*. Merrill Publishing Company.

Hanslow, D J & P Nielsen (1993): Shoreline setup on natural beaches. *J Coastal Res, Special Issue No 15*, pp 1-10.

Hegge, B J & G Masselink (1991): Groundwater-table response to wave runup: An experimental study from Western Australia. *J Coastal Res, Vol 7, No 3*, pp 623-634.

Kang, H-Y(1995): *Watertable dynamics forced by waves*. Ph D Thesis, Department of Civil Engineering, University of Queensland, 200pp.

Kang, H-Y, A M Aseervatham & P Nielsen (1994a): Field measurements of wave runup and the beach watertable. Research Report No CE148, Department of Civil Engineering, University of Queensland, Brisbane.

Kang, H-Y, P Nielsen & D J Hanslow (1994b): Watertable overheight due to wave runup on a sandy beach. *Proc 24th Int Conf Coastal Eng*, Kobe, A S C E, pp 2115-2124.

Katoh, K, S-I Yanagishima, S Nakamura & M Fukuta (1994): Stabilization of beach in integrated shore protection system. *Hydro-Port '94, Proc Int Conf Hydrotechnical Eng for port and harbour construction*. Yokosuka, Japan, pp 1077-1096.

Lewandowski, A & R Zeidler (1978): Beach groundwater oscillations. *Proc 16th Int Conf Coastal Eng*, Hamburg, A S C E, pp 2051-2065.

Li, L, D A Barry, J.-Y Parlange & C B Pattiarachi (1997): Beach watertable fluctuations due to wave runup. *Water Resources Research, Vol 33*, pp 935-945.

Longuet-Higgins, M S (1983): Wave setup, percolation and undertow in the surf zone. *Proc Roy Soc Lond, Vol A390*, pp283-291.

Martin, C S (1970): Effect of a porous sand bed on incipient sediment motion. *Water Resources Res,Vol 6, No 4*, pp1162-1174.

Nielsen, P (1990): Tidal dynamics of the watertable in beaches. *Water Resources Res, Vol 26, No 9*, pp 2127-2134.

Nielsen, P (1992): *Coastal bottom boundary layers and sediment transport.* World Scientific, Singapore, 324pp.

Nielsen, P, G A Davis, J M Winterbourne & G Elias (1988): *Wave setup and the watertable in sandy beaches.* Tech Memo 88/1, Coast & Rivers Branch, Public Works Dept, Sydney, 132pp.

Nielsen, P, J D Fenton, R A Aseervatham, P Perrochet (1996): Groundwater waves in aquifers of intermediate depths. *Advances in Water Resources, Vol 20, No 1*, pp 37-43.

Turner, I L (1993): Watertable outcropping on macrotidal beaches: A simulation model. *Mar Geol, Vol 115*, pp 227-238.

Turner, I L, B P Coates &R I Ackworth (1996): The effect of tides and waves on watertable elevations in coastal zones. *Hydrogeology Journal, Vol 4, No2*, pp 51-69.

Turner, I L & P Nielsen (1997): Rapid watertable fluctuations within the beach face: implications for swash zone sediment transport. Coastal Engineering, in press.

Philip, J R (1973): Periodic non-linear diffusion: an integral relation and its physical consequences. Australian J Physics, V26, pp 513-519.

Waddell, E (1973): *Dynamics of swash and implication to beach response.* Tech Report 139, Coastal studies Institute, Louisiana State University, Baton Rouge, 49pp.

DELTA'96: Surf-Zone and Nearshore Measurements at the Ebro Delta.

*Agustín Sánchez Arcilla [1], Andrés Rodriguez, [1] José Carlos Santás [2],
Vicente Gracia [1], Ruben K'osyan [3], Sergei Kuznetsov [3] and César Mösso [1]

Abstract

The Delta'96 large-scale hydromorphodynamic experiments and preliminary results are here presented. This campaign was carried out by the Laboratori d'Enginyeria Marítima (LIM) of the Technical University of Catalonia (UPC) and other Spanish and Russian research institutions within the framework of the CIIRC international projects and the European Union's (UE) FANS (*Flow Across Narrow Shelves "The Ebro Delta Case"*). The goal of these experiments was to study in an integrate manner the meteorological and hydromorphodynamical processes, their interrelations and dynamic impact on the nearshore (NS). With the obtention of high quality and 3D resolution data of the measured processes at different time and spatial scales, various NS hydromorphodynamic numerical models could be calibrated. Several weather conditions took place during the experiments, affecting the whole study area, causing some complex poorly-defined circulation patterns with superposition of waves and wind as driving mechanisms.

Introduction

DELTA'96 field campaign is the continuation of the NS hydromorphdynamcis investigation projects carried out by the LIM-UPC during the last 10 years. DELTA'96 presents important innovations with respect to the previous DELTA'93 field campaign (see Rodriguez et al. 1994 and 1995a). The main goal of DELTA'96 is an integral study and the obtention of high quality 3D resolution data of atmospheric conditions, hydrodynamic processes in the macroturbulence, wind waves, long waves and currents scales as well as suspended sediment transport in the NS, including the surf zone (SZ). With the obtained data, several numerical models that deal with hydrodynamics, irregular wave

* (1): **LIM-UPC**, D1 ETSECCPB, c/Gran Capitán s/n, 08034, Barcelona.
 (2): **CEPYC-CEDEX**, c/ A.López 81, Madrid.
 (3): **Shirshov Ocean Institute.**, Russian Acad. Of Scs.
 e-mail: arcilla@etseccpb.upc.es

ropagation and Q3D NS circulation in the study area will be calibrated. This project was done within the framework of the major-scale experiments FANS (*Flow Across Narrow Shelves "The Ebro Delta Case"*) of the EU which includes various types of oceanographic measurements in a much larger domain of the western Mediterranean Sea, covering a 75 x 50 km area, from Cape Salou, The Ebro Delta up to Columbretes Islands and including several sub-domains through the inner Shelf and Shoreface/Surf Zone. The global objective of FANS is to study the energy, water, sediments and nutrients fluxes through the shoreface and shelf. The NS campaign was carried out jointly by Spanish and Russian Research Institutions; LIM/UPC, ICM/CSIC, CEPYC/CEDEX, Gelendzik Oceanographic Institute and the Shirshov Oceanographic Institute, both belonging to the Russian Academy of Sciences. Different variables such as bathymetry, wheather conditions, hydrodynamics (waves and velocities at high frequency sample rate), suspended sediment concentrations and mean transport rate were measured during the experiments. Very complex circulation patterns, with typical wind induced structures superposed with wave breaking driven currents (longshore and weak undertow) were identified.

Study Area

The Ebro Delta, considered as a natural laboratory, was selected as the study area due to its nearly uniform longitudinal geometry and the availability of previous data. DELTA'96 experiments were carried out between 30/10/96 and 10/11/96 and were focused on the surf zone, from the shore line up to a depth of approximately 3 m at the Trabucador bar, southern hemidelta (see figure 1). The elected site corresponds to fine sand ($D_{50} \approx 0.2$ mm) cuasi longitudinally uniformly beach, with bars parallel to the shore line.

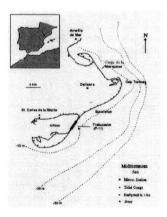

Figure 1: Study Area: the Ebro Delta.

Methodology

The general methodology included simultaneous and detailed measurements of the NS **topography**, **hydrodynamics**, **sediment transport** and **meteorology** outside (see figure 2) and inside the Surf Zone. The topo-bathymetric profiles in the surf zone, up to 2 m depth, were done before and after

the experiments with conventional topography (topographic station with infrared distance recor
and optic prisms) following and recording the sledge positions and depth and with aerial vie
images. Outside the surf zone, indirect methods were used, measuring depth from a zodiac boat a
following the positions from the beach. Figure 3 shows an aerial view of the study area to
bathymetric shape.

Equipment	Distance to shoreline (m)	Depth (m)	Sensors	Sample Freq, long. ST time interv (min.) / (Hs
NS-Tripods 1-2	1500-3000	8,5 - 12,5	3CEM,3OBS,1 SP	2 Hz, 20'/3Hr
SZ-Tetrapode	150	3.5	1 CEM, 1 s.pres.	2-4Hz, 20'-40'
DWR 1 y 2	1500-aprox10Km	8 - 50	3D acelerometers	1.78 Hz, 20'/3Hs
Sed. Traps	from 50 to 200	de 1 a 3	sampling recip.	1-2 days
Video IMG	de 0 a aprox.300	de 0 a 3-4	BN-color	50 Hz, 30'

Figure 2: Measurement equipments outside the Surf Zone.

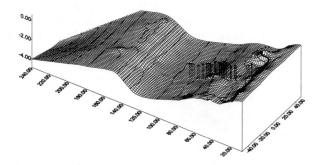

Figure 3: Trabucador's Bar bathymetry. The Sledge positions are marked with the vertical lines.

The hydrodynamic study inside and outside the surf zone included *Eulerian* measurements -hi
frequency rate sample (20 Hz) with electromagnetic currentmeters (EMC) in a movable sledge a
in the tetrapode-, and *lagrangian* measurements-spatial tracking of buoys and dye spots (milk w
fluorescein) done with topographic equipment and aerial video images, respectively.

i) Outside the surf zone, waves, water level (η) and horizontal velocities were measu
using a pressure sensor and a *Interocean S4* biaxial electromagnetic currentmeter -PUV.-
currents in the horizontal plane located at 1.5 m from the bed in a tetrapode moored at 3.5 m dep
and 150 m from the shoreline (see figure 4). The waves were measured at 4 Hz frequency in 35 m
continuous records every hour each. Tides and currents were obtained averaging 2 min recor
every half an hour. Velocities and pressure data were stored in the sensor's internal memory as
min-averaged vectors files that allow the obtention of measured mean currents and water leve
Directional wave measurements in open sea were done with a *Datawell Wave rider* direction

buoy; waves measurements, currents and suspension sediment measurements were available from ~ fixed tripodes, 3 spheric *2D Delft P-S* electromagnetic currentmeters, 3 Optical Back-Scatter (OBS with a pressure sensor, batteries and a *Campbell-CR10* Data logger each, moored between 10 and 2(m depth were available. The measuring period was approximately a month (see a detailer description in Jimenez et al. 1997 and Guillen et al. 1997)

Figure 4: Tetrapode with PUV (EMC and a PS) moored at 3.5 m. depth.

ii) In the surf zone, waves, water level (η) and velocities were measured with sensor assembled in a sledge (mobile data acquisition system guided from the shore line) at 45 position during the experiments (see vertical lines in figure 3). The velocities and suspended sedimer measurements were done using 8 electromagnetic currentmeters (6 Delft and 2 Shirsov) and turbidimeters (Shirshov) respectively, at a sampling rate of 20 Hz during the 45 -30 min each- tests allowing a detailed identification of the 3D vertical structure of the flux in the macroturbulence wind waves and mean flux hydrodynamic scales. The water levels were measured at a 4 H sampling rate using an *Etrometa* wave gauge, properly designed to be used in the severe condition inside the surf zone. Together with the sledge measurements, portable sediment traps, buoys an dye spots, and oblique and bird-view aerial video images of the surf zone were used. The topc bathymetric survey profiles in the surf zone -up to 2 m depth- were carried out with conventiona topography (a teodolite with infrared distance-meter and poles with optical prisms), recording th sledge position and depth. Outside the surf zone, indirect methods were used, with a zodiac boa measuring the depth at different sites. The atmospheric conditions were measured simultaneously i three near permanent stations (Sant Carles de la Ràpita, Deltebre and CasaBlanca; see figure 1) an (additionally) at the beach as follows, using a meteorological station (*Aandera*), located on a observation tower, approximately at 10 m above the sea level and at 800 m from the zone c experiments, countinously measuring temperature, atmospheric pressure, wind intensity an direction data in averaged 10 min series; and wind sensors on the sledge.

Innovations And Improvements Regarding DELTA'93 Field Campaign

At this point the authors would like to show some of the improvements and innovations of th present campaign with respect to the previous Delta'93 LIM/UPC field campaign.

More complete measures: Waves and currents were measured simultaneously inside and outside th
surf zone.
In this campaign, an EMC placed in a tetrapode at 3.5 m depth and 150 m from the shore line wa
available, thus obtaining continuous 2DH velocities and pressure time series. In the surf zone, th
3D structure of the flux and the bottom boundary layer were measured; emphasis was focused o
measuring the 3D structure of the flux. Therefore, 8 EMC were assembled on the sledge, four o
them as pairs in the same level, at 10 and 65 cm from bottom, to measure u-v and u-w, , allowin
the identification of the 3D structure of the flow. The rest, at 0.05, 0.10, 0.25 and 1 m from th
bottom in order to measure the horizontal component of velocity (see figure 5a), allowing to detec
the bottom boundary layer. Suspended sediment concentrations near the bed were also measured
using 2 EMC and 2 turbidimeters assembled in a cross-shore position on the sledge, separated 1 n
and at 10 cm from the bottom, measuring the fluctuations of the suspended sediment concentratior
related to the hydrodynamic flux (see figure 5b). With this setup, it is possible to follow step-by-step
the effect of the waves motion and to separate the 3D flux measured in the bottom boundary laye
into the different hydrodynamic scales, relating each individual contribution to the sedimen
suspension and further transport. Besides the portable surf zone sediment traps, a new set of 5 fixe
sediment traps were included, measuring the mean sediment transport rate (averaged in 24 Hs) nea
the bottom across the surf zone up to 3 m water depth. As mentioned before, the meteorologica
conditions were measured *in situ* at the beach and at the sledge.

Quality improved video images: Besides the B&W video camera assembled on a 40 m high crane, a
color videocamera, suspended from a 4 m diameter aerostatic balloon inflated with helium, was
used. With this set of cameras, oblique and bird-view video images of the dye spots were obtained
Most of the dye spots were released during the tests, in order to compare the dispersion coefficients
obtained with the analyses of video images and the hydrodynamic velocities measurements (see
figure 6; and Rodriguez et. al, 1997).

A longer field campaign: 45 tests were obtained during the 10 days long-DELTA'96 field campaign.
while 12 tests were obtained during the previous 6 days long-Delta'93 field campaign.

Figures 5a, b: Sledge with 8 EMC, 1 WG, 2 turbidimeters and wind sensors; Lateral view of the set of EMC
and turbidimeters at 0.10 m from bottom.

Figure 6: Aereal view of a dye spot released during a test, (03/11/96).

Preliminary Results

During Delta'96, a complete and harmonic data set distributed in 8 cases and 45 tests was acquired. *Sea conditions*: In general, during DELTA'96 there were weak wave conditions with a medium intensity wind distributed as daily breezes. In particular, during 05/11/96, a sudden change in meteorological and hydrodynamic conditions was observed, producing the situation for a detailed study and later numerical simulation. As seen in figure 7a, a decrease in atmospheric pressure leading to an increase of the water level, approximately of 0.10 m (of the same order as the typical astronomical tide of 0.15 m) was measured during the mentioned day. As shown in figure 7b, an important increase of the "external" current speed where the tetrapode was, outside the surf zone (3.5 m depth and 150 m from shore line) was also observed with speed values close to 0.35 m/s at m from the bed. Later, two different consecutive sea states, "wind dominant" at noon and "wave dominant" at the evening were recorded, with maximum mean wind speed values of ≈12 m/s and H of ≈ 0.7 m respectively. This sea state change can be seen in figures 7a, b and 8.

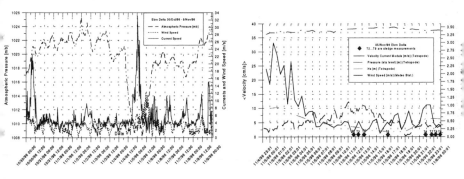

Figures 7a, b: Meteorology and mean sea level during DELTA'96; Wind, waves, mean water level and currents evolution during the selected day (05/11796).

During the selected day, 05/11/96, the different tests can be grouped together in two different sea states. The first one, named "*wind dominant*" lasted until noon (tests 1-4) followed by an intermediate case (test 5) and then the "*wave dominant*" (tests 6-9) during the night (see figure 8). The first state allows the study of wind driven mechanisms, particularly giving data for the sea friction roughness coefficients and the vertical distribution of momentum. The wind time series at two levels (the meteo station at 10 m above the sea level and the sledge at 1 m above) are shown in figure 8 and the corresponding horizontal current profile in figure 9a. On the other hand, the tests carried out at night during the *wave dominant* sea state are of interest for the study of the vertical structure of the long shore currents, showing a logarithmic shape, specially near the bottom (see figure 9b). Simultaneously with the waves and current data analysis, suspended sediment concentrations time series have been obtained from the turbidimeters, showing the close relationship existing between the instantaneous orbital velocity data ($\tilde{u},\tilde{w}$) and the suspended sediment concentration fluctuations. The intermittence analyses of the sediment suspension events related to wave breaking are now in process, using a methodology similar to that of Kos'yan et al. (1996).

Analysis Scales:

The obtained hydrodynamic data are being analyzed in 4 scales:

Macroturbulence, with frequencies higher than 1 Hz.

Oscillatory flow, corresponding to the wind wave orbital velocities, between 1 and 0.005 Hz ($T \leq 20$ s)

Long waves, from periods ranging form 20 to 300 s, and

Currents, for 5 to 30 min averaged values.

For various of these scales, the availability of new data, particularly the vertical component (e.g. w', $\tilde{w}$ y $\overline{w}$), will allow extended studies from those already done in the surf zone (Rodriguez et al. 1997).

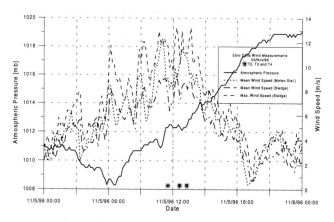

Figure 8: Wind speed velocity time series and vertical structure of longshore currents. during 05/11/96.

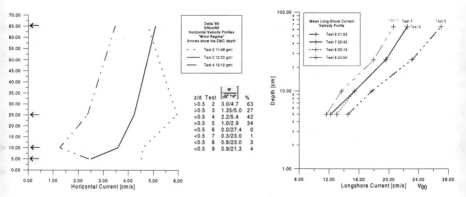

gure 9a, b: Vertical structure of longshore currents for "wind dominant" tests; Longshore currents vertical profiles, "wave dominant" case. (night measurements); tests 6,7,8 and 9.

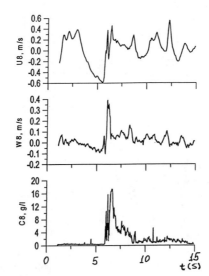

Figure 10: Suspended sediment concentrations signals and spatial components $\tilde{u}$, $\tilde{v}$ y $\tilde{w}$ of velocity.

urrent Scale: A feature to remark from these new data is their high resolution near the bottom (5,), 25 cm) allowing a detailed study of the bottom boundary layer vertical structure. Examples of garithmic fits of the longshore current profiles are shown in figures 11a, b. It can be seen that a tter fit (with higher correlation coefficient R^2) of the Delta'96 data profiles near the bottom, sults due to a higher measuring resolution. Finally, it was observed that the vertical component $\overline{w}$ the current was negligible (less than 1 cm/s, corresponding only to the 4% of the mean horizontal

ux) when the surf zone circulation was driven by wave breaking, whereas in the case of wind
riven circulation, the values of $\overline{w}$ range from 1 to 3 cm/s (between 27 and 63%) of the horizontal
mean flux (see figure 9a). To complete this vertical velocity study, it is necessary to analyze the
ests near the breaking point and swash zone where laboratory antecedents show the higher values
f $\overline{w}$ (Nadaoka, 1986)

DELTA'93 Test	$Log\ z'_0$	z'_0 (cm)	$K_w=30z'_0$ (cm)	κ/u_{*wc}	u_{*wc} (cm/s)	Corr.Coef.	DoF
5	-1,8368	0,015	0,44	0,055	7,4	0,96	2
6	-1,6798	0,021	0,63	0,037	11,1	0,91	2
7	-1,4459	0,036	1,07	0,039	10,3	0,89	2
8	-1,2626	0,055	1,64	0,123	3,3	0,96	4
9	-2,8718	0,001	0,04	0,076	5,3	0,985	1
10	-0,6726	0,212	6,40	0,031	13,4	0,999	1
11	-1,1867	0,065	1,95	0,054	7,5	0,99	2
Experimental values: $Log(z) = \kappa/u_{*wc}\ V(z) + Log\ (z'_0)$							

DELTA'96 Test	$Log\ z'_0$	z'_0 (cm)	$K_w=30z'_0$ (cm)	κ/u_{*wc}	u_{*wc} (cm/s)	Corr.Coef.	DoF
6	-0,568	0,271	8,12	0,087	4,7	0,997	2
7	-0,646	0,226	6,77	0,108	3,8	0,995	2
8	-0,803	0,157	4,72	0,115	3,6	0,996	2
9	-0,690	0,204	6,12	0,120	3,4	0,994	2
Experimental values: $Log(z) = \kappa/u_{*wc}\ V(z) + Log\ (z'_0)$							

Figure 11a, b: Logarithmic fits of longshore current profile measured in DELTA'93 and DELTA'96 .

Conclusions

ield campaigns, specially those in the surf zone, require great costs and effort. Therefore, to ensure
ne profitability, they must be carefully planned and faced by various research and public
nstitutions. DELTA'96 is an example of this kind of high scale experiment done in nature.

The very first results are a big set of meteo-hydro-morphodynamic data, useful to calibrate
redictive models of coastal processes.

The combination of numerical models and previous measurements (pre-campaign planning) is
dvisable to optimize costs, human effort and the variables to be studied and measured (modelling-
measuring interaction). This was done in the DELTA'96 case, where numerical simulations were
one (e.g. Rodriguez et al. 1994 y S.Arcilla et al. 1995) which permitted an improved distribution of
ne sensors on the transverse profile, with respect to the previous Delta'93 sensors array.

From the numerous tests of DELTA'96, most of the cases correspond to wind driven (induced)
irculation, complementing the previous DELTA'93 data. The test 6-9 done on the night of the
5/11/96 corresponds to breaking wave induced circulation.

ter these data analyses, it has been verified that in the near shore and surf zone of mediterranean ast, the waves, the wind and the meteorological tide are of the same relative importance as current ving mechanisms.

ie processes near to the bottom, affected by the oscillatory boundary layer (e.g. $z/d < 0.1$), are of eat importance regarding to the friction in the bottom boundary layer and sediment transport. The k of the corresponding field measures (detailed and available) makes the DELTA'96 data of great portance for the calibration of numerical models.

nally, it is remarked that the integral analysis of the obtained data and its utilization are still tasks progress.

cknowledgments

lese experiments were done within the framework of the FANS (MAS3-CT95-0037) projects of e UE's MAST program, CIIRC and MEC's DGICyT (CE94-0022) projects. The authors want to ank all of them who endured the field work, specially N.Pykhov, O.Pushkarev, I.Podymov, Rodriguez, B.Mercader, J.A.García, F.Nieto, P.Medina, C.Gallimani, E.Bahía, J.M.Redondo, Gómez, M.Diez, J.Pablo, M.Garrido, A.Caballero, and A.Chaler amongst others.

eferences

Garcez Faria,A., Thonton,E. y Stanton,T. (1996). Small-scale morphology related to wave d current parameters over a barred beach, IC Coastal Engineering, ASCE.

Guillén,J., Palanqués,A y Puig,P. (1997). Tripod measurements on the Ebro Delta shoreface-inter'96, Outer field site, MAST Workshop, Blanes.

Jiménez,J., Gracia,V. y S.Arcilla,A. (1997). Tripod measurements on the Ebro Delta oreface- Winter'96, Inner field site, MAST Workshop, Blanes.

Kos'yan, R.D., Kuznetsov, S.Yu., Kunz, H. and Pykhov, N.V. (1996). Sand Suspension vents and Intermittence of Turbulence in the Surf Zone. IC Coastal Engineering, ASCE.

Nadaoka, K. (1986). A fundamental study of shoaling and velocity field structure of water aves in the nearshore zone. Ph.D Thesis, Tokyo Inst of Tech. 125 pg.

Rodriguez, A., Bahia, E., Diez, M., Sánchez-Arcilla, A., Redondo, J. and Mestres, M. (1997) xperimental Study of mixing processes using video images. IC Coastal Dynamics'97, ASCE.

Rodriguez, A., Sánchez-Arcilla, A.,Collado, F., Gracia, V., Coussirat, M.G., y Prieto, J. 994). Waves and currents at the Ebro Delta surf zone: measured and modelled. IC Coastal ngineering, ASCE.

Rodriguez,A., Sánchez-Arcilla, A., Sospedra,J.; Gironella,X., Gómez,O. y Valdemoro,H. 995a). Delta'93: un estudio experimental de gran escala en la zona de rompientes, III JEICP, alencia.

Rodriguez,A., Sánchez-Arcilla, A., Gómez, J. y Bahia,E. (1995b). Experimental study of rf zone macroturbulence and mixing using Delta'93 field data, IC Coastal Dynamics'95, ASCE.

Rodriguez,A. (1997). Un estudio experimental de la hidrodinámica de zona de rompientes, esis Doctoral, Universidad Politécnica de Cataluña, Barcelona, 350 pg.

S.Arcilla,A.; Collado,F.; Coussirat,M. y Rodriguez,A. (1995). Q3D modelling of the nearshore rculation in the Ebro Delta; Coastal Dynamics'95, ASCE, Gdansk.

LITTORAL PROCESSES, SEDIMENT BUDGET AND COAST EVOLUTION IN VIETNAM

Ryszard B. ZEIDLER[1] & Hoang Xuan NHUAN[2]

Abstract

Almost entire Vietnam's coast exhibits evolution in various scales, with major contribution by river systems. Lithological indicators have been employed as a powerful tool for both explaining and predicting littoral processes. Sediment budget estimates are provided for Vietnam, along with other background (both qualitative and quantitative) for modelling large-scale littoral processes and coast evolution. Some features (e.g. equlibrium profile) remain unexplored. Numerous interactions of littoral processes, hinterland, flood control, coastal activities etc. have been identified for the Thua Thien Hue coastal cell (Central Vietnam).

1. Introduction: Background (VVA) and Motivation

About 23% of Vietnam's mainland coastline (estimated at 3800 km, depending on definition) is hard and rocky, some 22% being fronted by natural sand dunes; the remaining percentage consists generally of low coast with sandy or muddy deltaic beaches of the Red and Mekong Rivers (Fig. 1). – Since Vietnam has one of the most sensitive coastlines worldwide, a trilateral programme Vietnam Vulnerability Assessment (VVA) was sponsored by the Netherlands government, with emphasis placed on an extensive analysis of Vietnam's coast's vulnerability to climate change, first and foremost an accelerated sea level rise. Along with numerous analytical and synthetic tasks for the entire coast, three pilot studies were also undertaken to provide insight into particular problems of selected stretches of the coast. Some results stemming from the VVA project and the international cooperation it fostered are discussed here in the context of large-scale coastal processes and shore evolution.

If the 10-m line is taken as the uppermost elevation of Vietnam's coast vulnerable to 1-m sea level rise (as done under VVA), the impact will not be limited to a narrow coastal strip (Fig. 1) but will be even more serious further inland (backwater and riverbed rise effects). About 17 million people on 40,000 sq.km will be flooded annually, USD 17 billion of capital value being lost (ca 80% of yearly GDP).

[1,2] Polish Academy of Sciences' Inst. Hydro-Engineering (*IBW PAN*), Gdańsk 80 953, 7 Kościerska, & Institute of Physics, National Center Natural Sciences & Technology HCMCity, resp.

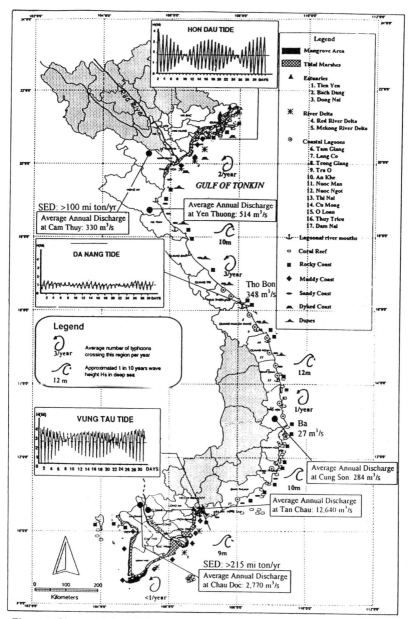

Figure 1. Characteristics of Vietnam's coast (SED = annual sediment load in rivers)

2. Large-Scale Overview of Vietnam's Coastal Zone

Figure 1 shows not only coast types but also some meteo and marine characteristics. By and large, the country is largely affected by monsoons and typhooons, and the associated phenomena, such as rainfall. They all vary along coast in some 'systematic' way, partly depicted in Figure 1. Inter alia, winter monsoons bring about air and water circulation from NE to SW, the summer counterparts being S-SW to NE-N. Rain falls mostly between October and January, earlier in the north, at the highest rate in the central provinces (3000 mm/yr) through 2000 mm/yr in the Red River and Mekong Deltas to 1000 mm/yr in the southern central region. An average of 6 typhoons cross Vietnam's coast per year, mostly from SE to NW, laden with high rainfall and coupled with extreme wind speeds, high waves and storm surges. The Mekong Delta is relatively free from typhoon attack.

Because of the seasonal climatic patterns and their time lags along coast, the latter is subject to large-scale evolution linked to the climatically affected chain rain-river discharge-sediment load. The entire coast is substantially controlled by the river load, mainly from the Red and Mekong Rivers, their annual input counted in hundreds of millions tons.

3. Littoral Processes: North Vietnam (Gulf of Tonkin)

3.1. Lithological/granulometric characteristics

The granulometric data collected for the entire Vietnam's coast, supported by other hydro- and morphodynamic evidence, have been employed to draw conclusions on coastal morphodynamics and littoral processes in various time scales, from annual to millenial. Adopted has been the Vietnamese classification (not the Wentworth one), in which sand is defined with grain diameters from 1 to 0.1 mm (medium from 0.5 to 0.25 mm, coarse and fine on top and below the medium fraction). An additional 'aleuritic' fraction is distinguished in the diameter range from 0.1 to 0.005 mm, clay being placed below it (hence combining part of the Wentworth very fine sand, silt and part of clay in the aleurite class).

The lithological pattern classes **(MP)** are shown in Figure 2 as:
MP1: Fluvial aleurite   *Original fluvial sediment load and material of the intertidal areas protected by sandy barriers or ridges;*
MP2: Aleuritic sand   *Bed material of poorly protected intertidal areas;*
MP3: Aleuritic clay   *Bed material of tidal channelss and mangrove marshes;*
MP4: Sand   *Sediment of sandy beaches, barrier islands and ridges;*
MP 5: Clay   *Bed material of offshore slope.*

For the sake of analysis, the following regions of coastal morphodynamics have been identified in the **cross-shore** direction:
Intertidal zone (+2m → -2m) consisting of MP1, MP2, MP3 and MP4; where MP3 on upper tidal flats of accretional muddy coasts often consists of three specific aleuritic clay layers: brown-gray, gray-brown and gray-blue;
Offshore slope (-2m → - 30m) consisting mainly of QIV(2-3) aleuritic clay;

Offshore shelf (-30m → - 120m), characterized by anomalous sediment distribution.

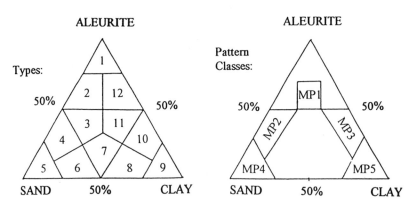

Figure 2. Lithological pattern classes **(MP)** distinguished

From the analysis of sediment samples taken from rivers and all over the coastal zone it may be inferred that the **transformation** of **river-borne sediment** load into **coastal** one occurs in two ways:

*   Differentiation/reduction to sand, as *continuous evolution MP1→ MP2→MP4, consisting in washing out clay and aleurite due to waves and currents;*
*   Differentiation/reduction to clay as *in two isolated stages:*
MP1→ MP3 deposited in the upper nearshore zone and
MP1→ MP5 deposited on the submerged shore slope.

3.2 Littoral processes in large scales

On the basis of the lithological data analysed with reference to the hydrology, physiography and geology of respective coastal environments, the following findings can be formulated for Vietnam's coast:

1. Nearly 95% of the total river sediment load is concentrated in two distinctive coastal segments nourished separately by the Red and Mekong Rivers. The cohesive coast of the Red River Delta is supplied with the river sediment load of 72,600,000 t/year, the composition of which is about 12% sand, 59% aleurite and 29% clay. Not more than 29% of this sediment load with about 41% sand, 44% aleurite and 13% clay contributes to maintain and develop sandy ridges and the intertidal zone. The remaining part (not less than 71%) with ~5% sand, ~62% and ~32% clay passes the intertidal plain to nourish the offshore slope.
2. Longshore sediment transport varies greatly and depends on shore orientation and wave climate. For the key locations along the coast of Bak Bo (Tonkin) Gulf one has Q_{RHS} = 352,000 → 1,028,000 m³/year, Q_{LHS} = -60,000 → 968,000 m³/yr and Q_{net} = -143,000 → 753,000 m³/yr. The typical convergent littoral drift is observed

on 16 km long Giao Lam section with the annual inflow of 1,053,000 m³/year. A typical divergent littoral drift is observed on 20 km long Hai Hau section with the annual outflow of about 800,000 m³/year.

3. The most important littoral processes observed in time scales of 10³ years are coastal changes associated with the secondary sea level variation and the recent development of the Red River Delta. Conspicuous littoral processes having the time scale of 10² years are dramatic erosion events along Ha Nam Ninh province and Ca Mau Peninsula. In the *first case*, the erosion is selective -- fine sediment is washed out from the intertidal plain and the sandy components remain in place. In the *second case*, the erosion could be caused by large-scale human interventions in the catchment area of the Mekong River.

4. The littoral processes on the decadal scale have not been sufficiently explored, while the important observed annual erosion events are linked to typhoons and storm surges.

3.3 Regional accretion/erosion and equilibrium profile

Estimates of erosion rates are provided, as collected from various sources, and are linked to different coastal factors and controlling phenomena. The rates vary from metres to episodic 200 m per year. The highest rates are given for Bac Cua Ganh Hao in Minh Hai as an average of 193 m/yr over 15 years between 1976 and 1991; 162 m/yr at Nghi Yen (Nghe An, between 1982 and 1991); 112 m/yr at Dien Du'ong (Quang Nam Da Nang, 1975-1995); ab. 100 m/yr at Hiep Thanh and Dan Thanh (both Tra Vinh, 1982-1992) and 100 m/yr at Nam Cua Ganh Hao (Minh Hai, 1945-1991). By and large, these figures are felt extremely high, and should certainly be verified by independent methods as they might represent local erosion events, not structural morphodynamics.

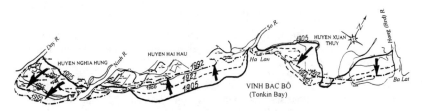

Figure 3. Decadal evolution of Hai Hau (Hai Trieu) coastline, both structural and anthropogenic (river closure)

Hai Trieu area (Fig. 3) is basically erosional, but its parts are accretional. Identification of such clear cases of regional erosion and accretion along Vietnam's coast in time scales of decades has shed light on the relative coverage of various sediment pattern classes (MP) in both types of environments. It has also been tempting to look at the cohesive bed profile (such as at Hai Trieu) and examine if it possesses any 'shore equilibrium' features, but this part of the study remains inconclusive.

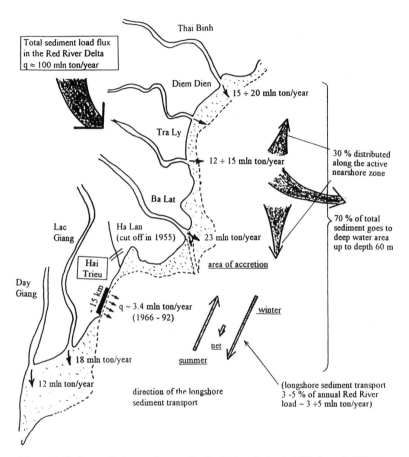

Figure 4. Sediment budget estimates for Red River Delta (cf. Vinh et al. 1996)

3.4 Sediment budget in the Gulf of Tonkin (Bak Bo)

The Gulf of Tonkin, to which the Red River conveys its sediments, is an extremely complex yet very interesting environment, where various lithological recognition techniques can and have been tried. On the background of the underlying geology and simplified lithology and some other estimates one arrives at the sediment budget given in Figure 4. Identification of large-scale sediment sinks is incomplete – the annual entrapment of sediment in Hoa Binh Reservoir (completed in 1987) is estimated at 70 million tons, sand mining figures for Red River remain undetermined and the impact of Cua Ha Lan (Song So) cut-off in 1955 is also unknown as to magnitude. The average decadal/centennial distribution of surficial sediments in the Gulf of Tonkin is illustrated in Figure 5.

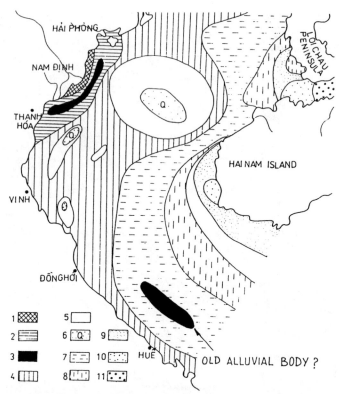

Figure 5. Surficial seabed sediment distribution in Gulf of Tonkin (Nguyen 1994)
1 - material of intertidal zone (MP1→MP4); 2 - aleuritic MP3 and old sediments; 3 - MP5
clay; 4 - QIII(2)-QIV(1) aleuritic sand Vinh Phuc layer?; 5 - QIII(2)-QIV(1) fine sand;
6 - QII(2)-QIV(1) medium, coarse sand; 7 - QIII(2)-QIV(1) aleuritic; 8 - QIII(2)-QIV(1)
aleuritic sand; 9 - fine sand; 10 - medium sand; 11 - coarse sand

4. Littoral Processes: Central Coast (Thua Thien Hue)

4.1. ThuaThien-Hue coastal cell: rivers, lagoons and coast

The Tam Giang - Cau Hai lagoon system in the Thua Thien *Hue* province is the
largest and most typical one among the twelve brackish tropical water bodies
stretching between 11 - 17° North of Central Vietnam. It was basically formed in
Holocene and late Pleistocene, keeping pace with glaciation and changing sea level. It
is about 68 km long and occupies 21,600 hectares, i.e. 4.3% of TTHue province. It is
nearly closed, with two narrow inlets, of which Thuan An in the north is 350 m wide,
5-6 m deep, while Tu Hien in the SE part is only 50 m wide, with average depth of
1 m.

The study area is largely controlled by the TTHue **river system**. Therefore the central issue around which coastal zone management (CZM) concepts for TTHue should be built consists in the flooding and its impact on the lagoon-oriented life (primarily lagoon salinity), i.e. both human activities and natural processes. The major objectives under VVA were then set forth as:

`Given that **flooding** creates most hazards for Thua Thien Hue on a large scale, with the city of Hue as the most conspicuous centre of vulnerability, and the Tam Giang - Cau Hai lagoon system on a regional scale, that the natural lagoon ecosystem attracts various human activities, and that both the natural ecosystem and the population of lagoon system users are equally affected by the flooding, outline as broadly as possible the processes and their interactions in the lagoon system, together with the effects due to climate change and socio-economic developments, and indicate the ICZM methodology applicable to solution of the natural, environmental and socio-economic problems so arising'.*

Lagoon salinity (from 1 to 33ppth, below 10ppth in the rain season, and above 20ppth in the dry season) is one of the most important features controlling the lagoon ecosystem and its users structure (e.g. agriculture vs aquaculture). On the other hand, salinity is affected by flood control measures, climate change and socio-economic developments. Littoral processes, lagoon sedimentation and lagoon circulation are found to contribute substantially to the lagoon morphodynamics, inlet stability in particular.

4.2 Littoral processes in CZM context

The sediment budget data for the TTHue cell include an input from rivers (total → 270,000 cu m per year, or 192, 450, 40 and 35 thousand tons from Rivers Bo, Huong, O Lau and Dai Giang, respectively). Sediment sinks encompass 60,000 cu m per year due to sand and gravel mining + rough quantities inferred for Ta Trach Reservoir under construction and other reservoirs when implemented (up to 200,000 cu m yearly). The sediment output from the lagoon to the sea remains to be explored – average inlet flow velocity has been measured as 4-5 m/s in rain season and below 1 m/s otherwise. Average sediment concentration in rain season is 50 mg/l; sedimentation patterns (i.a. lagoon circulation and granulometric variation) should be investigated. The littoral drift (300-400-500,000 cu m per annum net NW → SE), together with other budget components has been applied in our analysis of shoreline changes measured. Since the early estimates relied on standard software (such as the Dutch Unibest), more accurate computations are now being attempted. They must rely on improved wave figures, typical of at least the two predominant wave regimes in winter and summer; refined refraction plans for the study area; corrected data on coastal circulation (nonwave-induced currents such as tidal, inertial and wind-induced); and updated data for bedforms and bed roughness.

The coastal and lagoon morphodynamics are inherently coupled with the stability of the tidal inlets. During the Quaternary the beaches of Thuan Thien Hue were generated by the system of Rivers Huong, Bo, O Lau, which brought in sediment into an earlier gulf. During the first stage (unknown duration) the river bed was set in equilibrium but the coastline at that time was not affected. The second stage (medium Holocene) faced a system of dunes in Phong Dien and Phu Vang, consisting of a series of dunes 7--8 m high. At that time the sea level was higher by 4-5 m (4500 years ago) than at present, while 2--3 m higher 3000 years ago. Together with a system of dunes there was an ancient lagoon, which was later filled up with alluvial sediments to create the present day Hue plain. Finally, in the third stage (late Holocene), a system of very high dunes was generated from Quang Tri to Tu Hien inlet. -- At present, the sediment of the lagoon proper consists of coarse to fine sand, silt and clay. Samples with coarser silt and silty clay are found in shallow water of the lagoon while finer silt and clay are sampled in deeper water.

In recent forty years, the tidal inlet of Tu Hien was closed 3 times -- in 1953, 1979, 1994. In 1979, the salinity in Cau Hai was measured at 9 - 33ppth before closure and 11ppth after, thus proving the importance of lagoon flushing through the inlets. The last series of closing and opening events was initiated in 1989, with dramatic migration of Tu Hien in November 1994. Tu Hien is felt responsible for inundation, nutrition and flushing of the entire Cau Hai lagoon. It is quite obvious that, for the sake of the lagoon sanity, Tu Hien should be opened and measures should be taken to avoid its closure

The migration is also an inherent feature of the northern inlet, Thuan An. During this century, Thuan An has been moving 40 m/year to the north. At present the migration is slow and the old inlet at Phu Thuan is approached. Thuan An is directly affected by the Perfume River waters, which are flowing to the sea very dynamically, especially during the flood season. -- The history of the migration of both inlets, Thuan An and Tu Hien, from 1404 to date is well illustrated in Figure 6.

5. Major findings and conclusions

Almost entire Vietnam's coast exhibits evolution in various scales, to which major river systems, mostly Red R. and Mekong, contribute substantially. Interactions of seasonal (winter/summer) climatic features (waves and storm surges vs rain and flooding) bring about a diversification of littoral processes, North vs South. Lithological indicators can be employed as a powerful tool for both explaining and predicting littoral processes in various scales. Sediment budget estimates are provided for Vietnam (e.g. Red River) in various scales but mesoscales (1 km) and small scales (10 km) display regional and local patterns different from the large-scale ones (100 km) due to different exposure, distance from sources, lithological characteristics, climatic and longshore phase shifts and human intervention. Background (both qualitative and quantitative) has been provided for modelling large-scale littoral processes and coast evolution, although some features, e.g. equilibrium profile for (semi) cohesive beds, remain unexplored. Numerous interactions of littoral processes, hinterland, flood control, coastal activities etc have been identified for TTHue coastal cell (Central Vietnam).

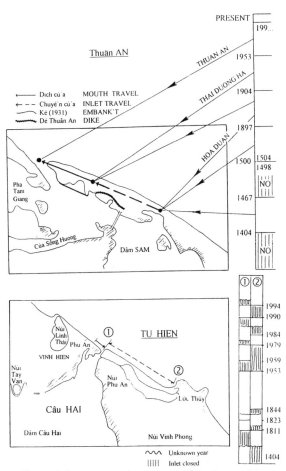

Figure 6. Long-term migration of TTHue lagoon inlets

Acknowledgements

This paper results in large part from the Vietnam Vulnerability Assessment programme carried out in the years 1994-1996. The data collected under that programme challenged the authors to formulate some concepts and findings which could contribute to the EU-sponsored PACE project (MAS3-CT95-0002), the support of which is gratefully acknowledged.

REFERENCES

Nguyen Chu Hoi (1994). Surface sediment in Tonkin Bay. Inst.Oceanology Haiphong
Vinh T.T., Z. Pruszak, N.N. Huan & G. Kant (1996). Sea dyke erosion and coastal retreat at Nam Ha, Vietnam. *Proceed. 25th Intnl Conf. Coastal Engng*, ASCE

FIELD STUDIES OF SHORE EVOLUTION

Alexander Khabidov[1], Lev Nikiforov[2], Leonid Zhindarev[3], Valeri Savkin[4],
Vadim Alymov[5] and Oleg Muraenko[6]

ABSTRACT

Long-term field observations of coastal dynamics were carried out on large
man-made lakes of Russia. These data offer a clearer view of the lake shores
evolution character and allow to reveal its trends. The noted tendencies of the lake
shores development and the tideless sea shore evolution behaviors were
compared. These trends turn out to be in excellent qualitative agreement; this being
so, it is felt that the data from man-made lake shores can be used to derive the
information giving better insight of long-term coastal dynamics of seas and to test
numerical models of coastal processes.

INTRODUCTION

It is common knowledge that direct observations of the long-term sea shore
evolution are extremely complicated or impossible in principle because of the
processes duration. Perhaps one of just a few sources of reliable information on its
features is field studies of large man-made lakes shores evolution. Under filling of
the man-made lakes the conditions of primary submersion and erosion of adjacent
lands are duplicated. In succeeding period of their existence the conditions of water
level rise, stabilization and subsidence are reconstructed over and over again, since
all of them carry out regulation of the inflow rivers discharge.

Observations of coastal dynamics on man-made lakes have been
conducting for a long time, therewith on some lakes since their filling. Because of
this, the analysis of such observation results is sufficient for revealing of
development tendencies, comparison of these tendencies with peculiarities of sea
shore evolution and defining key factors to be taken into account in numerical
models of morphological evolution of coastal areas.

---

[1] Associate Director and Head of Seas, Natural & Man-Made Lakes Geomorphology and
Sedimentology Lab, Institute of Water and Environmental Problems of the Russian Academy of
Sciences (IWEP), 105, Papaninsev St., 656099, Barnaul, Russia
2 Professor
[3] Senior Researcher, Geographical Department of the Moscow State University, Vorobyovy Gory,
118899, Moscow, Russia
[4] Head of the Hydrological Lab
[5,6] Post-Graduate Students, IWEP

STUDY SITES

To derive behaviors and quantitative analysis of the shore evolution the data of field observations on several large man-made lakes of Russia and on a few other reservoirs were used (Figure 1). As this takes place, monitoring of coastal processes on Novosibirsk Reservoir in West Siberia gave the most comprehensive information, since the man-made lake was formed in 1956 and the shore evolution has being watched here for about 40 years.

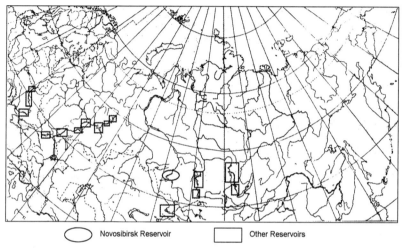

⬯ Novosibirsk Reservoir   ▭ Other Reservoirs

The monitored man-made lakes are distinguished by two main peculiarities. First, there are three main types of the dynamic relief-forming and depositional sedimentary environments presented within the basins of the reservoirs. These ones are wave-dominated and deltaic environments, separated by the transitional area. Second, all the man-made lakes show water level fluctuation amplitude close to future sea level rise value for the next century given in the most pessimistic forecasts of aftereffects of global warming, that is, not less than 3.5 m (Jacobson, 1990).

The primary emphasis in studies of the man-made lake shores was placed on the natural shores of sand, sandy loam, and sedimentary rock origin with coastal accretive forms of relief and without these ones. The coastal dynamics was monitored in all the three environments under water level rise at 0.01 to 0.3 m per day, periods of water level stabilization up to 150 days as well as under the periods of water level subsidence at 0.05 m per day and less. In present paper, however, only the data associated with the wave-dominated environment of the lakes (Table 1), where height of deep-water waves reached 3.5-4 m and the storm duration was up to 5 days, would be presented.

Table 1

GENERAL ENVIROMENTAL CONDITIONS OF THE STUDY SITES
(Wave-dominated environments)

Kind of the ground	Range of water level fluctuations					Storm duration, (days)	Range of h_{sig}, m
	Water level rise		Water level stabilizatio n	Water level subsidence			
	Duration, (days)	Rate, (m/day)	Duration, (days)	Duration, (days)	Rate, (m/day)		
Sand Sandy Loam Sandstone	10-100	0.01-0.3	14-150	120-250	≥ 0.05	0.3-5	0.1-3.5

MAIN RESULTS AND DISCUSSIONS

The evidences from soft ground shores turned out to be the most suitable for demonstration of shore evolution behaviors at initial filling of the man-made lakes. According to observations, wave-induced formation of bench and cliff starts just under the action of waves with height of 0.2-0.3 m. Prolonged erosion and the bluff retreat cause the developing of coastal platform. During the wave-cut platform formation, offshore directed sediment transport dominates in a complex of littoral processes. As a result, the sand particles are accumulated at the seaward edge of the bench where the superimposed aggradative terrace may appear (Figure 2). Such response of the coastal zone relief to water level rise is in qualitative agreement with Bruun Rule (Bruun, 1962, Schwartz, 1967).

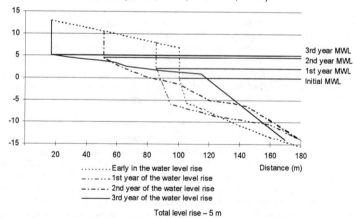

Fig. 2: Shore and beach profile response to initial water level rise: Observational data from man-made lakes

Like sea shores, the cross-shore profile evolution during stabilization of water level in the man-made lakes has a tendency to equilibrium (Figure 3). This way is conceptually identical with the familiar (Dean, 1990) beach response to increasing volume of sand added; the difference is that the source of loose material on the lakes is the eroded cliff.

A.

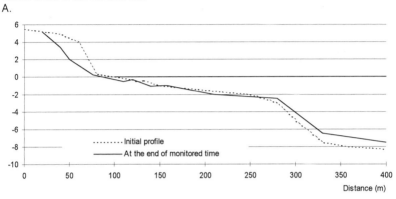

B.

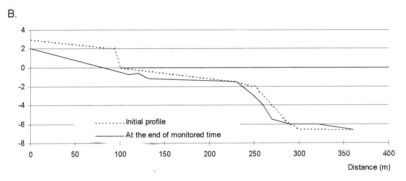

Fig. 3: Natural (A) and filled (B) beach profiles evolution
under water level stabilization of 130 day duration:
Observational data from man-made lakes

Water level subsidence on the man-made lakes results in shoreline advancement. As this takes place, such advancement of the shoreline is caused by both "passive" bottom drainage and rather coarse sediment accretion in the upper part of the cross-shore profile. The source of sediments (in the main coarse and medium sand) is the lake bottom at the nearshore periphery and, even, at the offshore area, where erosion intensifies after water level subsidence start. The finer sediments move mainly in offshore direction; accretion of such sediments at the outer border of coastal zone promotes development of bottom terrace (Figure 4). Similar tendencies occur on sea shores as well. Thus they were observed during the water level subsidence of Caspian Sea (Leont'ev, 1949) started in 1929 and lasted several decades.

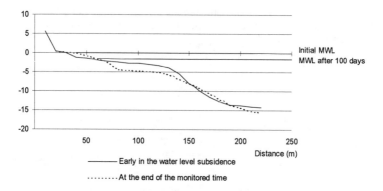

Fig. 4: Shore and beach profile response to water level subsidence:
Observational data for man-made lakes

As a result, because of negative changes in the mid and lower parts of the cross-shore profile, prerequsites for recommencement and, sometimes, for reinforcement of erosion emerged. It is safe to assume that one of a number of reasons why, despite the fact that man-made lakes have been existing for some decades already, their coasts are often far from equilibrium. Another important reason is progressive decrease of the mobile sandy particles content in bottom sediments as a result of sediment particles sizing along the cross-shore profile.

The latter can be illustrated by changes in distribution of percentage of particles of O.25-O.5 mm for 50 days of Novosibirsk Reservoir water level subsidence at the mean rate of 2 cm per day (Figure 5). Conceivably, mechanism of such selection is the onshore bedload transport due to wave asymetry and streaming versus the offshore directed suspended load transport in response to mean current. Of course, selective sediment transport in the nearshore zone takes place in other environments of the man-made lakes, but precisely within the wave-dominated one and under the water level subsidence its aftereffects appear in relief and sediment properties more distinctly. A large body of data from coastal studies on sediment transport in marine conditions reveals the same behaviors (e.g. Saf'ynov, 1978; Longinov, 1981; Antsyferov and Kos'yan, 1986; Kos'yan & Pykhov, 1991; Guillen & Hoekstra, 1996, etc.). The sediment selection by size in nearshore zone and step-by-step extraction of sediments for formation of coastal accretive relief forms and outer aggradative terrace finally results in the exhaustion of sand reserves on the bottom and first and foremost of mobile particles reserves. That is the reason why erosion may intensify at further stages of the coast development. It is felt, for one, that namely because of exhaustion of mobile sandy particles on the bottom barrier beaches, the formation of which is caused by cross-shore sediment transport, begin to experience sediment deficit and manifest signs of erosion. Man-made lakes show many examples of such type.

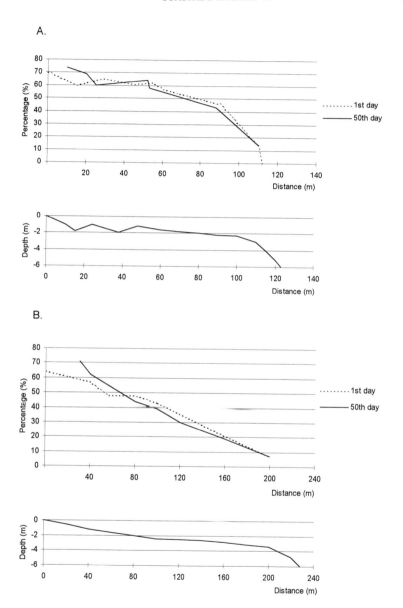

Fig. 5: Percentage distribution of particles of M_d=0.25 – 0.5 mm along the barred (A) and non-barred (B) cross-shore beach profiles under water level subsidence of 0.02 m/day, the 1st day vs the 50th day: Observational data from man-made lakes.

Abundant evidence of research stresses the fact that if one neglects the influence of non-wave character factors, so the individual features of storms, the storms duration and frequency are of prime importance at all stages of shore evolution. At the same time, the analysis of more than 200 storms observed during the period of man-made level change shows that the evolution tendencies of cross-shore profile depends highly not only on these parameters, but on the time derivatives of water level as well (Figure 6). In general, the upper slope erosion corresponds to positive value of the derivative, while sediment accretion corresponds to its negative one. Although Figure 6 also shows that this tendency is traced more clearly under coincidence of the signs of water level (dH/dT) and wave height (dh_{sig}/dT) derivatives, but it is abundantly clear that the rate of level change becomes one of the most essential factor of long-term coastal dynamics.

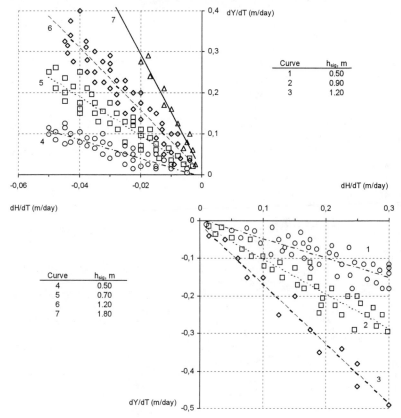

Fig. 6: Shoreline dynamics (dY/dT) vs water level gradient (dH/dT)
and significant wave height

CONCLUSIONS

1. The patterns of the shore evolution of man-made lakes were monitored at temporal scales from isolated storm and series of storms during the periods of water level rise, stabilization and subsidence up to 40 annual cycles fluctuations.

2. The data obtained testify the fact that in coastal area of man-made lakes the processes of sea shore evolution are reconstructed and, therefore, the results from coastal studies on man-made lakes can be used to derive comprehensive information and better insight of long-term coastal dynamics of seas and to test numerical models of coastal processes.

3. The effective numerical models of long term coastal dynamics should take into account the dependence of morphological evolution on the rate of water level change in time, peculiarities of wave height change during isolated storms and a series of storms, as well as consequences caused by selective sediment transport in the nearshore zone.

ACKNOWLEDGEMENT. The present work was supported by Russian Basic Research Foundation, RBRF grants # 93-05-12057, 96-05-64448 and 96-05-79087.

REFERENCES

Antsyferov, S.M. and Kos'yan, R.D., 1986. Suspended Sediments in the Upper Part of Shelf. Nauka Publisher, Moscow: 224, (in Russian).

Bruun, P., 1962. Sea-Level Rise as a Cause of Shore Erosion. J. Water ways and Harbor Div., ASCE 88: 117-130.

Dean, R.G., 1990. Equilibrium Beach Profiles: Characteristics and Applications, Technical report UFL/COEL-90/001, Coastal and Oceanographic Engineering Department, University of Florida, pp. 29-59.

Guillen, J. and Hoekstra, P., 1996. The "equilibrium" Distribution of Grain Size Fractions and its Implications for Cross-Shore Sediment Transport: A Conceptual Model. J. Marine Geology 135: 15-33.

Jacobson, J.L., 1990. Holding Back the Sea. State of the World: 79-97.

Kos'yan, R.D and Pykhov, N.V., 1991. Hydrogenous Sediment shift in the Coastal Zone. Nauka Publisher, Moscow: 280, (in Russian).

Leont'ev, O.K., 1949. Accretive Shore Profile Development Under Sea Level Subsidence. USSS Acad. of Sci. Reports 66 (3), Moscow: 578-583.

Longinov, V.V. (Edit.), 1981. Differentiation of Clastic Material Under Marine Conditions. Nauka Publisher, Moscow: 183, (in Russian).

Saf'ynov, G.A., 1978. Coastal Zone of Ocean in XX Century. Mysl Publisherÿ0, Moscow: 293, (in Russian).

Schwartz, M.L., 1967. The Bruun Theory on Sea Level Rise as a Cause of chore Erosion. J. Geol. 75: 79-92.

RIP CURRENTS

Graham Symonds[1], Robert A. Holman[2] and Barbara Bruno[2]

Abstract

Time exposure video images provide a means of observing rip currents under all incident wave conditions for an extended period of time. Time sequences of images illustrate the development and decay of rip currents in response to varying incident wave conditions over periods of days to months. The variability in the morphology as revealed by time exposures has been quantified using a fractal approach. High wave events typically form a shore parallel offshore bar with a low fractal dimension while the formation of the rip channels occurs during the subsequent period of lower waves. Rip formation appears to be associated with a gradual onshore migration of the offshore bar and a corresponding increase in fractal dimension. Once formed the rip channels remain relatively stable until the next high wave event. A relatively short period (1-2 days) of high waves can remove the rip channels and restore the initial shore parallel bar and the process repeats, though the location and length scales of the rip cells may be quite different. Converging longshore currents in the feeder channels either side of a rip current are revealed using video time stacks with estimated speeds of 0.5-0.6m/s.

Introduction

Australian surf life saving organizations report an average of 10,000 surf rescues per year and up to 50 drownings, most of which are associated with strong rip currents (Short, 1993; Short and Hogan, 1994). These currents also transport water and sediment cross-shore beyond the surf zone causing erosion of the adjacent beach and aiding the dispersal of storm water runoff which is a major source of coastal pollution on metropolitan beaches. Furthermore, the presence of rip cells provides potential for enhanced horizontal mixing within a few surf zone widths of the shore effectively increasing the nearshore horizontal eddy diffusivities necessary in larger scale models of the inner shelf region.

[1] School of Geography and Oceanography, University College, University of New South Wales, Australian Defence Force Academy, Canberra, Australia.
[2] College Oceanic/Atmospheric Sciences, Oregon State University, 104 Ocean Admin. Bldg., Corvallis, Oregon 97331-5503, U.S.A.

584

Rip currents are typically associated with longshore variations of wave induced setup driving longshore currents from regions of high setup to regions of low setup. Where these longshore currents converge the flow turns offshore in a rip current. Longshore variations in setup may result from longshore variations in water depth due to non-uniform topography (Sonu, 1972; Noda, 1974; Dalrymple, 1979), longshore variations in the wave field (Bowen, 1969; Dalrymple, 1975) or the presence of standing edge waves (Bowen and Inman, 1969). Rips are common on intermediate beaches and are often associated with rhythmic or transverse bars (Wright and Short, 1984). The morphology of rip currents on natural beaches has been described by Shepard et al (1941), Short (1979), Short (1985), Wright and Short (1984) and Huntley and Short (1992). However, the inherent intermittent nature of rip currents has hindered in situ observations. The nearshore circulation transports sediment, modifying the nearshore topography, which in turn affects the currents. Furthermore, the nearshore circulation and topography are continually adjusting to varying incident wave conditions and changing water depth due to tides and longer period motions. As a result the location and intensity of rip currents are highly variable in space and time making observations difficult, particularly on high energy beaches. Remote imaging of the nearshore using video allows long term monitoring of the nearshore topography and rip current cells and extends the previous morphological studies by providing potential for more detailed, quantifiable measurements of rip cell characteristics.

Site Description

In January, 1996 we installed an Argus video imaging station in Barrenjoey lighthouse, 110m above mean sea level, (shown in figure 1) overlooking Palm Beach, Sydney, Australia. The site was chosen because it is a medium to high energy, intermediate beach, and because rip currents consistently occur there throughout the year. The beach faces slightly south of east and is 2.5 km long with an average nearshore slope of .02. Two cameras were installed in the lighthouse, one with a 12mm lens imaging the whole beach and the second with a 16mm lens imaging the northern half of the beach. Each camera routinely acquires a single snap shot and twelve minute time exposure hourly.

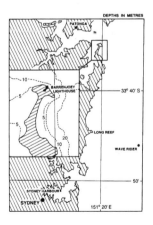

Figure 1. Site location with Palm Beach and the lighthouse shown in the inset.

At the time of installation the cameras and selected ground control points were accurately surveyed relative to a known bench mark which allows conversion from image

coordinates to world coordinates using the method described by Holland et al (1997). In addition to the video images water level data are obtained from a tide gauge at Patonga, shown in figure 1, and directional wave data from a buoy offshore of Long Reef in a water depth of 80m (also shown in figure 1). The tide is predominantly semi-diurnal with a tidal range of 1-1.5m. Time series of deep water significant wave height, period and direction are shown in

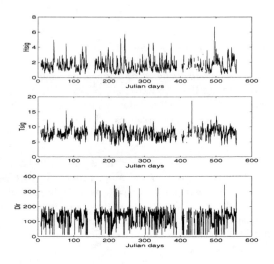

Figure 2. Time series of significant wave height (top), wave period (middle) and direction (bottom).

figure 2. The average significant wave height is about 1.5m increasing to 3-6m during high wave events. The dominant wave direction is from the SSE with occasional NE swells.

Morphology

In the time exposure images reduced breaking in the rip channels appears as regions of low intensity across the surf zone. Another signature of the rip currents in the time exposures is the offshore displacement of the breakpoint opposite the rip channels due to the possible presence of a rip head bar, wave breaking on the opposing current or foam, sediment and aerated water being transported offshore in the rip current. Shown in figure 3 are four oblique time exposure images showing the development of rip cells over a period of 12 days. The first image was obtained immediately after a period of large waves (H_{sig}~2.4m) on January 26, 1996 and shows a uniform, shore parallel bar with no obvious rip channels. By January 30 some longshore variability appears leading to fully developed rip channels one week later. Note the offshore bulge of the surf zone opposite the rip channels. The basic development sequence from the formation of a shore parallel bar under high waves to fully developed rip channels has been repeated numerous times over the past 15 months though the time scale of the development varies considerably. Once the rip channels form they appear relatively stable until the next high wave event. Shown in figure 4 is a sequence of images during April, 1996. The wave height was relatively low throughout this period with significant wave heights about 2m during the first

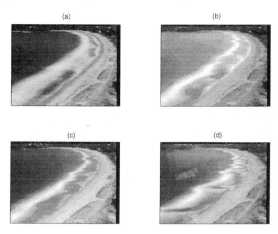

Figure 3. Rip cell development following a high wave event. (a) Jan. 26, (b) Jan. 30, (c) Feb. 3, (d) Feb. 7.

half of the month and falling to about 1m during the second half. The images shown in figure 4 have been rectified to a vertical perspective with the northern end of the beach to the right. The rip channel at the northern end of the beach persists

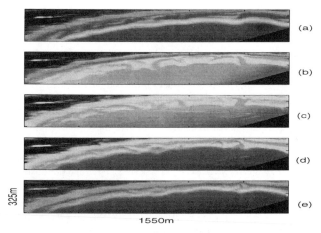

Figure 4. Rectified images with North to the right. (a)April 1, (b) April 7, (c) April 14, (d) April 21, (e) April 28.

throughout the period with little evidence of migration north or south. The rip channels near the middle of the beach are also relatively stable with perhaps some evidence of infilling towards the end of the period.

In contrast, the location and scale of the rip cells can change significantly before and after high wave events. Shown in figure 5 is a sequence of images from February through July, 1996, where there is at least one high wave event between each image.

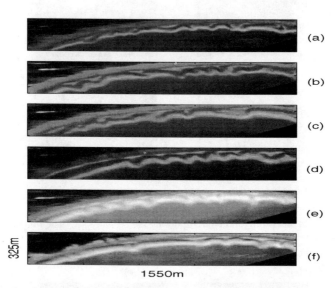

Figure 5. Sequence of rectified images spanning a 6 month period in 1996. (a) Feb. 7, (b) March 12, (c) April 3, (d) May 19, (e) June 23, (f) July 18.

There appears to be little correlation in the rip cell location from one image to the next. The longshore scale also varies between images and even within a given image the scale may vary along the beach for example on March 12. The May 19 image shows strong longshore periodic structure on the northern half of the beach with a sharp transition to a uniform, shore parallel bar on the southern part of the beach. In contrast, the rip cells on April 3 are more irregularly spaced alongshore tending towards a shore parallel bar at the northern end. While the rip channels remain relatively stable between high wave events the overall nearshore morphology continues to evolve during these periods. These changes can be illustrated using time series of intensity profiles along a shore normal transect. The bar between the rip channels can be identified in the profiles as a peak in intensity associated with wave breaking while seawards of the surf zone and between the bar and the shoreline are regions of low intensity. A series of intensity profiles along a shore normal transect between rip channels is shown in figure 6. Also shown in figure 6 are the corresponding significant wave height and period. The plot reveals periods of onshore migration of the bar for a few weeks separated by offshore displacements which occur over periods of order one day. The offshore displacements are associated with periods of high waves while the periods of onshore migration occur during periods of lower waves. The formation of shore parallel bars during high wave events followed by a slow onshore migration under low waves has been

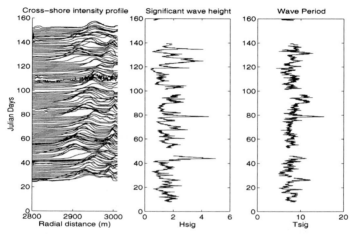

Figure 6. Daily cross-shore intensity profiles shown in the left panel. Time increases upwards from Jan. 24, 1996, at the bottom to May 31, 1996, at the top. The profiles extend from beyond the surf zone at the left to the shoreline on the right. Significant wave height (m) and wave period (s) are shown in the two panels on the right respectively.

reported by Greenwood and Davidson-Arnott (1979), Short (1979) and Wright et al (1979). The results presented in this section are only qualitative and we are currently investigating more rigorous, statistical analyses of the formation, location and persistence of the rip cells and the nearshore morphology.

Fractal Description

One technique to quantify the variability in the images is to use a fractal approach. If a curve is fractal (i.e., if it has the same shape at different scales), its shape can be described by the fractal dimension (D), which is a measure of convolution. A straight line has $D=1$, whereas a convoluted curve has $1 < D < 2$. Fractal dimension can be calculated using the "structured-walk" method (Richardson, 1961). This technique involves walking rods of various lengths along the curve. For each rod length (r), the length of the curve (L) is estimated according to the number of rod lengths (N) needed to approximate the margin; that is, $L=Nr$. Now suppose the rod length is halved and the process repeated. In this case, the number of rod lengths is increased by more than a factor of two as more of the peaks and troughs of the curve are resolved. In other words, L increases as r decreases. This process is repeated for a range of rod lengths and the results plotted as log(L) versus log (r), known as a Richardson plot.

Fractal behavior is characterized by a linear trend on the Richardson plot (i.e., a power law relationship). If the gradient of the linear regression is m, the fractal dimension (D) is given by $D=1-m$. Synthetic (or mathematical) fractals have the same D over an infinite range of scales. Natural objects are considered fractal if they have similar D values over a finite range of scales. In theory, the "structured-walk" method should be applied over a wide range or rod lengths, but in practice often the data set is too small to allow this.

We invoked the "structured-walk" method to examine the fractal properties of the outer edge of the surf zone using time-exposure images. We chose the outer edge because it was generally better defined than the inner-edge and also because the inner-edge sometimes merged with the shoreline. We analyzed 12 months of daily low-tide images from January through December, 1996. First, we auto-contoured the image so the surf zone margin would be objectively determined. We then selected only those images that had continuous intensity contours (185 in total); the structured-walk method cannot be applied to discontinuous curves. Each of these 185 images were analyzed over the same range of rod lengths(r=30-300m). The vast majority (152, or 82%) had a linear trend on the Richardson plot, as defined by a squared correlation coefficient exceeding 0.90. In other words, the vast majority appear fractal over the scale range 30-300m. However, this range of scale is too limited to confidently extrapolate the fractal behavior to larger & smaller scales. Thus, in the continuing discussion in which we discuss fractal dimension (D) and relate it to wave conditions, it's important to keep in mind that this fractal dimension may only describe the shape at the sand bar at the scale range 30-300m.

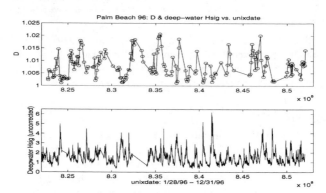

Figure 7. Time series of fractal dimension (top) and deep water significant wave height (bottom) for the period Jan 28 to Dec 31, 1996.

Shown in figure 7 is a time series of the fractal dimension (over the scale range 30 - 300m) with the corresponding wave heights for the period January 1 to December 31, 1996. These wave heights are the deep water wave heights obtained from a wave rider 20 km to the SSE, and do not necessarily correspond to the local wave heights experienced at Palm Beach. Nevertheless, it is reasonably clear that peaks in the wave height correspond to low D, which reflects the formation of a uniform, shore parallel bar during high wave events. The fractal dimension tends to increase during periods of low waves reflecting an increase in complexity as the rip cells develop. The fractal dimension therefore provides a more objective way of quantifying complexity though the present approach is limited in being one-dimensional and it relies on using images which have continuous intensity contours.

Longshore currents

A video technique for estimating longshore currents has been described by Leon and Holman (1994) and Holman and Frielich (1995). Foam left by breaking waves is used as a current tracer in a longshore video transect sampled at 1Hz and displayed as a vertical time stack. Longshore flow produces diagonal streaks in the time stack with the slope being equal to the mean velocity.

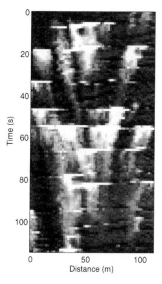

Figure 8. Time stack (left) from a longshore transect spanning two feeder channels either side of a rip channel shown in the time exposure on the right.

An example of this technique is shown in figure 8. In this case the shore parallel transect covers the two feeder channels either side of the main rip channel shown in the time exposure on the right. The converging streaks are consistent with the flow in the feeder channels converging towards the rip channel. In this case the estimated current speeds were -0.5m/s and 0.6m/s. Application of the technique to estimate cross-shore velocities has been less successful. In this case the orbital velocities of the waves and the actual wave fronts tend to dominate the signal obscuring the mean cross-shore flow. Further work is required to improve the robustness of the technique and in situ measurements are needed to assess the accuracy.

Conclusions

A qualitative analysis of hourly video time exposures over 15 months has been used to examine the morphological evolution of rip currents on a medium to high energy beach. During high wave events a uniform shore parallel bar develops and is characterized by a low fractal dimension. The subsequent period of low waves is characterized by an increase in complexity of the nearshore morphology and a corresponding increase in the fractal dimension. During this period the rip cells develop and their location remains relatively stable until the next high wave event. At the same time the bar between the rip channels migrates slowly onshore often becoming welded to the beach resulting in a series of transverse bars and rip channels.

This morphological sequence of offshore bar displacement during storms followed by a slow onshore migration under low waves has been observed by others (Greenwood and Davidson-Arnott (1979), Short (1979) and Wright et al (1979)). However, video provides the potential for more quantitative analysis with significantly improved temporal and spatial sampling. The location and spacing of rip cells can vary significantly before and after high wave events. In some cases the rip cells exhibit significant alongshore periodicity while in other cases a range of length scales can be present. The variation of length scales along the beach may be associated with changes in the wave field due to refractive effects along the beach which tends to be relatively sheltered at the southern end and more exposed at the northern end.

In addition to using time exposure images to examine the morphological evolution of rip currents rapid sampling (1Hz) of the video signal along an array of selected pixels provides a means of estimating surface current velocities. Converging longshore flow in the feeder channels either side of a rip channel could be discerned. Considerable more work is required to improve the robustness of the technique. The application to cross-shore flows has not been very successful to date due to the dominant orbital velocities of the waves.

References

Bowen, A.J., 1969. Rip Currents 1. Theoretical Investigations. *J. Geophys. Res.,* **74**(23), 5467-5478.

Bowen, A.J., and Inman, D.L., 1969. Rip Currents 2. Laboratory and Field Observations, *J. Geophys. Res.,* **74**(23), 5479-5490.

Dalrymple,R.A., 1975. A Mechanism for Rip Current Generation on an Open coast, *J. Geophys. Res.,* **80**, 3485-3487.

Dalrymple,R.A., 1979. Rip Currents and their causes, Proc 16th Int. Conf. Coastal Eng., ASCE, 1414-1427.

Greenwood, B. and Davidson-Arnott,R.G.D., 1979. Sedimentation and equilibrium in wave-formed bars: a review and case study, *Can. J. Earth Sci.,* **16**, 312-332.

Leon,P.F., and Holman,R.A., 1994 (abstract). The estimation of longshore currents from video imagery, *Trans. Amer. Geophys. Union,* EOS, **75**(44), 322.

Holland,K.T., Holman,R.A.,Lippmann,T.C.,Stanley,J., and Plant, N., 1997. Practical use of video imagery in nearshore oceanographic field studies, IEEE *J. Oceanic Eng.,* **22**(1), 81-92.

Holman,R.A.. and Frielich,M.H., 1995 (abstract). Estimation of longshore currents on complex topography, *Trans. Amer. Geophys. Union,* EOS, **76**(45), 293.

Huntley,D.A., and Short,A.D., 1992. On the Spacing Between Observed Rip Currents, *Coastal Eng.,* **17**, 211-225.

Noda, E.K., 1974. Wave-induced Nearshore Circulation, *J. Geophys. Res.,* **79**, 4097-4106.

Richardson,L.F., 1961. The problem of contiguity: an appendix to statistics of deadly quarrels, *General Systems Yearbook,* **6**, 139-167.

Shepard,,F.P., Emery,K.O. and LaFond,E.C., 1941.   Rip currents: a process of geological importance, *J. Geology,* **49**, 337-369.

Short,A.D., 1979. Three dimensional beach stage model, *J. Geology,* **87**, 553-571.

Short,A.D., 1985. Rip current type, spacing and persistence, Narrabeen beach, Australia, *Marine Geology,* **65**, 47-71.

Short, A.D., 1993. Beaches of the New South Wales Coast: a guide to their nature, characteristics, surf and safety, Australian Beach Safety and Management Program, Sydney, pp254.

Short, A.D. and Hogan,C.L., 1994. Rip currents and beach hazards: their impact on public safety and implications for coastal management, *J. Coastal Res.,* Sp. issue No **12**, 197-209.

Sonu,C.J., 1972.   Field observations of nearshore circulation and meandering currents, *J. Geophys. Res.,* **77**(18), 3232-3247.

Wright, L.D., and Short,A.D., 1984. Morphodynamic variability of surf zones and beaches: a synthesis. *Mar. Geology,* **56**, 93-118.

Wright,L.D., Chappell,J., Thom,B.G., Bradshaw,M.P., and Cowell,P., 1979. Morphodynamics of reflective and dissipative beach and inshore systems: Southeastern Australia, *Mar. Geology,* **32**, 105-140.

Rip Channels and Nearshore Circulation

Merrick C. Haller, R. A. Dalrymple, and I. A. Svendsen [1]

Abstract

Results from an experimental investigation into the effects of rip channels on the nearshore circulation system are presented. The results show that periodic channels in an otherwise longshore uniform bar induce longshore pressure gradients that drive pairs of counter-rotating circulation cells. Dense measurements of wave height, wave setup, and induced currents are presented. In addition, evidence of secondary circulation cells close to the shoreline is also given. Finally, the rip current is shown to contain low frequency oscillations probably due to jet instability and also tends to meander back and forth in the rip channel.

Introduction

Previous researchers have addressed the question of what mechanisms regulate the flows near a longshore bar. Dalrymple (1978) indicated that small longshore pressure gradients (neglected in most models of longshore currents) can drive strong longshore currents if the bathymetry is longshore varying. Putrevu et al. (1995) extended Mei and Liu's (1977) work to develop a semi-analytical solution to study the effect of alongshore bathymetric variability on longshore current predictions. These results again confirmed that longshore pressure gradients can substantially alter longshore current predictions. Recently, computational results (Sancho et al. , 1995) for the wave-induced flow over a rip-channel and barred beach showed that the longshore pressure gradient strongly contributes to the longshore momentum balance. However, it is still an open question as to which mechanisms govern how much water flows over the bar (Hansen and Svendsen, 1986), how the mean water level varies, and how rip currents in rip channels through the bar are influenced by these factors.

The concept of the rip current was first introduced by Shepard (1936) and since then many researchers have made qualitative observations of rip currents and their effects on nearshore morphology and circulation (e.g. Mckenzie, 1958; Short, 1985). Additionally, in the field, rip currents tend to be transient and they tend to elude investigators intent on measuring them with stationary instrument deployments, though some quantitative measurements do exist (Sonu, 1972; Bowman, 1988; Tang and Dalrymple, 1989). Therefore, the laboratory is very conducive

[1] all at: Center for Applied Coastal Research, Ocean Engineering Lab, University of Delaware, Newark, DE 19716, USA. Correspondence e-mail: merrick@coastal.udel.edu

to the study of nearshore circulation in the presence of rip currents since the environment is more easily controlled.

Physical Model

The results presented herein are from an ongoing experimental investigation being conducted in the 20m × 20m directional wave basin at the Center for Applied Coastal Research at the University of Delaware. The basin (Figure 1) contains a planar concrete beach of 1:30 slope, along with a steeper (1:5) toe structure. A discontinuous longshore bar made of molded plastic was attached directly onto the beach slope. The crest of the bar is 6cm above the planar beach and it has a parabolic shape in the cross-shore. The discontinuities result from the two gaps in the bar which act as rip channels and tend to fix the location of offshore directed rip currents depending on the wave conditions. These rip channels are located at 1/4 and 3/4 of the width of the basin and are each 1.8m wide with sloped sides.

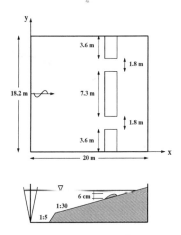

Figure 1: Plan view and cross-section of the experimental basin.

Separate arrays of ten capacitance wave gages and 3 acoustic-doppler velocity meters (ADV) were used to measure the generated wave fields, mean water levels ($\overline{\eta}$), and circulation patterns ($\vec{u} = (u, v)$). The measuring procedure consists of simultaneously initiating monochromatic wave generation and data collection and then collecting approximately 27 minutes of data sampled at 10 Hz. The relatively long recording time allows us to observe the development of the circulation system as it evolves to a steady state. Because of the limited number of measuring instruments available, the measuring procedure is to repeat the test for multiple runs of the same wave conditions in order to achieve a dense coverage of the nearshore region. Figure 2 shows the measurement locations for the 29 runs described in this paper. The location of wave measurements was chosen such that the mean water surface elevations governing the flow through one of the rip channels would be resolved. The location of ADV measurements spanned a wider area in the basin in order to evaluate any asymmetry in the circulation. The ADV measurements taken shoreward of x=10.85m were taken 3cm from the bottom and, for the present discussion, it is assumed that the measurements approximately represent the depth-averaged flow below wave trough level. The offshore ADV measurement lines at x=10m,9m were taken 4cm and 5cm from the bottom, respectively. The still water line was at x=14.89m, the wavemaker at x=0m, and the water depth at the bar crest (x=12m) was 3.6cm.

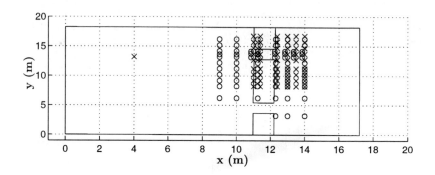

Figure 2: Measurement locations for wave gages (×) and ADVs (○).

Results - Wave transformation

The measured rms wave heights are shown in Figure 3a. The offshore wave height measured at the stationary gage (x=4m,y=13.2m) was H_{rms}=4.8cm with a frequency of 1 Hz and the waves were generated normally incident to the beach. The wave height at the offshore location was very repeatable and had a standard deviation of 0.6mm over the 29 runs. The wave heights measured in the nearshore were slightly more variable from run to run with most of the variability taking place in the region of the rip current. The spatial variation of wave heights indicates that as the waves approach the bar they are fairly uniform in the longshore direction except for near the rip channel (y=12.9-14.7m). Due to the presence of the opposing rip current, the waves are much higher through the rip channel and the breaking process is less dramatic as the waves that travel through the channel decay at a slower rate and remain higher than elsewhere until very near to the shoreline.

The spatial variation of $\overline{\eta}$ is shown in Figure 3b. The variation of $\overline{\eta}$ is essentially opposite to that of H_{rms} (note the change in figure orientation). As expected, the setup is highest shoreward of the bar due to the breaking process and, in the longshore direction, the setup is greatest away from the channels (except for very near to the shoreline). The induced longshore setup gradient shoreward of the bar drives the flow towards the rip channels where the flows converge and are driven out through the rip channel.

Figure 4 compares the cross-shore variations of H_{rms} and $\overline{\eta}$ measured through

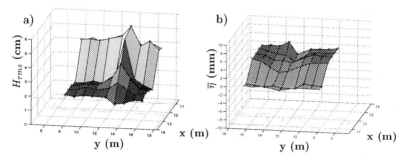

Figure 3: a) Measured rms wave heights b) measured setup/setdown. Note the change in horizontal orientation between the figures. Still water shoreline located at x=14.8m.

the channel and in the center of the basin. Notice the largest measured longshore gradients are at x=12.3m which is just shoreward of the bar in what would be considered the bar trough. It is also interesting to note that at x=14m, near the shoreline, the longshore gradients are reversed which drives a reverse flow away from the rip channel close to shore.

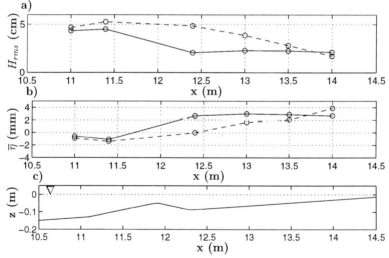

Figure 4: Cross-shore profiles of a)H_{rms} b)$\overline{\eta}$. Solid lines represent y=9.2m (bar), dashed line y=13.65m (rip channel); c)bathymetry profile.

Results - Circulation

Figure 5 shows the general circulation pattern induced in the basin. The current vectors represent the measured mean currents computed from a time average of the last 819s of each data record. It is evident from the measured flows that

there is a primary and secondary circulation and each circulation consists of a
pair of counter-rotating cells. The primary circulation consists of the converging
feeder currents, the rip current, the diverging flow at the exit of the channel, and
the return flux driven shoreward over the bar. Since the ADVs were deployed
below trough level, the mass flux brought shoreward over the bar in the wave
crests is not represented by these measurements. The maximum flow rate ($\vec{u}$)
was 17.5cm/s and occured in the rip channel at x=12.25m, y=14.15m. The
secondary circulation is driven by the breaking near the shoreline of waves that
have shoaled through the channels. Here the breaking process creates a local
maximum in the setup which drives a divergent flow away from the channel.
This flow travels along the shoreline and eventually turns and joins the feeder
currents of the primary circulation. As a consequence of the two circulations,
the longshore current behind the bar contains a strong shear near the channels
and is likely to be unstable.

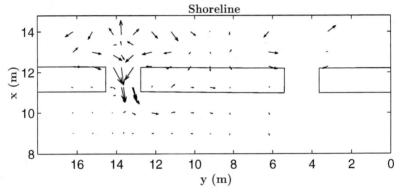

Figure 5: Induced circulation in basin. Vectors represent time average of last 819s of
each ADV record.

Figure 6 shows the evolution of the longshore current profile as it establishes
itself near its starting point in the center of the basin and grows towards the
channel. The center of the basin is in cross-shore balance which means the flow
carried shoreward by the waves is returned directly offshore in the undertow and
no longshore current exists. However, towards the channel the longshore current
gains in strength with the strongest longshore current located just shoreward
of the trough at x=13m. Correspondingly the reverse current located near the
shoreline also gains in strength with the strongest cross-shore shear in the current
profile located at y=12.2m.

Figure 7 shows the longshore variation of the cross-shore flow measured near
the shoreward edge of the central bar section. Negative u velocities indicate
offshore flow or undertow and the variation of u indicates that the strongest
offshore flow occurs in the center of the basin (y=9.2m) and decays to zero near
the channel (y=12.2m). Assuming that the flow below trough level is fairly
depth uniform and since the wave field is longshore uniform along the bar, the
decrease in undertow towards the channel implies that flow is being entrained in

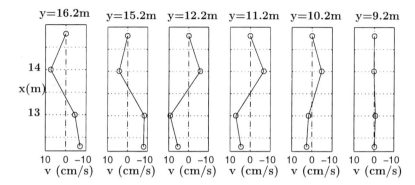

Figure 6: Longshore current profiles. Longshore current is assumed to be zero at the still water shoreline (x=14.8m).

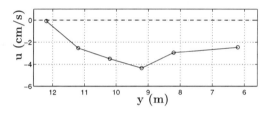

Figure 7: Longshore variation of cross-shore velocities along the central bar. ADVs located 3cm from bottom (x=12.3m).

the shoreward directed mass transport over the bar and further strengthens the longshore mass flux behind the bar.

Results - Low frequency motion

In addition to the general circulation pattern measured during the tests, another interesting aspect of the experiment was the presence of fluctuations and meanders in the rip current. Figure 8 shows three 1638s records of velocities (u, v) measured simultaneously by three ADVs spanning 1m in the longshore direction and centered near the channel exit. The time series show that the rip becomes relatively steady at about t=800s. As the offshore flow first begins to strengthen at y=13.15m (t=1050s), distinct fluctuations appear in both the u and v velocities. Notice that the strengthening of flow and the onset of oscillations occurs first at y=13.15m and then at y=13.75m and y=14.15m at successively later times. The strengthening of the flow from one sensor to the next represents a meandering of the entire rip in the positive y direction and this occurs at a slower time scale than the oscillations present within the rip itself. The oscillations seem to have a relatively regular frequency during t=1050-1638s and are more easily visible in the longshore velocities since they do not contain the wave induced oscillations.

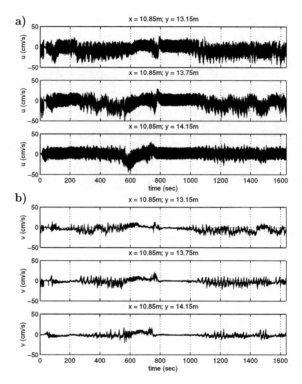

Figure 8: Time series of a)cross-shore velocity b)longshore velocity measured at exit of rip channel.

Spectral analysis of the longshore velocity records of the rip current are shown in Figure 9 (a and b). The spectra contain strong energy peaks near T=20s. In contrast, the longshore velocity spectra measured in the feeder currents and in the secondary circulation cells (Figure 9c,d) contain more energy at longer periods. Simultaneous visual observations and video recording of the rip current were made throughout the experiment with the aid of dye injected into the feeder currents. Besides following the dye patterns, it was also possible to track the location of the rip by watching the distinct breaking pattern (whitecapping) of the incident waves which was limited to the region of the strongest rip current (rip neck) (note: elsewhere wave breaking occurred at the bar crest). It was difficult to visually observe the 20s oscillations of the rip current. What was seen instead was the slower time meandering back and forth of the rip current in the channel. This supports the idea that the 20s oscillations are probably smaller scale vortices being advected with the current that are generated due to flow instability mechanisms. These oscillations are distinct from the larger scale, slower, side-to-side meanderings of the entire rip current which in turn cause the primary circulation cells to shrink and stretch along with it.

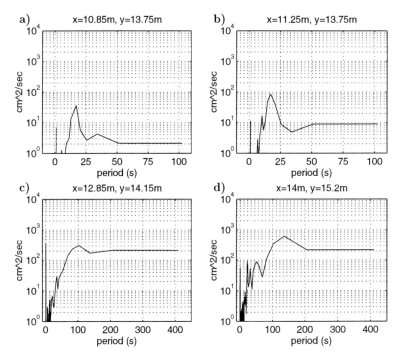

Figure 9: Measured spectra of longshore velocities from rip current a) and b) during t=1024s-1638s, D.O.F.=12, Δf=.00977Hz. Measured spectra of long-shore velocities from c)feeder current and d)secondary circulation, t=0-1638s, D.O.F.=8, Δf=.00244Hz.

Results - Wave-current interaction

The presence of strong currents in the region of the rip channel greatly affected the local wave field. The most obvious visual effects were refraction around the rip leading to criss-crossed wave crests shoreward of the rip, wave steepening/breaking on the rip, and subsequent diffraction of the relatively higher waves that had shoaled through the channel. In order to assess the effects of the rip on the incoming waves, the spectra of the waves entering the rip channel are compared to the spectra of the waves which travel over the bar in the center of the basin (Figure 10).

Figure 10a shows that as the waves encounter the strong opposing current ($u \sim O(10\text{cm/s})$) energy is transferred to the sidebands (f=.95,1.05Hz) of the spectral peak frequency. In addition, there is a low frequency peak located at f=.05Hz. This effect is not seen in the waves measured at the same cross-shore location (and

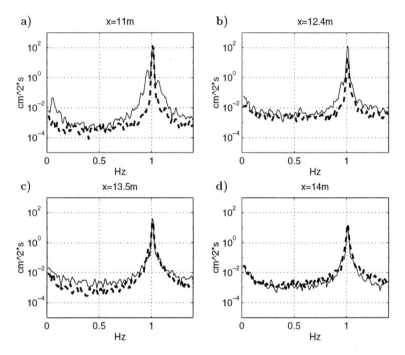

Figure 10: Comparison of wave spectra measured in the channel (y=13.65m, solid line) and in the center of the basin (y=9.2m, dashed line), t=819-1638s, D.O.F=16, Δf=.00977.

same depth) in front of the central bar. As the waves shoal through the channel the sidebands decay (Fig.10b,c) and near the shoreline (Fig.10d) the spectra have a very similar shape. It is also important to notice that the low frequency energy (f=.05Hz) present in Figure 10a is two to three orders of magnitude smaller than that shown in the rip spectra (Fig. 9a,b). Since the oscillations in the rip at f=.05Hz are much larger than the low frequency wave energy measured near the rip it is unlikely that the current oscillations are a manifestation of resonant tank modes (seiching/standing edge waves). In addition, wave spectra measured very near to shore (x=14m) do not show any significant energy in the f=.05Hz frequency band further indicating that the wave energy at this frequency measured near the rip is not due to basin resonance.

Discussion

The experimental results give a detailed description of the nearshore hydrody-namics existing on a barred shoreline with periodically spaced rip channels. The relatively dense set of measurements made during this experiment provide good resolution of the longshore variations in wave height and wave setup/setdown induced by the experimental bathymetry. A somewhat more extensive set of current measurements show that for the experimental wave conditions a complex circulation system exists. The circulation consists of primary and seondary circulation systems, each containing a pair of counter-rotating cells. It is expected that the data set will be very useful for the testing of computational models of nearshore circulation.

In addition, visual observations and data analysis of one of the rip currents present during the experiment shows that the rip tends to be unstable and contains oscillations of approximately 20s period. These oscillations are distinct from the larger scale, longer period meanderings of the entire rip current structure. These meanderings were observed to cause a shrinking and stretching of the primary circulation system associated with the rip. The feeder currents and secondary circulation cells are shown to contain long period oscillations related to the meandering of the rip current.

Finally, the experimental results indicate significant wave-current interaction occurs in the vicinity of the rip. Initially monochromatic waves propagating into the opposing rip current gain energetic sidebands. These sidebands disappear as the waves shoal past the rip neck. It is deemed unlikely that the low frequency energy ($f=.05$Hz) present in the waves near the rip and in the rip itself is due to resonant basin oscillations.

References

Bowman, D., D. Arad, D. S. Rosen, E. Kit, R. Goldbery, and A. Slavicz, Flow characteristics along the rip current system under low-energy conditions, *Marine Geology*, 82, 149-167, 1988.

Dalrymple, Robert A., Rip currents and their causes, *Proc. 16th Intl. Conf. Coastal Engrg*, ASCE, 1978.

Hansen, J. Buhr, and I. A. Svendsen, Experimental investigation of the wave and current motion over a longshore bar, *Proc. 20th Intl. Conf. Coast. Engrg.*, Taipei, ASCE, 1986.

McKenzie, P., Rip-current systems, *J. Geology*, Vol. 66, 103-113, 1958.

Mei, C. C., and P. L.-F. Liu, Effects of topography on the circulation in and near the surf-zone - linear theory, *Estuarine and Coastal Marine Science, 5*, pp.25-37, 1977.

Putrevu, U., J. Oltman-Shay, and I. A. Svendsen, Effect of alongshore nonuniformities on longshore current predictions, *J. Geophys. Res., 100*, 16119-16130, 1995.

Sancho, F. E., I. A. Svendsen, A. R. Van Dongeren, and U. Putrevu, Longshore nonuniformities of nearshore currents, *Coastal Dynamics '95, 425-436*, 1995.

Shepard, F. P., Undertow, rip tide or "rip current", *Science, Vol. 84*, 181-182, Aug. 21, 1936.

Short, A. D., Rip-current type, spacing and persistence, Narrabeen Beach, Australia, *Marine Geology, 65*, 47-71, 1985.

Sonu, Choule, J., Field observation of nearshore circulation and meandering currents, *J. Geophys. Res., 77, 18*, 3232-3247, 1972.

Tang, E. C.-S., and R. A. Dalrymple, Nearshore circulation: rip currents and wave Groups, in *Nearshore Sediment Transport Study* , R.J. Seymour ed., Plenum Press, 1989.

Coastal Evolution: Modeling Fabric and Form at
Successive Spatial Scales

Christopher W. Reed, Alan W. Niedoroda[1]
Donald P. Swift[2], Yong Zhang[2], John Carey[2]
Michael S. Steckler[3]

Abstract

An aggregated modeling approach for predicting
engineering scale coastal evolution is presented. The
approach is based on making extended simulations with
process-based models to generate long-term data sets
that substitute for data not generally available from
field measurements. The time-averaged simulated data is
used to determine the dynamic relationship between
hydrodynamic forcing and coastal system response at the
aggregated scale. The role of incomplete data sets in
calibrating and validating the models is reviewed.

Introduction

Predicting coastal and shelf morphology on time scales
of decades to centuries poses serious challenges to
coastal engineers and scientists because rigorous
quantitative approaches are only now being developed.
Meaningful predictions are needed by coastal managers
faced with evaluating the long-term merits of major
coastal projects (e.g. beach nourishment, inlet
dredging, induced subsidence). These analyses are also
critical to assessing the regional effects of global

---

[1] Woodward-Clyde-International-Americas, 3676 Hartsfield Road, Tallahassee, Florida 32303
[2] Old Dominion University, Department of Oceanography, Norfolk, Virginia 23529
[3] Lamont Doherty Earth Observatory, Columbia University, Palisades, New York 10964

sea level rise. Problems in developing predictive
models and techniques for these scales are described in
de Vriend (in press, this volume). Two of the most
prevalent issues are the inherent uncertainty in the
coastal systems brought on by the nonlinear nature of
the equations governing hydrodynamics and sediment
transport and the lack of reliable and sufficient data
for testing model predictions.

Most predictive morphodynamical models are based on the
conservation of mass (sediment), relating the time
evolution of profile elevation changes to spatial
gradients in the transport. The methods differ
enormously in both the time and space scales that are
represented and the means in which the sediment fluxes
are related to the system characteristics and external
forcing. Theoretically, a well calibrated detailed
process-based model representing small space and time
scales (seconds and centimeters) could be applied over
a large spatial range and integrated over decades or
centuries to predict the large scale coastal evolution.
Aside from the obvious computational limitations, the
inherent uncertainty associated with the governing
equations make the success of this approach suspect.
An alternate approach to predicting larger scale
coastal evolution is the aggregated-scale approach (see
de Vriend, this volume). In this approach, the system
dynamics at the desired scale is modeled directly, and
the smaller scales processes are aggregated and
manifested, for instance, as a net mean advection or
dispersion. The inherent difficulty, here, of course,
is determining the dynamical relationship between the
forcing and response, and just as important, how to
characterize the forcing.

Ideally, in any development of predictive models, a
complete data set for calibration and validation is
desired. A complete data set consists of well defined
system characteristics, system forcing and system
response. Complete data sets can be developed for
small scale processes which can be reproduced in a
laboratory experiment or in detailed field
measurements. However, the data sets become sparse and
less complete with increasing system scales. This can

be a serious limitation in the development of
aggregated-scale models since their development is
strongly dependent on empirical data.

We have adopted the aggregate-scale modeling approach
and developed a model which represents cross-shelf
morphodynamics and bed facies evolution on decadal to
century time scales. Critical to the model development
is the use of process-based models representing event
scale (i.e. storms) hydrodynamics, sediment transport
and morphologic changes. These models are used to help
develop the dynamical interaction between forcing and
response represented in the aggregated-scale model.
Inherent difficulties with data sets at the aggregated
scales are circumvented by use of the stratigraphic
data which represent the time-integrated morphologic
change on scales of decades to millennia. Thus, both
the event-scale and the aggregated-scale models contain
representations of bed composition and morphology.

Approach

The basis for the aggregated-scale modeling approach is
the behavior-oriented model developed by Niedoroda et
al.(1995). This model represents the cross-shelf
evolution of bed composition and elevation in response
to sediment supply, sea-level changes and changes in
the hydrodynamic climate. The model relates the time
average sediment flux to long-term average hydrodynamic
dispersion and residual advection. These quantities
vary across the shelf and depend on the local water
depth, slope and distance from the coast. It originally
operated on scales of millennia, has been used in
interpreting a number of geological data sets, and has
reproduced expected system behavior (Swift 1997, Carey
1997).

In order to apply the model for predicting engineering
scale (decades or centuries) cross-shore evolution, the
relationship between the time-averaged sediment fluxes
and the time averaged hydrodynamic climate needs to be
systematically determined. Furthermore, the model needs
to be verified against suitable data sets representing
these scales.

Two basic approaches to elucidating the underlying
relationships between the system forcing and response
can be considered. A formal averaging over time and
space can be applied to the process-based model
equations (DiSilvio 1991). This method is conceptually
similar to the turbulence modeling resulting from the
Reynolds-averaged Navier-Stokes equations. It
ultimately requires some means to model the additional
terms appearing in the time-averaged equations, and
some empirical data is required to determine the form
of these terms. A conceptually different approach is to
collect long term data sets and "back out" the
fundamental relationships, being guided by the data
itself. For instance, using measurements of the time-
and depth-averaged sediment flux and concentration at
points across the shelf, a suitable time- averaged
cross-shelf diffusion profile can be determined.
Unfortunately, no such data sets are available.
However, it is possible to create suitable data sets
using process-based models. In adopting this approach,
we have developed an efficient process-based model of
shelf hydrodynamics and sediment transport. The model
simulates hydrodynamic, sediment transport and
subsequent bed changes during storm events under the
influence of tidal and wind forcing. When calibrated
and then implemented to simulate a long succession of
events (numerous storms over one decade) it can produce
the required long term data sets.

Measured data is required to calibrate the event scale
model and to validate the aggregate-scale and the
extended simulations of the event scale model at the
larger scales. There are no complete data sets
available for calibrating the event-scale model, but
wind, current and wave data collected on the Texas
shelf LATEX and TABS projects and wind, current, wave
and sediment data collected on the Eel River shelf (N.
California) as part of the STRATAFORM project
(Nittrouer and Kravitz, 1996) provide sufficient data
to properly test the event scale model.

Validation of the aggregated-scale model (and the
extended simulations of the event scale model) relies
on detailed measurements of both morphology and

stratigraphy in shelf environments. A few suitable data
sets exists for morphological changes, such as those
for the Dutch, American and Australian coasts and have
been organized as part of the PACE project (de Vriend,
this volume). Stratigraphic and facies records
partially preserve the shelf response to hydrodynamic
forcing and sediment transport gradients from
individual events (upper few centimeters of the bed) to
decades and up to millennia (decimeters to 10s of
meters into the bed). Such data is currently being
collected as part of the STRATAFORM project. Prediction
of the fine scale vertical structure of the
stratigraphy and its variability at larger scales
across the shelf will provide a stringent validation of
both models.

The Aggregate-Scale Model

The governing equation for the aggregate scale model is
the conservation of mass equation

$$\frac{\partial h(x,t)}{\partial t} = \frac{1}{(1-e)} \sum_i \frac{\partial Q_i(x,t)}{\partial x}$$

where x represent distance in the offshore direction, t
is time, h is the shelf profile and Q_i is the time-
averaged sediment flux for the ith grain size component
and e is the bed porosity. The behavior oriented
aspect of the model is contained in the definition of
the sediment flux Q_i

$$Q_i(x,t) = U C_i(x,t) - W\left[e^{-\beta_i h(x,t)} \ e^{-d_i s(x,t)}\right] - \overline{D}_H \frac{\partial \overline{C}_i(x,t)}{\partial x} + G\left(s(x,t) - R_i\right)$$

Here s is the bed slope and C_i is the time averaged
depth integrated suspended sediment concentration. The
model consists of a finite volume numerical solution of
these two equations after specification of an initial
profile h(x), the sediment supply at the coast $Q_i(0,t)$,
the sea level variation over time and the model
parameters U, W, β_i, d_i, G, R_i and D_H. Detailed
discussion of the model and definitions of the
parameters can be found in Niedoroda (1995). Herein we
focus on the large-scale diffusion term, which is the
third term on the right side. The large-scale cross-
shelf diffusion term represents the aggregated effect

of many storms acting on the shelf sediments. Each individual storm may have significant advective component of transport. But so long as these individual excursions are small relative to the shelf width they can be averaged over many storms, and the transport can be represented as a small net or residual advection (i.e. U in the first term on the right hand side) and diffusion. In previous applications of the aggregated-scale model, the time-averaged diffusion D_H has been assumed to increase linearly with distance offshore and the time-averaged concentration C_i was specified as a function of the water depth (i.e. h). These prescription were based on experience and intuition, and their validity was founded on the model's ability to reasonably reproduce expected behavior.

It is the primary purpose of this effort to determine the appropriate form of these quantities and their dependence on hydrodynamic parameters. To this end, an event scale model is used to generate long term simulations of sediment transport across the shelf in response to a random sequence of storms. The instantaneous depth-integrated concentrations and fluxes at each point along the shelf are averaged over the many events to obtain the long term average cross shelf flux $F_i(x)$ and concentration $C_i(x)$. The associated time averaged diffusion coefficient D_H is determined from the definition of the flux:

$$F_i(x) = D_H(x)\frac{\partial C_i(x)}{\partial x}$$

This process is repeated for storm sequences with different mean levels of energy so that the dependence of D_H and C_i on the hydrodynamic climate can be elucidated.

The Event Scale Model

The cross-shelf event scale model is a direct extension of an existing vertical one-dimensional model of hydrodynamic, sediment transport and sediment stratigraphy. Details of the model can be found in Reed et al. (1997). The one-dimensional model consist of a

numerical solution of the conservation of horizontal
momentum and mass (sediment) equations throughout a
water column. The model calculates the time dependent
vertical velocity and concentration profiles for
multiple grain sizes in response to forcing such as
winds, tides and surface waves. Coupled to the model is
a bed armoring scheme which represents the effects of
limited supply on the suspended sediment.
Additionally, a bed tracking algorithm is included
which records the depth variation in bed grains size
class frequency during depositional periods.

A fundamental assumption in the one-dimensional model
is lateral homogeneity in both momentum and suspended
sediment concentration. However, lateral gradients in
suspended sediment concentration can be imposed to
simulate depositional or erosional events.

Figure 1 shows a stratigraphic record reproduced using
the one-dimensional version of the event scale model
with an imposed sediment accumulation. To obtain the
record, a random sequence of storms, (characterized by
mean currents and surface waves) was simulated.

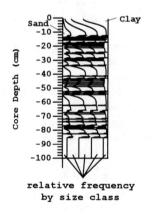

Figure 1 Simulated Core

Each successive storm cannibalizes part of the previous
deposited storm bed, and then deposits its own graded
bed. These beds range in thickness form millimeters to
centimeters.

The two-dimensional (cross-shelf) event scale model is a direct extension of the one-dimensional model, including time dependent hydrodynamic and sediment transport calculations, morphologic changes, and tracking of the developing stratigraphy. The fundamental difference is that the cross-shelf gradients are now an intrinsic part of the solution rather than an externally imposed condition. In this case, simulated stratigraphic cores arise due to calculated gradients in the cross-shore sediment fluxes. These gradients change across the shelf, and therefore the vertical structure of the deposits varies across the shelf, at a scale on the order of kilometers.

Large Scale Diffusion

The two-dimensional event-scale model has been used to generate a long sequence of storm events in order to test the proposed method for determining the time-averaged cross-shelf diffusion. A storm event was constructed based on data collected during the STRATAFORM project and consists of time profiles for wind speed and direction and wave height and period. A shelf profile and sediment distribution resembling those measured for the shelf were used as initial conditions. During the simulation, the duration, direction and intensity of each storm in the sequence was randomly varied and a sediment supply was introduced at the coastline. After the simulation the complete time series of depth-integrated concentrations and fluxes at each point along the shelf were averaged.

The time averaged depth-integrated concentrations and offshore fluxes are shown in Figure 2 for the three grain size classes represented in the simulation (sand, silt and clay). The sand concentrations and fluxes, not visible on the plots due to the scale, are similar to those for silt and clay. The concentrations show a decrease with distance offshore (or water depth). This is interesting given that the finer material in the bed had a larger bed fraction further offshore. The corresponding fluxes are all in the offshore direction

with peaks at about two kilometers offshore for all
grain size classes. Thus, for the conditions
simulated, there is a net erosion in the near shore
region and deposition further offshore, indicating that
the sediment supply, shelf profile and hydrodynamic
climate are not balanced. These results are likely due
to simplified representation of the storm events and
approximations in prescribing the initial conditions.
The time-average cross shelf diffusion, determined
independently from the concentration gradients and
fluxes for each of the grain size classes appears
fairly independent of the grain size. This indicates
that the dominant transport regime for all classes the
same, for instance, in the near bottom water layer.

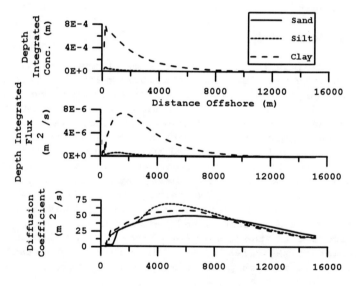

Figure 2 Time-averaged quantities

Summary and Conclusions

An approach for developing an aggregate-scale coastal
evolution model is presented and tested. A series of
process-based numerical models have been implemented to
cross successive space and time scales. They have
generated long term complete data sets that are used to

study the relationships between coastal system forcing and response at the larger aggregate scale.

The models require rigorous validation for this purpose. Ultimately, the long term data stored in the stratigraphic record across the continental shelf, combined with morphologic records will provide a basis for validation. Specifically, the large-scale across-shelf variation of fine-scale stratigraphic structure will be simulated.

References

Carey, J.S., Swift, D.J.P., Steckler, M., Reed, C.W., and Niedoroda, A., (1997) in review, High-resolution sequence stratigraphic modeling: effects of sedimentation processes: Watney, L.D., ed., Numerical Experiments in Stratigraphy: SEPM Special Publication.

DiSilvio, G., 1991, Averaging operations in sediment transport modeling: short-step versus long-step morphological simulations, International Symposium on Transport of Suspended Sediment Modeling

Niedoroda, A.W., Reed, C.W., Swift, D.J.P., Arato, H. and K. Hoyanami, 1995, Modeling shore-normal large scale coastal evolution, Marine Geology, v. 126, n. 1/4, p 181-201.

Nittrouer, C.A. and J. H. Kravitz, 1996, A Program to Study the Creation and Interpretation of Sedimentary Strata on Continental Margins, Oceanography, v.9, n.3, p 146-152.

Reed, C. W. Reed, Niedoroda, A. W. and Donald J.P. Swift, (1997) in press, Sediment Entrainment and Transport Limited by Bed Armoring, Marine Geology.

Swift, D.J.P., Carey, J.S., Zhang, Y., Reed, C.W., Niedoroda, A., and Steckler, M., (1997) in review, Quaternary stratigraphy and sedimentation on the northern California continental margin constrained by numerical modeling: preliminary results: Marine Geology.

Long Term Coastal Evolution and Regional Dynamics of a US Pacific Northwest Littoral Cell

George M. Kaminsky[1], Peter Ruggiero[1], Guy Gelfenbaum[2], and Curt Peterson[3]

Abstract

A regional study of sediment dynamics, long-term coastal evolution and shoreline response has been initiated for a 150 km long littoral cell along a tectonically active margin in the Pacific Northwest, USA. The study area includes a high-energy inner shelf, dissipative nearshore, and dynamic shoreline from Tillamook Head, Oregon to Point Grenville, Washington. Major goals of the study include determining and understanding the evolution of the regional sedimentary system with its natural and anthropogenic influences, and predicting coastal behaviour on a management scale of decades and tens of kilometers. This paper presents an overview of the study and the initial results from beach morphology monitoring, shoreline change analysis, and barrier-beach development analysis. Beach morphology monitoring is being conducted throughout the study area at many sites using real time kinematic, differential GPS survey techniques. Historical shoreline change analysis is being conducted with shoreline position mapping from late 1800 and early 1900-era topographic sheets, digitized aerial photography, and GPS. The long-term barrier-beach development processes are being studied with Ground Penetrating Radar (GPR), vibracores, and C-14 dating to obtain subsurface records of barrier accretion strata and episodic erosional scarps and to reconstruct ancient geomorphic processes and events. The study will also examine the effects of man-made influences (e.g. enhanced runoff, dredging operations, Columbia River dams and jetties) and natural processes (e.g. climate variability, co-seismic subsidence, coastal dune development) on sediment budgets and on the long-term shoreline change trends.

[1] Washington Department of Ecology, Coastal Monitoring and Analysis Program, P.O. Box 47690, Olympia, WA, 98504-7690, USA, gkam461@ecy.wa.gov (Kaminsky), prug461@ecy.wa.gov (Ruggiero)

[2] US Geological Survey, Center for Coastal Geology, 600 4th Street South, St. Petersburg, FL 33701, USA, guy@cfcg.er.usgs.gov

[3] Portland State University, Department of Geology, P.O. Box 751, Portland, OR, 97202-0751, USA, curt@ch1.ch.pdx.edu

Regional Setting

The Columbia River littoral cell extends from Tillamook Head, Oregon to Point Grenville, Washington and has four major concave-shaped, offset sections of prograded shoreline separated by the Columbia River and two large depositional estuaries of Willapa Bay (348 km^2 HW surface) and Grays Harbor (223 km^2 HW surface) (Figure 1). The Columbia River, the third largest river in the United States by discharge, governs the sediment budget for the region with a modern mean annual water discharge of 6,800 m^3 s^{-1}. Wide and gently sloped beaches throughout the littoral cell front barrier-beach plains that have developed over the past several thousand years with large linear dune ridges and broad dune fields and swales. The prograded barrier beaches along this tectonically active coastal margin have experienced episodic erosion (Meyers *et al.*, 1996) and sudden 1 to 2 m subsidence events associated with large earthquakes of approximately 500 yr recurrence intervals (Atwater *et al.*, 1995).

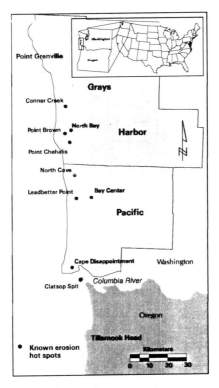

Figure 1. The Columbia River littoral cell from Tillamook Head, Oregon to Point Grenville, Washington.

Wide surf zones with multiple low-crested longshore bars are characteristic of this nearshore environment. Infragravity (low frequency) energy typically dominates the inner surf zone and rip currents and complex nearshore circulation patterns are common. Tides throughout the cell are mixed semi-diurnal with a 2 to 4 m range. The wave climate is large with annual mean H_{mo} heights of 2 m and peak periods of 10 s, and extreme winter storms producing H_{mo} heights of over 7 m and peak periods over 17 s. Strong El Niño events occur at decadal frequency, influencing wave direction and longshore sediment transport.

Background and Motivation

From a cursory historical perspective, the study area is known for being highly accretionary with shoreline progradation rates near 5 m yr^{-1}. This situation has been advantageous for the growth and development of coastal communities, with few societal problems to generate significant scientific research of the littoral region. However, a closer look at the present day conditions of the coast reveals many local exceptions to the accretion trend (Figure 1), as well as possible indicators of a regional-scale change towards erosion.

The only location in the cell where shoreline recession has been the historical trend is at Cape Shoalwater (North Cove), along the northern entrance to Willapa Bay (Figure 2). Terich and Levenseller (1986) reported a long-term erosion rate in excess of 30 m yr^{-1}, the most rapid and sustained erosion site on the Pacific Coast of the United States. The major cause of the erosion is attributed to a nearly continual migration of the deep natural channel along the north side of the entrance to Willapa Bay. A cyclical pattern (order 8-12 yr) of channel migration and outer bar-spit development followed by spit breaching and shoal migration has been documented by the annual bathymetric surveys conducted by the US Army Corps of Engineers, Seattle District. Erosion currently threatens homes and property, the previously relocated State Highway 105, and the potential for seawater intrusion into commercial cranberry bogs. Potential damages, both direct and indirect, have been estimated as high as $80 million if Highway 105 is lost to erosion.

The Columbia River is the major source of modern sediment for the Washington continental shelf, and 84 percent of the annual total sediment discharge accumulates on the mid-shelf, open slope and submarine canyons (Sternberg, 1986). Wright and Nittrouer (1995) assert that the high-energy coastal oceanographic regime precludes significant inner-shelf deposition and promotes deposition and accumulation of fines primarily on the mid-shelf. Most of the Columbia River sediment that enters the littoral system goes north into Washington, but seasonal reversals may be responsible for some southerly transport (Peterson et al., 1991). A trend of decreasing peak flows on the Columbia River since the 1880's has reduced the estimated Columbia River total sediment discharge from 15 million m^3 yr^{-1} to 5 million m^3 yr^{-1} (Sherwood et al., 1990). The construction of eleven major dams on the Columbia River between 1933 and 1968 and many smaller dams on tributary rivers and streams contributes to reducing the discharge of sand, potentially effecting shoreline change and nearshore morphology.

The Bonneville dam, constructed in 1938 on the lower Columbia River 234 km from the mouth, is believed to have the largest impact on regulating flows and sand supply to the coast.

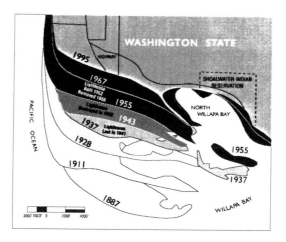

Figure 2. Historical erosion at Cape Shoalwater.

The ebb-tidal deltas on the inner-shelf have shown substantial erosion over the past century. The Grays Harbor delta and nearshore region have lost 120 million m³ of sediment since 1900, resulting in a deepening of 5 cm yr⁻¹ and an average depth increase of 4.5 m (Burch and Sherwood, 1992). The Columbia River ebb-tidal delta deepened 1.4 m between 1868 and 1958, and Peacock Spit on the north side of the Columbia River mouth, a modern source of sediments for northward beaches, has virtually disappeared at a rate of 1.5 million m³ yr⁻¹. These morphological changes are primarily in response to the construction of jetties at both inlets, but represent a large net volumetric loss of nearshore sediment supply.

A historically unprecedented erosion event occurred at the south entrance to Grays Harbor (Point Chehalis) in December 1993 when a breach formed across a narrow spit connecting the Grays Harbor south jetty with the adjacent land. The magnitude of the erosion and its implications caught coastal communities and governmental agencies by surprise, and there was much debate and controversy over the appropriate response. The breach grew from about 4 m in width to nearly 200 m in width over a six-month period, and caused accelerated shoreline recession up to 2 km south of the jetty. Eventually the breach was filled with 460,000 m³ of dredged material, and associated damages were repaired at a cost near $8 million until a long-term solution could be identified and implemented. This problem served as a wake-up call to state and local governments to improve on the existing knowledge base of regional coastal processes to help respond to and potentially prevent future shoreline erosion crises.

Other shoreline erosion problems have struck the communities at Conner Creek and Ocean Shores (Point Brown). The mouth of Conner Creek has migrated northward at approximately 300 m yr^{-1} over the last decade threatening coastal development and infrastructure. This migration is associated with longshore sediment transport and dune overwash processes. Ocean Shores, an area of most rapid accretion over the last 80 years, has experienced up to 60 m of shoreline recession over the last 2 years, threatening expensive shoreline development. Likewise, most of the current erosion hot spots (Figure 1) have historically been the location of the largest shoreline progradation and accumulation of coastal sediment. Nearly all of these sites appear to have either increasing erosion rates or an expanding spatial scale of erosion that is migrating away from the inlets, suggesting a new regional trend towards erosion. The recognition of escalating erosion combined with the awareness of other changes such as sediment source reduction, a slowing of widespread shoreline progradation rates, long-term erosion of ebb-tidal shoals and nearshore profile steepening, provided the motivation to begin this quantitative field investigation of regional coastal dynamics and sediment budget.

Study Elements and Initial Results

Primary goals of the study are to: predict coastal behaviour at management scale (i.e. decades and tens of kilometers); understand regional sediment system dynamics; and determine natural and anthropogenic influences on the littoral system. Specific objectives of the project are to: quantify the regional coastal sediment budget and transport rates; determine regional bathymetric and shoreline change trends; document baseline conditions and quantitatively monitor future coastal changes; and assist planners in identifying options to reduce hazards and impacts to natural resources and the economy.

There are several major study elements being performed to address the project goals and objectives including the following:

- Shoreline change analysis using historical aerial photography, topographic surveys and maps to document historical shoreline positions and rates of change from the 1870's to present.
- Barrier accretion and erosion studies using Ground Penetrating Radar (GPR) along with sediment cores and material dating to investigate the barrier-beach development over the last several thousand years.
- Beach morphology monitoring using real time kinematic differential global positioning systems (RTK DGPS) to measure seasonal and event-scale changes in beach elevation and topography.
- Bathymetric change analysis using historical seafloor surveys and determining elevation and topographic differences.
- Sediment source analysis using dredging records and sediment fill volumes in reservoirs to determine the sediment supply to the littoral cell.

Comparisons of the present shoreline position with shorelines derived from early US Coast and Geodetic Survey topographic maps reveal substantial changes. Shoreline change analysis for the Long Beach Peninsula shows that the shoreline has prograded on average 3.6 m yr^{-1} since the 1870's. The shoreline at Cape Disappointment had prograded most rapidly at 6 m yr^{-1} with diminishing rates northward over tens of kilometers alongshore. This alongshore gradient in progradation rate may be attributed to the construction of the Columbia River north jetty in 1917. The rates of average shoreline progradation for this region decrease from 4.9 m yr^{-1} during the period 1870's to 1926 to 4.1 m yr^{-1} during the period from 1926 to 1950's to 1.8 m yr^{-1} during the period from 1950's to 1995. In contrast with the earlier historical trend, the beach at Cape Disappointment has experienced erosion of up to 8 m yr^{-1} in recent years.

The historical shoreline change mapping and analysis from the 1870's to present is being combined with the barrier core sampling and dating techniques to compare historical and prehistoric shoreline changes. GPR is being used to map the barrier stratigraphy and analyze the development of the barrier-beaches over the last several thousand years. Initial results indicate an anomalous rate of accretion over the past 120 years compared with the late Holocene geological record. An example is shown in Figure 3 at the mid-section of Long Beach Peninsula, where shoreline progradation rates of 20 to 30 cm yr^{-1} over the past two thousand years is contrasted with historical rates near 4.8 m yr^{-1}. The reason for this sharp increase in shoreline progradation appears to be related to construction of jetties and potentially the redistribution of sediments from ebb-tidal deltas.

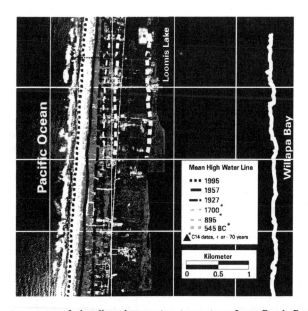

Figure 3. Long-term rates of shoreline change at a transect on Long Beach Peninsula. Shoreline progradation rates vary from 30 cm yr^{-1} during the period 545 BC-895 to 20 cm yr^{-1} during the period 895 - 1700 to 30 cm yr^{-1} during the period 1700 - 1927.   Rates of progradation then jump to 4.9 m yr^{-1} during 1927 - 1957 and 4.7 m yr^{-1} during 1957 - 1995.

Beach morphology monitoring is being implemented using RTK DGPS and a six-wheel drive all-terrain vehicle that collects beach position and elevation data along the coast. Parallel lines of data are collected to map beach surfaces over approximately 4 km reaches from low water to the toe of the primary dunes or bluff. Semi-annual surveys at several sites are being collected to document seasonal changes in beach morphology. Areas undergoing active erosion are being surveyed more regularly and after major storms to evaluate short-term variability.

Conceptual Model Development

Presently, a conceptual model of the Columbia River littoral cell is hypothesized whereby the supply of Columbia river sediment builds up the coastal barrier beaches from Tillamook Head, Oregon to Point Grenville, Washington. However, there is only a vague understanding of the time scale and pattern of sediment transport that leads to alongshore development of this region. Although it is expected that a decrease in sediment supply will slow or reverse the historical progradation of the coast, the time scale and spatial distribution of this effect is undetermined. It will be challenging to extrapolate the geological record as well as the historical change of geoindicators to predict the behaviour of this large-scale coastal system. For example, it appears that the most rapid shoreline progradation has occurred during a time of declining sediment supply, therefore lag times, anthropogenic impacts, and natural variability are important factors to be distinguished. The analysis approach in this study is conceptual understanding, model formulation, and finally the prediction of coastal evolution at scales of decades and tens of kilometers.

A framework for understanding and predicting large-scale coastal behaviour (LSCB) is provided in Figure 4. There are many classifications of analysis and modelling approaches to LSCB (e.g. Capobianco et al., 1993; Cowell and Thom, 1994). These classifications are considered here within the framework of three general categories, i.e. conceptual, quantitative, and combined analysis approaches. Each of these approaches have attributes that collectively form a process to acquire knowledge of LSCB. Regardless of the approach taken, the end-objective is to formulate a model that successfully explains the integrated fluid dynamics, sediment transport, and large-scale morphology in order to make medium-term and long-term predictions of coastal evolution. The model must be developed sufficiently to account for continuous morphodynamic processes as well as short-term variations and episodic extreme events. The development of a refined conceptual model is currently underway to constrain LSCB with boundary conditions, and provide insights to potential quantitative approaches that can evaluate the sensitivity of LSCB to fluctuations of various parameters. If the range of variability of the coastal system can be identified, then it should be possible to assess the susceptibility of the coast to adverse changes at management scale. A long-term monitoring program has been initiated to calibrate and refine modelling efforts to improve predictive capabilities and to enhance societal response to large-scale coastal changes.

Large Scale Coastal Behaviour Analysis

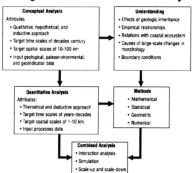

Figure 4. Categories and framework for performing LSCB analysis. **The conceptual approach provides the boundary conditions that constrain the evolution of the system. The quantitative approach employs modelling methods to refine conceptual understanding. The combined approach involves scaling-up through integration of physics-based processes and scaling-down through observational data and empirical relationships to simulate coastal behaviour.**

An example conceptual model is provided in Figure 5. This model is being used to develop sediment budgets and fluxes throughout the coastal sedimentary system. It is designed to incorporate short-term and event-scale fluxes that can be aggregated to obtain net fluxes at the decadal scale. Initial estimates of unit volumes are based on stratigraphic and topographic data that are inherently larger scale.

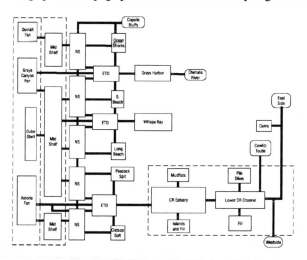

Figure 5. Conceptual model of the Columbia River sediment budget. Each box represents a sub-regional unit within the system that is connected with adjacent units by a limited number of transport pathways. (Figure courtesy of C. R. Sherwood)

Questions

A few of the most important questions fundamental questions that need to be addressed in this study include:

- How can a quantitative sediment budget be used to predict shoreline position?
- Can anthropogenic and natural influences on coastal change be distinguished?
- What are the effects of climate variability?
- Should the effects of long recurrence interval subsidence events be included in management scale predictive models?

As discussed above, a principal goal of the study is to predict coastal evolution at management scale, however, the shoreline appears to have prograded more slowly during prehistoric time when the sediment budget is assumed to have been much larger. Therefore, anthropogenic factors, such as jetty and dam construction, and the introduction of non-native dune grasses (Seabloom and Wiedemann, 1994) may be very significant in influencing historical sediment transport and accumulation patterns. Alternatively, very large scale events such as earthquake-induced subsidence that occurs instantaneously at approximately 500 year intervals clearly results in major erosion episodes that could account for the slow shoreline progradation over prehistoric time. The question of whether the coast rebounds vertically from these events over time scales of decades or centuries, and how this rebound may vary spatially is significant and remains to be answered. The distribution, fate, and transport of enormous quantities of the nearshore sediment following a subsidence event is unknown. The last major subsidence event occurred about 300 years before present, and it may be possible that the accelerated historical accretion, perceived to be primarily influenced by jetty construction, could also be an effect of a long-term recovery from the last major episodic subsidence and erosion event. Overlying these major questions are climate variability and many unknowns regarding processes-response relationships.

Despite these questions and lack of previous scientific investigation, the Columbia River littoral cell is situated where a variety of observational data can be collected to constrain the depositional history and enable stratigraphic reconstruction and interpretation verified by combined analysis and modelling. Nittrouer and Sternberg (1981) suggest that due to the distinct sediment accumulation patterns on the Washington continental shelf, the region is conducive to developing models of sedimentary strata formation. The use of GPR and radiocarbon dating of dune ridges is providing valuable data to interpret the Holocene development of barrier plains and the effects of major subsidence and erosion events (Jol et al., 1996). Various remote sensing techniques and field data collection programs such as sidescan-sonar, high-resolution seismic, bathymetric, and topographic surveys, drilling, and coring are expected to provide the essential sedimentary and stratigraphic data to understand and formulate prediction of LSCB for this system. GPS surveys and video imaging stations that are currently mapping short-term fluctuations in beach morphology over large spatial scales will help quantify coastal response to various hydrodynamic forcing conditions. The integration of these kinds of diverse data sets will be accomplished through conceptual model development, exploratory quantitative analysis, and combined scaling-up and scaling-down approaches. Continued monitoring and modelling programs are being pursued to advance management scale predictive capabilities of the behaviour of the Columbia River littoral cell.

Acknowledgements

This work was supported by the USGS and the Washington DOE. Additional support was provided by FEMA and NOAA. The participation and support of the local officials and citizens of Grays Harbor, Pacific, and Clatsop Counties is greatly appreciated.   Special thanks to Richard Daniels, Bob Huxford, Diana McCandless, Tim Schlender, and Brian Voigt who helped with the preparation of the manuscript.

References

Atwater, B.F., Nelson, A.R., Clague, J.J., Carver, G.A., Yamaguchi, D.K., Bobrowsky, P.T., Bourgeois, J., Darienzo, M.E., Grant, W.C., Hemphill-Haley, E., Kelsey, H.M., Jacoby, G.C., Nichenko, S.P., Palmer, S.P., Peterson, C.D., and Reinhart, M.A. 1995. Summary of coastal geological evidence for past great earthquakes at the Cascadia subduction zone, *Earthquake Spectra*, 11, pp. 1-18.

Burch, T .L. and Sherwood, C.R. 1992. Historical bathymetric changes near the entrance of Grays Harbor, Washington, report PNL-8414 prepared by the Battelle/Marine Science Laboratory for the US Army Corps of Engineers-Seattle District, 52 pp.

Capobianco, M., De Vriend, H.J., Nicholls, R.J., and Stive, M.J.F. 1993. Long term evolution of coastal morphology and its effects on the coastal environment. *Proceedings: 1st International Conference on the Mediterranean Coastal Environment*, pp. 889-903.

Cowell, P.J. and Thom, B.G. 1994. Morphodynamics of coastal evolution, In Coastal Evolution: Late quaternary shoreline morphodynamics, ed. R.W.G. Carter, and C.D. Woodroffe, Cambridge University Press, 517 pp.

Jol, H.M., Smith, D.G., and Meyers, R.A. 1996. Digital ground penetrating radar: an improved and very effective geophysical tool for studying modern coastal barriers (examples for the Atlantic, Gulf, and Pacific coasts, USA), *Journal of Coastal Research*, 12 (4), pp. 960-968.

Meyers, R.A., Smith, D.G., Jol, H.M., and Peterson, C.D. 1996. Evidence for eight great earthquake-subsidence events detected with ground-penetrating radar, Willapa Barrier, Washington, *Geology*, 24(2), pp. 99-102.

Nittrouer, C.A. and Sternberg, R.W. 1981. The formation of sedimentary strata in an allochthonous shelf environment: the Washington continental shelf, *Marine Geology*, 42, pp. 201-232.

Peterson, C.D., Darienzo, M.E., Pettit, D.J., Jackson, P.L., and Rosenfeld, C.L. 1991. Littoral-cell development in the convergent Cascadia margin of the Pacific Northwest, USA, SEPM Special Publication No. 46, From Shoreline to Abyss, p. 17-34.

Seabloom, E.W. and Wiedemann, A.M. 1994. Distribution and effects of Ammophila breviligulata fern (American beachgrass) on the foredunes of the Washington coast, *Journal of Coastal Research*, 10, pp. 178-188.

Sherwood, C.R., D.A. Jay, R.B. Harvey, P. Hamilton, and C.A. Simenstad. 1990. Historical changes in the Columbia River Estuary, *Progress in Oceanography*, 25, pp. 299-352.

Sternberg, R.W. 1986. Transport and accumulation of river-derived sediment on the Washington continental shelf, USA, *Journal of the Geological Society*, London, 143, pp. 945-956.

Terich, T. and Levenseller, T. 1986. The severe erosion of Cape Shoalwater, Washington, *Journal of Coastal Research*, Vol. 2, No. 4, pp. 465-477.

Wright, L.D. and Nittrouer, C.A. 1995. Dispersal of river sediments in coastal seas: six contrasting cases, *Estuaries*, 18(3), pp. 494-508.

Coastal Dynamics Applied to the Prediction of Beach Evolution: El Milagro Beach

J. Galofré [1], F. J. Montoya [2] and R. Medina [3]

Abstract

El Milagro beach (Spain). It is a 800 m long beach. Two nourishment works were carried out in 1986 and 1993, ten surveys have been carried out since 1986. A marina construction in 1995 has changed its stability.

Technical analysis of field data and a set of models are used in order to evaluated the coastal dynamics in the area and to understand how beach behaviour change and the prediction of beach evolution.

Introduction

The analysis of beach evolution, as a part of an integrated coastal zone management plan, is used as a coastal engineering tool in order to guarantee the functions of the beach: energy dissipation mechanism and useful free space for everyone. Coastal dynamics are needed in order to understand how the beach change.

The analysis of beach behaviour involves several processes that must be analysed in different space and time scales. Beach evolution can be studied in three time scale of variability: short term, less than 15 days, middle term, 15 days to 6 month and long term, years. The space scales, measured on shoreline length, that it can be divided are: microscale, less than 100 m, mesoscale, from 100 m to 10 km, and macroscale, more than 10 km. This case study are included on the meso space

1) Coastal Civil Engineer. Tarragona Coastal Service. Environment Ministry. Pl.Imperial Tarraco 4-4ª. 43005 Tarragona, Spain.
2) Coastal Civil Engineer. Head of Tarragona Coastal Service. Environment Ministry. Pl. Imperial Tarraco 4-4ª. 43005 Tarragona, Spain.
3) Associated professor. Ocean & Coastal Research Group. Universidad de Cantabria. Avda. de los Castros s/n. Santander 39005. Spain.

scale, the long-term is the timescale in which beach evolution is desirable. Unfortunately some processes and responses are often three-dimensional and only can be studied in the other two time scales, from these others it is possible to infer long-term behaviour. Important aspects of the coastal behaviour can be understood and predict on the bases of lower dimensional models. These models take advantage of the circumstance that the response of a beach often exhibits a different behaviour with essentially different length scales in three mutually orthogonal space directions (vertical, cross-shore and longshore), De Vriend (1992).

In order to understand beach performance it is necessary to analyse the beach forcings that are acting on the beach, these are the cause of all the processes. The beach responses are the consequence of the beach forcings, they are the phenomena that appear along the coast. Beach behaviour implies many methods and models working with the forcings and responses give a comprehension of beach performance. To improve predictive models for beach forcings and responses a wide range of them are available and it is necessary to select the most suitable in other to know beach evolution, Galofré et al (1995).

The goal of these paper is the analysis of coastal dynamics involved on beach behaviour, applied to prediction of beach evolution. This analysis must be made with the next methodology propounded:

- Morphological description
- Historical review
- Wave climate
    . Wave distribution
    . Sea and swell limits
    . Wave propagation
    . Wave driven current
- Morphodynamics
    . Shoreline evolution
    . Plantform analysis
    . Longshore transport
    . Profile analysis
- Diagnosis

This methodology will be shown applied to a case study, El Milagro Beach, in which a marina construction changed boundary conditions and its stability.

Case study description

The site of the field study is Tarragona , see Figure 1, a city located 100 km south of Barcelona, in Catalonia, on the Mediterranean coast of Spain. The morphological description and historical review is include in this section.

El Milagro is a half-opened beach located between El Milagro cape on the
east and Tarragona Harbour breakwater on the west. This harbour appeared from the
Roman time more than twenty centuries ago. This beach is graphed in several maps
from the beginning on the ninth century, that means that has been a stable beach for a
long time. The most recently history begins on 1986 when a nourishment work were
carried out consisting on 140.000 m^3. In 1994 a new nourishment work were carried
out, consisting on 165.000 m^3, and a new marina was built during 1995. The native
beach sand had a mean diameter had $D_{50}= 0.2$ mm and the borrowed sand $D_{50}= 0.6$ to
0.8 mm. The beach profile slope change from 2.0 % on the east, El Milagro Cape, to
4.0 % on the west, Harbour breakwater, from the shoreline to bathymetric - 5.0 m.
This value is considered the profile closure depth. Two monitoring programs were
carried out, one from the first nourishment project on 1986 to 1988, six bathymetric
surveys were made (jun-86, mar-87, sep-87, oct-87, jan-88, nov-93), and other from
the second nourishment project to 1997, six bathymetric and three sediment surveys
were made (aug-94, jun-95, feb-96, apr-96, oct-96,mar-97). In 1995 a new marina
was built and the data from the monitoring program can be used in order to evaluate
the change that has been produced. The nourishment works was done because of the
beach area reduction as a consequence sediment, brought by a stream in the middle
of the beach, reduction due to urbanisation work in the surroundings. This has been
theorised as a major factor in the erosion that has been witnesses in El Malign beach.

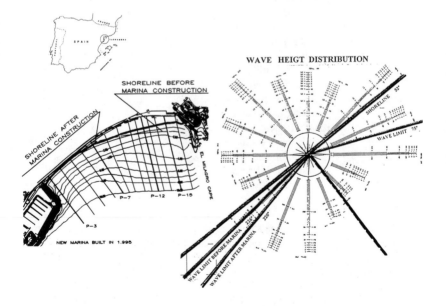

Figure 1. Site Location and Bathymetic Map   Figure 2. Wave Climate

Wave climate

There are two predominant directions of waves approach: SW and E. More than three quarter of deep water waves approach El Milagro beach from those sectors. The annual average significant wave height is about 0.5 m with typical winter storm waves of H_s of about 3.0 m. Tides at El Milagro are negligible. In figure 2 a visual wave distribution is made and the affected area is show, wave limits are defined before and after marina construction. The north limit is the same in both cases but the south limit change wit the marina construction, and the energy resultant is different.

In order to analyse the influence of the marina in coastal dynamics, that are responsible of beach stability. Qualitative and quantitative wave models calibration and verification have been made. Qualitative results have been obtained studying beach dynamics and comparing with aerial pictures and topobathymetrics. Wave propagation and wave driven currents models have been used considering pre and post marina construction behaviour. Different wave heights, periods and directions were used applying the REFDIF and COPLA programs, from Kirby and Dalrymple (1983), computed by the parabolic wave propagation model that combines refraction and diffraction phenomena. Figure 3 shows the wave induced currents determined from the wave field.

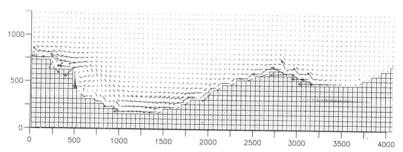

$H = 1.0$ m, $\alpha = -45°$, $T = 10$ s

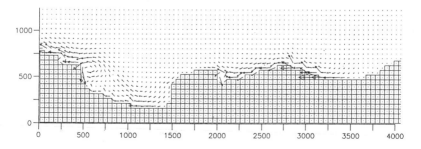

$H = 1.0$ m, $\alpha = -45°$, $T = 10$ s

Figure 3. Wave Induced Currents Before and After Marina Construction

It can be seen that the breaking wave induced currents direction decrease in the boundary of marina. It means that the sand in the west part of the marina is retained and accumulate in this area. The storms from the south-east transport the sand to this area and the storms from the south-west part can not return the sand to the beach, the dynamics in the area change the behaviour of the beach. Figure 4 shows the qualitative behaviour of the beach before and after the marina construction deduced from the wave climate models.

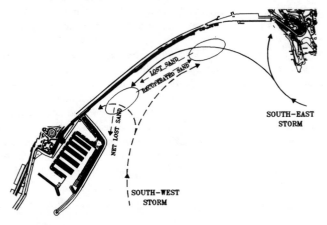

Figure 4. Beach analysis after marina construction

The recuperated sand, from the south-west storms is minor after the marina construction than before, these sand is accumulated on the shadow area of the marina.

Morphodynamics

Topobathymetric data are a good information in order to analyse morphodinamic characteristics, to understand beach behaviour and to predict beach evolution. Shoreline evolution, plantform analysis, longshore transport and profile analysis must be studied in this section.

On Table 1 a summary of events that have incidence on beach performance is shown. The most representative dates of beach incidences, the beach characteristics (including beach monitoring surveys, beach nourishment works and marina construction), beach area evolution measuring beach surface from the shoreline to landward limits, the sand volume lost/win (lost between monitoring surveys comparing the areas in each monitoring with the first one after nourishment works and win from filling projects), the longshore transport ratio deduced from the data obtained before and the grain size characteristic, can be found in it.

Date	Beach Characteristic	Beach area (m²)	Sand volume lost/win (m³)	Transport ratio (m³/year)	Grain size (mm)
jun-86	monitoring	15.069			0.2
nov-86	nourishment		+140.000		
jan-87	works		(5.8 m³/m²)		0.8
mar-87	monitoring	39.200			
sep-87	monitoring	33.085	-35.067	70.934	
oct-87	monitoring	27.170	-69.774	119.613	
jan-88	monitoring	31.299	-45.826	54.991	
nov-93	monitoring	23.596	-90.503	13.575	0.6-0.2
jan-94	nourishment		+165.000		
may-94	works		(6.8 m³/m²)		0.8
aug-94	monitoring	47.942			0.8-0.2
sep-94	marina				
may-95	construction				
jun-95	monitoring	45.784	-11.621	13.945	
feb-96	monitoring	45.952	-13.538	9.026	0.6-0.2
apr-96	monitoring	41.99	-40.412	24.247	0.6-0.2
oct-96	monitoring	40.113	-53.244	24.574	
mar-97	monitoring	39.800	-55.366	21.432	

Table 1. Historical beach evolution

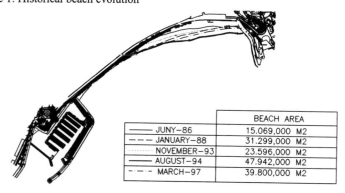

	BEACH AREA
———— JUNY−86	15.069,000 M2
− − − JANUARY−88	31.299,000 M2
·········· NOVEMBER−93	23.596,000 M2
———— AUGUST−94	47.942,000 M2
− · − MARCH−97	39.800,000 M2

Figure 5. Shoreline evolution

Figure 5 shows shoreline evolution taken from the most representative topobathymetric surveys, it is shown five shoreline position (jun-86, jan-88, nov-93, aug-94, mar-97) from the twelve it were taken. They represent the most significant date on beach behaviour the first one in before the first nourishment work, the second is after this works, the third is before the second nourishment works, the forth is afterthis works and before the marina construction and the fifth is after marina construction.

In Figure 6 the analysis of beach area evolution is made. It is compared the relation between the relation A_i / A_1 , A_i is the area corresponding to the i topobathymetric survey and A_1 is the corresponding to the post nourishment work, to the time. The figure shows the evolution before and after marina construction, A_1 in the first case correspond to march-87 and in the second case august-94. It can be inferred from this that comparing the graphic the medium slope is bigger after than before marina construction. It means that the erosion ratio, in El Milagro Beach, has been increasing since marina construction. A new beach appeared on the protected marina area, it is difficult to infer from the actual data about its stability, some consideration to profile slope must be considered in this analysis.

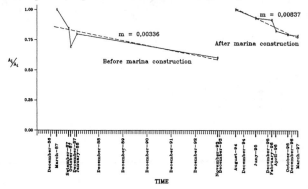

Figure 6. Beach area evolution

The analysis of beach profile must be made comparing different profiles in the beach and each one in different topobathymetric surveys. The most interesting aspect in this case is how the natural bottom submerged slope change from the east to the west part. In the morphological description it can be seen how this slope change from the 2 % on the east to 4 % on the west. In Figure 7 an overview of this changes is shown, four profiles have been selected for this analysis. Figure 7.a give a location map of the situation of this profiles, the profiles are numerated from the west to the east the analysis will be made in the opposite order from profile P-15 to P-3, intermiddle profiles will be studied too. Profiles P-15 and P-12 have similar behaviour profile changes has a range of variability from beach landward to bathymetric -5.0 m. The natural slope is around 2 % and beach nourishment works affect the slope of beach face and affect until bathymetric -2.5. Nourishment works were made from profile P-16 to P-7, it means that this profiles have the direct influence of the borrowed sand filling. The analysis of the three last bathymetrics give some information, after nourishment works a topobathymetric survey were made, august-94, in these profiles the shoreline has an accretion compared with prenourishment shoreline, november-93, but the survey that was made in march-97 an erosion phenomena occurs in both in P-15 shoreline rise more than the situation in November-93 and in profile P-12 was in front of it. It means that the shoreline

position is changing decreasing on the east part and accreting in the west.
The analysis of P-7 and P-3 show the sand accumulation that occurs in this

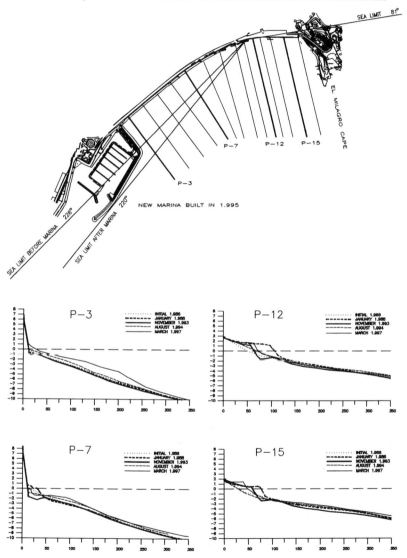

Figure 7. Beach profile location map and beach profiles evolution.

area this phenomena is specially important on P-3. In this point sand accumulation is really significant, but the profile form with borrowed sand is not compatible with the natural bottom. It appears two peaks in the profile that broke the equilibrium profile, and the sand is lost on the foot of the profile.

Diagnosis

The diagnosis of these beach is that marina construction change beach behaviour because less waves from the west side arrive and some sand are retained in the protected area of the marina. Beach profile in this area is not stable because of different bottom slope of natural beach and the profile formed by the borrowed sand.

Conclusion

1.- The analysis of beach evolution must be done with a methodology that includes:
              . Historical review
              . Morphology
              . Wave climate
              . Morphidynamics
              . Diagnosis

2.- Coastal dynamics in the methodology are included in wave climate and morphodynamics.

3.- The models that can be used on coastal dynamics must be chosen carefully, hypothesis and field data are the model bases.

4.- The case study shows how a marina construction changed the coastal dynamics in this area, historical evolution and the monitoring program that was carried out from 1986 to 1997 help to understand the coastal dynamics involved in beach evolution

5.- Computer models used in order to analyse coastal dynamics, wave propagation field and wave induced currents have been calculated using REFDIF and COPLA programs. Because of the difference on beach profile slope, from 2.0 % to 4.0 %, a one line model is not the most suitable in this case.

5.- The analysis of emerged beach area are agree with the interpretation from computer models. Comparing the evolution before and after marina construction it can be inferred that the erosion ratio is greater after than before it. It is necessary more data in order to calculate the value accurately.

6.- Profile equilibrium form and beach bottom slope are not compatible in the protected area of the marina. In this area some sand is accumulated from the south east storm and the south west storms can not return this to the beach. This sand has an equilibrium profile form that is not compatible with the natural bottom slope.

References

- Galofré J., Montoya F.J. and Medina R., 1995. "Study of the Evolution of a Beach Nourishment Project Based on Computer Models". Computer Modeling of Seas and Coastal Regions II. Computational Mechanics Publication, pp. 249-256.

- Kirby J.T. and Dalrymple R.A. 1983. " A Parabolic Equation for the Combined Refraction-Diffraction of Stokes Waves by Mildly Varying Topography" Journal of Fluid Mechanics vol 136, pp. 453-466.

- de Vriend, H.J., 1992. "Mathematic Modelling of 3D Coastal Morphology". Proc. Short Course on Design and Reliability of Coastal Structure. 23rd I.C.C.E. Venecia, Chapter 1 pp. 10-25

EVALUATION OF LONG TERM DUNE RECESSION DATA

Hans-H. Dette[1], M.ASCE

ABSTRACT The recession of dunes and cliffs along the west coast of Sylt is known to vary significantly in space and time along the coastline. The dunes, cliffs and the beaches from the dune's crest to low water have been surveyed at distances of 500 m (= approx. 70 profiles) since 1870 but at considerable time intervals and not always directly after severe storm surges. The number of existing surveys of individual profiles between 1870 and 1984 varies between 12 to 18.

It has been shown that the storm surge activity in the North Sea has increased appreciably since 1950 compared to the first half of the century. This affects both the occurrence of peak storm surge levels and the residence time of raised water level above a certain sensitive horizon. As a consequence the dune and cliff erosion on the west coast of Sylt became critical and repeated beach nourishment schemes to balance the yearly losses were commenced in 1985. The analysis of the data shows a direct long term relationship between environmental impact and coastline recession.

INTRODUCTION

The design life of man-made coastal structures and even that of regular beach re-nourishment schemes is typically decades. The ability to predict the changes of coastal morphology over these time spans is therefore an important requirement. Knowledge of coastal behaviour on this time scale is limited both in respect of the driving forces as well as the nature of long term processes. The analysis of long term

---

[1] Academic Director, Department of Hydrodynamics and Coastal Engineering, Leichtweiss-Institute for Hydraulics, Technical University Braunschweig, Beethovenstr. 51a, 38106 Braunschweig, Germany.

morphological data sets is a key element in the understanding and modelling of long term coastal morphodynamics (SOUTHGATE and CAPOBIANCO, 1997). The data analysis allows for practical engineering purpose the forecasting (estimation) of expected recession over a period of decades if it is assumed that the environmental parameters (water levels and waves) remain unchanged.

RELATIONS BETWEEN WATER LEVELS, BEACH RECESSION AND DUNE EROSION

It is common to all receding coastlines that beach and cliff - irrespective whether native beach material or that from beach nourishment is involved - retreat landward in form of an overall profile shift of a characteristic longterm mean profile. The longterm behaviour of such a coastline related to the applied forces is illustrated in Fig. 1 and 2.

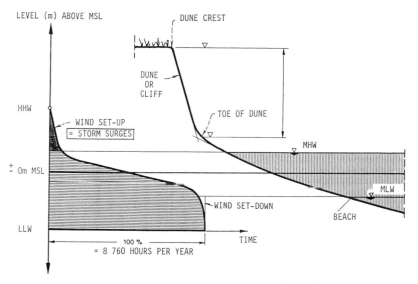

Fig. 1: Relations between beach, dune and water level (FÜHRBÖTER and DETTE, 1986)

Fig. 1 shows a dune and beach relative to the MHW- and MLW-line, above MHW is the "dry" beach, which is subject only to raised water levels (storm surges), wave run-up and longshore current. Above that level is the toe of the dune. The toe elevation at each beach is a characteristic parameter for the local beach conditions and is related to the predominant severe wave conditions. The position of the toe cannot always be located clearly because the eolian sand transport can in the meantime lead to a sand accumulation in front of the dune. However, after a storm surge the toe is well defined. Finally the crest of the dune marks the boundary to which the recession has proceeded but since the dune shapes vary the plan view of the crest is often irregular.

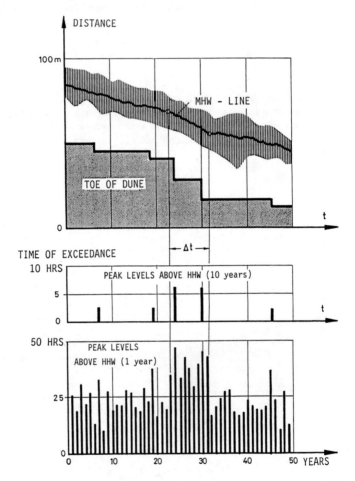

Fig. 2:   Continuous beach recession and discontinuous (stepwise) dune erosion (FÜHRBÖTER and DETTE, 1986)

The idealized profile in Fig. 1 is related to the mean annual water level curve. The latter indicates for each level the hours in the year (or in percentage of time) the beach is subject to the given water level, associated attack by waves and wave-induced currents. This curve in an approximation to an inverted sine function starting with values of 100 % = 8760 hours/year (with exemption of those events due to wind set-down below the MLW-line), reaches the value 50 % at about the MSL-line and decreases rapidly towards values of only a few percent above the MHW-line. The beach above MHW is reached by waves only during storm surges. It is a function of the elevation of the toe of the dune or cliff (and also on the width and height of the beach in front) when wave attack leads to dune erosion. The water level frequency curve indicates the average number of hours during the year that the toe is reached. It is apparent that the time is limited to only few hours in the year but can be very different from year to year according to the storm surge frequencies.

In the schematic illustration in Fig. 2 (below) the water level HW (1 year) is marked as that level which in a time series (here of 50 years) is reached or exceeded in each individual year. The duration of exceedance (residence time) as that number of hours per year during which the water level during a storm surge stays above that level can be regarded as a measure for the overall impacts on the whole beach. Moreover it is assumed, that a dune erosion occurs only when the still water level exceeds a level which in the average is only reached or exceeded e.g. once in 10 years. This water level is called HW (10 years). The number of hours, during which within one year this level is exceeded can be considered as a measure for the intensity of dune erosion. Fig. 2 also illustrates how beach and dune react to the storm surges over a longer time span. The HW line and its recession (distance) are regarded as typical for the development of the beach and the landward retreat of the toe of the dune, i.e. a measure for dune erosion.

Contrary to beach changes, which in short-term sense can be positive or negative and only in long term trend show a negative trend, the resultant changes at a cliff or dune are only negative and their characteristics are that they appear as discontinuities (stepwise recession), due to certain storm surges. The continuous negative beach changes are superimposed as a "white noise".

The differences in the dune recession due to equal storm surge parameters (water level, residence time and waves) and equal beach conditions occur due to the special conditions in front of the toe. During favourable seasonal weather conditions and even over years, with negligible storm surge activity eolian transport can build a sand depot in front of the dune. The waves of a storm surge have to remove this sand first before dune erosion is initiated and the damage to the dune can be considerably reduced.

ANALYSIS OF LONG TERM RECESSION

The Island of Sylt/North Sea which is a typical mesotidal receding coastline has been displaced more than 10 kilometers eastward over the last 7000 years. From plane-table surveys in 1793 and 1929 a mean annual recession of approx. 1 m over more than 130 years can be estimated (Fig. 3). The magnitude of recession of less than 100 m and more than 500 m during this time along the 40 km long coastline also illustrates a significant variability in space. This observation is more or less characteristic for all receding coasts.

Since 1870 comprehensive data sets on dune recession along the west coast of Sylt are available. Level surveys were carried out from the dune crest to the LW-line at spacing of 500 m. Fig. 4 shows the locations of the individual profiles between 35 S at the south and 35 N at the north end of the west coast. Unfortunately the dune surveys were carried out in relatively large time intervals and not always in connection with certain storm surges. Thus it is not possible to relate recessions to the storm surges (Fig. 2) in the period before 1950. Fig. 5 illustrates selected profile developments between 1886 till 1956 (LAMPRECHT, 1957).

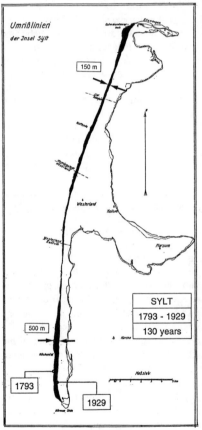

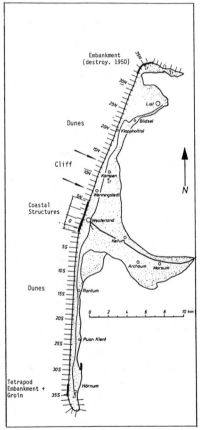

Fig. 3:   Long term recession of the west coast of the Island Sylt (MÜLLER and FISCHER, 1938)

Fig. 4:   Location of individual profiles (35 S till 35 N)

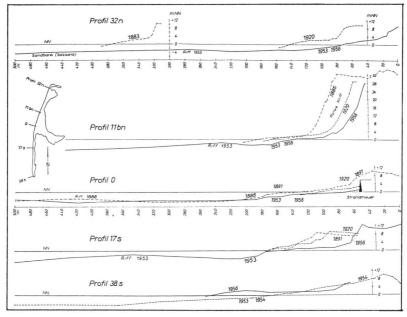

Fig. 5: Selected dune recession profiles from the west coast of Sylt (LAMPRECHT, 1957)

It can be assumed, that in case of dune recession the distance between crest and toe (Fig. 1) is maintained. So it can be expected that the analysis of longer time series results in similar recession rates for both parameters. By analysing the toe and crest position from all profile data the results provide a certain check on the survey i.e. both indicator points should lead to comparable results.

Another question has to be raised on the time dependent resolution of the data. In case of the recession behaviour along the west coast of Sylt it had been observed that with the increase in storm surge activity after 1950 an increase in recession was also present. The question is which time span for the apparent changes in meteorological conditions and coupled increased storm surges can be selected for an extrapolation in time. A detailed study of all the available data showed that for the determination of the average recession rate, the data should be divided in two periods:

- First measurements (1870) till 1950 resp. 1952
- 1952 till last levelling of undisturbed conditions in 1984 (before start of regular beach and dune nourishments). The rounded off data series (1950 - 1984) of 35 years could be used to extrapolate over the same time span to the year 2020.

Finally it has to be decided how the data points shall be analyzed. The first method is just to connect the initial and final point of a series, and to find from the slope of the line the average recession for that period. This method is called 'virtual regression' (Fig.6). Another possibility is the method of least mean squares applied to all data points of a period ('linear regression', Fig. 6).

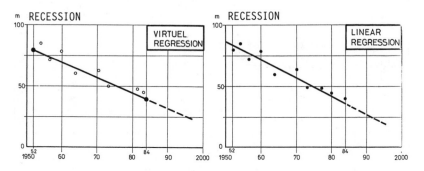

Fig. 6:   Methods for data analysis, virtual regression (left) and linear regression (right)

ANALYSIS OF LONG TERM DATA FROM SYLT

The methods of "virtual" and "linear" regression were applied for the profile data from ALW Husum and the recessions before and after 1950 determined (Fig. 7). In Table 1 the results of data analysis are summarized by determining the average recessions in 4 typical sections along the 35 km long coastline. Both methods, which were applied as well to the crest as to the toe data, show reasonably good agreement. From all plots it is evident that after 1950 the dune erosion has increased, especially for the middle part of the island (22 S till 22 N). It is apparent from Table 1 that the recession in the years since 1950 compared to the previous year has nearly doubled, this is in good agreement with the increase of storm surge activity represented by the cumulative annual residence time of raised water level above MHW + 1.2 m (Fig. 8).

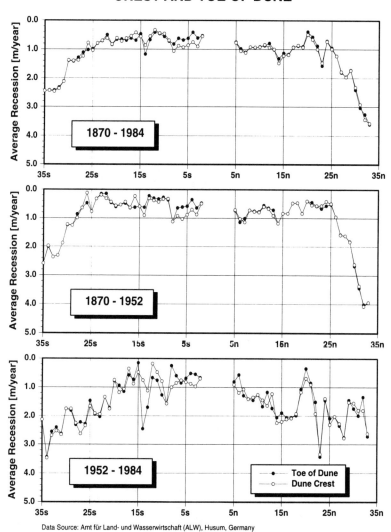

Fig. 7: Analysis of dune crest and toe data for overall period (1870 - 1984), and split periods (1870 - 1952 and 1952 - 1984), here: average recession from a baseline in m/year

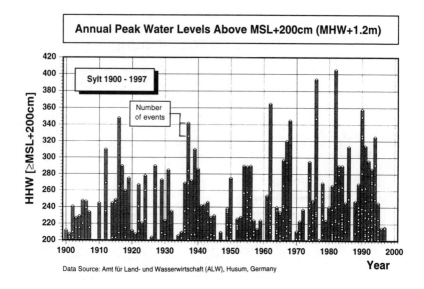

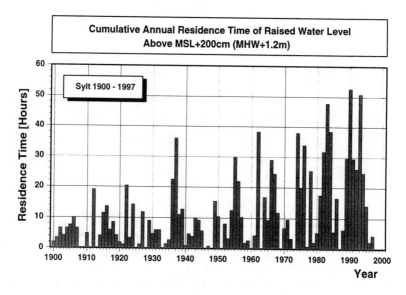

Fig. 8: Increase in storm surge activities in the period since 1950, here: yearly highest high water levels and number of yearly events above MHW + 1.2 m

Table 1:   Results of analysis based on virtual and regression method applied to dune crest and toe, here: average recessions with sections of coastline

Parameter	Regression Method	Average recession (m/year)			
		(35 S - 23 S) Hörnum	(22 S - 2 S) Rantum - Westerland	(5 N - 22 N) Westerland - Kampen	(23 N - 33 N) List - Weststrand
Crest	· virtual	1.2	0.4	0.7	1.9
	linear	1.3	0.4	0.7	1.9
Toe	virtual	1.3	0.4	0.7	1.9
	linear	1.3	0.4	0.8	1.9
Average	**1870 - 1952**	1.3	**0.4**	**0.8**	1.9
Crest	virtual	2.3	0.9	1.4	2.1
	linear	2.0	0.8	1.3	1.8
Toe	virtual	2.3	0.9	1.5	2.1
	linear	2.1	0.9	1.4	1.9
Average	**1952 - 1984**	2.2	**0.9**	**1.4**	2.1

ACKNOWLEDGEMENT

The study was made possible by the support of 'Amt für Land- und Wasserwirtschaft Husum' (Coastal Authority Husum) by giving access to archive material and allowing the results of data assessment to be published. The work was carried out in the framework of SAFE research program. It is funded by the Commission of the European Communities, Directorate General for Science, research and Development, under contract no. MAS3-CT95-0004.

REFERENCES

Führböter, A. and Dette, H. H. Zur Entwicklung der Düne Helgoland, Die Küste, Heft 43, S. 47 - 114

Lamprecht, H. O. (1957) Uferveränderungen und Küstenschutz auf Sylt. Die Küste, Jahrg. 6, Heft 2, S. 39 - 93

Müller, F. and Fischer, O. (1938) Das Wasserwesen an der schleswig-holsteinischen Nordseeküste, 2. Teil, Die Inseln, 7. Sylt, Verlag von Dietrich Reimer, Adrews & Steiner, Berlin, S. 1 - 301.

Southgate, H. N. and Capobianco, M. (1997) The Role of Chronology in Longterm Morphodynamics - Theory, Practice and Evidence, Abstracts, Int. Conf. on Coastal Research through Large Scale Experiments, Coastal Dynamics'97 Plymouth, UK.

Prediction of Aggregated-Scale Coastal Evolution

Huib J. de Vriend[1]

Abstract

The prediction of coastal behaviour at space and time scales which are much larger than the inherent dynamic scales of the system considered, is shown to be a non-trivial task, in spite of the recent increase of process knowledge and medium-term modelling expertise. The EU-funded project *Prediction of Aggregated-Scale Coastal Evolution* (PACE), in the framework of the Marine Science and Technology programme (MAST-III), has the objective to investigate and develop prediction methods for this large-scale behaviour. The paper summarizes the motivation of this work and outlines the approaches taken in the project.

Introduction

Coastal morphological changes take place at a variety of space and time scales, from small-scale ripple formation through to the evolution of entire coastal systems at historical and geological time scales (Fig. 1). On the other hand, the need to hindcast or predict these changes for practical purposes is usually restricted to a relatively narrow scale band: sedimentologists are interested in small-scale processes and the resulting stratification of deposits, coastal engineers are interested in scales which match those of the works they are involved in, and geologists are interested in the century- and millennium-scales at which large coastal systems change.
Yet, the modelling of coastal behaviour at this variety of scales is not a trivial task. The basic knowledge which has been incorporated in our present mathematical models concerns mainly small-scale processes (waves, currents, sediment transport) and their dynamic interaction with the bed topography at small and intermediate scales. To what extent such morphodynamic models can be integrated through to essentially larger scales is not yet clear. Data on larger-scale coastal changes, which are essential to asses the validity of these models, are scarce and

---

[1] Professor / Co-ordinator PACE-project, University of Twente, Civil Engineering & Management, P.O. Box 217, 7500 AE Enschede, The Netherlands

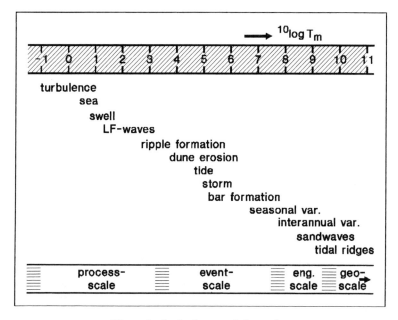

Figure 1. Scales in coastal dynamics

often inaccurate, and, by definition, it is too late to start monitoring at the moment the information is needed. Black-box empirical modelling techniques for data-abundant systems are therefore seldomly applicable here.

This leaves us with a number of research questions, such as:
- To what extent can medium-term models be used for long-term applications?
- How can these models be aggregated to models which operate at large scales?
- What comprehensive datasets on large-scale coastal behaviour are available?
- To what extent can these be used to derive empirical models of this behaviour?
- Are there alternative approaches of long-term modelling?

<u>Why not let the computer do the work?</u>

In recent years, significant progress has been made in the understanding and modelling of the physical processes on sandy coasts. This has led to modelling software and expertise which are now more or less ready for use (for instance, see Stive et al., 1995, and De Vriend, 1995). Although these models have been developed for medium-term applications - a few years of real-time simulation for coastal profile models, a few days of real-time simulation for coastal area models - extending their predictions to larger scales might be just a matter of computer power. If so, why should we bother about special large-scale models? There are a few reasons to have at least some doubts at this point.
(1)   The most practical reason is, that medium-term models work at the scale of the constituent processes, i.e. minutes or hours, and would need a prohibiti-

ve amount of computer time to cover decades or more. Brøker et al. (1996), for instance, describe an application of a 2-D depth-integrated medium-term coastal area model, the running of which took about half of the prototype time (Brøker, 1996, private comm.). In view of the rapidly increasing power of computers, however, this argument is bound to be temporary.

(2) Another reason for doubt is the possible build-up of errors in the numerical process. A long-term application with a medium-term model would require a very large number of time steps, in each of which some inaccuracy could build up (cf. De Vriend, 1987). If there are not enough "sinks" of inaccuracy, this will ultimately lead to nonsensical results. Whether or not this is the case remains to be investigated.

(3) Even if the computer power would be sufficient and the build-up of numerical errors would not be a problem, the results may be nonsensical. Medium-term process-oriented models are based on the description of processes which are important to small- and medium-term coastal behaviour. They have been validated against short- and medium-term data, usually concerning the coastal response to a major disturbance. To what extent the model is able to reproduce the equilibrium state is seldomly an issue, since the input conditions are changing all the time (equilibrium is a "moving target").

In the case of large-scale coastal behaviour, however, the system is continuously in a near-equilibrium state, i.e. the short-term processes almost balance and morphological changes are dominated by small residual effects, which are less important at smaller scales. One example is the bottom-slope effect on the sediment transport (sediment moves easier downhill than uphill). This is a minor effect in the instantaneous sediment motion, but it plays a major role in the morphological equlibrium state (cf. De Vriend et al., 1993a). Another example is the long-term behaviour of the lower shoreface and the inner shelf, in which relatively small 3-D tidal and wind-induced residual currents play a key role (cf. Wright et al., 1991). Many of these residual effects remain to be identified, and the present models cannot be claimed to include them all.

(4) A fourth reason not to rely exclusively on medium-term models refers to predictive-mode applications. Since the future input is inherently unpredictable, a predictive model requires a probabilistic approach, which involves a considerable increase of the number of runs. Even if a single long-term run would be feasible, the demand of computer time for this large number of runs would be problematic. Probabilistic approaches in morphodynamic modelling are still in an early stage of development and have hardly got beyond coastline models, but the first results seem to indicate that in some practical situations a simple probabilistic coastline model performs better than a sophisticated 3-D deterministic model (e.g. Reeve and Fleming, 1997). Probabilistic modelling is therefore worth pursuing in coastal morphology, be it with simple and aggregated models.

(5) Finally, theoretical arguments associated with the inherent uncertainty in coastal behaviour seem to favour aggregated large-scale models. This point will be elaborated in the following sections.

In summary, we can state that "letting the computer do the work" requires much further research before being a reliable approach. And, if it works, it will only do so under certain conditions. Therefore, this approach must not lead to the "black-box"-utilization these models, with the user is unable to explain the results and thus loses control.

Scale of interest vs. scale of phenomenon

A phenomenon can be viewed from different perspectives (scales of interest): the tidal stage, for example, is an extrinsic condition for small-scale processes, such as intra-wave sediment transport, whereas it is part of the dynamic process in the formation of tidal ridge, and noise in the coastal evolution at a geological scale. The latter does not mean that there are no net effects of the tide, but only that the time-variation of the tidal stage does not have to be described in detail in a model at this large scale.

Combined with the notion, that spatial and temporal scales are probably related (phenomena at larger spatial scales also involve longer time-scales), this leads to the conceptual picture of process-scale vs. scale of interest shown in Fig. 2.

Inherent uncertainty

Coastal hydrodynamics and sediment transport are strongly nonlinear, which means that the water and sediment motion may evolve into deterministic chaos.

A classical example is that of turbulence: even under a steady forcing, viscous flow evolves into chaotic motion, which is unpredictable in a deterministic sense[2]. Turbulent flow can be described with a high-resolution Navier-Stokes model, but every run of that model produces just one realization of the turbulent flow field. It would take many runs and a great effort to determine the mean flow. Therefore, a full Navier-stokes model is not a very efficient predictor of turbulence-averaged flows. This problem can be overcome by averaging the model *equations*, rather than the model *output*, over the turbulent motion. Because of the nonlinearity of the equations, this averaging operation yields residual terms (the Reynolds stress terms), which have to be related to mean flow properties (turbulence modelling) in order to close the model. This approach, which yields a deterministic mean-flow model, can be interpreted as stepping to a higher aggregation level, where the chaotic variation of the water motion is reduced to noise, though with a significant net effect.

Likewise, the motion of individual sediment grains is erratic and hence unpredictable in a deterministic sense. A model which describes the motion of individual grains is therefore not likely to be a very efficient transport predictor. In practice, this problem is overcome by modelling at a higher aggregation level, viz. that of the mean sediment flux. This has led to the well-known empirical transport formulae for all modes of transport and the diffusion-type transport models for suspended load transport. Once more, the problem of unpredictability is overcome by stepping to a higher aggregation level, where deterministic modelling is possible, again.

Under certain conditions, small-scale bedforms (e.g. ripples) develop on a mobile bed. Their pattern may seem regular at first sight, but it does exhibit irregularies. Although a formal proof is still lacking, bedform patterns must be expected to be unpredictable in a deterministic sense. Moreover, the prediction of individual

---

[2] Various models can claim to describe *statistical* properties of turbulence, such as the mean kinetic energy.

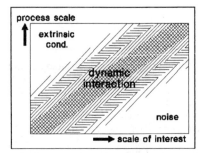

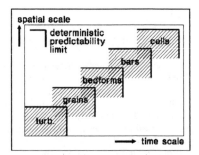

Figure 2.   Process-scales vs.   Figure 3.   Scale cascade
            scales of interest

bedforms is difficult, and models which might be able to do this in simple situations are laborious and time-consuming. In most cases, however, we are not interested in the bedforms, as such, but in their net effects on the water and sediment
motion, e.g. via the effective bed roughness. Therefore, (semi)-empirical bed
roughness predictors have been developed, instead of detailed morphodynamic
models at the bedform scale. This is another example of aggregation as a means to
overcome predictability limits, or to avoid inefficient sub-models.
Recent observations of nearshore bar systems (e.g. Lippmann et al., 1993) show
that these can jump from one dynamic state to another, probably under the impact
of major events. It looks as though the system is switching to another attractor,
which suggests that, under otherwise the same conditions, there can be more than
one dynamic state, each of which is more or less persistent under normal conditions. This seems to be in line with observations on the bar system along the
Holland coast (Wijnberg and Terwindt, 1995), where the occurrence of distinct
and sharply separated "provinces" in the bar behaviour cannot be related to spatial
variations of the input conditions. If bar systems can switch from one attractor to
another, their long-term behaviour will be unpredictable. Hence a morphodynamic
model which works at the time scale of the bar dynamics is probably not a very
efficient predictor of the coastal behaviour at scales of centuries or millennia.
Another example at this aggregation level refers to tidal inlets. Channels and
shoals in an inlet sytem evolve and migrate at a decadal scale, more or less
systematically, but not predictable in a deterministic sense. Resolving this
channel/shoal behaviour requires a sophisticated 3-D model (Wang et al., 1995).
So far, this type of model has not been shown to be able to reproduce the quasi-
equilibrium relationships between the tidal prism and e.g. the cross-sectional area
of the inlet, or the water volume in the embayment. In any case, the variability of
the channel/shoal pattern is not likely to play a key role in the evolution of the
system at a millennia-scale, and the efficiency of process-oriented 3-D medium-
term models to describe this evolution must therefore be doubted.
Fig. 3 shows the picture which arises from this: a scale cascade in space and
time, separated by aggregation steps which cope with limits of predictability or
computational efficiency. Its possible implication for the modelling strategy is
indicated in Figs. 4 and 5: one model which covers the entire scale range is not
likely to work, either because of the predictability limits which are inherent to the
system, or because of efficiency limits.

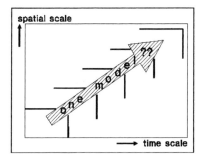

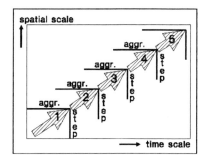

Figure 4. One model for all scales?   Figure 5. Or a cascade of models?

Research issues

In the light of the foregoing, large-scale modelling involves the following research issues:
(1) identification of the predictability horizons in coastal dynamics, via theoretical analysis, data analysis and possibly numerical simulations,
(2) identification of the limits of applicability of morphodynamic model concepts, from a predictability and/or an efficiency point of view,
(3) development and application of model aggregation techniques to overcome these limits, and
(4) development of a gamut of efficient models of large-scale coastal behaviour, each applicable within a certain scale range and to certain classes of phenomena.

The PACE-project

The EU-funded project *Prediction of Aggregated-Scale Coastal Evolution* (PACE), which runs from 1996 till 1999 in the framework of the Marine Science and Technology programme (MAST-III), addresses these issues for sandy coasts and attempts to develop and test prediction methods for large-scale coastal behaviour.
In order to limit the scope of work, the notion "large-scale" has been narrowed down to three scale-levels: (1) the interannual scale of nearshore bar behaviour, (2) the decadal scale, and (3) the century/millennium-scale. A separate sub-project on rhythmic features uses the more generic definition "at a much larger scale than that of the principal constituent processes". In the case of a nearshore bar system, for instance, this means that we attempt to predict the evolution of the bar *pattern*, rather than the behaviour of individual bars.
In the project, two different approaches are brought together:
- a "bottom-up" approach, which attempts to integrate small-scale process knowledge through to larger scales, and
- a "top-down" approach, which starts from observations and sticks to some basic conservation principles (e.g. sediment mass).
These approaches correspond more or less with what is sometimes called "upscaling" and "downscaling", or "deductive" and "inductive" modelling.

Bottom-up approach

The basis of the bottom-up approach is formed by a set of constituting equations
which describe the water/sediment motion and the sediment balance, at time-scales
of seconds to hours and at spatial scales of metres and less. As stated before, it is
not recommendable to take such small-scale model, put it on a big computer and
run it "blindly" in a long-term application. The reservations against this approach,
however, remain to be proven. Therefore, PACE is systematically investigating
these points, making use of the medium-term models developed in earlier MAST-
projects (cf. De Vriend et al., 1993a).

The project includes the development and tentative application of auxiliary
techniques to scale up small-scale process knowledge, and it attempts to form a
knowledge base to increase the robustness of "bottom-up" numerical model
utilization. One of these techniques is the analysis, linear and nonlinear, of the
inherent behaviour of known mathematical models. This concerns especially the
formation of rhythmic seabed features (free instabilities), such as tidal sandbanks,
sandwaves, shoreface bars, nearshore bars, and channel/shoal systems in tidal
inlets. Apart from explaining generation mechanisms, these analyses are expected
to lead to conclusions about pattern formation and predictability. Preliminary
results are described by Hulscher (1997) for storm ridges, Falqués et al. (1997)
for shoreface-connected ridges, Vittori et al. (1997) for cuspidal features, and
Schuttelaars and De Swart (1997) for channels and shoals in tidal inlets.

Keywords in PACE are reduction and aggregation. In the bottom-up context, they
are applied to models, i.e. sets of mathematical equations, as well as to their
stochastically varying inputs, i.e. time series of waves, tides, etc. De Vriend et al.
(1993) and Latteux (1995) describe examples of either approach. Since the system
is nonlinear, the effects of the chronology of forcing events (e.g. storms) deserve
special attention (e.g. Southgate and Capobianco, 1997; Millard et al., 1997).

Top-down approach

The basis of the top-down approach is formed by data and some general conservation principles, such as the conservation of sediment. In the case of nearshore bar
behaviour, video techniques provide a continuous and abundant flow of data from
various sites around the world (Holman, 1997). PACE includes research to
underpin this video technique, especially the translation from the observed
whiteness of the wave rollers to the actual seabed level (Aarninkhof and Plant,
1997). The video-data enable us to apply data analysis and data-driven modelling
techniques for data-abundant systems (e.g. Plant and Holman, 1997). PACE
involves the testing of a variety of these techniques, aiming at the identification of
chaotic or otherwise poorly predictable behaviour (e.g. Southgate and Beltran,
1995), and at empirical models of the interannual behaviour of nearshore bar
patterns (Wijnberg and Holman, 1997).

Another focus of research is the depth-of-closure concept. Given the time-scale of
interest, sediment exchange processes seem to be restricted to a zone near the
shore. If this is true, it would be a very important element in nearhore morphodynamic modelling, since it bounds the area over which the sediment budget is
closed (e.g. Nicholls and Birkemeier, 1997; Capobianco et al., 1997).

Scarcity of data, especially in the time domain, makes it usually impossible to
apply data-demanding techniques to larger-scale coastal behaviour. Purely

statistical extrapolation therefore seems to be not a very adequate modelling approach: the lack of data has to be compensated by physical knowledge. Yet, some datasets contain a wealth of spatial information. If it would be possible to find a sensible space-time transformation, there might be scope for a statistical approach. Novel data mapping and modelling techniques, such as PIP's and POP's (Hasselmann, 1993), are therefore being tested in PACE.

An alternative to stochastic extrapolation is "behaviour-oriented modelling", based on available knowledge and data, but not formally on small-scale process-descriptions. One example is the diffusion approach for decadal-scale coastal profile evolution, in fact a generalization of the cross-shore part of the n-line concept (De Vriend et al., 1993). This suggests that the approach can easily be extended with a longshore dimension, in order to include weak longshore variability, like in coastline models.

The one-line concept, which is probably the best established and robust modelling approach for large-scale coastal morphology at this moment, can also be considered as a behaviour-oriented model, in that it is not formally derived from a 3-D description of the constituent processes.

Another example of behaviour-oriented modelling is the hinged-panel approach, which was originally developed for large-scale coastal profile modelling (Stive and De Vriend, 1995), but which also allows for extension with a longshore dimension. An application of such an extended panel model to the Ebro Delta is described by Jiménez and Sánchez-Arcilla (1997).

Variations of this approach are also applied at scales of centuries and millennia (Niedoroda et al., 1995). PACE involves a special sub-project on very-large-scale modelling, in which this approach is compared with the hinged-panel approach, and with the geometrical approach developed by Cowell et al. (1995) for a number of reconstructed profile evolutions at geological scales. In order to have a better link with geologists, these models are made to produce the stratigraphy, in addition to space an time records of the seabed elevation. Auxiliary process-oriented research is done to improve the models at this point (Swift et al., 1997).

Prediction strategies

The above approaches lead to a range of possible prediction strategies for large-scale coastal behaviour, as summarized in Fig. 6.

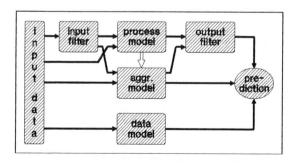

Figure 6. Prediction strategies for large-scale coastal morphology

Data

In view of the scarcity of data, PACE includes a worldwide collaboration network, which is meant to provide access to the world's most important long-term datasets. A special Task Force has been established to make these data accessible to the participants (see Hamm, 1997).

Conclusion

It can be concluded from the foregoing, that modelling of large-scale coastal behaviour (LSCB) is a relevant research issue, in which a number of challenging question have to be answered. The many papers related to LSCB in this volume indicate, that research is in full swing now, but also that it is too early to expect new, well-established and robust prediction methods. Moreover, the broad scatter of approaches indicates that this field is still in an early stage of development.

Acknowledgement

This paper was written on behalf of the Technical Committee of PACE, a project which is partly funded by the European Commission, as a part of the Marine Science and Technology Programme (MAST-III), contract no. MAS3-CT95-0002.

References

Aarninkhof, S.G.J. and Plant, N.G., 1997. Quantitative estimations of bar dynamics from video images. *This issue.*
Brøker, I., Zyserman, J. and Jakobsen, P.R., 1996. Thyborøn coastal investigation 1995: new lessons from an old coastal problem. In: B.L. Edge (Editor): *Coastal Engineering 1996*, ASCE, New York, p. 4703-4716.
Capobianco, M., Larson, M., Kraus, N.C. and Nicholls, R.J., 1997. Depth of closure: toward a reconciliation of theory, practice and evidence. *This issue.*
Cowell, P.J., Roy, P.S. and Jones, R.A., 1995. Simulation of LSCB using a morphological behaviour model. *Marine Geology*, 126: 45-61.
De Vriend, H.J., 1987. 2DH mathematical modelling of morphological evolutions in shallow water. *Coastal Engineering*, 11(1): 1-27.
De Vriend, H.J., 1995. G6/G8 Coastal Morphodynamics. In: M. Weydert et al. (Editors): *Marine Sciences and Technologies: Second MAST-days and EUROMAR-market*, Comm. of the European Communities, ISBN 92-827-5010-8, p. 631-650.
De Vriend, H.J., Zyserman, J., Nicholson, J., Péchon, Ph., Roelvink, J.A. and Southgate, H.N., 1993a. Medium-term 2-DH coastal area modelling. *Coastal Engineering*, 21(1-3): 193-224.
De Vriend, H.J., Capobianco, M., Chesher, T., De Swart, H.E., Latteux, B. and Stive, M.J.F., 1993b. Approaches to long-term modelling of coastal morphology: a review. *Coastal Engineering*, 21(1-3): 225-269.
Falqués, A., Montoto, A. and Iranzo, V., 1996. Coastal morphodynamic instabilities. In: B.L. Edge (Editor): *Coastal Engineering 1996*, ASCE, p. 3560-3573.
Hamm, L. (Editor), 1997. *Summary of available datasets*. PACE Task Force on Datasets, Sogreah internal report 51 8906 R1.
Hasselmann, K.H., 1993. PIP's and POP's: the reduction of complex dynamical systems using principal interaction and principal oscillation patterns. *J. Geoph.*

Res.: 11.015-11.021.

Holman, R.A., 1997. Bulk comparisons of the world's beaches. *This issue.*

Hulscher, S.J.M.H., 1997. Generation and migration of shoreface bars due to low-frequency waves. *This issue.*

Jiménez. J.A. and Sánchez-Arcilla, A., 1997. A conceptual model for barrier coasts behaviour at decadal scale. *This issue.*

Latteux, B., 1995. Techniques for long-term morphological simulation under tidal current action. *Marine Geology*, 126: 129-141.

Lippmann, T.C. Holman, R.A. and Hathaway, K.K., 1993. Episodic, nonstationary behavior of a double bar system at Duck, North Carolina, USA, 1986-1991. *J. Coastal Res.*, 15(SI): 49-75.

Millard, T.K., Southgate, H.N. and Bunn, N., 1997. Quantifying the uncertainty in beach planshape modelling to wave chronology. *This issue.*

Nicholls, R.J. and Birkemeier, W.A., 1997. Morphological and sediment budget controls on depth of closure at Duck, NC. *This issue.*

Niedoroda, A.W., Reed, C.W., Swift, D.J.P., Arato, H. and Hoyanagi, K., 1995. Modelling shore-normal large-scale coastal evolution. *Marine Geology*, 126: 181-199.

Plant, N.G. and Holman, R.A., 1997. Sandbar behavior: analysis of kinematic states. *This issue.*

Reeve, D.E. and Fleming, C.A., 1997. A statistical-dynamical method for predicting long term coastal evolution. *Coastal Engineering*, 30(3-4): 259-280.

Schuttelaars, H.M. and De Swart, H.E., 1997. Cyclic bar behaviour in a nonlinear model of a tidal inlet. *Proc. XXVIIth IAHR Congress*, San Francisco (in press).

Southgate, H.N. and Beltran, L.M., 1995. Time series analysis of long-term beach level data from Lincolnshire, UK. In: W.R. Dally and R.B. Zeidler (Editors): *Coastal Dynamics '95*, ASCE, New York, p. 1006-1017.

Southgate, H.N. and Capobianco, M., 1997. The role of chronology in long term morphodynamics; theory, practice and evidence. *This issue.*

Stive, M.J.F. and De Vriend, H.J., 1995. Modelling shoreface profile evolution. *Marine Geology*, 126: 235-248.

Stive, M.J.F., De Vriend, H.J., Hamm, L., Fredsoe, J., Soulsby, R.L., Teisson, C. and Winterwerp, J.C., 1995. *Advances in Coastal Morphodynamics: an overview of the G8 Coastal Morphodynamics Project*. Delft Hydraulics, Delft, 545 pp, ISBN 90-9009026-6.

Swift, D.J.P., Zhang, Y., Carey, J.S., Niedoroda, A.W., Reed, C.W. and Steckler, M., 1997. Coastal evolution: modelling fabric and form at successive spatial scales. *This issue.*

Vittori, G., De Swart, H.E. and Blondeaux, P., 1996. Crescentic forms in the nearshore region. *Submitted for publication.*

Wang, Z.B., Louters, T. and De Vriend, H.J., 1995. Morphodynamic modelling for a tidal inlet in the Wadden Sea. *Marine Geology*, 126: 289-300.

Wijnberg, K.M. and Holman, R.A., 1997. Cyclic behaviour of breaker bars viewed by video imagery. *This issue.*

Wijnberg, K.M. and Terwindt, J.H.J., 1995. Extracting decadal morphological behaviour from high-resolution, long-term bathymetric surveys along the Holland coast using eigenfunction analysis. *Marine Geology*, 126: 301-330.

Wright, L.D., Boon, J.D., Kim, S.C. and List, J.H., 1991. Modes of cross-shore sediment transport on the shoreface of the Mid-Atlantic Bight. *Marine Geology*, 96(1): 19-51.

Long-Term Observations of Migrating Shore-Normal Bars

Guy Gelfenbaum[1] and Gregg R. Brooks[2]

Abstract

A series of migrating shore-normal sandbars with wavelengths of 50-200 m and heights of 0.5-2 m have been identified off the northern tip of Anna Maria Island, a barrier island on the west-central Florida coast. Similar features have been described elsewhere since the 1930's and termed "transverse bars." The transverse bars identified off Anna Maria Island are found for about 3 km along the coast and extend 4 km offshore. No cusps or any other associated beach expression is evident despite the fact that the bars come to within about 75 m of the beach.

Historical aerial photographs from the early 1940's through the mid 1990's provide an excellent means of quantifying the migration of the bars for this time period. The historical photographs were orthorectified resulting in errors in geographic positions of 1-2 m. Analyses of the orthorectified photos clearly show movement or migration taking place in the bar field. In the forty year period from 1951 to 1991, the southern edge of the bar field moved 200-350 m to the south, with an average migration rate of 7.9 m/yr. A current-meter deployment suggests that southerly winds associated with the passage of cold fronts drives near-bed currents to the south that are strong enough to initiate sediment transport and cause the southerly migration of the transverse bars.

[1] Oceanographer, U.S. Geological Survey, Center For Coastal Geology and Regional Marine Studies, 600 4th Street South, St. Petersburg, FL 33701, USA.
[2] Associate Professor, Marine Science Department, Eckerd College, 4200 54th Ave. South, St. Petersburg, FL 33701, USA.

Introduction

Bars that are oriented steeply oblique or normal to the shoreline have been termed transverse bars (Shepard, 1952). Typically, transverse bars occur on wide, gently sloping foreshores under conditions of low to moderate wave energy and a large sediment supply (Niedoroda and Tanner, 1970). Short-crested near-normal bars have also been observed in higher energy nearshore environments such as off the coast of Duck, North Carolina (Konicki and Holman, 1996). The crest lengths of transverse bars have been reported to vary from tens of meters to several kilometers.

The processes responsible for the formation and maintenance of transverse bars are not well agreed upon. Niedoroda and Tanner (1970) suggest that convergent shoaling waves induce an along-crest circulation and sediment transport that is responsible for the maintenance of transverse bars. Barcilon and Lau (1973) argue that longshore tidal currents are responsible for transverse bar formation, and Falques et al. (1993) describe bathymetric resonance resulting from wave-induced longshore currents. The processes responsible for the formation and maintenance of transverse bars may not be responsible for any longshore migration of the bars.

The migration history of transverse bars has not been well studied. Niedoroda and Tanner (1970) found no migration of transverse bars off St. James Island, Florida over a 25-year period. Goud and Aubrey (1985), however, calculated a migration rate of 10-20 m/yr over a 10-yr period for shore-oblique bars off Popponesset Beach, Massachusetts. Since some transverse bar fields migrate whereas others do not, the processes responsible for the formation of transverse bars may not be the same as the processes responsible for their migration. This paper presents results from field observations of a field of transverse bars off the west coast of Florida, documents their migration over several decades, and hypothesizes that the waves and currents associated with the passage of cold fronts are responsible for their migration.

Description of Bars

A series of transverse bars have been identified off the northern tip of Anna Maria Island, a barrier island on the west-central Florida coast (Figure 1). Located just south of Passage Key Inlet, a large secondary inlet to the Tampa Bay estuary, the transverse bars are adjacent to an ebb-tidal shoal which may supply sediment to the bars. The regional cross-shore slope off of Anna Maria Island is small, approximately 0.002 (0.13°) from the shoreline to 9 m depth 4 km offshore. Wave and tidal energy in this region are also small (Tanner, 1960). Mean annual wave height is 25 cm and the mean tidal range along the coast is approximately 70 cm.

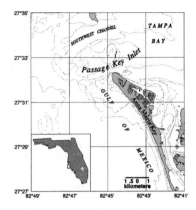

Figure 1. Location map of Anna
Maria Island, Florida.

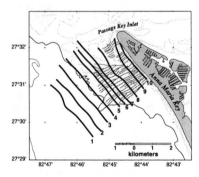

Figure 2. Oblique aerial photograph
of transverse bars off Anna Maria
Island.

The transverse bars off Anna Maria are somewhat unique in that they are
oriented normal to shore and extend a long distance offshore (Figure 2). The crests
of the bars are fairly straight and can be traced for several kilometers in the cross-
shore direction. The bar field extends approximately 3 km along the coast from
north to south, and nearly 4 km in the cross-shore direction. Some bar crests have
slight curvature, and a few stop or bifurcate (Figures 3 and 4). On the landward
side, the bars extend to within 75 m from the shoreline, however, they do not
appear to be associated with beach cusps or any other rhythmic beach structures.

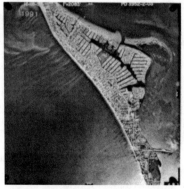

Figure 3. Vertical aerial photograph
of Anna Maria Island and transverse
bars from 1991. Bars extend farther
offshore than can be observed in
photograph.

Figure 4. Trace of transverse bar
crests from the 1991 aerial
photograph and bathymetry tracklines
from April, 1994.

A series of 10 bathymetric profiles were obtained across the transverse bars in April, 1994 (Figure 4). Examples of two of the bathymetric profiles are shown in Figure 5. Along the north side of the bar field is an ebb-tidal shoal that is shallower than the bars. Bed slopes normal to the bar crests are small, in the range of 2 -4 degrees. On average, the bars are asymmetrical with slightly steeper slopes on the south facing sides.

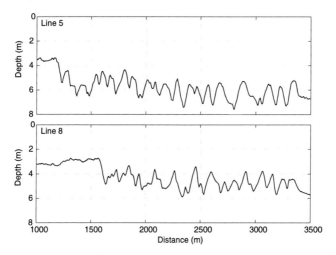

Figure 5. Bathymetric profiles across the transverse bars. North is on the left side of the figure.

Bar wavelength and amplitude were measured from the bathymetric profiles. The variations in bar wavelength and amplitude versus distance offshore are shown in Figures 6 and 7. Mean bar wavelength stays fairly constant across the bar field, varying between 75 m and 120 m. The full range of bar wavelengths varies between 30 m and 270 m. In contrast, mean bar amplitude varies significantly with distance offshore. Bar amplitude starts small at 0.4 m close to shore, increases to 1.7 m approximately 1.5 km from shore, and decreases to 0.5 m 3.5 km from shore.

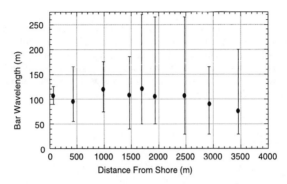

Figure 6. Average bar wavelength. Vertical bars represent range of wavelengths.

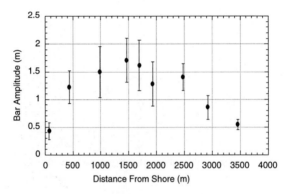

Figure 7. Average bar amplitude. Vertical bars represent the standard deviation in amplitude.

Migration of Transverse Bars

Historical Aerial Photographs

Historical aerial photographs from the early 1950's through the early 1990's provide an excellent means to quantitatively assess the migration of the transverse bars off Anna Maria. Vertical aerial photographs from 1951, 1965, 1977, 1980, 1984, and 1991 were scanned at 1 m resolution and orthorectified. The orthorectification process entails referencing known points on the photograph to their correct geographic position. This results in all the photographs being referenced to the same geographic datum, removing spatial and scale discrepancies between the photographs caused by differences in camera altitude, tilt, azimuth, or focal length. Using 8 reference points or more per photograph provided an accuracy of 1-2 m in position.

Analysis of the orthorectified historical photographs reveals significant movement in the position of the transverse bars. In fact, the time between photographs was too large to confidently identify bar crests from one photograph to another. Only along the southern edge of the bar field could the direction and magnitude of migration of the bars be determined with some certainty (Figure 8). Along the southern end of the bar field the bars migrated to the south at 5 - 12 m/yr. Over the 4 decades covered by these photographs the average rate of bar migration was approximately 8 m/yr.

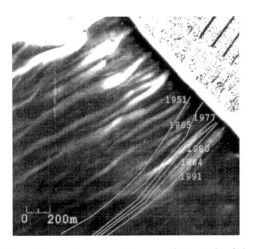

Figure 8. Position of the southern-most bar crest from each of the years shown superimposed on a 1991 image of the southern section of the bar field.

Repeated Bathymetric Surveys

Bathymetric surveys were repeated over the transverse bar field in May, 1996 and January, 1997. High-precision GPS was used to locate the survey lines, and a high-resolution fathometer and pitch-roll sensor were used to minimize vertical error. Horizontal and vertical resolution were both less than 10 cm. Repeated bathymetric surveys documented migration of the transverse bars over a shorter time scale than the photographic analysis and over the entire bar field (Figure 9). In general, individual bars migrated to the south. The average migration, as measured from 15 bar crests throughout the field was 23.3 m over the 17 month period between surveys. Bars along each line (except line 1 furthest from shore) showed migration to the south with no apparent growth or decay in bar height.

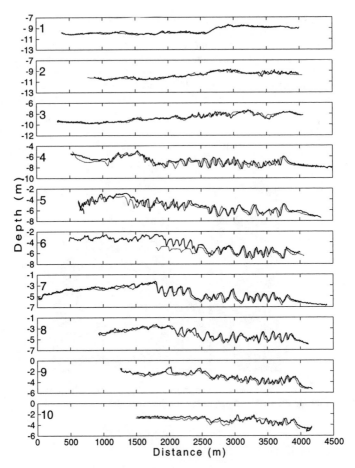

Figure 9. Repeated bathymetric surveys across the transverse bars. The thin line is from the May, 1996 survey and the thick line is from the January, 1997 survey. Distance increases from north to south. Line numbers correspond to transects shown in Figure 4.

Current Measurements

Two 1500 kHz *Sontek* ADP profiling current meters were deployed in an attempt to observe the processes responsible for the migration of the transverse bars. Currents were averaged over eight minutes and collected every 1/2 hour. The current meters were deployed on December 23, 1996 and positioned in the troughs between bars. They were recovered on February 13, 1997. At the shallow site (4.6 m water depth), a *Seabird* SBE-26 wave/tide gauge was deployed with the current

meter. The wave/tide gauge sampled tides every 20 minutes with waves measured
every two hours. The deeper site had just an ADP current meter deployed in 7.9 m
of water. The winds were collected from a NOAA weather buoy located
approximately 100 miles west of Tampa Bay.

Currents at both sites were characterized by low amplitude tidal currents
superimposed with infrequent but larger magnitude wind-driven currents associated
with the passage of cold fronts. Rotary spectral analysis revealed nearly rectilinear
semi-diurnal tidal ellipses with orientations of 27° west of north at the shallow site
and 17° west of north for the deeper site (Figure 10). The orientation of the tidal
currents was nearly shore parallel and approximately normal to the transverse bars.
Semi-diurnal tidal currents were small however, and unlikely to transport sediment.

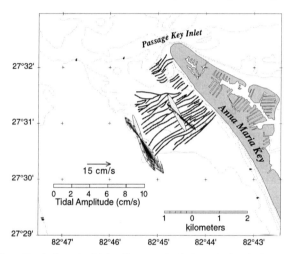

Figure 10. Semi-diurnal tidal ellipses (shaded) calculated from the entire
deployment, and current vectors (arrows) from the period January 9-11, 1997.

Wind-driven currents associated with the passage of two small cold fronts
moving north to south down the shelf were observed during the current-meter
deployment (Figure 11). The typical pattern of winds, currents, and waves can be
seen on January 8-11, and January 16-18, 1997. Ahead of the front, winds begin
blowing to the north and drive a weak flow in that direction. As the front passes,
winds shift to the south driving stronger southerly flow and building wave height
and period. The southerly flow lasts 1-2 days. Figure 11 also shows the estimated
value for the initiation of motion for the sediment that makes up the transverse bars.
During this deployment, the wind-driven currents near the bed during the northerly
flow phase ahead of the front may or may not be strong enough to move sediment.
During the southerly phase of the flow, the near-bed currents are strong enough to
move sediment, and in fact, the southerly sediment transport is augmented by the
enhanced bottom shear stress due to the combined waves and currents.

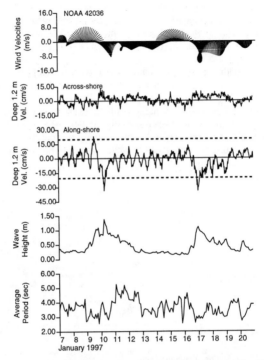

Figure 11. Time series of low-pass filtered wind, currents at 1.2 mab, wave height, and wave period. Winds are reported in the oceanographic sense, currents are 7.9 m site, and waves are from the 4.6 m site. Positive along-shore currents are in the northwestward direction and positive across-shore currents are onshore.

Conclusions

A series of transverse bars have been observed off Anna Maria Island, on the west-central coast of Florida. These bars are somewhat unique in that they are oriented normal to shore, extend a long distance across, and they have amplitudes which become a significant fraction of the water depth.

Through the use of historical aerial photographs and repeated bathymetric surveys, these bars have been shown to migrate to the south. The long-term rate of migration averaged over 40 years is about 8 m/yr, whereas the short-term rate as determined by the bathymetric surveys is substantially higher.

The migration of the transverse bars off Anna Maria Island seems to be associated with the passage of cold fronts as they move from north to south down the west Florida shelf and not related to tidal currents. The northerly directed winds ahead of the cold front drive a near-bed flow in a northerly direction normal to the transverse bar crests. This flow may or may not be strong enough to induce sediment transport. After the front passes, the wind reverses and drives a stronger flow to the south, again normal to the bar crests. This stronger flow, and the associated larger waves, force sediment transport to the south and results in the net southerly migration of the transverse bars. The enhanced sediment transport due to the combined waves and currents, along with the reversing nature of the sediment transport associated with the passage of the cold front may explain how the bars attain the large bar height to water depth ratios that were observed.

Acknowledgments

This work was funded by the U.S. Geological Survey's Coastal and Marine Geology Program as part of the West-Central Florida Coastal Studies Project. The authors would like to thank Mark Hansen, Nancy DeWitt, Sean Leatham, Bryan Black, and Matt Wallenstien for their assistance in the field and in preparing figures. This manuscript benefited from critical reviews by John Haines and Doug Wilson. Any use of trade names in this publication is for descriptive purposes only and does not imply endorsement by the USGS.

References

Barcilon, A. I., and Lau, J. P., 1973, A model for the formation of transverse bars. Journal of Geophysical Research, vol. 78, no. 15, 2656-2664.

Falques, A., Iranzo, V., and Montoto, A, 1993, Resonance of longshore currents under topographic forcing. Physics of Fluids, A5 (12), 3071-3084.

Goud, M. R., and Aubrey, D. G., 1985, Theoretical and observational estimates of nearshore bedload transport rates. Marine Geology, vol. 64, 91-111.

Konicki, K. M. and Holman, R. A., 1996, Transverse bars in Duck, North Carolina. Proceedings of the 25th international conference: Coastal Engineering '96, ed. Billy L. Edge, 3588-3599.

Niedoroda, A. W., and Tanner, W. F., 1970, Preliminary study of transverse bars. Marine Geology, vol. 9, 41-62.

Shepard, F. P., 1952, revised nomenclature for depositional coastal features. A.A.P.G. Bulletin, vol. 36, no. 10, 1902-1912.

Tanner, W. F., 1960, Florida coastal classification. Transactions Gulf Coast Assoc. Geol. Societies, vol. 10, 259-266.

Breaker Bar Dynamics Predicted by Kalman Filtering

W.T.Bakker[1], S.C.v.d.Biezen[1], A.W.Heemink[2], M.A.Knaapen[3]

Abstract

The morphodynamics of the seaward side of a barred coast is simulated: analytically by Airy waves and numerically by a diffusion model, in which the diffusion coefficient decays strongly in seaward direction. Theoretical results and prototype topographical data are merged with the aid of Kalman filtering.

Presumed that sufficient bathymetric data is available, in a dynamic equilibrium situation the model can describe, and to a certain extent forecast, coastal behaviour. A test case for the Terschelling (NL) coast yields promising results.

1. Introduction

This paper focuses on bar systems as found in the Netherlands. Here, beach and inshore often consist[4] of a system of 1 to 3 bars (depending on the site: wavelength$\approx$300 m; seaward propagation velocity 15 à 80 m/yr, i.e. periodicity 20 à 4 yrs resp.).

Complementary to most papers on breaker bars, which (logically) investigate de-stabilizing features, the present paper focuses on cross-shore diffusitivity[5] (ch.2), (dominating at the seaward end of the inshore) and its effect on bar propagation (ch.3). Still, prediction involves also de-stabilizing features, as those are implemented from nature itself. Merging of theory and nature is accomplished by Kalman filtering, as explained in ch.4 and demonstrated in ch.5. The present model applies in the first place for the lower side of the surf zone, where irregularities attenuate.

2. Cross-shore diffusitivity: the Van der Kerk theory

This chapter deals on the mean coastal profile[6] in dynamic equilibrium, $-h(y)$, and on the diffusitivity constant of the

---

1 Delft Univ. of Techn.; Fac.of Civil Engng.; P.O.Box 5048; 2600GA Delft

2 Delft Univ. of Techn.; Fac.of Techn.Math. & Comp.Science; P.O.Box 5031; 2600 GA Delft; Neth.

3 Univ. of Twente; Civ.Engng. & Management; P.O.Box 217; 7500 AE Enschede

4 Bakker & Joustra (1970);Bakker & De Vroeg (1988a,b), Wijnberg (1995); Roelvink (1993) concerning the physical aspects of bar systems

5 Longshore diffusitivity will not be treated in extenso, although in ch.4 (numerical model) it is implemented, using formulations according Pelnard-Considère (1954) and Bakker et al. (1988).

6 A stationary (2D-)wave climate is assumed with deep water wave number $k_0 = 2\pi/T$ and deep water rms-wave height $[H_{rms}]_0$. Cross-shore coordinate y positive in seaward direction.

momentaneous coastal profile $-h + z(y,t)$. It elucidates the relation between diffusitivity D_y and h. Use is made of the theory of Bailard (1981) with amendments of Van der Kerk (1987)[7], based upon investigations of De Vriend, Stive and Wind[8]. Van der Kerk derives (as part of his equilibrium profile theory) a linear relationship between local coastal slope and cross-shore transport; this determines D_y. His dimensionless profile model $k_0 h = f(k_0 y)$ has the constituents, wave dissipation and undertow (a), sediment transport (b) and equilibrium profile (c), being clarified below. V.d.Kerk finds that $f(k_0 y)$ only depends on deep-water wave steepness s_0 and dimensionless grain fall velocity $W^* = wT / [H_{rms}]_0$ [9].

a. wave propagation and dissipation; undertow:
Dimensionless local wave height H_{rms}/h is calculated as function of s_0 and of dimensionless depth kh [10]. Dimensionless expressions for the undertow are derived, following the method of Stive & Wind (1986). "Potential" landward current u_e is derived from Longuet Higgins' (1953) conduction solution and "potential" undertow u_s by breaking waves[11] from expressions, in which the local gradient of the cross-shore radiation stress and of the bottom level play a large role. Total undertow equals $Q_b u_e + (1 - Q_b) u_s$, where Q_b is the fraction of breaking waves.

b. sediment-transport and equilibrium slope:
Sediment transport is assumed to behave according to Bailard's (1981) paper, following Stive's (1986) amendments. Bailard extends Bagnold's (1966) theory to periodic motion plus resultant current. His result contains four components, which are proportional to the 3..5-power of the momentaneous velocity; wave-period averaged these can be translated into 0..5th velocity moments. Of those four components, two comprise the bed load and the other two the suspended load. Bed- and suspended load each have a component only depending on the momentaneous current velocity and another component linearly depending on the bottom slope. Thus, the

---

7 cf. Bakker et al.(1988);

8 Stive & Wind (1986); Stive (1986); De Vriend & Stive (1987);

9 w is the fall velocity of the grains. Parameter W^* is equivalent to: the "Dean parameter". Subscript "O" stands for "deep water"

10 k is the local wave number and can be related to deep water wave characteristic $k_0 h$. The simplicity of the formula is "paid" by abstaining from subtilies of modern wave theories: the relation has been derived from Battjes (1974), using $H_b/h = 0.48$. The model only holds for outer sides of breaker bars

11 u_e and u_s in dimensionless shape by dividing by gT

afore-mentioned diffusitivity constant D_y is the proportionality constant between slope and transport. V.d.Kerk
uses another application of D_y: if one calls S_{y0} the (combined
bottom- plus suspended-)transport for a horizontal bottom

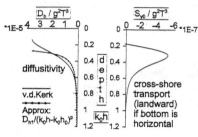

Fig.1a,b. D_h and S_{y0} as function of depth.
Valid for $s_0 = 0.02$ and $W^* = 0.1$

(addition of the two non-slope
depending components of Bailard's four components), the
local equilibrium slope dh/dy
is found equal[12] to $-S_{y0}/D_y$.
Fig 1a shows the dimensionless
diffusitivity D_h/g^2T^3 [13];
fig.1b the dimensionless
cross-shore transport[14]
S_{y0}/g^2T^3. The figure is valid
for $s_0 = .02$ and $W^* = .1$.

c. Local equilibrium slope integrated to (idem-) profile:
Bailard's (1981) paper results in a graph of equilibrium
slopes as function of k_0h and the Dean parameter. He finds
very high and rather unrealistic slopes for small water
depth, because he only takes non-breaking waves and Longuet
Higgins' conduction solution for the bottom transport
(cf."a") into account. The addition of Stive & Wind's (1986)
undertow by breaking waves, as done by V.d.Kerk, reduces
the calculated equilibrium bottom slopes to more realistic
values. At a certain depth the offshore transport by the
undertow is large enough to balance the onshore asymmetry
transport ($S_{y0} = 0$); this implies that the equilibrium profile
is horizontal at that depth. Higher in the profile the
increasing undertow (because of the higher percentage of
broken waves) dominates the offshore transport. Only slopes:

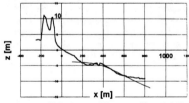

Fig.2 Equilibrium profile according V.d.Kerk
(1987); comparison with coastal profile

"downward in onshore direction"
could provide a balance. Under
those conditions V.d.Kerk's
assumptions do not hold anymore: this stops his equilibrium profile computation (an
integration, starting in deep
water, of the local equilibrium
slopes as calculated "ad b").
Fig. 2 shows an example of an

12 S_{y0} will be negative over the considered part of the profile

13 In fact, the V.d.Kerk theory considers the y-coordinate of the equilibrium profile as function
of z and thus gives diffusion as function of h. By taking the local equilibrium bottom slope into
account, D_h can be transferred to horizontal diffusion D_y

14 Seaward positive. Transport reverses at crest-level of breaker bar. From V.d. Biezen (1995)

equilibrium profile from V.d.Kerk's theory compared with a Dutch coastal profile[15] (V.Engelen, 1995).

Assigning to each water depth h an equilibrium distance y_e from the coast[16], the diffusivity constant D_y as function of y can be determined from D_h.

3. Sand, water, Airy: Analytic model for bar propagation

Consider the mathematical problem of finding the concentration of diffusant $z(y,t)$, given a harmonic variation $z_0(\omega_b t)$ at one boundary $y = y_0$; the concentration at the other boundary $y = \infty$ remains zero. In case of uniform diffusivity (fig.3a), a propagating wave, attenuating in positive y-direction results (fig.3b) (Wylie, 1960).

In the present application (breaker bars), diffusivity will be assumed, inversely proportional to a reference line at distance y_0 from the beach. Fig.1a shows, that the curve, resulting from V.d.Kerk's computations can be approximated by: $D_{h1}/(k_0h - k_0h_0)^p$, where D_{h1} is the value of D_h, at a dimensionless depth, 1 unit larger than k_0h_0 [17]. Airy waves result[18] from the assumed diffusivity (App.A), leading again to a propagating wave, decaying seawards (fig.3c).

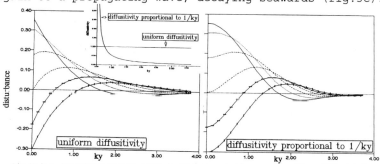

Fig.3a,b,c. Comparison between decay of sand waves in the case of uniform diffusivity and a diffusivity, inversely proportional to $y - y_0$

In the case of uniform diffusivity D propagation velocity would have been $\sqrt{2\omega_b D}$. In the present case (far from $y = y_0$) it is $\sqrt{2\omega_b D_y}$, where D_y is defined as $D_1/k_0(y - y_0)$; k_0 being the deep water wave number of the water waves, approaching

---

15 Delfland range 111.09 ($s_0 = .036$; $W^* = .06$; $T = 6$ sec; V.Engelen (1995))

16 if the site of the origin of the y-axis is defined at a given value of h

17 Approximation of D in fig.1: $k_0h_0 = .18$; $;D_{h1} = 4*10^{-7}g^2T^3$; $p = 1.9$

18 Under the condition that the equilibrium profile as calculated from the V.d.Kerk theory can be approximated by a horizontal shoal at depth h_0 plus a parabola $D_{h1}(k_0y - k_0y_0) = D_1(k_0h - k_0h_0)^p$, where y_0 is the y-coordinate of the seaward end of the shoal (Bakker et al.,1997)

the coast[19]. The diffusitivity D_1 is the value at the point where $k_0(y-y_0) = 1$ and can be derived from D_{h1}. Far from $y = y_0$ the decay of the wave is proportional to $\exp(-(y-y_0)^{3/2})$, strong compared with uniform diffusitivity $(\exp(-(y-y_0)))$. Near $y = y_0$, the site with infinitely high diffusion, the gradient of the diffusitivity plays a big role (appendix A). This leads to a very slow decrease of the amplitude of the wave motion in the vicinity of $y = y_0$ (fig.3c).

This conceptual Airy model leads to bar lengths proportional to the deep water wave length L_0 and proportional to $\sqrt[3]{T_b/T}$ (appendix A), where T_b is the period of bar motion. Furthermore, bar length, depending on W^*, becomes larger for finer grains.

4.Kalman filtering

In order to make the deficient[20] coastal diffusion model (ch.3) more adequate for prediction, a data assimilation technique as Kalman filtering (Jazwinski,1970, Maybeck,1979) is necessary. This provides a tool to merge measurements with theory.

Firstly, the physical model is embedded into a stochastic environment by introducing a system noise process. This in order to account for its inaccuracies and the systematic omission of important physical factors (ch.1).

By using a Kalman filter, the information provided by the resulting stochastic-dynamic model and the noisy measurements taken from the actual system are combined to obtain an optimal (least squares) estimate of the state of the system. In the last decade Kalman filtering has gained acceptance as a powerful framework for data assimilation in forecasting systems (Ghil et al., 1997).

Define the state of the system $\underline{X}(t_k)$ at time t_k as a vector, containing information on the coastal transport $(\underline{S}_x, \underline{S}_y)$ and on the bottomtopography $\underline{\zeta}$ (each of those constituing vectors having $m \times n$ elements; $m \times n$ indicating the dimensions of the two-dimensional spatial grid):

$$\underline{X}(t_k) = \left[\underline{S}_x^T(t_k), \underline{S}_y^T(t_k), \underline{\zeta}^T(t_k) \right]^T \tag{1}$$

---

19 Defining D as D_1/y would have violated considerations of dimensions. The choice of k_0 as a reference value with dimension [1/L] will be justified in Appendix A

20 As de-stabilization is neglected (ch.1)

Rewrite the physical model[21] described[22] in ch.3 formally as:

$$\underline{X}^{f}(t_{k+1}) = M(t_{k+1}, t_k)\underline{X}^{f}(t_k) + B\underline{U}(t_k) \tag{2}$$

where M and B are (non linear) operators and $\underline{U}(t_k)$ represents the boundary forcing at time t_k.

Kalman filter theory can now be applied, implementing the following stochastic elements in the state space model. It is assumed, that most inaccuracies are introduced by deficiencies in the transport equations ($\underline{S}_x, \underline{S}_y$). Thus a noise term[23] is added to eqn (2):

$$\underline{X}^{f}(t_{k+1}) = M(t_{k+1}, t_k)\underline{X}^{f}(t_k) + B\underline{U}(t_k) + \Lambda W(t_k) \tag{3}$$

Using the Kalman filter, measurements of $\underline{\zeta}$ are used to improve the model results.

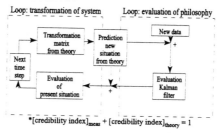

Loop: transformation of system   Loop: evaluation of philosophy

*[credibility index]$_{meas}$ + [credibility index]$_{theory}$ = 1

Fig.4.Loops, transforming the situation vector and the Kalman filter.

Fig. 4 indicates the various transformations (computational loops) to which the status vector is exposed, leading from a certain perception of reality at t_k to the perception at time t_{k+1}. Each time step the old situation vector is transformed by a transformation matrix (in the present case: based upon diffusion theory), which results in a new situation. This is compared with prototype data concerning the new time considered. This comparison results in an evaluation: the adaptation of the Kalman filter, which consists of an array of credibility indices. This index is

---

21 Superscripts "f" and "α" attached to $\underline{X}$ indicate the stage of the computation (before and after calibration with measurements)

22 Including longshore component $\underline{S}_x$ according Bakker et al.(1988) and making use of the cross-shore transport formulation from V.d.Kerk (1987): $S_y = S_{y0} - D_y \partial\zeta/\partial y$ with $\zeta = -h + z$

23 In the noise term, Λ is a matrix and $\underline{W}^{x(y)}(t_k)$ is a noise term, added to $\underline{S}_{x(y)}$. Noise processes in x- and y-direction are mutually independent. Expectation of $\underline{W}$ is zero and the covariances $E[W^x_{m,n}(t_k)W^x_{p,q}(t_k)]$ and $E[W^y_{m,n}(t_k)W^y_{p,q}(t_k)]$ are $\sigma_w^2 \exp\{-d_{i,j}/d_0\}$, where $d_{i,j} = \sqrt{(m-p)^2 + (n-q)^2}$ is the distance between point I (coordinates (m,n)) and J (p,q). The parameters d_0 and σ_w have to be specified. Continuity is still guaranteed.

a factor, giving the weight one should account to theory, compared to measurements. A more rigorous explanation is given by Jazwinsky (1970) and Maybeck (1979).

5.Application: The Terschelling coast.

With this model, including the Kalman Filter, breaker bar dynamics at the island of Terschelling between the ranges at RSP-line 18.00 and 20.00 is simulated. The inshore consists of two almost parallel bars, with a wavelength of ca. 300 m, migrating in offshore direction.
Measurements from the Jarkus-file are available. This data has a spatial frequency of 200 m longshore and 5 to 20 m cross-shore. Every range is measured once a year (since 1965). In figure 5a measurements of range 19.00 are displayed. Cross-shore the model extends from 300 to 2000 m from the "RSP"-main range. At the seaside of the domain the bottom slope is assumed constant, at the land side a periodic wave condition is posed. No longshore transport at the other boundaries is assumed.
The results of the model (cf.App.B) are promising. Figure 5b presents the reconstruction of the bathymetry at range 19.00. Mark the seawards moving breaker bar. It shows, that the model overestimates the decaying of the bar somewhat and under-estimates the propagation velocity.

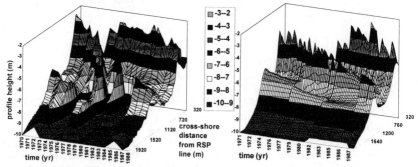

Figure 5a: Plot of Jarkus measurements from 1970 to 1988 for range 1900   Figure 5b: Model results for range 1900, hindcasted and reconstructed

The Kalman filter is able to improve the bathymetry estimations of the model. However, tuning of the model due to the adaptations is not yet visible. The tuning should result in better approximations after some time. A first solution is to improve the Kalman filter. Up till now, adaptations are only made to the bottom depth. The filter can easily be adapted to tune the parameters in the transport as well. Small changes could result in better approximations of the bathymetry.
The shape of the transport formulas might need review. The model has problem to combine the right propagation speed

with the correct negative growth of the bars. In advection/diffusion-like equations growth and propagation speed are heavily correlated.

6.Conclusions, evaluation and discussion

A. Kalman filtering provides a powerful tool for the prediction of coastal behaviour:

-   It is better than phenomenological extrapolation of nature data, as it takes the continuity between two successive checkpoints with nature strictly into account.

-   Physical considerations are implemented in an optimal way. Even a mathematical model, which does not contain all relevant features, imports those features by "listening to nature".

-   The model is self-learning by its successive evaluations of the credibility index. "Experience" is stored in a very efficient way.

B. The theory of V.d.Kerk (ch.2) predicts similarity of coastal shapes for equal $W^* = (wT/H_0)$ and deep water wave steepness s_0. Under similar circumstances the horizontal length of the deepest breaker bar should be[24] proportional to the deep water wave length and to $\sqrt[3]{T_b/T}$. Diffusitivity at equivalent sites should be proportional to T^3.

C. The analytical method (ch.2 and 3) leads to insight, but does not yield clear-cut values of diffusitivity as yet. One reason is the rough estimate of the wave climate, schematized by one kind of representative waves. Appendix B gives some discussion on orders of magnitude.

Discussion: The outer bar acts as an offshore breakwater for the higher bars. The wave attack on the higher parts of the shore is dominated by waves of shorter period. The same mechanism displayed in this paper can reproduce itself at a higher level, however at a smaller scale. This might explain the smaller bar lengths at higher levels.

Appendix A: Airy waves

Notation: in this Appendix, the usual mathematical notation of calling: "x" an independent variable, indicating a location (in one dimension) will be sustained. Harmonic periodicity is indicated by an (omitted) factor $\exp(i\omega_b t)$. The subscript "b" in ω_b reminds on the physical background (to be referred to) of "bar periodicity" instead of "water wave periodicity".

---

24 Comparing bar lengths in nature in areas with short and long bar periodicity T_b gave results of the right order of magnitude (Bakker et al.,1997).

Treated is the mathematical diffusivity problem, mentioned in ch.3.
The well-known diffusion equation (constant diffusivity) and its propagating-wave solution are shown below:

$$\frac{\partial z}{\partial t} = D\frac{\partial^2 z}{\partial x^2} \quad \rightarrow \quad z = z_0 e^{-k_b x} \quad with \quad k_b = \sqrt{\frac{\omega_b}{2D}}(1+i) \qquad (A1a,b,c)$$

If D is inversely proportional to x, (being D_1 for $k_0 x = 1$; (A2a)), the diffusion equation changes into (A2b), from which (A2c,d) result after elimination of the time factor[25]:

$$D_x = \frac{D_1}{k_0 x} \quad \rightarrow \quad \frac{\partial z}{\partial t} = \frac{\partial(D\partial z/\partial x)}{\partial x} \quad \rightarrow \quad (z'/x)' + \alpha z = 0 \quad with \quad \alpha = -\frac{i\omega_b k_0}{D_1} \qquad (A2a..d)$$

With appropriate boundary conditions, as solution is found:

$$z^* = -\sqrt{\pi}\{Ai'(-iu_r) - iBi'(-iu_r)\} \qquad (A3)$$

where Ai' and Bi' are the derivatives of the Airy functions Ai and Bi (Abramowitz & Stegun, 1970) and z^* and u_r are dimensionless z-, resp. x-coordinates:

$$z^* = \frac{z}{\sqrt[3]{D_1/\omega_b k_0}} \qquad\qquad u_r = \frac{x}{\sqrt[3]{D_1/\omega_b k_0}} \qquad (A4a,b)$$

Returning to physics $(x \rightarrow y - y_0)$:
According to ch.2 (V.d.Kerk theory), diffusivity should be proportional to $g^2 T^3$, i.e. D_1 in (A4a,b) can be written as $[D_1]_{dl} g^2 T^3$, in which $[D_1]_{dl}$ is a dimensionless "constant"[26]. Substitution of $[D_1]_{dl}$ into (A4b) shows, that u_r is proportional to $x/L_0\sqrt[3]{T_b/T}$, where $L_0 = gT^2/2\pi$.

Appendix B. Values of constants, used in the mathematical model (fig.5b).

The point of infinite diffusivity $(h = h_0, "y = y_0"; ch.3)$ has been situated at 200 m from the main range (fig.5b). $[D_1]_{dl}$ has been chosen $6*10^{-7}$, based upon measured bar propagation, compared with (cf.ch.3) $\sqrt{2\omega_b D_y}$; $T_b = 10$ yr; $T = 5$ s; resulting in $D_y = 1.5*10^6/(y_m - 200)$ m^2/yr, where y_m = distance from main range (fig.5a,b). From v.d.Kerk results $[D_1]_{dl} \approx 1.8*10^{-5}$ (cf.ch.6 and Bakker et al.,1997) for $s_0 \approx .02$ and $W^* = .1$ ($2.45*10^{-6}$ for $W^* = 1$). In ch.3 the following constants are applied: $D_1/D_{hl} = 44.83$; $p = 1.9$; $k_0 h_0 = .18$; $k_0 y_0 = 17$, when $s_0 = .02$ and $W^* = .1$.
A third method of calculation of diffusion coefficients (Bakker et al.,1997)[27] based upon an analysis of the NOURTEC supply at the Terschelling coast (ch.5) leads to some overall diffusion coefficient of 10^5 m^2/s at a depth $h = 4.5$ m. This supports the rather high diffusion rates from the V.d.Kerk theory.

References

Abramowitz, M. & Stegun, I.A. (ed),(1970), *Handbook of Mathematical Functions*, Dover Publ. Inc. New York.

Bagnold R.A.(1966), *An Approach to the Sediment Transport Problem from general Physics*, US Geol.Survey Paper 422-1

Bailard, J.A.,(1981), *An energetics total load sediment transport model for a plane sloping beach*. J.of Geoph.Res. 86-C11, p.10938-10954

Bakker, W.T. & D.Sj.Joustra (1970), *The History of the Dutch Coast in the last Century*, Proc. 12th Conf. on Coastal Engng., Washington

Bakker, W.T., C.v.d.Kerk & J.H.de Vroeg (1988), *Determination of coastal constants in mathematical line models.*, Second European Workshop on Coastal Zones, Loutraki

---

25 " ′ " means derivation to the independent spatial variable

26 Still depending on s_0 and W^* (ch.2.)

27 available on request

Bakker , W.T. & J.H. de Vroeg (1988a), *Coastal Modeling & Coastal Measurements in the Netherlands*, Proc. Seminar "Prototype Measurements to validate numerical models of Coastal Processes"; A.de Graauw & L.Hamm (Ed), May 1988

Bakker, W.T., S.C.v.d.Biezen, A.W.Heemink, M.A.Knaapen, *Kalman filtering as Tool for predicting the Behaviour of Breaker Bars.* (1997).Neth. Centre for Coastal Res.(to be publ.).

Battjes, J.A. (1974), *Computation of set-up, longshore currents, run-up and overtopping due to wind-generated waves.* Delft Univ. of Techn.; Fluid Mech.; Ph.D.Thesis

Biezen, S.C.v.d. (1995), *The dynamics of breaker bars, considered as a diffusion process.* Delft Univ. of Techn.; Coastal Engng.; M.Sc.Thesis

Engelen, H.J.v.(1995), *Sand-diodes for Delfland (in Dutch);* Delft Univ. of Techn.; Coastal Engng; M.Sc.Thesis

Ghil, M.(ed.) et al.(1997), *Data assimilation in meteorology and oceanography; theory and practice;* J. of the Met.Soc. of Japan

Jazwinski, A.H. (1970), *Stochastic Processes and Filtering Theory,* Academic Press, NY

V.d.Kerk, C.T.P. (1987), *The theoretical basis of the coastal constant sy derived with the aid of the Crostran concept (in Dutch);* Delft Univ. of Techn.; Coastal Engng.; M.Sc.Thesis

Knaapen, M.A.F. (1995), *Morphodynamic Modelling of the Coastal Zone;* Delft Univ. of Techn.; Fac. of Math. & Inf. (M.Sc.Thesis); Nat.Inst. for Coastal & Mar. Mgmnt.

Longuet Higgins, (1953) *Mass transport in Water Waves.* Phil.Trans.Royal Soc., vol.245; A903; pp.535-581

Maybeck, P.S.,(1979), *Stochastic Models, Estimation and Control,* Vol.1; Ac.Press, NY

Pelnard-Considère, (1954), *Essai de Théorie à l'Evolution des Formes de Rivages en Plages de Sables et de Galets.;* IVièmes Journées de l'Hydraulique; Paris; Question 3, les Energies de la Mer

Roelvink, J.A. (1993), *Surf beat and its effect on cross-shore profiles.* Ph.D.Thesis, Delft Univ. of Techn.

Stive, M.J.F.,(1986) *A model for cross-shore sediment transport;* 20th Int.Conf. on Coastal Engng., Taipei

Stive, M.J.F. & H.G. Wind (1986), *Cross-shore mean flow in the surf zone;* Coastal Engng; 10; no 4, p.235-340

De Vriend,H.J.& M.J.F.Stive (1987), *Quasi-3D nearshore current modeling: wave-induced secondary current.* ASCE Spec. Conference on Coastal Hydrodynamics, Delaware USA

Vroeg, J.H. de, E.S.P.Smit & W.T.Bakker (1988), *Coastal Genesis;* Proc. 21st Int.Conf. on Coastal Engng., Malaga

Wijnberg, K.M. (1995), *Morphologic Behaviour of a Barred Coast over a period of Decades.* Ph.D.Thesis, University of Utrecht

Wylie, C.R.,(1960), *Advanced Engineering Mathematics;* Mc Graw Hill Book Cy.,Inc.

WAVE TRANSMISSION AND SET-UP AT DETACHED BREAKWATERS

John Loveless[1] and Damian Debski[2]

ABSTRACT

The paper describes a large scale two dimensional flume study carried out at the University of Bristol on the performance of detached breakwaters. The object of the research was to determine how effective various rubble mound designs are in dissipating and reflecting the incident wave energy, to determine the inshore water level set-up generated by the different designs and to observe the limits to the stability of the rock under various conditions. The study has provided accurate information on the wave energy transmission and set-up to be expected with detached breakwaters. The performance on wave energy transmission was in line with previous estimates, but the authors made new discoveries regarding the set-up.

INTRODUCTION

In recent years the use of detached breakwaters for coastal defence has become widespread. Most are designed with crest levels well above the maximum mean sea level expected. However, recently there has been a move towards the design of breakwaters with lower crests. Information is needed on the performance of these low crest or submerged breakwaters with regard to wave transmission, their effect on inshore water levels and their general stability. This paper reports on some of the results obtained in a two year two dimensional flume study (Debski and Loveless, 1997) which was carried out to investigate these phenomena.

1)   Senior Lecturer, Dept. of Civil Eng. Univ. of Bristol, Bristol BS8 1TR, UK
2)   Asst. Engineer, Sir William Halcrow & Ptrs., Swindon SN4 OQD, UK

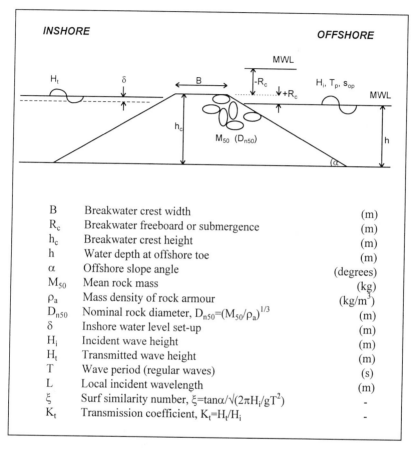

B	Breakwater crest width	(m)
R_c	Breakwater freeboard or submergence	(m)
h_c	Breakwater crest height	(m)
h	Water depth at offshore toe	(m)
α	Offshore slope angle	(degrees)
M_{50}	Mean rock mass	(kg)
ρ_a	Mass density of rock armour	(kg/m³)
D_{n50}	Nominal rock diameter, $D_{n50}=(M_{50}/\rho_a)^{1/3}$	(m)
δ	Inshore water level set-up	(m)
H_i	Incident wave height	(m)
H_t	Transmitted wave height	(m)
T	Wave period (regular waves)	(s)
L	Local incident wavelength	(m)
ξ	Surf similarity number, $\xi=\tan\alpha/\sqrt{(2\pi H_i/gT^2)}$	-
K_t	Transmission coefficient, $K_t=H_t/H_i$	-

Figure 1 : Major variables and definitions

DESCRIPTION OF THE MODELS AND VARIABLES

The major variables involved in the problem are listed and defined in Figure 1. The variable R_c has been defined as both the submergence of the crest (when negative) and the freeboard (when positive) below or above mean sea level. The rise in mean sea level or set-up inshore of the breakwaters has been defined as δ.

A total of six different model breakwater sections utilising three different grades of rock material were tested. The model sections are detailed in Table 1. The grading curves for the three grades of rock used are given in Figure 2. The experiments were all performed in a large random wave flume, 15m long, 1.5m wide and 1.1m high with a 5.0m long, 0.5m wide test section.

Model No	Offshore slope (tanα)	Inshore slope	Crest width B (cm)	Height h_c (cm)	Rock size M_{50} (g)
1	0.5	0.67	40	50	357
2	0.5	0.5	40	50	182
2a	0.5	0.5	60	50	182
2b	0.5	0.5	20	50	182
3	0.33	0.67	40	50	100
5	0.33	0.5	40	50	182

Table 1 : Models tested

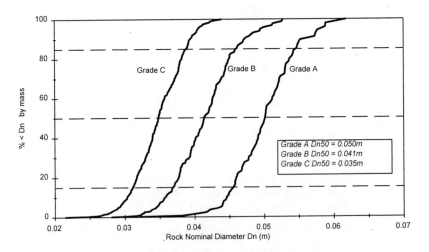

Figure 2 : Size distribution curves of breakwater rock

The range of wave and water level conditions tested are detailed in Table 2.

Variable	Values
H_i	5 to 20cm
T	1 to 3.33s
T_p	1.43, 2.00s
h	0.40 to 0.65m
R_c	-0.15 to +0.10m

Table 2 : Test conditions

WAVE ENERGY TRANSMISSION

Wave transmission past a submerged rubble mound breakwater occurs both over and through the structure. The relative contribution of each transmission mechanism depends on the wave conditions and sea level relative to the crest level. When the mean water level is close to or below the crest level, the permeability of the breakwater, determined by the rubble rock size and nature of the core zone, is an important factor in determining the transmission. It was assumed, following the findings of previous researchers that wave energy transmission is principally a function of the following non-dimensional variables:

$$K_t = f\left(\frac{H_i}{h_c}, \frac{R_c}{h_c}, \frac{D_{n50}}{h_c}, \frac{B}{L}, \xi, \frac{\delta}{h_c}\right)$$

The terms in brackets may be described as the relative wave height, relative submergence, a flow resistance parameter, relative crest width, surf similarity parameter and relative set-up respectively.

The principle governing the amount of wave energy transmitted inshore past a breakwater is the conservation of energy:

$$E_i = E_t + E_r + E_l$$

where $E_i =$ Energy of incident waves
          $E_r =$ Wave energy reflected seaward by breakwater
          $E_t =$ Wave energy immediately inshore of breakwater
          $E_l -$ Wave energy dissipated at breakwater

Figure 3 is a plot of one of the results obtained for transmission coefficient K_t for a wave period of 4.5s at the prototype scale (20x the model scale) for the four major models. It can be seen that the submergence or freeboard has a major influence on the energy transmission. The higher the freeboard the lower is the transmission and the greater the submergence the higher is the transmission.

The incident wave height also influences the energy transmission. For submerged conditions transmission is mostly influenced by wave breaking. The larger the wave height the lower is the transmission due to increased wave breaking and energy losses. For water levels below the crest level, transmission is mostly by overtopping and transmission through the breakwater. There is then a tendency for increasing transmission with increasing wave height as the amount of overtopping increases. The influence of stone size on transmission can be seen to be small for the range of sizes tested. The larger the stone size, the higher the transmission tends to be as the amount of transmission through the structure increases with the increasing permeability.

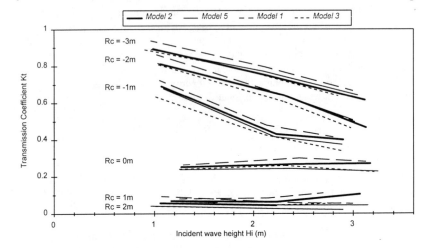

Figure 3 : Transmission Coefficient for T = 4.5s

The results obtained for transmission are generally in good agreement with existing empirical prediction methods such as that presented in the 'Manual on the use of rock in coastal and shoreline engineering' (CIRIA 1991) as illustrated in Figure 4.

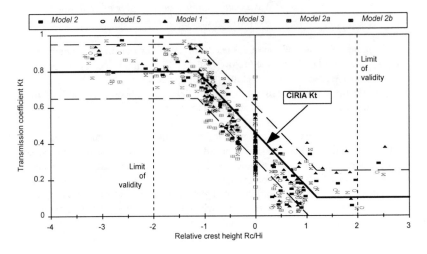

Figure 4 : CIRIA curve for K_t and test results

However, it can be seen that when the relative submergence is large the measured transmission is higher than predicted, tending towards 100% wave energy transmission, as may be expected for such conditions. For mean water levels near the crest level ($-1 < R_c/H_i < 1$) the measured transmission is generally less than that predicted by the formula whereas at larger freeboards ($R_c/H_i > 1$) the measured transmission is higher than predicted. These differences are due to the influence of the higher permeability of the tested models on the transmission mechanisms as compared to the test models from which the CIRIA formula was developed. Most of the test data falls within the 90% confidence bands of the CIRIA data curve (indicated by the dashed lines). There is considerable spreading of data around $R_c/H_i = 0$ where the effect of wave height cannot be accounted for by the formula.

A more comprehensive formula for predicting transmission was developed by van der Meer (van der Meer 1991). Figure 5 compares the measured values of K_t with those predicted by the van der Meer formula. The formula tends to underestimate the significant transmission of low steepness waves, probably due to the limited range of wave steepnesses from which the formula was derived.

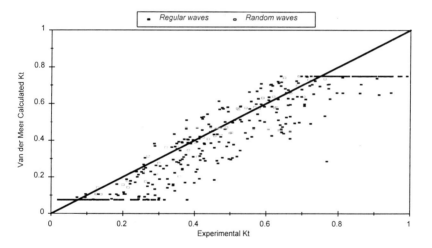

Figure 5 : Correlation between experimental K_t and K_t calculated by van der Meer equation

The formula limits the maximum transmission coefficient to 0.75 whereas values of K_t of up to 1.0 were measured experimentally for small wave heights and deep submergences.

SET-UP OF MEAN SEA LEVEL

Very little is known about the magnitude of the set-up to be expected inshore of a submerged breakwater. This two dimensional study has shown that the set-up can be very significant, giving rise to an increase in mean sea level of 1.0m with a 3.0m high wave under certain water level conditions. There is as yet no complete formula to predict set-up and the study has shown that the best existing prediction method (Diskin 1970) does not take account of all the important variables.

In this study it has been assumed that set-up δ is a function of the following variables:

$$\delta = f\left(\frac{H_i L}{T}, \frac{R_c}{h_c}, D_{n50}, B, \xi\right)$$

The terms in brackets may be described as the wave flow rate, relative submergence, a permeability factor, crest width and surf similarity parameter respectively.

Figure 6 shows the measured values of set-up as compared to those predicted by the Diskin formula. It can be seen that the formula predicts a much greater set-up for water levels at or below the crest level ($R_c \geq 0$) but is a reasonable estimate of set-up for water levels above the crest level ($R_c < 0$). This was assumed to be due to the difference in stone size and hence permeability of the Diskin models ($D_n 50 \approx 0.4$m) and the models tested in this study ($D_n 50 = 0.7$m to 1.0m).

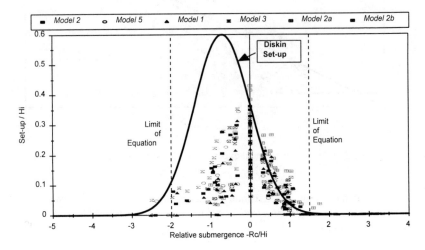

Figure 6 : Diskin curve for set-up and regular wave results

Figure 7 shows the influence of crest width on set-up where for Model 2 B=8m, for Model 2a B=12m and for Model 2b B=4m. It can be seen that the set-up increases with crest width.

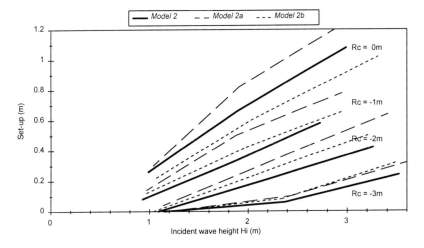

Figure 7 : Set-up for T = 11.2s

It was found that random waves, when defined by their mean wave height $\overline{H}$ rather than the significant wave height H_s ($\overline{H} \approx 0.6H_s$) generate the same amount of set-up as regular waves of equal wave height.

A new empirical method for predicting the set-up was developed from the data collected in this study:

$$\delta = \frac{1}{8} \frac{H_i}{T} \frac{L}{h} e^{-20(R_c/h_c)^2} \cdot f(D_{n50})$$

where $f(D_{n50})$ for the stone sizes tested in this study is approximately equal to 1. Figure 8 compares the measured values of set-up with the values predicted by the above equation.

CONCLUSIONS

This large scale model study of wave transmission and set-up at low crest rubble mound breakwaters has yielded high quality data with the minimum of scale effects. The research has confirmed the validity and detailed the limits of the van der Meer and CIRIA methods for the prediction of wave transmission at such breakwaters. The research has proposed a new formula for the maximum (unrelieved) set-up behind a detached breakwater.

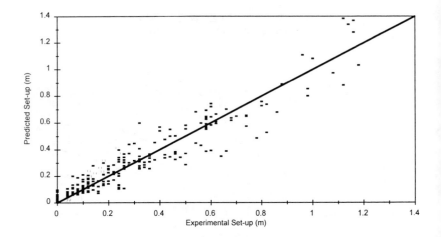

Figure 8 : Measured and predicted values of set-up (new prediction formula)

FURTHER RESEARCH

Although this research was conducted in two dimensions in a flume it was always realised that most detached breakwaters perform in a three dimensional environment. The advantages of working in two dimensions in this study have been those of greater control and increased scale and the use of a pumping system to control the actual amount of set-up. The research is due to be continued in a three dimensional model in the UK National Coastal Research Facility and the results obtained so far will provide a foundation for the interpretation of the results of the new study.

Further work will also be carried out on less permeable cross-sections in two dimensions in the University of Bristol flume to determine further the effect of stone size and permeability on the set-up.

ACKNOWLEDGEMENTS

The research described in this paper was funded by the UK Ministry of Agriculture Fisheries and Food.

REFERENCES

CIRIA, 1991 : 'Manual on the use of rock in coastal and shoreline engineering', CIRIA Special Publication 83, 1991.

Debski and Loveless, 1997 : 'The design and performance of submerged breakwaters', Dept of Civil Engineering Research Report, University of Bristol, February 1997.

Diskin, 1970 : 'Piling-up behind low and submerged permeable breakwaters', Journal of the Waterways and Harbors Division, ASCE 1970.

Van der Meer, 1991 : 'Wave transmission at low-crested structures', ICE Conference on Coastal Structures and Breakwaters 1991.

COMPOSITE MODELLING OF AN OFFSHORE BREAKWATER SCHEME IN THE UKCRF

S Ilic[1] , A Chadwick[1] , B Li[2] , C Fleming[2]

Abstract

A 3D physical model of an offshore breakwater scheme is to be constructed in the UK Coastal Research Facility at a scale of 1:28. Random, directional sea states and various tidal levels will be generated in the model based on field measurements of specific events. This experiment forms a part of a composite modelling. The paper describes the key aspects of the composite modelling approach with emphasis on the physical model design. Numerical models describing waves and wave induced currents have been used to design the optimum layout. The scaling problems associated with the offshore structures and mobile bed have been also investigated.

Introduction

The study of the effects of offshore breakwaters on beach evolution is a complex problem. Evaluation of the shoreline response is very complex because the beach response is governed by at least 14-15 different geometric, wave and sediment variables, which are still difficult to include in a single model. This complexity is exacerbated in a macrotidal environment. It is currently impossible to include all wave transformations, hydrodynamic and sediment processes in a physical or a numerical model. A composite modelling approach (Kamphuis, 1996) has therefore been chosen which integrates the benefits of both numerical and physical models. Incorporation of field measurements makes the composite model an even more powerful tool for investigating physical processes.

Numerical models do not suffer from the scaling problems associated with physical models in predicting prototype behaviour and can be applied, as a predictive tool, to both physical model and prototype scales. Current numerical models cannot describe the natural complexity of random directional sea states , wave non-linearities and long period motions. These phenomena are present in physical models however, and these results can then be used for numerical model validation. Field observations are limited to spot measurements of uncontrolled and uncontrollable physical events.

[1] Research Assistant; Reader - School of Civil and Structural Engineering, University of Plymouth, Plymouth, PL1 2DE, United Kingdom
[2] Senior Engineer; Director - Sir William Halcrow and Partners, Burderop Park, Swindon, UK

However, they are direct prototype measurements of random waves, non-linear wave characteristics and beach evolution.

Field measurements

Field measurements of wave transformation and beach evolution data were taken from September 1993 - January 1995. The chosen site, Elmer (Sussex, UK) is a renourished macro tidal beach comprising a shingle upper beach and lower sand beach overlying a chalk platform. The beach is protected by eight offshore breakwaters and a terminal groyne (see Figure 1). The scheme was designed using a lightweight physical model at a scale of 1:80. The construction of the breakwater scheme and beach fill along 2 km of frontage took place in 1992/1993 (described in Holland and Coughlan, 1993). Originally salients were predicted, however tombolos started to develop and interrupt the longshore transport along the coast.

Concurrent measurements of directional wave data offshore and in the lee of the breakwaters were taken at three hourly intervals almost continuously for a year. A summary of the results to date may be found in Chadwick et al (1994) and Ilic and Chadwick (1995). Additional fields measurements of wave transmission through the breakwater and currents around the breakwaters are currently in progress (Simmonds et al, 1997).

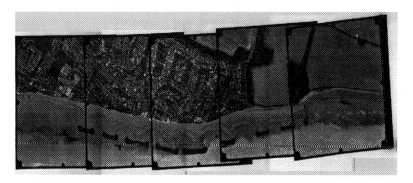

Figure 1   Aerial survey of the Elmer frontage (courtesy Arun District Council)

Six beach surveys using photogrammetry were taken in the period of wave data collection. This was arranged in collaboration with the local coastal protection authority - Arun District Council. Aerial surveys provided overlapping colour prints (Figure 1), at a contact scale of 1:3000, of the coastline up to 2 kilometres east of the scheme and 1 kilometre west.

Numerical models

A wave transformation numerical model based on the mild-slope equation (Li, 1994), has been evaluated using this field data (Ilic and Chadwick, 1995). Even with only the spot measurements, the agreement between measured and computed wave conditions, for directional seas, was within 8-12 %. To reduce this error, it was suggested that the influence of transmission through the breakwater, wave breaking and bed friction should be taken in account. It was also recommended, that wave current interactions should be further investigated. The new field measurements and physical model measurements should enable further model validation and improvements. Axe et al (1996) reported an evaluation of current design formulae for shoreline response behind multiple breakwater schemes, concluding that more comprehensive process based modelling is necessary. Currently a plan shape model is being validated. The data which will be collected in the UKCRF should help in the validation and development of more advanced 3D morphological models.

Physical model

One complete beach cell, between two breakwaters together with two further breakwaters will be modelled in the UKCRF (Simons et al, 1995) which determines the scale of 1:28. A physical study of the beach evolution will be modelled in two stages. The beach will be modelled first as a fixed bed during November/December 1997 and then as a mobile bed in May/June 1998. Two sets of tests will be carried out in the mobile bed model, using anthracite (currently used by HR Wallingford for modelling shingle beaches) and also coarse sand. The offshore bathymetry will be modelled as a fixed bed representing the chalk platform. The physical model studies will hopefully show the importance of the longshore and rip currents and their influences on the beach evolution. Also, the new data will be used for the validation of the numerical models and their further improvement.

The interaction between those three parts is incorporated in data and information exchange. The field data which were used for the numerical model validation, will also be used for the physical model simulation. Measurements taken during physical model tests will be compared with the field data. They may be used for a further study of the physical processes or for further model validation and developments.

However the physical model will include laboratory and scale effects. The design of the physical model requires careful planing of experiments and examination of scaling and laboratory effects. In the preparation phase the influence of the laboratory effects using numerical models has been studied. The scale effects influencing reflection, transmission and sediment transport have been studied in 2D physical models.

Physical model design using numerical modelling

In designing the physical model, a decision has to be made on the optimum positioning of the lateral boundaries, either mid way in the gap between breakwaters or at the mid point of a breakwater. One of the purposes of investigating this with a numerical model is to determine which configuration introduces the least boundary effects within the measurement area. A numerical model of wave propagation (Li, 1994) and a wave induced currents model (Li and Maddrell, 1994) have been used to define the optimum layout.

The chosen scale of 1:28 allows 25-28 cm of water offshore in front of the paddles. The required minimum depth of water to enable the paddles to work properly is 30 cm. Therefore, it was necessary to deepen the bathymetry in front of the paddles. Three different layouts were proposed - short steep slope 1:10, slope 1:20 and gentle long slope approximately 1:50. The mild-slope equation model was run for each of those proposed bathymetries with two different wave heights (prototype values, 2m and 1m), periods (prototype values, 10 s and 5 s) and three different directions (SW (12°), S (0°), SE (-12°)). It was found that the steeper slope in combination with the greatest wave height and larger period introduces significant bed reflection and thus changes the wave conditions in the lee of the breakwaters. Therefore, the gentle slope was accepted for further consideration.

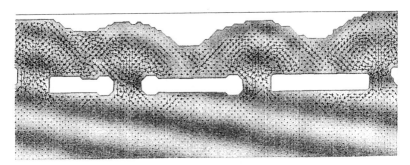

Figure 2   Wave-induced currents model results for the wave period of 10 sec, wave height of 2 m, SW direction, when the right lateral boundary was in the middle of the gap (only breakwaters 2, 3 and 4 are plotted here (counted from left to right))

Figure 2 show the wave-induced currents for a wave period of 10 sec and wave height of 2 m (prototype values) when the right lateral boundary was in the middle of the gap. The results of the wave-induced currents model is based on the results of the wave model. From the results it can be concluded that for this wave condition the relative gap width (ratio of the gap width to the incident wave length) is small. The diffraction occurs from the middle point of the gap and the longshore currents are not significant. There was no difference observed in bay 3 when the right lateral boundary was moved to the mid point of the breakwater. However, this solution allowed longshore currents to develop on the left part of the scheme. This might be useful for the feeding of sediment in the second stage of the experiment. Therefore, this proposed solution will be further considered for the bathymetry construction in the UKCRF.

2D studies of scale effects

Scale effects are defined as differences between prototype and model response that arise from the inability to simulate all relevant forces in the model at the proper scale, dictated by the scaling (similitude) criteria. Froude similarity is chosen to model wave propagation and hydrodynamics. It is important to check how the chosen similitude effects the flow through the permeable structures. It is also important to check, what influence it has on sediment scaling and hence on the modelled beach profile and longshore transport.

Structures

Wave interaction with permeable structures is complex due to the variable properties of the permeable breakwater - in particular reflection, transmission and friction. The wave transmission through the offshore breakwaters is thought to be an important process in salient/tombolo formation behind the breakwater (Hanson et al, 1989). To model properly the reflective and transmissive properties of the structure, it is important to model accurately the size, density and shape of the individual pieces of rock.

First, a check was made to determine whether turbulent flow through the model breakwater occurred. This proved to be the case. Next, a 2D process model with limestone rock prepared for the 3D model was constructed. The effect of rock placement and grading on the transmission and reflection characteristics of the breakwaters was investigated in terms of porosity, packing and void ratios. The model scale was 1:28 as in 3 D model. Two different gradings in terms of D_{85}/D_{15} were chosen, narrow grading (D_{85}/D_{15} =1.36) and wide grading (D_{85}/D_{15} =1.5 respectively. The difference was quite small because the material was graded from already prepared material for the 3D experiment. Also two different packings, loose and tight, were chosen. The water depth was kept constant and five different tests with five different periods were performed. The details are given in Edwards (1997) and Souster (1997).

The measured reflection coefficient (calculated as the ratio of the incident to reflected wave energy) values varied with different grading and packing showing an influence on the amount of reflection from the structure. The amount of reflection was higher for the tight packing than for the loose packing in the case of the same grading. Also the reflection coefficient was higher for the narrow grading than for the wide grading.

In Figure 3, the measured reflection coefficient is plotted versus nondimensional reflection number R

$$R = \frac{d_t \lambda_o^2 \tan \beta}{H_i D^2}$$

where d_t is mean water depth measured at the toe of the structure; D is characteristic diameter of rock armour; H_i is significant incident wave height; λ_o is deep water wave length (=$g/2\pi f_p^2$); and $\tan\beta$ is a structure gradient. The nondimensional reflection number was derived by Davidson (1996a) on the basis of field measurements in the front of the breakwater 4. The measured values in the 2D experiment are compared with the values calculated using Davidson's (1996a) formula:

$$K_r = 0.151 R^{0.11}$$

The calculated values are plotted by dashed lines in Figure 3. The results show rapid increase of the measured reflection coefficient for the lower R. No further increase in measured reflection is observed for a further increase in R. This is in accordance with the theoretical curve. This saturation can be related to surging wave conditions. However, the reduction in the reflected wave is primarily due to the wave transmission. The measured values differ slightly from values predicted by using Davidson's formula. During the 2D test only regular waves were simulated, thus irregularity and directionality of waves were not considered which might introduce a difference along with the scale and laboratory effects (Davidson, 1996b).

In the case of measured transmission (calculated as the ratio of transmitted to incident energy), the transmission coefficient values also vary with different grading and packing. The amount of transmission was larger in the case of loose packing than in the case of tight packing for both grading. Also for the wide grading the transmission was larger than in the case of narrow grading. It should be pointed out

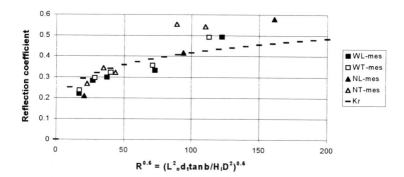

Figure 3 Measured reflection coefficient plotted versus reflection number $R^{0.5}$
(WL-mes, WT-mes, NL-mes, NT-mes represent measured values for wide
grading- loose packing, wide grading - tight packing, narrow grading -
loose packing, narrow grading - tight packing respectively, Kr represents
calculated values)

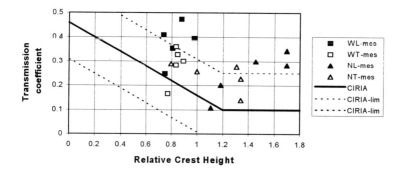

Figure 4 Measured transmission coefficient plotted versus relative crest height (
WL-mes, WT-mes, NL-mes, NT-mes represent measured values for wide
grading- loose packing, wide grading - tight packing, narrow grading -
loose packing, narrow grading - tight packing respectively, CIRIA
represents calculated values using Van der Meer expression given in
CIRIA manual (1991))

that the grading differed only slightly and that packing had an influence on the porosity and transmission. In the case of wide grading and loose packing the porosity had the highest value of 29%.

The transmission coefficient is plotted versus the relative crest freeboard Rc/Hs (crest height/ incident wave height) in Figure 4. Van der Meer's (1990) analysed several hydraulics model transmission results and suggested an expression for the transmission coefficient related to the relative crest freeboard Rc/Hs. The values calculated by Van der Meer's expression are plotted with a solid line in Figure 4. The upper and lower bounds of the calculated values are plotted with dashed lines. The 2D experiment results are very close to the proposed expression or lie within its confidence limits. It is evident that, for the same Rc/Hs, the transmission coefficient varied with packing and grading. This graph does not show the dependence of the transmission coefficient on wave shape, wave length, angle of approach. In the 2D study, it was confirmed that wave transmission through a breakwater is related to its permeability. Also, Kt decreases with decreasing wave length, breakwater porosity, increasing wave height and breakwater width.

An improved expression for wave transmission should be derived from current field measurements and analysis (Simmonds et al, 1997). The comparison between laboratory and field measured reflection and transmission will enable a choice to be made of the grading and packing which will minimise the difference between model and prototype.

Mobile bed

Mobile bed modelling is more complex. The review of previous work showed that it is still unclear how to properly scale sediment and the definition of the grain size for the mobile bed remains an art.

Kamphuis (1985) classified coastal models. Two of them are considered here, the best model (BM) and the light weight model (LWM). The best model accounts for the similitude of densimetric Froude number, relative density and length, resulting in a model sediment with the same density as the prototype and grain size reduced in accordance with the geometric scale. The Lightweight Model does not satisfy the relative density and consequently the grain size is not scaled geometrically. Usually those models satisfy the grain Reynolds number and the densimetric Froude number. Neither of these model satisfies the similitude of fall velocity. Thus they do not represent onshore/offshore motion correctly. However, they should simulate initiation of motion correctly which is more significant for the shingle beaches where bedload transport predominates.

There are still doubts, about the accuracy of the use of light weight material for beach modelling (Janssen, 1992; Loveless et all 1995). Therefore, the performance of two materials, sand and anthracite, will be investigated. Both materials will introduce some scale effects. Considering a shingle beach with D_{50}=10.22 mm, at a scale 1/28, using sand with a specific gravity of 2650 kg/m^3, and anthracite with a specific gravity of 1390 kg/m^3 , the best model and the lightweight model produce the grain sizes of 0.365 mm and 2.5 mm respectively. In Table 1 the different scale effects for both models are presented.

For the LWM, the HR approach, described by Powell (1988), was considered. Firstly the grain size was determined satisfying the same permeability in the prototype and in the model as given by Yallin (1965). Using the threshold of motion of the sediment particles, given by Komar and Miller (1973), the specific gravity is obtained.

Table 1   Scale effects for BM and LWM

Model	N_D	N_ρ	N_{M^*}	N_{R^*}	N_{G^*}	N_n	N_Q
BM	28	1	~1	148.15	1	1.14	0.9-1
LWM	4.36	1.91	0.39	11.49	6.42	3.18	~0.67

The grain scale ($N_D = D_p/D_m$) is 28 and 5.5 for BM and LWM respectively. The density scale ($N_\rho = \rho_p/\rho_m$) is 1 and 1.91 for BM and LWM respectively. The mobility ($N_{M^*} = n^{1/4}\, n_D^{-1/4}\, n_\Delta^{-1}$) is larger in LWM than in the prototype which can influence the model longshore transport. Grain Reynolds Number ($N_{R^*} = n^{3/2}$) is smaller in BM. According to Kamphuis (1985) if the Grain Reynolds Number in the prototype is larger than about 3 $n^{3/2}$ no serious problems should occur. This should be satisfactory for the shingle beach. When v_* goes to zero (twice each wave cycle), the sediment transport conditions will be distorted. The geometric link between waves and sediment ($N_{G^*} = n_a/n_D$) is not equal to 1 in the LWM, which means that the bottom friction will be distorted. Slope distortion ($N_n = n^{1/2}/n_B^{3/2}$) was calculated using Bruun/Dean expression as given by Kamphuis (1996) even though this expression is more relevant for sand than for a shingle beach. The slope in LWM will be larger than in the prototype which will also influence the longshore transport. The longshore sediment transport scale effect ($N_q = N_L^{5/2}\, N_{mb}^{3/4}(N_L/N_{D50})^{1/4}$) was calculated using Kamphuis' (1991) expression. The result shows that the longshore transport will be overestimated in the LWM.

Loveless (1994, 1995) considered the percolation force on the sediment at or near the bed. If the percolation force is larger than the submerged weight, it will have an effect on the sediment transport. In the case of LWM the percolation force will be 1.5-2 times larger than in the prototype which might cause a difference in the predicted sediment transport.

Table 1 shows that both models will introduce scaling effects. It is necessary to design a 2D model prior to the UKCRF study to compare the beach profile with the prototype profile. Also, some other considerations like angle of repose should be further investigated.

Conclusions

The composite modelling approach was chosen to study the shoreline response behind the offshore breakwaters. The approach which incorporates the advantages of numerical, physical modelling and field measurements can be a powerful tool to study the complex problem of beach evolution.

A wave transformation model and wave induced currents model were used to design the physical model. The numerical models results enable the bathymetry layout to be chosen which reduces the bed reflection and lateral boundary effects.

A 2D study of reflection and transmission and a comparison with the field measurements should determined the best grading and packing to be chosen to design the model structure with the same permeability characteristics as the prototype structure.

Some further investigations of sediment scaling is suggested. A 2D beach profile model might help to define the differences between best model (BM) and prototype and lightweight (LWM) and prototype.

Acknowledgements

This work was supported through funding by the Engineering and Physical Sciences Research Council (EPSRC). The authors would also like to acknowledge the contribution of Adrian Channel (HR Wallingford), Steve Edmonds, Rob James, Ian Morgan and Tony Tapp (University of Plymouth) in technical aspects of this work. Thanks are due to students Chris Edwards and Mike Souster who carried out the experiment and also to Sally Leng. We are very grateful to Roger Spencer and Ray Trainer at ARUN DC. Thanks to Phil Axe, Mark Davidson and David Simmonds for their collaboration, and to Tom Baldock for his helpful comments.

References

Axe P, Ilic S, Chadwick A J, 1997, Evaluation of Beach Modelling Techniques Behind Detached Breakwaters, *Proc. of 25th International Conference on Coastal Engineering*, Orlando 1996, ASCE, pp 2036- 2049.

Chadwick A J, Fleming C, Simm J, Bullock G N, 1994, Performance Evaluation of Offshore Breakwaters - A Field and Computational Study, *Proceedings of Coastal Dynamics '94*, ASCE, pp. 950-961.

CIRIA, Manual on the Use of Rock in Coastal and Shoreline Engineering, *Centre for Civil Engineering Research and Codes, Report 154*, London

Davidson M A, Bird P A D, Bullock G N, Huntley D A, 1996a, A New Non-dimensional Number for the Analysis of Wave Reflection from Rubble Mound Breakwaters, *Coastal Eng 28 (1-4) Sept*, pp 93-120.

Davidson M A, Bird P A D, Huntley D A, Bullock G N, 1996b, Prediction of Wave Reflection from Rock Structures: an Integration of Field and Laboratory Data, *Proc. 25th International Conf. Coastal Engineering*, Orlando 1996, ASCE, pp 2077-2086

Edwards C, 1997, An Investigation of Scaling Parameters for Rock in Physical Model Tests, *University of Plymouth, School of Civil and Str. Eng, Final Year Project*, PRCE 303

Hanson H, Kraus N C, Nakashima L, 1989, Shoreline change behind transmissive detached breakwaters, *Proceedings of coastal zone '89*, ASCE, pp 568-582

Holland B, Coughlan P, 1993, The Elmer Coastal Defence Scheme, *Proceedings of MAFF Conference on River and Coastal Engineering*, Keele, UK

Ilic S, Chadwick A J, 1995, Evaluation and Validation of the Mild Slope Evolution Equation Model for Combined Refraction-Diffraction Using Field Data, *Proceedings Coastal Dynamics 95*, ASCE, pp 149 -160.

Kamphuis J W, 1985, On Understanding Scale Effect in Coastal Mobile Bed Models, *Physical Modelling in Coastal Engineering*, R. Dalrymple (Ed), Balkema, pp 141-162.

Kamphuis J W, 1991, Alongshore Sediment Transport Rate, *Journal of Waterway, Port, Coastal and Ocean Engineering*, Vol 117, No 6, ASCE, pp. 624-640.

Kamphuis J W, 1996, Coastal Hydraulic Models, *Physical Modeling Workshop, 25th International Conference on Coastal Engineering*

Komar P D, Miller M C, 1973, The Threshold of Sediment Movement under Oscillatory Water Waves, *J. Sediment Petrol.*, 43, pp 1101-10.

Janssen C M, 1993, Evolution of Beach Profiles Under Random Waves, *Unpublished Msc Thesis,* Imperial College of Science, Technology and Medicine

Li B, 1994a, An Evolution Equation for Water Waves, *Coastal Engineering*, 23, pp 227-242

Li B, Maddrell R, 1994b, A three-dimensional Model for Wave Induced Currents, *Proceeding of 24th International Conference on Coastal Engineering*, Kobe 1994, ASCE, pp 2297-2310.

Loveless J H, 1994, Modelling Toe Scour at Coastal Structures, *2nd International Conference on Hydraulic Modelling: Development and Application of Physical and Mathematical Models,* Stratford upon Avon, June.

Loveless J H, Grant G T, 1995, Physical Modelling of Scour at Coastal Structures, *Proceedings HYDRA 2000 Vol. 3,* Thomas Telford, London, pp. 293-298.

Powell K A, The Dynamic Response of Shingle Beaches to Random Waves, *Proc. 21st International Conference on Coastal Engineering*, ASCE, pp. 1763-1773.

Simons R R, Whitehouse R J S, MacIver R D, Pearson J, Sayers P B, Zhao Y, Channell A R, 1995, Evaulation of the UK Coastal Research Facility, *Proc. Coastal Dynamics 95,* ASCE, pp. 161-172.

Simmonds D J, Chadwick A J, Bird PA D, Pope D, 1997, Field Measurements of Wave Transmission Through a Rubble Mound Breakwater, *Proc. Coastal Dynamics 97,* ASCE, (this publ).

Souster M, 1997, An Investigation of Scaling Parameters for Rock in Physical Model Tests, University of Plymouth, *School of Civil and Str. Eng, Final Year Project*

Van der Meer J W, 1990. Data on Wave Transmission Due to Overtopping, *Delft Hydraulics Report No. H 986,* Delft, The Netherlands

Yallin M S, 1963, A Model Shingle Beach with Permeability and Drag Forces Reproduced, *Proceedings IAHR Congress,* London 1963, pp 169

CRF Study of Wave Kinematics in Front of Coastal Structures

James Sutherland[1] and Tom O'Donoghue[2]

Abstract

Experiments carried out in a large coastal facility to measure water surface elevations and water particle velocities in front of different coastal structure models using a wide range of normally-incident and obliquely-incident, long-crested random seas are described. The chosen experimental parameters resulted in a wide range of wave-structure interaction, varying from almost full reflection to heavy breaking on the structure. Comparison of predicted and measured velocities in the frequency and time domains, have shown that a linear theory-based prediction method, developed previously for normally-incident waves, can be extended to obliquely-incident waves to yield good estimates of the velocity field in front of the structure. In practice, errors in the predicted velocities are more likely to result from poor definition of the reflection coefficient spectrum than from the limitations of linear wave theory.

Introduction

Beach response in front of reflective coastal structures is dependent on the near-bed hydrodynamic regime generated by the interaction of incident and reflected waves. It is important therefore to be able to predict the near-bed flow-field for such conditions. O'Donoghue and Goldsworthy (1995, 1996) compared measurements of velocity at many positions in front of impermeable structures with various slopes with predicted velocities based on simple superposition of incident and reflected wave components and linear wave theory. They found that the relatively simple prediction method yields encouragingly good results at least as far as u_{rms} values are concerned. Hughes (1992) also obtained good agreement between measured and predicted velocities in front of a vertical, impermeable wall using a prediction method similarly based on simple superposition and linear wave theory and later

---

[1] Research Fellow, University of Aberdeen, Department of Engineering, King's College, Aberdeen AB24 3UE, Scotland
[2] Lecturer, University of Aberdeen, Department of Engineering

applied the method to estimating velocities in front of sloping structures (Hughes and Fowler, 1995).

The work of O'Donoghue and Goldsworthy and of Hughes and Fowler was concerned with waves approaching with normal incidence to the structure. The present paper is concerned with extending this previous work to the case of obliquely-incident random waves. Oblique wave attack results in a complicated short-crested sea being generated in front of the structure. The angle of attack influences the phase and magnitude of reflected wave components which need to be known in order to determine the kinematics. The main emphasis of the present paper is to see how well the extended theory performs in predicting the near-bed velocity-field through time in front of the structure given the incident waves at some location seaward of the structure. To this end a series of experiments were carried out in the UK Coastal Research Facility (CRF) to measure surface elevations and velocities at a number of locations in front of various coastal structure models and to measure reflection coefficient and reflection phase spectra. Velocities predicted using the extended theory are compared with the measured velocities. The measured incident wave condition, the measured reflection coefficient spectrum and an empirical formulation developed during the course of the study for the phase shift on reflection are used as input for the prediction.

Problem Definition and Linear Theory

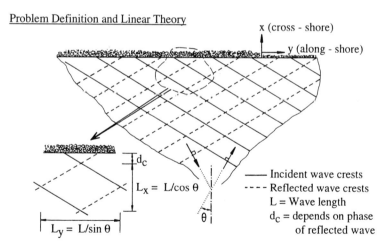

Figure 1: Definition sketch

Figure 1 presents a definition sketch for the wave field generated by the interaction of obliquely-incident and reflected waves. For simplicity the sketch shows the situation for the case of a single frequency wave with angle of incidence θ and wavelength L approaching the structure over a uniform water depth h. The x axis corresponds to the cross-shore direction with $x=0$ at the toe of the structure; the y axis corresponds to the along-shore direction; the z axis is vertically upwards with $z=0$ at the still water level. The short-crested sea and associated water particle velocities which result from the interaction of the incident and reflected waves for this single frequency case is very well described by Hsu and Silvester (1989).

For the case of long-crested, multi-frequency or random waves, it is assumed here that the incident and reflected wave components can be linearly superimposed. The surface elevation spectrum at any cross-shore position x in front of the wall is independent of the along-shore position and is given by

$$S_{\eta\eta}(x,f) = S_{ii}(f)\left[1 + C_r^2(f) + 2C_r(f)\cos(2k_x(f)x + \gamma(f)\right] \qquad ...(1)$$

where f is linear frequency, $S_{ii}(f)$, $C_r(f)$ and $\gamma(f)$ represent the incident wave spectrum, the reflection coefficient spectrum and the reflection phase spectrum respectively and k_x is the cross-shore wave number given by $k_x = 2\pi\cos\theta / L$, with L and θ dependent on x and f if the bed is sloping. Corresponding expressions for the along-shore and cross-shore velocity spectra are given by

$$S_{vv}(x,f) = S_{ii}(f)\left(TF\right)^2 \sin^2\theta\left[1 + C_r^2(f) + 2C_r(f)\cos(2k_x(f)x + \gamma(f)\right] \quad ...(2)$$

and

$$S_{uu}(x,f) = S_{ii}(f)\left(TF\right)^2 \cos^2\theta\left[1 + C_r^2(f) - 2C_r(f)\cos(2k_x(f)x + \gamma(f)\right] \quad ...(3)$$

respectively, where TF is the linear wave velocity transfer function given by

$$TF = \frac{2\pi f \cosh k(z+h)}{\sinh kh} \qquad ...(4)$$

Equations (1), (2) and (3) show that the surface elevation spectrum and the along-shore and cross-shore velocity spectra can be calculated for any position in front of the structure if the incident wave spectrum, the reflection coefficient spectrum, the reflection phase shift spectrum and the incident wave angle are known. Frequency-domain statistical quantities such as rms values can be determined directly from the spectra.

To generate time series of surface elevation or velocity for a location in front of the structure, an IFFT is applied to the spectrum with an appropriate phase spectrum. The appropriate phase spectrum must account for the random phase ε associated with each component of the incident wave spectrum at $(x,y,t)=(0,0,0)$, the location (x,y) in front of the structure and the phase shift on reflection γ. The phase associated with a component of the surface elevation spectrum and the along-shore near-bed velocity spectrum at (x,y) in front of the wall is

$$-k_y y - \tan^{-1}\left[\frac{\sin(k_x x + \varepsilon) - C_r \sin(k_x x + \gamma - \varepsilon)}{\cos(k_x x + \varepsilon) + C_r \cos(k_x x + \gamma - \varepsilon)}\right] \qquad \ldots(5)$$

where $k_y = 2\pi \sin\theta / L$ is the along-shore wavenumber. The corresponding phase associated with the cross-shore near-bed velocity spectrum is

$$-k_y y - \tan^{-1}\left[\frac{\sin(k_x x + \varepsilon) + C_r \sin(k_x x + \gamma - \varepsilon)}{\cos(k_x x + \varepsilon) - C_r \cos(k_x x + \gamma - \varepsilon)}\right] \qquad \ldots(6)$$

Therefore, if the incident wave spectrum and random phase spectrum are known, the corresponding time series for surface elevation and velocity can be calculated for any position in front of the reflective structure, provided of course the reflection coefficient spectrum and the reflection phase spectrum are known.

CRF Experiments

Experiments were carried out in the UK Coastal Research Facility (CRF) in which surface elevations and water particle velocities were measured at a number of locations in front of reflective coastal structures for long-crested random waves normally- and obliquely-incident to the structure. The CRF is a 57m by 27m coastal wave basin containing a central 36m wide by 22m long section for wave generation, testing and dissipation. In the present study the facility was used to generate long-crested, random waves which propagated from a depth of 0.5m at the wave paddles to a 1:20 beach slope 8.7m from the paddles and then along the slope to the structure aligned parallel to the paddles. Experiments were conducted with no structure, a vertical, smooth, impermeable wall, a 1:2 (v:h) sloping, smooth impermeable wall and a rubble mound breakwater. However, the results from the rubble mound breakwater experiments are not available for the present paper. Both impermeable structures had a total length of 8.4m and a toe water depth d_t of 240mm.

The wave conditions were synthesised from JONSWAP spectra with the following nominal f_p(Hz), H_s(mm) combinations: (0.5,50), (0.7,50), (1.0,50), (0.5,100), (0.7,100) and (1.0,100), i.e. a combination of 3 peak spectral frequencies and 2 significant wave heights, one relatively low and one relatively high. For each sea, tests were carried out for $\theta = 0^0$, 15^0 and 30^0. A wide range of wave behaviour was produced at the structure, ranging from heavy breaking onto the structure to almost perfect reflection.

Measurements of surface elevation and water particle velocity were made along one cross-shore line. A 2.4m long aluminium beam, suspended from the facility's bridge, supported 6 conductance wave gauges and 3 acoustic Doppler velocimeters (ADV) which measure 3 components of velocity at a point. The inshore ADV was located in a water depth of 248mm, the middle ADV was 695mm away in a water depth of

283mm and the offshore ADV was a further 1554mm away in a water depth of
361mm. The measurement volume associated with each ADV was located
approximately 50mm above the bed. Three conductance wave gauges were co-
located with the 3 ADVs with the other 3 gauges located between the offshore and
middle ADV. The complete set of wave gauges constituted a variable spacing array
of gauges from which reflection could be determined.

Figures 2(a) and 2(b) show examples of the measured surface elevation and
cross-shore velocity spectra for the inshore and offshore locations respectively for a
particular experiment. At the inshore location the surface elevation spectrum shows a
strong peak at the incident wave peak spectral frequency of 0.5Hz while the velocity
spectral density is relatively low in the incident wave frequency range. For this
particular sea therefore, the inshore location corresponds to an "anti-nodal" location
with high wave activity and low horizontal velocities. In contrast, the offshore
location is seen to be "nodal" (Figure 2(b)) with high velocity spectral energy and
low wave activity.

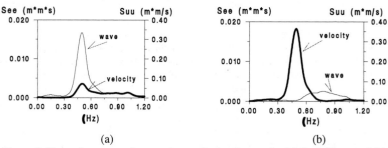

(a)   (b)
Figure 2. Example wave and cross-shore velocity spectra for (a) the inshore and (b)
the offshore locations for the same experiment.

Wave Reflection

In order to predict the wave kinematics in front of the structure, the incident
waves, the reflection coefficient spectra and the reflection phase spectra need to be
known. Using data from regular and irregular wave experiments conducted in the
CRF and in the wave flume at Aberdeen University, as well as data from other
sources, Sutherland and O'Donoghue (submitted) show that the wave phase shift on
reflection is uniquely determined by a non-dimensional parameter χ_3 defined by
structure slope, toe depth, wave period and angle of incidence. Based on the large
experimental dataset, they recommend the following empirical equation for the phase
shift on reflection:

$$\gamma = -8.84\pi\chi_3^{1.25} \qquad\qquad ...(7)$$

where $$\chi_3 = \frac{1}{\tan\alpha}\sqrt{\frac{d_t\cos\theta}{gT^2}} \qquad\qquad ...(8)$$

and α is the structure slope. The phase shift is independent of structure permeability and wave height. For a given wall slope, toe depth and angle of incidence, equation (7) can be used to calculate the phase shift on reflection for any wave period or frequency, i.e. the phase shift spectrum can be determined.

The present paper uses the above result of Sutherland and O'Donoghue for determining the wave reflection phase shift spectrum in the prediction of the wave kinematics in front of the structure. However, determining the reflection coefficient spectrum is much more problematic, mainly because reflection coefficient depends on wave height and structure permeability. There is therefore no formulation similar to equation (7) for reflection coefficient. (Hughes and Fowler (1995) do suggest an empirical equation for reflection coefficient as a function of frequency but, as argued by O'Donoghue (1996), the equation is over-simplistic.) Therefore, in the present absence of a method for predicting reflection coefficient spectra with any real accuracy, measured reflection coefficient spectra are used here as input to the prediction method. A least squares method using simultaneous measurements of surface elevation from an arbitrary number of wave gauges in any shape of array was developed for the present study to determine the incident wave and phase spectra and the reflection coefficient spectrum within a limited frequency range around the incident wave spectral peak.

Comparison of Measured and Predicted Velocities

Frequency domain comparison
Using the measured incident wave spectrum and reflection coefficient spectrum obtained using data from 3 of the wave gauges and a wave phase shift on reflection spectrum determined using equation (7), the water particle velocity spectra corresponding to the offshore, middle and inshore locations were calculated and compared with the corresponding measured velocity spectra: rms velocities, peak spectral period, average spectral period and spectral width parameter can all be determined from the spectra for comparison.

Figure 3 shows the predicted rms cross-shore velocities plotted against the corresponding measured rms cross-shore velocities for the no structure, vertical wall and sloping wall cases. There are 54 data points in each case, corresponding to 6 seas, 3 angles of incidence and 3 locations in front of the structure. The trends in the results for different angles of incidence were the same and so the angles are not identified in Figure 3. Velocities for the no structure case reach a maximum rms value of 0.12m/s while rms velocities in front of a structure reach 0.26m/s, the higher velocities corresponding to "nodal" positions. The results in Figure 3 show that the predicted velocities are generally higher than the measured velocities: the average difference is just over 8% and nearly 90% of the predicted values are within 15% of the measured values. The over-prediction is a feature of the no-structure results as well as the with-structure results indicating that the over-prediction is inherent in the linear wave theory used and that it is not related to the reflection processes.

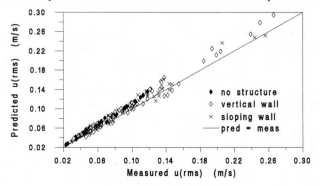

Figure 3. Predicted versus measured rms cross-shore velocity

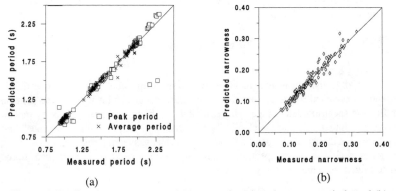

(a)   (b)

Figure 4. Predicted versus measured (a) spectral peak and average periods and (b) spectral narrowness parameter

Figure 4 shows predicted peak spectral period, average spectral period and spectral width parameter plotted against the corresponding measured values. Agreement between the predicted and measured values is excellent indicating excellent agreement between the predicted and measured spectral shapes. Because measured reflection coefficient spectra are used in the predictions, the agreement shown in Figure 4 means that the reflection phase spectra are well predicted using the empirical equation of Sutherland and O'Donoghue (submitted).

Time domain comparison
The predicted velocity time series for the 3 ADV positions were then compared with the measured velocity time series: peak velocity statistics and non-dimensional peak statistics have been compared and the rank order correlation between each measured and predicted time series calculated. Figure 5 shows 3 examples of measured and predicted cross-shore velocity time series at the inshore location corresponding to the same sea with no structure present (top), with the vertical wall present and with the sloping wall present (bottom). The corresponding Spearman rank order correlation coefficients for the 40s of data shown are are 0.98, 0.82 and 0.93.

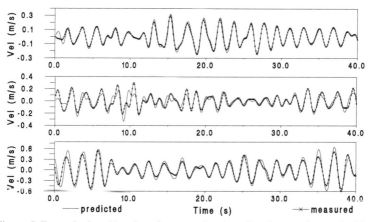

Figure 5. Example time-histories of measured and predicted cross-shore velocities

Figure 6(a) presents the predicted mean of the highest one third of the peak cross-shore velocities ($U_{\frac{1}{3}}$) plotted against the corresponding measured values. As for the rms velocities, the predicted $U_{\frac{1}{3}}$ velocities are generally higher than the measured values and this is true for both positive and negative peak velocities: again, the average difference is just over 8% and over 85% of the predicted values are within 15% of the measured values. Figure 6(b) presents the predicted versus measured $U_{\frac{1}{3}}$ values normalised with respect to the rms velocity. There is generally very good agreement between the predicted and measured values as evidenced by the very large percentage of the results falling on the predicted=measured line. This implies that the overpredictions seen earlier are a result of general overprediction rather than overpredictions of the extreme values only. The non-dimensional results are seen to cluster around ±2.0, the expected value for a Gaussian probability density function with zero mean.

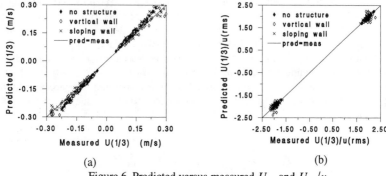

(a)   (b)

Figure 6. Predicted versus measured $U_{1/3}$ and $U_{1/3}/u_{rms}$

Figure 7 presents the Spearman rank order correlation coefficients between predicted and measured cross-shore velocity time series plotted against rms cross-shore velocity for the no-structure, the vertical wall and the sloping wall tests. The correlation values are normally very high and always very high in the case of the no-structure tests. Low values of correlation tend to correspond to low values of rms velocity. This is understandable as small errors are relatively more significant in these cases.

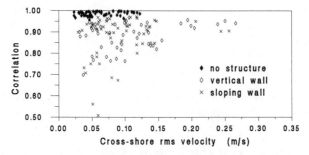

Figure 7. Spearman rank order correlation coefficients between predicted and measured cross-shore velocity time series

Conclusion

Water surface elevations and water particle velocities have been measured at a number of locations in front of different coastal structure models in the CRF using a wide range of normally-incident and obliquely-incident, long-crested random seas. The chosen structure types, toe depth and incident wave conditions resulted in a wide range of wave-structure interaction, varying from almost full reflection to breaking on the structure. Comparison of predicted and measured velocities in the frequency and time domains, have shown that the linear theory-based prediction method used previously for normally-incident waves can be extended to obliquely-incident waves to yield good estimates of the velocity field in front of the structure given the incident waves, the incident wave angle, the reflection coefficient spectrum and the reflection phase shift spectrum. Results from the rubble mound breakwater experiments are expected to confirm the general conclusions. The results of Sutherland and O'Donoghue (submitted) provide a good basis for determining the reflection phase shift spectrum but determination of the reflection coefficient spectrum is more problematic. In practice therefore, errors in the predicted velocities are more likely to result from poor definition of the reflection coefficient spectrum than from the limitations of linear wave theory. Current work on the CRF and other data is focussed on the reflection coefficient spectrum problem.

Acknowledgements

The research was carried out with funding from the UK Engineering and Physical Sciences Research Council (EPSRC). The authors also acknowledge HR Wallingford Ltd. as co-sponsors of the UK Coastal Research Facility.

References

O'Donoghue,T. and Goldsworthy,C.J., (1995), *Near-bed velocities in front of sea walls,* XXVIth IAHR Congress, HYDRA 2000, London, 1995, Vol 3, pp311-316.
O'Donoghue,T. and Goldsworthy,C.J., (1996), *Random wave kinematics in front of sea walls,* in Advances in Coastal Structures and Breakwaters, J.E.Clifford (ed.), Inst Civil Engineers, pp241-252 (ISBN: 072772509).
O'Donoghue,T., (1996), *Estimating wave-induced kinematics at sloping structures,* ASCE, J. Waterway, Port, Coastal and Ocean Engng., 122(6), pp303-305, Discussion of paper 07357.
Hsu,J.R.C. and Silvester,R., (1989), *Model test results of scour along breakwaters,* ASCE, J. Waterway, Port, Coastal and Ocean Engng., 115(1), pp 66-85.
Hughes,S.A., (1992), *Estimating wave-induced bottom velocities at vertical wall,* ASCE, J. Waterway, Port, Coastal and Ocean Engng., 118(2), pp175-192.
Hughes,S.A. and Fowler,J.E., (1995), *Estimating wave-induced kinematics at sloping structures,* ASCE, J. Waterway, Port, Coastal and Ocean Engng., 121(4), pp209-215.
Sutherland,J. and O'Donoghue,T., (submitted Aug.'97), *Wave phase shift at coastal structures,* submitted to ASCE, J. Waterway, Port, Coastal and Ocean Engng.

SUSPENDED SEDIMENT RESPONSE TO WAVE REFLECTION

Jonathon R. Miles[1], Paul E. Russell[1] and David A. Huntley[1].

Abstract

A field experiment has been carried out to examine the nature of sediment suspension and transport in front of a seawall. Synchronous high frequency measurements of surface elevation (using pressure transducers), water velocity (using electromagnetic current meters) and suspended sediment concentrations (using optical backscatter sensors) have been taken from in front of the Brunel seawall at Teignmouth, South Devon, UK. Measurements were also made simultaneously on an adjacent natural beach to provide control data. Wave reflections from the seawall lead to an increase in suspended sediment concentrations, and alter the phase of sediment suspension. The results emphasise the importance of wave reflection on the dynamics of beaches backed by seawalls.

Introduction

Measurements of sediment suspension and transport processes have been made on a variety of different types of beach, from macrotidal high energy dissipative beaches (e.g. Russell, 1993) to steeper reflective beaches (Davidson *et al.*, 1993). Sediment transport processes vary depending on the incident wave conditions, the sediment characteristics and the steepness (reflectivity) of the beach. An impermeable vertical seawall at the head of the beach increases the reflection coefficient of the shoreline considerably, and would therefore be expected to alter the nature of sediment suspension and transport both in the cross-shore and longshore directions. Initial results of this field investigation of the suspended sediment response to wave reflection from a coastal structure using *in-situ* measurements were presented by Miles *et al.* (1996). In the present paper results are presented which illustrate that there are significant differences between the suspension of sediment in front of a seawall and on an adjacent natural, undefended beach.

---

[1] Institute of Marine Studies, University of Plymouth, Drake Circus, Plymouth, Devon, PL4 8AA

Field site

　　　　To compare sediment suspension on a natural beach with suspension in front of a wall, a location was required with a wall adjacent to a natural beach. The site chosen was Sprey Point at Teignmouth in South Devon, U.K., which has a 7m high seawall fronted by a macrotidal beach. To the south of Sprey Point is a natural beach which receives the same incident wave conditions as the Sprey Point wall.

A rig designed to minimise interference with the hydro and sediment dynamic conditions whilst providing a stable base for instrumentation was built and bolted to the wall for a period in June 1995. Two EMCMs, (electromagnetic current meters) two OBSs (optical backscatter sensors) and a PT (pressure transducer) were attached to this rig and suspended 1.2m seaward of the wall in a vertical array. A similar array of instruments was dug in to the adjacent beach on approximately the same depth contour to provide an experimental control. The layout of the site is shown in Figure 1. The base of the wall at Sprey Point slopes at an angle of $53°$ and the average beach slope in front of the wall is approximately $3.3°$. The beach to the south is of similar slope to the beach in front of the Sprey Point wall, having an average slope of $3.9°$. The grain size of the sand is that of medium quartz sand with $D_{50} = 0.24$mm.

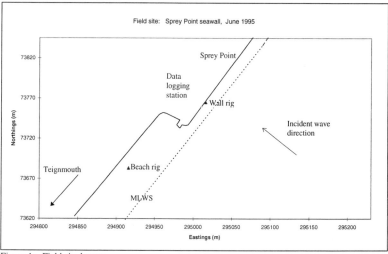

Figure 1: Field site layout.

Data was logged at 8Hz and pre-filtered at 2 or 4Hz. Surveys and sediment samples were obtained daily to establish beach morphology. The beach profiles and the position of the instruments on the beach are shown in Figure 2.

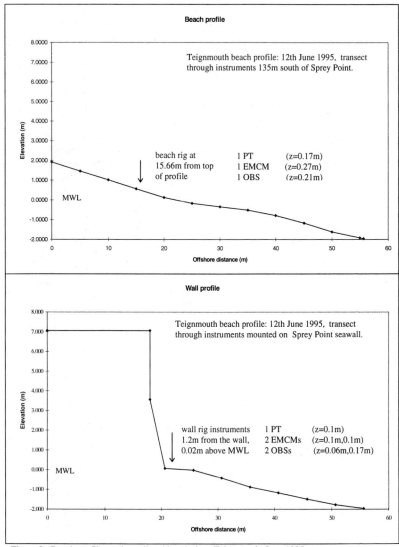

Figure 2: Beach profiles at the wall and beach sites, Teignmouth, June 1995.

<u>Results</u>

        Data analysed in this paper were collected on the afternoon tide of 13th
June, 1995. Water depth was taken as the normalising factor for the comparison of
data between the two rigs, as the wave height was approximately constant between

runs when similar water depths covered the two rigs. The incident wave heights measured at the beach rig decreased gradually during the tide from 0.22m to 0.14m.

<u>The nature of sediment suspension</u>

Major differences in the hydrodynamics of the natural beach and seawall environments result in different sediment dynamic conditions for these two cases. In order to illustrate the major differences several sections of data are presented. The first two examples demonstrate that more sediment is suspended in front of the wall (Figure 3) than on the natural beach (Figure 4) in similar incoming wave conditions.

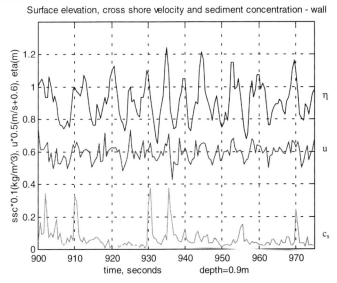

Figure 3: Time series of surface elevation, velocity (scaled by 0.5, offset by 0.6m/s), sediment concentration (scaled by 0.1) at wall.

Maximum wave elevations are greater at the wall rig (Figure 3) than at the beach rig (Figure 4), while the beach rig measured greater cross-shore orbital velocities (note scale of axes used). Typical values of wave period (T=4s) and water depth (h=1m) give a wavelength of λ=12.5m, placing the node at $\lambda/4$=3.1m from the wall. The differences in velocities and elevations are therefore to be expected, as the instruments are closer to the elevation antinode at the wall than the node. Traditional models of sediment suspension suggest that velocity gradients in the boundary layer lead to a shear stress which is responsible for suspending sediment (Bagnold, 1946). Despite smaller velocities at the wall rig, a greater number of suspension events resulted in a considerably larger amount of entrainment at the wall than was observed at the beach rig. In the examples shown, instruments were at similar heights and incident wave conditions were identical. Even so, maximum

concentrations at the wall reach 3.9kg/m^3, while at the beach rig suspension events
are absent, and the concentration fails to rise above 0.1kg/m^3.

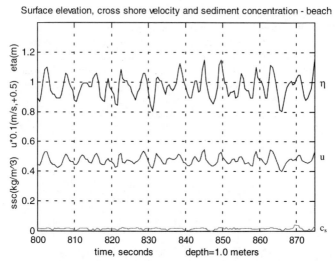

Figure 4: Time series of surface elevation, velocity (scaled by 0.1, offset by 0.5m/s), sediment
concentration at beach.

Wave heights at the breakpoint during this tide were approximately 0.3m on the
natural beach. The relationship $H_b/h = 0.78$ indicates the breakpoint water depth
should therefore be approximately 0.4m. The closest the beach rig instruments
were able to get to this before becoming intermittently dry was when the water
depth was 0.48m. Measurements made at this time are shown in Figure 5. They
differ considerably from those made at the wall in a similar water depth (Figure 6).

In Figure 5, wave elevation and velocity traces exhibit noticeable asymmetry as the
waves are shoaling and are about to break. Suspension events recorded during this
run reached a maximum of 1.2kg/m^3. In a similar water depth at the seawall, with
the same incident wave conditions, maximum concentrations were recorded of just
over 9kg/m^3 (Figure 6). The complicated elevation and velocity traces in the
signals were typical of runs where the water was shallow and waves passed the
sensors as both incoming and outgoing 'solitary' waves. The resulting sediment
suspension traces are complex.

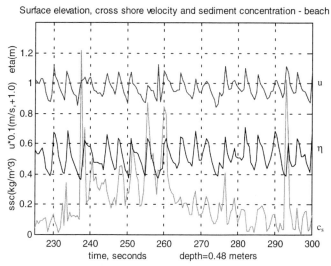

Figure 5: Time series of velocity (scaled by 0.1 and offset by 1.0m/s), surface elevation and sediment concentration at beach.

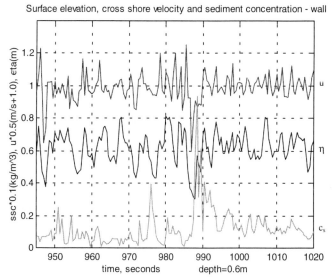

Figure 6: Time series of velocity (scaled by 0.5 and offset by 1.0m/s), surface elevation and sediment concentration (scaled by 0.1) at wall.

The phase of suspension

At the beach rig, the time-series show suspension events occurring in phase with onshore velocity (e.g. Figure 5, peak at 251 seconds). Suspension events which occur at this phase of the wave may be driven by the velocity shear in the boundary layer (Bagnold, 1946), which is maximum when the velocity outside the boundary layer is maximum. Shoaling waves become skewed, and their onshore velocities become larger than the offshore velocities. When the offshore velocity fails to rise above the threshold for suspension, suspension events only occur at the wave crest. In this case, the suspension events often appear to develop during the acceleration phase of the wave, and reach their maximum just before the crest of the wave passes (e.g. Figure 5, peak at 294 seconds). Generally however, the majority of sediment suspension peaks observed at the beach rig reach their maximum at or approaching the maximum velocity. Huntley and Hanes (1987) show that when suspension events occur in phase with onshore velocity, this gives rise to onshore sediment transport at the frequency of the waves.

At the seawall however, the phase of suspension is less clear. Suspension events are observed to occur most frequently with the maximum elevation. However, the complicated nature of the velocity field means that this may be at times of maximum velocity or at times of flow reversal. When the velocity is maximum, a plausible explanation is that the sediment peaks are driven by the velocity shear in the boundary layer (e.g. Figure 3, peak at 910 seconds). However at times of flow reversal, the velocity (and shear stress) is 90^0 out of phase with the elevation for the standing wave case, and the suspension events which occur at this time are therefore occurring at times of minimum shear stress (e.g. Figure 3, peak at 955 seconds).

The suspension of sediment at flow reversal has also been documented by Foster *et al.* (1994) who suggested that the mechanism may result from the release of turbulence from the bed during the reversal of the velocity. Murray (1992) also documented a flow reversal peak from a flume experiment and found that in some cases, the flow reversal peak was in fact larger than the shear peak.

Data simulation

A simple data simulation has been applied in order that the processes driving the time series results can be conceptualised. In the simulation, a wave is advected past some 'instruments', reflected from a wall 1.2m away, and the wave passes the instruments on its way out. Instrument heights in the data simulation are the same as in the field experiment. Solitary wave theory was applied because of its simplicity and similarity to waves in shallow water (Bagnold, 1947).

The equations used were as follows:

$$\eta_{in,out} = H \sec h^2 \left(\sqrt{\frac{3}{4} \frac{H}{h}} \frac{x}{h} \right)$$
  Boussinesq (1872)

$$u = c N \frac{1 + \cos(Mz/h)\cosh(Mx/h)}{\left[\cos(Mz/h) + \cosh(Mx/h)\right]^2}$$
  McCowan (1891)

where $c = \sqrt{g(h+H)}$, and both M and N are functions of H/h, calculated by Munk (1949). Typical values of H=0.2m, h=1.0m give values of M and N as 0.65 and 0.3 respectively.

For a progressive solitary wave, the elevation is in phase with the velocity. The total elevation and velocity for the incident and reflected wave were obtained by adding the incoming and outgoing components of η and u. A reflection coefficient (r) of 0.9 was determined from the field data (Miles *et al.*, 1996) and this was used in the simulation to obtain the amplitude of the outgoing wave. A mean onshore velocity was added of similar magnitude to that observed in the field ($\overline{u} = 0.1$m/s) Suspension was assumed to be a function of the u_{tot}^n where $u_{tot} = u_{in} + u_{out} + \overline{u}$. Various values of n have been suggested from n=2 (Inman and Bagnold, 1963), n=3 (Owen and Thorn, 1978), to n=5 (Madsen and Grant, 1976). The magnitude of the suspension peak was considered less important than the phase at which it occurred, so for the purposes of the simulation a value of n=3 was used, and the relationship $c_s = u_{tot}^3 = (u_{in} + u_{out} + \overline{u})^3$ was applied. The simulation results are shown in Figure 7.

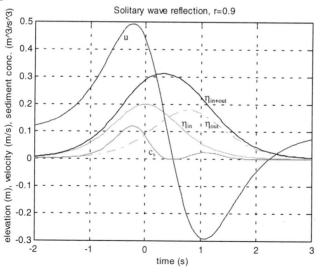

Figure 7. Simulation of solitary wave reflection past the instruments and the associated suspension.

The incoming and outgoing waves (η_{in} and η_{out}) are separated by the time it takes to travel to and from the wall. The difference in the direction of the velocity which is associated with each of these waves, combined with the lag between them results in the rapid change in cross-shore velocity (u). The reflection coefficient of 0.9 and the superimposition of the oscillatory flow on a mean onshore flow results in two sediment peaks, the second of which is smaller than the first.

The structure in the time series of the simulation above was also found to exist in the field data. An example is given in Figure 8. Of particular interest in this example is the suspension of sediment which occurs by both the incoming and outgoing waves, as in the simulation. Maximum suspension of sediment occurs at the same time as the velocity maximum in each case, and as in the simulation, the onshore directed sediment peak is smaller than the offshore peak.

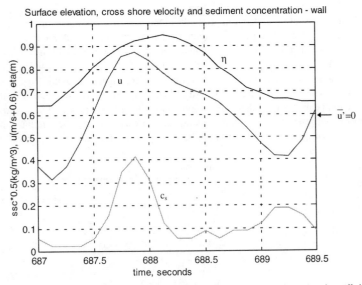

Figure 8. Time series of elevation (η), velocity (u) and sediment concentration (c_s) at the wall rig.

While the data simulation above is a gross simplification of the processes driving suspension in front of the wall, it is a useful tool in understanding the contributions of the incoming waves, the outgoing waves and the mean cross shore flow to sediment suspension. Wave reflection from a seawall radically alters the hydrodynamics in front of the wall, and this experiment has shown that this in turn affects the nature of sediment suspension considerably.

Conclusions

1. The observations have shown that near-bed suspended sediment concentrations on a beach backed by a seawall are increased by an order of magnitude relative to near-bed suspended sediment concentrations on an adjacent undefended beach.

2. The suspension of sediment in front of the wall may occur at times of maximum shear *or* at times of flow reversal.

3. Sediment suspension in front of the seawall results from the complex interaction of incoming waves, outgoing waves, mean currents and turbulence. Further work is required both in the observation and modelling of these processes.

References

Bagnold, R.A., 1946. Motion of waves in shallow water; interactions between waves and sand bottom. *Proceedings of the Royal Society*, London, series A, 187, 1-15.

Bagnold, R.A., 1947. Sand movement by waves: some small scale experiments with sand of very low density. *Journal of the Institution of Civil Engineers*, 27(4), 447-69.

Boussinesq, J. 1872. Theorie des ondes et des remous qui se propagent le long d'un canal des vitesses sensiblement paralleles de la surface au fond. *J. Math. Pures et Appliquees*, Lionvilles, France, 17, 55-108.

Davidson, M.A., Russell, P.E., Huntley, D.A., Hardisty, J. and Cramp, A., 1993a. An overview of the British beach and nearshore dynamics programme. *Proceedings of 23rd International Conference on Coastal Engineering '92*, 1987-2000.

Foster, D.L., Holman, R.A. and Beach, R.A., 1994. Sediment suspension events and shear instabilities in the bottom boundary layer. *Proceedings of Coastal Dynamics '95*, 712-726.

Huntley, D.A. and Hanes, D.M., 1987. Direct measurement of suspended sediment transport. *Proceedings of Coastal Sediments '87*, American Society of Civil Engineers, 723-737.

Inman, D.L. and Bagnold, R.A., 1963. Beach and nearshore processes. Part II. Littoral processes. In Hill, M. N. (Ed), *The Sea*: Vol. 3, Wiley Interscience, New York, 529-553.

Madsen, O.S. and Grant, W.D., 1976. Quantitative description of sediment transport by waves. *Proceedings of 15th Coastal Engineering Conference*, American Society of Civil Engineers, 1093-1112.

McCowan, J., 1891. On the solitary wave. *Phil. Mag., series 5, 32*, 45-58.

Miles, J.R., Russell, P.E., Huntley, D.A., Sediment transport and wave reflection near a seawall. *Proceedings of the 25th International Conference on Coastal Engineering '96*, 2612-2624.

Murray, P.B., 1992. *Sediment pick up in combined wave-current flow*. Unpublished Ph.D. Thesis, University of Wales, 189pp.

Owen, M.W. and Thorn, M.F.C., 1978. Effect of waves on sand transport by currents. *Proceedings of 16th Coastal Engineering Conference.*, 1675-1687.

Russell, P.E., 1993. Mechanisms for beach erosion during storms. *Continental Shelf Research 13(11)*, 1243-1265.

A Field Investigation of Functional Design Parameter Influence on Groin Performance: The Great Lakes Groin Performance Experiment

Guy A. Meadows[1], William L. Wood[2], Brian Caufield[3],
Lorelle Meadows[4], Tom Bennett[5], Marc Perlin[1],
Hans Van Sumeren[4], Don Carpenter[6], and Robert Mach[6]

Abstract

A five week long field experiment was conducted on the U.S. tideless, coastline of one of the North American Great Lakes. Four individual groins, varying in two distinct design parameters, height and length, were installed and monitored during this experiment. The response of the shoreline and the nearshore and short term beach evolution are evaluated to determine the net impact on sediment dynamics. An even/odd analysis method is utilized to examine shoreline change. A composite beach profile model is used to modify the cross-shore profile of longshore current and sediment transport and hence, to determine effective groin length. This analysis shows that groin height influences the degree of bathymetric change, while the groin length determines the dynamic responsiveness of the system.

Introduction

For centuries, groins have been built for the purpose of protecting beaches along the world's coasts. During this extensive period of time, design changes have

----

[1] Associate Professor, The University of Michigan, Ann Arbor, Michigan, USA
[2] Professor, Purdue University, West Lafayette, Indiana, USA(deceased)
[3] Graduate Student, Purdue University, West Lafayette, Indiana, USA
[4] Research Associate, The University of Michigan, Ann Arbor, Michigan, USA
[5] Great Lakes Specialist, Michigan Department of Environmental Quality, Lansing, Michigan, USA
[6] Graduate Student, The University of Michigan, Ann Arbor, Michigan, USA

resulted primarily from "practical experience." While research has been directed at the problem of effective groin design a great deal of controversy exists both in literature and in practice over the use of groins as shore protection structures. Ostensibly, groins are built in an attempt to stabilize a length of beach where erosion is primarily due to net longshore sediment transport. To this end they are intended to increase beach width, extend the performance life of beach nourishments, and prevent the loss of sand from beaches to longshore sinks such as inlets and navigation channels. As pointed out by Kraus et. al. (1994), the performance of a characteristically simple groin structure is actually governed by a large number of parameters. In fact, Kraus et. al. (1994) argue that the response of the shoreline to groins is expressed by a functional relation involving at least 27 parameters.

A great deal of controversy remains over the influence of various design parameters on the performance of these structures. Kraus et. al. (1994) suggest that no guidance on groin functional design is available, which results in the loss of a potentially valuable shore-protection structure that can function effectively and economically under certain conditions. In addition, there is a public perception, reflected in the coastal zone management policy of many countries, that the placement of groins necessarily results in the destruction of beaches. It is within this context that the Great Lakes Groin Performance Experiment(GLGPE) was conceived and carried out.

Field Experimental Site and Climatic Conditions

The site chosen for GLGPE is the Ludington State Park north of Ludington, Michigan, USA (Figure 1). The Ludington State Park is located on the eastern shore of Lake Michigan, one of the five North American Great Lakes. This is a temperate zone lake experiencing prevailing westerly winds. The site is centrally located along the east coast of Lake Michigan with a maximum fetch of approximately 250 km to the northwest or southwest. The maximum significant waves incident upon this site are about 6.9 m in height. Over the majority of the year, the waves are locally generated wind waves with no swell. The field site is essentially a tideless environment, however, other factors can cause significant water level fluctuations. Both storm setup and the annual hydrologic cycles can cause fluctuations in mean water level on the order of 0.5 m. In addition, long term decadal climatic changes can result in water level fluctuations on the order of 2.0 m. The beach and nearshore region is composed of fine to medium sand and is free of man-made and natural barriers to longshore sediment transport. The nearshore region is a multi-barred profile throughout the year.

The experimental goals achieved as a part of the GLGPE were: 1) to instantaneously place four typical groin structures on a natural beach with straight

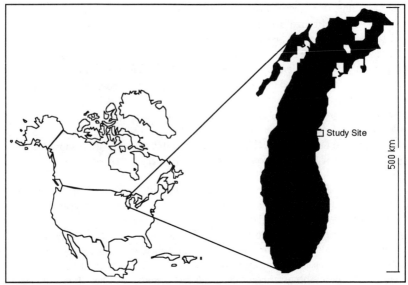

Figure 1: Location Map

and planar bathymetry, 2) to characterize the short term beach and nearshore response through multiple storm events, 3) to quantify up and down drift effects of typical groin structures, 4) to evaluate the effects of functional groin design parameters, and 5) to characterize the post-placement recovery of the beach.

The experimental design involved the monitoring of the effects of four groin types. The groins varied in two primary design parameters, groin length and groin height. Two long (30 m) and two short (15 m) impermeable groins were placed perpendicular to the shore (Figure 2). One long and one short groin were constructed as "high-profile" groins, (crest elevation totally above still water level), the other set of long and short groins were constructed as "low-profile" groins, (crest elevation totally below still water level). The groins were constructed of steel I-beams and were installed over a six hour quiescent period. The groins were spaced such that each groin acted as a singular unit. Therefore, the experiment consisted of four individual groins.

Field work consisted of periodic bathymetric surveys, daily offset measurements and daily still photography. In addition, environmental data was obtained from the National Oceanic and Atmospheric Administration (NOAA) buoy 45002, located approximately 65 km offshore; and the National Ocean Service (NOS) water level station 9087023, located 1.5 km south.

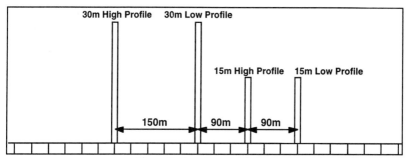

Figure 2: Experimental Setup

Over the duration of the experiment, the wave conditions incident at the site consisted of distinct storm seas with no swell component (Figure 3). The direction from which the storm waves originated varied with successive storms. The maximum storm wave was approximately 1.6 m. On average, the waves were less than 1 m in height.

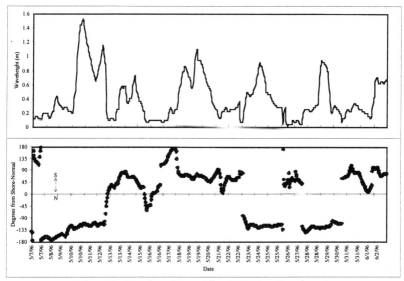

Figure 3: Nearshore Wave Climate During Experiment

Shoreline and Nearshore Response

As the incident wave direction shifted, the shoreline responded with changes in the offset magnitudes. Comparing the shoreline from May 7, 1996, and June 1,

1996, a general buildout of shoreland is apparent immediately adjacent to most of the groins (Figure 4). However, summed over the total control volume, the shoreline has receded between the groins, resulting in an overall net loss of shoreland.

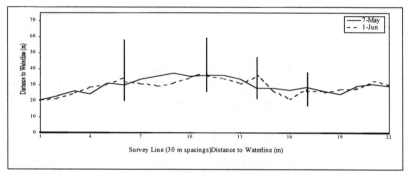

Figure 4: Shoreline Response

A similar pattern is found in an evaluation of the bathymetric response of the study site over the course of the experiment (Figure 5). There was an overall loss of approximately 452 cubic meters of beach and nearshore sediment between May 7 and June 1. The primary loss of material occurred near the shoreline between the structures.

Even/Odd Analysis

Shoreline change at any site consists of a complex response of the bathymetric contours to the incident environmental forces. The complexity of this interaction can cloud the analysis of a cause-effect relationship between the response and any single force. In the quest for a determination of the effect a groin structure induces on its surroundings, it is of use to somehow separate its effect from that of other factors influencing the site. One method for such a "separation" is the Even/Odd analysis utilized by Work and Dean (1990 a,b). By separating the shoreline response into even (symmetric) and odd (anti-symmetric) changes centered about the groin axis, one may singularly evaluate the effect of the structures.

For example, the even component of shoreline change, symmetric about the groin, reflects the influences of water level changes and the erosive impact of storms. The odd component of shoreline change, anti-symmetric about the groin, reflects the interruption of longshore sediment transport. For the purposes of this work, a shoreline change parameter, the position of the waterline on each side of the groin, was used instead of a shoreline change function. The shoreline change parameter for the duration of the experiment was split into its even and odd components (Figure 6).

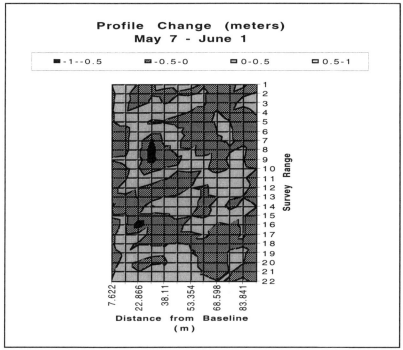

Figure 5: Bathymetric Response

The higher groins induced a greater initial and sustained response in both the even and odd parameter analysis. This indicates that the high-profile structures more severely impact the shoreline for a given set of environmental conditions. The fact that the even parameter behaves in such a manner is indicative of the potential for high-profile groins to trap material on both sides and build shoreland immediately adjacent to the structures quickly. However, it should be kept in mind that this build up is not without some cost, specifically, that of the shoreland outside of the immediate structure vicinity (Figure 5). The short groins induced faster response to changes in the incident wave direction and also had more dramatic offset magnitudes than their long counterparts. In essence, the shorter the structure, the quicker and more dramatic the reaction to changes in incident climate.

Profile Change Analysis

The effectiveness of a groin structure at trapping material and thus increasing beach width is a result of its interaction with the complex hydrodynamics in operation in the nearshore zone. One of the most important driving forces is the

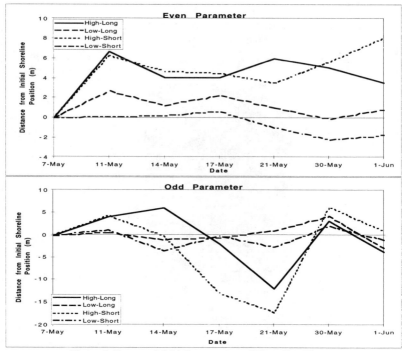

Figure 6: Even and Odd Components of Shoreline Change

longshore current distribution. Several theories exist as to the structure of this current over various nearshore profiles. As coastal engineering structures tend to modify the nearshore profile, they act also to modify one of the major forces which drive longshore transport. McDougal and Hudspeth (1989) propose a longshore current distribution based on a "composite beach profile."

The composite beach profile consists of a nearshore planar beach profile and a non-planar, concave up offshore beach profile. The intersection between the two profiles is denoted by a dimensionless parameter known as the R-factor.

$$R = \frac{\text{distance to intersection of planar and non - planar beach profiles}}{\text{distance to breaker line}}$$

The R-factor has been used to modify the Longuett-Higgins (1970 a,b) longshore current profile. As the value of R decreases, the point of maximum velocity shifts onshore and the velocity decreases in magnitude. Therefore, it would be expected that on a beach with a low R-value, the placement of a shorter groin would result in more cost-efficient sediment trapping and allow for faster bypassing.

Since the GLGPE study region hosts a composite beach profile, this approach to groin length design was evaluated. Based on an average profile from a line adjacent to the short high-profile groin for the beginning of the experiment and using a breaker of 0.6 meters, the R-factor was determined to be 0.08. During the course of the experiment there was an offshore movement of the location of the intersection between the planar and non-planar beach profiles (Figure 7). At the same time, the location of the breaker line did not change. Therefore, with time, the groin caused an increase in the value of R. Given the McDougal and Hudspeth longshore current distribution, the higher value of R implies a shift in the maximum velocity of the longshore current offshore. Thus, use of the initial R-factor in the determination of effective groin length may be misleading.

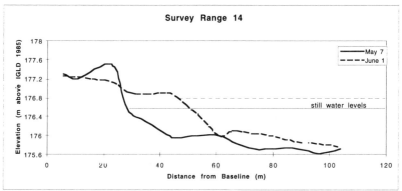

Figure 7: Profile Change from May 7 to June 1

Conclusions

The purpose of this investigation was to determine the influence of initial design parameters on groin performance on a natural sand beach. Two design parameters were chosen as the focus of this field investigation; groin height and groin length. The results of this analysis show that high-profile groins have a more marked impact on the shoreline offsets adjacent to the structures, responding more dramatically to changes in incident conditions. Also, the high-profile groins were more efficient at building shoreland immediately adjacent to the structures. In terms of groin length, the short groins responded quicker to changes in incident wave conditions with more dramatic results than the long counterparts. Also, the short groins were more efficient at building shoreland adjacent to the structures. However, this buildup is at the cost of shoreland outside of the structure boundaries.

An evaluation of the use of an R-factor profile analysis in designing effective groin length was undertaken. It was determined that the use of the preconstruction R-factor can be misleading. Because of the dynamic nature of the coastal zone, the R-factor is constantly changing. Thus, the initial estimate for groin length may result in an effective groin after the region undergoes bathymetric readjustment to the presence of the structure. These results could render the structure less effective than original designed.

Overall, the experiment region underwent drastic adjustment to the placement of the groin structures. Although the intention of such a structure in most settings is to build or maintain beach, it was found that this deployment resulted in a net loss of shoreland, as well as a net volumetric loss of bottom material. Clearly, further evaluation of structural parameters for proper groin design is needed in order to assure that structures perform in a positive, cost-effective manner.

Acknowledgment

The Great Lakes Groin Performance Experiment (GLGPE) was conducted with support of the U.S. Department of Commerce NOAA/Coastal Zone Management Program through the Michigan Department of Environmental Quality. The authors would like to thank the staffs of The University of Michigan's Ocean Engineering Laboratory; Purdue University's Great Lakes Coastal Research Laboratory; Michigan's Department of Environmental Quality; Michigan's Department of Natural Resources, Parks and Recreation Division; Hardman Construction, Ludington, Michigan; the United States Army Corps of Engineers, Detroit District; and the United States, National Oceanic and Atmospheric Administration's, Great Lakes Environmental Research Laboratory.

References

Berek, E. P. and R. G. Dean, "Field Investigation of Longshore Transport Distribution," Proceedings, 18th International Conference on Coastal Engineering, 1982, pp. 1620-1639.

Dean, J. L. and J. Pope, "The Redington Shores Breakwater Project: Initial Response," Coastal Sediments '87, 1987, pp. 1369-1384.

Kraus, N. C., Hanson, H., Blomgren, S. H., "Modern Functional Design of Groin Systems," Proceedings, 24th International Conference on Coastal Engineering, 1994, pp. 1327-1342.

Longuet-Higgins, M. S., "Longshore Currents Generated by Obliquely Incident Sea Waves, 1," Journal of Geophysical Research, Vol. 75, No. 33, 1970a, pp. 6778-6789.

Longuet-Higgins, M. S., "Longshore Currents Generated by Obliquely Incident Sea Waves, 2," Journal of Geophysical Research, Vol. 75, No. 33, 1970b, pp. 6790-6801.

McDougal, W. G. and R. T. Hudspeth, "Longshore Current and Sediment Transport on Composite Beach Profiles," Coastal Engineering, Vol. 12, No. 4, 1989, pp. 315-338.

Work, P. A. and R. G. Dean, "Shoreline Changes Adjacent to Florida's East Coast Tidal Inlets," Coastal and Oceanographic Engineering Department, University of Florida, Gainesville, Florida, January, 1990a.

Work, P. A. and R. G. Dean, "Even/Odd Analysis of Shoreline Changes Adjacent to Florida's Tidal Inlets," Proceedings, 22nd International Conference on Coastal Engineering, 1990b, pp. 2522-2535.

Interaction of Floating Breakwaters with the 2DH Hydrodynamic Processes in the Coastal Zone

E.E. Kriezi[1], Th. V. Karambas[1], C. Koutitas[2] and P. Prinos[3]

Abstract

A numerical model for the prediction of wave and wave induced current field behind a floating breakwater is presented in this work. An existing nonlinear breaking wave model is extended in order to include such structures. The model predicts well the transmission coefficient in comparison with analytical solutions and experimental data. It is concluded that the current field behind the structure (due to breaking waves) is very similar to that predicted in the case of a detached breakwater.

Introduction

The purpose of the present work is to investigate the effects of floating breakwaters on wave propagation and nearshore circulation. Floating breakwaters can be used for coastal protection and restoration in regions with limited tidal mean water variations (Mediterranean sea). Their deployment and withdrawal in different periods of the year offer dynamic opportunities for full control of the hydromorphodynamic conditions in the coastal zone. They can offer a satisfactory - not high - level of wave attenuation and protection from external wave disturbance. The main advantages of floating breakwaters are:

[1] Research Associate, [2] Professor, [3] Ass. Professor, Aristotle University of Thessaloniki, Department of Civil Engineering, Division of Hydraulics and Environmental Engineering, Thessaloniki, 54006, GREECE.

- Low cost construction independent of water depth and sea bed geological conditions because they do not need to be founded.

-Easy transportation, essential for temporary facilities.

-Ecologically advantageous since the sea water exchange beneath the structure is allowed.

There are numbers of performance studies reported on the transmission, reflection and energy dissipation due to a fixed or freely moved floating breakwater (Drimer et al., 1992, Fugazza and Natale, 1988, Black et al., 1971, Svendsen and Madsen, 1981, van Daalen, 1993).

Stoker (1957) gave an analytical solution for long waves, using a simplified assumption for the velocity potential beneath the structure. Other authors (Svendsen and Madsen, 1981, Drimer et al., 1992) also proposed solutions in the case where the floating breakwater width and the incident wave length are much larger than the gap between the breakwater and the sea bed. In the present work the wave length L is assumed to be much larger than the width b_o of the breakwater.

The present paper is divided in two parts. In the first part an existing time domain non linear dispersive wave propagation model is extended in order to include such structures. In the second part the effects of the structures on the wave induced nearshore circulation are studied.

Wave model

In the region in front and behind the breakwater a Boussinesq type model is used. Although the water depth can generally be considered as shallow, the waves produced by the breakwater movement are not long (their period is about 1-2 secs). Thus a dispersive wave model with improved linear dispersion characteristics is required in order to be valid in deep waters. Here, the type of equations proposed by Schaffer and Madsen (1995) is adopted.

$$\zeta_t + ((d+\zeta)U)_x + ((d+\zeta)V)_y + (\alpha-\beta+1/3)\ d^3\ (U_{xxx}+V_{xxy}) +$$
$$+ (\alpha-\beta+1/3)\ d^3\ (U_{xyy}+V_{yyy}) - \beta d^2\ \zeta_{xxt} - \beta d^2\ \zeta_{yyt} + \text{slope terms} = 0$$

$$U_t + UU_x + VU_y + g\zeta_x + (\alpha-\gamma)d^2\ (U_{xxt}+V_{xyt}) - g\gamma d^2\ (\zeta_{xxx}+\zeta_{xyy}) +$$
$$+ \text{slope terms} = 0$$

$$V_t + VV_y + UV_x + g\zeta_y + (\alpha-\gamma)d^2\ (U_{xyt}+V_{yyt}) - g\gamma d^2\ (\zeta_{xxy}+\zeta_{yyy}) +$$
$$+ \text{slope terms} = 0$$

where d is the water depth, ζ the surface elevation U, V are the horizontal velocities at an arbitrary level and α, β and γ are constants:

$\alpha = -0.02865$, $\beta = 0.40528$ and $\gamma = 0.01052$

Wave breaking is simulated in a similar way to Nowgu   (1996) introducing a dispersion term in the momentum equations:

$v_\tau U_{xx} + v_\tau U_{yy}$

$v_\tau V_{xx} + v_\tau V_{yy}$

where v_τ is the eddy viscosity coefficient, calculated by the solution of the turbulent kinetic energy equation:

$k_t + Uk_x + Vk_y = v_\tau (k_{xx} + k_{yy}) + B D - C_{dk} k^{3/2}/l_t$

where k is the turbulent kinetic energy, $C_{dk} = 0.08$, l_t is the turbulence length scale, $l_t = 0.3d$, D is the production term of the turbulent kinetic energy (D = $0.015 (c-u_o)^3$, as in Karambas et al., 1995), u_o is the bottom velocity and B is a parameter which ensures that turbulence is produced only when the surface velocity u_s exceeds the celerity:

$$B = \begin{cases} 0 & \text{in the region where } |u_s| < \text{celerity} \\ 1 & \text{in the region where } \qquad |u_s| > \text{celerity} \end{cases}$$

The eddy viscosity is given by:

$v_\tau = 3.0 \, k^{1/2} \, l_t$

The mean current is calculated by a numerical integration over the time of the horizontal velocities U and V.

The model is applied for the simulation of diffraction, refraction, partial reflection and transmission, shoaling, breaking, dissipation after breaking and wave induced current.

Effects of a floating breakwater on wave height distribution in the nearshore region

A rectangular cross section body is considered as a floating breakwater. Two cases are examined: first, when the body is fixed partially immersed and the second when the body is moved with only one degree of freedom.

In the case of a fixed body we consider the assumption that the water flux beneath the floating breakwater can be simulated like a flux in a tube. The continuity condition is applied on the boundaries 1-1 and 2-2, i.e. the left and right edge of the floating body and linear variation of the pressure is considered beneath the floating breakwater.

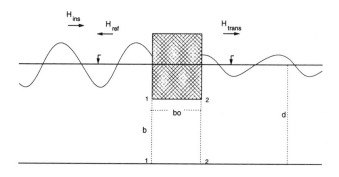

Figure 1. Definition sketch.

The velocity is calculated by the momentum equation between the two ends of the structure.

$$\frac{\partial v_o}{\partial t} + \frac{1}{\rho} \cdot \frac{\partial P}{\partial x} = 0$$

or

$$\frac{\partial v_o}{\partial t} = \frac{P_r - P_l}{\rho \cdot b_o} \tag{1}$$

where v_o is the velocity beneath the floating breakwater, b_o is its width, and P_r, P_l are the pressure at the edges (right and left respectively), calculated from the Boussinesq theory.

The above assumption is considered valuable when the rate of breakwater's width b_0 to wave length L is small.

The numerical solution is compared with the analytical solution of Stoker (1957), given for zero draft (b=0) and referred to long waves in shallow water:

$$C_r = \frac{\pi\theta}{\sqrt{1 + \theta^2\,\pi^2}}$$

$$C_t = \frac{1}{\sqrt{1 + \theta^2\,\pi^2}} \tag{2}$$

where $\theta = b_0/L$ and C_t and C_r are the transmitted and reflected coefficient respectively. The same assumptions, as in the numerical model, were adopted.

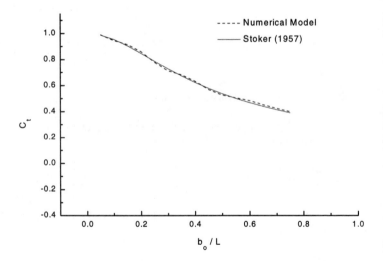

Figure 2. Transmission coefficient versus b_0/L in comparison with the analytical solution given by Stoker (1957) for zero draft (b=0).

The results of the model are comparable with the analytical solution.

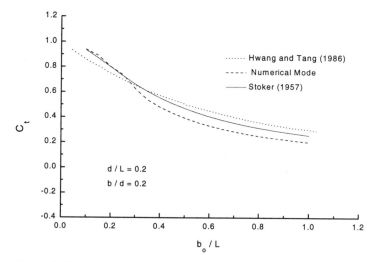

Figure 3. Transmission coefficient versus the rate b_o/L in comparison with analytical solutions (Stoker, 1957 and Hwang and Tang, 1986).

In figure 3 the numerical solution is compared with the analytical solution given by Hwang and Tang (1986) for $d/L = 0.2$ and body draft equal to $b=0.2d$. In this figure, the Stoker solution has been modified in order to include the effect of finite draft (given by Wiegel, 1964, p. 137).

Let's consider now that the body is floating with one degree of freedom (heave motion only). The dynamic equation of the motion of the floating body is written:

$$\sum F = m \frac{\partial^2 h_b}{\partial t^2} \tag{3}$$

where m is the mass of the body, h_b the vertical displacement and ΣF are the vertical forces applied on the body, i.e.:

- Body Weight
- Resistance R_{st}
- Buoyancy B_u

$$R_{st} = C_d \, \rho S w |w| / 2 \tag{4}$$

where w is the vertical velocity, S the body surface and C_d is the drag coefficient, $C_d=2$.
and

$$B_u = \int_S P \, n \, dx \tag{5}$$

Momentum and continuity equation are valid for calculating the pressure below the breakwater:

$$\frac{\partial v_o}{\partial t} + \frac{1}{\rho} \cdot \frac{\partial P}{\partial x} = 0 \tag{6}$$

$$\frac{\partial h_o}{\partial t} + \frac{\partial h_o v_o}{\partial x} = 0 \tag{7}$$

where $h_o = b + h_b$, the depth under the body and h_b the vertical displacement.

The analytical solution of the above equations is easily obtained by deriving the equations with respect to x and t, respectively, and by applying the boundary conditions (pressures P_r and P_l). The procedure leads to the determination of the pressure variation under the structure:

$$\frac{P}{\rho} = f \frac{x^2}{2} + \left(\frac{P_r - P_l}{\rho \cdot b_o} - f \frac{b_o}{2} \right) x + \frac{P_l}{\rho} \tag{8}$$

where

$$f = \frac{h_{ott}}{h_o} - \left(\frac{h_{ot}}{h_o} \right)^2$$

The integration of the pressure variation (8) gives the buoyancy B_u:

$$B_u = \frac{P_r + P_l}{2} b_o - \rho \frac{f}{12} b_o^3 \tag{9}$$

In figure 4 the model results   concerning the prediction of the transmission and reflection coefficient are presented.

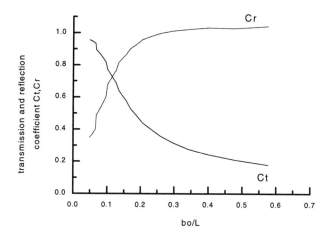

Figure 4. Transmission and reflection coefficients versus b_o/L.

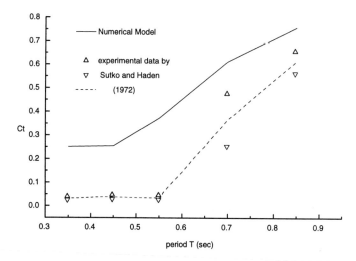

Figure 5. Transmission coefficient in comparison with experimental data (Sutko and Haden, 1974).

In figure 5, a comparison between the model results and experimental data by Sutko and Haden (1974) is presented, for a floating breakwater with the freedom of the heave movement. The depth of the tank was 0.3 m and the breakwater had a square cross section 0.1x0.1 m^2. Here the effects of the energy loss due to vortices are not taken into account and hence the model overestimates the transmission coefficient.

Effects of a floating breakwater on nearshore circulation

In the following numerical experiment waves are diffracted and transmitted behind the floating breakwater in a basin with slope 1/50 and initial depth 0.33 m. The breakwater (along x-axis) is located at a depth d=0.187 m and its length is 13.2 m. The paddle generate waves (with period T=1.7 s and height H=0.075 m) normal to the breakwater and the beach.

The velocity field (Figure 6) is very similar with the case in which there is a wall instead of a floating breakwater. However, as it is expected, the present case predicts smaller velocities. The current velocities were calculated by integrating over the time the horizontal wave velocities U and V.

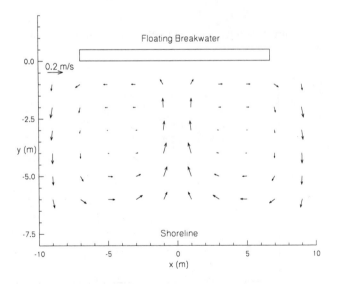

Figure 6. Breaking wave induced velocity field behind a floating breakwater.

Conclusions

The model is applied for the simulation of diffraction, refraction, partial reflection and transmission, shoaling, breaking, dissipation after breaking and wave induced current. The assumption made for the floating breakwater and the numerical model which describe the interaction with waves gives satisfactory results. The breaking wave induced current field behind the breakwater is of a similar pattern with that produced by a detached breakwater.

References

Black J.L.B., Mei C.C. and Bray M.C.G., 1971, "Radiation and scattering of water waves by rigid bodies", *Journal of Fluid Mechanics*, vol.46, part 1, pp. 151-164.

van Daalen E.F.G., 1993, Numerical and theoretical studies of water waves and floating bodies, Ph.D. Thesis University of Twente, Enschede.

Drimer N., Agnon Y., Stiassnie M., 1992, "A simplified analytical model for a floating breakwater in a water of finite depth", *Applied Ocean Research*, 14, pp. 33-41.

Fugazza M. and Natale L., 1988, "Energy losses and floating breakwater response", *Journal of Waterway Port Coastal and Ocean Engineering*, vol.114, pp. 191-205.

Hwang C. and Tang F.L.W., 1986 "Studies on rectangular surface barrier against short waves", *Proc. 20th Int. Conf. on Coastal Engineering*, ASCE, pp 1915 - 1928.

Nwogu, O. G., 1996, "Numerical prediction of breaking waves and currents with a Boussinesq model", to appear in *Proc. 25th Int. Conf. on Coastal Engineering*, ASCE.

Stoker J.J., 1957, *Water Waves,* Interscience Publishers Inc., New York.

Sutko A.A. and Haden E.L., 1974, "The Effect of Surge, Heave and Pitch on the Performance of a Floating Breakwater", *Proceeding of Floating Breakwater Conference*, Rhode Island, pp 41 -53.

Svendsen A. and Mandsen P.A., 1981, "The dynamics of wave induced ship motion in shallow water", *Ocean Engineering*, vol.8, no 5, pp. 443-479.

Wiegel R. L., 1964, *Oceanographical Engineering*, Predice-Hall, N. J.

Field Measurements of Wave Transmission through a Rubble Mound Breakwater

D.J. Simmonds[1], A.J. Chadwick[1], P.A.D. Bird[1] & D.J. Pope[2]

Abstract

A field experiment to measure transmission of wave energy through a porous rock island breakwater at Elmer on the South coast of the UK is described. Preliminary estimates of transmission, reflection and dissipation are discussed with reference to the statistical parameters of the incident wave field, and with respect to gravity band and low frequency response. For the wave conditions observed during the measurement period, results indicate that transmission is primarily affected by wave steepness and energy flux magnitude, but may also be changed by the existence of standing waves within the breakwater structure. Examination of the low frequency spectrum shows that seiches are excited in the bays between salients. The geometry of the bays appears to support such modes.

Introduction

Rubble mound breakwaters are widely used as cost–effective solutions for the protection of coastlines and harbours. The transmissivity characteristics of these structures can be a very important design consideration. The performance of island structures, in particular, can be radically affected by energy transmission through the interstices since overtopping may be insignificant. Transmission alters the degree of shelter offered by a breakwater and affects current circulations in the lee which, in turn, controls sediment dynamics. The wavelength dependence of the transmission coefficient (*Kt*) can also be important for design. In particular, it is recognised that long period motions preferentially propagate through porous structures and can induce dangerous seiching in "protected" harbours.

Whilst there have been many laboratory and numerical studies of energy transmission through porous structures relatively few full–scale measurements have been published. Dickson *et al* (1995) reported measurements of transmission through the Monterey Harbour breakwater. They note that transmitted energy

---

[1] School Civil & Structural Eng, Univ. Plymouth, Palace Court, PLYMOUTH PL1 2DE, UK
[2] Dept Civil Eng, Univ. Brighton, Moulsecoombe, BRIGHTON BN2 4GJ, UK.

decreases with increasing incident energy flux. Troch et al (1996) have reported field measurements of pore pressures gradients within a breakwater at Zeebrugge and found that steeper and shorter period waves were more rapidly attenuated.

At the simplest engineering design level, van der Meer (1993) has presented a widely recognised diagram which relates laboratory derived measurements of *Kt* to a single design parameter, the relative crest height[†]. This empirical approach provides no information about the relationship between *Kt* and wave shape, angle of approach or wave length. It simply represents transmission through overtopping and is of little use to the design of island breakwaters. Van der Meer therefore developed two more sophisticated but still empirical relationships for calculating *Kt* for submerged and island breakwaters using four dimensionless "primary parameters". These were *relative crest height, relative wave height, wave steepness, relative breakwater width*. Kondo & Toma (1972) remarked that thickness effects caused by standing waves within the structure can also change *Kr*. These comments were made in relation to tests performed on an idealised porous structure made from a lattice of cylinders. Their published data appears to indicate that *Kt* is also affected.

Theoretically, Solitt & Cross (1972) have predicted a change in the transmissivity of a porous structure at the point at which the scale of the fluid motions begins to exceed the armour diameter and hence the structure roughness. The structure behaviour passes through a transition after which the bulk properties of the porous structure can be considered in place of interactions of the waves with individual armour units.

There remains a question of how representative laboratory measurements of *Kt* are. There is a dilemma in choosing whether to scale to reproduce the effects of gravity or of turbulence within models. Both of these phenomena can significantly alter porous transmission. In addition, laboratory tests of schemes tend to employ extreme wave conditions, but when looking at the overall performance of a design it is important to assess performance during a range of conditions, representative of the day to day response of the system.

Hence there is a clear requirement for comparison with "full–scale models".

A Field Experiment

Although rubble mound breakwaters have been used in many countries since the beginning of this century, their adoption for use in Britain has been only recent with several schemes built since 1985. A composite model study (Ilic et al, 1997) of one of these schemes is currently in progress. Numerical modelling and laboratory tests in the UK Coastal Research Facility, Wallingford, will be compared to field measurements in order to evaluate state of the art coastal engineering practice.

The field site chosen for this study is at Elmer, on the South coast of UK and represents one of the most successful British schemes. The coast is macrotidal (*2.9–5.3m*, Neap–Spring) and subject to South Easterly storms from across the English Channel. The scheme consists of eight rubble mound island breakwaters with a

---

[†] R_c/H_s — crest height divided by incident wave height

single groyne to the east and has been augmented by a beach nourishment, completed in 1993. The scheme has performed well though there is some indication that excessive sediment has been trapped within the system (Simm et al, 1996) leading to possible starvation downstream.

Previous studies of this site have addressed wave field modelling around the structures and morphological development (Ilic & Chadwick (1995), Axe *et al.*,(1996)). Field measurements of the reflection properties of these breakwaters have been reported by Davidson *et al.* (1996a, 1996b). These studies have pointed to the need to quantify *Kt* in order to enhance understanding.

Measurement

A joint field measurement programme involving the Civil Engineering departments of the Universities of Plymouth & Brighton was started early in February *1996* and finished in July *1997* with the aim of studying energy transmission

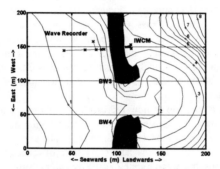

Figure 1: Plan of field site & equipment

through the breakwaters and the current circulations set up behind them.

Figure 1 shows the breakwater under study and the instrument layout. Measurements were made either side of the third breakwater (**BW3**) from the Western end. BW3 stands around *7m* above the chalk platform to seaward but the sea bed is up to *3m* shallower on the landward side due to accumulation of sediment. The seaward and landward slopes were estimated at $\approx 7/11$ and $\approx 5/9$ respectively with breakwater width between *20m* at the base to *4m* at the crown. The estimated D_{n50} was *1.2m* for the section of interest. The original irregularly shaped armour blocks are limestone but these have been reinforced with additional irregular syo*nite* blocks on the seaward faces.

The incident (& reflected) wave climate was monitored using the *P*lymouth *W*ave *R*ecorder *S*ystem (PWRS) a distributed pressure transducer array (*Bird et al.* (1994)). The wave climate inshore of the breakwater was observed using the *I*nshore *W*ave *C*limate *M*onitor (IWCM), a close packed star–array of wave–staffs reading surface elevation directly (Chadwick et al. 1995(2)).

At Elmer, the 3D topography of the bays, the macrotidal environment and the sloping rubble mound geometry combine to complicate measurements of *Kt*. When designing the instrument layout the first concern was to enable isolation of the transmitted signal from the measured wave field. Wave components reach the shadow of the structure through diffraction around the breakwater heads, refraction over the topography, reflection from the beach, as well as transmission and refraction through the structure. The contribution of all these varies with tidal elevation.

The site of the IWCM was chosen to minimise reflection from the beach behind the breakwater (the salient reflects energy obliquely away) and to minimise incidence of diffracted energy by siting on a relatively level area close up behind the breakwater. The star–array is compact which helps to ensure homogeneity of the wave field and water depth over the base area, but does not permit resolution of the directional wave field spectrum at long wavelengths.

The PWRS was installed seaward of the breakwater from the IWCM above the spring low water mark. Deployment was combined with a survey of both the beach and instrumentation using a *Leica* total station.

Time series from the IWCM were logged direct to PC via a buried beach cable. Control of and data download from the PWRS was achieved employing a *150m* RS–232 link operating at a reduced transfer rate. This was routed through the breakwater and buried in the salient. The simultaneous data sets from the PWRS and IWCM were stored to floppy disk and then archived onto compact disk. each of these are of *17* minutes duration, recorded every *40* minutes during intervals of inundation. PWRS records were acquired at *2Hz* and are approximately *2k* points in length. The IWCM acquired data at *4Hz* producing records of twice the length.

Currently, over *1,500* records have been collected between March 1996 and July 1997. This paper concentrates on data collected between Sept.–Dec. 1996 which includes several storm events (total *400* records).

Analysis

PWRS voltage files were converted to pressure equivalent using laboratory calibrations These files are corrected to equivalent water surface elevation by removal of atmospheric pressure and application of an inverse hydraulic filter. A judicious cut–off of *0.2Hz* was subsequently employed to prevent unrepresentative amplification of noise by the filter.

The *2D* method of Gaillard, Gautier and Holly (1980) was used to estimate the incident and reflected spectra. Davidson (1996a) has found that for the PWRS array geometry, the reflection analysis is insensitive to wave angles of incidence $\pm 30°$ of normal and previous field studies have shown that the observed value rarely exceeds $\pm 20°$ of normal. Only three time series are required for this technique and those from the transducers nearest to the structure excluding the most nearest were selected by the virtues of highest coherence and stability. Identification of the transmitted signal is achieved by directional analysis of the *4* IWCM channels using a maximum likelihood technique (Chadwick et al, 1995(1)).

The two above techniques both require cross–spectral estimates from the time series. In view of their different digitisation rates, estimates from the PWRS files were calculated using Welch method with Hanning window of width 256 points and 50% overlap and those from the IWCM files were obtained similarly but using a window of length 512 pts. This generated spectra with approximately 30 degrees of freedom, spectral bandwidth 0.0146Hz and spectral interval of 0.0078Hz.

Wave statistics of the incident field were derived following the *IAHR* guidelines (Darras *et al.*, 1987) from the incident PWRS spectra using a gravity band defined as 0.05–0.2Hz. Significant values of the spectra were selected by identifying those frequencies for which coherence was in excess of a threshold of 0.8 between all transducers used in the directional analysis and for which the power density was in excess of 5% of the peak value within the gravity band.

Parameter	Range
H_{sig}	0.06—1.57m
f_p	0.06—0.2Hz
d_t	0.6—5m
L_0	41—520m
L_{local}	15—120m
Ir	4—180
Ur	< 1
$\tan\beta$	0.636

**Table 1 Summary of data
collected Sept.–Dec. 1996**

A summary of the data collected over the period September–December 1996 is given in Table 1. Several periods of stormy weather were monitored with the largest waves (H_{sig}~1.6m) occurring over 27–29 October.

Estimation of *Kr, Kt & Kd*

Kr was calculated using the ratio of reflected to incident signal variances. Due to the difference in depth of the seabed either side of the structure the transmission coefficient *Kt* was calculated from the square–root of the ratio of transmitted to incident energy *fluxes* (see Dickson *et al.*, 1995).

The gravity–band averaged values *Kr_av, Kt_av & Kd_av* were calculated using the significant values of the discrete spectra in the band 0.05—0.2Hz.

$$K_{r\_av} = \sum_{0.05}^{0.2} \sqrt{\frac{E_r}{E_i}} \Bigg/ N$$

$$K_{t\_av} = \sum_{0.05}^{0.2} \sqrt{\frac{E_t . c_{gt}}{E_i . c_{gi}}} \Bigg/ N$$

E_i and E_r spectral estimates are generated by the 2D reflection analysis of the PWRS records. E_t is estimated by performing a summation of the directional spectrum produced by analysis of the IWCM over ±45°. Values of the coefficients, *Kr_p, Kt_p & Kd_p*, were calculated at the peak frequency of the incident spectrum.

Dissipation coefficients, **Kd** were inferred from conservation of energy.

$$K_d = \sqrt{1 - K_r^2 - K_t^2}$$

Where there was significant energy at long periods, the infra–gravity band averaged coefficients **Kr_ig, Kt_ig** were calculated using significant values in the band $f<0.05$. Summation was performed over 360° to calculate **Kt_ig** since the IWCM cannot resolve directionally at these scales.

A spreadsheet database of statistics and estimates of these coefficients was assembled. Some preliminary results are now presented.

Results — Gravity Band

The relationships between **Kr_av**, & **Kt_av** with Iribarren number, calculated using local wavelength are shown in figure 2. This effectively shows the dependence of the coefficients upon the wave steepness for the constant structure slope.

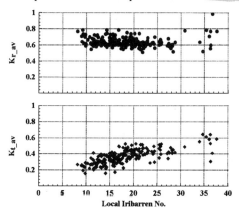

Figure 2 : Values of Kr_av **and** Kt_av **vs. local Ir.**

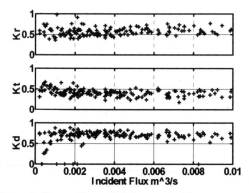

Figure 3: $Kr\ av, Kt\ av$ & $Kd\ av$ **vs. incident energy flux**

Reflection appears to be insensitive to *Ir* and takes higher values than those measured by Davidson (1996a) on an adjacent breakwater. This is likely to have been caused by the noticeable accretion of sediment within and behind the breakwater in the intervening period. The *Kt* values show a clear increase with decreasing wave steepness (larger *Ir*) caused by preferential dissipation of steeper waves on the face of the breakwater. This relationship is less obvious if the deep water wavelength is used to calculate *Ir*.

Kt values were found to increase with van der Meer's relative crest height parameter, (not shown). This conflicts with the general trend indicated within the limits of applicability of the van der Meer diagram, since most of the results from this field experiment lie outside those limits. This strengthens the argument that the relative crest height parameter is inappropriate for characterising *Kt* in cases where transmission is dominated by porous flow through the structure rather than overtopping.

Figure 3 shows *Kr_av*, *Kt_av* & *Kd_av* plotted against energy flux. Values of *Kt_av* decrease with increasing flux. This is similar to the trend reported by Dickson *et al.* (1996). The more energetic and steeper waves are more readily reflected & dissipated leading to lower transmission. Closer examination reveals a corresponding increase in *Kr_av* with energy flux, though there is a great deal of scatter and obviously other dependencies are in play.

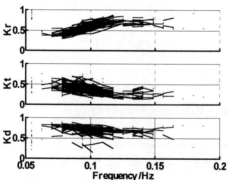

Figure 4: Discrete values of *Kr*, *Kt* & *Kd*

Whilst the most obvious relationship appears to be with wave steepness there is also a remarkable dependence on local wavelength (figure 4). The largest values of *Kr_p* occur for a local wavelength of approximately *50m*. There is a corresponding increase in values of *Kt_p* as the wavelength exceeds this value. The first observation supports the comments made by Kondo & Toma (1972). When the breakwater width is comparable to the scale of a *1/4*-wavelength, resonance can occur, whereby the wave-front reflected from the landward face of the breakwater reinforces that reflected at the seaward face. For a wavelength of *50m* this would equate to a breakwater width of *12m*. Although complicated by the beach geometry, this value corresponds well with the approximate width of the breakwater at around the water level for the majority of these measurements. The behaviour of **Kt_p** with wavelength shows a rapid rise at this point. This demonstrates the easier passage of the longer period and less steep waves through the structure.

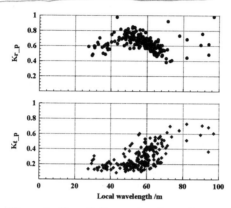

Figure 5 : Kr_p **&** K_pt **vs. local wavelength.**

All selected values of the coefficients from the data set within the gravity band are presented in figure 5. These show a remarkable consitency over the data set. The reflection function **Kr(f)** indicates a maximum in the middle of the gravity band possibly related to the quarter wave resonance. **Kt(f)** decreases at higher frequencies indicating that wave shape is softened (low–pass filtered) by transmission through the structure.

Results — Infra–Gravity band

Values of the infra–gravity coefficients show no clear relationship with wavelength, depth or Iribarren number. Values of **Kr_ig** are generally large as

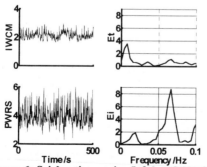

**Figure 6: Seiche: time series (left hand graphs)
and energy spectra (right)**

expected but difficult to interpret given the topography and geometry of the structure (sometimes *Kr_ig* exceeding unity). At these periods the long wavelength motions are strongly reflected by the beach behind the structure, whilst the transmission coefficient is predictably larger. The IWCM is not able to resolve direction at low frequency but one feature of the records is of note. During energetic periods, a low frequency signal can be clearly seen in the time series collected by the IWCM behind the breakwater (figure 6). Comparison of the expanded spectrum from any of the PWRS transducers with that from any of the IWCM poles shows that under these conditions a very low frequency peak occurs at around *0.014Hz*. Figure 6 shows that energy at this frequency is not obviously related to the incident spectrum. It is suggested that this corresponds to a *seiche* mode in the bays.

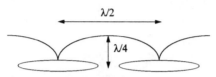

Figure 7: for λ/2=200m , d=3m, f_0=0.014

Figure 7 is a schematic of the bay geometry which has a spacing of around 200m between salients. Two possible resonant modes are that a half–wavelength *seiche* could be set up between the salients, or a quarter–wavelength *seiche* could occur between the beach and the breakwater gap. In fact the dimensions of the bay appear to satisfy both these conditions simultaneously. No variation of this frequency is evident with depth. However, for a parabolic bay geometry the resonant frequency would be nearly independant of depth.

Conclusions

Field measurements of transmission through an island breakwater situated on a macrotidal coast have been made using two instrument arrays. An initial review of the extensive data set indicates that transmission is primarily affected by wave steepness and local wave length. Other parameters such as water depth and angle of wave approach need further investigation. The relative crest height parameter is not appropriate for describing wave transmission through these island breakwaters under even moderate storm conditions.

Two interesting phenomena have been observed in the field. Firstly the values of *Kt* and *Kr* show a strong dependence on local wavelength, and this is believed to be related to standing wave resonances in the breakwater structure. Further justification of this process is required from numerical model comparisons. Secondly, evidence of seiching modes have been observed under storm conditions in the bays behind the breakwaters. It is interesting to consider that just as the coupling of long–period motions to short wave groups is recognised as an important mechanism for transport of sediment on natural beaches, seiching could also be an important mechanism for shaping the bays behind breakwaters.

Acknowledgements

This work was funded by the Engineering and Physical Sciences Research Council (EPSRC). and through the continued support of ARUN District Council, West Sussex. The authors would also like to thank those involved with the field experiment, especially Mr. Ian Black, Bob Gayler, Ian Morgan and Brian Worley. We are especially grateful for the continued enthusiasm of Roger Spencer and Ray Trainer at ARUN DC. We are grateful to Suzana Ilic and Mark Davidson for their expert advice on data analysis. Finally we are indebted to Rosemary Enticknap for loan of her field station (garden shed).

References

P.Axe, S.Ilic & A.Chadwick. 1996. Evaluation of beach modelling techniques behind detached breakwaters. *Proc. 25th International Conf. Coastal Engineering Orlando 1996* ASCE, pp 2036–2049

P.A.Bird , M.A.Davidson, G.N.Bullock & D.A.Huntley, 1994. Wave measurements near reflective structures. *Proceedings of Coastal Dynamics '94.* ASCE

A.J.Chadwick, D.J.Pope, J.Borges & S.Ilic, 1995. Shoreline directional wave spectra. Part 1: An investigation of spectral and directional analysis techniques. *Proc. Instn Civ. Engrs Wat. Marit. & Energy* **112** Sept, pp198–208.

A.J.Chadwick, D.J.Pope, J.Borges & S.Ilic, 1995. Shoreline directional wave spectra. Part2: Instrumentation and field measurements. *Proc. Instn Civ. Engrs Wat. Marit. & Energy* **112** Sept, pp 209–214.

M.Darras, 1987, IAHR List of sea state parameters: a presentation. *Proc. XXII IAHR Congress, Lausanne, Switzerland.*

M.A.Davidson, P.A.D.Bird, G.N.Bullock & D.A.Huntley, 1996a. A new non-dimensional number for the analysis of wave reflection from rubble mound breakwaters **28**(1-4) Sept *Coastal Eng* pp93–120.

M.A.Davidson, P.A.D.Bird, D.A.Huntley & G.N.Bullock, 1996b. Prediction of wave reflection from rock structures: an integration of field & laboratory data *Proc. 25th International Conf. Coastal Engineering Orlando 1996* ASCE, pp 2077–2086.

W.S.Dickson, T.H.C.Herbers & E.B.Thornton, 1995. Wave reflection from breakwater. *J. Waterway, Port, Coastal & Ocean Eng* Sept/Oct pp262–268.

P.Gaillard, M.Gauthier & F.Holly, 1980. Method of analysis of random wave experiments with reflecting coastal structures. *Proc. 17th International Conf. Coastal Engineering, Sydney* ASCE, pp 204–220.

S.Ilic & A.Chadwick 1995. Evaluation & validation of the mild slope evolution equation model using field data. *Proc. Coastal Dynamics '95, Gdansk* pp149–160 ASCE.

S.Ilic A.J.Chadwick, B.Li & C.A.Flemming, 1997. Composite modelling of an offshore breakwaterscheme in the UKCRF, *Proc. Coastal Dynamics '97, Plymouth* ASCE *(this pub).*

H.Kondo & S.Toma, 1972, Reflection and transmission for a porous structure, *Proc.13th International Conf. Coastal Engineering, Vancouver 1972* ASCE pp1847–1866

J.D.Simm, A.H.Brampton, N.W.Beech, J.S.Brooke, 1996. Beach Management Manual, *CIRIA Report No 153*. 6 Storey's Gate, Westminster, LONDON SW1P 3AU, UK.

C.K.Solitt & R.H.Cross, 1972, Wave transmission through permeable breakwaters, *Proc.13th International Conf. Coastal Engineering, Vancouver 1972* ASCE pp1827–1846

P.Troch, M.De Somer, J.De Rouck, L.Van Damme, D.Vermeir, J-P.Martens & C.Van Hove, 1996, Full scale measurements of wave attenuation inside a rubble mound breakwater, *Proc. 25th International Conf. Coastal Engineering Orlando 1996* ASCE, pp 1916–1929.

J.W.Van der Meer, 1993 Conceptual design of rubble mound breakwaters. Delft Hydraulics Publication No. 483, Netherlands, December 1993.

Beach Variability Behind Detached Breakwaters

P.G. Axe[1], and A.J. Chadwick[2]

1 Abstract

Beach profile data collected around detached breakwaters in a macrotidal environment were analysed using Principal Components (empirical orthogonal functions) to demonstrate longshore changes in profile variability updrift of, within, and downdrift of the breakwater scheme. The resulting eigenfunctions were gridded to show their spatial characteristics.

The breakwater scheme was found to reduce profile variability when compared to the updrift (beach backed by a vertical seawall), and the downdrift (groyned) coastlines. From gridding eigenfunctions, the downdrift coastline, and to a lesser extent, the updrift coastline, was found to follow the seasonal bar-berm variation described by Winant et al (1975). Within the scheme, material was exchanged between the western sides of the tombolos, and eastern sides of the bays (behaving as the summer berms) and the floors and western sides of the bays.

2 Introduction

Beach changes occur as a result of the interaction of waves, currents and tides, with the sedimentology of the beach. Forcing factors may vary spatially along a beach, and also temporally, from individual wave periods through seasonal and annual changes, to long term changes due to variability of sediment supply and sea level variability due to tectonic or post-glacial changes.

Principle component analysis allows a data set consisting of a large number of correlated variables to be reduced to a set with fewer, uncorrelated components that retain most of the information of the original data. These uncorrelated

---

[1] Proudman Oceanographic Laboratory, Bidston Observatory, Birkenhead, Merseyside, UK, L43 7RA

[2] School of Civil and Structural Engineering, University of Plymouth, Plymouth, UK, PL1 2DE

components are derived from the eigen analysis of a correlation matrix of the initial data. Each function represents a certain proportion of the total variance of the original sample, indicated by the magnitude of that eigenfunction's associated eigenvalue.

In this paper, principal component analysis is used to analyse the temporal and spatial variability of a renourished shingle beach on the U.K. south coast, and to compare the behaviour of that beach with that of the immediately adjacent coastline.

2.1 Eigenfunction analysis

Principal component (or 'empirical orthogonal function') analysis for the study of beach profile data was first presented by Winant et al (1975), in a study of three profiles at Torrey Pines beach, California. Most profile variance was found to be accounted for by the three largest eigenvalues. The spatial eigenfunction associated with the largest eigenvalue was called the *'mean beach function'*, and was related to the time -averaged beach profile. The spatial function associated with the next largest eigenvalue had strong maxima at the position of the summer berm, and a minimum at the winter bar position, and so was named the *'bar-berm function'*. The third function had a maximum at the position of the low tide terrace, and was called the *'terrace function'*. In Winant et al's study, the temporal eigenfunction associated with the largest eigenvalue showed no seasonal variation, but second largest eigenvalue exhibited strong annual periodicity, reaching a maximum in the late summer (August-September), and a minimum in the spring (March).

This method of analysis has since been applied to a wide variety locations, including the U.K. east coast (Aranuvachapun and Johnson, 1979), the Baltic and Black Sea (Pruszak, 1993) and to renourished beaches on the Atlantic coasts of Spain and the United States (e.g. Medina et al, 1992; Larson et al, 1997). The technique has also been extended to two spatial dimensions, and has also been used as the basis for a predictive model of beach changes (Hashimoto and Uda, 1980; Hsu et al, 1994). This work follows the one dimensional method described by Winant et al and Aranuvachapun and Johnson.

To calculate the eigenfunctions and corresponding eigenvalues for this data, each profile line was interpolated to give beach elevations at 1 metre (horizontal) intervals along it's length (so that each profile was made up of exactly coincident points). For each profile, a matrix H was then constructed. Each row of H consisted of values of beach elevation along the profile at a particular time, while each column of H showed the time series of beach elevation for a particular point on that profile. From this matrix, two square symmetric correlation matrices A and B are then formed, where

$$A = \frac{H'H}{tx}, \quad B = \frac{HH'}{tx}.$$

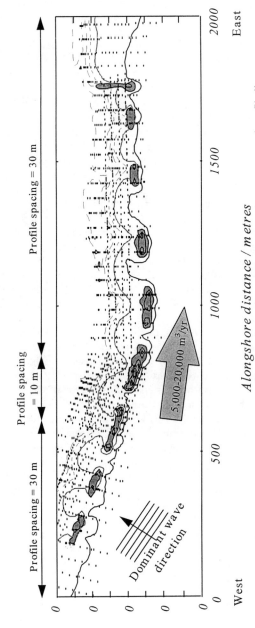

Alongshore distance / metres

Figure 1. Contour plot of Elmer frontage, showing positions of breakwaters, terminal groyne and profile lines. Contour lines are at 1.5 metre intervals, from 1.5 m below UK Ordnance datum, to 4.5 m above Ordnance Datum. Filled arrow shows net longshore sediment transport rate and direction. Dominant wave direction (180=192°) is also shown.

These matrices each have a set of eigenvalues λ_n and corresponding eigenfunctions e_{nx} and e_{nt} respectively, defined by the following equations:

$$Ae_{nx} = \lambda_n e_{nx} \ , \ Be_{nt} = \lambda_n e_{nt}$$

where the subscript x labels the spatial eigenfunction derived from matrix A, and subscript t labels the temporal eigenfunction calculated from B. The eigenvalues may then be ranked by percentage of variance (where the variance is defined as the *mean square value* of the data), such that the first eigenfunction explains most of the variance. The magnitudes of the eigenvalues associated with beach profile data decrease rapidly. The proportion of the total variance described by each eigenfunction is found from its corresponding eigenvalue, and therefore the relative importance of the eigenfunctions in determining the beach profile configuration can be observed.

2.2 Site Description

The Elmer field site is a renourished shingle beach on a chalk platform, with a median grain size of 11 millimetres, and poor sorting. The beach is stabilised by eight detached breakwaters and a terminal groyne. The spring tidal range is 5.3 metres, and at low water the chalk platform at the beach toe is exposed. Longshore transport rates are of the order of 16000 cubic metres/annum, with a predominant drift direction from west to east. Updrift of the scheme, the beach is of sand and shingle, backed by a vertical seawall. Beach levels are lower than within the renourished area. Downdrift of the breakwater scheme, the beach is also sand and shingle, and is stabilised by low wooden groynes.

Profile data were collected within the renourished area along 43 cross shore profile lines with a longshore spacing of between 10 and 30 metres. Outside of this area, a further 22 profile lines were sampled at 50 metre intervals. Figure 1 is a contour plot of the study area, showing the profile spacing within the scheme, the incident wave direction, and the average longshore sediment transport rate. During the experiment (described by Chadwick *et al*, 1994), one survey was carried out by the contractors on completion of the scheme and aerial surveys were flown at four monthly intervals thereafter. Since the end of the wave recording exercise, surveys have been taken at 3 monthly intervals by the local coast protection authority. In addition to beach profile data, wave conditions were recorded both offshore of the scheme, using an array of six pressure transducers, and within the scheme, using six resistance gauges. Monthly averages of wave conditions showed them to be strongly seasonal.

3 Results

3.1 Eigenvalues

Figure 2 shows the percentage of variance associated with the first eigenfunction (the *mean beach* function), as it varies alongshore from updrift of the breakwater scheme, to 1500 metres downdrift of it. Figure 2 parts *a* and *c* are contour plots

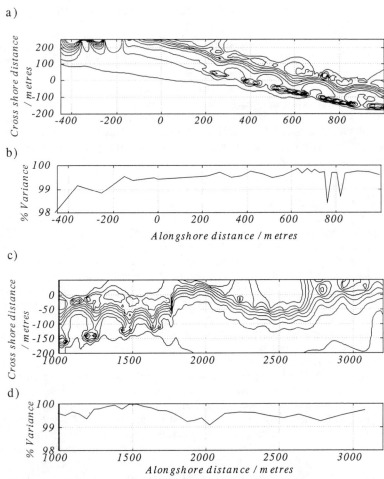

Figure 2. a) Contour plot of western (updrift) part of Elmer frontage showing four western breakwaters. b) Magnitude of first eigenvalue (as a percentage of the total variance of the data), showing variation along the western part of the coast. c) Contour plot of eastern (downdrift) frontage, showing last four breakwaters, terminal groyne and downdrift coastline. d) Magnitude of first eigenvalue for downdrift coastline.

of the Elmer topography (contour interval 1 m). The eigenvalues are plotted against longshore distance in figure 2 parts *b* and *c*. The proportion of variance associated with the first eigenfunction is between 98 and 100% throughout the study site. Updrift of the renourished area, in an area of narrow beach fronting a vertical seawall, the associated variance is lower than that observed further east. In the western half of the scheme, the proportion of variance is generally greater than 99.5%, with the exception of two points at around +800 metres alongshore. Continuing eastwards through the scheme, the proportion of variance continues to increase, to a maximum of almost 100%, around the position of the seventh breakwater (around +1450 metres alongshore). For the next 600 metres, the proportion of variance decreases steadily, reaching a minimum of 99% in the embayment down-drift of the terminal groyne. Between this point and the end of the surveyed area (+2000 to +3200 metres alongshore), the proportion of variance remains fairly constant, at around 99.5%.

3.2 Eigenfunctions

Spatial eigenfunctions were gridded and contoured. Figure 3 shows the second eigenfunction (the *bar berm* function) gridded for the western and eastern ends of the study site. The parts *a* and *b* show the western end of the scheme. Figure 3a shows the positive values of the bar berm function. Figure 3b shows negative values of this function. From these, berm (maxima) and bar (minima) positions may be inferred. Figure 3b and 3c show the same information for the eastern end of the study site. The dashed lines in each panel represent the actual beach morphology from the first (*mean beach*) eigenfunction.

Starting to the west of the scheme, figure 3a shows the strong maximum that occurs in the bar berm function immediately in front of the seawall. This maximum extends alongshore towards the start of the breakwater scheme. Additional local maxima occur offshore of the wall, and extend for 100-200 metres alongshore up to 100 metres from the beach crest. Figure 3b shows the position of minima in this function which, following Winant *et al*'s (1975) interpretation, represent winter bar features. These occur offshore of the seawall, and also at the toe of the beach slope. The minimum at -200 to 0 metres alongshore in figure 3b lies inshore of a maximum in the function. If Winant et *al*'s classification is correct, it suggests that the seasonal movement of material in front of the seawall is more complex than a simple onshore-offshore summer winter movement.

Looking at the eigenfunction structure within the scheme, behind the two westernmost breakwaters, there is a simple configuration, with maxima in the bar-berm eigenfunction at the beach crest, and minima at the beach toe, particularly in the bays shorewards of the breakwater gaps. The seasonality associated with these eigenfunctions indicate a building of the berm in summer, and a transfer of material to the toe of the beach and the floor of the bays in winter to flatten the beach profile.

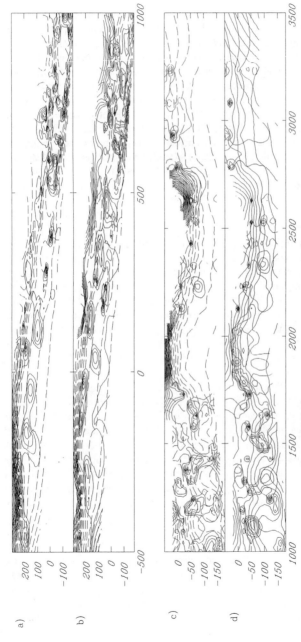

Figure 3. Gridded bar-berm eigenfunction. Panels *a)* and *b)* show the western half of the study area. Panel *a)* is the positive values of the bar berm function, while panel *b)* shows negative values. Panels *c)* and *d)* show the eastern study area. Panel *c)* is the positive values, and panel *d)* are negative values. Dashed lines on all plots show the underlying bathymetry (from the mean beach function). The top *x* axis gives alongshore distance for the top two panels. The *y* axis shows cross shore distance. The lower *x* axis gives alongshore distance for the bottom two panels. All distances in metres

Further into the scheme, the patterns of the eigenfunctions was more complex. This could be due to the decrease in profile line separation in this area, rather than an increase in complexity of the beach behaviour. Behind breakwaters 3,4,5,6 and 7 (numbering from west to east), maxima in the bar berm function occurred on the western side of the tombolos, and the eastern (exposed) sides of the embayments. Minima occurred on the eastern side of the tombolos, in the middle of the bays, and on the western (sheltered) sides of the bays. This could be explained as follows: In summer, the berm is built up on the exposed (eastern) side of the bays, and western sides of the tombolos. The winter (barred) configuration moves material across shore from these areas to raise the level of the bay floor (thus flattening the profile), and also alongshore, building the lower parts of the western side of each bay.

The downstream limit of the scheme is maintained by the last breakwater and a terminal groyne. The bay immediately west of the breakwater has the same configuration as the others, with a maximum in the bar berm function over the tombolo and eastern sides of the bay, and a corresponding minimum in the bottom, and western side, of the bay. To the east of the breakwater, the maximum extends alongshore to the groyne, and also offshore. The coast beyond this point is marked by a strong peak in the bar berm function extending around crest of the downdrift bay (1800-2000 metres alongshore in figure 3c). The floor of this bay is covered with an extensive minimum, which is strongest close to the terminal groyne. This indicates a seasonal movement of sand from the berm, to just downdrift of the groyne in winter, and a return movement in the summer. Further alongshore (2000-2500 metres) the situation is similar, with a berm maximum, and an offshore minimum, although there is some longshore variability in this pattern.

4 Discussion and conclusions

The effectiveness of the breakwater scheme in reducing profile variability is demonstrated in figure 2. Updrift of the scheme, in the area of beach backed by a seawall, a lower proportion of variance was associated with the mean beach eigenfunction, indicating that the bar-berm and terrace functions are relatively important in controlling beach levels. Beach levels vary more in this area (although there may be no net loss of material). Downdrift of the scheme, the mean beach function accounted for a higher proportion of the variance than on the updrift beach, but was still more variable than the protected, renourished beach. Gridding the eigenfunctions demonstrated the variability in the bar berm function in longshore, as well as cross shore directions. Winant et al (1975) described the locations of the summer berm and winter bar by looking at the variation in eigenfunctions in the cross shore. In this study, longshore variations in the second (bar berm) eigenfunction were apparent, indicating a more complex response of the beach to seasonal changes in wave climate. Within the breakwater scheme, two main beach configurations were observed. The winter beach has beach material stored as raised beach levels in the bay floors, and as more material on the western slope of the embayments. In summer, the berm is rebuilt primarily on the western sides of salients and tombolos, and on the eastern sides of the embayments. Downdrift of the scheme, the beach is more two dimensional, with a maximum in the bar berm function at the position of the berm crest, and a minimum at the beach toe.

5 Acknowledgements

The fieldwork described in this paper was funded by the U.K. Engineering and Physical Sciences Research Council. Data processing was supported by the Standing Conference on Problems Associated with the Coastline (SCOPAC). Additionally, the authors would like to thank David Green, Roger Spencer and Ray Trainer of Arun District Council for their help in supplying aerial survey data, and the U.K. Natural Environment Research Council for supporting the first author's attendance at this conference.

6 References

Aranuvachapun S. and Johnson J.A.,1979, Beach profiles at Gorleston and Great Yarmouth, in: Coastal Engineering vol 2., pp 201-213

Chadwick A.J., Fleming C.A., Simm J., and Bullock G.N., 1994, Performance evaluation of offshore breakwaters, a field and computational study. in Proceedings of the Conference on Coastal Dynamics '94, ASCE, Barcelona, pp 701-711

Hashimoto H., and Uda T.,1980, An application of an empirical prediction model of beach profile change to the Ogawara coast, in: Coastal Engineering in Japan, vol 23, pp 191-204

Hsu T-W., Ou S-H., and Wang S-K., 1994, On the prediction of beach changes by a new 2D empirical eigenfunction model, in: Coastal Engineering, vol 23, pp 255-270

Larson M., Kraus N., Hanson H., and Gravens M., 1997, Beach topography response to nourishment operations at Ocean City, Maryland, in: Proceedings on the conference on Coastal Dynamics '97, ASCE, Plymouth (this volume)

Medina R., Vidal C., Losada M.A., and Roldan A.J., 1992, Three-mode principal component analysis of bathymetric data, applied to 'Playa de Castilla' (Huelva, Spain), in: Proceedings of the 23rd International Conference on Coastal Engineering, Venice, ed: Edge B., pub by ASCE, pp 2265-2278

Pruszak Z., 1993, The analysis of beach profile changes using Dean's method and empirical orthogonal functions, in: Coastal Engineering vol 19, pp 245-261

Winant C.D., Inman D.L., and Nordstrom C.E., 1975, Description of beach changes using empirical eigenfunctions, in: Journal of Geophysical Research, vol. 80 no. 15, pp 1979-1986

7 Notation

H   Matrix of beach profile data
A   Spatial correlation matrix

B   Temporal correlation matrix
'   Transpose operator
t   Number of beach profiles in time
x   Number of interpolated survey points per profile

LARGE-SCALE HYDRODYNAMIC EXPERIMENTS ON SUBMERGED BREAKWATERS

Francisco J. Rivero, Agustín S.-Arcilla, Xavier Gironella and August Corrons[1]

Abstract

This paper presents some preliminary results of a set of hydrodynamic experiments on submerged breakwaters conducted in the large-scale wave flume CIEM of LIM/UPC. Experimental values of the free surface elevation and simultaneous (horizontal and vertical) velocity components were collected at 12 stations for a number of (regular and irregular) wave conditions and breakwater geometries. The obtained results are finally compared with state-of-the-art formulations and a mathematical model for wave transformation.

1. Introduction

In the last decades there has been an increasing interest on the use of submerged breakwaters for coastal protection. Yet most of the laboratory experiments with this type of structures have focused mainly on determining wave transmission and reflection coefficients for various wave conditions and breakwater geometries, and little research has gone into a detailed description of the hydro-morphodynamics associated with submerged structures.

The "Dynamics of Beaches" project (see Prinos *et al*, 1995), funded by the Human and Capital Mobility programme of the European Commission, is an experimental research project, conducted at several wave flumes and basins of the 6 participating universities (Catalonia University of Technology, Delft University of Technology, University of Liverpool, University College Cork, Aristotle University of Thessaloniki, and Ghent University), aiming at providing high-quality experimental data on the hydro-morphodynamic impact of submerged breakwaters on the beach.

---

[1] Maritime Engineering Laboratory (LIM), Catalonia University of Technology (UPC) Campus Nord (mód. D-1), Gran Capitán s/n, 08034 Barcelona, Spain

This paper will concentrate on the set of experimental tests conducted at the large-scale wave flume CIEM of LIM/UPC within the "Dynamics of Beaches" project, where simultaneous measurements of free surface elevation and (horizontal and vertical) velocities were collected for a number of (regular and irregular) wave conditions at 12 stations with a (fixed) submerged breakwater on a 1:15 rigid sloping bottom. Additional (similar) experiments to investigate the influence of the breakwater freeboard that were conducted at LIM/UPC under a national research project (CYTMAR programme) will also be presented.

The outline of the paper will be as follows: the experimental setup is given in #2, some preliminary results are presented in #3, and a comparison with a wave transformation model is shown in #4; some final remarks are finally summarised in #5.

2. Experimental Setup

For its dimensions (100m length, 3m width, 5m depth) and performance (capable to generate individual waves up to 1.7 m height and to absorb the reflected waves), the large-scale wave flume CIEM is one of the largest in the world. The submerged breakwater, with 1:2 slopes, impermeable core and armour layer of quarrystones with mean weight of 25 kg (Fig. 1), was constructed on a 1:15 concrete sloping bottom (Fig. 2).

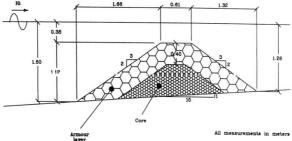

Figure 1. Cross-section of the submerged breakwater.

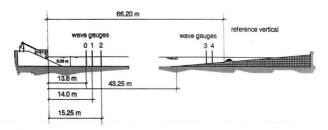

Figure 2. Sketch of the wave flume and the submerged breakwater model.

A moving carriage, mounting 2 twin-wire wave gauges, 6 pressure sensors and 5 electromagnetic currentmeters (Fig. 3), was placed at 12 stations along the wave flume: 3 before, 2 over and 7 behind the breakwater. The currentmeters were evenly distributed along each vertical and coincide horizontally with the position of one wave gauge to enable the separation of incident and reflected wave components using the method of Hughes (1993). Video records were also taken to investigate wave breaking features and to determine the fraction of breaking waves.

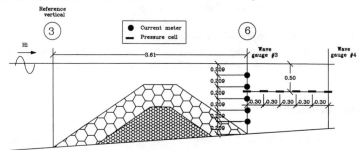

Figure 3. Sketch of the moving carriage with the measuring equipment (station #6).

Ten different wave conditions (5 irregular and 5 regular) were generated, that correspond to 6 cases (3+3) with same wave heght and varying wave period and to 6 cases (3+3) with same deep-water equivalent wave steepness and varying wave height (Tables 1 and 2).

Test	Hs (m)	Tp (s)	F / Hs	Hs / Lo,p
A	0.375	4.0	-1.00	1.5 %
B	0.375	3.0	-1.00	2.6 %
C	0.375	2.5	-1.00	3.8 %
D	0.500	3.5	-0.75	2.6 %
E	0.250	2.5	-1.50	2.6 %

Table 1. Irregular wave conditions (at the toe of the structure).

Test	H (m)	T (s)	F / H	H / Lo
F	0.375	4.0	-1.00	1.5 %
G	0.375	3.0	-1.00	2.6 %
H	0.375	2.5	-1.00	3.8 %
I	0.500	3.5	-0.75	2.6 %
J	0.250	2.5	-1.50	2.6 %

Table 2. Regular wave conditions (at the toe of the structure).

In addition to the tests defined in the "Dynamics of Beaches" project, already described, some experimental tests were also conducted with identical wave conditions (tests A to J) but less resolution (5 measurement stations instead of 12) for 3 smaller freeboards (F = -0.250m, F = -0.125m and F = 0m), which correspond to water depth values at the seaward toe of the breakwater of 1.375m, 1.250m and 1.125m, respectively.

3. Experimental Results

In the following some preliminary experimental results are presented and compared with state-of-the-art formulations.

3.1. Transmission Coefficient

The complete set (for the irregular wave tests only) of experimental results for the transmission coefficient K_T, defined as the ratio Hs (transmitted) / Hs (incident), are compared with the formulations of van der Meer (1990) and Daemen (1991). The comparison is made using 2 different positions for the measurement of the transmitted wave height: position 1 at the shoreward toe of the structure, and position 2 at some distance behind it. In general, the obtained Hs values at position 1 are smaller than at position 2 due to wave reformation and shoaling behind the breakwater.

The graphs (Fig. 4) clearly show that both formulations (especially that of Daemen) tend to underpredict the measured K_T values. The main differences between the datasets used by both researchers for their analyses and the present one are the gently sloping bottom (1:15) and large-scale conditions in our tests.

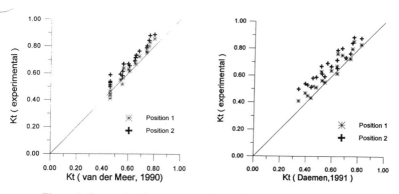

Figure 4. Comparison between experimental values of the transmission coefficient K_T and formulations of van der Meer (1990) and Daemen (1991).

3.2. Reflection Coefficient

The values of the reflection coefficient K_R for the irregular wave tests were obtained using collocated measurements of the cross-shore velocity and free surface elevation (Hughes, 1993). This method showed important discrepancies depending on the currentmeter applied in the calculation of K_R, especially in stations close to the breakwater. This could be explained by the fact that such method is based on linear wave theory for finite water depth and neglects sloping bottom effects. Fig. 5 shows a plot between the obtained values for K_R and the widely used dimensionless freeboard $F/Hs_{,inc}$, where $Hs_{,inc}$ is the incident significant wave height. From the graph it can be clearly seen that such parameter is not appropriate to characterise wave reflection from a submerged breakwater.

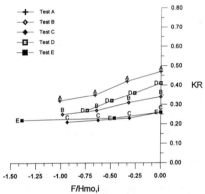

Figure 5. Reflection coefficient K_R *versus* dimensionless freeboard $F/Hs_{,inc}$.

3.3. Wave Spectra Transformation

Fig. 6 shows the cross-shore transformation of non-dimensional spectra of the free surface elevation and the horizontal velocity (at mid depth) over the submerged breakwater for test A (case with freeboard $F = 0$ m). These plots show the generation of superharmonics (energy transfer to higher harmonics) due to the presence of the submerged structure (see e.g. experiments of Battjes and Beji (1992) on wave transformation over a bar). From observation of the complete datasets it was found that this energy transfer increases with smaller freeboard and larger wave period.

An interesting feature found only in tests with irregular waves and small freeboard (in particular, with $F = 0$ m; see Fig. 6) is the presence of low frequency motions between the submerged breakwater and the beach. This phenomenon, which was also found in similar experiments by Petti and Ruol (1992), is, according to preliminary calculations, associated to mean water level oscillations induced by breaking of wave groups in the leeside of the breakwater.

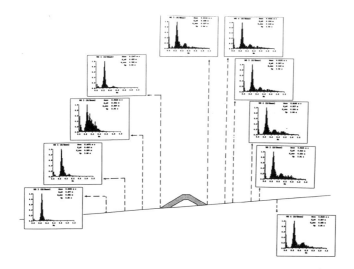

(a) Free surface elevation

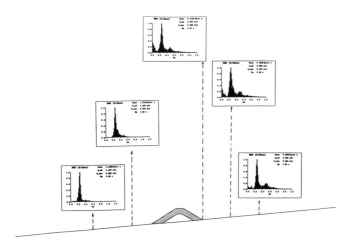

(a) Horizontal velocity at mid depth

Fig. 6. Cross-shore transformation of non-dimensional wave spectra
(Test A: Hs = 0.375 m; Tp = 4.0 s; freeboard F = 0 m)

3.4. Vertical Circulation

As shown in Fig. 7, the vertical circulation pattern is mainly governed by a strong return flow over the submerged breakwater, which causes the formation of a large cell in front of the structure. This could have implications in the development and evolution of scour at the toe of the breakwater. A new set of morphodynamic experiments with emphasis on scour processes will soon be conducted at LIM/UPC within a MAST-3 project (SCARCOST). In the leeside of the structure the flow pattern is similar to that found in the surf zone, with a return flow compensating the mass flux driven shorewards by the waves over the trough level.

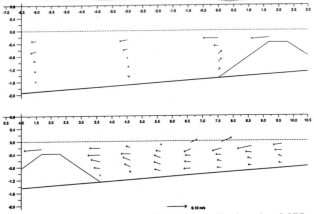

Figure 7. Time-averaged velocity field (Test A; Freeboard = -0.375 m)

4. Mathematical Modelling

A simple numerical model has been developed to calculate wave transformation on a beach in the presence of a submerged breakwater. Wave kinematics is defined based on linear wave theory and the (complex) dispersion relation for a porous flow given by (Sawaragi and Deguchi, 1992). Wave dynamics is governed by the wave energy balance equation taking into account the following mechanisms of energy dissipation: wave breaking for regular (Dally et al, 1985) or irregular waves (Battjes and Janssen, 1978), bottom friction (e.g. Nielsen, 1992) and due to porous flow, associated to the imaginary part of the wavenumber (e.g. Sawaragi and Deguchi, 1992). Wave reflection is taken into account after calculating the phase shift between the incident and reflected wave trains, and correcting the associated expressions for the modulated wave height and radiation stress tensor with the corresponding reflection coefficient, which is now complex to account for the 1:2 slope of the submerged breakwater. The starting point of the reflection process is assumed to be at the seaward edge of the breakwater crest.

The cross-shore distribution of the mean water level, $<\eta>$, is calculated from the time-averaged depth-integrated cross-shore momentum balance equation:

$$\text{\textit{Where is it?}} \tag{1}$$

where S_{xx} is the radiation stress tensor, ρ is the fluid density, g is the gravity acceleration, d is the still-water depth, u is the (cross-shore) x-velocity component, and $<...>$ is the time-averaging operator. In this equation the advection term (third on the LHS), usually neglected in the modelling of beach hydrodynamics, will be shown to play an important role in the prediction of the set-up/down gradients over the breakwater. This term is approximated assuming a uniform distribution of the undertow that compensates the wave mass flux in the upper layer, calculated with the formulation of Stive and De Vriend (1987). S_{xx} is computed using linear wave theory and incorporates transition zone effects due to the wave roller (Rivero et al, 1994).

Fig. 8 shows a typical comparison between numerical results and measured values of H (or Hs) and $<\eta>$ for test F (with freeboard F = -0.375 m). After tuning the parameters of the wave breaking dissipation model (K=0.13, Γ=0.40, in this case) to fit the wave height data, it can be clearly seen the difference between the predicted mean water level (MWL) variation neglecting the advection term (dashed line) and taking it into account (solid line), especially in the prediction of the large set-down over the crest of the breakwater. This is reasonable after noting the large gradients in the undertow velocity over the breakwater (approximately, mass flux divided by the fluid density and water depth). The model, however, cannot predict the large values of the set-up at the shoreward toe of the structure and its further decrease shorewards.

5. Final Remarks

The following relevant conclusions may be drawn from a preliminary analysis of the experimental results and the calibration of the numerical model:

- In wave flume experiments, the hydrodynamics associated with breaking waves on a submerged breakwater is mainly governed by a strong return flow over the structure. This situation does not necessarily exist in a general 3D situation, however, where the mass flux over the breakwater is diverted alongshore. Caution must therefore be taken when interpreting results from 2DV tests.
- The obtained values of the transmission coefficients K_T are found to be larger than those predicted by state-of-the-art formulations based on available experimental data. This could maybe suggest influence of the bottom slope (1:15 in our tests).
- Both sub- and superharmonic wave generation in the frequency domain take place over the submerged breakwater. These phenomena appear to be more relevant for smaller breakwater freeboards (especially the bound-long-wave generation).

- The calibration parameters of the wave breaking models found to fit the experimental data differ from those recommended in beach modelling applications.
- The calibrated values of the phase, ψ, of the reflection coefficient ($R=K_R e^{-i\psi}$) increases, for a given submerged breakwater geometry, with smaller wave periods.
- The magnitude of the reflection coefficient, K_R, is mainly dependent on the wave period and the breakwater freeboard; little dependence was found on wave height.

Acknowledgements

The main part of these experiments were funded by the Human and Capital Mobility programme of the European Commission (contract n° CHRX-CT93-0392). The execution of additional experiments and the analysis of the collected data was funded by the CYTMAR programme of the Comisión Interministerial de Ciencia y Tecnología (reference n° MAR95-1915). The authors would also like to acknowledge the collaboration of Dr. Michael Høgedal, Jan Petersen, Jesús Gómez and Quim Sospedra in the preparation and execution of the experiments.

References

Battjes, J.A. and Beji, S. (1992). Breaking waves propagating over a bar. *Proc. 23rd Int. Conf. on Coastal Eng.*, Venice, pp. 42-50..

Battjes, J.A. and Janssen, J.P.F.M. (1978). Energy loss and set-up due to breaking of random waves. *Proc. 16th Int. Conf. on Coastal Eng.*, Hamburg, pp. 569-587.

Daemen,I.F.R. (1991). Wave transmission at low-crested breakwaters. *MSc thesis*, Delft University of Technology, The Netherlands.

Dally, W.R., Dean, R.G. and Dalrymple, R.A. (1985). Wave height variation across beaches of arbitrary profile. *Journal of Geophys. Res.*, 90, C6, pp. 11917-11927.

Hughes, S.A. (1993). Laboratory wave reflection analysis using collocated wave gauges. *Coastal Engineering*, 20 (2,3), pp. 223-247.

Nielsen, P. (1992). *Coastal bottom boundary layers and sediment transport*. World Scientific, Singapore, 324 p.

Petti, M. and Ruol, P. (1992). Laboratory tests on the interaction between nonlinear long waves and submerged breakwaters. *Proc. 23rd Int. Conf. on Coastal Eng.*, Venice, pp. 792-803.

Prinos, P., Arcilla, A.S.-, Christiansen, N., van de Graaff, J, Høgedal, M., Lewis, A., O'Connor, B.A., Rivero, F.J. and de Rouck, J. (1995). The "Dynamics of Beaches" project. *Proc. Coastal Dynamics'95*, Gdansk, pp. 571-582.

Rivero,F.J.,Arcilla, A.S.- and Beyer, D. (1994). Comparison of a wave transformation model with LIP-11D data. *Proc. Coastal Dynamics'94*, Barcelona, pp. 518-532.

Sawaragi, T. and Deguchi, I. (1992). Waves on permeable breakwaters. *Proc. 23rd Int. Conf. on Coastal Eng.*, Venice, pp. 1531-1544.

Stive, M.J.F. and De Vriend, H.J. (1987). Quasi-3D nearshore current modelling: wave-induced secondary current. *Spec. Conf. Hydrodyn.*, Delaware, pp.356-370.

van der Meer, J.W. (1990). Data on wave transmission due to overtopping. *Report H-986*, Delft Hydraulics, The Netherlands.

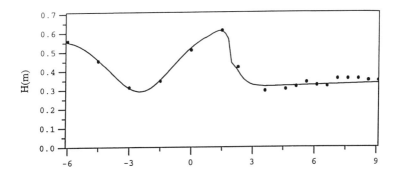

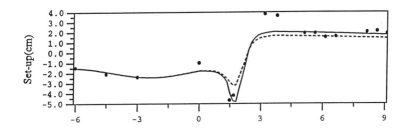

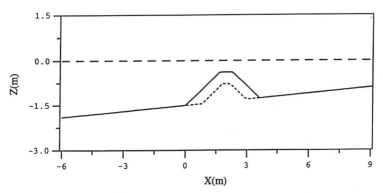

Figure 8. Comparison between numerical results (lines) and measurements (circles): bathymetry, wave height and MWL variation (Test F, freeboard F = -0.375 m). The dashed line in the set-up distribution solves Eq.(1) neglecting the advection term.

Modelling of Hydrodynamic and Morphological Effects of Submerged Breakwaters on the Nearshore Region

Th. V. Karambas[1], P. Prinos[2] and E.E. Kriezi[1]

Abstract

This paper describes the modelling of the wave breaking propagation over submerged breakwaters as well as the morphological effects of such structures on the cross-shore beach profile. A new breaking criterion for nonlinear time domain models, valid for regular and irregular waves is introduced. The cross-shore sediment transport rate is estimated using model results and the energetics approach. Model predictions for wave breaking, dissipation after breaking, nonlinear interaction between harmonics and bed evolution behind the structure are validated against experimental data.

Introduction

A submerged breakwater is an offshore structure with its top at or below still water level. Submerged breakwater can only be used when the tidal range is relatively small (Mediterranean sea) in order to be effective at all times. Due to the presence of the structure a part of the wave energy is reflected offshore while a part of it is transmitted shoreward. A significant amount of wave energy is dissipated due to friction and breaking on the structure. In this way a small part of the energy is transmitted behind the breakwater and hence the structure can be used to protect the shoreward area from wave action.

[1] Research Associates, [2] Ass. Professor, Aristotle University of Thessaloniki, Division of Hydraulics and Environmental Engineering, 54006 Thessaloniki, GREECE

A Boussinesq type model can be successfully used for the simulation of nonlinear breaking wave propagation (Karambas and Koutitas, 1992, Schaffer et al., 1993, Nwogu, 1996). In addition, the model, with improved linear dispersion characteristics, is capable to describe nonlinear wave transformations (i.e. nonlinear interaction between higher harmonics) over and behind submerged bars and breakwaters (Beji and Battjes, 1994). Karambas et al. (1995) presented an extension of this type of models for the prediction the cross-shore sediment transport based on the energetics approach.

In the present work the above types of Boussinesq models are combined and extended in order to describe nonlinear breaking wave propagation over a submerged breakwater and to predict the effects on cross-shore evolution.

Wave Model

A new time domain form of Boussinesq equations, with improved linear dispersion characteristics, proposed by Schaffer and Madsen (1995) is used. Wave breaking is introduced as in Karambas et al. (1995). The equations are written:

$$\zeta_t + ((d+\zeta)U)_x +$$

$$((z_a^2 - d^2/6)dU_{xx} + (z_a + d/2 - \beta_1 d)d\,(dU)_x + \beta_2\,d^2\,(dU)_x - \beta_1 d^2\zeta_{tx})_x = 0$$

$$U_t + g\,\zeta_x + (\frac{1}{h}\int_{-d}^{\zeta} u^2 dx)_x - U/h\,(Uh)_x + (z_a^2/2 - \gamma_1 d^2)\,U_{txx} +$$

$$(z_a + \gamma_2 d)((dU_t)_{xx} - \gamma_1 g d^2 \zeta_{xxx} + \gamma_2 g d\,(d\zeta_x)_{xx} - 1/h\,(v_\tau h U_x)_x + \tau_b/\rho h = 0 \tag{1}$$

where d is the water depth, ζ is the surface elevation, $h = d + \zeta$, U the velocity at an arbitrary level $z = z_a(x)$, v_τ the eddy viscosity coefficient, τ_b is the bottom shear stress and

$z_a = -0.02907\,d$
$\beta_1 = 0.2043$
$\beta_2 = -0.20097$
$\gamma_1 = 0.01196$
$\gamma_2 = 0.00144$

The integral in the above momentum equation (1) is estimated from the horizontal velocity distribution in a bore (or hydraulic jump) according to Madsen and Svendsen (1983) (used also in Karambas et al., 1995, and Karambas, 1996).

The eddy viscosity coefficient is estimated from the solution of a turbulent kinetic energy model (k-model).

$$k_t + Uk_x = \nu_\tau \, k_{xx} + D - C_{dk} \, k^{3/2}/ \, l_t \tag{2}$$

where k is the turbulent kinetic energy, $C_{dk} = 0.08$, l_t is the turbulence length scale, $l_t = 0.3d$, D is the production of turbulence kinetic energy.

The production term D is taken equal to the dissipation of the wave energy and is given by (Madsen and Svendsen, 1983, Karambas et al., 1995):

$$D = 0.015 \, (c-u_o)^3 \tag{3}$$

where c is the wave celerity and u_o the bottom velocity (Peregrine, 1972, $u_o = U + h^2/6 \, U_{xx}$).

The roller region, where dissipation D takes the above value, is detected as in Schaffer et al. (1993). Outside the roller region D=0.

The eddy viscosity is given by:

$$\nu_\tau = k^{1/2} \, l_t \tag{4}$$

The submerged structure is simply introduced as a stepwise variation of the water depth. A partially standing wave is formed offshore, due to reflection from the depth variation, while the wave is propagated as a progressive one behind the structure. In monolithic structures with sharp corners, the flow separation and the formation of eddies of high vorticity cause significant energy loss. In this case the energy dissipation is simulated as in Karambas and Kriezi (1997).

Breaking criterion

A new breaking criterion for the simulation of irregular and regular wave breaking in a time domain non linear wave model is presented. The criterion is a function of the ratio of the local wave height over a mean depth d_m proposed by Seyama and Kimura (1988). The latter is related to the still water depth and the vertical wave asymmetry :

$$d_m = d + (\zeta_c + \zeta_t)/2 \tag{5}$$

where ζ_c and ζ_t are the crest and trough elevation respectively.

Hansen (1990) proposed a semi-empirical relation for ζ_c and ζ_t:

$$\frac{\zeta_t}{H} = -0.5 \tanh \frac{4.85}{\sqrt{U_b}}, \tag{6}$$

with

$$U_b = 10.1 S^{0.2} \left(\frac{H_{o,rms}}{L_{o,rms}} \right)^{-1}$$

with S the slope, $H_{o,rms}$ and $L_{o,rms}$ the deep water height and length respectively.

Invoking the above (and the definition $\zeta_c/H = \zeta_t/H + 1$), the criterion is written:

$$\frac{H}{d_m} = \frac{H}{d} \left(\frac{1}{1 + \frac{H}{d}(0.5 - B_h)} \right) \tag{7}$$

with $B_h = 0.5 \tanh \dfrac{4.85}{\sqrt{U_b}}$,

where H/d is the well known breaking criterion.

The purpose of the above approach is to improve the simulation of breaking for irregular waves (for different values of wave steepness H_o/L_o, H/d is very different, but H/d_m is almost constant taking values near 0.65, $H/d_m = 0.65$).

This criterion is used in the next to determine the break-point of each wave, whilst the surface roller region is detected geometrically as in Schaffer et al. (1993).

Comparison with experimental data

The validity of the present model is tested against laboratory experiments by Beji and Battjes (1994) concerning non linear irregular wave propagation over a submerged bar. A time-domain nonlinear breaking wave Boussinesq model with improved dispersion characteristics it seems to be the most appropriate to include the above type of structures since it can describe wave reflection from the breakwater, nonlinear interactions and energy dissipation due to breaking.

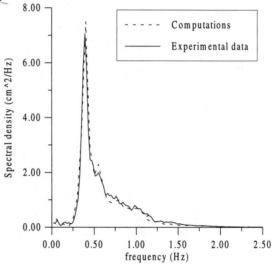

Figure 1. Measured and calculated spectral density functions at Station 2.

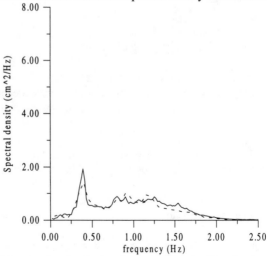

Figure 1. Measured and calculated spectral density functions at Station 6.

As mentioned in the introduction, the non linear interaction between frequencies (spectral energy is transferred to higher frequencies) and the generation of higher free harmonics dominate the phenomenon. A trapezoidal bar (submerged breakwater) is taken into consideration, in which the offshore slope is 1:20 and the shoreward slope 1:10, while the depth in the horizontal depth parts is 40 cm and on the horizontal part of the bar (of extent 2 m) the depth is 10 cm. A JONSWAP (with peak frequency f_p=0.4 Hz) is used as input. In this experiment waves were breaking on the breakwater. A comparison between experimental data and the results from the model are shown in figure 1 and2. Station 2 were placed over the 1:20 slope   and   station 6 on   the downslope 1:10. It is concluded that the model is capable of reproducing the generation of high frequency energy and its transfer among nearly harmonic wave components due to the non linear interactions taking place in the deeper water region behind the bar. Wave energy dissipation due to breaking on the bar is also predicted well.

In the next the experimental data from the EU Project "Dynamics of Beaches" (TU Delft data, Prinos et al. , 1995, Dekker, 1996) are reproduced. In that experiment the effects of a submerged breakwater on beach profile changes under the wave action are studied. The water depth in front of the structure was 0.4 m (height of structure=0.3m   and freebroad=0.1). The generated wave spectrum was of JONSWAP type with significant wave height H_{sig} =0.098 m and peak period T_p=1.55 s (experiment B). Tests were done without and with breakwater with an 1 in 15 slope and a movable bed. The bed consisted of sand with D_{50}=95 μm and fall velocity w=0.01 m/s. The measurements are compared with the computed results in figures 3 (significant wave height H_{sig}), 4 and 5 (bed evolution).

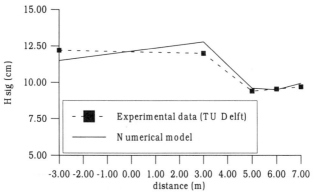

Figure 3. Significant wave height H_{sig}: comparison between experimental data (experiment B) and numerical model.

Model results (near bottom velocity moments) are used as input in the sediment transport model based on an energetics approach (Karambas et al., 1995).

In an energetics approach the submerged weight transport i_t rate is given by the Bailard formula:

$$< i_t >= \rho c_f \frac{\varepsilon_b}{\tan\varphi} \left(<|u_o|^2 u_o > - \frac{\tan\beta}{\tan\varphi} <|u_o|^3 > i \right)$$

$$+ \rho c_f \frac{\varepsilon_s}{w} \left(<|u_o|^3 u_o > - \frac{\varepsilon_s}{w} \tan\beta <|u_o|^5 > i \right) \qquad (8)$$

where u_o represents the near bottom instantaneous vector velocity field, angled brackets represent time-averaging, i is the direction of u_o, $\tan\beta$ is the bed slope, φ is the angle of internal friction, w is the fall velocity, c_f is the friction coefficient and ε_b and ε_s are the bed and suspended load efficiency factors respectively (with ε_b=0.10 and ε_s=0.02).

Inside the surf zone the suspended load is increased for the inclusion of the turbulence effects due to wave breaking (Roelvink and Stive, 1989, Morfett, 1995, Karambas et al., 1995):

$$i_{sb} = \frac{\varepsilon_s}{w / u_o - \varepsilon_s \tan\beta} \omega_b \qquad (9)$$

where ω_b is the local rate of energy dissipation due to breaking wave induced turbulence inside the surf zone ($\omega_b \equiv D$).

The model is not extended in the run-up region. It stops at a small depth h_s. In the swash zone an extrapolation of the transport rate is adopted (Larson, 1996):

$$i_{swash-zone} = i_s ((h-h_r) / (h_s-h_r))^{1.5} \tan\beta/\tan\beta_s \qquad (10)$$

where h_s, i_s and $\tan\beta_s$ is the depth, transport rate and the slope at the shoreward end of the (numerical) surfzone and h_r is the run-up height.

Bed profile changes with and without the presence of the structure are shown in figures 4 and 5 respectively. From the comparison of the two figures it seems that this type of structure can be used to protect the beaches against erosion due to attack by storm waves.

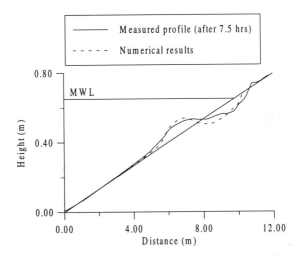

Figure 4. Bed evolution without breakwater: comparison between measurements and simulation.

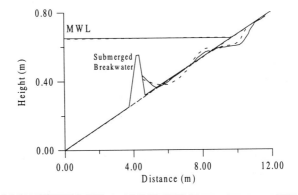

Figure 5. Bed evolution with breakwater: comparison between measurements and simulation.

The model describes well the accumulation of the sediment behind the breakwater as well as its effects on the longshore bar formation.

Conclusions

A breaking wave time domain nonlinear model with improved linear dispersion characteristics and appropriate breaking criterion was successfully applied for the prediction of:
- energy dissipation due to wave breaking over a submerged breakwater.
- energy transfer among wave components due to the nonlinear interactions behind the structure.
- breakwater effects on bed evolution.

Acknowledgement

Part of the work in the present paper was carried out in the framework of SAFE research program, funded by the Commission of the European Communities, Directorate General for Science, Research and Development, under contract No MAS3-CT95-0004.

References

Bailard J. A. (1981), "An energetics total sediment transport model for a plane sloping beach", *J. of Geophysical Research*, vol. 86, No C11, pp 10938-10954.

Beji S. and Battjes J.A. (1993) "Experimental investigation of wave propagation over a bar", *Coastal Engineering*, 19, pp 151,162.

Dekker R., (1996), Submerged breakwater influence on beach hydrodynamics. A 2DV approach, Master Thesis, UPC, Barcelona, TU Delft.

Hansen, J. B. (1990), "Periodic waves in the surf zone: Analysis of experimental data", *Coastal Engineering*, 14, pp 19-41.

Prinos P., Arcilla et. al., (1995) "The Dynamic of Beaches Project", *Coastal Dynamics '95*, Gdansk, Poland, eds W.R. Dally and R.B. Zeidler, pp. 571-582.

Karambas Th.V., H.N. Southgate and C. Koutitas, (1995) "A Boussinesq Model for Inshore Zone Sediment Transport using an Energetics Approach", *Coastal Dynamics '95*, Gdansk, Poland, eds W.R. Dally and R.B. Zeidler, pp. 841-849.

Karambas Th. V., (1996) "Nonlinear Wave Energy Modelling in the Surf Zone", *Nonlinear Processes in Geophysics*, 3, pp 127-134.

Karambas Th. V. and E. E. Kriezi, (1997) "Wave Reflection and Transmission for Submerged Monolithic Breakwaters", *Computer Methods and Experimental Measurements*, eds P. Anagnostopoulos, G.M. Carlomagno and C.A. Brebbia, Computational Mechanics Publications, pp 543-550.

Larson M. (1996), "Model of beach profile change under random waves", *J. of Waterway, Port, Coastal and Ocean Engineering*, ASCE, 122, no 4, pp 172-181.

Madsen P. A. and I. A. Svendsen, (1983) "Turbulent bores and hydraulic jumps", *J. of Fluid Mechanics*, vol 148, pp 73-96.

Morfett J. C. (1995) "Computational modelling of sediment transport by breaking waves", *2nd Int. Conference on Computer Modelling of Seas and Coastal Regions*, Mexico, vol II, pp 241-248.

Peregrine D. H., (1972) "Equations for water waves and approximation behind them", *Waves on Beaches*, ed. R. E. Meyer, Academic press.

Schaffer H.A., Madsen P.A. and Deigaard R., (1993) "A Boussinesq model for waves breaking in shallow water", *Coastal Engineering*, 26, pp 185-202.

Schaffer H.A. and Madsen P.A. (1995) "Further enhancements of Boussinesq-type equations", *Coastal Engineering*, 26, pp 1-14.

Seyama A. and Kimura A. (1988), "The measured properties of irregular wave breaking and the wave height change after breaking on the slope", *Proc. 21st ICCE ASCE*, pp 419-432.

MONITORING AND MODELLING GROUND WATER BEHAVIOUR IN SANDY BEACHES AS A BASIS FOR IMPROVED MODELS OF SWASH ZONE SEDIMENT TRANSPORT

Andrew J. Baird[1], Travis E. Mason[2], Diane P. Horn[3], and Thomas E. Baldock[4]

ABSTRACT
An examination is made of two aspects of beach ground water behaviour: (i) water table fluctuations in response to tidal forcing, and (ii) pressure propagation through a saturated sand bed under swash. Models and measurements of each are presented. It is found that a 1-D Boussinesq model of water table behaviour provides good predictions of observed water table fluctuations. Further field work is needed before a simple model of pressure propagation can be properly tested.

INTRODUCTION
The link between beach ground water behaviour and sediment transport processes in the inter-tidal zone has long been recognised by coastal researchers (see Baird and Horn 1996). Two linking processes seem to be of most importance:

Infiltration
When water tables in the swash zone are below the sand surface there is the potential for runup and backwash infiltration to occur. Because runup and backwash are usually shallow, a small change in water volume due to infiltration will significantly decrease the energy available for sediment transport. Work by Packwood (1983) has suggested that backwash infiltration, in particular, may be important in affecting sediment movement in the swash zone. A limitation of Packwood's analysis is that the water table under an advancing swash was assumed to be at the still water level. In reality, the water table will often be much closer to the sand surface causing a reduction of swash infiltration. The position of the water table is, therefore, of critical importance.

---

[1] Lecturer, Dept. Geography, University of Sheffield, Sheffield S10 2TN, UK
[2] Research Associate, Southampton Oceanography Centre, Southampton SO14 3ZH, UK
[3] Lecturer, Dept. Geography, Birkbeck College, University of London, London W1P 2LL, UK.
[4] Research Associate, Dept. Civil Engineering, Imperial College, London SW7 2BU, UK

Fluidisation

It has been suggested by a number of authors going back to Grant (1946, 1948) that ground water outflow from a beach during the ebb tide may enhance the potential for fluidisation of sand, and therefore the ease with which sand can be transported by swash flows. However, ground water outflow from a beach in response to a falling tide is unlikely to be sufficient to induce fluidisation since hydraulic gradients under the beach surface will tend to be relatively small. It is likely that fluidisation is only generally possible in the presence of swash on a seepage face. As a swash flow advances over the saturated beach surface there will be a rapid increase in pore water pressures below the beach surface. When under swash flow, the beach sediment behaves like a confined aquifer. The sediment is saturated and movement of water into the beach is extremely limited since changes in porosity due to expansion and contraction of the mineral 'skeleton' will be minimal. However, water pressures will propagate rapidly through the sediment. As the swash retreats there will be a release of pressure on the beach face giving large hydraulic gradients acting vertically upwards immediately below the surface. The resultant seepage force associated with rapid ground water outflow could be sufficient to induce fluidisation of the sand grains at the surface. This process may occur during the latter stages of backwash, providing readily entrainable material which is then available for transport, either seawards by the backwash or landwards by the following uprush. To our knowledge no attempts have been made to model or measure such effects under swash action. Madsen (1978) and Okusa (1985) examined pressure propagation in sand beds under waves but neither considered the situation under the different hydrodynamic conditions of swash.

It is clear from the foregoing that in order to model sediment transport in the swash zone it is first necessary to understand and hence measure and model: (i) water table dynamics and the extent of the seepage face, and (ii) pressure propagation through the sand bed under the action of swash. Both are addressed in this paper.

MODELLING BEACH GROUND WATER BEHAVIOUR

Water table behaviour

Ground water flow in a shallow aquifer can be described using the 1-D Boussinesq Equation given by

$$\frac{\partial h}{\partial t} = \frac{K}{s} \frac{\partial}{\partial x}\left(h \frac{\partial h}{\partial x} \right) \qquad (1)$$

where h is water table elevation above an arbitrary datum (L), t is time (T), s is specific yield (1), x is distance (L), and K is hydraulic conductivity (LT⁻¹). A problem in using the Boussinesq Equation is that no provision is made in the equation for the development of the seepage face. The seepage face occurs when ground water flows out of the beach giving the sand a glassy appearance. This surface is different from the water table in that its shape is dictated by beach topography. However, water on the seepage face is under atmospheric pressure as is water on the water table. This basic property can be exploited in numerical models so that the Boussinesq Equation can be used to simulate seepage face development. In a numerical model in which a sloping beach is simulated, a seepage face can be assumed to be present if a computational cell 'fills' with water, i.e. if

$$h_{i,t} + \left(\frac{(Q_{i-1,t} - Q_{i,t})\Delta t}{\Delta x s} \right) > e_i \qquad (2)$$

where $Q_{i-1,t}$ is the rate of seaward ground water discharge per unit longshore distance ($L^2 T^{-1}$) into cell i during Δt (from cell i-1), $Q_{i,t}$ is the rate of discharge out of cell i during Δt, Δx is the cross-shore width of each cell (L), Δt is the timestep (T), and e is the cell height (L).

The numerical solution to the Boussinesq Equation incorporating seepage face development used in this study has been called the GRIST (GRound water Interaction with Swash and Tides) model (see Baird and Horn 1996 and Baird et al. in review). The current version of the model simulates ground water behaviour in response to tides and wave-induced setup. The model has been coded in Visual Basic™ 3.0 and uses a simple explicit block-centred finite-difference solution to equation (1). Water on the seepage face is assumed to be 'lost' from the solution domain, i.e. it cannot enter another cell down beach. The model is very flexible in that it can be used to simulate sloping beaches (topographic data are entered directly into the model) and a non-horizontal impermeable barrier at the base of the beach. The model can have one of three inland boundary conditions (Dirichlet, von Neumann, or Cauchy), and tide levels and wave conditions (root mean square wave height and wave length for the calculation of setup) can be entered directly into the model as time series. In this paper only the tidal part of the model is discussed. The predictions of the model using an estimate of wave-induced setup are discussed in Baird et al. (in review).

Pressure propagation through the sand bed

Pressure propagation through a saturated or partially saturated sand bed (i.e. one that has occluded gas bubbles) is a complex process. Under swash, pressure propagation will occur in three dimensions and will involve contraction and expansion of the porous bed. To our knowledge no full solution to this problem has been presented although it is possible that sophisticated numerical thermo-elastic models (if they exist) could be used, since the governing equations will be the same as for the current problem. As a first stage in modelling pressures in the sand bed under swash, the 1-D parabolic or diffusion equation can be used:

$$\frac{K}{\gamma} \frac{\partial^2 p}{\partial z^2} = (\alpha + n\beta) \frac{\partial p}{\partial t} \qquad (3)$$

where K is the hydraulic conductivity of the sediment (LT^{-1}), γ is the specific weight of water (FL^{-3}), p is the incremental fluid pressure (FL^{-2}), z is the Eulerian vertical coordinate from bottom to top of the sand bed (L), α is the classical vertical compressibility of the sediment (L^2F^{-1}), n is the porosity of the sediment (dimensionless), β is the compressibility of water (L^2F^{-1}), and t is time (T). Use of the diffusion equation means that the movement of the sediment under stress changes is partially ignored. However, in a comparison with rigorous theory (Gambolati 1973a, b) it is clear that, for sand beds of all grain sizes under typical 1-D loading and release of pressure under uprush/backwash, the use of the diffusion equation introduces negligible error.

Trials with a numerical solution to the equation, using values of K, α, and n typical of fine to medium sand (0.001 cm s^{-1}, 3 x 10^{-6} cm^{-1}, and 0.3 respectively - see Okusa 1985), have indicated that seepage forces in the sand bed under swash may be sufficient to induce fluidisation. Example results from the trial are shown in Figure 1 where an assumed pattern (based on measurements by Hughes 1992) of uprush and backwash depths over time is shown. Packwood and Peregrine (1980) noted that for many sands and fine gravels, fluidisation occurs when the upward-acting dynamic pressure gradient (expressed in head units and therefore dimensionless) is greater than about 0.6 to 0.7. In loosely packed sediments, gradients of less than 0.6 may also induce fluidisation while for densely packed sediment, gradients of greater than 1.0 may be required. Fluidisation occurs when the effective stress becomes zero. It should also be noted that downward seepage has the effect of increasing the effective stress; hence it may increase the stability of the sediment. Thus, the importance of the seepage force may be different for uprush and backwash.[5]

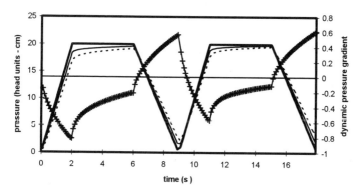

Figure 1. Simulated changes in incremental pressure in a sand bed, over time. The reference datum is the bed surface. Thick solid line = swash depth; thin solid line = pressure at 2 cm below surface; broken line = pressure at 4 cm. Crosses represent the hydraulic gradient acting between 2 cm depth and the bed surface (expressed in head units per unit distance and therefore dimensionless). During uprush the pressure gradient acts downwards (expressed as -'ve values); during backwash it acts upwards (+'ve values). The maximum gradient acting upwards is between 0.5 and 0.6. The total thickness of the simulated bed is 60 cm. The initial condition of the simulation was a hydrostatic pressure distribution.

---

[5] It should be noted that values of $\gamma(\alpha + n\beta)$ (i.e. the specific storage of hydrogeology) given in Baird et al. (1996) were unrealistically high. However, with correct values for the sediment properties the diffusion equation still predicts large upward acting pressure gradients in a sand bed under swash.

FIELD TESTING OF THE MODELS
Water table predictions

To test the Boussinesq model, measurements of water tables, tides, and waves were made on a microtidal beach at Canford Cliffs, Poole, Dorset, England (50° 42'N, 1° 57'W) (see Figure 2). The beach was backed by a concrete sea wall set in underlying poorly permeable silts giving no-flow conditions on the inland and lower boundaries. This stretch of the south coast of England experiences a distorted tidal regime, with double high water conditions. Canford Cliffs beach is a groyned beach with a mean slope of 3.2° (tan β = 0.055) on the upper foreshore and a mean slope of 1.5° (tan β = 0.027) on the lower foreshore. Field measurements from 21st - 24th October 1995 were undertaken in a period leading from a neap to a spring tide, with a tidal range of 0.9 to 1.7 m. No rainfall was recorded during the period of measurement or in the preceding two weeks.

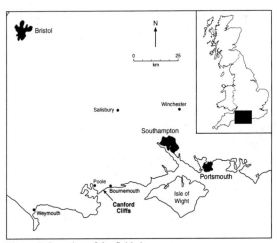

Figure 2. Location of the field site.

Water table elevations were measured with Druck™ PDCR 830 pressure transducers in ten screened 'dipwells' placed at distances of 1.9, 3.8, 5.8, 7.8, 9.3, 10.8, 13.1, 15.6, 18.7, and 21.7 m seawards from the sea wall (see Figure 3). Measurements were made on 22nd and 23rd October 1995. The dipwells were lined with polyvinyl chloride (PVC) tubing with an outside diameter of 28 mm. The liners were drilled with 5 mm holes in four lines throughout their length with each hole protected against sand ingress with a Terram™ geotextile.

The hydraulic conductivity of twelve surface sand samples (10.2 cm in length and 3.8 cm in diameter) was measured using a constant head laboratory permeameter. The mean K of the sediment cores was 0.225 cm s^{-1}, with a range from 0.036 to 1.179 cm s^{-1}, and a coefficient of variation of 143 per cent, which is typical for this parameter.

The GRIST model was run for Canford Cliffs using 40 computational cells with a width of 1 m. Measured values of tidal elevation were used in the simulation. The seaward cell was assumed to be permanently saturated, i.e. the water table was assumed to be always at the cell surface. A value of K/s of 0.75 cm s^{-1} was used in the GRIST model simulations. This value was calculated using the arithmetic mean of the permeameter measurements of K (0.225 cm s^{-1}) and an assumed value of s of 0.3.

Good agreement between modelled and observed water tables was found for all of the dipwells (see Figure 4). Thus the model was able to predict with some accuracy water table rise and fall at different positions in the beach, giving a good representation of changes in the cross-shore water table profile over time. We suggest, therefore, that the model can be used to provide satisfactory boundary conditions for swash infiltration and pressure propagation models. Even better predictions were obtained when setup was accounted for in the model; these predictions and other aspects of model testing are discussed in much more detail in Baird et al. (in review).

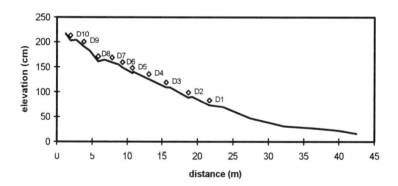

Figure 3. Location of the dipwells used in the study. The open diamonds show the insertion point of each dipwell.

The position of the water table relative to the beach surface

Figure 5 shows the measured position of the water table relative to the sand surface for two positions of the tide. At low water, the water table intersected the beach face at a distance of about 32 m from the sea wall and tended to approach the beach face at a tangent. The water table in dipwells 1 and 2 fell little during the period between low water and subsequent immersion by the rising tide (see Figure 4). Thus

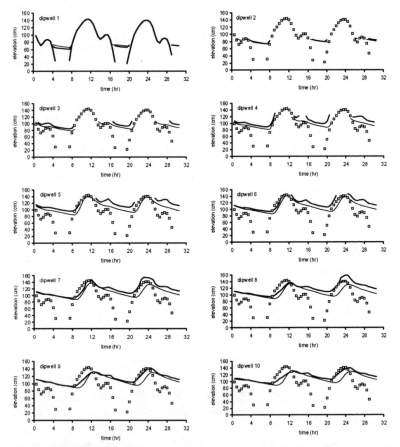

Figure 4. Measured and modelled water table elevations for each of the dipwells. In each case: medium solid line - observed water table; thin solid line - simulated water table using tide-only inputs (i.e. setup ignored); thick solid line (dipwell 1 only) and open squares - observed tide. Elevations are given above a datum which represents the position of impermeable material below the beach surface. Time is from 20.00 GMT on 22nd October 1995.

during the advance of the tide over the seepage face and later across the zone between 18 and 32 m from the sea wall, where the water table remained close to the beach surface, there would have been little opportunity for large amounts of swash infiltration. Any infiltration that did occur is unlikely to have had an effect on swash dynamics because the beach would have behaved like the impermeable case analysed by Packwood (1983). The water table at high water shows a pronounced landward slope and is superelevated in dipwells 6 and 7. This superelevation can be attributed to setup and runup, and indicates that infiltration into this part of the beach may have the potential to affect swash and backwash hydraulics in the way described by Packwood (1983) for a permeable beach.

It would appear from the above reasoning that uprush and backwash infiltration are only likely to be important in affecting swash hydraulics on a site such as Canford Cliffs during the latter part of the rising tide when the uprush and backwash move over an unsaturated beach. However, it is important to stress that even small amounts of swash infiltration into the seepage face followed by rapid ground water outflow could induce fluidisation on middle and lower parts of the beach, and that the interaction between beach ground water behaviour and sediment dynamics may be more subtle than suggested by existing uprush/backwash infiltration models such as that of Packwood (1983).

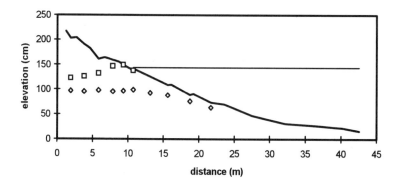

Figure 5. Measured water table elevations along the cross-shore profile at low water after an elapsed time of 6.00 hours (open diamonds), and high water at 11:30 hours (open squares) (see Figure 4). The still high water level is shown as a thin solid line. The still water level at low tide was below the level of the x axis and is not shown.

Pressure propagation through the sand bed

Pressures in the sand bed under swash were measured on a separate deployment at Canford Cliffs on 25 March 1997. Pressures were measured using the device of Baldock and Holmes (1996). Briefly, this consists of a series of 2 mm ID stainless steel probes. Each probe is connected to a 0.1 Bar pressure transducer via a three way ball valve and short length of 3 mm ID semi-rigid nylon tubing. Air may be flushed from the tubing and probe by opening the valve to a small header reservoir mounted above the transducer/valve system. The pressure variation is transmitted into each probe though twelve 0.4 mm holes drilled into the tip. In order to measure pressure variations in the sand bed at Canford Cliffs four probes were arranged in an array such that the vertical pressure gradients could be obtained for a single point. For logistical reasons it was not possible to take measurements on the lower part of the beach profile. Measurements were, therefore, taken inland of the break of slope on the beach as the tide was rising over the exit point. The sand surface was not saturated for every swash that passed the measuring point, although the water table was never far from the sand surface (it was always within 1 to 2 cm of the surface). The four probes were placed at depths of 0.5, 1.5, 2.5 and 3.5 cm below the surface and were adjusted regularly to take account of bed level changes.

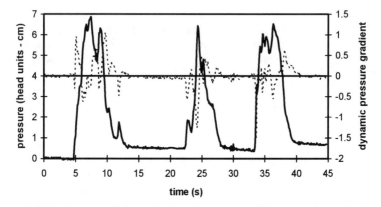

Figure 6. Measured changes in dynamic pressure gradient over time for the sand bed at Canford Cliffs. Medium solid line is incremental pressure at a depth of 0.5 cm (probably broadly similar to swash depth). The thin broken line is the pressure gradient acting between a depth of 0.5 cm and 1.5 cm. Negative values indicate a downward acting pressure gradient, positive values an upward gradient.

The results from the deployment were somewhat equivocal. Pressures at depth were seen at times to respond more quickly than pressures near the sand surface. This gave, simultaneously, gradients in the sand bed acting both downwards and upwards for several seconds which seems counter-intuitive. It is not clear whether this behaviour was apparent, and therefore a function of the measuring device, or real and a function of heterogeneity and complex three dimensional flows in the sand bed. Some observed behaviour was, however, consistent with the 1-D diffusion equation and is shown in Figure 6 where it can be seen that the upward acting gradient was at times relatively large and possibly close to inducing fluidisation. Large upward acting gradients were also observed during the complex behaviour described above. However, these results should be treated with caution.

The 1-D diffusion model is obviously an over-simplification of pressure propagation under swash and further measurements will be needed to elucidate flow processes in sand beds. However, the model suggests a mechanism for fluidisation in the swash zone apparently not considered by other researchers, and, as noted in the discussion of Figure 5, it is one which may exert a greater control on sediment movement in the swash zone than infiltration of uprush and backwash into the unsaturated part of the beach. It is important to emphasise that, although the pressure propagation mechanism described here involves infiltration into the sand bed, the amount of water lost from uprush and backwash is negligible since the specific storage of saturated sand is extremely small. Therefore, the hydrodynamics of uprush and backwash are unlikely to be affected. Thus, it appears that infiltration *sensu* Grant (1948) (*i.e.* infiltration into unsaturated sand above the water table) may not be the most important mechanism by which ground water affects swash zone sediment dynamics on fine and medium sand beaches.

ACKNOWLEDGEMENTS

The work on beach water tables was funded by the UK Natural Environment Research Council (Grant number GR9/01726). Poole Borough Council kindly allowed us to use the Surf Lifesaving Club building at Canford Cliffs for housing logging equipment.

REFERENCES

Baird, A.J. and Horn, D.P. 1996. Monitoring and modelling ground water behaviour in sandy beaches. Journal of Coastal Research 12: 630-640.

Baird, A.J., Mason, T.E., and Horn, D.P. 1996. Mechanisms of beach ground water and swash interaction. Proceedings 25th ICCE, ASCE, pp. 4120-4133.

Baird, A.J., Mason, T.E. and Horn, D.P. (in review). Validation of a Boussinesq model of beach ground water behaviour. Submitted to Marine Geology.

Baldock, T.E. and Holmes, P. 1996. Pressure gradients within sediment beds. Proceedings 25th ICCE, ASCE, pp. 4161-4173.

Gambolati, G. 1973a. Equation for one-dimensional vertical flow of ground water. 1.Rigorous theory. Water Resources Research 9: 1022-1028.

Gambolati, G. 1973b. Equation for one-dimensional vertical flow of ground water. 2.Validity range of the diffusion equation. Water Resources Research 9: 1385-1395.

Grant, U.S. 1946. Effects of ground water table on beach erosion. Bulletin of the Geological Society of America 57: 1252 (abstract).

Grant, U.S. 1948. Influence of the water table on beach aggradation and degradation. Journal of Marine Research 7: 655-660.

Hughes, M.G. 1992. Application of a non-linear shallow water theory to swash following bore collapse on a sandy beach. Journal of Coastal Research 8: 562-578.

Madsen, O.S. 1978. Wave-induced pore pressures and effective stresses in a porous bed. Geotechnique 28: 377-393.

Okusa, S. 1985. Wave-induced stresses in unsaturated submarine sediments. Geotechnique 35: 517-532.

Packwood, A.R. 1983. The influence of beach porosity on wave uprush and backwash. Coastal Engineering 7: 29-40.

Packwood, A.R., and Peregrine, D.H. 1980. The propagation of solitary waves and bores over a porous bed. Coastal Engineering 3: 221-242.

SWASH HYDRODYNAMICS ON A STEEP BEACH.

Tom E. Baldock[1] and Patrick Holmes[2]

Abstract

New data on swash hydrodynamics on a steep laboratory scale beach are presented. Wave grouping is found to be particularly important and directly induces large amplitude low frequency shoreline motions which describe the run-up due to individual bores. Consequently, run-up spectra may be dominated by energy at low frequency harmonics, even in the absence of free incident long waves. A new semi-empirical swash kinematics model is developed which is based on the physical constraints on the swash motion and a non-dimensional shape function to describe the cross-shore swash profile. The model shows good agreement with laboratory data obtained over fixed sediment beds during both the uprush and backwash phases of the swash cycle.

Introduction

The swash zone covers the region of wave run-up and run-down on beaches and coastal structures. It may therefore extend over, and influence, the whole of the intertidal zone. Swash hydrodynamics therefore play a particularly important role in the accretion of beaches and, consequently, the longer term stability and evolution of the coastal zone. However, numerical modelling of beach processes in the swash zone has yet to prove successful, largely because of the uncertainties in the hydrodynamics. Previous work (e.g. Huntley et al., 1977; Holland et al., 1995; Raubenheimer et al., 1995) has shown that on mildly sloping beaches the swash is often dominated by low frequency infragravity motions, which frequently have a cross-shore standing wave structure. However, an alternative hypothesis is that the swash motion is principally driven by bores which collapse at the shoreline and then propagate up the beach face (Shen and Meyer, 1963; Hibberd and Peregrine, 1979; Yeh et al., 1989; Hughes, 1992).

---

1) Research Associate. 2) Professor.
Dept. of Civil Engineering, Imperial College, London, SW7 2BU, UK.

The present paper addresses this point and compares new experimental data with a semi-empirical numerical model that can provide a good description of the swash kinematics. The effects of wave grouping on swash run-up spectra are highlighted and show that, although the spectral characteristics of swash on steep and mildly sloping beaches may frequently be very similar, the swash hydrodynamics appear to be driven by two physically distinct mechanisms. As a result, sediment transport over the two types of beach is likely to be very different.

Experimental setup

The experiments were carried out in the large wave flume in the Civil Engineering Department at Imperial College (figure 1). A flat bottom-hinged wave paddle was used to generate regular or random waves up to frequencies of 2Hz, with simultaneous absorption of short waves reflected from the far end of the flume. A slope (β) of gradient 0.05 starts 32m from the paddle, rising to 0.1 in the surf and swash zones. The last 6m of the flume are subdivided into three sections, providing two working sections of width 0.9m and a central section for access to instrumentation. Within the swash zone, fixed and mobile beds of sediment (d_{50}=0.5mm; d_{50}=1.5mm) were used to investigate frictional effects on run-up.

WAVE FLUME

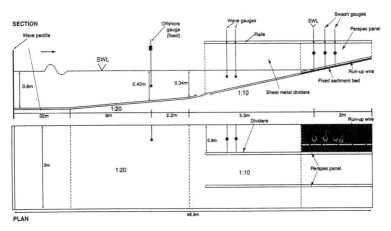

Figure 1. Wave flume and instrumentation layout.

The vertical elevation of the water surface was measured using surface piercing resistance type gauges with an absolute accuracy of order ±1mm. Surf and swash zone kinematics were measured with an Acoustic Doppler Velocitimeter (ADV), which was also co-located with a wave gauge in the outer surf zone to enable wave reflections to be estimated (figure 2). A run-up wire was also installed above the fixed and mobile sediment beds. This comprised of two 1.5mm diameter rigid wires held parallel and 10mm apart by six 5mm high spacers. The wires passed through the centre of the spacers, fixing the wire at a constant elevation above the bed of 3mm. The run-up wire measurements have a relative accuracy of about 1mm, but due to the finite thickness of the component wires, the absolute accuracy was estimated to be of order 20mm parallel to the bed. The origin of the horizontal co-ordinate is taken at the SWL, positive onshore.

The present investigation considers the swash motion induced by a regular wave train and a random wave (Jonswap spectrum) simulation (table 1). In each case, the wave heights shown in table 1 are the measured wave heights at the offshore gauge (x=-5.5m, d=0.45m), denoted by a subscript $_o$, where H_{rms} is taken as $\sqrt{(8m_o)}$. Other cases, including wave groups, may be found in Baldock et al. (1997). Note that the random wave spectrum may be considered to be deterministic, rather than a single representation of a random process. Consequently, the spectral plots presented later do not require the confidence limits associated with stochastic processes. All waves were generated using a linear representation of the water surface at the wave paddle. Although this may induce some spurious free long waves, previous work has shown that generating the waves in deep water reduces their magnitude to a minimum. Furthermore, tests indicated that the phase relationship between the wave groups and the bound long waves was consistent with second order wave theory, which would not be the case if spurious free long waves were present.

Case	T (s), f_m (Hz), T_p (s)	H_o , H_{rmso} (mm)	H_o/L_o	Surf similarity, ζ
R4	2.5	50	0.01	1.39
J1	1.5	90	0.033	.62

Table 1. Experimental wave conditions.

Swash model

Previous field data obtained by Hughes (1992) suggests that the analytical model proposed by Shen and Meyer (1963) describes the shoreline motion during the uprush phase of the swash cycle with a fair degree of accuracy. However, Hughes (1992) found that the swash depth across the beach profile was significantly overestimated. Similar results were found during the present study and the cross-shore shape of the swash lens was observed to be convex rather than concave as proposed by Shen and Meyer (1963). Therefore, in order to model the swash kinematics, a semi-empirical model has been formulated as follows:

Taking X_s as the shoreline position and x as the position of a point within the swash zone (with the origin of x at the point of bore collapse, positive shorewards) then from Shen and Meyer (1963):

$$X_s = U_o t - \frac{1}{2} g t^2 \sin \beta \qquad (1)$$

where t is the time elapsed since bore collapse, U_o is the initial speed of the shoreline, g is the acceleration due to gravity and β the beach slope. The natural swash period, T_s, from the start of the uprush to the end of the backwash is given by:

$$T_s = \frac{2U_o}{g \sin \beta} \qquad (2)$$

The bore collapse involves the rapid conversion of potential energy to kinetic energy, with an associated acceleration of the fluid velocity at the bore front (Yeh et al., 1989). U_o is therefore given by $U_o = 2\sqrt{(gH_b)}$, where H_b is the height of the incident bore. Although a bed friction coefficient, f_r, may be readily incorporated into (1) and (2) (see Hughes, 1995), inviscid conditions have been used for clarity. In order to describe the water depth, h, across the swash zone, the model is then based on the following physical constraints:

i) $(X_s-x)=0$; $h=0$,
ii) $(X_s-x)=X_s$, $t=0$; $h=H_b$,
iii) $(X_s-x)=X_s$, $t=T_s$; $h=0$,

i.e. the water depth is equal to the height of the incident bore at the start of the uprush, reducing to zero at the end of the backwash and at the run-up tip. Both field (Hughes, 1992) and laboratory data (see section 4) show that the swash front tends to be convex and in order to obtain the swash kinematics an integrable function is required. A non-dimensional shape function of the following form is therefore used to describe the swash lens profile:

$$h(x,t) = \left[\frac{(X_s - x)}{X_s} \right]^a \left(\frac{(T_s - t)}{T_s} \right)^b H_b \qquad (3)$$

The coefficients a and b are based on laboratory data and might be expected to vary slightly with beach slope. The data in section 4 suggests that for $\beta \approx 0.1$, a value for $a=0.75$ and $b=2$ provides a good approximation to the shape of the swash lens. The cross-shore shape of the swash lens used in the model is shown in figure 3, where the time interval between profiles is 0.2s. Equation (3) maybe integrated w.r.t. x to find the volume shoreward of a specific point and the kinematics are then obtained from a finite difference scheme by considering the change in the swash volume over time. However, because the bore is collapsing during the uprush, the depth averaged fluid velocity is not equal to the velocity close to the bed. For example, at the start of the uprush, the depth averaged velocity will only equal the speed of the shoreline if the bore has a vertical front and the swash depth is uniform over space. For the shape of the swash lens in figure 3, or the parabolic shape proposed by Shen and Meyer (1963), the depth averaged velocity will always underestimate the velocity close to the bed. The

swash kinematics are therefore calculated by only considering the change in the
swash volume below a certain depth, z, given by:

$$z = h(x,t)\frac{X_s}{X_{sm}} \tag{4}$$

where X_{sm} is the maximum uprush distance. This ensures that at the start and end
of the swash cycle the fluid velocity close to the bed tends to the shoreline speed.
This appears to be a realistic physical constraint and is in good agreement with the
data presented in the next section.

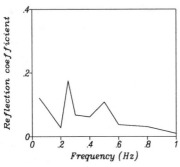

Figure 2. Reflection coefficient, $R^2(f)$,
during at x=-1.8m, case J1.

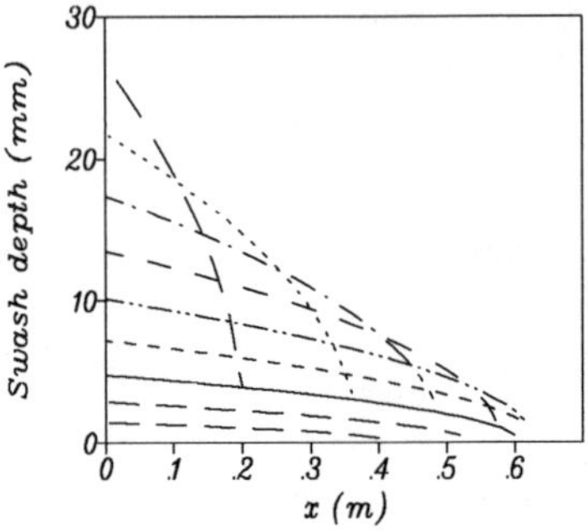

Figure 3. Shape of the swash lens
used in the model (equation 3).

Results
Wave grouping

Figure 4a shows the shoreline motion and swash depth, S_o, at the seaward
limit of the swash zone ($x\approx0$) for part of the random wave case, J1. The data have
been plotted so that the run-up height may also be read off the right hand scale of
the figure. Significant wave grouping is still apparent at the seaward limit of the
swash and large amplitude shoreline motions are clearly driven by a combination of
two or three incident bores. The low frequency components ($f<0.2$Hz) of these
two data sets (figure 4b) show that although the shoreline motion and swash depth
are in phase, the amplitude of the low frequency motion (LFM) in the run-up is
about three times that present in the inner surf zone. This difference is much larger
than that expected for a low frequency standing wave. The low frequency motion
in the run-up is therefore best regarded as the slowly varying mean shoreline
position induced by the incident wave grouping and, if wave setup is taken into
account, simply describes the envelope of the run-up due to individual bores of
different amplitude (figure 5a). The amplitude of the envelope is large because it
describes the difference in the run-up due to the largest and smallest incident bores
and therefore has a magnitude similar to that of the mean swash zone width. The
spectra of the run-up height and the offshore wave envelope (which describes the

amplitude of the incident wave groups) are found to be very similar at wave group frequencies (f<0.2Hz), which shows that, on a steep beach, much of the incident wave groupiness may be transmitted into the swash zone (figure 5b). Further examples of this effect, including those due to well defined individual wave groups, may be found in Baldock et al. (1997).

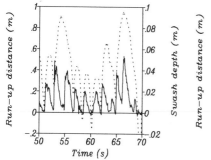

Figure 4a. Run-up and swash depth, at x=0, case J1.
······ Run-up wire, ——— S_o (rhs).

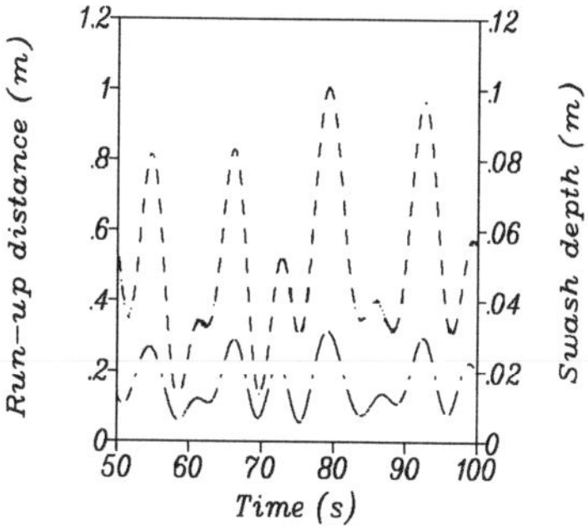

Figure 4b. Low frequency run-up and swash depth at x=0, case J1.
– – – Run-up wire, — · — S_o (rhs).

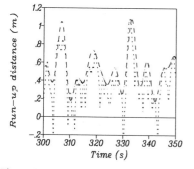

Figure 5a. LFM of shoreline and bore run-up, case J1.
– – – Run-up wire (LFM), ···· run-up - $\overline{\eta}_{So}$

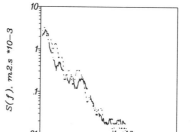

Figure 5b. Power spectra of run-up height and offshore envelope, case J1.
······ Run-up wire, – – – – envelope.

The effect of wave grouping on the swash may also be easily simulated. For example, figure 6a shows a simulation of the shoreline motion using (1), where U_o is obtained from the height of the bores at the seaward limit of the swash. Although there is no low frequency input to the simulation, the low frequency components again appear in the run-up and describe the run-up envelope. Furthermore, low frequency components also dominate the spectrum (figure 6b) and the spectrum is very similar to that obtained from the measured data. Consequently, in an

unsaturated or partially saturated surf zone, any remaining wave grouping will
induce an additional low frequency motion of the swash. This might help explain
why linear standing long wave theory tends to underestimate the motion of swash
on mildly sloping dissipative beaches (e.g. Holland et al., 1995).

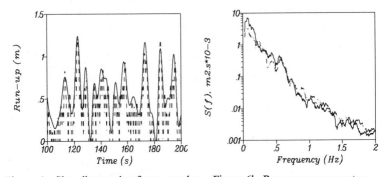

Figure 6a. Shoreline motion from a random   Figure 6b. Run-up power spectra -
wave simulation of the swash run-up.   simulation and measured data.
——— Run-up (LFM), – – – – run-up - $\overline{\eta}_{So}$ ——— Simulation, – – – – case J1.

Swash profile

Typical examples of the cross-shore shape of the swash lens over the fixed
beds are shown in figures 10a&b. The front of the swash lens is generally found to
be convex upwards, rather than concave as proposed by Shen and Meyer (1963).
The water depth during the backwash also appears to be fairly uniform over much
of the swash zone, inconsistent with standing long waves. In contrast, the inviscid
theoretical solution of Shen and Meyer (1963) significantly over-estimates the
water depth within the swash zone. The swash model outlined above provides
good estimates of the maximum water depth across the swash zone for both cases
R4 and J1 (figures 11a&b), although, for case R4, the data suggest that the bore
collapse is slightly slower than that estimated by the model. The theoretical
solution again over-estimates the water depths, particularly at the seaward limit of
the swash zone.

Swash kinematics

Examples of the swash kinematics at the seaward limit of the swash zone
are shown on figures 12a&b. However, the ADV could not provide data during the
initial stage of the uprush, due to the presence of bubbles in the leading edge of the
swash lens, or at the end of the backwash when the depth decreased below 5mm.
Additional kinematic data was therefore obtained from the swash probes by
calculating the rate of change in the measured swash volumes. Diverging flow is

evident within the swash zone for both the regular and random wave cases, since the velocity at the seaward limit of the swash turns offshore prior to the time of maximum uprush. The data suggest that the physical constraint imposed on the model, i.e. that the fluid velocity should equal the speed of the shoreline at the start and end of the uprush, is realistic, at least for the steep beach slope used here. The model predictions are generally in good agreement with the data, even in the latter stages of the backwash. Both the model and data also show that the swash kinematics are asymmetric, even in the absence of infiltration. Note that no evidence of backwash bores was observed during this study, although a hydraulic jump was frequently formed just seaward of the swash zone and this appeared to be the region in which the majority of the energy dissipation occurred.

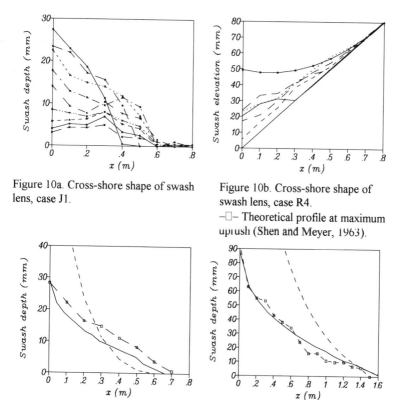

Figure 10a. Cross-shore shape of swash lens, case J1.

Figure 10b. Cross-shore shape of swash lens, case R4.
–□– Theoretical profile at maximum uprush (Shen and Meyer, 1963).

a) case R4.   b) case J1.

Figure 11. Maximum depth across the swash zone, fixed bed, $d_{50}=0.5$mm
— □ — data points, —— predicted (swash model), – – – Shen and Meyer (1963).

Two points of further interest are evident from the random data, figure 12b. Firstly, the large amplitude low frequency motion of the shoreline is clearly driven by two incident bores and is not due to a single standing long wave. Secondly, the horizontal velocity changes from offshore to onshore very rapidly as the second bore arrives at the measurement position, giving rise to very large fluid accelerations in the uprush. The implications of such high accelerations on the sediment dynamics is not clear at present, but it seems likely that treating the sediment dynamics as purely drag dominated might neglect significant inertial forces on larger particles. High fluid accelerations might also lead to the saltation of sediment particles out of the front of the swash lens (T. E. Baldock, M. G. Hughes, personal observation at Slapton Sands, Coastal Dynamics, 1997), although the effect on sediment transport rates is likely to be minimal.

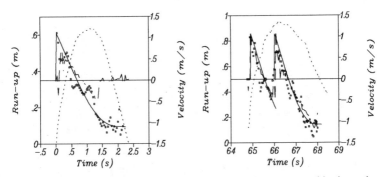

Figure 12a. Run-up and horizontal velocity at $x=0$, case R4.

Figure 12b. Run-up and horizontal velocity at $x=0$, case J1.

····· Run-up, ▯▯▯▯ U (swash probes), ── ── U (ADV), ───── U (swash model).

Conclusions

On a steep beach, where the surf zone is frequently unsaturated, significant wave energy may remain at the shoreline in the form of short wave bores, and, in a random sea state, bores of differing amplitude will directly induce low frequency swash motions. These low frequency motions describe the run-up due to individual bores and therefore describe the run-up envelope. Spectra of the run-up and offshore envelope are also found to be very similar in terms of both magnitude and shape, showing that a large proportion of the incident wave groping may be transmitted into the swash zone. Frictional effects over fixed sediment beds appeared to be small, with little difference observed in the data obtained over the two sediment sizes ($d_{50}=0.5$mm; $d_{50}=1.5$mm). In contrast, considerable differences were found over mobile beds, suggesting that frictional effects were much greater. However, infiltration of fluid into the swash zone will also affect the data and further work is required to distinguish between the two processes.

A semi-empirical swash model based on laboratory data to describe the shape of the swash lens is found to give good agreement with swash kinematics during both the uprush and backwash. The model is formulated so that it may be readily incorporated into cross-shore sediment transport models, and for this reason more exact solutions (e.g. Hibberd and Peregrine, 1979) have been avoided. Further work is in progress to identify the effects of fluid infiltration on sediment stability and swash kinematics, which is considered to be of crucial importance for the modelling of accretionary beach processes. Preliminary findings are discussed in Baird et al. (1997, this issue).

Acknowledgements

The authors gratefully acknowledge the support of the UK Engineering and Physical Sciences Research Council (EPSRC).

References

Baird, A. J., Mason, T. E., Horn, D.P. and Baldock, T. E., 1997. Monitoring and modelling ground water behaviour in sandy beaches as a basis for improving models of swash zone sediment transport. *Coastal Dynamics '97* (this issue).

Baldock, T. E., Holmes, P. and Horn, D. P., 1997. Low frequency swash motion induced by wave grouping. *Coastal Engineering* (in press).

Hibberd, S. and Peregrine, D. H., 1979. Surf and run-up on a beach: a uniform bore. *J. Fluid Mech.*, **95**, 323-345.

Holland, K. T., Raubenheimer, B., Guza, R. T., and Holman, R. A., 1995. Run-up kinematics on a natural beach. *J. Geophys. Res.*, **100** (C3), 4985-4993.

Hughes, M. G., 1992. Application of a non-linear shallow water theory to swash following bore collapse on a sandy beach. *J. Coastal Res.*, **8**, 562-578.

Hughes, M. G., 1995. Friction factors for wave uprush. *J. Coastal Research*, **11**, 1089-1098.

Huntley, D. A., Guza, R. T. and Bowen, A. J., 1977. A universal form for shoreline run-up spectra? *J. Geophys. Res.*, **82**, 2577-2581.

Nielsen, P., 1992. Coastal bottom boundary layers and sediment transport. World Scientific, Singapore.

Raubenheimer, B., Guza, R. T., Elgar, S. and Kobayashi, N., 1995. Swash on a gently sloping beach. *J. Geophys. Res.*, **100**, 8751-8760.

Shen, M. C. and Meyer, R. E., 1963. Climb of a bore on a beach 3: run-up. *J. Fluid Mech.*, **16**, 113-125.

Yeh, H. H., Ghazali, A. and Marton, I., 1989. Experimental study of bore run-up. *J. Fluid Mech.*, **205**, 563-578

MEASURING SWASH HYDRODYNAMICS ON A LABORATORY BEACH

Matthew Foote[1] and Diane Horn[2]

ABSTRACT: This paper reports on experiments designed to provide high-resolution video images of the time-varying water surface elevation and shoreline position for individual swash events. The swash from a series of regular waves was measured in a 50 metre wave flume on a steep beach (gradient 0.1). Water surface elevations were measured with wave gauges and video. Uprush and backwash events were filmed against the side wall of the flume, painted fluorescent red to allow a strong spectral differentiation with the water. An automated procedure has been developed to convert analogue video images into digital imagery formats and rectify the images. Spectral classification techniques, such as clustering and temporal differencing, are used to classify the images, and edge detection algorithms are then used to identify the water surface and to create a 2D representation of the water surface change over time. This allows the measurement of water surface elevations and the calculation of 2D water volumes to a vertical and horizontal spatial resolution of 1 cm or less. Results from this work have shown that video imagery can be a valuable data capture technique in the swash zone.

INTRODUCTION

Although the region above the still water level is the most critical in terms of shoreline management, this is the region about which least is known. As Birkemeier (1992) commented, shoreline change is by definition a swash process, yet the swash zone is one of the least understood areas of the nearshore environment. The importance of beach water table and swash interaction to accretion and erosion above the still water level has long been recognised by coastal researchers, suggesting that swash processes play an important role in the accretion of beaches. Swash processes therefore provide an important control on beach recovery in response to storm erosion. Existing cross-shore sediment transport models have demonstrated some success in predicting eroding beach profiles, but have not been found to be particularly successful at predicting accretionary episodes (Schoones and Theron 1995). Since an accretionary event is defined by the deposition of sediment above mean sea level, the lack of realistic models of swash zone hydrodynamics and sediment transport is probably the principal factor in the

---

(1) Postgraduate research student; (2) Lecturer
Department of Geography, Birkbeck College, University of London, 7-15 Gresse Street, London W1P 2LL, England.

inability of current numerical models to simulate accretionary events accurately. At present, numerical models of shoreline change do not include sediment transport processes in the swash zone. This failure to model the swash zone correctly means that sediment transport at the coastline will not be adequately represented. Recent observations indicate that flows in the swash zone can affect the beach profile seaward of the intertidal profile, influencing sediment transport in the bar region (Oh and Dean 1994, Sato et al. 1994), suggesting a possible backwash contribution to undertow. Laboratory experiments also show that the different evolution of fine and coarse-grained beaches is strongly influenced by sediment transport in the swash zone (Holmes et al., 1996).

Accurate predictions of swash zone sediment transport depend on accurate predictions of swash and backwash hydrodynamics. However, predictive models of swash and backwash hydrodynamics have had only limited success, particularly for backwash, which has been shown to be significantly different from swash (Hughes 1992, Packwood 1983). Ogawa and Shuto (1981) distinguished three types of sediment transport in the swash zone: transport in bore-like waves advancing up a dry bed, transport in bore-like waves advancing in shallow water, and transport in receding waves. This suggests that any research on swash and backwash sediment transport will need to be able the characterise the hydrodynamics of each of these different transport environments more accurately before it is likely to succeed in predicting sediment transport.

USE OF VIDEO TECHNOLOGY IN THE SWASH ZONE

The swash zone has been a problematic environment for undertaking process measurements. High energy conditions, shallow depths, and complex, rapidly reversing flows all conspire to make measurement of the time-varying hydrodynamic and morphodynamic character of the swash zone extremely difficult. A particular need at present is for high-resolution measurements of time-varying swash lens geometry, including depth and precise lens shape, from which a greater understanding of the detailed swash hydrodynamics, particularly the differences which exist between uprush and backwash, can be made (Hughes et al. 1997). Even the most detailed measurements available to date (Holland et al. 1995, Raubenheimer et al. 1995) only provide data at six relatively widely spaced positions in the swash lens from five vertically stacked resistance wires and a video of the shoreline as the sixth (beach surface) position. Most deployments have tended to favour the use of electronic probe technology. However, Holman et al. (1993) noted that while probes have provided good quality data, they have a number of limitations. Fragility to wave and swash energy, even in relatively benign conditions, plus the poor horizontal or vertical resolution gained through limitations on the number of probes which can realistically be deployed at any one time, may impose significant barriers to their sole use within the swash zone. These problems can be magnified in the very shallow flows of the backwash, where water depths can typically be less than 1 mm and sediment transport often occurs as a slurry of sediment and water, and where interference between the water body and the instruments can result in significant measurement error.

The use of video technology, and before that cine film, to record swash and other nearshore hydrodynamics is not new. Indeed, most recent deployments have used video as a complementary record of the field conditions, and in some cases, as a data collection system in its own right (see Hughes 1992, Raubenheimer et al. 1995, Holland et al. 1995). Video techniques can be used to measure both morphology and processes in the coastal environment, and a valuable review of the uses of videography in the measurement of nearshore processes and morphology has been given by Holman et al. (1993).

For morphological measurements, video techniques have the advantage over ground survey in that temporal error is reduced since the morphology of the area can change over the duration of a ground survey. The technique uses oblique photogrammetric techniques to correct the imagery geometrically. Video techniques are particularly appropriate for long-term, large-scale measurements of beach morphological evolution. A series of unmanned installations, known as Argus stations, are currently operating on at least 8 coastal sites. Each Argus station is a PC-based image processor which is programmed to collect two types of images: snapshot images of the beach or cliff and 12-minute time exposure images, collected on an approximately hourly schedule. The images are automatically downloaded to a central computer each night and stored in a central database. Video measurements have also been used for measurements of nearshore fluid processes, and have proved particularly useful in the swash zone, where video identification of the run-up edge is equivalent to a measurement very close to the bed (Holland et al. 1995; Raubenheimer et al. 1995; Raubenheimer and Guza, 1996).

It should be pointed out at this stage that video data is not presently able to completely replace electronic probes. Indeed, it is likely that future field deployments will use both electronic and video methods. However, there is a great potential for video data to become an important data capture technique. We argue here that video data used in conjunction with state of the art image processing and GIS technology may well provide a previously ignored avenue for the solution of the swash zone data resolution problem. With commonly available camcorder technology it is possible to capture imagery of microscale swash zone phenomena to extremely high temporal and spatial resolutions, typically 25 frames per second and potentially to the nearest millimetre. Video cameras are relatively cheap, employ standardised technologies, and have the potential to record phenomena and events which may otherwise be missed by non-visual means. Once data are captured, they are readily converted into standard digital image formats and can then be treated in the same way as any other digital images within image processing software. A major and until now relatively unrealised advantage of this is the similarity of such data to other geographic images such as those captured at much smaller scales by airborne and satellite sensors. Techniques designed to enhance and manipulate such complex data using GIS concepts are therefore applicable to video records of swash zone processes.

Camcorders are two separate subsystems: the sensor and the recorder (VCR). Standard camcorders using CCD sensors continuously sample electromagnetic energy reflected from real world objects in the visible and very near infrared portion of the spectrum (0.3 to 1.0 microns) in three sections corresponding to the red, green and blue bands. A matrix array of silicon capacitance tiles converts the incident energy into voltages, which are amplified and 'pushed' one after the other to a continuously recording VCR mechanism. An analogue record of voltage is thus created, with each row of the array followed by a timing signal, or synchronisation pulse. In most cases, the recording mechanism is in the form of an analogue tape, such as S-VHS or Hi-8, although digital video is becoming more available. The reflected energy is focused onto the tile array using standard optical lenses which allow the camera image to be treated as any other optical image (Holland and Holman 1997), allowing the 2D camera image to be reproduced relative to the real world objects focused onto the sensor. Filters commonly stop the infrared portion of incident energy from reaching the sensor. The time taken for a full frame image to be captured can be assumed to be the nominal sample rate of the system. In Europe this is 0.04 second (Trundle 1996, Watkinson 1995).

The final stage of the imaging process converts the analogue signal, as recorded by the VCR, into a digital form of discrete pixels of information through a second sampling (or resampling) process. This frame grabbing has until recently only been achieved using high-end graphics workstation technology. Eight-bit arrays allow for 256 discrete values of intensity (digital numbers) to be recorded for each of the three colour bands (Watkinson 1995). Such information richness therefore allows relatively subtle variations in spectral signature to be identified. A cartesian raster, or pixellated array, is therefore generated for each selected frame. The timing signals recorded onto the analogue signal allow the frame grabber to recreate the correct image geometry through a synchronisation of the computer timer and the image signal.

LABORATORY EXPERIMENTS

A series of field and laboratory experiments have been conducted with the aim of capturing high-resolution 2D water surface measurements. This paper concentrates on the work carried out between August 1996 and March 1997 in the 50 metre wave flume in the hydrographic laboratory of the Department of Civil Engineering at Imperial College. Baldock et al. (1997) give a detailed description of the experimental set up. The flume is split into three sections, two sides of 90 cm and 86 cm respectively, and a central access channel of 105 cm. Perspex panels running alongside the beach section allowed cameras to be positioned in the access section of the beach area, focused onto the far wall. The swash from a series of computer-generated random and regular wave patterns was measured on a steep beach (gradient 0.1), using both fixed and mobile beds with two different mean sediment diameters (0.5 mm and 1.5 mm).

The flume SWL was set to intersect the relevant beach at an x axis position 150 cm from the landward end, with the SWL raised when necessary to accommodate the increased height of the mobile beds. The SWL/beach intersection was taken as the horizontal (x) origin of the beach. A series of repeatable wave programs were used to generate random and regular wave trains. As a result of the pre-determined nature of the wave programs, it was not possible to match exactly the natural conditions of bore development and shoreline collapse noted by Hughes (1992, 1995). However, it was assumed that the swash events generated by the available wave programs would possess sufficiently wide ranging characteristics to adequately simulate those occurring on natural beaches. To allow a good assessment of the applicability of the video technique, a regular wave train 120 seconds long was generated with a period of 2.5 seconds and a deep water wave height of 50 mm, giving the most uninterrupted swash cycle possible at the beach angle, therefore allowing video measurement from collapse to the end of the backwash.

Water surface elevation was measured using both video and conventional resistance wave wires, with a sampling frequency of 25 Hz. Five wave wires were placed in a cross-shore array at 30 cm intervals from the SWL, with a sixth wave gauge measuring offshore conditions. The wave programs have been shown to generate the same wave fields and swash each time they are run (Baldock et al. 1997), making it possible for a higher density data collection to be achieved by moving all five wires upslope in 10 cm increments. In this way, a synthetic measurement array with a density of one measurement every 10 cm was created from probes spaced 30 cm apart. It must be pointed out that such a technique is of little use in field experiments, but was deemed to be valid in this environment.

Two standard waterproof Hi-8 CCD video cameras were used to record the swash zone optically. The cameras were positioned against the perspex sheets of one of the dividing walls, focused on the wall at the back (the target wall). Camera 1 was positioned 20 cm from the SWL origin. Camera 2 was positioned 60 cm from the SWL origin. The trade-off between the amount of side wall imaged by each camera lens (the field of view, FOV) and the spatial resolution of the resulting images (the instantaneous field of view, IFOV) is controlled by the distance between the focused object and the lens. In this case, positioning the cameras with the lens hard against the perspex gave a FOV of approximately 80 cm, while also allowing IFOV resolutions of 1 mm. As a result, problems of reflection in the perspex were also minimised. These camera positions allowed a small overlap of 10 - 15 cm, ensuring that none of the swash event was missed. Swash events from the regular wave run described above were used to test the technique whose maximum extent fell within the FOV of camera 1. The target wall was painted fluorescent red to allow strong spectral definition. A grid of control lines was drawn over the paint to allow the images to be geo-rectified (Figure 1). In order to optimise the spectral difference between the water and the wall, potassium permanganate was used to dye the water in the swash zone.

DATA CONVERSION AND IMAGE PROCESSING

Once the video was captured, analogue to digital conversion was carried out using the IRIS suite of video software on a 64 Mb Silicon Graphics Indy workstation. The camera is linked to the workstation via a standard S-video lead, and IRIS Capture used to convert analogue video to digital images in a JPEG movie format. At this stage, the sampling quality of the A/D converter is controlled by the RAM memory of the workstation. The result is a trade-off between digital pixel resolution, radiometric resolution and frame sampling rate. In this case, it was decided to reduce sampling frequency to ensure that both radiometric and spatial image resolution were maximised. As a result, sampling rate was reduced from 25 frames per second to 5 frames per second. In future experiments, a machine with larger RAM will eliminate this problem. A single image of the beach before the start of the wave run was captured to serve as the control image against which the time varying water surface was to be measured.

At this point, the images, while in JPEG format, are still stored as an IRIS movie. This is non-standard within image processing and GIS packages. Stage 2 uses IRIS MovieConvert to save each JPEG image as individual RGB format graphics files. Stage 3 uses version 3 XV in a graphics package available on the INDY to convert the RGB files to TIFF. At this stage, the images are readable by image processing software. The next conversion stage is to rectify the images to cartesian co-ordinate space. This was done in Arc/Info using the register and rectify commands. Control points were chosen at intersections of the control grid and converted to x,y positions relative to the SWL. An affine transformation of the images to the control points resulted in a series of geo-rectifed images (Figure 1).

At this stage, each image can be regarded as a digital representation of 2D cartesian space in the same way as any other raster map. A final conversion process from rectified TIFF to ERDAS Imagine raster format allows the images to be manipulated using image processing techniques. Each image at this stage is composed of three layers corresponding to the red, green and blue spectral bands. Nominal spatial resolution at this stage is still approximately 1 pixel, equal to 1 millimetre. As a large proportion of the image area will be static over time, (this is particularly true of the area of backboard which does not get wet during swash events, but which is still imaged on the video), a subset of the image relating to the area of swash extent was made at this stage. As well as optimising data storage, this also reduced the potential for poor differentiation between the board and beach spectral signatures due to variations in spectral response over large areas of board.

ERDAS Imagine allows for a large number of processing operations to be carried out on images. In this case, the first procedure was to classify the control image to identify beach and board pixels. In spectral classification, a multi-dimensional feature space allows pixels with similar spectral signatures to be grouped together (Schowengert 1983). Training areas of known pixels were used to define the beach and board spectral signatures. A maximum likelihood classifier was used to partition the whole image. Figure 2 shows the resulting image, which was then used as a mask to remove the redundant foreground beach pixels from the swash event images before classification. This was particularly useful in reducing the amount of classification confusion between true board pixels and pixels which gave similar responses due to reflection in the shallow backwash flows. Another mask was generated to remove the ADV probe from the image.

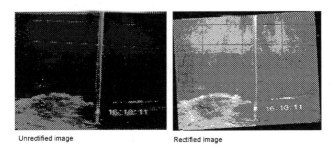

Unrectified image   Rectified image

Figure 1. Comparison between unrectified and rectified images

RESULTS

Before classification of the swash images, small variations in the spectral signature ranges between each image (due to small changes in lighting conditions) were reduced through a histogram normalisation process. Here, the histogram of each image band was shifted in order for the mean digital number to be equal to that of the corresponding control image histogram. Classification, using the same training area signatures as for the control image, resulted in images as seen in Figure 3.

As can be seen from Figure 3, errors of commission, where board pixels were mis-classified as non-board, occurred with increasing regularity as the water surface elevation reduced. In most of the images, however, a good differentiation between board and non-board was made at the water surface. In some areas omission errors occurred where water pixels were mis-classified as board. This is most noticeable in image 6, at the thinnest part of the backwash. This is most likely to be the result of board reflection in the water, or to the board being imaged through the more transparent thin water.

Classified bare beach control image

Swash image with bare beach mask

Figure 2. Classified control image and masked swash image

DISCUSSION

The video technique as deployed here allows for the capture of relatively high-resolution water surface measurements in a laboratory swash zone. However, a number of logistical and technological problems were present which reduced the overall effectiveness of the technique as outlined below.

Classification error. All classified images will contain errors of commission or omission. In this case, mis-classified pixels can make large differences to the accuracy of the data, particularly in the thinner backwash flows. The degree of mis-classification can be reduced through a number of actions, such as the use of other classification techniques or achieving greater spectral differences between the board and water pixels of each image. In the first case, simple parametric supervised classification procedures were used to classify the image. These rely on training area pixels being chosen which fully describe all other pixels in terms of multi-dimensional feature space. It was noticed that even with a strong fluorescent red background, the resulting board images showed a large variation of saturation over the extent of the whole image, and that in some areas, the spectral signatures of the board matched those of the water too closely to allow accurate classification. It may be that unsupervised (isodata) classifiers, which use statistical clustering methods to partition the image feature space without the need for operator input of training areas, might produce a more accurate classification. However, it may be that the lack of significant variation between water and board pixel signatures in some areas of the images would not allow any standard classification technique to improve significantly on the results obtained here. Greater spectral differences between the board and water pixels of each image could be accomplished by increasing the recorded opacity of the water relative to the board. The use of infrared video cameras has also been considered as a potential means of ensuring that the greatest opacity is achieved, as very near infra red wavelength electromagnetic radiation is absorbed by water. Other dye colours rather than purple might also allow for a better differentiation.

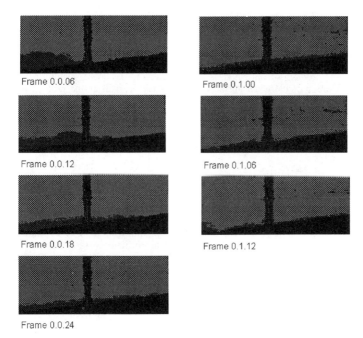

Figure 3. Series of classified swash images

Poor temporal synchronisation. In these experiments, synchronisation between camera clock and the computer clock was difficult due to the lack of an easy link between the two systems. A stopwatch was imaged at the start of each run in an attempt to allow for this link to be recorded to the nearest second. However, it was found that it was difficult to relate this to the probe records after data capture. In future deployments it will be necessary to construct a more rigorous means of time synchronisation. A potential solution to this problem would be to genlock the cameras (i.e. time synchronise them using the timing signal of the VCR system) using a GPS signal as the parent time pulse (see for example Holland and Holman 1997). A continuous time record could then be added to the video data, which is extractable at the A/D stage. It is not possible to achieve this easily with the available Hi-8 cameras, although S-VHS cameras have this capability.

Reduced temporal resolution. It has already been mentioned that the low memory capability of the computer used to perform the A/D conversion resulted in a much lower frame capture rate than that of the cameras. A larger RAM capacity would nullify this problem to some extent, although the increasing availability of digital camera technology may soon make the A/D conversion stage obsolete. However, while logistical difficulties have reduced the effectiveness of this particular data conversion process, the original data recorded at 25 frames per second is still held on the original video tapes, and can be converted at a later date to full resolution.

Problems of field deployment. While relatively useful data has been collected in the controlled environment of the laboratory, the technique as it is used here has a number of problems when transferred to the field. Fragility of the backboard and the cameras themselves to even relatively low wave energy conditions reduces their effectiveness as field measurement instruments. The use of a backboard also introduces the problem of uprush striking at an angle to the board, resulting in a spurious elevation of the water surface. Water splash, as well as the rain on the camera lenses can also reduce the usefulness of data captured using this technique. However, it is possible to reduce the impact of these factors through careful deployment design. At present, non-intrusive video techniques are being developed which will remove the need for a backboard.

CONCLUSION

This series of experiments has shown that video data capture has the potential to be a valuable measurement technique within the field of swash zone hydrodynamics. The data produced through the deployment of standard Hi-8 camcorders in a laboratory environment have allowed high-resolution water surface measurements to be made at a potential temporal resolution of 25 frames per second. Problems stemming from logistical difficulties may be solved relatively easily, making this technique more useful both in the laboratory, as well as in the field in good conditions.

REFERENCES

Baldock, T.E., Holmes, P. and Horn, D.P. 1997 in press. Low frequency swash motion induced by wave grouping. Coastal Engineering.
Birkemeier, W.A. (Editor). 1992. SandyDuck: a field study of sediment and bathymetric response to fluid forcing. From SandyDuck web site: http://www.frf.usace.army.mil/duck94/sandy.html
Holland, K.T. and Holman, R.A. 1997. Video estimation of foreshore topography using trinocular stereo. Journal of Coastal Research 13(1): 81-87.

Holland, K.T., Raubenheimer, B., Guza, R.T., and Holman, R.A. 1995. Runup kinematics on a natural beach. Journal of Geophysical Research 100(C3): 4985-4993.

Holman, R.A., Sallenger, A.H., Lippmann, T.C. and Haines, J.W. 1993. The application of video image processing to the study of nearshore processes. Oceanography 6(3): 78-85.

Holmes, P., Baldock, T.E., Chan, R.T.C. and Neshai, M.A.L. 1996. Beach evolution under random waves. Proceedings of the 25th International Conference on Coastal Engineering, 3006-3019.

Hughes, M.G. 1992. Application of a non-linear shallow water theory to swash following bore collapse on a sandy beach. Journal of Coastal Research 8: 562-578.

Hughes, M.G. 1995. Friction factors for wave uprush. Journal of Coastal Research 11(4): 1089-1098.

Hughes, M.G., Masselink, G. and Brander, R.W. 1997. Flow velocity and sediment transport in the swash zone of a steep beach. Marine Geology 138: 91-103.

Ogawa, Y. and Shuto, N. 1981. Field measurements of hydraulics and sediment movements in a swash zone. Proceedings of the 28th Japanese Conference on Coastal Engineering, pp. 212-216 (in Japanese, quoted in Horikawa 1988).

Oh, T-M. and Dean, R.G. 1994. Effects of controlled water table on beach profile dynamics. Proceedings of the 24th International Conference on Coastal Engineering, 2449-2460.

Packwood, A.R. 1983. The influence of beach porosity on wave uprush and backwash. Coastal Engineering 7: 29-40.

Raubenheimer, B., Guza, R.T., Elgar, S. and Kobayashi, N. 1995. Swash on a gently sloping beach. Journal of Geophysical Research 100(C5): 8751-8760.

Raubenheimer, B. and Guza, R.T. 1996. Observations and predictions of run-up. Journal of Geophysical Research 101(C10): 25575-25587.

Sato, M., Hata, S. and Fukushima, M. 1994. An experimental study on beach transformation due to waves under the operation of coastal drain system. Proceedings of the 24th International Conference on Coastal Engineering, 2571-2582.

Schoones, J.S. and Theron, A.K. 1995. Evaluation of 10 cross-shore sediment transport/ morphological models. Coastal Engineering 25: 1-41.

Schowengerdt, R.A. 1983. Techniques for image processing and classification in remote sensing. New York: Academic Press.

Trundle, E. 1996. Newnes guide to tv and video technology. Oxford: Butterworth Heinemann.

Watkinson, J. 1995. An introduction to digital video. Oxford: Butterworth Heinemann.

Toward a Better Understanding of Swash Zone Sediment Transport

Michael Hughes[1], Gerhard Masselink[2], David Hanslow[3] and David Mitchell[1]

Introduction

At present the inclusion of swash zone sediment transport into numerical beach models is rarely attempted. This highlights our limited understanding of the swash zone compared with other parts of the beach system. Nevertheless, decades of theoretical research and the data emerging from recent field experiments points to the key elements that a successful swash zone sediment transport model must contain. The purpose of this paper is to discuss implications of recently reported data (Hughes et al., 1997; Masselink and Hughes, in review) in the context of developing a complete physical description for swash zone sediment transport. The data presented here is intended only to illustrate points of discussion. A complete description of the data set collected by the authors, field sites and methods can be found in the abovementioned references. Briefly, the data were obtained from two Australian Beaches: Palm Beach, NSW and Myalup Beach, WA. Palm beach displayed a beach face gradient of 0.12 and was composed of sediment with a mean grain diameter of 0.3 mm, whereas Myalup beach was marginally steeper (gradient of 0.14) and coarser (mean grain diameter of 0.5 mm). The waves arriving at the seaward edge of the swash zone were either plunging breakers or fully developed surf zone bores.

Observations of flow behaviour and sediment transport in the swash zone are described in the following two sections. The important issue to emerge from these observations is that sediment transport model calibration coefficients for uprush are larger than for backwash. The paper concludes with a brief discussion of how we might proceed in gaining a physical explanation for this result and ultimately

[1] Dept of Geology & Geophysics, University of Sydney, NSW 2006, Australia
[2] Centre for Water Research, University of Western Australia
[3] Coastal Management Program, NSW Dept of Land and Water Conservation

develop a successful swash zone sediment transport model. It appears that this future model will be significantly different to existing sediment transport models for waves in the surf zone.

Flow Kinematics

The shoreline motion and geometry of the swash lens have been measured in both the laboratory and the field (Yeh et al., 1989; Hughes, 1992; Hughes et al., 1997; Masselink and Hughes, in review). The salient features of swash on steep beaches are listed below.

1. The shoreline (front of swash lens) position is generally parabolic through time, although, the occurrence of a retrogressive (backwash) bore can cause a departure from this pattern in the latter stages of the backwash.
2. The water depth at a given position on the beach face increases rapidly, immediately after the shoreline has moved past, but begins to decrease before the shoreline reaches its maximum height on the beach. This causes the swash lens to thin as it advances. Throughout the backwash the water depth continues to decrease until zero water depth is reached, when the shoreline recedes past the point of interest. Again, this situation can be modified in the latter stages of the backwash by the occurrence of a retrogressive bore, which causes a localised increased in water depth on the lower beach face. Irrespective of the presence of a bore, the largest water depths occur during the uprush.
3. The shoreline generally has a constant negative acceleration during the uprush and the same constant positive acceleration during the backwash, so that its velocity is symmetric through time (i.e. the duration of landward advance and seaward retreat are equal).
4. In contrast the flow velocity at a given position on the beach face is asymmetric through time. Locally the flow direction will switch from landward to seaward before the shoreline reaches its maximum height on the beach, thus the duration of landward flow is less than the duration of seaward flow. It is this divergence of flow that produces the thinning of the swash lens during uprush (see point 2).

It is apparent from Point 4 that some ambiguity exists in the definition of uprush and backwash, since the flow in the interior of the swash lens can be seaward when the shoreline on the upper beach face is still moving landward. Here uprush and backwash are defined in terms of the local flow direction, rather than the direction of shoreline movement. Notably the above list of observations for flow kinematics in the swash zone of steep beaches is well described by solutions of the non-linear shallow water equations that include bores (see Ho et al., 1963; Hibberd and Peregrine, 1979).

The water depth and flow velocity within the swash lens are of particular relevance to the local sediment transport rate on the beach face. Measurements of these parameters over a single swash cycle are shown in Figure 1. The asymmetric nature of the flow field is clearly evident in this example. The maximum landward (positive) velocity occurs near the beginning of the uprush whereas the maximum seaward (negative) velocity occurs near the end of the backwash. Moreover, the uprush duration is clearly smaller than the backwash duration. The effect of a secondary maximum in water depth on flow velocity during the backwash is indicated by the arrows on the figure. The increase in water depth is associated with a decrease in the flow acceleration (i.e. change in slope of the velocity time series). The origin of this secondary maximum in water depth cannot be determined from this data set, but two possibilities exist: (1) a retrogressive bore located immediately up-beach or (2) an incoming bore overriding the backwash. Either scenario has the effect of slowing the backwash flow. Field measurements showing the development of a retrogressive bore on similar beaches to the example shown here have been presented by Hughes (1992).

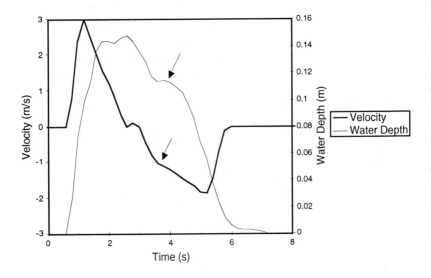

Figure 1: Field data from Palm Beach showing time series of water velocity and depth for a single swash cycle. The instruments were located approximately at the mid swash position (After Hughes et al., 1997).

Two non-dimensional parameters relevant to later discussion are the Froude Number, F_r, and the Shields Parameter, θ. Calculated time series of these parameters for a single swash cycle are shown in Figure 2. The calculations were made using

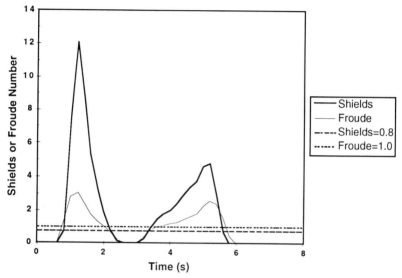

Figure 2: Calculated time series of the Froude Number and Shields Parameter for the data shown in Figure 1. The parameters were calculated using Equations 1 to 3.

$$F_r = \frac{u}{\sqrt{gh}} \tag{1}$$

and

$$\theta = \frac{0.5\rho f u^2}{gD(s-1)} \tag{2}$$

where u is the flow velocity, g is the gravitational acceleration, h is the water depth, ρ is the water density, f is a friction factor, D is the grain diameter of the sediment and s is the ratio of sediment to fluid density. The friction factor was calculated using a formulation for sheet flow under waves (Wilson, 1989):

$$f = 0.114\left(\frac{\hat{a}}{(s-1)gT^2}\right)^{0.4} \tag{3}$$

where $\hat{a}$ is the peak wave orbital excursion given by $\hat{a} = T\hat{u}/2\pi$, T is the wave period and $\hat{u}$ is the peak flow velocity. In this example $f=0.012$. For most of the swash $F_r>1$ indicating supercritical flow (Allen, 1985), and $\theta>0.8$ indicating sheet flow conditions (Wilson, 1989).

Sediment Transport

Field measurements of sediment transport in the swash zone collected to date support a relationship between the total (bedload+suspended) load immersed mass transport rate and the mean flow velocity cubed (see Hardisty et al., 1984; Hughes et al., 1997; Masselink and Hughes, in review). One model for sediment transport that is consistent with this observed relationship is the energetics-based bedload model developed by Bagnold (1963; 1966):

$$I_u = \frac{k_u \bar{u}^3 T_u}{\tan\phi + \tan\beta} \tag{4}$$

$$I_b = \frac{k_b \bar{u}^3 T_b}{\tan\phi - \tan\beta} \tag{5}$$

where the subscripts u and b denote uprush and backwash respectively, I is the immersed mass sediment transport per unit metre beach width (kg m^{-1}), k is a calibration coefficient (kg m^{-4} s^2), $\bar{u}$ is the mean flow velocity (m s^{-1}), T is the flow duration, ϕ is the friction angle of the sediment and β is the angle of the beach face. In this formulation $k = \frac{1}{2g}\rho f e_b$, where e_b is Bagnold's bedload efficiency factor (Bagnold, 1966). The uprush sediment transport data contained in Hughes et al. (1997) and the joint uprush and backwash data contained in Masselink and Hughes (in review) are shown together here in Figure 3. The calibration coefficient k has the following values:

k_u=1.37±0.17 (Palm Beach)
k_u=2.60±0.52 (Myalup Beach)
k_b=1.50±0.20 (Myalup Beach)

It is not immediately clear why k_u is different for the two beaches. Intuitively it is expected that both f and e_b, and hence k depend on grain size (see Hardisty, 1983), but many more measurements from a variety of beaches are necessary in order to confirm this. Unfortunately there is insufficient knowledge of either f or e_b in the swash zone to speculate further on their direct role.

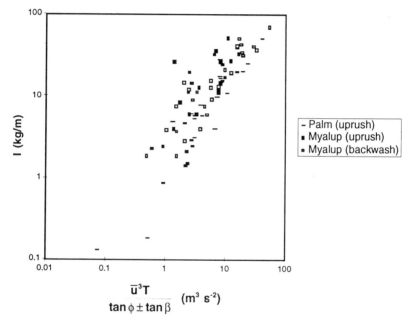

Figure 3: Field data from Palm Beach and Myalup Beach showing separately the total immersed mass transported during the uprush and backwash as a function of Bagnold's model. Data are from Hughes et al. (1997) and Masselink and Hughes (in review).

The fact that $k_u > k_b$ (statistically significant at the 95 % confidence level) on Myalup Beach is an important result, as it suggests something about the completeness of Bagnold's model. The existing formulation of the model (Eqs. 4 and 5) accounts for the fluid shear stress causing sediment motion and the bed slope effects influencing this. If these were the only physical processes relevant to sediment transport by swash then the model coefficients for uprush and backwash should be equal. The result that $k_u > k_b$ for a given beach indicates, therefore, that the transport model is incomplete. The "measured" value of k must involve the net influence of unaccounted for processes. It could be argued that the difference between k_u and k_b reported here has arisen because Bagnold's bedload model does not account for the suspended load component of the total load that was measured. The magnitude of the Shields Parameter throughout most of the swash (Figure 2) is indicative of sheet flow conditions. Interestingly, a Shield's type transport model proposed by Wilson (1989) to describe the total load (bedload+suspended load) transport in sheet flow yields a similar result to the above - i.e. the proportionality coefficient for the uprush is significantly larger than that for the backwash (Masselink and Hughes, in review).

Discussion

The important issue to emerge from the preceding field observations is that conventional sediment transport model calibration coefficients for uprush are larger than for backwash. Interestingly, attempts to combine the non-linear shallow water theory with these sediment transport models in order to simulate beach face morphodynamics have only been successful if the model coefficients assigned to the uprush are arbitrarily larger than those assigned to the backwash (e.g. Hughes, 1989; Turner, 1995; Ling et al, in review). It seems that the most important lesson learnt so far is that existing sediment transport models do not adequately account for the processes governing swash zone sediment transport. There are at least two unaccounted for processes believed to be important: turbulence/sediment advection processes and infiltration/exfiltration processes.

Turbulence/sediment advection

Plunging breakers and fully developed bores arriving at the beach face are highly turbulent, and will probably transfer some of this turbulence to the swash lens. Turbulence intensity during the uprush is therefore expected to be larger than during the backwash. Moreover, incoming waves are carrying sediment, and will probably transfer some of this sediment to the uprush. Intuitively we expect that turbulence and sediment advection from the inner surf zone into the swash zone will influence the mechanics of uprush transport. At present there are precious few measurements of the sediment load carried by plunging breakers and bores near the beach face (see James and Brenninkmeyer, 1977) and no measurements of turbulence intensity in the swash zone.

Infiltration/exfiltration

It has long been argued that landward of the saturated part of the beach face where the water table outcrops, the uprush and backwash flows are expected to experience a loss of volume due to fluid infiltrating into the "dry" beach (e.g. Grant, 1948). Moreover, within the saturated part of the beach the flow is expected to gain additional volume due to fluid exfiltrating from the saturated beach. In a more general sense, it should be expected that infiltration will occur anywhere on the beach where the pressure head provided by the swash lens thickness is larger than the pore water pressure in the beach whereas exfiltration will occur anywhere on the beach where it is less. So in addition to changing swash volumes the loading and unloading of pore water pressure is also likely to impact on swash kinematics and sediment transport.

The changes in swash volume due to infiltration/exfiltration on a sandy beach are probably not sufficient to significantly change the swash behaviour directly (see Packwood, 1983). It is more likely that vertical drag forces and boundary layer thinning/thickening resulting in modified bed shear stresses is of greater importance. Turner (1995) provides a detailed description of these processes. Their likely effects are summarised in the following:

1. Infiltration (exfiltration) is likely to result in a vertically downward (upward) directed fluid drag, providing a stabilising (destabilising) effect on the bed sediment and producing a corresponding impact on the mechanics of sediment entrainment.
2. Infiltration (exfiltration) is likely to result in a thinning (thickening) of the boundary layer, causing an increase (decrease) in bed shear stress and producing a corresponding impact on the mechanics of sediment transport.
3. Loading and unloading of pore water pressure due to rapid infiltration and exfiltration over a swash cycle has the potential to cause bed fluidisation and produce a corresponding impact on the mechanics of sediment entrainment.

The relative importance of each of these effects during either the uprush or backwash has not been measured to date. There has been recent attempts to assess the relative importance of the first two effects listed above (Turner and Masselink, in review) and assess the individual importance of the third (Baird et al., elsewhere in this volume) through heuristic modeling. However, unequivocal interpretation of results is hampered by the lack of direct measurements of these processes.

Key elements of a swash zone sediment transport model

The preceding discussion suggests that a physically complete swash zone sediment transport model requires at the very least an extension to the range of mechanisms implicit in existing surf zone sediment transport models. In particular the effects of (1) turbulence advection and (2) infiltration/exfiltration on sediment stability and bed shear stress need to be incorporated. Nielsen (elsewhere in this volume) has already begun addressing (2) by proposing a modified Shields parameter that accounts for these effects. In order to proceed further with developing existing transport models the following measurements are considered to be of highest priority: turbulence intensity and boundary layer structure within the swash, pore water pressures within the beach and sediment loads carried by plunging breakers and bores.

Until we have some knowledge of the boundary layer structure in the swash we can only persist with conventional sediment transport models originally developed for waves or unidirectional flows. There is no reason to expect, however, that the boundary layer structure in the swash will be similar to either, and a completely different approach to sediment transport might be required. Some qualitative observations of sediment transport are relevant in this regard. It has been observed that both bedload and suspended load transport occurs during the uprush, whereas the backwash tends to be dominated by bedload transport (Horn and Mason, 1995). The entire uprush and the early part of the backwash looks like a two-phase flow of water transporting sediment, whereas the latter part of the backwash looks like a single-phase flow or sand/water "slurry" (Hughes, 1992). The two-phase flow is consistent with the concept underpinning conventional surf zone sediment transport models, that a fluid flow transports the sediment, but the single-phase flow is not. The single-phase flow, whose upper surface is the free surface, may be more consistent with transport models that describe debris flows.

Summary

It appears that particular solutions of the non-linear shallow water equations involving bores provide a reasonably accurate description of swash on steep beaches. Moreover it appears that conventional sediment transport equations of the type proposed by Bagnold (1963, 1966) and Wilson (1989) provide a gross description of swash zone sediment transport. However the application of these models has masked some fundamental aspects of swash zone sediment transport behind calibration coefficients. It is apparent from our observations to date that a successful swash zone sediment transport will need to either extend the processes implicit in existing surf zone formulations or be completely novel. Either way the implication is that existing surf zone sediment transport models are inadequate for the swash zone.

References

Allen, J.R.L., 1985. *Principles of Physical Sedimentology*. George Allen and
        Unwin, London, 272 pp.
Bagnold, R.A., 1963. Mechanics of marine sedimentation. In M.N. Hill (Ed.), *The
        Sea*, Volume 3, Wiley-Interscience, 507-528.
Bagnold, R.A., 1966. *An Approach to the Sediment Transport Problem from
        General Physics*. U.S. Geological Survey Paper 422-I, 37 pp.

Grant, U.S., 1948. Influence of the water table on beach aggradation and degradation. *Journal of Marine Research*, 7, 655-660.

Hardisty, J., 1983. An assessment and calibration of formulations for Bagnold's bedload equation. *Journal of Sedimentary Petrology*, 53, 1007-1010.

Hardisty, J., Collier, J. and Hamilton, D., 1984. A calibration of the Bagnold beach equation. *Marine Geology*, 61, 95-101.

Hibberd, S. and Peregrine, D.H., 1979. Surf and runup on a beach: A uniform bore. *Journal of Fluid Mechanics*, 95, 323-345.

Ho, D.V., Meyer, R.E. and Shen, M.C., 1963. Long surf. *Journal of Marine Research*, 21, 219-230.

Horn, D.P. and Mason, T., 1994. Swash zone sediment transport modes. *Marine Geology*, 120, 309-325.

Hughes, M.G., 1989. *A Field Study of Wave-Sediment Interaction in the Swash Zone*. Ph.D. thesis, University of Sydney, 243 pp. (Unpubl.).

Hughes, M.G., 1992. Application of a non-linear shallow water theory to swash following bore collapse on a sandy beach. *Journal of Coastal Research*, 8, 562-578.

Hughes, M.G., Masselink, G. and Brander, R.W., 1997. Flow velocity and sediment transport in the swash zone of a steep beach. *Marine Geology*, 138, 91-103.

James, C.P. and Brenninkmeyer, B.M., 1977. Sediment entrainment within bores and backwash. Geoscience and Man, 18, 61-68.

Ling, L, Barry, D.A., Pattiaratchi, C.B. and Masselink, G., in review. Sediment transport and beach profile changes in the swash zone (part A): Model formulation and testing. *Journal of Geophysical Research*.

Masselink, G. and Hughes, M.G., in review. Field investigation of sediment transport in the swash zone. *Continental Shelf Research*.

Packwood, A.R., 1983. The influence of beach porosity on wave uprush and backwash. *Coastal Engineering*, 7, 29-40.

Turner, I.L., 1995. Simulating the influence of groundwater seepage on sediment transported by the sweep of the swash zone across macro-tidal beaches. *Marine Geology*, 125, 153-174.

Turner, I.L. and Masselink, G., in review. Swash infiltration-exfiltration and sediment transport. *Journal of Geophysical Research*.

Wilson, K.C., 1987. Analysis of bedload motion at high shear stress. *Journal of Hydraulic Engineering*, ASCE, 113, 97-103.

Yeh, H.H., Ghazali, A. and Marton, I., 1989. Experimental study of bore run-up. *Journal of Fluid Mechanics*, 206, 563-578.

SURF ZONE AND SWASH ZONE SEDIMENT DYNAMICS
ON HIGH ENERGY BEACHES: WEST AUCKLAND, NEW ZEALAND

Philip D. Osborne[1] and Geraldine A. Rooker

Abstract

Suspended sediment concentration (SSC), fluid velocity and morphological response were measured on the foreshore of a high energy dissipative beach west of Auckland, New Zealand. Re-suspension events exhibit a distinct temporal structure associated with both the uprush and backwash phases of infragravity swash. An initial peak in SSC of > 50 g l^{-1} occurs under the turbulent, accelerating flow of the shoreward propagating swash bore.   SSC decreases rapidly from the initial peak and then more gradually as water depth increases to a maximum at the end of the uprush. Nearly all suspended sediment settles back to the bed before the start of the backwash cycle. SSC increases gradually during the backwash acceleration phase, and reaches a peak under the super-critical (Froude number >>1), accelerating flows near the end of the backwash. Hydraulic jumps occur and anti-dune bedforms often develop near the end of backwash events and these are associated with large increases in SSC.   Net transport is potentially sensitive to small variations in the phase angle between SSC and velocity in the uprush and backwash cycle as well as to the variability in bed conditions during swash cycles. These measurements suggest that foreshore equilibrium is constrained by relatively small differences in the velocity - SSC phase relationship.

---

[1]Lecturer, Department of Geography, Division of Science & Technology, Tamaki Campus, University of Auckland, Private Bag 92019, Auckland, New Zealand; e-mail: p.osborne@auckland.ac.nz; phone: 64 9 373 7599 ext 6606; fax: 64 9 373 7042

Introduction

Sediment transport in the swash zone and inner surf zone is important to the overall nearshore (littoral) sediment budget and to the sorting of beach sedimentary deposits, since swash/backwash and inner surfzone hydrodynamic processes determine whether sediment is stored onshore or carried into the outer surf zone with potential to be lost offshore. In particular, long-period swash motions associated with high energy incident waves have been shown to be a major factor resulting in foreshore erosion (see review by Komar and Holman, 1986). Unfortunately, quantitative understanding of the sediment transport processes in the swash zone and inner surf zone remains rudimentary due to a lack of knowledge concerning the relative importance of various transport mechanisms and processes. In particular, observations of sand transport on beaches indicate that episodic suspension events occur in response to a variety of mechanisms, including long period infragravity and far-infragravity waves (Beach and Sternberg, 1992; Aagaard and Greenwood, 1994), short period waves and wave groups (Hanes, 1991; Dick et al., 1994). The nature, causes and relative importance of episodicity in producing net sediment transport are not fully understood largely due to a lack of adequate field observations. This problem is particularly acute with respect to the swash zone and inner surf zone of high energy beaches where infragravity waves dominate the velocity field and where it has been particularly difficult to obtain process measurements.

This paper presents results from an experimental programme designed to increase understanding of the sediment dynamics and morphological response in the swash zone and inner surf zone of high energy dissipative beaches on the west coast of New Zealand's North Island. These rather unique beaches are composed of relatively dense black ironsands and are exposed to almost continuous high energy swell originating mainly from the Southern Ocean to the southwest. The primary objective was to examine the temporal structure of swash zone suspension events in relation to the local hydrodynamics in order to gain an improved understanding of the controls on net sediment transport and morphological response in this region.

Study Site and Environmental Conditions

Experiments were conducted at Bethells Beach near Auckland on the west coast of the North Island of New Zealand over four days in May and July, 1995. Bethells Beach is typical of the relatively undisturbed west coast beaches which are exposed to almost continuous high energy swell (common breaker heights of 2-5 m) originating mainly from the southwest particularly in winter and spring. Bethells Beach is a multi-barred dissipative beach with a broad surf zone (200-400 m) and a meso-macro tidal regime (approximately 3.5 m at spring tides). The wide swash zone (30-60 m), dominated by cross-shore infragravity motions, permits frequent observations of bed elevation relative to sensors and observations of foreshore response between major swash events.

The beach is composed of the dense black "ironsands" which are common on the west coast of New Zealand's North Island. The "ironsand" consists mainly of heavy minerals (ρ_s~4.0-5.0 g cm^{-3}) including titanomagnetite (15-40%;), and ilmenite (17%) as well as small quantities of lighter minerals (ρ_s<3.0 g cm^{-3}) including quartz and feldspars. The sand in the vicinity of the instrumentation was well sorted fine-medium sand with a standard deviation of 0.26 ϕ (0.150 mm). There is a tendency for higher density grains to accumulate on the upper foreshore and backshore regions of these beaches particularly in the wake of storms and for mean density to decline offshore (e.g. Komar, 1985).

Process measurements were taken over four high tide cycles with moderate to high energy incident swell present (H_{sb}~2 m). Results presented here focus on the measurements taken on July 28, 1995 when 23 time-series were collected.

Field Methods

Measurements of 3-D velocities within 2-5 cm of the bed were obtained using a Sontek Acoustic Doppler Velocimeter (ADV). Measurements of suspended sediment concentrations (SSC) were made using 2 optical backscatter sensors (OBS-3 by D&A Instruments) deployed at nominal elevations of 5 and 10 cm above the bed. Water surface elevations were measured with a 1.5 m teflon coated capacitance wave gauge (WG-1 by Richard Brancker Research, Ltd.). The instruments were collocated in the longshore direction, separated by no more than 30 cm horizontally and mounted on a single steel frame. Sensor position was recorded before and after each burst of data collection and the vertical position was adjusted frequently between bursts to maintain a consistent elevation for measurements.

Sensors were hardwired to a shore-based power supply and a high speed data acquisition system controlled by laptop computer. Bursts of 4100 data points were acquired at rates of 10 Hz. OBS sensors and wave gauges were calibrated according to standard laboratory procedures.

The foreshore was surveyed at low tide before and after each set of process measurements was taken at high tide. Small scale changes (1-2 m^2) in beach response were monitored with a grid of depth of disturbance rods equipped with washers and located around the sensor frame. The instruments were situated on the lower portion of the foreshore in order to be fully immersed at high tide and to be in the swash zone for up to 2 hours before and after high tide (Fig. 1).

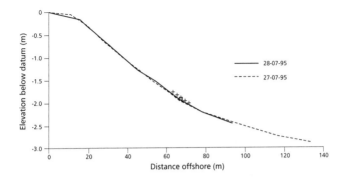

Figure 1. Beach profiles showing location of instrumented station
and grid of depth-of-disturbance rods.

Near-bed velocities and Sand Re-suspension under Infragravity Swash

Time-series of horizontal velocities (U), total 3-D velocity variances
(TKE), ADV signal-noise ratios (SNR), water surface elevations and SSC typical
of mid-tide conditions on 28 July 1995 are illustrated in Fig. 2. The series in Fig.
3 represent the transition region between the swash zone and inner surf zone where
the sensors were almost continuously immersed. Segments of the velocity record
in which the ADV SNR dropped below 25 dB have been excluded; segments of
low SNR represent periods when the velocimeter probe was out of the water.
Typically, despite the highly turbulent, often foamy and high suspended loads in
the water column, the ADV exhibited extremely rapid response with SNR >> 25
dB and also high signal correlations when immersed in the swash zone.

Total 3-D velocity variances (TKE) were estimated from ADV velocity
components using:

$$TKE = \tfrac{1}{2}\rho \overline{k^2} = \tfrac{1}{2}\rho \left(\overline{u'^2} + \overline{v'^2} + \overline{w'^2} \right)$$

where u', v' and w' are the velocity fluctuations about the mean in the three
coordinate directions and ρ is the fluid density. In this case, TKE is an estimate of
the total variance and not strictly the turbulent kinetic energy, as the flow
kinematics are dominated by infragravity wave orbital motions which have been
included in the estimates.

Cross-shore currents dominate the velocity field and are characterized by
maximum speeds of 200 cm s^{-1} in the swash zone. Significant onshore current

speeds of 145 cm s^{-1} with periods of approximately 50 s occur in the inner surf zone. Smaller waves with periods of 12-15 s are superimposed on the larger scale motion but these constitute a minor proportion of the total variance. Mean shore-parallel and shore-normal currents reach speeds of up to 25 cm s^{-1} on the foreshore.

SSC events in Fig. 2 exhibit a distinctive temporal structure associated with both uprush and backwash portions of the infragravity cycles. Peak SSC (>100 g l^{-1}) is associated with high shear stresses (high velocities, accelerating flow and shallow water) early in the uprush and towards the end of the backwash phases. SSC decreases rapidly from the initial peak and then more gradually as water depth increases to a maximum at the end of the uprush. SSC increases gradually at first, as the backwash accelerates, and then reaches a peak under the rapidly thinning and accelerating flow near the end of the backwash. Very high SSC are also observed in association with super-critical flows, hydraulic jumps and backwash ripples generated as incoming swash interacts with the end of a backwash phase. These processes appear to be responsible for the secondary peaks in SSC near the end of the backwash phase. High SSC close to the end of the backwash may lead to considerable advection of SSC into the inner surf zone and also to enhanced SSC during subsequent uprush events.

Although the frequency of events increases, SSC declines as water depth increases in the swash-surf transition (Fig. 3). A conceptual diagram which illustrates the distinctive temporal variations in swash zone re-suspension and hydrodynamics is illustrated in Fig. 4.

Cross-shore sediment Transport and Foreshore Dynamics

A number of contemporary physically based models for sediment transport in the nearshore are based on the assumption of a direct dependence between SSC, sediment transport rates and the instantaneous bed shear stresses based on local fluid velocity field (eg. Bagnold, 1966; Bailard, 1981). In the swash zone, major suspension and transport events coincide closely with maximum velocity variances derived from local 3-D velocity measurements. Fig. 5 illustrates that cross-shore transport rates computed from instantaneous products of U and SSC are well correlated with the Bagnold velocity moments (U^3 and $U^3|U|$) though there is considerable scatter in the data particularly associated with low values of the velocity moments. This implies that significant amounts of sediment are being advected during accelerating and decelerating flows. Time series (Fig. 2 and 3) reveal that major onshore transport events coincide with accelerating flows as peak flow speed is approached and with maximum SSC whereas offshore transport maxima coincide with maximum deceleration and high SSC near the end of the wave cycle. Net cross-shore fluxes therefore represent a balance between large opposing transports separated by relatively small differences in phase angle.

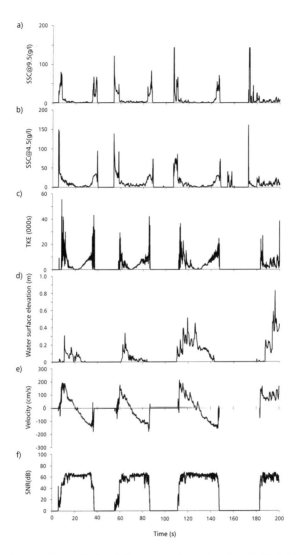

Figure 2. Time series of SSC at 9.5 cm and 4.5 cm elevation (a, b), total velocity variance (TKE, c), water surface elevation (d) cross-shore velocity (e) and ADV signal-noise ratio (f) from the mid swash zone on a rising tide (28-07-1995 8:24).

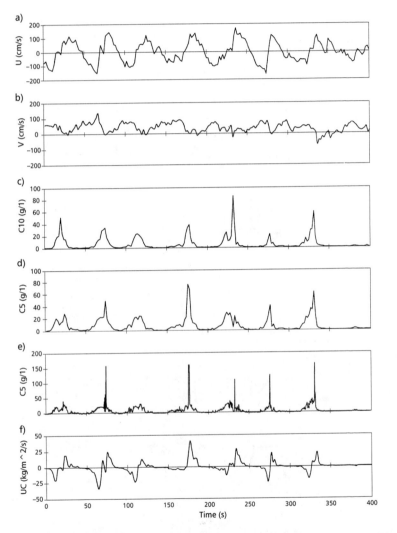

Figure 3. Time series of cross-shore velocity (a), shore-parallel velocity (b), SSC at 10 cm and 5 cm elevation (c, d, e) and instantaneous suspended sediment flux (f) from the swash zone/inner surf zone transition (28-07-1995 9:30). Note: time series in a, b, c, d, f have been block-averaged to 2 Hz while the time series in e is a 10 Hz time series.

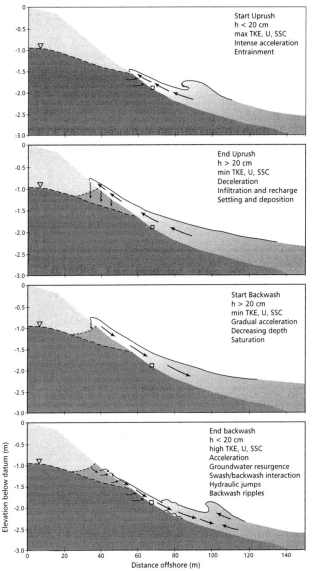

Figure 4. Schematic illustrating suspension and hydraulic conditions during an infragravity swash event.

Beach profiles measured at low tide and after the process measurements at high tide did not differ significantly indicating that the foreshore was in stable equilibrium over the half tidal cycle. On the other hand depth of activity measurements indicated that short term fluctuations in local bed level of up to 5-10 cm occurred over much shorter time scales (< 1-3 swash cycles) in response to the large instantaneous fluxes associated with the cross-shore infragravity motions and high concentration events. The measurements analysed here suggest that foreshore equilibrium may be brought about by relatively small differences in the velocity - SSC phase relationship.

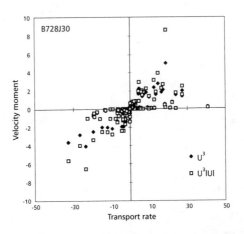

Figure 5. Transport rates computed from instantaneous products of U and SSC with velocity moments (U^3 and $U^3|U|$).

Conclusions

Re-suspension events in a high energy swash zone exhibit a temporal structure which is characterised by two major peaks in SSC coinciding with shallow water, super-critical flows, and highly turbulent conditions under the accelerating flows very near the start of the uprush and the decelerating flows near the end of the backwash. Additional peaks in SSC may also occur in association with the development of anti-dune bedforms under the backwash (Fig. 4).

These major suspension and transport events coincide closely with near-bed total velocity variance from a high resolution 3-D velocimeter implying that velocity-based transport models should perform reasonably well in the high energy infragravity swash environment.

However, it is clear that net transports represent a small difference between two large quantities and therefore prediction is likely to be complicated by relatively small differences in the phase angle between velocity and SSC and also by the presence or absence of bedforms which may develop during the backwash cycle leading to enhanced SSC levels.

Acknowledgements

The authors gratefully acknowledge funding from the University of Auckland Research Committee in support of this research. Thanks also to Dr Kevin Parnell, Shane Thomson, Janine Pritchard, Libby Boak, and Alex Lee for their assistance while in the field. Jonette Surridge provided expert cartographic and document production skills.

References

Aagaard, T. and Greenwood, B. (1994) Suspended sediment transport and the role of infragravity waves in a barred surf zone. Marine Geology, 118: 23-48.

Bagnold, R.A., 1966, An approach to the sediment transport problem from general physics. U.S.G.S. Professional Paper, 422-I.

Bailard, J.A., 1981, An energetics total load sediment transport model for a plane sloping beach. J. Geophys. Res., 86: 10938-10954.

Beach, R.A. and Sternberg, R.W. (1992) Infragravity driven suspended sediment transport in the swash, inner and outer surf zone. Proc. Coastal Sediments '91, A.S.C.E., pp. 114-128.

Dick, J.E., Erdman, M.R. and Hanes, D.M. (1994) Suspended sand concentrations due to shoaled waves over a flat bed. Marine Geology 119: 67-73.

Hanes, D. (1991) The structure of events of intermittent suspension of sand due to shoaling waves. In: B. LeMehaute and D.M. Hanes (eds), The Sea, Ocean Engineering Science. Wiley, New York, pp. 941-951.

Komar, P.D. and Holman, R.A. (1986) Coastal processes and the development of shoreline erosion. Annu. Rev. Earth Planet. Sci., 14, 237.

Komar, P.D. (1985) Physical processes of waves and currents and the formation of marine placers. CRC Critical Reviews in Aquatic Sciences, Vol. 1 (3): 393-423.

SUSPENDED SEDIMENT TRANSPORT AND MORPHOLOGICAL EVOLUTION ON AN INTERTIDAL BEACH

Troels Aagaard[1], Jørgen Nielsen[1], Niels Nielsen[1] and
Brian Greenwood[2]

Abstract

Suspended sediment transport was measured on the intertidal beach over a 72-hour long storm event. The transport direction was found to depend critically upon the pre-existing topography. During the first part of the event, a swash bar and runnel complex existed landward of the instrument position. Mean currents were directed onshore at times when the bar was submerged, leading to a net onshore transport of suspended sediment with a concomitant landwards and upwards migration of the swash bar. When the runnel infilled, the mean flow reversed as longshore drainage of onshore mass transport was inhibited, and the beach eroded. Morphodynamic feed-back mechanisms would appear to be instrumental in the morphological evolution of the intertidal beach.

Introduction

Sediment transport and morphological behaviour in the intertidal zone have received considerable attenttion in the literature. Classic models of beach behaviour state that beaches erode under storm conditions (e.g. Hayes and Boothroyd, 1969; Komar, 1976), when the water level is high, e.g. during storm surges when the inner surf zone extends to relatively high elevations. During such events, a pre-existing berm or swash bar is predicted to erode when it is overtopped by waves and the sediment is transported offshore to form a nearshore bar. Subsequent low energy conditions lead to onshore bar migration and restoration of the berm/swash bar (Davis et al., 1972; Hayes, 1972). Similarly, recent field observations from the beach/inner surf zone as well as numerical models suggest that suspended sediment transport is directed offshore during storms (e.g. Svendsen, 1984; Beach and Sternberg, 1991; Smith et al., 1992; Russell, 1993).

---

[1] Institute of Geography, University of Copenhagen, Øster Voldgade 10, DK-1350 Copenhagen K. e-mail: taa@geogr.ku.dk.
[2] Scarborough College Coastal Research Group, The University of Toronto at Scarborough, 1265 Military Trail, Scarborough, Ontario M1C 1A4, Canada.

In this paper, we document the behaviour and evolution of a swash bar, located within the intertidal zone which was subjected to a moderate but prolonged storm surge. At this time, the beach experienced a cycle of accretion and erosion. Measurements of suspended sediment transport are correlated with morphological evolution, and the tidal modulation of the hydrodynamics and sediment transport pattern is discussed.

Experimental design

The field experiment was part of the Morphodynamics Of Storm-surge Shorelines (MOSS) Programme (Aagaard et al., 1995) and was conducted during October 1995 at Skallingen, a barrier spit in the northern part of the Danish Wadden Sea. The barrier is exposed to the moderately-high energy, storm-wave dominated North Sea and is subjected to semi-diurnal tides with a spring tidal range of 1.8 m. Storm surges due to low barometric pressure and wind set-up may cause water levels to exceed +4.0 m DNN (Danish Ordnance Datum).

The nearshore profile is gently sloping, on the order of 0.008 with a somewhat steeper intertidal slope (β = 0.02-0.03; Figure 1). The tidal cycle therefore results in significant horizontal displacements of the shoreline, on the order of 50-90 m. The intertidal zone, which has a mean grain size of 200-300 microns, often contains a swash bar backed by a runnel and dissected by small, regularly spaced ($\lambda \approx$ 200-300 m) rip channels. At times of high water level, wave energy is dissipated by breaking across this bar. Visual observations and topographic surveys conducted over a 3-year period indicate that the upper beach and dunes do not erode until this swash bar has been eroded. The bar thus serves as a buffer against significant erosion and shoreline recession.

Six sediment transport stations (S1-S6) were deployed across the intertidal and subtidal zones (Figure 1). Each were equipped with a Marsh McBirney electromagnetic current meter and three optical backscatter sensors (OBS-1P). A wave recorder was deployed seaward of the bars, in a water depth of approximately -5.0 m DNN.

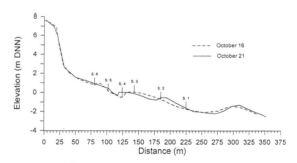

Figure 1. Cross-shore profile of the instrument transect.

In this paper, results from station S5 in the intertidal zone, which was also equipped with a pressure transducer, will be discussed.

The current meter at S5 was initially set at an elevation of 0.20 m above the bed and adjusted at each low tide to compensate for changes in bed elevation. It was oriented using a compass and level to record positive onshore and northward flows. The optical backscatter sensors were deployed at z=0.05, 0.10 and 0.20 m, and similarly adjusted at each low tide. The sensors were cable-linked to a shore-based data acquisition system and sampled at 4 Hz for 34 minutes, yielding 8192 data points per record. Individual data runs were spaced 1-1½ hours apart. Sensor outputs were screened to check data quality and records which contained noise or errors (usually due to sensor exposure or burial) were discarded.

Net suspended sediment flux at a point was determined from

$$<q_s>_n = 1/N \sum uc$$

where u is instantaneous velocity (m s^{-1}) and c is instantaneous suspended sediment concentration (kg m^{-3}). N is the number of data points in each record.

Morphological changes were documented using 28 half-inch diameter survey rods embedded in the beach surface and spaced at 5 m increments across the intertidal (and subtidal) zones. The rods were displaced approximately 10 m north of the instrument transect. The top of the rods were surveyed relative to a known datum and bed elevation changes were recorded at each low tide throughout the experiment.

Waves, currents and tides

The focus of this paper is a prolonged, moderate-energy storm event which occurred on Oct. 17-20, 1995 and comprising 6 tidal cycles (Figure 2), denoted cycles 5 through 10. Significant offshore wave heights peaked at 1.86 m with accompanying peak spectral periods of 8.0 seconds (Figure 2). The event occurred during neap tides although a ≈ 0.6 m storm surge persisted throughout the storm. Figure 3 illustrates the water level at S5, as well as the mean cross-shore and longshore current velocities. Maximum water level on the beach reached an elevation of +1.35 m DNN during tidal cycle 6. Cross-shore mean currents were initially negligible, but significant onshore directed currents developed during tidal cycles 6 and 7 ($\dot{u}$ = 0.05-0.12 m/s) at times when the crest of the swash bar

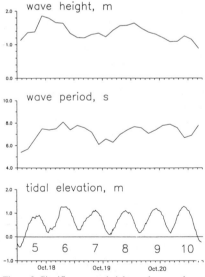

Figure 2. Significant wave height, peak spectral period and tidal elevation, measured seaward of the bar zone. Tidal cycles are numbered in the lower panel.

was being overtopped. An onshore directed mass transport of water drained alongshore in the runnel and eventually offshore through a small rip channel dissecting the swash bar, approximately 150 m south of the transect. This flow resembled a horizontally segregated cell circulation. During tidal cycle 9, the cross-shore mean current reversed and was directed offshore throughout the remainder of the event, peaking at -0.30 m/s. The longshore current was predominantly directed south and peaked at -0.79 m/s when the wind direction became parallel to the shoreline. Cross-shore and longshore currents were tidally modulated, with maxima at high tide.

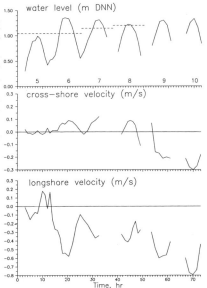

Figure 3. Water level and cross-shore and longshore current velocities measured at S5. Currents are positive onshore and to the north. The dashed line in the upper panel shows the elevation of the swash bar crest. Data gaps reflect instrument exposure. Tidal cycles are numbered in the upper panel.

Morphological evolution

Figure 4 illustrates the morphological changes over the entire experimental period. The axes of this time-distance diagram represent respectively the cross-shore distance from the baseline in the dunes, time given in number of low tides (tidal cycles) from the beginning of the experiment and elevation relative to Danish Ordnance Datum (DNN). Each line extending cross-shore in the diagram represents a low-tide survey at the termination of a tidal cycle; a total of 28 surveys were conducted over the 2-week period.

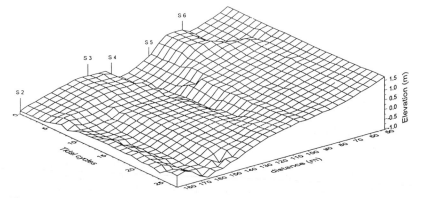

Figure 4. Morphological development of the beach and inner surf zone over the experimental period.

There was continuous erosion of the seaward slope of the swash bar during tidal cycles 5-8, with sediment being deposited at the step (x=105-110m) and in the runnel, which gradually infilled (see also Figure 5). The swash bar migrated landward and upward and the crest elevation increased from +1.04 m DNN to +1.23 m DNN. At S6, the bed level increased by 0.35 m. Early in tidal cycle 9, visual observations indicated that the runnel behind the swash bar was finally infilled by onshore sediment transport in low swash bores; the upper/mid parts of the beach then eroded continuously throughout cycles 9-10 until the swash bar had disappeared and the beach became planar. At S6, the bed was eroded by 0.20 m. The eroded sediment was deposited in the lower intertidal zone and seaward of S5, gradually building a berm (Figure 5) which served as the nucleus for a new swash bar which appeared towards the end of the experiment. However, landward of S5, a planar sloping beach persisted through the remainder of the experiment.

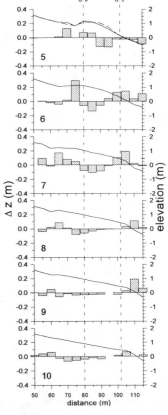

Figure 5. Bed elevation change (Δz) and the resulting cross-shore profile at the ends of tidal cycles 5-10. The dashed line in the upper panel is the profile prior to cycle 5.

Suspended sediment transport

The suspended sediment transport measured by the sensors during this event is illustrated in Figure 6. Net transport has been determined by integrating measurements at all three OBS-elevations over the interval z=0.025-0.275 m (nominally). Data gaps (Figure 6) indicate periods when the sensors at S5 were exposed, except during the latter half of tidal cycle 7, when the sensors were buried by sand.

Initially (tidal cycle 5), suspended sediment transport was negligible probably because water depth over the inner nearshore bar was small and dissipation of wave energy over this bar was more intense compared to later tidal cycles. Cycles 6-8 displayed a characteristic cyclic pattern with maximum onshore sediment transport at high tide and a small onshore transport, trending progressively more offshore, at low tide; a topographic control on the sediment transport pattern clearly existed. The presence of a swash bar/runnel system induced a net landward transport of sediment integrated over individual tidal cycles. Further evidence for this topographic control is provided by the fact that transport rates were small and/or offshore directed at low tide when the swash bar was not submerged. Furthermore, the offshore transport at low tide became progressively more important, especially during cycle 8 when the swash bar had migrated further up the beach. Thus, the crest was only submerged for very short periods of time around high water (Figure 6). An almost perfect correspondance between the transport direction and the water depth over the bar is evident; whenever the water level exceeded the bar crest level, sediment transport was directed onshore.

Early in cycle 9, the runnel behind the swash bar infilled, the beach became planar and the transport pattern shifted radically. Maximum net transport rates were again attained at high tide but they were directed offshore. At low tide, smaller offshore transport rates were observed.

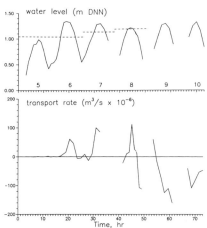

Landward of the survey rods, changes in elevation were small (Figure 2) and assuming that the inner rod represents a closed landward boundary to the active profile, sediment transport volumes and directions at S5 (ΔQ) can be compared to volumetric changes landward of S5 (ΔV), determined from the survey rods (Table 1). There was a perfect correspondence between transport directions at S5 and volumetric change in the profile. Net onshore sediment transport over individual tidal cycles was correlated with

Figure 6. Water level (top) and net suspended sediment transport rate recorded at S5 (lowest optical sensor). Positive transport is onshore. The number of the tidal cycles are given in the upper panel. The dashed line is the level of the swash bar crest.

volumetric gain; net offshore transport corresponded to volumetric loss.

Quantitatively, it is very difficult to relate ΔV and ΔQ as several sources of error exist in the determination of the sediment transport. Errors include: Longshore transport gradients; the assumption of a vertically homogeneous velocity field; shifts in sensor calibration, and textural segregation within the water column and/or between the bed sediment and the suspended sediment (Greenwood and Jagger, 1995; Sheng and Hay, 1995). Transport in the high-concentration layer very close to the bed, as well as transport above the level of the OBS-sensors was not measured. Finally, changes in bed level over individual tidal cycles will affect measured sediment transport. The most significant error source in this case was probably that the bed at S5 accreted throughout most tidal cycles. This will tend to cause artificially increased suspension due to increased turbulence around the sensors, an effect which was observed particularly during the latter half of tidal cycles 7, 9 and 10 when one (or more) of the backscatter sensors became buried. Therefore, transport rates during the ebbing tide of these cycles are expected to be overestimated. Accordingly, it appears that measured sediment transport volumes are comparable to or exceed (cycle 9-10) volumetric changes.

Tidal cycle no.	ΔV, m³	ΔQ, m³
5	- 0.105	- 0.014
6	+ 0.830	+ 0.638
7	+ 1.025	(+ 1.034)
8	- 0.555	- 0.476
9	- 1.055	- 2.267
10	- 0.530	- 1.782

Table 1. Volumetric change (m³/m shoreline; ΔV) landward of S5 and net suspended sediment transport rate and direction at S5 (m³/m; ΔQ) over tidal cycles 5-10. Positive values indicate accretion and onshore sediment transport, respectively.

However, it is encouraging that discrepancies between measured and observed transport volumes are generally within a factor of two and indications are that the measured transport was a realistic representation of the actual transport rates.

The suspended sediment flux was also separated into a mean and an oscillatory component:

$$<q_s>_n = \bar{u}\bar{c} + <u'c'>$$

where $\bar{u}$, $\bar{c}$ are the mean current velocity and the mean sediment concentration, respectively, while u'c' are the oscillatory components of velocity and concentration. The oscillatory term was further separated into an infragravity and an incident wave contribution using cospectral analysis (Huntley and Hanes, 1987) with a frequency separation at 0.067 Hz which was marked by consistently small variances in the wave spectra.

Figure 7 illustrates the sediment flux through time at the level of the lower OBS at S5. Transport by mean currents was clearly the most important contribution to the net flux and the mean transport characteristics closely resemble the pattern seen in Figure 6. During cycles 5-8, the mean component was directed landward at times when the swash bar was submerged and was at a maximum at, or shortly after, high tide in a manner similar to the net sediment transport (Figure 6). During the later part of the storm (cycles 9-10), the mean component was consistently directed seaward but again with a tendency towards a transport peak at or shortly after high tide. Averaged over the event, the mean flux constituted 56% of the gross flux (which is found by adding the mean, the incident wave and the infragravity wave contributions without regard to sign).

The sediment flux due to incident wind and swell waves was generally directed landward at all tidal stages and contributed approximately 26% to the gross sediment flux over the event. The infragravity wave contributions were more variable in direction with a tendency towards maximum transport rates at low tide (small water depths); at times it was equal to or exceeded the mean transport. This agrees with earlier reports by e.g.

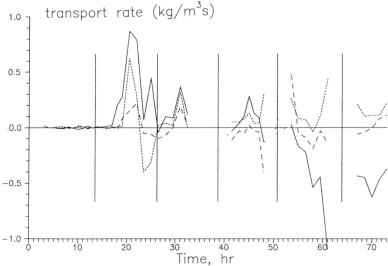

Figure 7. Suspended sediment transport rates due to mean currents (solid line), incident wind waves (dots) and infragravity waves (dashed line) measured at S5. Vertical lines separate tidal cycles and mark the time of low tide.

Beach and Sternberg (1991) and Russell (1993). However, when integrated over entire tidal cycles infragravity fluxes were secondary to the mean flux, comprising only 18% of the gross flux.

Discussion and conclusions

The response of the intertidal beach at Skallingen to a storm event appears to be somewhat more complicated than predicted by existing models as bar forms in this zone are not immediately destroyed with the onset of a storm. In the present case, a swash bar migrated landward and upward from the beginning and through the peak of the storm. As the storm waned, the bar was eroded, sediment was transported offshore and a berm-like feature was built around the storm low-tide water level. This feature gradually evolved into a convexity, *i.e.* a new swash bar slightly seaward of S5 (Figure 4) towards the termination of the field experiment. Swash bar behaviour thus does appear to be cyclic.

The key mechanism driving this cyclic morphological behaviour was sediment transport by the mean current. Under the conditions monitored (a moderate storm event with moderate but prolonged surge elevations), mean currents accounted for almost 60 percent of the gross transport in the lower intertidal zone. Mean current velocities peaked at high tide. However, contrary to existing nearshore circulation models (e.g. Svendsen, 1984; Smith at al., 1992; Svendsen and Putrevu, 1995), these currents were directed landward as long as the swash bar was in existence, probably reflecting the presence of a wave-induced mass transport. At such times, a horizontally segregated cell circulation was set up with landward directed mean flows across the swash bar, alongshore (southward) currents in the runnel and a seaward return flow in a small rip channel dissecting the bar.

The onshore directed mean flow depended critically upon the existence of a pathway to evacuate the onshore discharge along- and offshore, i.e. the runnel. When the runnel finally infilled as the storm was easing, an offshore directed flow (probably an undertow) was generated, eroding the remnants of the bar and leaving a gently sloping planar beach.

Sediment transport on the intertidal beach therefore may be a function primarily of the mean water level and the pre-existing topography and involving a feed-back between morphology and hydrodynamics. The lower the swash bar relative to the MWL, the more pronounced will be the tendency towards onshore migration of the form. As the bar migrates landward and upward, the length of time it is submerged will decrease and increasing amounts of sand can be transported offshore at relatively low tidal stages. In the present case, at least, wave energy level was of secondary importance to sediment transport.

The implications from this study are that if a storm does not last long enough, or if the MWL is not sufficiently high, the swash bar may remain more or less intact. On the other hand, it is likely that a critical threshold exists between onshore mass transport of water and the cross-sectional area of the runnel. If the water level becomes very high with ensuing large local wave heights (e.g. during a major surge), the runnel will probably not be able to accomodate the longshore flow necessary to balance the onshore mass transport and the bar will erode very rapidly, leaving the backshore and the dunes open to severe erosion.

Oscillatory waves were important secondary contributors to the overall sediment transport. Incident wind waves and swell transported sediment landward and augmented the onshore transport due to mean flows early in the event, and offset the mean transport at later stages. Infragravity wave activity was variable and these waves did not appear to be critical to swash bar/ridge and runnel behaviour as proposed by Simmonds et al. (1995). In small water depths (low tidal stages), infragravity wave transport equalled or at times exceeded the sediment transport due to mass transport/undertow. However, in the lower intertidal zone their effect averaged over tidal cycles was small due to shifting sediment transport directions and magnitudes caused by variations in phase coupling between velocity and concentration and varying spectral composition of these waves.

Acknowledgements

This study was supported by the Danish Natural Sciences Research Council, grant no. 11-0925. Field assistance was provided by Ulf Thomas, Rasmus Nielsen, Carlo Sørensen and Torsten Christensen.

References

Aagaard, T., Nielsen, J., Greenwood, B., Christiansen, C., Lund-Hansen, L. and Nielsen, N., 1995. Coastal morphodynamics of Skallingen, SW Denmark. 1.Low and moderate energy conditions. *Danish Journal of Geography,* 95, 1-11.

Beach, R.A. and Sternberg, R.W., 1991. Infragravity driven suspended sediment transport in the swash, inner and outer surf zone. *Proc. Coastal Sediments '91,* ASCE, 114-128.

Davis, R.A., Fox, W.T., Hayes, M.O. and Boothroyd, J.C., 1972. Comparison of ridge and runnel systems in tidal and non-tidal environments. *Journal of Sedimentary Petrology,* 42, 413-421.

Greenwood, B. and Jagger, K., 1995. Sensitivity of optical sensors to grain-size variation in the sand mode: implications for transport measurements. *Proc. Canadian Coastal Conference 1995,* CCSEA, 383-397.

Hayes, M.O., 1972. Forms of sediment accumulation in the beach zone. In: R.E.Meyer (ed) *Waves on Beaches and Resulting Sediment Transport,* Academic Press, New York, 297-356.

Hayes, M.O. and Boothroyd, J.C., 1969. Storms as modifying agents in the coastal environment. SEPM Field Guide, Coastal Environments NE Massachusetts and New Hampshire.

Huntley, D.A. and Hanes, D.M., 1987. Direct measurement of suspended sediment transport. *Proc. Coastal Sediments '91,* ASCE, 723-737.

Komar, P.D., 1976. *Beach Processes and Sedimentation.* Prentice-Hall, Englewood Cliffs, 429pp.

Russell, P.E., 1993. Mechanisms for beach erosion during storms. *Continental Shelf Research,* 13, 1243-1265.

Sheng, J. and Hay, A.E., 1995. Sediment eddy diffusivities in the nearshore zone, from multifrequency acoustic backscatter. *Continental Shelf Research,* 15, 129-147.

Simmonds, D., Voulgaris, G. and Huntley, D.A., 1995. Dynamic processes on a ridge and runnel beach. *Proc. Coastal Dynamics '95,* ASCE, 868-878.

Smith, J.M., Svendsen, I.A. and Putrevu, U., 1992. Vertical structure of the nearshore current at DELILAH: measured and modelled. *Proc. 23rd Coastal Engineering Conference,* ASCE, 2825-2838.

Svendsen, I.A., 1984. Mass flux and undertow in a surf zone. *Coastal Engineering,* 8, 347-365.

Svendsen, I.A. and Putrevu, U., 1995. Surf-zone Hydrodynamics. In: P.L.-F. Liu (ed) *Advances in Coastal and Ocean Engineering Volume 2,* World Scientific, Singapore, 1-78.

The erosion response of estuarine mudflats
to tidal currents and waves

Christie, M.C[1]., Dyer, K.R., & Huntley, D.A.

Abstract

Successful environmental management of intertidal regions depends upon in-situ measurements, which help specify the interactions between the active processes and so allow the development of predictive models. Three experiments, located in the Severn, Humber and Dollard estuaries, examined the response of mudflats to local waves and tidal currents. Three dominant physical processes are defined as erosion by wave action, erosion due to tidal currents, and the resuspension of a thin surface biofilm. The effects of wave and current action are quantified by simple expressions describing the relationships between velocity, bed shear stress and concentration. The results show that the highest sediment fluxes occur in the shallow water periods at the beginning and end of tidal coverage.

Introduction

Shallow water mudflats can experience rapid morphological changes, responding to changes in local waves and tidal currents. The interactions between the mudflat and shallow waters (i.e. depth, h <1.0 m) at the beginning and end of tidal immersion are particularly important but not well understood. High concentrations of suspended sediment are often found in the waters edge and the tidal movement of the shallow water over the intertidal area can potentially erode or deposit large quantities of sediment. Surface waves are extremely important at these times and can cause significant erosion in only a few centimetres of water. However, there are few field measurements of the physical processes that occur in shallow water, due to both the difficulty of working in these locations and a lack of instrumentation.

Equipment description

The Profiles Of Sediment Transport (POST) system (described in detail by Christie *et al.*, 1997) was developed to measure velocity and suspended sediment concentration (SSC) profiles in very shallow water over an intertidal region. The system is based around four miniaturised electro-magnetic current meters (EMCM's) and eight miniaturised optical turbidity sensors. The EMCM's and turbidity sensors sample velocity and concentration respectively from approximately 1 cm^3 of water, and can be deployed to within about half a centimetre of the sediment surface. The sensor array provides vertical profiles of velocity and concentration in a few centimetres depth of water. Data is continuously recorded at high frequency (18 Hz.)

---

[1]Post-doctoral research fellow, Institute of Marine Studies, University of Plymouth, Plymouth, Devon, PL4 8AA, U.K. tel. 44 1752 232430 fax. 44 1752 232406.

and is low pass filtered (-3dB at 5 Hz.) prior to storage on a computer. Additional data is recorded simultaneously from a CTD and pressure transducer.

Methodology

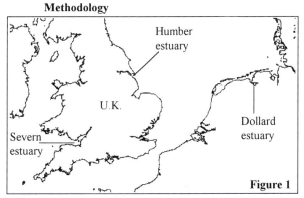

Figure 1

Three deployments of POST, two in the U.K. and one in the Netherlands (see figure 1) investigated the hydrodynamic and physical processes affecting the bed in very shallow water, with the aim of defining the dynamics of the high concentration zone near the waters edge. The first experiment examined wave dominated processes over the Portishead mudflats in the Severn estuary. These mudflats are considered to be undergoing net erosion from historical records. A second experiment investigated tidally dominated processes across the Skeffling mudflats of the Humber estuary, a site experiencing net annual accretion. The Dollard estuary was the location for the third experiment. This site was tidally dominated and there was significant biological activity which appeared to stabilise the upper sediments.

Portishead Severn Estuary

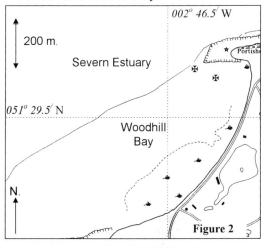

Figure 2

POST was deployed in June 1994, at different locations across the mudflat depicted by ✠ in figure 2. This region is a macrotidal environment with a spring tidal range in excess of 12 m, which produces tidal currents over $0.5~ms^{-1}$ in the deep water offshore. During the study, prevailing south westerly winds often produced a wave dominated environment and significant wave heights H_s reached 0.6 m.

LISP UK '95 experiment

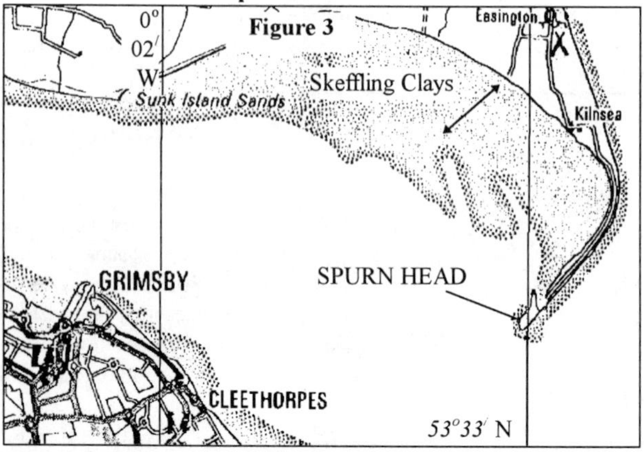

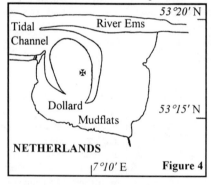

Figure 3

The LISP UK study of April 1995 examined the Skeffling mudflats of the Humber estuary. The mudflat in figure 3, was sheltered, and tidally dominated. A tidal range of about 6 m produced currents over 0.5 ms^{-1} as the tide flowed across the 5 km expanse of mudflat. The equipment was deployed along a shore normal transect depicted by ←——→ in figure 3.

Dollard INTRMUD experiment

Figure 4

POST was deployed in May 1996 at a mid-tide level, in the middle of the Dollard Heringsplaat mudflat (see ✠ in figure 4). Sea conditions were relatively calm during the deployment with H_s of centimetre order, The tidal range was 3.0 m, and peak velocities of about 0.15 ms^{-1} were measured over the mudflat during both the flood and the ebb phases. There were significant amounts of organic matter in the sediment which appeared to strengthen the mudflat surface.

Results

Typical burst mean data (from 3 min 48 s of time series) are described, with emphasis placed on the first hour of immersion by the advancing flood tide. The data are used to highlight the main physical differences between the three study locations. All data shown in the following figures were obtained from sensors while completely covered by water. The measurement heights are noted in centimetres above the bed by the graph legends. Depths are plotted alongside the concentration data. The mean $(\overline{uc})$ and oscillatory $(\overline{u'c'})$ fluxes of suspended sediment are also plotted. The $(\overline{u'c'})$ values were greatest at the beginning and end of tidal immersion, particularly when wave activity was significant, forming as much as 10% of the total sediment flux value. Flow conditions were classified by an approach of Soulsby and Humphrey (1990) who used a wave:current dominance factor to identify when wave action enhanced the current related drag coefficient. A factor W_x/U_x was defined, such that for a sample height $_x$, W_x was a wave activity parameter defined as $\sqrt{2}$ times the standard deviation of the velocity time series, and U_x was the total horizontal current speed. For a measurement height of 15 cm above the sediment, flow conditions were considered to be current dominated when $W_{15}/U_{15} < 0.18$ and wave dominated when $W_{15}/U_{15} > 0.54$.

Shear stresses due to mean currents (τ_c) were calculated using a logarithmic velocity profile approach. However, the measured velocity profiles did not always fit the law of the wall. In these cases the calculations produced erratic and unreasonable estimates of bottom shear stress. Glenn and Grant (1987) note high concentrations of suspended material may effect the shape of the velocity profile by altering the nature of the boundary generated turbulence and the eddy viscosity. For this work there were also problems obtaining accurate zero flow offsets for the EMCM's, which could have produced errors of ± 0.05 ms^{-1} in the mean flow value. This size of error in a velocity profile could cause significant deviation from the logarithmic profile. Because it was not possible to obtain reliable values for z_0 from the in situ velocity data, and values for D_{50} were not recorded; in order to calculate τ_c, an assumed logarithmic profile was fitted the mean velocity value from one of the near bed sensors (taking $z_0 = 0.02$ cm, from Soulsby 1997). Bed shear stresses due to the combined effects of waves and currents (τ_{wc}) were calculated using a parameterisation (see Soulsby, 1997) of Huynh-Tanh and Temperville's (1991) model of the rough turbulent boundary layer. The calculated shear stresses τ_c and τ_{wc} were considered to provide a useful indication of the comparative forces acting on the bed during both current and wave dominated conditions.

Regression analysis was used to examine relationships between the measured and calculated variables. The "least squares" method was used to calculate an expression (95% confidence) that best fitted the data, where concentration was a function of velocity, τ_c or τ_{wc}.

Mean velocity, concentration and flux data

The Portishead data in figures 5 to 8 show spring tide data obtained from a mid-tide level on the mudflat. The data were obtained following a period of calm weather, and it appeared that a 1-2 cm thick layer of sediment deposited during the preceding tidal cycle was resuspended by increasing wave activity. Data recording started at 06:40 hrs after the flood tide had covered the sensors, and continued through high water, finally ending with uncovering by the ebbing tide at 12:15 hrs. Mean velocities in figure 5, ranged from 0 - 0.22 ms^{-1}, with the current direction reversing at high water. Sediment concentration variations are shown in figure 6. Maximum concentrations in excess of 3.0 gL^{-1} were typically recorded in the shallow water

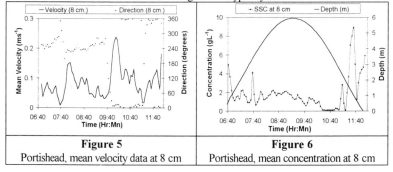

Figure 5	**Figure 6**
Portishead, mean velocity data at 8 cm	Portishead, mean concentration at 8 cm

periods at the beginning and end of mudflat immersion. These high concentrations were attributed to the erosive effects of the shallow water waves. Concentrations exceeded 8 gL⁻¹ around 11:00 hrs due to surface water draining off the mudflat down a shore normal channel. Figures 7 and 8 show the ($\overline{uc}$) and ($\overline{u'c'}$) values respectively. Oscillatory fluxes were about 5% of the mean fluxes, illustrating the increased effects of wave activity in the first and last half hour of tidal immersion. However ($\overline{u'c'}$) values were at a minimum around high water.

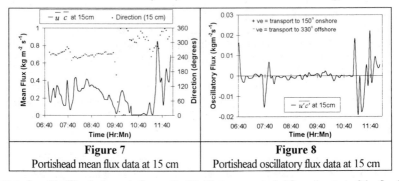

| **Figure 7** | **Figure 8** |
| Portishead mean flux data at 15 cm | Portishead oscillatory flux data at 15 cm |

Data from Skeffling is shown in figures 9 - 12. Data was recorded from the start of the flood tide until water depth was about 2.5 m. Velocities exceeded 0.4 ms⁻¹ at the beginning of covering and decreased to about 0.1 ms⁻¹ near high water (figure 9). The very large mean speeds resulted from the 6 m tide flowing over the very shallow sloping mudflat (gradient about 1:1000). H_s values ranged from 5 to 15 cm. The flow direction was directly onshore throughout the flood period. Concentrations were greatest (≈ 0.8 gL⁻¹) during the first half hour of immersion and values decreased to a background level of about 0.4 gL⁻¹ (in figure 10). Mean fluxes closely followed the trends for mean velocity and the oscillatory flux values were less than 1% of the mean values (figures 11 and 12). Wave effects were relatively unimportant in terms of the sediment transport during the first few minutes of immersion, and W_{15}/U_{15} values indicated the flow conditions were current dominated. It appeared on site that the strong tidal flows were able to erode surface sediment during the first hour of immersion, and so form a high concentration band at the advancing waters edge.

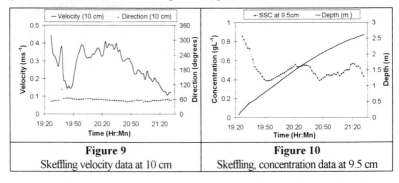

| **Figure 9** | **Figure 10** |
| Skeffling velocity data at 10 cm | Skeffling, concentration data at 9.5 cm |

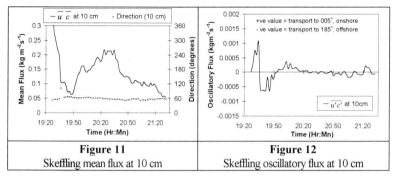

Figure 11
Skeffling mean flux at 10 cm

Figure 12
Skeffling oscillatory flux at 10 cm

Figures 13 to 16 show data from the Dollard estuary, recorded from initial covering until the depth reached about 1.0 m at high water. Surface waves during this study were very small with H_s of order a few centimetres. The onshore flood flow velocities (in figure 13) were less than at Portishead and Skeffling because of the reduced tidal range, and peak speeds of about 0.13 ms^{-1} were reached about 45 minutes after immersion. Concentrations exceeded 1.0 gL^{-1} at the start of water coverage, and decreased rapidly as the depth increased. The suspended particles

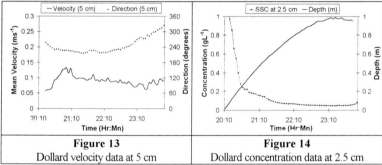

Figure 13
Dollard velocity data at 5 cm

Figure 14
Dollard concentration data at 2.5 cm

consisted largely of organic material. Suspended sediment transport was dominated by the mean fluxes and the oscillatory fluxes were negligible, as shown in figures 15 and 16. Compared to Portishead and Skeffling, the slower flow speeds and smaller waves seemed

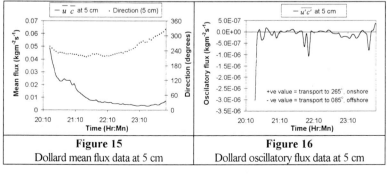

Figure 15
Dollard mean flux data at 5 cm

Figure 16
Dollard oscillatory flux data at 5 cm

unable to cause significant erosion during immersion. In this case, the high concentrations in the leading waters edge were thought to be due to resuspension of a thin (order μm) biofilm that formed on the mudflat surface during exposure.

To summarise; at Portishead shallow water wave activity appeared to erode a surface sediment layer deposited by the previous tide, and as a result oscillatory fluxes were significant at the beginning and end of water coverage. At Skeffling, very strong onshore flows seemed to cause sediment resuspension during the first hour of immersion by the flood tide. Erosion processes appeared negligible at the Dollard site as biological activity served to strengthen the bed. However, high concentrations of organic material were observed in the leading waters edge. The following sections attempt to quantify these observations.

Bed shear stresses

The large velocity gradients associated with the wave boundary layer produced much larger shear stresses than for currents alone. Thus, when waves were significant, τ_{wc} values were greater than the τ_c values. Figures 17 to 19 show the shear stress results from the three different sites. Wave action noticeably enhances the bed shear stress at Portishead when $h <$ 1.0 m. For both the Skeffling and Dollard data the hydrodynamics were dominated by the mean currents such that $\tau_c \approx \tau_{wc}$ at all depths. Critical shear stresses for erosion (τ_{crit}) of the

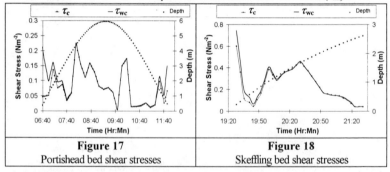

Figure 17
Portishead bed shear stresses

Figure 18
Skeffling bed shear stresses

surface sediment were also measured during each of the three experiments. Values of $\tau_{crit} \approx$ 0.06 to 0.17 Nm^{-2} at Portishead, while at Skeffling τ_{crit} values ranged from 0.20 to 0.23 Nm^{-2} (Mitchener, *pers comm.*). Measurements indicated τ_{crit} values were much higher at the Dollard

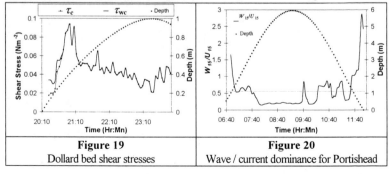

Figure 19
Dollard bed shear stresses

Figure 20
Wave / current dominance for Portishead

site, ranging from 0.55 Nm^{-2} to 0.60 Nm^{-2} (see Mitchener and Feates, 1996). Using figures 17 to 19, and making the assumption that erosion could occur when either τ_c or $\tau_{wc} > \tau_{crit}$, it can be seen that erosion was possible during both the majority of the Portishead record, and the early part of the Skeffling record. But, following this approach, no erosion could have occurred at the Dollard despite the implications of the high concentrations in the first half hour.

Regression analysis

An attempt was made to correlate the nearbed Portishead SSC data with bed shear stresses, when either τ_c or τ_{wc} exceeded τ_{crit} (i.e. during periods of erosion). Only nearbed concentrations were used for this analysis as any relationship with the bed shear stress would become less significant with increasing height off the bed. In practice, such a distinct relationship was not found, largely because advective processes could produce an increase in SSC when $\tau_{crit} > \tau_{wc}$. Advective processes seemed most significant during periods of low flow speed, especially around the time of slack water. Periods of wave dominance for the Portishead record are shown by figure 20, such that when $W_{15}/U_{15} > 0.54$ the hydrodynamic conditions were considered to be wave dominated. Relationships were found between τ_{wc} and SSC data at Portishead when $h < 3.0$ m and the flow conditions were wave dominated. Figure 21 shows a polynomial relationship obtained between concentrations at 10 cm above the bed (C_{10}) and τ_{wc}. This was an empirical expression and does not follow any particular transport law.

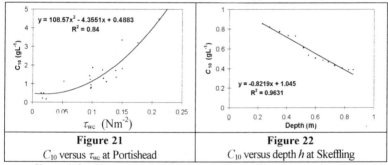

Figure 21	Figure 22
C_{10} versus τ_{wc} at Portishead	C_{10} versus depth h at Skeffling

Clear relationships were obtained for all the Skeffling records between nearbed concentration and h for the Skeffling data during the flood phase, when $h < 1.0$ m. A typical relationship is illustrated in figure 22, noting that there are only 14 data points because of the

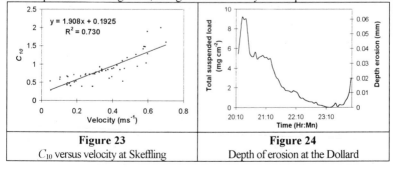

Figure 23	Figure 24
C_{10} versus velocity at Skeffling	Depth of erosion at the Dollard

rapid rise in water depth. A physically more meaningful linear relationship was that between nearbed concentration and mean velocity, as shown in figure 23. The correlation coefficient was increased slightly to $R^2 = 0.800$ when a 2nd order polynomial expression was fitted to the data. Good relationships were necessarily obtained between C_{10} and either τ_c or τ_{wc} due to the dependence of τ_c and τ_{wc} upon the mean velocity value (remembering for the Skeffling data $\tau_c \approx \tau_{wc}$ as flow conditions were largely current dominated).

The Dollard flow regime was also predominantly current dominated, however in this case there was not a clear relationship between nearbed concentration and either velocity or the bed shear stress. It was considered unlikely that substantial erosion could have occurred as the bed shear stresses were less than the measured value of τ_{crit}. However as at the Skeffling site, there were high concentrations measured at the start of the record, indicating that some form of erosion occurred as the leading waters edge progressed across the mudflat. There is some evidence that the turbid waters edge at the Dollard was largely due to erosion of a thin surface biofilm, formed by the action of micro-algae during emmersion. Analysis of suspended material indicated a high biological content (Mitchener and Feates, 1996). Measurements showed the concentration profile to be homogeneous, allowing a value for the total suspended load to be obtained by depth integrating the concentration data. Dividing this total suspended load by an assumed sediment bulk density of 1500 kgm^{-3} gives an estimate of the actual depth of erosion required to produce the local SSC profile. Figure 24 shows the variation of the suspended load and the depth of erosion with time. Concentrations (shown in figure 14) were greatest for the first 20 minutes of the record before decreasing rapidly to reach "background" levels of about 0.1 gL^{-1}. This behaviour implies erosion was limited to the first 20 minutes of the Dollard record. Integrating the depth of erosion over this period gives a total depth of erosion at the start of the record to be about 250 μm, which is slightly larger than the thickness of a typical surface biofilm which is of order 50 - 100 μm (Paterson, *pers comm*). However, the value obtained for the depth of erosion is likely to be an upper estimate. In-situ bulk density values of between 1480 - 1820 kgm^{-3} were measured at the Dollard site by Mitchener and Feates, (1996) and the 1500 kgm^{-3} value was considered to be typical of the very surface layer. But, this approach assumes all the suspended load is from erosion of the local bed and does not allow for material advected shorewards with the advancing tide, which may account for the over estimation of the erosion depth. It seems likely that the high concentrations measured at the start of covering at the Dollard were predominantly the result of resuspension of surface biological material. This biofilm layer was unlikely to picked up by the in-situ shear strength device of Mitchener and Feates (1996), which means biofilm erosion could occur despite the calculated bed shear stress being less than the measured τ_{crit} value.

Concluding remarks

We have identified three different modes of erosion during the early part of immersion at the three study sites, specifically wave enhanced resuspension, erosion by mean tidal currents and removal of a thin surface biofilm.

Some simple expressions were obtained relating nearbed concentrations to flow velocity and bed shear stress. Multiple regression analysis did not yield such clear results. At Portishead, when $h < 3.0$ m and flow conditions were wave dominated, concentrations within 10 cm (C_{10}) of the bed were related to the bed shear stress due to waves and currents τ_{wc} by:

$$C_{10} = 108.5 \; \tau_{wc}^2 - 4.355 \; \tau_{wc} + 0.488 \qquad R^2 = 0.840 \qquad \textbf{(1)}$$

At Skeffling, for current dominated flow conditions while $h < 1.0$ m, concentrations within 10 cm (C_{10}) of the bed were related to velocity at 10 cm (U_{10}) by either:

$$C_{10} = 1.908 \; U_{10} + 0.1925 \qquad R^2 = 0.730 \qquad \textbf{(2)}$$
$$C_{10} = 3.40 \; U_{10}^2 - 0.554 \; U_{10} + 0.539 \qquad R^2 = 0.800 \qquad \textbf{(3)}$$

At the Dollard site, the calculated bed shear stresses were less than the measured values for the critical shear strength of the bed (τ_{crit}) and erosion seemed unlikely to occur. But analysis of concentration results suggest that a thin biofilm was eroded off the mudflat surface during the first 20 minutes of water coverage.

The results support observations from other studies, that the edge of the advancing tide is a particularly important factor in the fluxes of sediment across the mudflat. This work illustrates the importance of the shallow water period ($h < 1.0$ m), at the start of immersion by the flood tide, in the resuspension of surface material deposited during the previous tidal cycle.

Because of the sensitivity of the results to local hydrodynamics (in particular the wave climate) future studies need to take into account the important effects of the very shallow flood tide period when considering the overall intertidal zone sediment budget.

Acknowledgements
This work was partly funded by the LISP UK experiment (a Special Topic 122 of the LOIS RACS (C) programme), as part of MAST II G8M Project 4, and the MAST III contract MAS3-CT95-0022 INTRMUD.

References

Christie, M.C., Quartley, C. P., & Dyer, K. R., (1997). *The development of the POST system for in-situ intertidal measurements.* In proceedings of the 7th International Conference on Electronic Engineering in Oceanography, I.E.E. publication nos. 439.

Glenn, S. M. & Grant, W. D., (1987). *A suspended sediment stratification correction for combined wave and current flows.* J. Geophys. Res., 92 (C8): p 8244-8264.

Huynh Tanh, S. & Temperville, A., (1991). *A numerical model of the rough turbulent boundary layer in combined wave and current interaction,* Euromech 262-Sand Transport in Rivers, Estuaries and the Sea, Soulsby & Bettes (eds). ISBN 90 6191 186 9

Mitchener, H. J. & Feates, N. G., (1996). *Field measurements of erosional behaviour and settling velocities at the Dollard, Netherlands 21-23 May 1996.* HR Wallingford Report TR 16, November 1996.

Soulsby, R. L., (1997). *Dynamics of marine sands.* HR Wallingford Report SR 466, February 1997.

Soulsby, R. L. & Humphery, J. D., (1990). *Field observations of wave -current interaction at the sea bed.* In: Water wave mechanics, ed. A Tormun and O T Gudmestad, p 413-428. Kluwar Academic Publishers, Netherlands.

BEACH TOPOGRAPHY RESPONSE TO NOURISHMENT
AT OCEAN CITY, MARYLAND

Magnus Larson[1], Hans Hanson[1], Nicholas C. Kraus[2], and Mark B. Gravens[2]

ABSTRACT: Beach topography data collected along the northern
Atlantic Coast of Maryland in a monitoring program of two large beach fill
operations were analyzed to determine the response of the beach to the
placement of the fill material. The data set comprised profile surveys taken
along 24 lines at 16 different dates extending over a period of 5 years (not
all lines were surveyed on every date). Different data analysis techniques,
such as empirical orthogonal functions and canonical correlation analysis,
were employed to identify spatial and temporal patterns that characterize
the beach topography and how the fills appeared in these patterns. The
long-term adjustment of the fill material could be distinguished from the
signal due to individual storms and offshore shoals.

INTRODUCTION

In the design and performance evaluation of beach fills, both short- and long-term
responses of the fill to the forcing conditions (waves, water levels, currents, winds, etc.)
have to be considered. Short-term responses involve initial adjustment of the fill and the
redistribution of the material during storms. More long-term responses are mainly related
to the evolution of the fill topography towards a new equilibrium state and the effect of
gradients in the longshore sediment transport. The relative importance of short- and long-
term responses for the project design depends on the purpose of the nourishment
operation, whether it aims at increasing the recreational value of a beach, protecting
against flooding, or combating erosion and shoreline retreat. Another purpose for
studying beach fill responses is to obtain information for the development and

---

[1] Department of Water Resources Engineering, University of Lund, Box 118, S-
221 00 Lund, SWEDEN. [2] U.S. Army Engineer Waterways Experiment Station,
Coastal and Hydraulics Laboratory, 3909 Halls Ferry Road, Vicksburg, MS 39180-
6199, USA.

validation of numerical models, which are important tools in designing nourishment schemes and selecting optimal nourishment strategies.

The main objective of this study was to investigate the three-dimensional (3D) topographical response of a beach to the placement of nourished material. A high-quality data set collected during monitoring of a beach-nourishment project at Ocean City, Maryland, was analyzed to determine the short- and long-term responses of the placed material. Several different data analysis techniques were used to quantify the effect of the nourishment on the topographic evolution in the area. First, simple geometric predictors, such as the standard deviation in depth, depth change, and contour movement, were calculated to quantify the local topographic response in time and space. Then, more sophisticated techniques were used to determine large-scale temporal and spatial patterns in the data, and how the fill modified these patterns. Empirical Orthogonal Functions (EOF) were employed to calculate spatial eigenvector maps that gave information on the cross-shore and longshore variability in the fill response. Also, Canonical Correlation Analysis (CCA) was applied to investigate the covariability of the profile response at different survey lines and to determine the associated dominant modes of variation. During the measurement period several major storms passed through the Ocean City area (Kraus 1993), providing an excellent opportunity to evaluate the performance of the beach fills during severe storms. Thus, the results provide insights both to the beach response during short-term events and to the long-term adjustment of the topography.

OCEAN CITY BEACH NOURISHMENT PROJECT

In order to provide beach erosion control along the northern Atlantic Coast of Maryland and to protect the town of Ocean City (see Figure 1), a large beach nourishment project was jointly undertaken by the State of Maryland and the US Army Corps of Engineers (CE) in the late 1980's (Stauble et al. 1993). The project was carried out as two separate beach fill operations; the State of Maryland placed one fill during the summer of 1988 (state fill) and the other fill was placed by the CE during the summer of 1990 (federal fill). The state fill consisted of approximately 2 million m³ of sand and the federal fill of almost 3 million m³. About 13 km of beach were nourished involving an area from just north of Ocean City Inlet to the Maryland-Delaware state line. Fill material was taken from two borrow areas located off the coast in about 10 m water depth. The representative grain sizes for the two borrow areas were 0.25 and 0.35 mm (numbered 2 and 3, respectively, in Figure 2), to be compared with a composite grain size of 0.37 mm for the native beach determined by Anders and Hansen (1990).

The long-term shoreline recession in the area is related to a divergence in the longshore transport in the vicinity of Bethany Beach on Fenwick Island (north of the fill area; see Figure 1), where there is a change in the orientation of the coast, and the net transport to the south causes an impoundment on the north jetty at Ocean City Inlet (Dean and Perlin 1977). The sediment trapping capacity of the north jetty has effectively reached its maximum, and most of the sediment transported southward is presently bypassing the jetty (Stauble et al. 1993). This sediment deposits in the entrance channel of the inlet, settles on the ebb-tidal shoal, or bypasses the inlet and supplies northern Assateague Island with material. Sediment periodically dredged from the entrance channel are placed on the northern end of Assateague Island. Previous studies indicate that the annual net longshore transport rate to the south varies between 100,000 and 250,000 m^3. Stauble et al. (1993) reported an average shoreline recession rate for Fenwick Island from Ocean City Inlet to the state line of 0.6 m/year, ranging from 0 to 1.2 m/year.

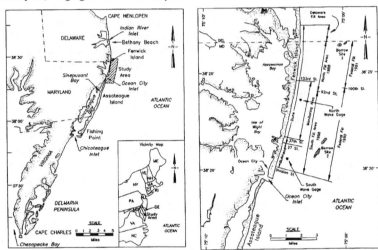

Figure 1. Location map for the Ocean City area.   Figure 2. Detail of the area.
(from Stauble et al. 1993).

FIELD MEASUREMENTS

The State of Maryland, the U.S. Army Corps of Engineers (CE) District in Baltimore, and the Coastal Engineering Research Center of the CE carried out a detailed monitoring program in connection with the placement of the two beach fills (Stauble et al. 1993). Profile surveying started in June 1988 immediately before the state fill was placed and has continued to the present, although only data collected between June 1988 and July 1993 were analyzed in this study. The surveys were made from the dune region out to a typical water depth of about 8 m, and data were available from 24 survey lines encompassing 16 different dates (complete spatial coverage did not exist for all dates). The surveys were performed with a towable sled and a total survey station, except on the subaerial portion of the beach where manual surveying with a rod was employed. The typical resolution in the cross-shore direction for the survey was about 6 m near the shoreline and 12 m in deeper water (with higher resolution around distinct morphological features), and in the alongshore direction the spacing between the survey lines was typically 300-600 m. The survey lines were located close to local streets that run perpendicular to the shoreline orientation and the lines were named after these streets. In this study focus was on the lines located from 37th to 103rd street (see Figure 2), which were located in the central portion of the fills and had the most dense coverage in time.

RESULTS

Geometric Profile Properties

Figure 3 displays a typical contour map of the beach topography (elevation in m) surveyed in April 1989 less than one year after the placement of the state fill. The location of the survey lines used to generate the map is also shown (12 lines in total). Two shore-attached shoals, commonly occurring features on the inner Atlantic continental shelf on the US East Coast (Duane et al. 1972), are clearly seen in Figure 3 (at about $y=6250$ and 9250 m, close to lines 56 and 92, respectively, where y is the longshore direction; the origin of the coordinate system was arbitrarily placed somewhat south of the study area with the y-axis pointing along the trend of the shoreline), and these features tended to modify changes in the beach topography. For example, Figure 4 shows the calculated standard deviation in elevation along three of the survey lines, namely 52, 63, and 74. Line 52 is located in the vicinity of where one of the shoals attaches to shore. A larger standard deviation is observed here at greater depths (> 7 m) than for lines 63 and 74 because of movement on the shoal. However, it is difficult to quantitatively determine how much of the material in the nearshore (including the fill material) participates in the material exchange with the shoal. The similarities in the standard deviation between the lines directly shoreward of the 6-m depth seem to indicate that the long-term fill adjustment primarily takes place down to this depth. The variation in the standard deviation above the zero-contour is mainly related to differences in where the fill material was placed. Figure 5 shows the mean absolute elevation change between consecutive surveys as a function of the mean elevation for lines 52, 63, and 74. This figure basically reveals the same behavior as Figure 4; however, the quantity might be a better measure of the profile activity than the standard deviation because the influence of single events is less (Larson and Kraus 1994).

Empirical Orthogonal Function (EOF) Analysis

EOFs were used to describe the variation in the beach topography in order to find characteristic temporal and spatial patterns in the data (Aubrey 1979). The analysis was carried out for time series of measured topographies so that spatial EOF maps (functions of x and y, where x is the cross-shore direction) could be derived. The subset from the center of the fill area for which good survey coverage existed during the measurement period was selected for this analysis. Linear interpolation was employed to obtain elevations at the same cross-shore locations for all surveys taken at a particular line. The elevation data encompassed the region from the dune out to a water depth of about 6 m covering the area of significant sediment movement. The data matrix D, which contained the elevations surveyed at a specific date in the rows (the values from the different lines were concatenated), was represented in terms of the eigenvectors according to $D=ALE^T$, where A and E are matrices with the temporal and spatial EOFs as column vectors (orthonormal), respectively, L a matrix containing the square-root of the eigenvalues, and T denotes transpose.

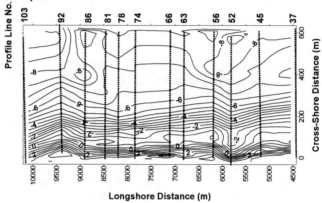

Figure 3. Bottom topography surveyed at Ocean City in April 1989.

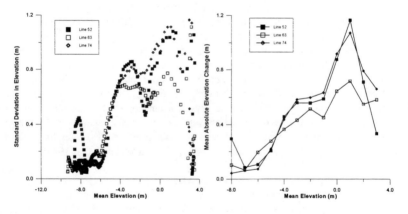

Figure 4. Standard deviation in elevation.   Figure 5. Mean absolute elevation change.

Figure 6 displays a contour map of the mean beach topography, which was subtracted out before the EOF analysis. The presence of the two shore-attached shoals is clearly seen again. Figures 7-9 show the spatial EOF maps corresponding to the three first eigenvectors (E_1-E_3), respectively, whereas the associated temporal EOFs are displayed in Fig. 10 (A_1-A_3). The first three eigenvectors explained almost 90% of the variation in the data, with E_1, E_2, and E_3 individually contributing 59.6, 21.2, and 7.3%, respectively, to the total variation.

Eigenvector E_1 (Figure 7; shaded areas denote negative values in the contour maps) reflects the general increase of material in the area as a result of the fills. With respect to the mean topography there is less material in the area in the beginning of the period compared to at the end, which is reflected in A_1 (Figure 10). After the state fill (summer of 1988), A_1 increases rapidly, but after about a year a constant value is attained. This indicates a time scale for the adjustment of the beach towards a new equilibrium state in response to the fill placement. A similar response is seen in connection with the federal fill (summer of 1990): A_1 increases at first and then falls back towards a near-constant value after approximately a year. E_2 (Figure 8) reflects the exchange of material between the berm and the bar area that exhibits a distinct seasonality (compare A_2). The largest offshore accretion of material (eroded from the inshore; vice versa for negative A_2-values) typically occurs at about 3-m water depth. The areas of accretion and erosion are not completely uniform alongshore, indicating that the shore-attached shoals might modify the pattern of erosion and accretion.

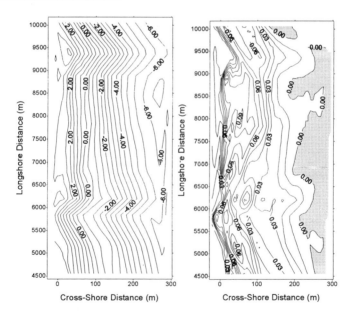

Figure 6. Mean bottom topography.   Figure 7. First spatial EOF map.

Eigenvector E_3 (Figure 9) primarily characterizes the exchange of material between the dune region and the more shallow portion of the beach (< 2 m). Thus, the response of the topography to the most severe storms occurring during the measurement period may be seen in A_3 (the peaks around March 1989 and January 1992). During these storms large amounts of material were eroded from the dune and some was deposited in fairly shallow water, especially in the vicinity of the shoals. For example, Figure 11 illustrates the change in elevation between the two topographies surveyed just after and before the March 1989 storm (surveys done in January and April 1989). Four storms were observed in the area between 23 February and 25 March 1989 that generated significant wave heights in excess of 2.25 m and in two cases wave heights close to 3.0 m (Stauble et al. 1993). A_3 also has a peak just after the federal fill was placed that is associated with the addition of material in the dune region. The higher-order EOFs could not be given any physical interpretation and mainly represent noise.

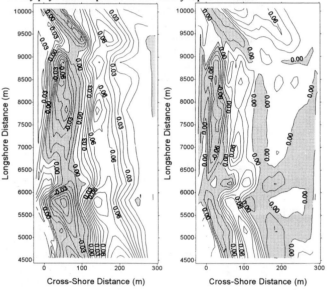

Figure 8. Second spatial EOF map.   Figure 9. Third spatial EOF map.

Canonical Correlation Analysis (CCA)

CCA may be used to investigate if there are any patterns that tend to occur simultaneously in two different data sets and what the correlation is between associated patterns (Barnett and Preisendorfer 1987). The main idea is to form a new set of variables from the original data sets so that the new variables are linear combinations of the old ones and maximally correlated. Thus, if the two original data sets are denoted Y and Z (in the present analysis Y is the profile data matrix from one survey line and Z the profile data from another), the new transformed variables in matrices U and V have maximally correlated column vectors for the same index and zero correlation for differing indices.

Furthermore, the column vectors in U and V are orthonormal. The desired weights for transforming Y into U are given by the solution to the eigenvalue problem (Graham 1990),

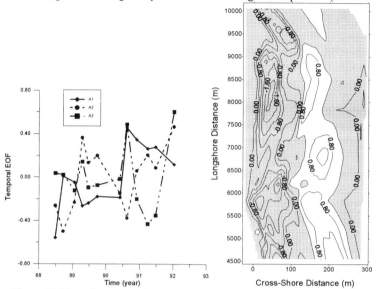

Figure 10. Three first temporal EOFs.   Figure 11. Elevation change during storms.

$$\left[(Y^T Y)^{-1}(Y^T Z)(Z^T Z)^{-1}(Z^T Y) - \mu^2 I\right] = 0$$

where μ^2 denotes the eigenvalues and the subscript T is transpose. The eigenvalue gives the squared correlation between the corresponding temporal amplitudes of the canonical modes (column vectors in U and V), and the associated eigenvectors R yield the transformation $U=YR$. A similar eigenvalue problem as the above equation defines the transformation of Z into V having the same μ^2 values and eigenvectors Q ($V=ZQ$). The spatial amplitudes (g and h) of the canonical modes are given by $g=Y^T U$ and $h=Z^T V$ (thus, the original data sets are expressed as $Y=Ug$ and $Z=Vh$).

CCA was applied to investigate if there were any patterns that tended to occur simultaneously in the profile response along different survey lines. Thus, time series of measured profiles at two different locations were used in the CCA and modes of profile variation that displayed maximum correlation between the locations were determined. As an example, Figure 12 displays the two first CCA spatial modes emerging in the analysis of lines 63 (g_1, g_2) and 74 (h_1, h_2), whereas Figure 13 shows the corresponding CCA temporal modes (U_1, U_2, and V_1, V_2, respectively). The squared correlation between U_1 and V_1 was 0.98 and between U_2 and V_2 0.95. The first modes (g_1, h_1) reflect the adjustment of the fills and the material is deposited mainly in 2-6 m water depth (approximately corresponding to $x=120$-240 m), although some material remains in the subaerial portion of the profile (see Figure 12). In contrast, the second modes (g_2, h_2) are associated with the placement of the fills showing that more material was placed further offshore along line 74 compared to 63. The CCA temporal modes (Figure 13) show a increase from negative to positive values for U_1 and V_1 as the fill material is deposited offshore, whereas U_2 and V_2 fluctuates between negative and positive values related to cycles of fill placement and seaward transport due to short- and long-term fill adjustment. The CCA analysis did not show any clear modes that indicated exchange of material between the bar and berm region (as could be seen in the EOF maps; compare Figure 8). This seems to indicate that the profile adjustment varies alongshore, possibly due to the presence of the shoals.

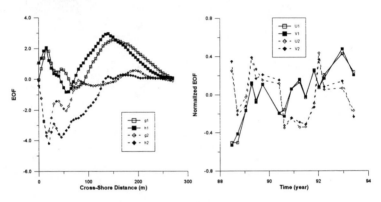

Figure 12. CCA spatial modes.   Figure 13. CCA temporal modes.

CONCLUDING REMARKS

The present analysis showed that it is possible to distinguish different mechanisms, such as fill adjustment (short- and long-term), storms, and the presence of offshore shoals, on the topographic evolution using advanced statistical techniques (for example, EOF and CCA). Placement of fill material disturbs the existing state of a beach, and it takes about one year (a seasonal cycle) before the material has been redistributed and a new state attained that is in balance with the forcing conditions. The offshore shoals attaching to shore at several locations along the Ocean City coastline clearly induced a beach response, although it was difficult to determine the cross-shore material exchange through these features. This exchange is probably limited, judging from how the standard deviation in elevation changes with depth, but the main influence is on the nearshore hydrodynamics (waves and currents). The second and third spatial EOF maps displayed the influence of the shoals on the longshore variability in the cross-shore exchange of material, where the second EOF map characterized the bar-berm exchange and the third EOF map the impacts of severe storms. CCA analysis showed the spatial modes that exhibited the largest degree of correlation between profile lines. These modes represented the placement and long-term adjustment of the fill material alongshore.

ACKNOWLEDGMENTS

The research presented in this paper was partly conducted under the SAFE Project (Contract No. MAS3-CT95-0004) and the PACE Project (Contract No. MAS3-CT95-0002) of the Marine Science and Technology Program funded by the Commission of the European Communities, Directorate for Science, Research, and Development, and partly under the Coastal Inlets Research Program of the U.S. Army Corps of Engineers. Permission was granted to N.C.K. and M.B.G. by the Chief of Engineers to publish this information.

REFERENCES

Anders, F. and Hansen, M. 1990. "Beach and Borrow Site Sediment Investigation for a Beach Nourishment at Ocean City," Technical Report CERC-90-5, USAE Waterways Experiment Station, Vicksburg, MS.

Aubrey, D.G. 1979. "Seasonal Patterns of Onshore/Offshore Movement," Journal of Geophysical Research, Vol 84, No. C10, pp 6347-6354.

Barnett, T.P. and Preisendorfer, R. 1987. "Origins and Levels of Monthly and Seasonal Forecast Skill for United States Surface Air Temperatures Determined by Canonical Correlation Analysis," Monthly Weather Review, American Meteorological Society, Vol 115, pp 1825-1850.

Dean, R.G. and Perlin, M. 1977. "Coastal Engineering Study of Ocean City Inlet, Maryland," Proceedings of Coastal Sediments '77, ASCE, pp 520-542.

Duane, D.B., Field, M.E., Meisburger, E.P, Swift, D.J.P, and Williams, S.J. 1972. "Linear Shoals on the Atlantic Inner Continental Shelf, Florida to Long Island," Shelf Sediment Transport (Eds. Swift, Duane, and Pilkey), Dowden, Hutchinson & Ross Inc., Stroudsburg, PA.

Graham, N.E. 1990. "Canonical Correlation Analysis," Report, WMO Review of Climate Diagnostic Methods.

Kraus, N.C. (ed.) 1993. "The January 4, 1992 Storm at Ocean City, Maryland," Shore and Beach, Vol 61, No. 1.

Larson, M. and Kraus, N.C. 1994. "Temporal and Spatial Scales of Beach Profile Change, Duck, North Carolina," Marine Geology, Vol 117, pp 75-94.

Stauble, D.K., Garcia, A.W., Kraus, N.C., Grosskopf, W.G., and Bass, G.P. 1993. "Beach Nourishment Project Response and Design Evaluation: Ocean City, Maryland. Report I. 1988-1992," Technical Report CERC-93-13, USAE Waterways Experiment Station, Vicksburg, MS.

ON BEACH NOURISHMENT DESIGN

Karsten Peters[1], Jürgen Newe[1] and Hans-H. Dette[2], M.ASCE

ABSTRACT: In order to develop recommendations for coastal protection by beach nourishment, data are required on beach profile changes as a function of waves and water level. Experiments were carried out in the Large Wave Flume (LWF) in Hannover to investigate the development of beach profiles. For this purpose the slope of the beach (1:5, 1:10, 1:15 and 1:20) above the normal water level (dry beach) is defined as the variable. The first part of the experiments is focussed on beach profile development under normal wave and water level conditions. The second part concentrates on changes of the profiles due to a storm surge at raised water levels with the aim to study erosion of the initially dry beach by storm surges.

INTRODUCTION

Coastal protection measures, especially beach nourishments, involve a significant and continuous expenditure. In some instances nearly all of sand placed to protect a coastline for several years, can be lost in a severe storm. In order to achieve an adequate protection of the coastline the beach fill must have a form that does not lead to large sand losses in severe storms. The renourishment intervals have to be measured for the ongoing profile changes and sand losses.

The most stable shorelines in the nature display a broad beach and surf zone. Hence the aim of beach protection should be not only to create a sand reservoir as a dune, but also to create a profile seawards from the dune that provides at all water levels a surf zone profile approximating an equilibrium profile. On such a profile the dissipation of wave energy per unit area is minimized and with it the erosion

---

[1] Research Assistant [2] Academic Director, Department of Hydrodynamics and Coastal Engineering, Leichtweiss-Institute for Hydraulics, Technical University Braunschweig, Beethovenstr. 51a, 38106 Braunschweig, Germany.

potential. This implies that any shore protection concept has to incorporate the nearshore underwater profile as well as the usually dry beach profile. These have a major influence on beach and dune erosion.

OBJECTIVE OF EU-MAST PROJECT 'SAFE'

Periodic artificial nourishment is now regarded as an environmentally acceptable method of beach and dune protection. The durability of the initial beach fill and subsequent nourishment requirements are of major importance in the planning of beach nourishment. The concepts of design are as yet far from being generally accepted and means for prediction of morphodynamic behaviour of nourished beaches still have a low confidence level.

A group of participants from 11 institutes and authorities in 8 European countries in 1996 within the EU-MAST III Research contract 'SAFE' (Performance of soft beach systems and nourishment measures for European coasts) joined their efforts to improve the methods of design and performance assessment.

For attainment of this the three existing sources of knowledge namely field observations, large scale physical modelling and numerical models are used and integrated. Analysis of high quality field data leads to improved empirical knowledge, which is a key element for the development of behaviour orientated long-term numerical models. It should also lead to a better scientific evaluation of existing or ongoing nourishment schemes and dune restoration programs in the European Union. In particular, agreed and clear definitions of performance indicators like the volume of sand loss, rate of loss, coastline changes need to be defined (SAFE, Technical Annex, 1996).

SCOPE OF LARGE WAVE FLUME EXPERIMENTS

The studies were carried out in the Large Wave Flume (LWF) in Hannover from November 1996 to August 1997. The experiments within the SAFE project were focussed on beach and dune profile changes as a function of waves and water level.

In order to provide experimental data for numerical modelling and to develop recommendations for practical coastal protection by beach nourishment methods details on beach profile changes as a function of waves and water level are needed. In the LWF-SAFE tests the slope of the beach above the normal water level (dry beach) was defined as the variable. For each test series the profile was first established under normal wave and water level conditions using a TMA spectrum, $H_{mo} = 0.65$ m, $T_m = 5$ s and h = 4 m. The initial test lasted until the profile reached an equilibrium for the selected wave parameters. This profile was then the initial profile for the profile change due to a 1 m raised water level and storm surge (TMA spectrum $H_{mo} = 1.20$ m, $T_m = 5$ s). The dry beach slopes used were 1:20, 1:15, 1:10 and 1:5 in order to determine the sand loss as a function of beach slope (Fig. 1).

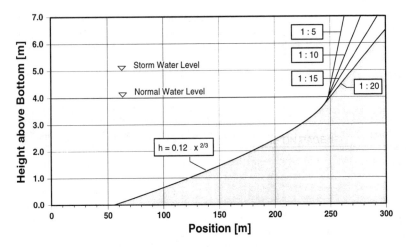

Fig. 1:   Profile geometries in SAFE-LWF experiments

INSTRUMENTATION

The different instruments used for the experiments are summarized in Table 1.

Table 1:   Instrumentation for SAFE-LWF experiments 1996/97

Variable	Instrument	Accuracy
Free surface elevation (waves)	Resistance type gauges	∓ 1 cm
Water velocity	ADV ocean probes (3 components of flow, sampling volume 3 cm³), 3 levels (see Fig. 3)	∓ 1 % of velocity range
Suspended load concentration	OBS probes, 6 levels (point measurements)	∓ 0.05 g/l
	Multi-frequency acoustic concentration device (instantenous concentration measurement over 50 cm of water column)	∓ 0.1 g/l
Bed levels, bed forms	Profile measuring device mounted on mobile carriage (speed: 10 m/min)	horizontal: ∓ 3 mm vertical: < 1 mm

27 wave gauges were deployed along the flume (Fig. 2). The arrangement in measuring harps (no. 1 - 4, 5 - 8, 9 - 23, 14 - 17) allows the analysis of reflection characteristics. The measuring instruments for water velocity and suspended load were mounted on a mobile carriage (Fig. 3). Measurements at one location were taken for 15 min (cycle of the spectrum), and the carriage moved to the next location (distances 2.5 m to 5 m). Fig. 4 shows the wave gauges along the flume.

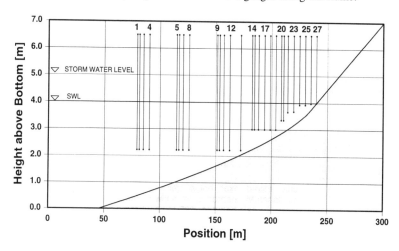

Fig. 2:   Position of the 27 wave recording locations along the flume

Fig. 4:   Wave gauges along the flume

Fig. 3: Current and concentration measuring instruments on the instrument carriage

An online absorption control system installed in the LWF enables the compensation of reflections directly at the wave paddle. Fig. 5 shows a record of a TMA spectrum of wave heights over 3.5 hours and illustrates the stability of wave pattern.

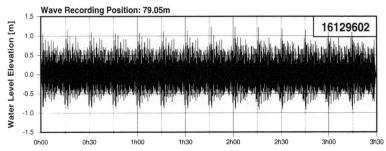

Fig. 5:   Time series of wave heights (TMA spectrum, 15 min cycle, $H_{m0} = 1.20$m, $T_m = 5$ s) over 3 hours 30 min

PRELIMINARY RESULTS

In Table 2 lists all the test runs of profile changes with the slope of the initial dry beach as the variable.

Table 2:   Summary of test runs for beach slopes 1:20, 1:15, 1:10 and 1:5

Test Series	Water Depth h	Wave Height H_{m0}	Wave Period T_m	Generated Wave	Total Test Duration	Initial Beach Slope
ID	**[m]**	**[m]**	**[s]**	**[]**	**[hh:mm]**	**[1:n]**
A8	4,00	0,65	5,0	TMA	11:30	1:20
A9	5,00	1,20	5,0	TMA	25:00	1:20
B1	4,00	0,65	5,0	TMA	17:00	1:10
B2	5,00	1,20	5,0	TMA	22:51	1:10
C1	4,00	0,65	5,0	TMA	05:30	1:5
C2	5,00	1,20	5,0	TMA	23:00	1:5
H1	4,00	0,65	5,0	TMA	05:00	1:15
H2	5,00	1,20	5,0	TMA	25:15	1:15

Fig. 6 illustrates the measured wave parameters ($H_{1/3}$ and H_m) along the profile, at raised water level for the test series A9 (Table 2) after 30 minutes and 25 hours of profile development.

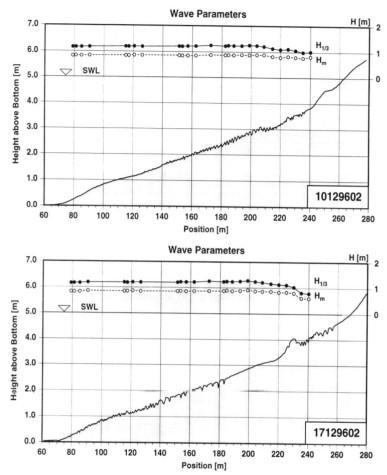

Fig. 6: Measured wave parameters along the profile for test series A9 at 1 m raised water level after 30 minutes (above) and 25 hours of test duration (below) (Dette et al., 1997)

The profile developments (initial and final profile after 25 hrs) are shown for all 4 beach slopes in Fig. 7 to 10. It is apparent that with steep slopes 1:5 and 1:10 the profiles developed an erosion cliff whereas the slopes 1:15 and 1:20 maintained a continuous slope. Thus, the transition from slope to erosion cliff type profile evolution lies approximately at an initial dry beach slope of 1:15. This is a significant result for quantitative numerical modelling as well as for practical engineering solutions.

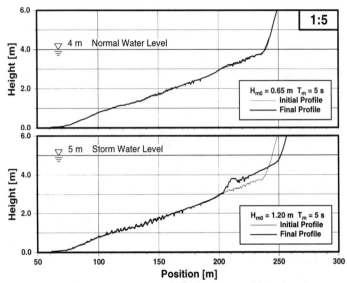

Fig. 7: Profile change for the 1:5 beach slope case under normal wave conditions (above) and under storm conditions (below)

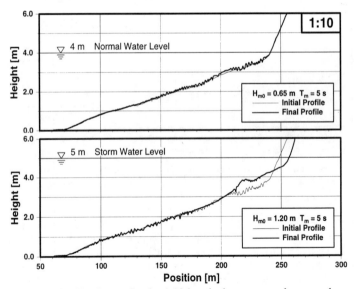

Fig. 8: Profile change for the 1:10 beach slope case under normal wave conditions (above) and under storm conditions (below)

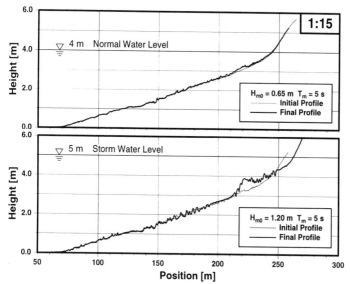

Fig. 9:   Profile change for the 1:15 beach slope case under normal wave conditions (above) and under storm conditions (below)

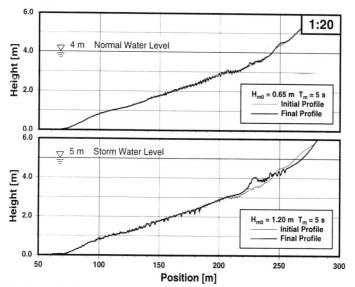

Fig. 10:   Profile change for the 1:20 beach slope case under normal wave conditions (above) and under storm conditions (below)

The final profiles from the investigated slopes after more than 20 hours of test are compared in Fig. 11.

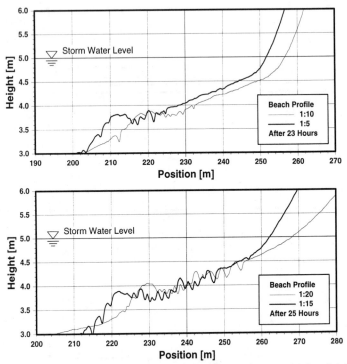

Fig. 11:   Comparison of final beach profiles for the cases of 1:5 and 1:10 (above) and for the cases of 1:15 and 1:20 (below)

ACKNOWLEDGEMENT

This work is undertaken as part of the MAST III 'SAFE'-Project supported by the Commission of the European Communities, Directorate General for Science, Research and Development, under contract no. MAS-CT95-0004.

REFERENCES

DETTE, H.H.; PETERS, K.; NEWE, J. (1997) Large Wave Flume Experiments '96/97 - Equilibrium Profile with Beach Slope 1 : 20 -, Report No. 819, Leichtweiss-Institute, Technical University Braunschweig (unpublished).
MAST III - SAFE (1996) Technical Annex.

Shoreface Nourishment at Terschelling, the Netherlands: Feeder Berm or Breaker Berm ?

R. Spanhoff[1], E.J. Biegel[2], J. van de Graaff[3] and P. Hoekstra[4]

Abstract

A 2 Mm^3 shoreface nourishment has been placed in 1993 at the island Terschelling to maintain the local coastline position. The nourishment was designed to feed sand to higher parts of the profile through net cross-shore transports. The observed gain of sand landward of the nourishment is largely attributed to induced gradients in the longshore sand transport.

Introduction

Human activities are concentrated in coastal zones, which thus may exhibit high population densities and capital investments. Coasts are often eroding, e.g. due to sea level rise, which calls for measures to protect not only lives and goods but also other functions of the coastal system like its ecological values. Nourishing beaches with sand is nowadays commonly applied in replacing the classic approach of building hard structures, like sea walls and groynes that not only are expensive to construct and maintain but also often show adverse side effects. Shoreface nourishments may offer an attractive alternative for beach nourishments. The former can be performed under a wider range of conditions, without interfering with any on-going recreation, and are

---

[1] National Institute for Coastal and Marine Management/*RIKZ*, P.O. Box 20907, 2500 EX The Hague, The Netherlands. e-mail: R.Spanhoff@rikz.rws.minvenw.nl

[2] SEPRA BV, Leidschendam, The Netherlands.

[3] Delft University of Technology, Stevinweg 1, 2628 KN Delft, The Netherlands.

[4] Institute for Marine and Atmospheric Research, Utrecht University, P.O.Box 80.115, 3508 TC Utrecht, The Netherlands.

potentially cheaper by avoiding some handling like sand transfer to smaller vessels, coupling to pipelines, pumping and/or rainbowing, and bulldozing.

In the Netherlands the Dutch coastline is being prevented from receding beyond the position it had in 1990, a national policy effective since that year. Yearly circa 6 Mm^3 sand is being nourished, on beaches so far. To assess the advantages of (future) shoreface nourishment schemes, in 1993 (May-November) 2 Mm^3 sand has been dumped on the foreshore of an eroding part of the island Terschelling. Subsequently the forcing conditions (water levels, waves, currents) and the adaptation of the coastal system have been monitored in the MAST-2 project NOURTEC (Innovative Nourishments Techniques Evaluation), that was carried out in the years 1993-1996 and involved similar activities in Denmark and Germany (Niemeyer et al., 1995). Besides the monitoring programme, at Terschelling extensive so-called process measurements have been carried out in NOURTEC, with e.g. 6 instrumented frames at the sea bottom in 4 campaigns of circa 6 weeks each (Hoekstra et al., 1997). This paper describes some phenomenological observations of the coastal system, based on echo sounding and beach levelling.

Study area at Terschelling

Terschelling is a barrier island that fringes the Wadden Sea (Figure 1). Its seaward part is fully exposed to the Northwest, and thus to swell from the North-Atlantic and to large-fetch wind waves from that direction. Prevailing winds are southwesterly. Tidal range is mesotidal (between 1.2 m and 2.8 m). Tidal and residual currents are basically shore parallel (order 0.3 and 0.03 m/s, respectively), with occasionally relatively important wind-driven currents (order 0.5 m/s). The morphology is characterized by a cyclic, seaward moving three-bar system (Ruessink and Kroon, 1994). The level of the dune foot is +3.0 m NAP (NAP: Dutch ordnance level that virtually coincides with Mean Sea Level). Part of the centre of the island, notably to the east, retreats with order 2 m/yr, without any obvious link to the bar behaviour.

The bathymetry is yearly monitored (JARKUS data base) since 1965 with conventional echo sounding in cross-shore transects: from the reference beach poles, called RSP, lines extended 800 m off-shore till late eighties and 1500 m since, and have a spacing of typically 200 m. Beach and dune levels are determined from aerial photographs. In NOURTEC the study area (boundaries transects RSP 10.0 and RSP 22.0) was surveyed circa 4 times a year (Figure 1), with extra lines being added to a spacing of typically 100 m, and the lines often being longer, (i.e. 2 km in the centre of the study area and 1.5 km at the sides). Levels

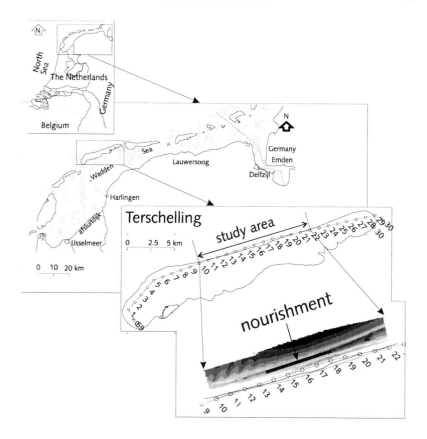

Figure 1. Locations of the island Terschelling, the study
          area and the shoreface nourishment. The 1-km
          spaced reference beach poles RSP are indicated.
          The lower figure shows the monitored area and
          the bathymetry (interpolated on a 20 m grid)
          with e.g. shore-parallel breaker bars and sand
          waves. Depths are indicated with grey tones
          (max. depth 11 m).

of beach and part of the dune (at least including the dune
foot) were obtained with levelling. Data have been
analysed per transect and with GIS techniques in an
ARC/INFO environment.

     The coastline position in the Netherlands, called
momentary coastline position MKL, is defined by a (cross-
sectional) "BKL" control volume stead of by a level

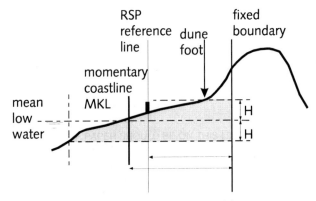

Figure 2. Definition Momentary Coastline MKL.

crossing of the profile (Figure 2). This BKL volume is
bounded by (i) the beach profile, (ii) a horizontal line
at a distance 2H below the dune foot level, (iii) a
vertical reference line at a fixed position and (iv) a
horizontal line through the dune foot, with H = the
vertical distance between the dune foot and the elevation
of MLW (Mean Low Water). The dune foot level is considered
constant (+3 m NAP at Terschelling) in the period 1965-
present, as is MLW level (- 1.12 m NAP), in line with the
observations. Dividing the control volume by 2H gives the
MKL distance from the fixed vertical reference line and
thus the MKL position with respect to RSP. [In practice
the coastline position is yearly assessed from
extrapolating the linear regression to the MKL values in
the previous ten years. The basal coastline BKL, the value
for the year 1990 that should not be receded, has been
derived in this way as well.] When recession occurs,
nourishments are applied with as design objective to keep
the MKL for a selected number of years (typically 5)
seaward of the BKL. Nourishments may have different
objectives, e.g. warranting safety of the hinterland
against flooding or sufficient beach width for recreation;
then other design parameters are appropriate (NOURTEC,
1997) than the MKL discussed here.

The shoreface nourishment at Terschelling has been
designed to compensate the natural erosion at the
nourished coastal section during a period of 8 years. The
average erosion in the area (2.9 m/yr ==> 24 m³ per lineal
metre, since H= 4.12 m) has been derived from linear
regressions per transect to the MKL values over the period
1982-1991. The nourishment was placed between km 13.6 and
km 18.2, thus the 8-year erosion amounts ca. 1 Mm³
(8x24x4600) sand. The shoreface nourishment was placed in
the trough between the outer two bars, largely below the

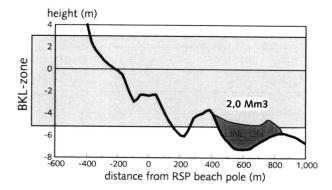

Figure 3. Position shoreface nourishment with respect to
          the profile and the BKL zone. 2 Mm^3 sand has
          been dumped in a 4.6 km long coastal stretch.

BKL zone (Figure 3), and it was anticipated (Mulder *et
al.*, 1994) that circa half of the dumped sand would enter
that zone, therefore 2 Mm^3 was dumped. The expected
response of the system, in terms of sand volume in the BKL
control volume, is depicted in Figure 4. The nourishment
is dumped at T=0, with a small fraction (top layer)
directly in the BKL zone. Next, the profile will adapt
according to a relaxation process, with a net transport
of nourished material into the zone, higher up in the
profile, until 50% has entered it. The rate of gain
initially will be relatively high, subsequently decreasing
monotonously to zero with time. The natural erosion will
thus prevail again in later stages.

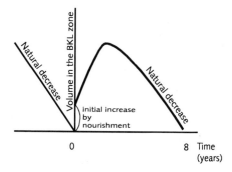

Figure 4. Anticipated response of the sand volume in the
          BKL zone to the shoreface nourishment placed at
          t=0. The nourishment has a design life time of
          8 years.

Observations

MKL values were derived from soundings and levelling for the periods before NOURTEC (annual JARKUS data 1965-1992) and during NOURTEC (1993-1996, several surveys per year). Figure 5 shows averaged values for the western part, eastern part and the whole of the nourished coastal sector, respectively. The absolute value of the vertical axis lost its meaning in the averaging. The data scatter around the linear regressions to them, despite of averaging over many transects. These fluctuations are partly due to systematic errors in the soundings, partly to differences in annual profile lengths, partly to the bar behaviour in relation to the BKL zone and partly to other volume changes. The data do not demonstrate trend changes in the period 1965-1993. A seeming positive trend in 1990-1993 is deemed an artefact, the annual fluctuations being in the range of the previous ones. Thus we made regressions over the whole period 1965-1993.

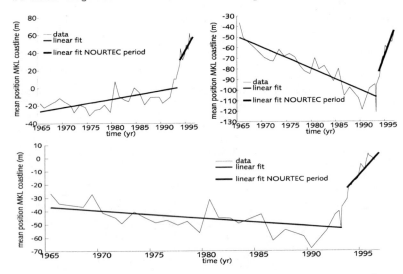

Figure 5. Coastline positions as a function of time. Average values are shown for the total nourishment area (bottom), its western half (top left) and its eastern half (top right). Linear regressions for the pre- and post-nourishment periods, respectively, are shown.

It can be seen that the eastern half and the whole of the nourished sector are eroding, albeit less than follows from a ten-year regression 1981-1992. The western half rather is accreting, in contrast with its trend in

1982-1991. In the three years since the nourishment the trends are strongly positive, order 10-15 m/yr, and hardly any saturation is noticed yet.

The response of the coastal system to the nourishment was further investigated with GIS tools. Depth and height data were interpolated in a 2D grid (20 m grid size) and colour coded, providing synoptic views of the bathymetric features in the study area (compare Figure 1). Similar views of differences between 2 surveys clearly showed the developments in time. With the system, volumes of sand and differences therein were determined quantitatively for selected areas. Some results for the monitored area (compare Figure 1) between transects RSP km 10.0 and km 20.0 are shown in Figure 6, with as subareas the dump site, the areas directly shoreface and off-shore of it, and two control volumes at either side.

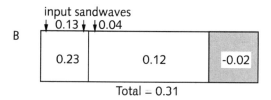

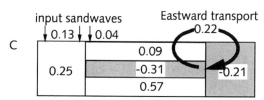

Figure 6. Sand budgets in Mm³/yr for selected areas (A) in the (1965-1993) pre-nourishment (B) and 2-year post-nourishment (C) period. The shore-parallel eastward migration of the nourishment in the control area marked with an asterisk (A) is indicated seperately (C).

In the eastern control volume a separate rectangular shore-parallel control volume (400x1200 m^2) was defined to quantify the observed significant shore-parallel eastward transport of the nourishment. The results are expressed as observed (linear) trends of the volume changes in the ten soundings after the nourishment in the period November 1993 - December 1995. Also the trends in the period 1965-1993 before the nourishment are indicated; the latter were derived by horizontally integrating the available profiles.

Discussion

The shoreface nourishment has been designed as a feeder berm. Sand dumped at -5 m to -7 m NAP in the trough of the outer two bars was supposed to be transported shoreward. Longshore transports, or better gradients therein, were deemed less significant, and the system was considered homogeneous in longshore direction. At most some head effects near the western and eastern ends of the nourishment should occur. Furthermore, it was assumed that the added amount of sand was not that large that it can push the system to a basically other state. Thus the autonomous erosion should virtually remain the same. In this approach the whole active zone will be shifted seaward at infinite time, the shift depending on its height and on the amount of sand. Assuming an active zone height of 8 m (+3 m NAP till -10 m NAP, the adopted closure depth), the BKL zone (circa 8 m heigh) benefits for circa 50% of the nourished sand, which justifies the applied factor 2 that led to the applied 2 Mm^3 sand. The relevant question is whether the above picture holds and whether the response time of the system matches the desired 8-year period.

The MKL gain of 10-15 m/yr already in the first three years (Figure 5) after the shoreface nourishment at Terschelling by far exceeds the expectations. A net total MKL gain in 8 years of circa 25 m superimposed on the unchanged autonomous erosion was designed. This would involve 1 Mm^3 extra sand. Since twice the amount has been dumped, in theory 50 m MKL gain is possible when ignoring the autonomous erosion which turned out to be relatively small anyway. However, the extrapolated continuing MKL gain after 1996 rules out this solution. Obviously the actual response deviates from the anticipated one (Figure 4), indicating that the original assumptions are not or only partly met.
Figure 6 shows that the autonomous behaviour is virtually unaffected by the nourishment, at least in the total area and in the western control area. The

nourishment loses 0.31 Mm^3/yr due to its eastward displacement (0.22 Mm^3/yr) and maybe due to a small (0.09 Mm^3/yr) off-shore transport. From the balance (Figure 6) on-shore transport is small (between zero and the order of the uncertainties) compared to the gain (0.57 Mm^3/yr) in the coastal stretch landward of the nourishment. The latter rather should be attributed to trapping of longshore transported sand behind the berm that reduces the wave energy responsible for the littoral drift. According to the CERC formula and to an integration over the profile of the longshore transport rates calculated with UNIBEST-TC (compare Hoekstra et al., 1997), both for the wave climate at Terschelling, this longshore transport amounts circa 1 Mm^3/yr, so trapping of the order 0.5 Mm^3/yr is conceivable. The eastern control sector then should show an extra erosion with the same amount. Indeed erosion has increased with 0.19 Mm^3/yr compared with the pre-nourishment situation (Figure 6), but in fact amounts 0.41 Mm^3/yr when taking into account the eastward shift of the nourishment (Figure 6). Thus it is confirmed that most gained sand landward of the nourishment is trapped in a salient due to the breaker berm effect. Part of it, order 0.1-0.2 Mm^3/yr, still might be due to cross-shore transport in view of the uncertainties in the numbers. Such a value would be in line with the expectations (1 Mm^3 in 8 years) and with recent calculations with UNIBEST-TC fine-tuned to the natural bar behaviour and the initial (150 days) response of the system to the nourishment (Roelvink et al., 1996). No error margins have been attributed to the above rates yet, since the sources of the fluctuations are hard to interpret (see OBSERVATIONS). The autonomous behaviour in the pre-nourishment period of the western part of the nourished area differs markedly from that of the eastern part (Figure 5). This by itself raises doubts on the assumed long-shore homogeneity. In a cross-shore description, one is mainly concerned with the ditto transport components and (small) longshore gradients are not necessarily prohibitive. However, also the GIS analysis has revealed pronounced 3D phenomena. Obviously the area has to be analysed in more detail, both in cross- and long-shore directions.

Conclusions

From a practical point of view the nourishment already now has more than satisfied its goal. The downdrift negative effect happens to be harmless in this case.
Most of the effect of the nourishment on the coastline is in the form of a salient due to the breaker berm effect.

It can not be ruled out that the anticipated net cross-shore sand transports from the nourishment do occur.
A more detailed (2D) analysis of the data, and a longer series of data after the nourishment may resolve some remaining questions.
Trends in sand volumes have to be determined with great caution in view of the present scatter in the echo sounding data.

Acknowledgements

This work was carried out as part of the NOURTEC and SAFE projects. It was funded jointly by the Ministry of Transport, Public Works and Water Management, Directorate General Rijkswaterstaat, National Institute for Coastal and Marine Management/RIKZ and the Commission of the European Communities, Directorate General for Science, Research and Development, under Contracts MAS2-CT93-0049 and MAS3-CT95-0004.

References

Hoekstra, P., Houwman, K.T., Kroon, A., Ruessink, B.G., Roelvink, J.A. and Spanhoff, R., 1997. Morphological development of the Terschelling shoreface nourishment in response to hydrodynamic and sediment transport processes. Proc. 25th Int. Conf. on Coastal Eng., ASCE New York, 2897-2910.

Mulder, J.P.M., J. van de Kreeke and P. van Vessem, 1994. Experimental Shoreface Nourishment, Terschelling (NL). Proc. 24th Int. Conf. Coastal Eng. Vol. 3, ASCE New York, 2886-2899.

Niemeyer, H.D., E. Biegel, R. Kaiser, H. Knaack, C. Laustrup, R. Spanhoff, and H. Toxvig Madsen, 1995. General Aims of the NOURTEC project - effectiveness and execution of beach and shoreface nourishments. Proc. 4th Conf.on Coast. & Port Eng. in Develop. Countr., Rio de Janeiro, Brazil.

NOURTEC, 1997. Innovative nourishment techniques evaluation, Final report. Coord. National Institute for Coastal and Marine Management/RIKZ, The Hague, The Netherlands, 105 pp with figures.

Roelvink, J.A., Meijer, Th.J.G.P., Houwman, K.T., Bakker, R. and Spanhoff, R., 1996. Field validation of a coastal profile model. Proc. Coastal Dynamics '95, ASCE New York, 818-828.

Ruessink, B.G. and A. Kroon, 1994. The Behaviour of a Multiple Bar System in the Nearshore Zone of Terschelling, the Netherlands: 1965-1993. Marine Geology, 121, 187-197.

SAND BAR DYNAMICS AND OFFSHORE BEACH NOURISHMENT

B. Boczar-Karakiewicz[1], W. Romańczyk[1] and J.L. Bona[2]

ABSTRACT

Our work concerns beach protection methods restoring the wave-dominated dynamic equilibrium of eroding coasts. The approach is environmentally-sound, innovative and cost-effective. The recommended offshore nourishment involves placing sediment in the offshore, on artificial bars.

The proposed work applies mathematical models describing coastal processes in a wave-dominated environment. In the present stage, we analyze seabed evolution using a two-dimensional model that predicts the evolution of longshore bars under nonlinear waves from an initially featureless bed. Model predictions explain that in contrast to universal properties of the process of bar formation, the dynamics of coastal morphology are site-specific. Coastal stability and erosion depend on sequences of storm events in the local wave climate. Model simulations reproduce the observed quasi-periodicity of beach erosion and beach recovery that result from a cyclic exchange of sediment between offshore bars and the beach above the waterline.

The mode of model application is explained by an example of offshore beach nourishment in the Gold Coast, Australia.

1. INTRODUCTION

Our work considers the dynamic evolution of nearshore topography under wind-generated waves. The objective of our analysis is to develop efficient protection methods for eroding coasts where multiple sand bars are common. Observations show that erosion and retreat of unconsolidated coastlines are at present a worldwide phenomenon. Global warming will increase wave activity, raise sea level, and further degrade the present catastrophic state of many beaches under

---

[1]INRS-Océanologie (Université du Québec), Rimouski, Québec, Canada
[2]University of Texas, Applied Math, Austin, Texas, USA

social developmental pressure. In Canada, fragile Arctic coasts of the Beaufort
Sea, beaches of the Great Lakes and of the Gulf of St. Lawrence already display
signs of deterioration (Boczar-Karakiewicz, Forbes and Drapeau 1995, Boczar-
Karakiewicz, Hill and Romańczyk 1995). Severe erosion is likely to spread
to many coastal regions within half a century (Vellinga and Perdomo 1992,
Committee on Beach Nourishment and Protection 1995).

The leading idea of our work is to justify a method of protection restoring
the dynamic equilibrium of coastal systems. This view competes with methods for establishing static beach equilibrium profiles. The proposed procedure
recommends offshore nourishment by placing sediment in the offshore, on artificial bars. The interaction of wave-induced forces in the nourished coastal
environment with the underlying seabed leads to sediment redistribution and
to the restoration of site-specific dynamic processes (Boczar-Karakiewicz, Bona
and Jackson 1993). In some situations, the proposed offshore mechanism competes in efficiency with established but more expensive nourishment of beaches
above the waterline (Dean 1996, Committee on Beach Nourishment and Protection 1995). The proposed approach is environmentally-sound, innovative and
in many cases cost-effective.

2. MODELING WAVE-SEABED INTERACTION

The proposed work is an attempt to understand coastal processes acting in
a wave-dominated environment and to describe seabed evolution by nonlinear
mathematical models.

In the present stage, we apply a simple two-dimensional (2D) model which
considers forces under incident wind-generated waves. The structure of our
model is modular and comprises: (1) a description of surface waves over topography, using scaled Boussinesq equations with modified dispersive terms
(Boczar-Karakiewicz, Bona and Cohen 1987, Restrepo and Bona 1997, Bona
and Chen 1997); (2) a description of boundary-layer dynamics using the idea of
drift velocity, and a description of sediment transport (Longuet-Higgins 1953,
Chapalain and Boczar-Karakiewicz 1992); and (3) the bed treated as a continuum (Boczar-Karakiewicz, Bona and Cohen 1987). The modular structure of
this model allows us to take advantage of disparate time scales for bed and fluid
evolution. The two time scales are treated independently.

This 2D model applies in nearshore regions with regular bathymetry where it
predicts the evolution of longshore bars under nonlinear waves from an initially
featureless bed. The model also shows how longshore bars affect the dynamics of the coastal environment (Boczar-Karakiewicz and Davidson-Arnott 1987,
Boczar-Karakiewicz, Forbes and Drapeau 1995, Boczar-Karakiewicz and Jackson 1990, Boczar-Karakiewicz, Hill and Romańczyk 1995).

Quantitative results obtained with this model have shown that sandbar formation under shallow-water waves is controlled by parameters of incident waves
and mean slopes of the underwater beach. The process of bar formation is in-

dependent of geometric scales or climatological conditions. In contrast to these universal properties of bar formation, 2D model predictions explain the site-specific dynamics of the foreshore morphology controlled by sequences of events in the local wave climate. Model simulations reproduce the observed quasi-periodic beach erosion and beach recovery that result from cyclic exchanges of sediment between offshore bars and beaches above the waterline. It also has been shown that wave-exposed coasts are more resistant to erosion when sand bars are part of a permanent equilibrium morphology (Boczar-Karakiewicz, Forbes and Drapeau 1995).

The 2D model was tested extensively with laboratory and field data. Wave-channel experiments in three different scales were carried out in three labo-ratories (Boczar-Karakiewicz, Bona and Cohen 1987, Chapalain, Cointe and Temperville 1992). Field data for 2D model tests stem from fifteen different coastal zones (see, e.g., Osborne and Greenwood 1992, Boczar-Karakiewicz and Davidson-Arnott 1987, Forbes, Frobel, Heffler, Dickie and Shiels 1986, Amos and Nadeau 1988). Comparisons of predictions and observations have shown that the 2D model applies in coastal environments in a wide range of geometric scales: in Lake Huron, Canada, crest-to-crest distances of outer bars are about 50 m; in the Gulf of St. Lawrence, Canada, these distances are 100-400 m; and at open oceanic coasts such as Sable Island, the Scotian Shelf and Gold Coast, Australia, the spacing of bars is about 500 m. Our experience also derives from studies in coastal zones located in a variety of climatic regions (e.g., subtropical, moderate, Arctic).

3. APPLICATION IN A BEACH NOURISHMENT PROJECT

The mode of model application is presented using data from an offshore beach nourishment project in the Gold Coast, Australia (Boczar-Karakiewicz and Jackson 1991). The Gold Coast seafront comprises 42 km of beaches facing the South Pacific (Fig. 1).

In the **preliminary stage** of design of the nourishment project available field data and observations were analyzed to identify the site-specific erosion problem. In the Gold Coast area, beach erosion and related damage caused by flooding resulted from:

- a deficit of sediment supply to the beach system caused by the construction and extension of Tweed River training walls (evaluated to be of some 250.000 m^3/yr, see Fig. 1),

- a diminution of buffer zones consisting of natural dunes due to resort development,

- waterfront constructions (boulevards, seawalls, etc.) associated frequently with elimination of the visible beach at these locations, and

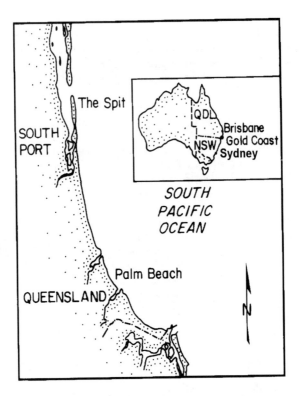

Figure 1: Location map.

- intense recreational use of Gold Coast beaches (stimulating further urbanization of the waterfront area).

Under such conditions, beach nourishment is an alternative to the threat of losing the beach, as observed in the Capricorn area, Queensland. The lost beach is located 600 km north of the Gold Coast.

The **first stage** of the design includes an analysis of site-specific wave and topography data. These data are used in model simulations reproducing the pre-erosion state of the chosen site presented in Figure 2. For the Gold Coast, the pre-erosion situation is characterized by a non-deficit sediment supply and a fully dissipative beach system. These two additional assumptions were introduced into model simulations in which the local wave climate in the Gold Coast area is simplified to a sequence of an extreme cyclonic event (lasting a couple of days, with a peak period $T_p = 17$ s), followed by several months of moderate storm activity (18 months, $T_p = 10$ s). (For evaluation of time scale τ in Fig-

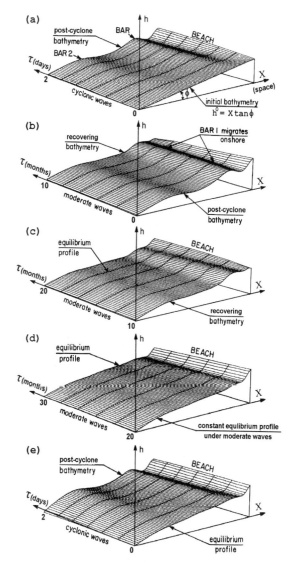

Figure 2: Figure 2. Dynamic cycle of beach transformation: (a) formation of post-cyclone bars, (b) and (c) modification of post-cyclone bars by moderate waves, (d) equilibrium bathymetry under moderate storm waves, (e) reformation of post-cyclone bars.

ure 2, see Boczar-Karakiewicz and Jackson 1990, and McGrath and Patterson 1972). The featureless initial bed profile in Fig. 2a (at $\tau = 0$) represents mean bed characteristics at Kirra (Fig. 1). The mean slopes in the Gold Coast area are of the order of 0.01 (Gold Coast City Council 1990).

The results shown in Figure 2 reproduce an observed cycle of seabed response to the ambient wave climate (Jackson, Smith and Piggot 1988). A prominent two-bar system is formed in a couple of days by cyclonic waves (Fig. 2a). The formation of these bars is associated with erosion of the upper beach. Moderated storm activity during 18 months leads to the formation of a nearly uniform and featureless bed profile (Fig. 2b-2c) which reaches equilibrium with the incident moderate storm wave (Fig. 2d). Moderate wave activity redistributes sediment of the post-cyclonic bars in the nearshore and partially returns it to the eroded visible beach (Chapman 1980, Smith and Jackson 1990). In the following extreme cyclonic event, the equilibrium profile established by milder weather conditions (Fig. 2d) reforms into a two-bar post-cyclone system shown in Figure 2e. The initial post-cyclone bed profile generated from the featureless bed (Fig. 2a) and the final post-cyclone bed profile generated from the equilibrium profile (Fig. 2e) are virtually identical. The simple simulation presented in Figure 2 correctly reproduces the main features of the observed site-specific recurrent cycle of beach response to the incident wave climate in the Gold Coast area (Boczar-Karakiewicz and Jackson 1991).

In the **second stage** of design, recommendations for offshore nourishment are formulated. At this stage, technical parameters of the nourishment procedures are chosen by specifying the localization and volume of sand placement, and the shape of sand bodies to be formed.

For the Gold Coast area, the suggested nourishment procedures are:

1. to reconstruct the observed and simulated post-cyclonic two-bar bed profile, or alternatively

2. to reconstruct only one among two bars of the post-cyclonic profile.

The first version of offshore nourishment (solution 1) is costly and time-intensive. This version provides the most effective protection of the visible beach, assures its stable condition and prevents flooding of urbanized areas next to the beach. The alternative version of partial nourishment (solution 2) is less expensive and simpler in its realization.

An additional model simulation presented in Figure 3 has been conducted to justify the choice of location of sediment to be placed on the inner bar (solution 2a) or, alternatively, on the outer bar of the post-cyclone profile (solution 2b). The results show that in both cases of the prenourished underwater beach (solution 2a: construction of the inner bar, Fig. 3a, $\tau = 0$; solution 2b: construction of the outer bar, Figure 3b, $\tau = 0$), the incident cyclonic wave forms a two-bar system after two days (see bed configuration at $\tau = 2$ days in Fig. 3a and 3b, respectively).

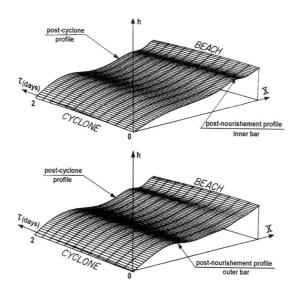

Figure 3: Figure 3. Dynamics of the foreshore under cyclonic waves: (a) for nourished inner bar, and (b) for nourished outer bar.

However, nourishing the outer bar (solution 2b, Fig. 3b) is more efficient in assuring the stability of the visible beach:

- by supplying a volume of sediment reconstructing the outer bar that approximately covers the estimated deficit caused by Tweed River training walls, and

- by constructing a movable underwater breakwater causing offshore breaking of the incident cyclonic wave and related dissipation of some 50% of its energy at a safe distance from the beach and shoreline (Smith and Jackson 1990).

Therefore, solution 2b has been justified and recommended for beaches in the Gold Coast area.

In contrast, nourishment of the inner bar would be a less effective measure of protection because its volume only partially covers the estimated sediment deficit. The inner bar is already partially formed under moderate storm activity (see Fig. 2b - 2d), and observations confirm its permanent presence in all stages of beach transformations over the whole stretch of the 42 km beach (aerial photography and current beach surveys, Gold Coast City Council 1990).

4. DISCUSSION

Our calculations have confirmed the efficiency of protection measures applied to earlier beach nourishments in the Gold Coast area and justified beach restoration procedures applied in 1990 by the Gold Coast City Council and the Beach Protection Authority (Boczar-Karakiewicz and Jackson 1991).

Application of offshore nourishment has lowered the total cost of operations from A$18 M to A$14 M (for 10 km of beaches). Dumping of sediment in the offshore area is some 50% cheaper compared to the cost of the same volume placed on the beach above the waterline (Jackson and Tomlinson 1990).

In further stages of the design (not considered in the previous section), model simulations may also provide estimations for follow-up beach-maintenance operations comprising the frequency of future nourishments and recommended surveys of the protected area.

Model predictions may also estimate the long-term local coastal response to increased wave activity coupled with rising sea levels such as that caused by global change.

Our work answers the call for 'soft' protection methods that would enable the sustained survival of beaches in situations where application of other measures failed or, where alternative nourishment of the visible beach is no longer possible.

At present, we are using a simple 2D model that has serious limitations. A three-dimensional model allowing for some effects of wave breaking and longshore current activity in sediment dispersal is now under development.

5. REFERENCES

Amos, C.L. and O.C. Nadeau. 1988. Surficial sediments of the outer banks, Scotian Shelf, Canada. *J. Earth Sci.*, 25(12): 1923-1944.

Boczar-Karakiewicz, B., J.L. Bona and D.L. Cohen. 1987. Interaction of shallow water waves and bottom topography. In: *Dynamical Problems in Continuum Physics. Mathematics and its Applications (ed. J.L. Bona).* Springer Verlag, New York, 4: 131–176.

Boczar-Karakiewicz, B., J.L. Bona and L.A. Jackson. 1993. On 'soft' protection of sandy beaches in wave-dominated environments. *Int. Conf. Hydro-Science & Eng.*, ASCE, Washington (ed. S.S. Wang), Vol. 1: 1689-1696.

Boczar-Karakiewicz, B. and R.G.D. Davidson-Arnott. 1987. Nearshore bar formation by non-linear wave processes – a comparison of model results and field data. *Marine Geol.*, 77: 287-304.

Boczar-Karakiewicz, B., D.L. Forbes and G. Drapeau. 1995. Nearshore bar development in the Southern Gulf of St. Lawrence. *J. Waterways, Port, Coast. & Ocean Eng.*, ASCE, 129: 49–60.

Boczar-Karakiewicz, B., P. Hill and W. Romańczyk. 1995. Sand bar formation on an Arctic coast: Beaufort Sea, Canada. *Proc. Can. Coast. Conf.*, Dartmouth, Vol. 1: 58-65.

Boczar-Karakiewicz, B. and L.A. Jackson. 1990. The analysis and role of bars on the protection of a beach system, Gold Coast, Queensland, Australia. *Proc. 22th Int. Conf. Coast. Eng.*, Delft, The Netherlands: 2265-2278.

Boczar-Karakiewicz, B. and L.A. Jackson. 1991. Beach dynamics and protection measures in the Gold Coast area, Australia. *Proc. 10th Australian Conf. Coast and Ocean Eng.*, Auckland, New Zealand, Water Quality Center Publication No. 21: 417-422.

Bona, J.L. and M. Chen. 1997. A Boussinesq system for two-way propagation of nonlinear dispersive waves. *Phil. Trans. Roy. Soc. London* (to be submitted)

Chapalain, G. and B. Boczar-Karakiewicz. 1992. Modelling of hydrodynamics and sedimentary processes related to unbroken progressive shallow water waves. *J. Coast. Res.*, 8(2): 414-441.

Chapalain, G., R. Cointe and A. Temperville. 1992. Observed and modeled resonantly interacting progressive water-waves. *Coast. Eng.*, 16: 267-300.

Chapman, D.M. 1980. Beach nourishment as a management technique. *Proc. 17th Int. Conf. Coast. Eng.*, Sydney, Australia: 86-92.

Committee on Beach Nourishment and Protection, National Research Council. 1995. Beach Nourishment and Protection. *National Academy Press*, Washington, USA.

Dean, R.G. 1996. Beach nourishment: theory & practice. *World Scientific Advanced Series & Ocean Eng.*, World Scientific Publishing Co. Pte. Ltd., Singapore.

Forbes, D.L., D. Frobel, D.E. Heffler, K. Dickie and C. Shiels. 1986. Surficial geology, sediment mobility and transport processes in the coastal zone at two sites in the southern Gulf of St. Lawrence: Pte Sapin (N.B.) and Stanhope Lane (P.E.I.). *Report C2S2-20*, Canadian Coastal Sediment Study, National Research Council, Canada. Ottawa, Canada.

Gold Coast City Council. 1990. *Coastal Engineering Works, Internal Report*, Gold Coast, Australia.

Jackson, L.A., A.W. Smith and T. Piggot. 1988. Data for coastal management in a beach zone. *Coast. Zone Manag. 88 Workshop Proceedings*, Lismore: 102-110

Jackson, L.A. and R.B. Tomlinson. 1990. Nearshore nourishment implementation, monitoring and model studies of 1.5M^3 at Kirra Beach. *Proc. 22th Int. Conf. Coast. Eng.*, Delft, The Netherlands: 2241-2264

Longuet-Higgins, M.S. 1953. Mass transport in water waves. *Phil. Trans. Roy. Soc. London*, A245: 535-581.

McGrath, B.L. and D.C. Patterson. 1972. Wave climate at Gold Coast, Qld, Australia. *Inst. Eng. Aust., 8 Div. Tech. Papers*, 13(5):1-17.

Osborne, P.D. and B. Greenwood. 1992. Frequency dependent cross-shore suspended sediment transport. 2. A barred shoreface. Marine Geol., 106: 45-51.

Restrepo, J.M. and J.L. Bona. 1997. Model for the formation of longshore structures on the continental shelf. *J. Geophys. Res.* (to be submitted)

Smith, A.W. and L.A. Jackson. 1990. The siting of beach nourishment placements. *Shore and Beach*, 58(1): 17-24.

Vellinga, P. and M. Perdomo. 1992. Global climate change and the rising challenge of the sea. *Report of the Coastal Zone Management Subgroup*, Ministry of Transport, Public Works and Water Management, Directorate General Rijkswaterstaat, The Netherlands.

Morphodynamic modelling of shoreface bars

Suzanne J.M.H. Hulscher *

Abstract

In this paper the behaviour of near-shore large-scale bars is studied using a morphological model. The model describes the interaction between the sandy sea bed and propagating weakly-dispersive, weakly-nonlinear free low-frequency waves, which are incident at the offshore boundary. A linear stability analysis is performed to study the formation and migration of bed patterns. The fastest growing sea-bed wave in this model has a wavelength close to the recurrence length; it moves slowly shorewards. These results are compared with the ridges on the North American, Atlantic shelf and show good agreement. The model shows that the bars grow due to wave-skewness differences between uphill and downhill moving surface waves and the bars migrate due to differences in the shape (peakedness and skewness) between the surface waves straight above the top and the ones straight above the trough. The bar pattern in the present analytical model agrees with the results of related numerical studies; however, in the present study the bars migrate, whereas this was impossible in the numerical studies because the boundary conditions were fixed.

Introduction

The sandy bed of many shelf seas shows a variety of submarine rhythmic sea-bed patterns. This paper focuses on a specific model for shoreface bars, a pattern having multiple ridges whose crests are oriented (nearly) parallel to the shoreline.

The fields of sand ridges in the North Atlantic shelf [*Swift et al., 1972a*] occur at a typical mean depth ($H \sim 15$ m) on a very gentle mean slope (order 0.0005) towards the coast. The mean crest line spacing in all these systems shows a variation between 1.4 km and 6.1 km. The mean ridge length is typically 10 times the ridge spacing. The height of these ridges is of the order of 4 m.

*University of Twente, Civil Engineering & Management, P.O. box 217, 7500 AE Enschede, The Netherlands.

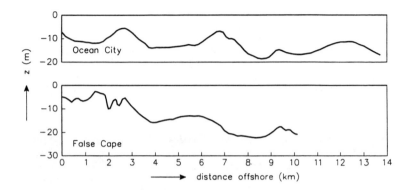

Figure 1: *Cross-section of shore-face ridge systems of False Cape, near the Virginia inner shelf region, and Ocean City area. After Duane et al. [1972]*

On the North Atlantic shelf tidal currents are medium to low. At these locations far more energy is supplied by wind waves, storm currents and the associated wave trains [*Duane et al., 1972, Swift et al, 1972b*], which are expected to play a leading role in the morphodynamical processes. Two specific fields of shoreface ridges on the North Atlantic shelf are shortly discussed now: those located near False Cape and Ocean City; representative cross-sections are presented in figure 1.

Comparisons of the depth surveys in the False Cape area [*Swift et al., 1972a*] of 1922 and 1969 indicate landward migration of 140 m for the offshore ridges system, although the authors themselves have serious doubts about the accuracy of the earliest observation. Given an average ridge spacing of 5 km, it will take 1.5 millennia to move up one complete bar. The bars near Ocean City have shown an offshore migration of 150-450 m over 80 years [*Duane et al., 1972*], indicating a period of 1 to 3 millennia for the migration of one complete bar offshore.

The numerical nonlinear morphological model of *Boczar-Karakiewicz & Bona* [1986] simulates the nonlinear evolution of multiple bar systems due to the harmonic wave-wave interaction. For low-frequency waves this model simulates successfully the bottom topography near False Cape and Ocean City. This shows that the low-frequency waves are indeed capable of forming such ridge systems, but the formation mechanism is still not understood in detail. Yet, there are some points of further discussion:

- The model does not predict bar migration, which does occur in nature

- The influence of initial and boundary conditions on the model results is unclear

- It is not clear how the bed-wave interaction process works which underlies the formation and migration of the bed waves

T(s)	H(m)	A(m)	$\lambda_w(m)$	ϵ	μ
47	12.5	0.15	518	0.02	0.15
80	15	0.15	969	0.01	0.097
80	20	0.2	$1.10 \, 10^3$	0.01	0.11

Table 1: *Model parameters for free low-frequency waves at various depths, for the North-Atlantic, North American shelf.*

The model here is studied using linear stability analysis, which has been found successful in the study of bar formation due to tide-topography interactions [*Hulscher*, 1996]. Such an analytical technique yields the initial behaviour (growth and migration) of small-amplitude bed forms. In this way the initially fastest-growing bottom pattern is found. As this investigation method explicitly yields the initial feedback between water waves due to the bottom perturbation, it will contribute to a better understanding of the underlying bar processes.

Model description

The morphological model, derived in [*Hulscher*, 1996], describes the interaction between the sandy sea bed and propagating weakly-dispersive, weakly-nonlinear free low-frequency waves, which are incident at the offshore boundary. The model consist out of

- The Boussinesq flow model in one horizontal co-ordinate x, including weak friction and weakly forcing of low-frequency waves due to the modulating character of wind waves.

- Bedload sediment transport model. The sand is transported due to the near bed streaming of the nonlinear low-frequency waves. The transport is stirred by wind waves, a slope effect in included.

- Conservation of sand on the model domain

By using the Boussinesq model is required that the dispersion, measured by $\mu = k^*H$, and the nonlinear effects of the free surface elevation, measured by $\epsilon = A/H$ are such that $\epsilon = \mathcal{O}(\mu^2)$. Herein k^* is the wave number of the low-frequency waves, A is the amplitude of the surface waves. Table 1 shows examples of the values of the small parameters ϵ and μ.

In this study the reference values $\epsilon=0.01$, $\mu=0.1$ and $H = 15$ m are chosen. As this study focuses on a length scale larger that the wavelength of the low-frequency waves, a new spatial variable χ for the morphological length scale is introduced, the ratio between these scales is chosen here as $\epsilon_L=0.01$. Furthermore a friction parameter r is involved, the value of the Jonsson friction coefficient $f_w = 0.24$ leads to the value $r = 1.0$. Sediment transport values are taken as usual, see *Hulscher* [1996].

Averaging over the period of the low frequency waves leads here to the morphological time scale yields

$$T_m = 930 \pm 10 \quad \text{yr}.$$

Stability Analysis

This paper examines whether the presence of near-shore multiple bar systems (crests parallel to the shoreline) can be associated with instabilities in the morphological system. Due to the system the nondimensional velocity of the low-frequency waves is described by its first and second harmonic component as follows:

$$u(x, \chi, t) = C_{\beta 1} \beta_1(\chi) \cos \varphi + C_{\beta 2} \left[-B_s(\chi) \sin (2\varphi) + B_c(\chi) \cos (2\varphi) \right], \quad (1)$$

in which β_1 gives the rescaled amplitude of the first harmonic component and B_s, B_c the 90° shifted and the in phase component of the second harmonic component of the water wave. The constants $C_{\beta 1, \beta 2}$ are scaling constants which are necessary to express the mathematical problem in terms of equal order. The phases are expressed in

$$\varphi(x, \chi, t) = kx - t + \phi(\chi), \quad (2)$$

in which ϕ is the (real, not continuous) phase of the first harmonic component. These surface waves are asymmetric in two ways: firstly, they can have more peaked tops than troughs (or vice versa), and, secondly, they have slightly steeper fronts than backs (or vice versa). The first effect depends on the ratio $\breve{B}_c / \breve{\beta}_1$ and is indicated as peakedness; larger positive values mean more peaked wave tops than troughs. The second effect depends on the ratio $-\breve{B}_s / \breve{\beta}_1$ and in this paper is indicated as skewness. The skewness increases for larger differences in the steepness of the front and back slopes of the water waves.

The basic state The stability analysis can be carried out by first defining a basic state which describes water waves over a sea floor without a rhythmic topography. The flow is characterised by $\psi = (\beta_1, B_s, B_c)$, the basic flow is indicated as $\breve{\psi}$. Here the basic flow corresponds to an asymmetrical (positive peaked and positive skewed), permanent surface wave which propagates over a flat sea bed, see for details *Hulscher* [1996].

Bed perturbations The next step is to introduce sine-like bed perturbations h with arbitrary wavenumbers κ in the horizontal direction χ, shorter than the wavenumber k of the incoming water waves. In the case of a small bed perturbation the solution is expanded as follows

$$(\psi, h) = (\breve{\psi}, 0) + \int \left(\tilde{\psi}, \tilde{h} \right) e^{-i\kappa\chi} d\kappa + c.c., \quad (3)$$

which means that harmonic boundary conditions have been taken. This type of boundary condition has the advantage that the free behaviour in the system can be investigated.

Growth rate   Subsequently, one studies the initial interaction between bed forms and waves and determines the growth rate ω which shows whether the bed perturbations are amplified or reduced. A basic state is said to be stable if arbitrary bed form disturbances with nonzero wavenumbers decay exponentially in time. This implies that all bed perturbations κ have negative growth rates. Conversely, if there is at least one bed perturbation κ with a positive growth rate ω, the basic state is unstable. After some algebra, see *Hulscher* [1996] the growth rate ω can be written as

$$\omega(\kappa) \;=\; i\kappa \frac{(i * D_1\kappa + D_2)}{(\lambda_1 + i\kappa)(\lambda_2 + i\kappa)(\lambda_3{'} + i\kappa)} - \kappa^2 \hat{\lambda}, \tag{4}$$

in which D_i are real coefficients. The λ_i are eigenvalues of the hydrodynamic-stability problem; λ_1 is real and negative, the remaining two are complex conjugates $\lambda_2 = \tilde{\lambda}_3$. These eigenvalues are connected with the so-called reccurence length $\kappa = |Im(\lambda_2)|$.

The first term on the right-hand side of (4) describes the complex growth (amplification and migration) of the sea-bed perturbation κ due to the feedback given by the fluid. The last term of (4) is always real and negative; it decreases quadratically in κ. This term hinders the development of small-wavenumber bed patterns.

Results

Growth rate of bed perturbations   The resulting real and imaginary parts of the growth rate $\omega(\kappa)$ as a function of the wavenumber κ are shown in figure 2.

Due to the scaling and rescaling as shown in *Hulscher* [1996] the bed mode κ is related to the dimensional sea-bed wavelength λ_m^* following:

$$\lambda_m^* = \frac{\lambda_w}{2\pi\kappa S_{12}\sqrt{2S_{22}\epsilon}} = \frac{H}{\mu\kappa S_{12}\sqrt{2S_{22}\epsilon}}, \tag{5}$$

herein, the factor $S_{12}\sqrt{2S_{22}} \simeq 2.7$ [*Hulscher*, 1996].

The imaginary part of the growth rate is related to the migration of a particular bed perturbation κ. The imaginary part of the growth rate $Im(\omega)$ denotes the number of bed waves which pass over on a certain position in the morphological time scale. This means that the sea-bed wave migration period yields $T_{migr} = T_m/Im(\omega)$.

In order to discuss the results dimensionally, the reference values as discussed in section model description have been chosen. The dimensional bed wavelength

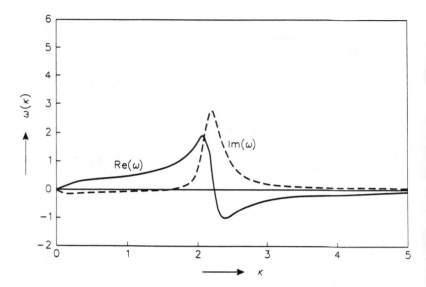

Figure 2: *Growth rate $Re(\omega)$ (solid) and $Im(\omega)$ (dashed) as a function of the wavenumber κ. Parameters are $\mu=0.10$, $\epsilon = \epsilon_L = 0.01$ and $r=1.0$. For more information, see text.*

then becomes $\lambda_m = 5.5\,\text{km}/\kappa$. This means that sea-bed patterns which have wavelengths down to 2.4 km ($\kappa \geq 2.25$) initially grow. The most unstable mode has a wavelength of $\lambda_m = 2.6\,\text{km}$ ($\kappa \simeq 2.1$). The growth rate factor is related to the morphological time scale, which means that the typical time scale for morphological changes of the fastest growing bed forms is 520 years, and even longer for larger bed forms.

Bed waves which have crest-spacing of 3.3 km or longer ($\kappa < 1.65$) show small negative values, $-0.2 \leq Im(\omega) \leq 0$, which corresponds to a migration in the seaward direction of one bed wave in a period of at least 4.5 millennia. An initially shoreward movement is found for sea-bed waves shorter than 3.3 km ($\kappa \geq 1.65$). However, note that wavelengths shorter than 2.4 km are not amplified but decay exponentially in time. The most unstable mode gives a positive value for the imaginary part, $Im(\omega) = 1.4$, which means a shoreward migration period of 670 years. Note that the mode having the smallest migration period is marginally stable: it neither grows nor decays. Note also that the fastest growing mode migrates shoreward.

The spatial and temporal characteristics of the bed forms in this model are comparable to the ridge systems on the North-American North-Atlantic shelf, as reported in *Swift et al.* [1972a]. The model shows slow migration of the bed features in either direction, which is in agreement with the observations.

However, the fastest growing mode always migrates shorewards, one order of magnitude faster than the estimates derived from the observations. Since these estimates themselves give rise to doubts, this aspect deserves further study.

Flow-topography feedback Here the shape of the surface waves is investigated, under conditions which lead to growth and migration of the perturbed bottom.

The sea-bed perturbation $\epsilon \hat{h} \cos{(\kappa \chi)}$ leads to spatially periodic changes in the ratio of the harmonic components, given by the complex values of $\tilde{\beta}_1, \tilde{B}_s, \tilde{B}_c$. As a function of the bed wavenumber κ figure 3 shows the real and the imaginary part of the perturbed harmonic components $\tilde{\beta}_1/\epsilon, \tilde{B}_s/\epsilon, \tilde{B}_c/\epsilon$. The rescaled form is chosen here as an order ϵ change in topography, it balances order 1 changes in the harmonic components, which subsequently leads to order ϵ changes in the residual velocity near the sea bed. The perturbations in the basic shape of the surface waves which are caused by the bed-topography perturbation are thus known exactly here.

Figures 2 and 3 imply that growing modes correspond to $Im(\tilde{B}_s) > 0$, whereas decaying modes show $Im(\tilde{B}_s) < 0$. Physically, $Im(\tilde{B}_s)$ corresponds to spatial differences in surface-wave skewness at the sides of the bar, such that $Im(\tilde{B}_s) > 0$ gives steeper fronts for uphill moving waves than for the front which move downhill. So in this model a growing bar has steeper surface wave-fronts on the seaside than on the shore side. If these surface-wave differences produces bars in the real world will, this will indicate whether the presently studied mechanism works and is therefore an interesting topic to investigate.

From the previous considerations it is clear that the real part of the flow perturbations is generated indirectly due to friction. Figures 2 and 3 imply that in the case of seaward migration $Re(\tilde{\beta}_1) < 0$ and $Re(\tilde{B}_c) > 0$, which means more peaked waves over the crests than in the troughs. In the case of shoreward migration of the bars, this effect is dominated by $Re(\tilde{B}_s) > 0$, which means steeper fronts at the crests than in the troughs. In this model therefore the migration orientation of the bars is determined by differences in surface-wave peakedness (seaward) or skewness (shoreward) between top and trough. Again it still has to be verified, whether these results agree with morphological behaviour in reality.

Comparison of results with related models In order to compare the results with those of the numerical evolution model of *Boczar-Karakiewicz & Bona* [1986] special choices have to be made. The model in the present paper is limited to small bed amplitudes and thus it is only comparable to the earliest stage of the evolution of the finite amplitude bars in the numerical model. Furthermore, the small parameters in the present model $\epsilon, \epsilon_L, \mu$ have to be used differently, in order to create similar circumstances here, ϵ in equation (5) has to be replaced by $\mu/(2\pi)$ [*Hulscher*, 1996]. Under these conditions equation (5) gives the following

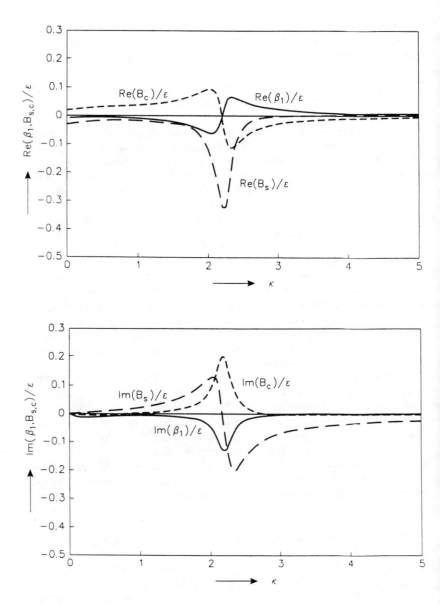

Figure 3: *The (a) real and (b) imaginary part of the wave components*
$\tilde{\beta}_1/\epsilon, \tilde{B}_s/\epsilon, \tilde{B}_c/\epsilon$ *as function of the bed wave number* κ. *For parameters see text.*

formula to estimate the dimensional wavelength of the most unstable bar κ_f beyond critical conditions

$$\lambda_m^* \simeq \frac{\lambda_w}{\kappa_f S_{12}\sqrt{2S_{22}}\mu} = \frac{2\pi H}{\kappa_f S_{12}\sqrt{2S_{22}}\mu^2}. \tag{6}$$

From the previous results it is estimated that $\kappa_f \simeq 2.1 \pm 0.1$.

Boczar-Karakiewicz & Bona [1986] estimated the relative wave amplitudes being $\epsilon = 0.05$; the typical wavelength of the incoming low-frequency waves at False Cape and Ocean City are $\lambda_w^* = 2.5$ km; which mean periods of some $T = 80$ s. Substitution of these values into equation (6), taking realistic depth-values, that is between $H=30$ m and $H=15$ m, yields bars with wavelengths between 1.7 and 11.3 km. This agrees with the model results as presented in *Boczar-Karakiewicz & Bona* [1986], in which the recurrence length of the surface waves is of the order of magnitude of the bar spacing.

However, some critical remarks are called for here. The previous derivation shows the sensitivity of the bar spacing to the wavelength of the incoming waves: for every sea-bed wavelength a suitable low-frequency wave can be selected which is able to form it. A model result which is so dependent on the (variable) extrinsic conditions is not satisfactory.

A second comment concerns the boundary conditions. The present analysis has shown that migration of bars, either seawards or shorewards, occurs naturally in the present morphological model. As the sea bed at the seaward side in the *Boczar-Karakiewics*-model is fixed, the migration of the bed features is simply not able to occur. As the migration and the growth of sea-bed features are both due to divergences in the sediment transport and are therefore closely connected processes, a physically realistic simulation of bars requires that movement is permitted.

Discussion/Conclusion

This study confirms the conclusion of *Boczar-Karakiewicz & Bona* [1986] that the ridge systems on the South Eastern North American Atlantic Shelf [*Swift et al.*, 1972a] may be formed by weakly dispersive and weakly nonlinear low-frequency waves. The most preferred bed wavelength equals the recurrence length of the surface wave amplitudes. The present study quantifies the strong relation between the sea-bed wavelength and the wavelength of the incoming surface waves. The present analysis shows that both onshore as offshore movement of bars under conditions of an overall onshore oriented sediment transport are intrinsic processes in this system. Because the solution technique is analytical, the model enables detailed investigations to be made in order to discover the underlying processes.

According to this model the topographically induced growth of sand bars is connected to wave-skewness differences at the sides of the bar, in such a way that the propagating surface waves show steeper fronts when going uphill than when going downhill. Bar migration is connected to the differences in wave shapes between the surface waves at top and trough. In this model more peaked waves at the crests cause bars to move seawards; steeper fronts at the crests coincide shoreward migration. It would be interesting to find out whether such wave-shape differences actually occur in nature. It may be that the validity of the processes established here even extends to different types of waves (wind- or storm-waves).

The various papers on the numerical evolution model, the basis of the morphological model in this paper [Boczar-Karakiewicz et al, 1995 and references herein] show striking agreement between numerical simulations and observations of barred profiles. However, the strong relation between the resulting sea-bed wavelength and the low-frequency wavelength tends to undermine evidence for the validity of the model. Moreover, a fixed sea bed at the seaward boundary keeps the bars from moving and so the inherent sea-bed behaviour in this model cannot develop.

Acknowledgements This paper is based on work in the PACE-project, in the framework of the EU-sponsored Marine Science and Technology Programme (MAST-III), under contract no. MAS3-CT95-0002. It was co-sponsored jointly by Delft Hydraulics and RIKZ, within the framework of the Netherlands Centre for Coastal Research. The work was started during the author's stay at the INRS-Océanologie Rimouski, Canada. The author thanks Prof. B. Boczar-Karakiewicz for introducing her to her morphological evolution model. The model has been a continuous source of inspiration for the study which has led to this paper.

References

Boczar-Karakiewicz, B. and Bona, J.L., Wave-dominated shelves: a model of sand ridge formation by progressive infragravity waves, in *Shelf sands and sandstones*, edited by R.J. Knight and J.R. McLean, pp 163-179, Can. Soc. Petr. Geol. Memoir II, 1986.

Boczar-Karakiewicz, B., Forbes D.L. and Drapeau, G., Nearshore bar development in southern gulf of St. Lawrence, *J. of Waterw., Port, Coast. and Oc. Engineering*, 121, 1, 49-60, 1995.

Duane, D.B., Field, M.E., Meisburger, E.P., Swift, D.J.P. and Williams, S.J., Linear shoals on the Atlantic inner continental shelf, Florida to Long Island, in *Shelf sediment transport: process and pattern*, edited by D.J.P. Swift, D.B. Duane and O.H. Pilkey, pp 447-498, Dowden, Hutchinson & Ross, Inc., 1972.

Hulscher, S.J.M.H., Formation an migration of large-scale rhythmic sea-bed patterns, PhD-Thesis IMAU, Utrecht University, 1996.

Swift, D.J.P., Kofoed, J.W., Saulsbury, F.P., Sears, P., Holocene evolution of the shelf surface, central and Southern Atlantic Shelf of North America, in *Shelf sediment transport: process and pattern*, edited by D.J.P. Swift, D.B. Duane and O.H. Pilkey, pp 499-574, Dowden, Hutchinson & Ross, Inc., 1972.

MODELING OF SEASONAL VARIATIONS BY CROSS-SHORE TRANSPORT USING ONE-LINE COMPATIBLE METHODS

Hans Hanson[1], Magnus Larson[1], Nicholas C Kraus[2] and Michele Capobianco[3]

ABSTRACT: A procedure for representing seasonal variations in shoreline position as caused by on-offshore sediment transport is presented. The study was based on an analysis of an 11-year long time series of waves and shoreline location obtained at Duck, North Carolina. Preliminary calculations show that the method may be used for analyzing the seasonal behavior of a beach. At the same time, the impact of individual storms are not represented to any significant degree.

INTRODUCTION

Over the past decade, one-line shoreline response models have demonstrated their predictive capabilities in numerous projects (Hanson *et al* 1988), and their usefulness as a planning and design tool has been confirmed. The shoreline-change model predicts shoreline position changes that occur over a period of several months to several years. Changes in shoreline position are assumed to be produced by spatial and temporal differences in the **longshore** sand transport rate (Hanson 1989, Hanson and Kraus 1989). The model is best suited to situations where there is a systematic trend in long-term change in shoreline position, such as recession down drift of a groin. Cross-shore transport effects, such as storm-induced erosion and cyclical movement of shoreline position as associated with seasonal variation in wave climate, are assumed to cancel over a long simulation period or are accounted for through external calculation.

Cross-shore models, on the other hand, predict beach change as a result of cross-shore transport produced by storms (Larson and Kraus 1989). This type of models is

[1]Assoc. Prof., Dept. of Water Res. Engrg., Univ. of Lund, S-22100 Lund, Sweden. [2]Res. Scientist, U.S.A.E. Waterws. Exp. Sta., Coastal and Hydraulics Lab., 3909 Halls Ferry Road, Vicksburg, MS 39180-6199, USA.
[3]Senior Systems Engineer , Tecnomare SpA, 3584 San Marco, 30124 Venezia, Italy.

simplified by omitting longshore transport processes, i.e., constancy in longshore processes is assumed. In principle, the two types of models could be combined to obtain both long- and short-time changes in shoreline position. Such attempts are being done (e.g. Larson *et al.* 1990), but these models are still in a research stage and require significant computer resources and labor to be used in practical applications. Previous attempts have been made to combine longshore and cross-shore transport in a schematic way (Bakker 1969, Perlin and Dean 1979), but these approaches have never found their way into engineering practice.

Thus, it seems that there are no models available for simulating long-term cross-shore related coastal evolution for engineering applications. At the same time, it is recognized that seasonal variations can play a significant role in the interpretation of long-term shoreline evolution. Therefore, it is important to be able to represent the main features of cross-shore related long-term coastal evolution.

In this work, a standard one-line response model was extended to include cross-shore processes. A precondition for the formulation of the cross-shore contribution was that it should be compatible with the one-line formulation in terms of independent variables and level of sophistication. Due to the complexity of profile response to variations in wave height and water level, a relatively simple model cannot be expected to represent the impact of individual storms, but rather to capture the more long-term effects. The objective of this study was, thus, to represent seasonal variation due to cross-shore transport using a one-line model.

FIELD DATA

As an application of the proposed procedure, simultaneously collected data on waves and beach profiles from the US Army Field Research Facility at Duck, North Carolina (Figure 1) (Howd and Birkemeier 1987, Lee and Birkemeier 1993) were analyzed to investigate the relationship between the incident waves and the seasonal shoreline variations over a longer time period. The data set comprised shoreline positions extracted from surveys taken biweekly in profile lines 58, 62, 188, and 190 (Figure 2) during 11 years. Spectral wave properties (significant height and peak period), recorded at least every 6 hours, were available for the same period. The waves were measured in 18-m water depth, here considered as deep water. The beach at Duck is one of the most well-documented field sites in the world

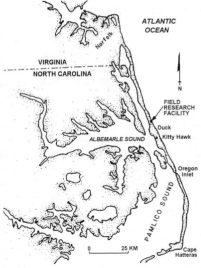

Figure 1. Field site at US Army Field Research Facility at Duck, North Carolina.

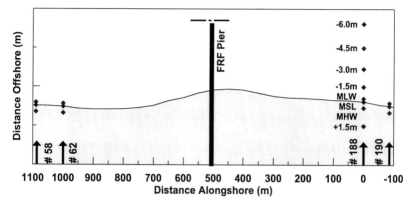

Figure 2: Profile lines and extracted contours at Duck.

with collected time series on waves and profiles that are unique in their length and quality. Thus, it was a natural choice to employ data from Duck in the development and validation of the cross-shore transport algorithm. However, coarse material is often found around the shoreline at Duck (Larson 1991) creating a steep beach face and an armoring effect that could significantly reduce the shoreline response to changes in the wave conditions. Because of this the beach response at Duck might not be as well-behaved as on other beaches. In spite of this, the Duck data was chosen.

The long-term variation of different contour lines was extracted by linear interpolation from the measured profiles. In all lines, MSL (+0.08 m NGVD), MHW (+0.62 m NGVD), and MLW (-0.41 m NGVD) were analyzed. In addition, for line 188, the temporal variation of the contour lines from MSL + 1.5 m to MSL - 6.0 m were extracted at every 1.5 m (Figure 2). As an illustration, Figure 3 shows temporal variation of the obtained shoreline position in profile line 188 over the 11 years. In this study, the long-term trend is assumed to be associated with longshore processes and will be subtracted before performing further analyzes.

In order to investigate the temporal scale in the shoreline variation, an FFT analysis was performed on the data series after the linear trend was removed. Figure 4 (solid line) shows that there is a strong annual variation that most likely is associated with seasonal variation in the wave climate. There is also a significant low-frequency component associated with a non-linear long-term trend.

Figure 5 shows the shoreline change relative to the linear trend at MSL in all four profile lines. Lines on the same side of the FRF pier show very similar behavior ($r_{188,190}$ = 0.87, $r_{58,62}$ = 0.82). However, lines on opposite sides show much less correlated behavior ($r_{188,58}$ = 0.50, $r_{190,62}$ = 0.34).

As an illustration of the profile behavior in line 188, Figure 6 shows the relative change of MSL together with that of the MSL - 3.0 m, which is a good indicator of the location the nearshore bar. Both contour lines were smoothed using a 5-point moving

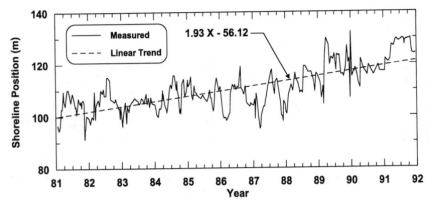

Figure 3: Temporal variation of MSL in profile line 188.

average for easier viewing. During most of the period, the two contours show anti-symmetric behavior, indicating that the main part of the changes is associated with cross-shore processes.

PROCEDURE

Transport magnitude

Because the one-line model does not provide any information about the actual shape of the bottom profile, the cross-shore transport must be calculated using a relation that is independent of the profile shape. Numerous formulae for the cross-shore sand transport rate may be found in the literature (Horikawa 1988, p. 196 ff.). Many of these

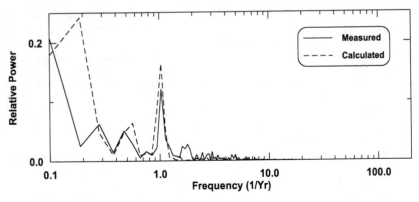

Figure 4: FFT analysis of periodicity in shoreline fluctuations.

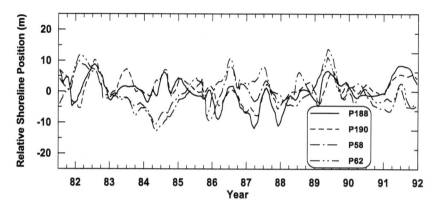

Figure 5: Temporal variation of MSL relative to linear trend in all profile lines.

relations may be written in the generic form:

$$\frac{q_o}{wd} = K_q \, (\Psi - \Psi_c)^a \tag{1}$$

where q_0 = cross-shore transport rate per unit width, w = sediment fall speed, d = grain size, K_q = transport coefficient, Ψ = Shield parameter, Ψ_c = critical Shield parameter, and a = exponent. The value of K_q varies quite significantly between the different formulas. In the present study, the value of this coefficient will be determined in a calibration procedure. The value of the exponent a varies between 1 and 3 in the formulae reported in Horikawa (1988). Here the value a = 1.5, as proposed by Watanabe (1982), will be

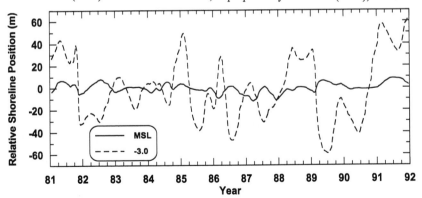

Figure 6. Temporal variation of MSL and the -3 m contour in profile line 188.

used. For simplicity, the value of Ψ_c is set to zero. Strictly, Eq. (1) represents a point value, in this case at wave breaking. To be applicable to the one-line concept, the transport rate is assumed to be uniformly distributed across the surf zone.

The Shield parameter may be written on the form

$$\Psi = \frac{f_w \, u^2}{2sgd} = K_\psi \frac{u^2}{gd} \tag{2}$$

where f_w = Jonsson's (1966) wave friction factor, u = maximum horizontal wave-induced fluid velocity at the bottom, s = sediment specific density in water, g = acceleration due to gravity, and K_ψ = friction coefficient. In addition, at wave breaking, the maximum fluid velocity may be approximated as $u^2 = g \, H_b \approx g \, H_o$ (Kaminsky and Kraus 1994), where H_b = breaking wave height and H_o = deep water wave height. Thus, the cross-shore sediment transport rate per unit length alongshore, q_o, here regarded as a potential rate, may be calculated, with $K = K_q \, K_\psi$:

$$q_o = K_q K_\Psi w d \left(\frac{u^2}{gd} \right)^{3/2} = K w d \left(\frac{H_o}{d} \right)^{3/2} = K w \left(\frac{H_o^3}{d} \right)^{1/2} \tag{3}$$

Transport direction

Equation (3) gives, however, only the potential magnitude of the transport rate, not the direction. Kraus et al. (1991) examined the capability of several criteria for predicting the direction of cross-shore sediment transport. One of these criteria was based on the non-dimensional fall speed $N_o = H_o / (wT)$, also known as the Dean number. The study identified a critical non-dimensional fall speed, $N_c = 3.2$, for which the following limiting values were found:

$$N_o < 0.75 \, N_c = 2.4 \quad \Rightarrow \quad \text{onshore transport}$$
$$\tag{4}$$
$$N_o > 1.25 \, N_c = 4.0 \quad \Rightarrow \quad \text{offshore transport}$$

where T = wave peak period. For values between the two limits, the direction of transport was ambiguous.

In the present study, the same criterion is used, although the critical value, N_c, that separates onshore from offshore transport, will be determined in the calibration procedure. Based on the wave time series, the potential cross-shore sediment transport rate, q_o, is calculated at each time step according to Eq. (3). At the critical value $N_o = N_c$ the cross-shore transport rate is set to zero. From here, the actual transport q is assumed to decrease linearly to $-q_o$ at $N_o = 0.75 \, N_c$, where the minus sign indicates offshore transport or erosion. For lower values of N_o, q remains at $-q_o$. On the opposite side of N_c,

q is assumed to increase linearly to $+q_o$ at $N_o = 1.25\ N_c$, where the plus sign indicates onshore transport or accretion. For higher values of N_o, q remains at $+q_o$.

Shoreline change

The shoreline location y is calculated based on the continuity relation:

$$\frac{\partial y}{\partial t} + \frac{1}{D_c}\left(\frac{\partial Q}{\partial x} - q\right) = 0 \tag{5}$$

where t is time, Q is the longshore sediment transport rate, and x is the alongshore coordinate. By assuming no longshore transport gradients ($\partial Q/\partial x = 0$), the shoreline change Δy during a time step Δt is given by:

$$\Delta y = \pm\ q\ \frac{\Delta t}{D_c} \tag{6}$$

where a positive sign corresponds to onshore transport. The best fit value of N_c was determined to be 3.8, which is in good agreement with the data presented in Kraus et al. (1991). Figure 7 shows a comparison between the calculated shoreline variation and the actual change relative to the linear trend filtered with the moving average. Although there are - as expected - some differences, the general behavior of the shoreline seems to be quite well reproduced. Through the main part of the simulation period the two curves show similar behavior although with some phase lag. However, during the period 1989-90 there is quite poor agreement between calculated and measured shoreline movement. Figure 8 shows an enlargement for this period, where the temporal variation

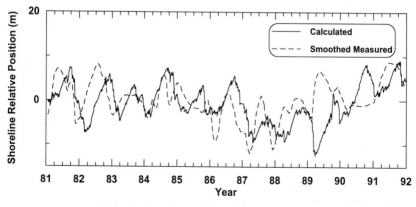

Figure 7. Comparison between calculated and measured shorelines in profile line 188.

of the Dean number was added to the shoreline evolution. The calculated shoreline change responds quite closely to the variation in the Dean number, where erosion takes place as soon as the Dean number exceeds the critical value. Figure 6 shows that not only the shoreline (MSL) but also the nearshore bar (-3.0 m contour) shows an accretive behavior in the beginning of 1989, as the bar moves onshore when the shoreline moves off-shore. It is not known at this point why the measured shoreline does not respond similarly to the calculated one.

As for the actual shoreline change, the calculated shoreline variation was analyzed by FFT. The temporal behavior of the oscillations is shown in Figure 4. Similar to the measured variations the calculated changes show a clear annual or seasonal component. They also have a low-frequency (long-term) component, whereas there are virtually no higher frequencies represented. Thus, the calculations show that the method may be used for analyzing the seasonal behavior of a beach fill. However, the impact of individual storms is not represented to any significant degree.

MODEL SENSITIVITY

The classical one line concept have proven to be amazingly robust and stable with respect to variations in input parameter values (Hanson and Kraus 1989). Thus, the variations in simulated results are smaller than the variations in the input parameters. However, cross-shore models are typically more sensitive to changes in the input conditions or the parameter values used. This is often related to the fact that the profile evolution is dependent on critical limits (or similar formulations) that determine the direction of sediment transport. Thus, small changes may cause the transport direction to change inducing offshore transport and bar formation or the opposite, onshore transport and berm formation. A small error in the predictions close to the critical limit will produce a response that is contrary to the one observed. This may be one reason for the disagreement between measured and calculated changes during the period 1989-90.

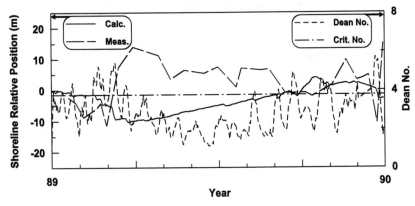

Figure 8. Shoreline change and Dean number variation in profile line 188.

CONCLUSION

The present analysis showed that it is possible to reproduce seasonal cross-shore transport related shoreline variation using a simple one-line model approach. Similar to measured shoreline changes at Duck, N.C., model calculations over 20 years showed clear seasonal variations. FFT analysis of measured and calculated shoreline changes also showed strong similarities in long-term variations. As expected, however, the impact of individual storms that are present in the measurements, are not represented in the simulated shoreline behavior to any significant degree.

ACKNOWLEDGMENTS

The research in this paper was partly conducted (HH, ML, and MC) under the PACE project of the European Union's MAST research program (Contract No. MAS3-CT95-0002) and partly (NCK) under the Coastal Inlets Research Program of the US Army Corps of Engineers. Permission was granted to NCK by the Office of Chief of Engineers to publish this information.

REFERENCES

Bakker, W.T. 1969. "The Dynamics of a Coast with a Groin System," *Proc. 11th Coast. Engrg. Conf.,* ASCE, 492-517.

Jonsson, I.G. 1966. "Wave Boundary Layers and Friction Factors," *Proc. 11th Coast. Engrg. Conf.,* ASCE, 492-517.

Hanson, H. 1989. "GENESIS - A Generalized Shoreline Change Numerical Model," *Journal of Coastal Research,* Vol 5(1), pp. 1-27.

Hanson, H., Gravens, M. B., and Kraus, N. C. 1988. "Prototype applications of a generalized shoreline change numerical model," *Proc. 21st Coast. Engrg. Conf.,* ASCE, 1265-1279.

Hanson, H. and Kraus, N. C. 1989. "GENESIS - Generalized model for simulating shoreline change," Vol. 1: Reference Manual and Users Guide, *Tech. Rep. CERC-89-19,* US Army Eng. Wtrwy. Exp. Sta., Coast. Engrg. Res. Ctr.

Howd, P.A. and Birkemeier, W.A. 1987. "Beach and Nearshore Survey Data: 1981-1984 CERC Field Research Facility," *Tech. Rep. CERC-87-9,* US Army Eng. Wtrwy. Exp. Sta., Coast. Engrg. Res. Ctr.

Horikawa, K. 1988. *"Nearshore Dynamics and Coastal Processes,"* Univ. of Tokyo Press, Tokyo.

Kaminsky, G. and Kraus, N.C. 1994. "Evaluation of Depth-Limited Breaking Wave Criteria," *Proc. Ocean Wave Measurement and Analysis,* ASCE, 180-193.

Kraus, N.C., Larson, M., and Kriebel, D.L. 1991. "Evaluation of Beach Erosion and Accretion Predictors.," *Proc. Coastal Sediments'91,* ASCE, 572-587.

Larson, M. and Kraus, N. C. 1989. "SBEACH: Numerical Model for Simulating Storm-Induced Beach Change," Report 1: Empirical Foundation and Model Development, *Tech. Rep. CERC-89-9,* US Army Eng. Wtrwy. Exp. Sta., Coast. Engrg. Res. Ctr.

Larson, M., Kraus, N.C., and Hanson, H. 1990. "Decoupled Numerical Model of Three-Dimensional Beach Change," *Proc. 22nd Int. Conf. on Coastal Eng.*, ASCE. pp. 2173-2185.

Lee, G. and Birkemeier, W.A. 1987. "Beach and Nearshore Survey Data: 1985-1991 CERC Field Research Facility," *Tech. Rep. CERC-93-3*, US Army Eng. Wtrwy. Exp. Sta., Coast. Engrg. Res. Ctr.

Perlin, M. and Dean, R.G. 1979. "Prediction of Beach Planforms with Littoral Controls," *Proc. 16th Coast. Engrg. Conf.*, ASCE, 1818-1838.

A Behaviour Oriented Model for the Evaluation of Long-Term Lagoon-Coastal Dynamic Interaction along the Po River Delta

Marcel J.F.Stive[1,2], Piero Ruol[3], Michele Capobianco[4] and Maarten Buijsman[1,2]

Abstract

We are further developing an aggregated-scale behaviour-model (Stive et al., 1996) describing the interaction between a tidal lagoon or basin and its adjacent coastal environment, including the ebb-tidal delta. While the model is based on general sediment transport equations, the behavioural aspect concerns the a priory assumption that the equilibrium state of each element of the system is known or at least can be defined. Here, we are able to rely on a vast body of reported experience for each of these system elements. The importance of the present work lies in the application of the model to the microtidal environment of the Po Delta, while the original model was validated for the mesotidal environment of the Wadden Sea.

Introduction

The sediment accommodation space that is provided by relative sea level rise in tidal basins and lagoons causes, in general, a net influx of sediments. It is generally assumed that this influx goes to the cost of the coastal stretches adjacent to the tidal entrance. Also, sediment may be delivered by the ebb tidal delta, which is, however, considered as a temporary source.

The idea behind these lines of thinking is strongly derived from the empirical findings (since O'Brien) that the morphological state of tidal basin elements (flats,

---

[1] Delft Hydraulics, PO Box 177, 2600 MH Delft, The Netherlands
[2] Netherlands Centre for Coastal Research (NCK), Dept. of Civil Engineering, Delft Univ. of Technology, PO Box 5048, 2600 GA Delft, The Netherlands
[3] University of Padova, Istituto di Costruzioni Marittime e di Geotecnica, Via Ognissanti, 39 - 35129 Padova, Italy
[4] Tecnomare SpA, 3584 San Marco, 30124 Venezia, Italy

channels, ebb-tidal deltas) seems to be in quasi-static equilibrium with the tidal prism. However, for some time now there has been the notion that morphological interactions take place between the adjacent coastal stretches, the tidal basins and their submerged deltas at different spatial and temporal scales, according to intrinsic dynamics and to external forcing factors. We are interested in developing approaches to account for these interactions on decadal and on centennial time-scales.

The ASMITA approach

A first modelling approach (ASMITA - Aggregated Scale Morphological Interaction between a Tidal basin and the Adjacent coast) has been developed, focusing on the "residual" interaction occurring on long term scales. Because of the time-scales of interest, the geophysical elements of the coastal fringe are considered at an aggregated scale, e.g. single-inlet lagoons are schematised into two or at most three spatial units such as channel area, and (high and low) flats area. The new element is that we introduce the nonlinear dynamic interaction between the lagoon elements, the submerged delta and the adjacent coastal sections.

These principal elements are morphodynamic systems themselves, consisting of interacting sand bodies and water masses. They can be considered as being mutually linked by sediment "conveyors", one of which goes through the tidal entrance.

The approach is an aggregation and an extension of a model formulation for tidal basins (Wang et al., 1996). The aggregation concerns the fact that we characterise the system elements by only one state variable (viz. a total wet or dry volume (V)). The extension concerns the incorporation of formulations for the ebb tidal delta and directly adjacent coast as well, without modifying the basic concepts.

We assume that the morphological state of each of these elements can be described by a small number of *aggregate state variables* (e.g. the total sand volume of the outer delta, or the total water volume below mean sea level and the area of the intertidal zone in the lagoon basin). The reason is that we want to use the model on very simple and representative variables for which we can "control" the accuracy on the long term. Conservation of sediment requires sediment exchange with other elements, or with the open sea, if these state variables change.

The most important hypothesis used in the model concept is that an equilibrium state can be defined for each element depending on the hydrodynamic condition, e.g., tidal volume (P) and tidal range (H). An empirical relation is required for each element to define the morphological equilibrium state. These relations are derived from literature (cf. Eysink, 1990).

A summary of the model equations is as follows. For the equilibrium volume we assume:

$$V_e = f(P, H) \tag{1}$$

A disturbance of these volumes causes changes in the local equilibrium concentration (c_e), which depends on the actual volume V. The equation determining the local equilibrium concentration is equal to:

$$c_e = c_E \cdot \left(\frac{V_e}{V}\right)^n \tag{2}$$

The overall concentration c_E, which is the long-term concentration of the system in equilibrium, also applies to the concentration at the boundaries and is assumed constant. The coefficient n equals 2, based on the usual assumption in sediment transport formulas that the sediment transport is proportional to the third power of the flow velocity. The difference between the actual concentration (c) and the local equilibrium concentration results in changes in the morphology. Morphological changes are governed by:

$$\frac{\partial V}{\partial t} = w_s \cdot A \cdot (c_e - c) \tag{3}$$

Here w_s equals the vertical exchange coefficient and A the horizontal area. As a result of the disturbance, sediment is exchanged between the various elements. Considering the neighbouring elements "a" and "b", the diffusion based sediment exchange yields:

$$\delta_{ab} \cdot (c_a - c_b) = w_s \cdot A_a \cdot (c_{ae} - c_a) \tag{4}$$

The exchange of sediment continues until the overall equilibrium concentration and the equilibrium volumes are reached again. Every element has a morphological time-scale, depending on the magnitude of the disturbance. For a single element the morphological time-scale τ equals:

$$\tau = \frac{1}{c_E \cdot n} \left(\frac{V_e}{w_s \cdot A} + \frac{V_e}{\delta}\right) \tag{5}$$

For a more complete description we refer to Stive et al. (1996).

Simplified "process-based" computation of the equilibrium concentration

The problem of definition of equilibria for the internal lagoon model can be translated into the definition of one "equilibrium concentration", one single empirical constant, with reference to the inlet. This is particularly important in microtidal environments like the Po Delta where both wave climate and tidal flux play a significant role, on the long term, in the determination of its values.

Similarly to the concept of equilibrium profile in longshore sediment transport modelling, in the modelling of the evolution of the internal lagoon we use the concept of "equilibrium concentration". Such a concept was also used extensively, even though still in a very experimental way, for the morphological evolution of the inlets in the Lagoon of Venice (Di Silvio, 1989). The equilibrium concentration represents the fundamental link between hydrodynamics, sediment transport and morphology at the entrance of the lagoon. Conceptually, c_E is the average annual concentration induced by waves and currents. In principle it would be possible to obtain a formulation for c_E by integrating over a long period of time any sediment transport formula, and obtaining something like:

$$c_E = c_E \text{ (local wave climate, tidal currents, water depth, grain diameter)} \quad (6)$$

Further it should be noted that if a constant c_E is used[5], than c_E by itself does not have any influence on the final equilibrium state of the system but it is an important parameter determining the time scale of the morphological development together with the dispersion coefficients and the fall velocity. Therefore c_E could be used as one of the calibration parameters in the model.

Model application results

The ASMITA model was applied to two lagoons of the Po River Delta (Italy): the Scardovari and Barbamarco lagoon. In the Po River Delta area a large number of tidally flushed lagoons is present (the tidal range is about 1.0m in this part of the Adriatic Sea) and there is an increasing need of predicting their evolution, focusing on the impact of management measures. The most important characteristics of the lagoon inlets are presented in Table 1.

Lagoon	area at MLW (km^2)	fishery valley (km^2)	n. of inlets	inlet name	inlet section (m^2)	exchanged volume ($m^3 * 10^6$)	Q_{max} in (m^3/s)	Q_{max} out (m^3/s)
Caleri	10.5	30.0	2	Pozzatini	284	2.95	146	128
				P. Caleri	605	6.27	313	272
La Vallona	11.0	37.0	2	P. Levante	542	16.90	610	610
				Bocchetta	104	1.42	83	63
Barbamarco	6.9	18.0	2	Nord	330	totally	68	47
				Sud	203	7.4	250	150
Scardovari	29.0	0.0	1	---	3295	15.7	1625	1250

Tab. 1. *Main characteristics of the most important Po River Delta lagoons*

We present the details of the application to the lagoons Barbamarco and Scardovari for which the inner channel network and the recent morphological evolution of the inlets were derived from bathymetric information. In Table 2 we summarise the characteristics for the Scardovari and Barbamarco lagoon.

---

[5] This is not a logical incongruence; simply, as we will further state in the following, the equilibrium concentration can be made dependent on variables that are subject to change.

	Scardovari	*Barbamarco*
BASIN DATA:		
lagoon area at MWL A_b (km^2)	29.0	6.74
exchanged volume V per tidal cycle (tidal prism or flood volume) (10^6 m^3)	15.7	7.4
tidal range R_t (m)	1.28	1.34
CHANNEL DATA:		
total length (km)	18.0	7.5
average depth (m)	1.75	3
mean width (m)	150	79
channel area A_{ch} (km^2)	2.7	0.59
channel volume V_{ch} (10^6 m^3)	4.7	1.78
FLAT DATA:		
lagoon area at HWL (km^2)	30.0	7
lagoon area at LWL (km^2)	28.0	6.5
tidal flat area (km^2)	27.3	6.38
tidal flat volume (above LWL) (10^6 m^3)	22.7	1.94
DELTA DATA:		
active base over a slope of 1:100 (m)	7.0	7.0
longshore extension (km)	2.0	0.6
delta volume V_d (10^6 m^3)	4.9	1.47
COAST DATA:		
active base over 500 m cross-shore (m)	5.0	5.0
longshore extension (km)	2.0	2.0
coast volume (10^6 m^3)	7.5	7.5

Tab. 2. *Characteristics of Scardovari and Barbamarco lagoon*

Applying the Wadden Sea relations for Barbamarco we acquire the following equilibrium data:

- for the equilibrium channel volume: $V_{ch} = 65*10^{-6}*V^{1.5} = 1.31*10^6$ m^3;
- for the equilibrium delta volume: $V_d = 6.57*10^{-3}*V^{1.23} = 1.85*10^6$ m^3.

Applying the Wadden Sea relations for Scardovari we acquire the following equilibrium data:

- for the equilibrium channel volume: $V_{ch} = 65*10^{-6}*V^{1.5} = 4.04*10^6$ m^3;
- for the equilibrium delta volume: $V_d = 6.57*10^{-3}*V^{1.23} = 4.66*10^6$ m^3.

We thus can conclude that the Wadden Sea relations may well be applicable to the Po Delta region.

A plan view of all the lagoons of the Po Delta is shown in Fig.1.

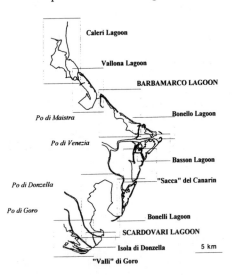

Fig.1. *Plan view of the Po River Delta*

Lacking validated information about the equilibrium flat volume we assume that the flats are in equilibrium in the case of the Barbamarco lagoon. Applying the equilibrium Wadden Sea relations we find the following model behaviour (Fig.2):

Barbamarco

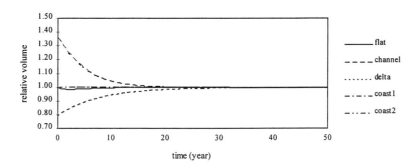

Fig.2. *Applying the Wadden Sea relations to the Barbamarco lagoon*

The volumes are taken relative to the proposed equilibrium volumes. Because the lagoon is small, the responses of the system are very rapid. The channel and the delta are subsequently too large and too small compared to their equilibrium states.

It is management practice to dredge the channels for navigation and flushing purpose. To maintain its initial volume we adapt a scenario in which the channel is dredged with an interval of about 4.1 years (Fig.3).

Barbamarco

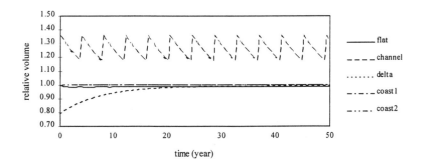

Fig.3. *Dredging the channel of the Barbamarco lagoon with an interval of 4.1 years*

Due to dredging much sand is withdrawn from the system. As can be seen all elements develop to a new equilibrium. Much sand is imported through the boundaries to compensate for the loss of sediment. The time response scales are relatively short, indicating the need for regular dredging.

A management scenario which is investigated concerns the creation of tidal flats with the material from the channel. In Fig.4 we see the effects of dredging sand from the channel, and dumping this volume of sand onto the flat. The system's responses are rapid, and much sand is directly transported back to the channel.

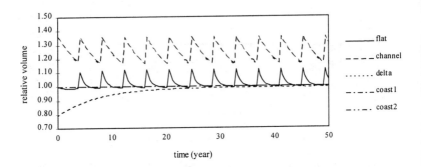

Fig.4. *Dredging and dumping of sediment in the Barbamarco lagoon*

A similar analysis was done for the Scardovari lagoon (Fig.5, Fig.6 and Fig.7). In this case we assumed also that the flats are in equilibrium. The channel and delta volumes in the initial states are a little larger than in their equilibrium states (Fig.5).

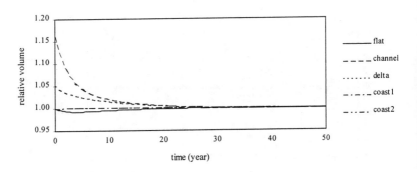

Fig.5. *Applying the Wadden Sea relations to the Scardovari lagoon*

In Fig 6 it can be seen that the flat suffers mostly from the dredging. Much sand is transported directly from the flat to the channel. Because of the import of sand across the boundaries the small ebb-tidal delta starts to grow. As can be seen all elements develop to a new equilibrium. The Scardovari lagoon is larger than the Barbamarco lagoon, resulting in larger morphological time-scales.

Scardovari

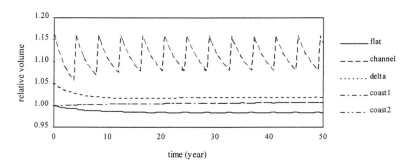

Fig.6. *Dredging the channel of the lagoon Scardovari with an interval of 4.1 years*

As we can see in Fig.7 the delta is negatively affected by the dumping of sediment onto the flats. Because the import of sediment equals zero the demand of sand of the channel affects also the delta.

Scardovari

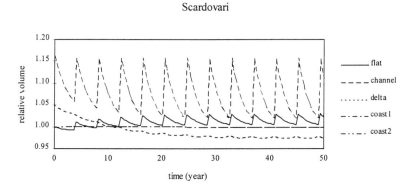

Fig.7. *Dredging and dumping of sediment in the Scardovari lagoon*

Conclusions

We have applied our mesotidal model ASMITA to microtidal lagoons of the Po Delta. ASMITA is a tool to evaluate with large scale intervention scenarios. There is a need for verifications concerning the evolution of the adjacent coastline and the extension of the flat area, however we can conclude that the first results look promising.

The adoption of ASMITA to the Po Delta lagoons is justified by the need to investigate the effects (on safety and on morphology) of possible accelerated sea-level rise, the effects of local subsidence and the effects of "reclamation" or "renaturalisation" interventions currently being planned.

Such questions need an integrated approach for a suitable answer and we recognise that we can integrate *if and only if* we proceed to a simplification of the available long term lagoon evolution and long term coastal evolution models.

Our analysis shows that the time-scales of the Barbamarco lagoon are shorter than the time-scales of the Scardovari lagoon. This indicates that dredging in the Barbamarco lagoon would have to be more frequent than in Scardovari lagoon.

A management scenario aimed at enlarging the tidal flat volume is more successful for the larger Scardovari lagoon.

Acknowledgements

Part of this work is based on work in the PACE-project, in the framework of the EU-sponsored Marine Science and Technology Programme (MAST-III), under contract no. MAS3-CT95-0002.

References

Bruun, P. (editor), 1978. Stability of tidal inlets. Elsevier Sc. Publ. Cy, Amsterdam.

Di Silvio G., 1989. Modelling the Morphological Evolution of Tidal Lagoons and their Equilibrium Configuration. In Proc XXIIIrd IAHR Congress, Ottawa, Canada, pp 169-175.

Eysink, W.D., 1990. Morphologic response of tidal basins to changes. Proc. Int Conf Coastal Eng., ASCE, New York, pp 1948-1961.

Stive, M.J.F., Capobianco M., Wang, Z.B. and Ruol, P., 1996. Morphodynamics of a tidal lagoon and the adjacent coast. Manuscript submitted for the 8th International Biennial Conference on Physics of Estuaries and Coastal Seas, The Hague, Sep.1996.

Wang, Z.B., Karssen, B., Fokkink, R.J. and Langerak, A., 1996. A dynamic/empirical model for estuarine morphology. Manuscript submitted for the 8th International Biennial Conference on Physics of Estuaries and Coastal Seas The Hague, Sep.1996.

A conceptual model for barrier coasts behaviour at decadal scale. Application to the Trabucador Bar

José A. Jiménez[1, 2] and Agustín Sánchez-Arcilla[1,2]

Abstract

A conceptual model for barrier coasts evolution a decadal scale is presented. The model considers all the dynamic processes acting on a barrier system, and in this sense, the RSLR is only considered as a passive factor. The model includes the concept of critical width for overwash contribution to backbarrier deposition. A first simplified version of the model has been used to simulate the landward rollover experienced by the Trabucador Bar during the last decades. The main identified agent driving such behaviour was the overwash transport and reasonable results were obtained. The effect of man actions (such as artificial dunes) preventing overwash on barrier evolution was also assessed.

Introduction

Traditionally, barrier coasts evolution has been largely related with mean sea level fluctuations. Although when long term evolution is studied, RSLR has to be considered as an element of the dynamic system, this is a relatively simplistic approach. In fact, when RSLR is used to explain some observed barrier type of evolution, e.g. landward rollover, most of the real dynamic processes controlling such behaviour are implicitely included in it but in a parametric and hidden way.

In order to solve this, and to explain and model the Ebro delta coast behaviour, where a relatively large barrier-spit complex (The Trabucador Bar-La Banya spit, Figure 1) does exist, a more comprehensive approach is presented.

This approach follows the large number of previous works done in

---

[1] Laboratori d'Enginyeria Marítima, ETSECCPB, Catalonia University of Technology, c/ Jordi Girona 1-3, Campus Nord, Ed. D1, 08034 Barcelona, Spain (jimenez@etseccpb.upc.es).
[2] International Centre for Coastal Resources Research (CIIRC), UPC, c/ Jordi Girona 1-3, Campus Nord, Ed. D1, 08034 Barcelona, Spain.

geomorphology of barrier systems (mainly in the US) but adapted to the specific environment of the Mediterranean Sea. The final objective is to derive a specific barrier evolution model to be included in a general large scale coastal evolution model being developed for the Ebro delta coast (a "mixed-type" model in the sense that it makes use of the underlying physical processes and "observed" morphological evolution, from which governing equations can be derived in an heuristic manner, see Sánchez-Arcilla and Jiménez, 1996 for further details).

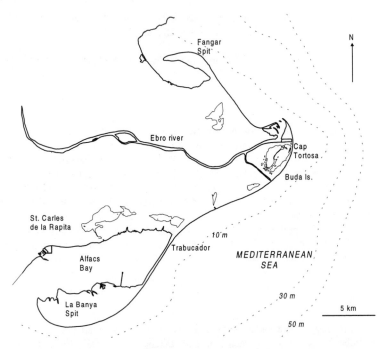

Figure 1. The Ebro delta.

Processes affecting barrier behaviour

Recently, McBride et al. (1995) proposed a classification for barrier coasts by analysing the large scale evolution of Louisiana, Mississippi and Georgia barrier systems during the last 150 years. Their classification is based on the observed response (although most of the selected types were previously identified and explained by other researchers) and, at the same time, they associated the trend in response-type with the existing relative-sea-level-rise (RSLR) rates as the primary controlling factor. However and although these authors did the analysis at the maximum aggregated level, other authors as e.g. Leatherman (1983, 1985), have demonstrated that some of the analysed response-types cannot be explained by

RSLR alone even at very long time scales.

Thus, in this work a bottom-up approach is followed in which the main processes determining barrier evolution considered in the model are (Figure 2):

- *aeolian transport*, which transfers sediment between bayside and seaside barrier shorelines, being the direction of the net sediment transfer controlled by the dominant wind action (landwards -*Eon*- or seawards -*Eoff*-);

- *overwash transport -Ovw-*, which mainly transfers sediment from the outer shoreline towards the backbarrier, although depending on the lagoon dimensions some episodes of overwash from the lagoon is also possible (in principle they would be less significant);

- *breaching or inlets formation -Breach-,* which are genetically related to overwash being the main difference their corresponding intensity and morphological response (e.g. Pierce, 1970; Leatherman, 1979);

- *RSLR*, which is here considered as a passive factor affecting both barrier shorelines and that can produce bayside submergence (e.g. Leatherman, 1983) and not only a landward migration of the backbarrier as the model of Dean and Maurmeyer (1983) among other predicts;

- *wave-induced cross-shore transport*, mainly acting in the outer coast but also in the bayside depending on the lagoon dimensions. It includes: net onshore transport at the shoreface level -*Stsho*-, offshore transport during storms of very long return period -*Stoff*- and, wave-induced crosshore transport in the backbarrier, -*Wav*-;

- *longshore sediment transport gradient -ΔSl-*, acting in the outer coast and potentially inducing erosion or accretion of the seaside shoreline depending on the sign of the gradient.

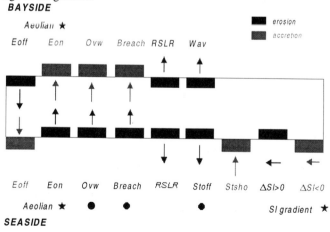

Figure 2. Processes contributing to barrier evolution at decadal scale (→: main direction of sediment flux for each process; •: genetically related processes; ★:

processes able to contribute either landwards and seawards but with only one "exclusive" net contribution).

Figure 2 shows the potential contribution of each process to the barrier evolution (in terms of local sediment budget and shoreline evolution for both bayshore and seashore). For processes transferring sediment between seaside and bayside, i.e. aeolian and overwash transport, one of the points to be considered is that the emerged part of the barrier can act as a neutral system (i.e. sediment eroded from one side is "conservatively" transported towards the other side), as a sedimentary sink (i.e. the barrier behaves as a temporary sediment storage) or a sediment source (i.e. emerged barrier erosion). This sediment transfer between both shorelines can be strongly affected by human activities in the emerged part of the barrier, e.g. construction of dunes along the barriers, etc. (see Dolan, 1972, among others).

Model development

The morphological result of the processes before mentioned depends on their intensity and also on the existing barrier morphology. The main morphological variables to be considered are: barrier width, barrier height, depth of closure at both sides (it depends on wave climate) and lagoon dimensions (Figure 3).

The *barrier width* controls the effective contribution of overwash to the bayside shoreline evolution. Following Leatherman (1979), a *critical width* -depending on the wave and water level climates- has been considered. Barriers wider than this critical value will not experience overwash deposition in the bayshore under "normal conditions". This has been used to explain why overwash deposition in the bayshore is mainly associated with eroding barriers. This critical width has been estimated in about 200 m for the Trabucador Bar.

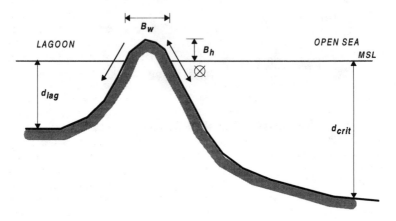

Figure 3. Morphological variables representing the barrier system.

The *barrier height* will also affect the effective contribution of overwash to the bayshore. The higher the barrier is, the lower the overwash will be (for a determined wave and water level climates). This variable is generally the most easily altered by e.g. construction of artificial dunes. Usually, it is possible to find an equilibrium height for barriers subjected to overwash action and experiencing landwards rollover.

The *critical depth* at the seashore and bayshore will control the "amplification" of overwash deposits. In many field studies of overwash events it has been detected a larger bayshore advance than the observed seashore retreat (e.g. Ritchie and Penland, 1989). This can be due to the deposition of sediment eroded from the outer coast in a shallower water column in the bay. However, other factors can also contribute to this difference as e.g. an additional release of sediment from the emerged part of the barrier.

Finally, the *lagoon dimensions* will control the lagoonal hydrodynamic regime, e.g. fetch limitation for wave generation, etc.

This proposed model is being built considering all the before mentioned processes as well as the morphological variables able to affect the barrier evolution. As a first validation of the model concepts, barrier evolution without considering RSLR has been analysed. The objective was to assess the model capability to adequately simulate barrier response considering only dynamic processes and to stress their contribution to barrier evolution. In this sense, the main test was to simulate the barrier landward rollover at a decadal time scale without the effects of SLR, which is usually considered as the main agent driving such behaviour.

One of the main problems is how to evaluate the magnitude of each one of the contributing processes appearing in Figure 2. Before to apply this to the Ebro delta some considerations were done. Firstly, the longshore sediment transport gradient along the outer coast of the Trabucador Bar is known and, even it has been successfully included in a long-term coastline model (see Sánchez-Arcilla and Jiménez, 1996). On the other hand and although usually, breaching is considered as one of the main processes transferring sediment from the outer coast to the backbarrier, this process is not so common in the Ebro delta at least during the last decades (e.g. Sánchez-Arcilla and Jiménez, 1994). Thus, a period of several decades has been considered in which no significant breaching event took place.

With respect to the wave action on both shorelines, it will not be considered at this first approach. Although this implies to avoid one process that can be relevant, the lagoon dimensions behind the Trabucador permits to assume that wave action on the backbarrier will be of second-order with respect to the remaining processes. On the other hand, as the main objective of this work is to simulate the observed behaviour of the Trabucador Bar during the last decades in which rollover has been detected (see Figure 4), the wave-induced processes considered in Figure 2 (i.e. cross-shore transport at the shoreface level and offshore transport due to erosion during big storms) are of secondary importance.

Thus, two processes emerge as the main agents controlling barrier rollover: overwash and aeolian transport.

The remaining problem, i.e. the evaluation of the magnitude of both processes, is not a trivial task, and although at a microscale the problem has been solved at a certain level of detail (see e.g. Sánchez-Arcilla and Jiménez, 1994), at a larger scale there remains some questions on how to evaluate it. In this work, an integrated approach has been followed. Thus, the transport rates associated with these two processes have been evaluated through the integral response of the analysed barrier, by estimating the amount of sediment deposited in the backbarrier during several decades and assuming that this sediment has been transported by the net contribution of aeolian and overwash transport during that period. With this approach, we are assuming that this estimated amount is due to the combined action of the wave and water level climates typical of the Ebro delta during that period. bThis also implies that a hypothesis of steadiness in such climates is also imposed.

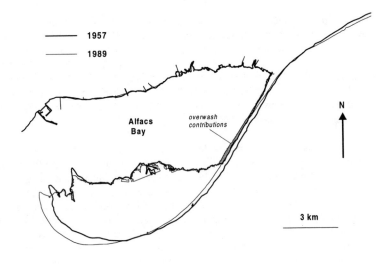

Figure 4. Trabucador Bar evolution at the decadal time scale (from 1957 to 1989).

Thus, taking into account all these hypotheses, the seaside, X_s, and bayside shoreline, X_b, movements will be given by the continuity equation applied to both shorelines

$$\frac{\partial X_s}{\partial t} = -\frac{1}{d_{crit}}\left(\frac{\partial Sl}{\partial y} + Q_{out}\right)$$

$$\frac{\partial X_b}{\partial t} = -\frac{1}{d_{lag}} Q_{in}$$

where Q_{out} and Q_{in} are the net contribution of overwash and aeolian transport at the outer and inner coasts of the barrier.

In this work, Q_{in} is considered to be a function of the overwash transport rate at the outer coast in the following form (depending on the actual barrier width, Bw, and the critical width, $Bcrit$)

$$Bw > Bcrit \qquad\qquad Q_{in} = 0 \qquad\quad \text{(no overwash contribution)}$$

$$Bw \leq Bcrit \qquad\qquad Q_{in} = Q_{out}\,(Bcrit\,/\,Bw)^{\alpha}$$

where α is a parameter greater than 1 and it takes into account the amount of material eroded from the emerged part of the barrier that contributes to the backbarrier displacement. Differences between Q_{in} and Q_{out} will determine the role played by the emerged part of the barrier (by-pass system, sediment storage or sediment feeding).

As it can be deduced, even in the case that both transport rates were of the same order of magnitude ($Q_{in} = Q_{out}$) and without considering any other agent, for shallow lagoons, i.e. $d_{lag} < d_{crit}$, the backbarrier rate of displacement would be larger than the outer shoreline displacement. This explains why barriers suffering outer coast erosion (e.g. due to a longshore sediment transport gradient) but also experiencing landward rollover are able to cope with the erosion without a significant loss of emerged area.

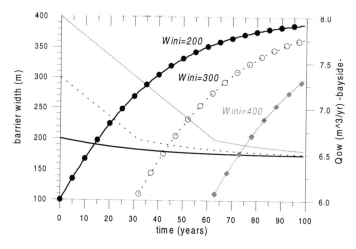

Figure 5. Barrier width evolution and associated overwash transport rates in the

bayside coastline under the same "hydrodynamic scenario" for different initial barrier
widths (critical barrier width for overwash contribution to the bayside is 200 m).

With the assumed relationship between overwash transport rates at the inner and
outer coasts, the findings of Leatherman (1979, 1985) about the concept of critical
width for overwash contributions are incorporated into the model. Thus, the barrier
will only experience backbarrier displacement due to overwash when the outer
shoreline begins to erode and the barrier width becomes narrower than this value.
This critical width depends on specific conditions such as local hydrodynamic
climate (wave and water level climates) as well as the own geomorphological
characteristics of the barrier. According to the evolution experienced by the
Trabucador Bar (see Figure 4), this parameter has been estimated in about 200 m for
the actual wave and water level climates (see review in Jiménez et al. 1997).

Figure 5 shows a sensitivity analysis of the model concepts for typical conditions
of the Trabucador Bar in the Ebro delta coast. Thus, three runs were made in which
three different barrier configurations were subjected to the action of the same
hydrodynamic climate. Thus, for the widest barrier configuration (400 m), it can be
seen that during the first 60 years, the barrier width decreases (due to the existing
longshore sediment transport gradient along the outer coast) but as during this time
the barrier was wider than the critical width (200 m), no overwash at the backbarrier
occurred. However, as the barrier width decreases to the critical width, the overwash
transport began to act and the barrier reaches an "equilibrium width". Because the
outer coast was experiencing erosion, the constant value of the barrier width
indicates the landward displacement of the backbarrier. Moreover, in the same figure
is also seen that as width decreases, the overwash contribution increases.

The same behaviour is also observed for an initial barrier configuration of 300 m
width. In this case, the overwash begins to act after 30 years of continuous erosion of
the outer coast but the final width is the same.

When the initial barrier width is equal to the critical value, the barrier experiences
rollover from the beginning (in this test the outer coast is also eroding), and it can be
also seen that overwash transport rates show a more or less asymptotic value at the
end of the analysed period, which is the equilibrium value between the incident
wave and water level climate and a barrier configuration given by the "equilibrium
width".

This test serves to stress the importance of overwash contributions to maintain the
barrier stability even in the case of large erosion rates along the outer coast.

Discussion and conclusions

As it has been shown through this work, it is possible to explain barrier coast
evolution without use of RSLR. In this case, this implies to consider the dynamic
processes acting on the barrier system and, at the same time, it helps to explain or to
adequately model some of the reported observations that are not able to justify

considering the RSLR as the main driving agent.

Moreover, with this approach it is also possible to predict as the man action can influence the barrier evolution. Thus, Figure 6 shows a simulation of the effect of preventing overwash contribution to the backbarrier on barrier evolution. For the hydrodynamic conditions and barrier morphology considered, when overwash is not affected, the barrier width for 100 years of evolution is about 135 m which is reached after 40 years. After that time, the barrier width is maintained through the backbarrier deposition and the subsequent rollover.

If the overwash contribution to the backbarrier is prevented by any means (e.g. an artificial dune) in such a way that the magnitude is decreased to 50% (0.5) of the original value, the final width will decrease to about 90 m. This has important implications since narrower the barrier is, higher the probability of breaching will be (see e.g. Sánchez-Arcilla and Jiménez, 1994).

If the overwash is fully prevented, the model predicts the barrier disappearance after 35 years. However, this is not a real scenario and it is only indicating that after 30 years, the probability of breaching will be extremely high because its width will be about 70 m.

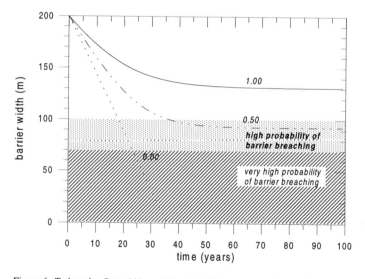

Figure 6. Trabucador Bar width evolution for different overwash scenarios (*1.00*: present situation; *0.50*: overwash reduced to the half; *0.00*: no overwash allowed).

In other words, even with a simplified model such as the here presented, it is possible to assess the barrier evolution as well as the influence of the variation of the dynamic processes acting on the barrier system.

Acknowledgements

This work was carried out as part of the *PACE* project. It was supported by the EU Marine Science and Technology (MAST-III) Programme under Contract No. MAS3-CT95-0002. This work was also co-sponsored by the Spanish Ministry of Education and Research through DGICYT (UE95-0033). The first author also thanks to Departament de Política Territorial i Obras Públiques de la Generalitat de Catalunya and to the ETSECCPB by the support to attend to the conference.

References

Dean, R.G. and Maurmeyer, E.M. 1983. Models of beach profile response. In: Komar, P.D. and Moore, J. (Eds.), *CRC Handbook of Coastal Processes and Erosion*, CRC Press, 151-165.

Dolan, R. 1972. The barrier dune system along the Outer Banks of North Carolina. *Science*, 175, 286-288.

Jiménez, J.A., Sánchez-Arcilla, A., Valdemoro, H.I., Gracia, V. and Nieto, F. 1997. Processes reshaping the Ebro delta. *Marine Geology*, (in press).

Leatherman, S.P. 1979. Migration of Assateague Island, Maryland, by inlet and overwash processes. *Geology*, 7, 104-107.

Leatherman, S.P. 1983. Barrier dynamics and landward migration with Holocene sea-level-rise. *Nature*, 301, 415-417.

Leatherman, S.P. 1985. Geomorphic and stratigraphic analysis of Fire Island, N.Y. *Marine Geology*, 126, 143-159.

McBride, R.A., Byrnes, M.R. and Hiland, M.W. 1995. Geomorphic response-type model for barrier coastlines: a regional perspective. *Marine Geology*, 126, 143-159.

Pierce, J.W. 1970. Tidal inlets and washover fans. *Journal of Geology*, 78, 230-234.

Ritchie, W. and Penland, S. 1989. Erosion and washover in Coastal Louisiana. *Proc. 6th Symp. on Coastal and Ocean Management*, ASCE, 253-264.

Sánchez-Arcilla, A. and Jiménez, J.A. 1994. Breaching in a wave-dominated barrier spit: the Trabucador bar (NE Spanish coast). *Earth Surface Processes and Landforms*, 19, 483-498.

Sánchez-Arcilla, A. and Jiménez, J.A. 1996. A morphological mixed-type model for the Ebro delta coast. *Proc. 25th Coastal Eng. Conf.*, ASCE, 2806-2819.

Sediment Transport Model and Its Application for Assessment of Sedimentation in Navigable Channels

Leszek M. Kaczmarek & Rafał Ostrowski[1]

Abstract

The sediment transport model, taking account of bedload and the sediment suspended in a thin layer above sea bed, is proposed for the evaluation of accumulation processes in water-ways. The model has been tested against available laboratory and field (radio-tracer) results and its practical applicability is now checked using the long-term field data associated to dredging works in the navigable channel of Łeba harbour (Poland).

Introduction

The maintenance of navigational depth of water-ways is a serious problem in operation of harbours. The accumulation processes, as a result of sediment transport, cause the necessity of periodic dredging works. Sometimes, after heavy storms, the navigation can locally be impossible due to occurrence of shoals in water-ways. Therefore, extensive studies are undertaken to optimise the layouts of harbour entrances and navigable channels. The prediction of sedimentation rate in designed water-ways plays an important role in the optimisation schemes. However, there are few predictive methods in use and they are most often simplified and inaccurate.

Goals

The theoretical model has previously been tested using a number of available published laboratory data, radio-tracer field data obtained at *IBW PAN* Coastal Research Facility of Lubiatowo (Poland) and the most recent bedload *IBW PAN* laboratory results. Now, following Kaczmarek et al. (1997), the model is applied in a case study, in order to carry out the validation of the proposed approach in engineering practice.

---

[1] both Polish Academy of Sciences' Institute of Hydro-Engineering, *IBW PAN*, 7 Kościerska, 80-953 Gdańsk, Poland

Site description

The small fishing harbour of Łeba is located in the mouth of the Łeba river, with the entrance at 50°46' N and 17°33' E, some 100 km north-west off Gdańsk, see Fig. 1. The breakwaters were rebuilt and re-arranged for a few times in the harbour history. The present layout of the harbour is shown in Fig. 2.

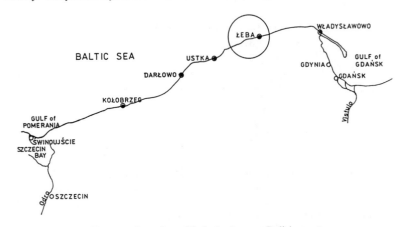

Figure 1. Location of Łeba harbour at Polish coast

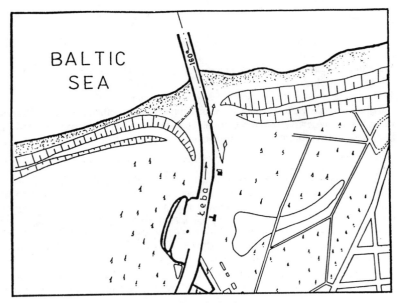

Figure 2. Łeba harbour

Water depth at the harbour entrance and in the approach channel is maintained by relatively rare dredging works. Good operation of the harbour requires the navigable channel to be 35 m wide and 5 m deep. However, the irregularity of dredging activities results in a certain variability of the navigable channel depth. The coast nearby Łeba harbour is classified as stabilised one and has not revealed any distinct erosive or accumulative trends recently. The nearshore bathymetry is shown in Fig. 3.

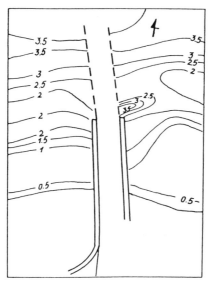

Figure 3. Łeba harbour entrance, navigable channel and nearshore bathymetry

Assumptions and methodology

The cross-shore profile characteristic for the vicinity of the harbour is depicted in Fig. 4. The points A and B denote the location of the harbour entrance and the offshore end of the water-way, respectively.

The analysis of the bottom profile geometry has led to the determination of Dean's parameter A, describing an equilibrium profile represented by the formula:

$$y = Ax^{2/3} \tag{1}$$

In the region of the harbour, the parameter A amounts to 0.0758. The equilibrium profile has been used in the determination of generalised geometrical features of the navigable channel. In the calculations it has been assumed that the geographic azimuth of the seaward directed cross-shore profile equals to 347.5° while the navigable channel azimuth is 160°. Hence, the angle between the cross-shore profile and the water-way is $\alpha = 7.5°$. The projection area of the navigable channel on the cross-shore profile is denoted as F_r' in Fig. 4. Following the harbour maintenance standards, the nominal navigable depth has been assumed as $h=5$ m.

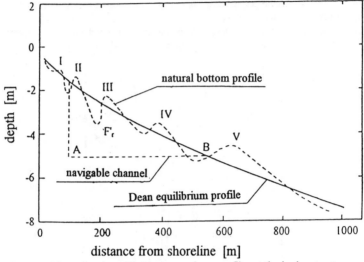

Figure 4. Natural and equilibrium cross-shore profiles at the harbour entrance

The navigable channel crosses three bars: II, III and IV (see Fig. 4). The projection of the navigable channel on the cross-shore profile yields three representative segments of its total length, corresponding to the three underwater bars.

In the analysis of sedimentation, the cross section of the navigable channel has been assumed to have rectangular shape. Further assumption has been made that sand moves as bedload and suspended load in a thin layer above the bed in the direction of wave propagation due to oscillatory motion of water, falls into the navigable channel where it is trapped and distributed evenly across the channel's width. In computations the wave climate of a statistic year has been taken into account. Sediment transport rate and accumulation in the water-way have been calculated for a number of wave conditions occurring at the analysed site. The results have then been integrated over one-year time and spatially - along the water-way and compared thereafter with the long-term dredging data.

Theoretical bedload model

The nearbed dynamics is examined and modelled for the flow regions above and beneath the original static bed line, see Fig. 5. The collision-dominated granular-fluid region I stretches below the nominal static bed while the wall-bounded turbulent fluid region II extends above it. Since both water and sand grains are assumed to move in both regions, there must be a certain transition zone between I and II, in which the velocity profiles of I and II would merge and preserve continuity of shape.

The velocity distribution about a porous rough bed is controlled by various features of roughness and bed permeability. At first it is assumed that the velocity is determined by roughness geometry and outer flow parameters, such as the free-stream wave velocity.

The iteration procedure permits matching of velocities in the regions I and II. The velocity profile in the upper turbulent layer, which is linked with identification of roughness k_e, is determined first and then passed to the lower collision-dominated layer. The intersection of the two velocity profiles is marked as point A in Fig. 5.

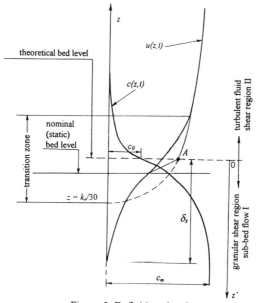

Figure 5. Definition sketch

The flow in the turbulent upper region for regular waves is described by the integral momentum model based on the solution proposed by Fredsøe (1984):

$$\frac{dz_1}{d(\omega t)} = \frac{30\kappa^2 U}{k_e \omega e^{z_1}(z_1 - 1) + 1} - \frac{z_1(e^{z_1} - z_1 - 1)}{e^{z_1}(z_1 - 1) + 1} \frac{1}{U} \frac{dU}{d(\omega t)} \qquad (2)$$

where $U(t)$ is the free stream velocity.

The solution of Eq. (2) is achieved by the Runge-Kutta second-order method. As a result for a particular case, the function $z_1(t)$ is obtained and the time distributions of the friction velocity $u_f(t)$ and boundary layer thickness $\delta(t)$ are calculated thereafter using the following equation:

$$z_1 = \frac{U\kappa}{u_f} \qquad (3)$$

$$\delta = \frac{k_e}{30}\left(e^{z_1} - 1\right)$$ (4)

The solution of Eq. (2) enables the value of u_{fmax} to be determined, if the quantity k_e is specified. To evaluate the roughness parameter k_e an iterative procedure is proposed for finding the matching point A.

In the sub-bed flow region, the sediment concentration is high and chaotic collisions of grains are the predominant mechanism. Particle interactions are assumed to produce two distinct types of behaviour. The Coulomb friction between particles give rise to rate-independent stresses (of the plastic type) and the particle collisions bring about stresses that are rate-dependent (of the viscous type). The use of the mathematical description by Sayed & Savage (1983) for determination of the stress tensor was made and the balance of linear momentum according to Kaczmarek & O'Connor (1993) leads to the following equations:

$$\alpha^0 \left(\frac{c - c_0}{c_m - c}\right) \sin\varphi \sin 2\psi + \mu_1 \left(\frac{\partial u}{\partial z'}\right)^2 = \rho u_f^2$$ (5)

$$\alpha^0 \left(\frac{c - c_0}{c_m - c}\right)(1 - \sin\varphi \sin 2\psi) + (\mu_0 + \mu_2)\left(\frac{\partial u}{\partial z'}\right)^2 =$$

$$\left.\left(\frac{\mu_0 + \mu_2}{\mu_1}\right)\right|_{c = c_0} \rho u_f^2 + (\rho_s - \rho)g\int_0^{z'} c \, dz'$$ (6)

in which:

ρ_s and ρ are the densities of the solid and fluid, respectively;

α^0 is a constant;

c_0 and c_m are the solid concentrations corresponding to fluidity and the closest packing, respectively;

μ_0, μ_1 and μ_2 are functions of the solid concentration c:

$$\frac{\mu_1}{\rho_s d^2} = \frac{0.03}{(c_m - c)^{1.5}}$$ (7)

$$\frac{\mu_0 + \mu_2}{\rho_s d^2} = \frac{0.02}{(c_m - c)^{1.75}}$$ (8)

The value φ in Equations (5) and (6) is the quasi-static angle of internal friction, while the quantity ψ is equal to:

$$\psi = \frac{\pi}{4} - \frac{\varphi}{2}$$ (9)

For the calculations the following numerical values are recommended:

$$\frac{\alpha^0}{\rho_s g d} = 1 \qquad c_0 = 0.32 \qquad c_0 = 0.32 \qquad \varphi = 24.4°$$ (10)

where d denotes the diameter of grains.

The apparent bed roughness parameter k_e is an important quantity in the theoretical model. Both the turbulent and sub-bed velocity profiles depend on k_e, which is not known a priori. Therefore an iteration procedure is proposed for finding the matching point A between these profiles. The matching is assumed to take place at the phase of maximum shear stress, although at any other phase of oscillatory motion there must be some transition between the two profiles. Under monochromatic waves the maximum shear stress is the maximum value of shear stress during a wave period.

For engineering purposes it is useful to approximate the theoretical results by a curve expressed in terms of skin friction. To determine the skin friction one can follow Nielsen's (1992) description:

$$\theta_{2.5} = \frac{1}{2} f_{2.5} \; \psi_1 = \frac{1}{2} f_{2.5} \frac{(a_{1m}\omega)^2}{(s-1)gd} \tag{11}$$

$$f_{2.5} = \exp\left[5.213\left(\frac{2.5d_{50}}{a_{1m}}\right)^{0.194} - 5.977 \right] \tag{12}$$

in which $s = \rho_s/\rho$.

The iteration procedure proposed for finding the matching point A between the velocity profiles (Fig. 5) was run for a wide range of small and large scale conditions and for a few diameters of sandy bed. Then the results were approximated yielding the following formula:

$$\frac{k_e}{d} = 47.03\theta_{2.5}^{-0.66} \tag{13}$$

Once the roughness k_e is determined and the friction velocity $u_f(t)$ is obtained from Eq. (2), the velocity and concentration distributions in bedload layer can be computed from Eqs. (5) and (6). Knowing these quantities one can calculate the instantaneous bedload rate:

$$Q_B = \int_0^{\delta_s} u(z',t) \cdot c(z',t)\, dz' \tag{14}$$

and the dimensionless bedload rate defined as:

$$\phi_B = \frac{Q_s}{d\sqrt{(s-1)gd}} \tag{15}$$

The theoretical results averaged over half a wave period ($T/2$) can be approximated by the following formula:

$$\phi_{T/2} = 1.3 \cdot \theta_{2.5}^{1.5} \tag{16}$$

Experimental evidence and model extension

The bedload theoretical model has been incorporated in bottom profile modelling by Kaczmarek et al. (1995) and tested successfully against radio-tracer field data obtained at *IBW PAN* Coastal Research Facility of Lubiatowo (Poland). The exemplary results of comparison between measured and computed sedimentation are presented in Fig. 6.

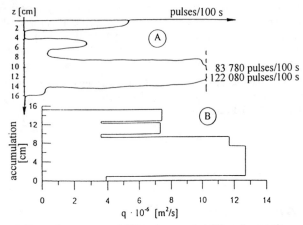

Figure 6. Exemplary measured (A) and computed (B) sedimentation profiles

The model has also been tested using a number of laboratory data presented by Nielsen (1992) and the most recent bedload *IBW PAN* laboratory results. The comparison is shown in Fig. 7. A good conformity between the theoretical curve and the laboratory data is visible for small sediment transport intensity, i.e. for $\theta_{2.5}<1$. The sediment transport rates $\phi_{T/2}$ measured for high values of $\theta_{2.5}$ are much bigger than the theoretical bedload values.

It is supposed that this is due to the increasing importance of suspended transport in a thin layer above the bed. In order to check the above, the system of equations proposed by Deigaard (1993) has been run, combined with the present bedload approach which has provided the bed boundary conditions. The governing momentum and sediment exchange equations of Deigaard (1993) have the following form:

$$\left[\frac{3}{2}\left(\alpha\frac{d}{w_s}\frac{du}{dz}\frac{2}{3}\frac{s+c_m}{c_d}+\beta\right)^2 d^2c^2\left(s+c_m\right)+l^2\right]\left(\frac{du}{dz}\right)^2 = u_f^2 \tag{17}$$

$$\left[3\left(\alpha\frac{d}{w_s}\frac{du}{dz}\frac{2}{3}\frac{s+c_m}{c_d}+\beta\right)^2 d^2\frac{du}{dz}c+l^2\frac{du}{dz}\right]\frac{dc}{dz} = -w_sc \tag{18}$$

in which w_s denotes settling velocity of grains, s is a relative sediment density, c_m and c_d are the added mass and drag coefficients, respectively, and l is a mixing length defined as $l=\kappa z$. It should be noted that the iterative procedure to match the solutions of Eqs. (17) and (18) with the solution in bedload layer, including the determination of the calibration coefficients α and β, has been used.

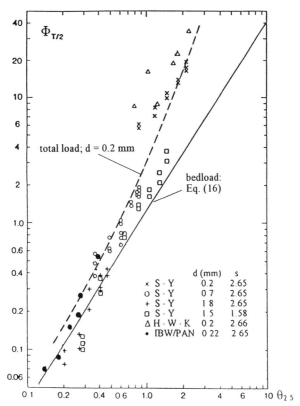

Figure 7. Theoretical bedload rate vs. lab. data of Sawamoto & Yamashita, Horikawa et al. (as given by Nielsen (1992)) and *IBW PAN* lab. data

The preliminary results shown in Fig. 7 indicate that the increase of laboratory data for $\phi_{T/2}$ with respect to the theoretical bedload values, which reflects the suspended load contribution, can be modelled quite well using Eqs. (17) and (18). Further studies are in progress and their completion requires the verification using other laboratory and field data. At present, however, for the engineering purpose of this study, it has been assumed that for high shear stresses the suspended sediment transport is 5 times bigger than bedload.

Model results vs. dredging data

For small $\theta_{2.5}$ the sediment transport as bedload has been determined using Eq. (16). For $\theta_{2.5} > 1$ the presence of suspended load has been included and the total transport - in accordance to the previous quantitative estimation - has been determined using Eq. (16) results multiplied times 6.

Using the above assumptions, the volume of sand trapped in the navigable channel has been calculated for the sequence of wave conditions occurring at the site in a statistic year. The results integrated over one-year time and along the water-way have yielded the value of about 83000 m³ which corresponds well to the dredging data, see table below.

Table: Annual dredging data

Period	Number of years	Sediment dredged [$*10^3$ m³]	Source of data
1915-1916	2	42	Szopowski, Z. (1962a)
1949	1	70	Szopowski, Z. (1962b)
1952-1960	9	47*	Szopowski, Z. (1962b)
1949-1970	22	64	Michowski, A & M. Skurczyński (1971)
1964-1974	11	88	Huszcza, A. & M. Pachowski (1974)

*approximately

References

Fredsøe, J. (1984). Turbulent boundary layer in combined wave-current motion, *J. Hydraulic Eng.*, ASCE, Vol. 110, No. HY8, 1103-1120

Deigaard, R. (1993). Modelling of sheet flow: dispersion stresses vs. the diffusion concept, *Prog. Rep.* 74, Inst. Hydrodyn. and Hydraulic Eng., Tech. Univ. Denmark, 65-81

Huszcza, A. & M. Pachowski (1974). Study on dredging works in Łeba harbour (in Polish), Gdańsk, *IBW PAN*

Kaczmarek, L.M. and B.A. O'Connor (1993). A new theoretical approach for predictive evaluation of wavy roughness on a moveable flat/rippled bed, Parts I/II, *Reps. CE*/14-15/93, Dept. Civ. Engng., Univ. Liverpool

Kaczmarek, L.M., R. Ostrowski & R.B. Zeidler (1995). Boundary Layer Theory and Field Bedload, *Proc. International Conf. Coast. Res. in Terms of Large Scale Experiments (Coast. Dynamics '95)*, ASCE, 664-675

Kaczmarek, L.M., A. Mielczarski & R. Ostrowski (1997). Application of a new sediment transport model for assessment of sedimentation in navigable channels (in Polish), *Maritime Engineering and Geotechnics*, No. 2

Michowski, A. & M. Skurczyński (1971). Dredging works, in: „Investigations on sedimentation of Łeba harbour entrance" (in Polish), Gdańsk, Instytut Morski, Rep. No. 610

Nielsen, P. (1992). Coastal bottom boundary layers and sediment transport. *Advanced Series on Ocean Engineering*, Vol. 4, World Scientific, Singapore

Sayed, M. and Savage, S.B. (1983). Rapid gravity flow of cohesionless granular materials down inclined chutes, *J. Applied Mathematics and Physics* (ZAMP), Vol. 34, 84-100

Szopowski, Z. (1962 a). Small harbours of west Pomerania before II World War (in Polish), Warsaw-Poznań, PWN

Szopowski, Z. (1962 b). Dredging works at navigable channels, in: „Materials of Polish shore monograph" (in Polish), No. 2, Gdańsk-Poznań, *IBW PAN*

Equilibrium Beach Profiles: Effect of Refraction

Mauricio González[1], Raúl Medina[2] and Miguel A. Losada[3]

Abstract

A beach equilibrium model for beaches in which refraction effects are important is presented. The model assumes uniform energy dissipation per unit volume and considers the convergence/divergence of wave-rays within the surf-zone. The resulting profile shape is given as the expression $h = A_r A x^{2/3}$ which resembles Dean's (1977) formulation. Although A_r is a function of the cross-shore coordinate, it is shown that for typical field refraction coefficients this dependence can be neglected and A_r can be determined as a function of the refraction coefficient, K_r. The proposed profile shape formulation is validated using several profiles along the Spanish coast. It is concluded that the profile shape is mainly related to the way the energy is transformed (dissipated/refracted) along the profile and is almost independent on the total amount of energy reaching the beach. It is also concluded that the shape parameter is not only a function of the sediment grain size or fall velocity, as proposed by Dean (1977), but the wave propagation within the surf zone.

Introduction

The concept of equilibrium beach profile has been a matter of great interest to numerous investigators for the last few decades. The hypothesis behind the equilibrium beach profile is that beaches respond to wave forcing by adjusting their form to an "equilibrium" (or constant) shape attributable to some physical parameters (e.g. grain size, wave steepness...). Several expressions have been proposed over the years (Keulegan and Krumbein, 1919; Sunamura and Horikawa, 1974; Vellinga, 1983; Bodge, 1992; ...). The most widely used formulation, probably due to its simplicity, is the one proposed by Bruun (1954) and Dean (1977).

[1] Researcher, Ocean & Coastal Research Group. Universidad de Cantabria. Dpto. Ciencias y Técnicas del Agua y del Medio Ambiente. Avda. de los Castros s/n. 39005 Santander, Spain
[2] Associate Profesor, Ocean & Coastal Research Group. Universidad de Cantabria. Dpto. Ciencias y Técnicas del Agua y del Medio Ambiente. Avda. de los Castros s/n. 39005 Santander, Spain
[3] Professor' Ocean & Coastal Research Group. Universidad de Cantabria. Dpto. Ciencias y Técnicas del Agua y del Medio Ambiente. Avda. de los Castros s/n. 39005 Santander, Spain

Bruun (1954) assumed that the beach profile shape could be given as:

$$h = A x^{\frac{2}{3}} \qquad (1)$$

where h is the total water depth, A is a dimensional shape parameter and x is the horizontal distance measured from the shoreline. This equation was found by fitting beach profiles from California and the Danish North Sea coast. Dean (1977) adjusted the same expression to 504 beach profiles collected by Hayden et al (1975), along the east coast of the U.S.A. and the Gulf of Mexico.

Moore (1982) analyzed the shape parameter, A, and concluded that this parameter can be determined by means of the beach grain size (D_{50}). Dean (1987) modified Moore's relationship between A and D_{50} using the fall velocity, ω, and found a linear fit between A and ω in a log-log plot.

Pilkey et al (1993) reanalyzed Moore's relationship and suggested that the shape parameter, A, must be a function of some other variables, since different A values are found in natural beach profiles which have the same grain diameter. Vellinga (1983) studied the influence of the wave height and wave steepness on the shape parameter, A, and found a weak dependence on this latter parameter. Suh and Dalrymple (1988), using Sunamura and Horikawa's (1974) laboratory data, showed that both storm and normal beach profiles can be adequately represented by equation (1) using Moore's relationship and neglecting the dependency on incident wave characteristics.

Although only a weak, or even negligible, relationship is found in laboratory studies between wave characteristics and beach profile shape, investigators have long recognized that there is a strong relationship between the beach slope and the degree of wave exposure of a beach. Bascom (1959), using data from more than 40 beaches on the Pacific coast of the United States, concluded that the slope of a beach face is related to the median diameter of the sand and the amount of wave energy reaching that point. The amount of energy is a function of the refraction conditions; consequently, beaches that are protected are steeper for any given sand size than exposed beaches. In this paper, the dependence of the beach profile shape on wave refraction is further analyzed.

Modified Equilibrium Beach Profile

According to the balance of destructive and constructive forces and assuming wave energy dissipation per unit volume to be the dominate destructive force, Dean (1977) proposed that a sediment of specific size, d, would be stable in the presence of a particular level of wave energy per unit water volume, D_*, leading to the following equation for equilibrium beach profiles:

$$\frac{1}{h} \frac{\partial}{\partial x} (E \ C_g) = D. \tag{2}$$

in which, E is the local wave energy density and C_g is the local wave group velocity. Equation (1) can be easily obtained by integrating equation (2) if shallow water linear wave theory and constant breaker-to-depth ratio are assumed.

If refraction effects are important, the distance between wave-rays, b, can not be considered constant along the x-coordinate and equation (2) must be modified resulting in:

$$\frac{1}{h} \frac{\partial}{\partial x} (E \ C_g \ b) = D. \ b \tag{3}$$

In order to integrate equation (3), an expression for $b(x)$ must be specified. Two cases will be analyzed: (1) linear and (2) exponential.

Case 1: Linear

A simple linear function of the form:

$$b = \alpha x + \beta \tag{4}$$

can be used to analyze the influence of wave propagation in beach profile shape. This linear relationship can properly represent a wide range of natural situations such as the wave refraction behind a breakwater or the wave propagation in a bay with a headland. In these cases, and assuming that wave diffraction phenomena are concentrated in the vecinity of the breakwater or headland, wave crests can be approximated by circles and the distance between wave rays is defined as:

$$b = (R - x) \ \theta \tag{5}$$

where R is the distance between the headland and the coast and θ is a given angle in radians. Notice that equation (5) and equation (4) are the same for $\alpha = -\theta$; $\beta = R\theta$.

Plugging equation (4) into equation (3) and after some straightforward algebra, a solution for the beach profile shape can be obtained as:

$$h = A \left\{ \frac{5}{8\alpha} \ \frac{(\alpha x + \beta)^{\frac{8}{5}} - \beta^{\frac{8}{5}}}{(\alpha x + \beta)^{\frac{3}{5}}} \right\}^{\frac{2}{3}} \tag{6}$$

Equation (6) can be written in terms of the refraction coefficient, K_r. From figure 1, K_r can be determined as:

$$K_r^2(x) = \frac{b_0}{b(x)} \tag{7}$$

Since K_r is a function of the x-coordinate, a particular value of x must be selected. For convenience, the refraction coefficient at the beginning of the profile, $x = 0$, is chosen. This refraction coefficient, herein denoted as K_{r0}, takes into account the total convergence/divergence that occurs within the surf zone, W, and can be written as (see figure 1):

$$K_{r0}^2 = \frac{\alpha W + \beta}{\beta} \tag{8}$$

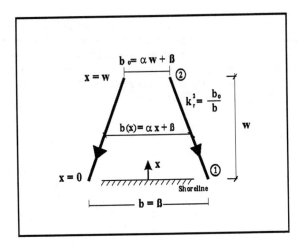

Figure 1. Sketch of wave-rays and definition refraction coefficient

Deriving an expression from equation (8) and plugging it into equation (6), it yields:

$$h = A_{rl}(x) \cdot A \, x^{\frac{2}{3}} \tag{9}$$

where:

$$A_{rl}(x) = \left[\frac{5}{8} \left(\frac{1}{\phi} \left[1 - (1 + \phi)^{-\frac{3}{5}} \right] + 1 \right) \right]^{\frac{2}{3}} \tag{10}$$

and:

$$\phi = \frac{\left(K_{r0}^2 - 1 \right) x}{W} \tag{11}$$

It can be easily shown that if no refraction effects are present, the value of K_{r0} is equal to one, and is also the value of A_{rl}. In this case, equation (9) is the same as equation (1).

Notice that since A_{rl} depends on the x-coordinate, the beach profile given by equation (9) is no longer represented by a parabolic shape. In order to analyze this dependence, the variation of the refraction shape parameter, A_{rl}, along the dimensionless cross-shore distance, x/W, is presented in figure 2 for different K_{r0} values.

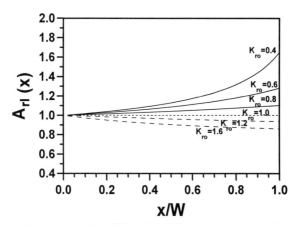

Figure 2. Variation of the refraction shape parameter, A_{rl}, along the dimensionless cross-shore distance, x/W, for different refraction coeficients, K_{r0}. Case 1, $b(x)$=linear

Case 2: Exponential

An exponential function of the form:

$$b = b_0 \, e^{-\Gamma x} \tag{12}$$

where, Γ, is a parameter which determines the rate of convergence/divergence of the wave rays, is also used to analyze the influence of wave propagation in beach profile shape.

Plugging equation (12) into equation (3) and after some algebra, a solution for the beach profile shape is obtained as:

$$h = A_{re} \cdot A \, x^{\frac{2}{3}} \tag{13}$$

where:

$$A_{re} = \left[\frac{5}{3\Gamma x} \left(1 - e^{\frac{3}{5}\Gamma x} \right) \right]^{\frac{2}{3}} \tag{14}$$

Notice that, again, A_{re} is a function of the x-coordinate and consequently, equation (13) is not a parabolic profile. Equation (14) can be written in terms of the refraction coefficient, K_{r0}, definded as:

$$K_{r0}^2 = e^{-\Gamma W} \tag{15}$$

In figure 3, the variation of the refraction shape parameter, A_{re}, along the dimensionless cross-shore distance, x/W, is presented for different K_{r0} values.

From figure (2) and (3) it can be observed that for K_{r0} values in the range of $0.7 < K_{r0} < 1.5$ the values of A_{rl} and A_{re} are very similar. Furthermore, if K_{r0} is assumed to be close to one, A_{rl} can be approximated by:

$$A_{rl} \sim K_{r0}^{-\frac{4}{5}} \tag{16}$$

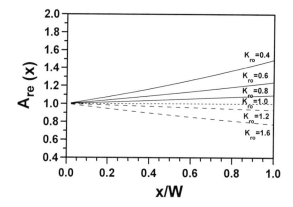

Figure 3. Variation of the refraction shape parameter, A_{re}, along the dimensionless cross-shore distance, x/W, for different refraction coeficients, K_{r0}.
Case 2, $b(x)$ = exponential

at $x = W$. It is worth noting that the hypothesis of $K_{r0} \sim 1$ is fullfiled in most natural beaches within the surf zone. Under the above mentioned assumptions, and for engineering purposes, equation (9) can be written as:

$$h = K_{r0}^{-\frac{4}{5}} \cdot Ax^{\frac{2}{3}} \qquad (17)$$

Field Data

Over 50 profiles from 10 beaches from the Spanish coast have been collected to verify equations (9, 13). In figure 4, beach profile fitting for Plencia Beach is shown.

Plencia Beach is a pocket-beach, located between two breakwaters, which that was nourished in 1986. The breakwaters shelter part of the beach and consequently, a different degree of wave exposure can be found along the beach. Sand size is, however, almost uniform all along the beach.

When comparing exposed and protected profiles, it can be clearly observed that protected profiles are steeper. In figure 4, a beach profile from the sheltered area is compared with equations (1, 9, 13). It is concluded that the equation (1) formulation can not properly represent the profile which is steeper than the one predicted using Moore's relationship. On the other hand, equations (9, 13) are doing a good job.

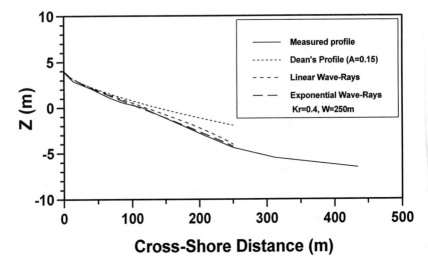

Figure 4. Sandy beach profile measured in Plencia (Spain) and profile fitting equations (1, 9 ,13)

Discussion and Conclusion

The main hypothesis behind the heuristic equilibrium beach profile model presented by Dean (1977), see equation (2), is that the wave energy dissipation per unit volume in the surf zone is uniform. Assuming that this hypothesis is valid, the correct application of equation (2) needs to take into account the variation of the available wave-energy, E, along the profile due to processes, such as wave refraction or shoaling. Shoaling effects are taken into account in Dean's (1977) model once a breaker-to-depth relationship is assumed. However, if there is an increase in the wave energy flux due to a wave-ray concentration, the beach profile will need a longer distance (a greater volume) in order to dissipate that energy. This effect is not included in Dean's (1971) model.

It is worth noting that, from the engineering point of view, the profile shape can still be represented by a parabolic function similar to that proposed by Dean (1977), with a different shaped parameter. However, the parameter that must be used in this case, is not only a function of the sediment grain size or fall velocity but the refraction coefficient as well, $A_T = A_r \cdot A$, where A_T is the total shape parameter, A_r is the refraction shape parameter and A is the grain shape parameter.

The above mentioned conclusion is not in contradiction with Suh and Dalrymple's (1988) or Vellinga's (1983) results which show a weak or negligible dependency of the profile shape on the wave height. Both works were conducted in a 2D lab where refraction did not exist. Vellinga (1983) in very clearly in his lab tests, that an increase in wave height results in an increase in the profile length (or closure depth) with minor changes in the profile shape. Consequently, it can be concluded that the profile shape is mainly related to the way the energy is transformed (dissipated/refracted) along the profile and is almost independent of the total amount of energy reaching the beach.

Sediment sorting is also an important factor that must be taken into account in beaches where wave refraction phenomena are important. Waves take the available supply of clastic material and sort and redistribute the components so that the finer materials are ultimately deposited at points of lesser energy. Consequently, and due to sediment sorting, exposed areas may be steeper than protected ones. This trend will work against the refraction one. The final equilibrium depends on the available grain sizes.

Acknowledgements

This study has been partly funded by *Comisión Interministerial de Ciencia y Tecnología* under contract MAR96-1845. The field data have been kindly provided by the *Dirección General de Costas (Ministerio de Medio Ambiente)*, whose support is hereby gratefully acknowledged.

Appendix - References

Bascom, W., (1959). "The relationship between sand size and beach-face slope". *Am. Geophys. Union Trans.* Vol. 32, N° 6, pp. 866-874.

Bodge, K., (1992). "Representing equilibrium beach profiles with an exponential expression". *Journal of Coastal Research.* Vol. 8, N° 1, pp. 47-55.

Bruun, P. (1954). "Coast erosion and the development of beach profiles". Beach Erosion Board, *Technical Memorandum, N° 44.*

Dean, R.G. (1977). "Equilibrium beach profiles: U.S. Atlantic and Gulf coasts". Department of Civil Engineering, *Ocean Engineering Report N°. 12*, University of Delaware, Newark, Delaware.

Hayden, B., Felder, W., Fisher, J., Resio, D., Vincent, L. and Dolan R. (1975). "Systematic variations in inshore bathymetry". Department of Environmental Sciences, *Technical Report N°. 10*, University of Virginia, Charlottesville, Virginia.

Keulegan, G.H. and Krumbein, W.C. (1919). "Stable configuration of bottom slope in a shallow sea and its bearing on geological processes". *Transactions American Geophysical Union*, 30(6) 855 - 861.

Moore, B. (1982). "Beach profile evolution in response to changes in water level and wave height". M. S. Thesis, University of Delaware.

Pilkey, O., Young, R., Riggs, S.R., Sam Smith, A.W., Wu H. and Pilkey, W.D. (1993). "The concept of shoreface profile of equilibrium: A critical review". *Journal of Coastal Research*, Vol. 9, N°. 1, pp. 255 - 278.

Suh, K. and Dalrymple, R. (1988). "Expression for shoreline advancement of initially plane beach". Jour. of Waterway, Port, Coastal and Ocean Eng. ASCE Vol. 114, N°. 6.

Sunamura, T. and Horikawa, K. (1974)." Two-dimensional beach transformation due to waves". *Proceedings, Fourteenth International Conference on Coastal Engineering* (ASCE), pp. 920-938.

Vellinga, P. (1983). "Predictive computational model for beach and dune erosion during storm surges". *Proceedings, (ASCE), Specialty Conference on Coastal Structures '83*, pp. 806-819.

Zheng J. and Dean, R.G. (1997). "Numerical models and intercomparisons of beach profile evolution". *Coastal Engineering.* Vol. 30, pp. 169-201.

The Role of Chronology in Long-Term Morphodynamics. Theory, Practice and Evidence

Howard N Southgate[1] and Michele Capobianco[2]

Abstract

In the field of modelling of long-term coastal morphodynamics, one of the fundamental questions that still remains to be answered, and to a large extent also properly formulated, is that of the role of chronology, or sequencing, of (effectively) random hydrodynamic forcing events. Chronology can refer to the effects on morphology on the timescale of groups of storms, but also to longer term "events" such as entire seasons. Wave conditions are the main type of random hydrodynamic forcing events considered here, but the same ideas apply to water level and wind conditions. This paper introduces a theoretical framework for understanding chronology in terms of non-linear dynamical systems, describes a methodology for detecting chronology effects from field data, and discusses the implications for using computer models for making long-term predictions of coastal morphology. Uninterrupted sandy coastal systems are considered from a cross-shore and longshore point of view, but the general principles apply to more complex grain types and coastline geometries. In this paper, "medium-term" refers to timescales up to several weeks and thus covers individual storms and small groups of storms, while "long-term" refers to periods between several weeks and a few decades and thus covers seasonal and annual "events".

Introduction

Coastal morphodynamics can be regarded as a strongly non-linear dynamical system, and generally the response of such a system will be sensitive to the sequence of forcing events. These effects of chronology, however, are system-dependent and difficult to generalise. For coastal morphodynamics, some observational features can be identified which could be attributable to the occurrence of chronology effects (whether they are requires further data analysis or modelling, which will be discussed later in the

[1]HR Wallingford Ltd, Howbery Park, Wallingford, Oxfordshire, OX10 8BA, UK
[2]Tecnomare SpA, R&D Division, 3584 San Marco, 30124 Venezia, Italy

paper).

Concerning coastal profiles, it could be argued that a given subsequence of storms can create conditions to protect the upper profile from erosion by further storms (e.g. by creating a breaker bar to dissipate incoming large waves). On the other hand, a different set of storm conditions may make the upper profile more vulnerable to further attack (e.g. if the bar created is lower). On the timescale of seasons, it can be expected that the summer and winter profiles will be affected by different sequences of strong and mild winters (and the same for summers). The behaviour of nearshore bars can also be expected to respond differently to different sequences of forcing conditions. Different dynamical states can be identified in the evolution of nearshore bar systems, often with relatively abrupt transitions between them. The timing of these transitions (or even whether they occur at all) could depend on the chronology of forcing events.

Similar considerations can be applied to the dynamics of longshore processes. It may be anticipated, however, that chronology effects may be more significant in complex morphological systems than for uninterrupted coastal stretches. In such situations the number of possible dynamic attractor states, and hence the number of possible transitions between them, are likely to be higher.

In a general sense, chronology effects could be substantially different on different timescales. Over periods of weeks, months and years, observations commonly show a large variability of nearshore levels, compared with longer term trends, which may indicate that chronology effects are significant. Over longer timescales, perhaps many decades, nearshore responses may be insensitive to the sequencing of events at the storm or seasonal timescales, and depend only on the overall statistical characteristics of the forcing conditions and the chronology of events in the last period.

The ability to answer, or even to formulate properly, some of these questions will make an important contribution to the wider problem of predicting long-term coastal morphodynamic behaviour. This is especially relevant in present-day engineering practice in which it is common to run morphodynamic models using "representative waves" or "synthetic time series of wave data" or "measured wave sequences" without considering the effects that the range of possible alternative wave sequences would have.

Theoretical Framework

A definition of chronology, as it relates to the effects on coastal morphology, can be given as follows:

"The effects on coastal morphology of different sequences of waves (or, in general, any random forcing conditions) with the same overall statistical properties"

The phrases "random forcing conditions" and "overall statistical properties" require some clarification. Some types of forcing conditions, such as tides, are very largely deterministic and predictable over long periods of time into the future (at least several decades, the longest time periods of concern in this paper). In these cases, chronology is not an issue. In contrast, other forcing conditions such as wind, waves and processes induced by wind and waves (e.g. water level variations and nearshore currents) are to a large extent random and therefore unpredictable into the future. Nevertheless, there is some (limited) information about these forcing conditions that can be predicted. Some examples for wave forcing conditions are:

1)   An initial period (about one or two weeks) corresponding to the period of predictability of meteorological events and the travel time of distantly generated waves.
2)   The statistical properties of waves over cycles smaller than the full wave sequence (e.g. the annual cycle).
3)   The sequence of wave conditions within individual storms or other meteorological events.
4)   The statistical properties of grouping of storms

In a study of chronology effects on morphology, these types of information should be treated as constraints on the construction of otherwise random sequences of forcing conditions.

Returning to the definition of chronology, the "overall statistical properties" refer to properties which pertain to the whole wave sequence, but not to the ordering of wave data in that sequence. Such properties would include the average wave height, period and direction, and their variances and higher moments. However, these properties would not include the autocorrelations or cross-correlations of the above parameters, since these depend on the sequencing of data values.

These "overall statistical properties" can often be predicted at a future time to a reasonable level of accuracy. For example, the frequency of occurrence of wave heights, periods and directions can be estimated and is usually expressed in tabular form or diagrammatically as a wave rose.

Detecting Chronology Effects from Field Data

There are two fundamental difficulties with detecting chronology effects from field data. The first is to distinguish between a direct response of bed levels to the forcing conditions, and an inherent dynamical behaviour. Both are generic types of response which can show apparently random variations of bed levels. The second difficulty is that nature provides only one actual sequence of forcing and response; it is not possible to go back in time and "rerun" the forcing conditions in a different order. Methods to address these two difficulties will be considered in turn.

Distinguishing "Chaotic" Responses from Random Forced Responses

Coastal morphodynamic data sets which cover a long-term period (a few decades) but which are sampled at medium-term intervals (a few weeks) are quite rare. The data sets which do exist usually show gradual long-term trends in bed levels with much larger medium-term variations about these trends. This is evident in the data sets from Duck, North Carolina, USA (Lee et al, 1996; Möller and Southgate, 1997), and Lincolnshire, UK (HR Wallingford, 1991; Southgate and Beltran, 1995). The effect is most pronounced in and around the surf zone, and is less strong further offshore. Successive bed level differences are typically an order of magnitude larger than measurement errors.

Time series of these bed level variations appear to the eye and to simple statistical tests to be essentially random. One possibility is that they represent a direct response to the effectively random sequence of wave conditions. Another possibility is a chaotic type of behaviour which would give rise to random-appearing sequences of bed levels, but which would not be related to the sequence of forcing conditions. There is strong *prima facie* evidence for the latter possibility from present understanding of the non-linear processes and feedback mechanisms involved in coastal morphology (Southgate, 1997; Southgate and Beltran, 1996). In general, it would be expected that both the chaotic and direct forcing mechanisms would be present to some degree, although one or the other may be stronger over different timescales and in different regions of the coastal zone.

In order to assess the importance of wave chronology on morphodynamic development, it is essential to be able to distinguish the relative importance of the two mechanisms. If chaotic behaviour predominates, the sequence of wave conditions is relatively unimportant, and therefore the effects of alternative sequences will also be unimportant. However, if direct forced responses predominate, then the possibility exists that the chronology of the forcing conditions is a significant effect.

Möller and Southgate (1997) have used morphodynamic data from Duck to attempt to identify the timescales and coastal regions over which each of these two generic mechanisms predominate. The techniques make use of the fact that chaotic behaviour of the dynamic variables (especially in high-dimensional systems such as pertain in coastal morphology) is often revealed by fractal statistics of these variables. An approximate summary of the findings is that chaotic-type behaviour was stronger over timescales longer than 12-24 months in regions of wave breaking, and shorter than 12-24 months outside wave breaking regions. One would therefore look for evidence for chronology effects in the regions and timescales where chaotic behaviour was relatively weak.

Figures 1 and 2 show an example from Duck where possible effects of chronology can be detected by visual inspection. The bathymetry surrounding the research pier at Duck is shown in Figure 1, together with four profile lines along which

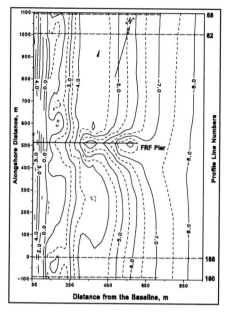

Figure 1. Bathymetry at Duck, NC, USA showing the Four Profile Transects

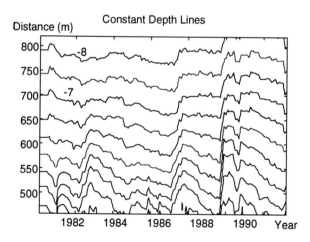

Figure 2. Depth Values along Profile 188 at Duck

bed level measurements were made. These measurements were made over a 10.5 year period between 1981 and 1992. Figure 2 shows the locations of depth values along one of these profiles, labelled Profile 188 (located 0.5 km south of the research pier). The distances (on the vertical axis) are measured relative to a fixed datum on the shore. Depth values are relative Mean Sea Level, and are shown at 0.5m intervals, the furthest depth offshore being at 8m. In Figure 2 these profile measurements are "stacked" side-by-side through time so the time history of depth values can be seen by following particular "contours" from left to right.

Figure 2 reveals strong, short-term accretionary events in 1982, 1987 and 1989 with periods of gradual erosion between them. Comparison with simultaneous wave data shows that the accretionary events coincide with large storms. These erode the upper profile, and the material is deposited further offshore, in the region covered by the figure. Under more moderate conditions, onshore transport mechanisms bring this material back onshore. The chronology of these storms can be seen to be significant. After the 1982 event, the depth values have returned to their original position by the time of the next event in 1987. However, after this event the depth values have not had sufficient time to return to their previous values (at some offshore distances) before the next event in 1989.

Nature Provides Only One Realisation

An obvious difficulty in detecting chronology effects from field data is that the data pertains to only one realisation of all the possible time sequences that could occur satisfying the overall statistical properties and any predictability constraints. However, evidence for chronology effects can be found, in principle, by searching the morphology and wave data sequences for two (or more) cases where the initial morphology is nearly the same and where the subsequent wave statistics over a given interval are nearly the same. If the wave sequences in these intervals are substantially different, and the morphological development in these intervals is also substantially different, this will provide evidence for a significant effect of chronology. There is some similarity with nearest-neighbour searching used in the method of "forecasting signatures" for analysing time series to determine the best of a set of trial models of the system dynamics (see Sugihara, 1994; Rubin, 1992 and Southgate, 1997).

There are some substantial practical difficulties with such a method. The most obvious is that long time series of data will be needed to find two or more cases with similar starting morphology and subsequent wave statistics. In fact, there will be a trade-off between the accuracy in determining what are "similar" initial morphologies and subsequent wave statistics, and the length of the time series needed.

An initial attempt along such lines is described in Capobianco and Southgate (1997) where the 10.5 years of profile evolution data, together with 6 hourly time series of wave data at Duck, have been considered. Classes of significant wave forcing events (on the timescale of meteorological events) have been defined and morphodynamic responses of the profiles to such events have been identified. The results however cannot be considered to be statistically significant due to insufficient numbers of these classes. A possible method of analysis with limited data is to identify initial starting morphologies in terms of initial attractor states and study the effects of chronology on transitions between such states. Such states could be defined in quite broad terms, for example, they may relate to mean levels of profiles, or to the presence and locations of one or more nearshore bars.

Although in-situ measurements do not provide sufficient quantities of morphodynamic data for the procedures outlined above to be carried out in the detail that would be desired, much larger morphological data sets can potentially be derived from remote sensing. An example is the video-based ARGUS system from which it is possible to derive data on the locations of nearshore bars (Holman, 1997).

Investigating Chronology Effects by Modelling

The problems of detecting chronology effects from only one realisation of a time sequence of data do not apply to computer modelling of chronology effects (apart perhaps from the sheer computational effort involved). Repeated running of a computational model with identical starting morphology, identical overall wave statistics, but different wave sequences can be readily done. It would also be possible to follow this approach in physical models, although the time involved may make it impractical in some cases.

One requirement for a modelling study of chronology effects is to be able to generate wave sequences with the same overall statistical properties. Starting from these overall statistical properties, it is possible to generate different time series using techniques such as those described in Scheffner and Borgman (1992) and Le Mehaute et al (1983). These methods will need to respect any predictable properties such as those mentioned earlier under "Theoretical Framework". Furthermore, it may be desirable to combine different wave parameters into more useful parameters for a given application. For example, in studies of longshore transport, it is more appropriate to use single time series of potential longshore drift rates (which are a combination of wave height and direction) rather than separate series of heights and directions (Millard et al, 1997).

A drawback with using wave time series generated from a given statistical distribution, is that the overall statistical properties of the series do not exactly match the original statistics (or agree exactly with each other). For example, in Scheffner and Borgman (1992), there were typically differences of 2-4 per cent in mean wave height and period, and 10-15 per cent in the standard deviation of these parameters, based on generation of yearly data sequences. For monthly data sequences, differences were in some cases as much as 50 per cent. This illustrates that the reproduction of the wave statistics, especially of higher order moments and extreme events, is a difficult problem. One practical solution is to generate just one time series (or to use a measured series if available) and to reorder wave values within this series. This ensures that the overall statistical properties of each wave series are exactly the same. Southgate (1995) has indicated some guidelines for taking account of predictable properties as part of the procedure for reordering the wave series.

In Southgate (1995), the procedure was applied to a four-month set of measured wave height data with a time interval of 3 hours. The data sequence was split into five portions, respecting as far as possible the predictability constraints. These five portions were arranged in every possible combination, giving a total of 120 wave sequences. Each of these sequences was run through a morphodynamic profile model (COSMOS-2D, see Southgate and Nairn, 1993, and Nairn and Southgate, 1993). The initial bathymetry consisted of a uniformly-sloping beach, and was identical for each run. Results were processed into probability of exceedance envelopes of bed levels along the profile. The results indicated that the variations in bed levels between runs could be as significant as trends common to all the runs.

The procedure can be applied in principle to any type of model or coastal geometry appropriate to the model being used. However, the computational effort to perform repeated model runs may render the method impractical for detailed 2DH or 3D models operating over many morphodynamic timesteps. Profile models and one-line (or N-line) planshape models are examples of morphodynamic model types where the computational effort is sufficiently low for the procedure to be readily carried out. Millard et al (1997), Hanson (1988) and Le Mehaute el al (1983) describe applications to one-line planshape models, focusing on longshore transport processes.

In this general approach to modelling chronology effects, care has to be taken to determine whether the variability between runs is a genuine 'forced' morphodynamic response, and is not the result of an inherent chaotic behaviour or a simple accumulation of numerical errors. One test that could be carried out is to make slight variations to the input data sequence and investigate how strongly the morphodynamic response varies between the runs. It may also be possible to detect chaotic behaviour by fractal analyses of the morphodynamic response.

Conclusions

Given our present knowledge of the strongly non-linear relationship between hydrodynamic forcing and coastal morphodynamic response, there is a strong possibility that chronology effects will be an important factor, at least over some timescales and for some coastal situations. The detection of such effects from morphodynamic field data is tricky, and may require higher resolution measurements over longer time periods than are presently available for definitive conclusions to be drawn.

In contrast, chronology effects can be readily studied in computational morphodynamic models. It should be possible to perform a wide range of computational investigations of chronology effects using different model types, coastal situations and timescales. These would determine the conditions under which chronology effects are (and are not) likely to be important. Where chronology effects are shown to be important, it is recommended that they are investigated, as a standard procedure, in practical studies using methodologies such as those outlined above.

Acknowledgements

This paper is based on work in the PACE project, in the framework of the EU-sponsored Marine Science and Technology Programme (MAST-3), under contract no. MAS3-CT95-0002

HNS acknowledges additional funding from the UK Ministry of Agriculture, Fisheries and Food (Flood and Coastal Defence with Emergencies Division) as part of the CAMELOT Commission (FD1001).

MC acknowledges additional funding from the SAFE project in the EU-sponsored Marine Science and Technology Programme (MAST-3), under contract no. MAS3-CT95-0004.

References

Capobianco M and Southgate H N (1997), "Investigation on the Role of Chronology in Long-Term Morphodynamics. Looking for Field Evidence", PACE Report, First Overall Workshop, Barcelona, March 3-7, 1997

Hanson H (1988), "GENESIS - A Generalised Shoreline Change Numerical Model", J. Coastal Research, Vol.5, No. 1, pp1-27

Holman R A (1997), "Bulk Comparison of the World's Beaches", This Issue

HR Wallingford Ltd (1991), "Mablethorpe to Skegness, Lincolnshire. Strategic Approach Study. Topic 1. Understanding Beach Level Trends and Predicting Future Erosion Rates", HR Report EX 2333

Lee G H, Nicholls R J, Birkemeier W A, and Leatherman S P (1995), "A Conceptual Fairweather-Storm Model of Beach Nearshore Profile Evolution at Duck, North Carolina, USA", J. Coastal Research, Vol. 11, No. 4, pp1157-1166

Le Mehaute B, Wang J D and Lu C-C (1983), "Wave Data Discretization for Shore Line Processes", J. Waterw. Port Coastal Ocean Eng., ASCE, Vol. 109, No. 1, pp63-78

Millard T K, Southgate H N and Bunn N P (1997), "Quantifying the Uncertainty in Beach Planshape Modelling to Wave Chronology", Coastal Dynamics '97 (abstract)

Möller I and Southgate H N (1997), "Fractal Properties of Beach Profile Evolution at Duck, North Carolina", J. Geophysical Research (submitted)

Nairn R B and Southgate H N (1993), "Deterministic Profile Modelling of Nearshore Processes. Part 2. Sediment Transport and beach Profile Development", Coastal Engineering, Vol. 19, pp57-96

Rubin D M (1992), "Use of Forecasting Signatures to help distinguish Periodicity, Chaos and Randomness in Ripples and other Spatial Patterns", Chaos, Vol. 2, No. 4, pp525-535

Scheffner N W and Borgman L E (1992), "Stochastic Time-Series Representation of Wave Data", J. Waterw. Port Coastal Ocean Eng., ASCE, Vol. 118, No. 4, pp337-351

Southgate H N (1995), "The Effects of Wave Chronology on Medium and Long Term Coastal Morphology", Coastal Engineering, Vol. 26, pp251-270

Southgate H N (1997), "Non-Linear Methods for Analysis of Beach Level Data", Proc. 27th IAHR Congress, Theme A, San Francisco, USA, pp46-51

Southgate H N and Beltran L M (1995), "Time Series Analysis of Long-Term Beach Level Data from Lincolnshire, UK", Proc. Coastal Dynamics '95, Gdansk, Poland, ASCE, pp1006-1017

Southgate H N and Beltran L M (1996), "Self-Organisational Processes in Beach Morphology", Proc. Conference on Physics of Estuarine and Coastal Seas, The Hague, The Netherlands

Southgate H N and Nairn R B (1993), "Deterministic Profile Modelling of Nearshore Processes. Part 1, Waves and Currents", Coastal Engineering, Vol. 19, pp27-56

Sugihara G (1994), "Non-Linear Forecasting for the Classification of Natural Time Series", Phil. Trans. R. Soc. Lond. A, Vol. 348, pp477-495

Measurements at a Cusped Beach in the UKCRF

Alistair G L Borthwick[1], Yolanda L M Foote[2] and Andrew Ridehalgh[2]

Abstract

Wave-induced currents at cusped beaches are important for large-scale horizontal mixing processes. In this paper, data are presented on the nearshore currents due to a sinusoidal multi-cusped beach installed at the U.K. Coastal Research Facility (UKCRF). Horizontal spatial patterns of the currents have been determined by digital image analysis of the motions of neutrally-buoyant markers. For normally incident regular waves, combinations of rip and longshore currents gave rise to a steady system of multiple circulation cells. Oblique incident waves generated a stable meandering longshore current. Detailed information on the depth profiles of rip and meandering nearshore currents was obtained using Sontek acoustic Doppler velocimeters (ADVs). The profiles indicate that the outflowing rip current is nearly-horizontal with low frequency oscillations evident. The meandering current is steady and has a logarithmic profile near the bed.

Introduction

At multi-cusped beaches, waves do not break uniformly, and so the excess mass and momentum fluxes lead to the production of nearshore currents. The pattern of nearshore currents is extremely important in the nearly-horizontal advection and large-scale mixing of suspended sand and buoyant pollutants (such as oil slicks which cover several surf-zone widths). For normally incident waves at a multi-cusped beach, longshore currents are generated parallel to the beach contours, and directed from cusp crests to troughs where they feed into rapidly flowing offshore jets (rip currents). In turn, these are balanced by inflowing currents at the cusp crests. Thus a system of primary nearshore circulation cells is established at a multi-cusped beach by normally-incident regular waves. For oblique

---

[1] Reader, Dept of Engng Science, Oxford University, Oxford OX1 3PJ, U.K.

[2] Research Assistant, Dept of Engng Science, Oxford University, Oxford, U.K.

incident waves, the excess fluxes predominate in one alongshore direction, and so a meandering longshore current occurs instead of a primary nearshore circulation system. For all regular waves, whether normally incident or oblique, a system of secondary circulation cells is generated in the shallowest water, inshore of the breaker line and encompassing the swash zone. At erodible beaches, the cusp spacing is due to spatial variations in bed topography, the sediment fall velocity and the breaker height (Huntley and Short, 1992). Mechanisms proposed for rip current generation include the uneven distribution of breaking wave heights, edge waves, hydrodynamic instability and intersecting wave trains (Arcilla and Lemos, 1990).

Although field and laboratory measurements have been made of multiple circulation cell systems (e.g. Sonu, 1972, Horikawa and Sasaki, 1972, and Ozaki *et al.*, 1977), the majority of investigations have considered single circulation cells (e.g. Dalrymple *et al.* (1977), da Silva Lima (1981) and Hamm (1992)). There is a need for detailed information on the wave-induced currents near cusped beaches; Arcilla and Lemos (1990) note the lack of data on the vertical structure of a rip current.

This paper describes experimental data obtained at the U.K. Coastal Research Facility (UKCRF) on nearshore wave-induced currents due to a sinusoidal multi-cusped beach. Horizontal spatial patterns of the wave-induced currents have been determined by digital image analysis of video film of neutrally-buoyant markers. Wave gauge data have given information on the wave height field. Detailed descriptions of the three-dimensional vertical profiles of rip and meandering longshore currents have been determined from point velocity measurements using Sontek acoustic Doppler velocimeters (ADVs). The data obtained should be useful for the verification of nearshore wave-current interaction numerical models.

The UKCRF Multi-cusped Beach

The UKCRF has plan dimensions 27 m (cross-shore) x 36 m (alongshore) with a working area of 20 x 15 m. Waves are generated using a 72 paddle wave-maker; longshore currents may be recirculated using 4 pumps. The mean water level at the paddles was 0.5 m.

The multi-cusped beach had overall dimensions of 12 m alongshore and 5 m across-shore. It consisted of three sinusoidal cusps situated on a 1:20 plane beach, with the still water depth given by

$$h(x,y) = s\left[\left(x_L - x\right) - A \sin\left(\frac{\pi \left(x_L - x\right)}{x_L} \right)\left(1 + \sin\left(\phi - \frac{2\pi y}{R} \right)\right) \right] \quad \text{(1)}$$

where x is the distance measured onshore from the cusp's toe (located 5m onshore from the toe of an underlying plane beach of slope, $s = 0.05$), y is measured alongshore from the edge of the cusps, $x_L = 5$ m is the cross-shore length of the cusps, $A = 0.75$ m is the amplitude of the sine wave used to generate the cusped profile above the plane beach, $R = 4$ m is the longshore wave length of a single cusp and $\phi = 3\pi/2$ is a phase angle. The three cusps were therefore located within $0 \leq x \leq 5$ m and $-12 \leq y \leq 0$ m. The product of A and s represented the maximum height of cusp possible for the above configuration; a larger value would have resulted in a shoal being created. The choice of formula was based on earlier laboratory-based work by da Silva Lima (1981) and Borthwick and Joynes (1992) who modelled a single circulation cell in the vicinity of a half-sinusoidal beach. Figure 1 depicts the still water depth contours and outline of the multi-cusped beach within the UKCRF basin.

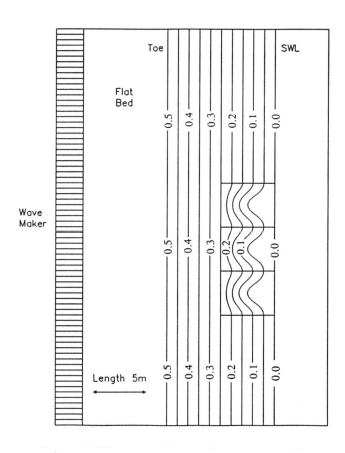

Figure 1   Still water depth contours (all values in m).

The cusped beach design was aided by numerical predictions of the wave-induced current patterns using a depth-averaged 2DH wave-current interaction model based on Yoo and O'Connor's (1986) scheme. The model was used to predict the nearshore circulation pattern in the vicinity of the multi-cusped beach due to regular offshore waves of height $H = 0.125$ m and period $T = 1.2$ s. For waves at zero incidence, the simulations predicted that symmetric counter-rotating primary and secondary circulation cells with rip currents of velocity above 30 cm/s would be produced at the multi-cusped beach. It should be noted that the grid used for the simulations was rather coarse (a mesh spacing of 0.5 m) and so the magnitude of the rip current may have been underpredicted.

Although the cusp geometry was not fitted *a priori* to the incident wave parameters, calculations of the Iribarren number (for the plane beach) gave a typical value of $Ir_o = 0.2$, corresponding to spilling breakers and indicating that the rip spacing on an erodible plane beach would be dictated by infragravity waves, with values ranging from 4 to 5 m (using the empirical formula given by Ozaki *et al.*, 1977). The designed cusp spacing was slightly narrower in order to fit three cusps into a budgeted 60 m^2 plan area, allowing for 5 m cross-shore length (half the distance across the plane beach).

The sinusoidal cusps were fabricated using a cement mortar skim moulded over granular fill placed on an existing concrete plane beach. Beach construction was carried out over a two week period by builders sub-contracted from HR Wallingford Ltd. An as-constructed survey indicated that the multi-cusped beach profile was accurate to within ± 0.1 cm. The final surface of the cusps was smooth, though the surrounding plane beach had an average roughness height of 1.7 cm.

After the tests were completed, the moulded cusped beach was removed and the concrete surface of the plane beach reinstated.

Experimental Measurements

The tests took place in March 1996. During the first phase, two overhead cameras were used to record the movement of 10 cm diameter neutrally-buoyant markers distributed throughout the test zone. The video images were digitised using a 768 x 572 RGB pixel frame grabber on a Pentium computer system. After enhancement by thresholding and lens distortion calibration, the images were processed wave-by-wave to give the spatial distribution of velocity vectors corresponding to the depth-averaged currents. The corresponding wave height fields were determined from an array of point measurements of the free surface elevation using one mobile and 7 fixed wave gauges. In the second phase, 3-D velocity component measurements were taken using the ADVs, with particular emphasis on the vertical profiles in rip and meandering nearshore currents at selected intervals along a cusp trough and a cusp crest. Four wave conditions were considered, with the mean water level at the wave-maker fixed at 0.5 m. The wave parameters were as follows: (i) Case A, regular waves, period 1.0 s, height 0.1 m, 0° incident angle; (ii) Case B, regular waves, period 1.2 s, height 0.125 m, 0° incident angle; (iii) Case C, oblique regular waves, period 1.2 s, height 0.125 m, 20° incident angle; and (iv) Case D, random waves, peak period 1.2 s, significant wave height 0.125 m, incident angle 0°. The results presented below concentrate on Cases B and C.

Results

The velocity vectors obtained from the digital image analysis were scattered, but tended to be concentrated in regions where the current was reasonably strong. In each case, approximately 5,000 wave-averaged velocity vectors were estimated; these were then interpolated onto a uniform grid using an inverse weighted interpolation scheme. In order to help identify the flow patterns (in particular the circulation cells) the uniform field of velocity vectors was multiplied by the local depths and the resulting fluxes integrated from the centre of the domain of interest to give an estimate of the depth-averaged stream function. Figure 2 shows the wave-averaged velocity vectors for Case B; the corresponding interpolated vectors and depth-averaged stream function contours are presented in Figure 3. Examination of these data indicates that there are nearshore circulation patterns formed from fan-shaped inflowing currents running up the peak of each cusp and narrow jet-like outflowing rip currents flowing rapidly offshore at the cusp troughs. The outflowing rip currents last a short distance after the wave breakers.

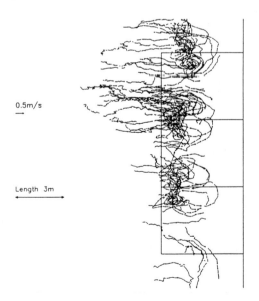

0.5m/s

Length 3m

Figure 2   Scattered current vectors: Case B

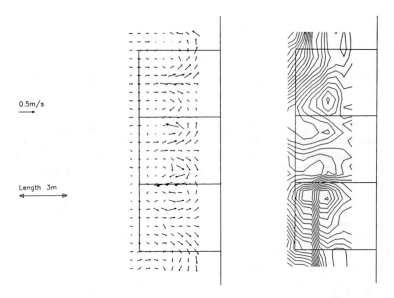

Figure 3   Interpolated current vectors and stream contours: Case B

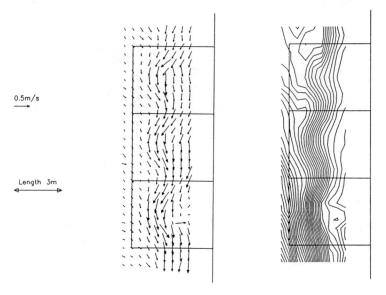

Figure 4   Interpolated current vectors and stream contours: Case C

The interpolated wave-averaged velocity field and depth-averaged stream function contours for Case C presented in Figure 4 clearly show the presence of a meandering nearshore current.

Figures 5 and 6 illustrate the time-varying horizontal current vectors at 0.02s intervals obtained at two locations in the rip current along the central cusp trough using the Sontek ADVs, at 10% of the water depth. Locations P1 and P2 are 2 m and 1 m inshore, respectively, from the toe of the cusp. It is clear that, although the vector directions are reasonably constant (i.e. always pointing offshore), there is some low frequency oscillation evident. The vectors at point P1 (at the heart of the rip current) are of greater magnitude than P2 where the rip current is decaying. Figure 7 shows the equivalent (meandering longshore) current vectors obtained for Case C at point P1 (again at 10% of the water depth). In this case, current vectors are much more constant indicating the stability of the oblique wave-induced meandering current.

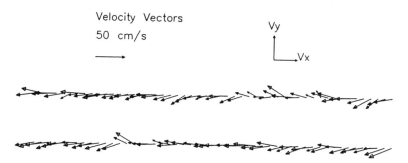

Figure 5   Time variation in horizontal current components: Case B, location P1

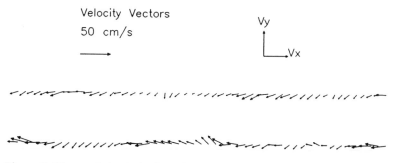

Figure 6   Time variation in horizontal current components: Case B, location P2

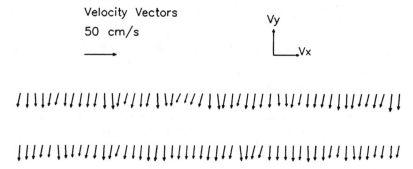

Figure 7   Time variation in horizontal current components: Case C, location P1

Figure 8 presents the vertical profile of rip current magnitude (with the bars representing the standard deviation in the measurements) for Case B at location P1. Although the vertical profile further offshore was found to be logarithmic close to the bed, it is more complicated at P1 where the current is strongest.

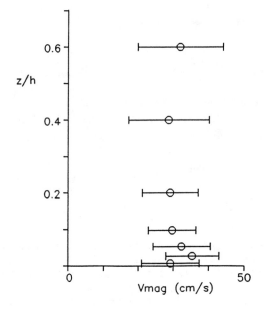

Figure 8   Vertical profile of current (magnitude): Case B, location P1

Figure 9 shows the vertical profile of the meandering longshore current for Case C at point P1. The profile is logarithmic at the bed, with a maximum value at about 10% of the water depth, and then gradually decreases towards the free surface. The very small standard deviations confirm the stability of the meandering longshore current. In all cases, the vertical velocity component was near zero.

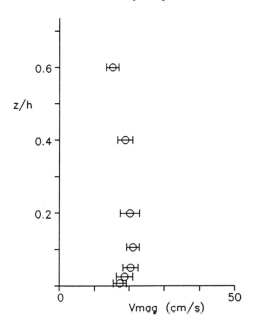

Figure 9   Vertical profile of current (magnitude): Case C, location P1

Conclusions

Measurements have been presented of the large-scale horizontal mixing currents which occur in multiple circulation cells at a multi-cusped beach installed in the UKCRF. Sontek acoustic Doppler velocimeters gave detailed information on the depth profile of rip and meandering nearshore currents.

The measurements demonstrated that regular waves normally incident to the multi-cusped beach caused nearshore circulation cells to develop. The primary cells consisted of fan-shaped inflowing currents which run up the cusp peak, longshore currents from cusp peak to trough, and narrow jet-like rip currents which flow offshore at the cusp troughs. The outflowing rip currents are nearly horizontal (especially at the region of maximum velocity) with low frequency oscillations evident. The high velocity rip currents last a short distance offshore of the wave breakers. Within the range of tests considered, peak current velocities were in the order of 50 cm/s. Oblique incident waves produced an extremely stable meandering nearshore current, with a logarithmic profile near the bed.

Acknowledgements

The work described herein was funded by the Built Environment Committee of the Engineering and Physical Sciences Research Council through EPSRC grant GR/K04125. The authors would like to acknowledge the UK EPSRC and HR Wallingford Ltd for their co-sponsorship of the UKCRF. Mr W Wooster (Oxford University), Mr A Channell (HR Wallingford) and Professor R Eatock Taylor (Oxford University) made significant contributions to managing, and implementing the test programme.

References

Arcilla, A.S., and Lemos, C.M. (1990) *Surf-zone hydrodynamics*. Pineridge Press.

Borthwick, A.G.L., and Joynes, S.A. (1992) Laboratory study of oil slick subjected to nearshore circulation. *J. of Environ. Engrg.*, ASCE, 118(6), 905-922.

Dalrymple, R.A., Eubanks, R.A., and Birkemeier, W.A. (1977) Wave-induced circulation in shallow basins. *J. Waterway, Port, Coastal and Ocean Div.*, ASCE, 103(WW1), 117-135.

da Silva Lima, S.S. (1981) *Wave-induced nearshore circulation*. PhD Thesis, Department of Civil Engineering, University of Liverpool, U.K.

Hamm, L. (1992) Directional nearshore wave propagation over a rip channel: an experiment. *Proc. 23rd Int. Conf. on Coastal Eng.*, ASCE, 226-239.

Horikawa, K., and Sasaki, T. (1972) Field observations of a nearshore current system. *Proc. 13th Int. Conf. on Coastal Eng.*, ASCE, 635-672.

Huntley, D.A., and Short, A.D. (1992) On the spacing between observed rip currents. *Coastal Engineering*, 17: 211-225.

Ozaki, A., Sasaki, M., and Usui, Y. (1977) Study on rip currents. *Coastal Engineering in Japan*, 20: 147-158.

Sonu, C.J. (1972) Field observations of nearshore circulation and meandering currents. *J. Geophys. Res.*, 77(18), 3232-3247.

Yoo, D., and O'Connor, B.A. (1986) Mathematical modelling of wave-induced nearshore circulations. *Proc. 20th Int. Conf. on Coastal Eng., ASCE*, 1667-1681.

WAVE-INDUCED CURRENTS IN MULTIPLE-BAR NEARSHORE ZONE

Marek SZMYTKIEWICZ, Zbigniew PRUSZAK & Ryszard B. ZEIDLER[1]

Abstract

Recent field campaigns at IBW PAN's CRF Lubiatowo, with its multiple bars, have provided insight into the complex three-dimensional structure of wave-induced circulation. Tracer experiments (mostly bedload) point to importance of longshore transport between bars (effect of transport parallelisation by bars) and confinement of transport between bars, even at high waves. Longshore velocities are almost identical at bar crests and troughs and display a fairly uniform vertical profile. Cross-shore velocities are vertically non-uniform but their mean values are also similar at crests and troughs. The cross-shore circulation displays a vortex (whirl) between bars. Free-surface elevation gradients may become as important as radiation stresses in generation of nearshore currents. Reliable linkage between Lagrangian and Eulerian characteristics must be worked out to shed more light on nearshore velocities and sediment transport

1. Introduction: site, objectives and scope

The data analysed and discussed in this paper stem exclusively from the Coastal Research Facility (CRF) Lubiatowo. The facility is situated on the Polish coast of the Baltic Sea, some 75 km from Gdańsk (see insert in Figure 1). The seabed topography consists of multiple longshore bars (Fig. 1). The mean bed slope is 1%, and the mean grain diameter equals 0.22 mm. The wave energy flux is usually directed at an angle to the shoreline, and so is the resulting sediment transport. Two basic directional sectors can be distinguished – more frequent westerly winds (waves) and the eastern incidence, of lower occurrence but much more energetic, due to longer fetches. The net annual littoral drift at Lubiatowo is directed from west to east.

Recent field studies on coastal hydrodynamics and morphodynamics at Lubiatowo have provided insight into the complex three-dimensional structure of

---

[1] Res. Assoc., Assoc. Prof. & Professor, resp.; Polish Academy of Sciences' Institute of Hydro-Engineering (*IBW PAN*), Gdańsk 80 953, 7 Kościerska, PL

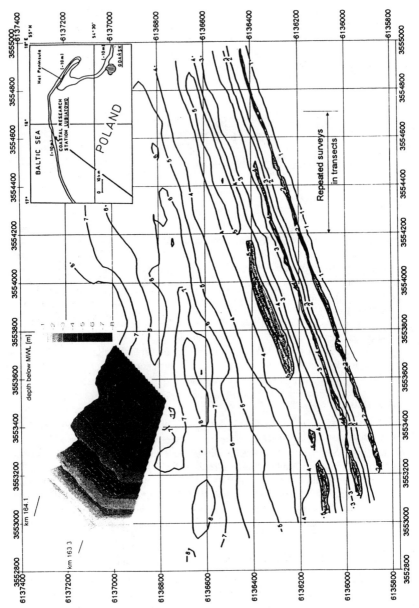

Figure 1. Multiple bars at Coastal Research Facility Lubiatowo (IBW PAN)

wave-induced circulation, in particular some peculiar features of longshore currents in the nearshore zone, with its multiple bars. Measurements of both currents (with the aid of electromagnetic current meters) and sand grain speed (by tracking radiotracers), along with the coupling of sand and water velocities, have implied that the velocities of both media in troughs between bars could be much higher than those predictable by models based on commonly adopted assumptions stemming from the concept of radiation stress.

Two basic sets of data are presented: one on 2-year tracer studies and Lagrangian characteristics of bedload and water flow and another resulting from measurements of water velocities. The radiotracer studies were carried out in the years 1993 and 1994. These findings and other data stemming from radiotracer studies (Pruszak & Zeidler 1995) and some earlier measurements of wave-induced currents have inspired us to look more closely at the hydrodynamics of the water velocity field in the vicinity of bar crests and troughs. This intention was implemented during the Lubiatowo'96 field campaign.

2. Tracers '93+94 and Lubiatowo'96 field campaigns

The field tests with radioactive tracers at CRF Lubiatowo are described extensively i.a. by Pruszak & Zeidler (1995). Figure 2 presents some results of the studies carried out in 1993 and 1994, from which it can be seen that the longshore sediment transport rate (vertical axis) does not differ substantially between bar crests and troughs. The transport rate measured between bars, and the longshore sand speed producing that transport as well, are comparable in magnitude with their counterparts at bar crests. This remains true for any wave climate observed in the radiotracer studies (classes 1-7).

The field campaign Lubiatowo'96 was undertaken to elucidate the phenomena and effects outlined in introduction, in addition to shore evolution investigations (under PACE) and other accompanying studies. The primary objective of the campaign in terms of this paper was to shed light on the peculiarities of longshore currents at sand bars. The arrangement of the measuring equipment is shown in Figure 3, on the background of bed topography measured all over the campaign, from August to December 1996. Five electromagnetic current meters were deployed; waves were measured by a wave rider on a depth of 20.4 m and wave gauges in the vicinity of bars. Wind, waves and currents are recorded over a time span of seven weeks of 1996. The waves measured in November'96 are illustrated in Figure 4. For later purposes of analysis three characteristic time spans have been selected as shown with arrows - days 310 to 315, 321 to 322, and 325 to 328.

3. Findings from Lagrangian description

As pointed out, the flow intensity in troughs may even exceed that at bar crests, if the wave incidence angle is small (Fig.2). It would suggest a `nonenergetic' genesis of the flow in bar troughs, linked rather to inertial and gradiental effects of wave breaking at adjacent bars than to transfer of the radiation stress. This can be put forth especially for bars II and III, where wave breaking occurs almost permanently in the situations studied. While cross-shore velocities become significant seawards of bar IV, primarily in terms of bedload, the zone between bar IV and the water line is strongly controlled by the bars, which bring about intensive reorientation of the water flow and enhance its shore-parallel component.

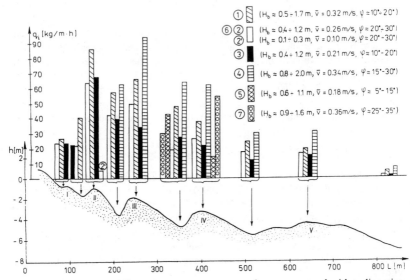

Figure 2. Longshore sediment transport rate across shore, measured with radioactive sand tracers, in various classes of waves and currents

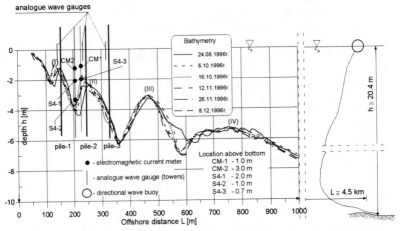

Figure 3. Layout of the measuring apparatus in field campaign Lubiatowo '96

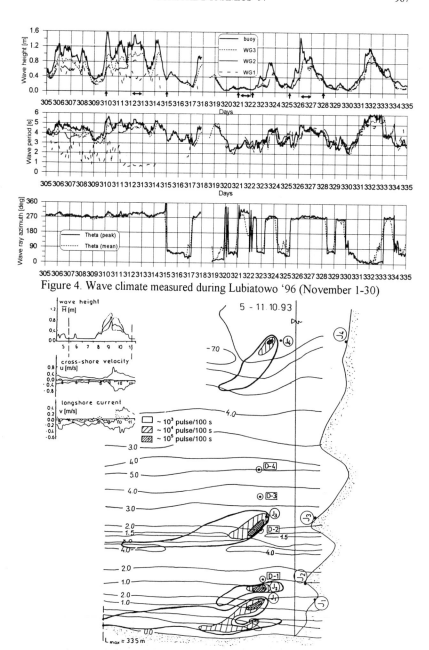

Figure 4. Wave climate measured during Lubiatowo '96 (November 1-30)

Figure 5. Radiotracer studies October 1993

Figure 5 illustrates a typical situation measured in radio-tracer studies. One can note the occurrence of a strong longshore current in bar troughs. It is only far away from the zone of conspicuous bars that bedload in the onshore direction prevails, compare injection J4 beyond the outermost bar. On the contrary, in the vicinity of the other bars the onshore movement predominates over a short distance, and is followed by longshore transport once the bar crest is passed by the tracers. Basically, the same picture is observed in all experiments of 1993 and 1994, in various situations with larger and smaller wave incidence angles, different wave parameters, etc. The occurrence of pronounced littoral drift between bar crests is somehow surprising and does not find a lot of support in the earlier studies, or common belief among coastal researchers. Yet it is evidenced by the Lagrangian characteristics found in our tracer studies and can be attributed to the "nonenergetic" origin of longshore currents in bar troughs. In other words, the gradiental components of free surface elevation become as important as to bring about longshore water velocities comparable in magnitude with those resulting from radiation stresses and wave breaking.

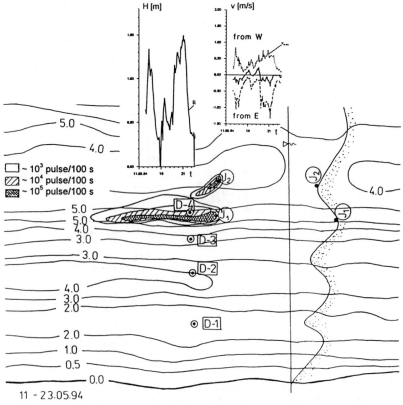

Figure 6. Radiotracer studies May 1994

From the perspective of sediment transport itself, it appears that sand grains are confined to subregions between bars and seldom pass over the boundaries created by bar crests, at least in the region of bars I, II and III (counted from the shoreline, the ephemeral bar excluded). This finding is clearly confirmed by the measurements illustrated in Figure 6, where the radiotracers deployed on the seaward slope of bar 3 move shorewards only to overtop bar crests and then travel exclusively in the longshore direction between bars II and III.

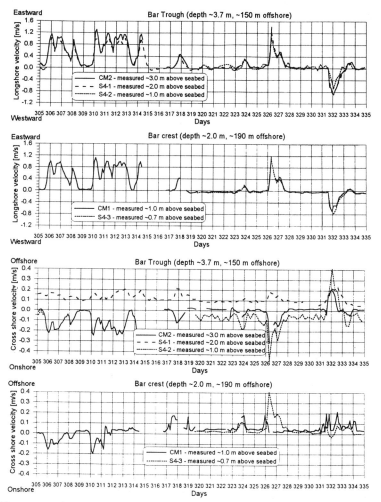

Figure 7. Mean longshore (top) and cross-shore (bottom) velocities at bar crest and trough measured from 1st to 30th November 1996 (days 305 to 335)

4. **Findings from Eulerian description** (Lubiatowo'96)

Results of the measurements of water velocities during the Lubiatowo'96 field campaign are illustrated in Figure 7, for both longshore and cross-shore velocities. One can postulate the following general findings. First, as seen from their temporal behaviour, both longshore and cross-shore velocities at both crests and troughs are very well correlated with surface waves (compare Figure 3). Second, the mean longshore velocities are obviously greater than mean cross-shore velocities. Thirdly, the vertical nonuniformity of longshore velocities is much lower (if any) than that of cross-shore velocities, compare Figure 7a vs. Figure 7b. In the case of cross-shore velocities, one also observes a certain reversal of direction in the vertical profile, from onshore to offshore and onshore again, compare onshore CM2 at the highest elevation above bed (mid-depth), through offshore S4-1 below it, to onshore S4-1 positioned most closely to the seabed. Finally, the absolute velocities at bar crests and troughs are almost identical for longshore velocities on nearly all occasions, while the differences between cross-shore velocities at bar troughs and crests are quite conspicuous, although the mean cross-shore velocities are generally small.

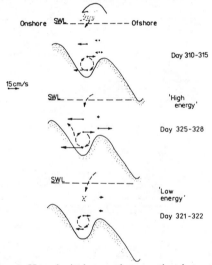

Figure 8. Hypothetical cross-shore vortices between bars

Subject to a slightly closer analysis have been water velocities for the three time spans identified in Figure 4, that is days 310 to 315 and 325 to 328 referred to as „high energy" events and days 321 to 322 labeled „low energy" events. The mean cross-shore velocities are indicated in Figure 8, on the background of a slightly distorted bed topography. The velocity vectors shown in both high and low energy situations suggest the presence of a whirl (vortex) between bar crest and trough. It seems to be more intensive in high energy situations, but its occurrence is quite persistent over the entire period of measurements. Because of the limited quantity of current meters, one cannot claim that the vortex sketched in the drawing is the only possible explanation of the velocity field measured. Some other circulation patterns are also likely to arise. Nevertheless, it seems to be clear that one experiences an offshore outflow of water close to the bar crest and onshore circulation at the very trough, close to its bottom along with mid-depth layers. It should be stressed that the picture results from continuous measurements but does not reflect the velocity field close to and above still water line, where considerable fluxes of mass and momentum occur during the passage of breaking or nonbreaking waves. Hence, the conservation of mass and momentum is secured if one takes into account the entire water column, not only the part indicated in the drawing with velocity vectors.

It should be added that one experiences a substantial variation of instantaneous longshore and cross-shore velocities and the deviation from the mean values shown in Figure 7. For instance, during the days 310 to 315, for the mean longshore velocity about 1 m/s the maxima and minima are respectively above 2.4 m/s and -0.3 m/s (negative values in the westward direction). This is even more pronounced for the onshore velocities, which oscillatey between 1.2 m/s and -1.2 m/s, while the mean value is roughly 0 m/s. The above figures do not differ very much between bar crests and troughs. The oscillatory motion seems to be slightly more pronounced at bar crests, which is due to the shallower water depth above crests vs. troughs.

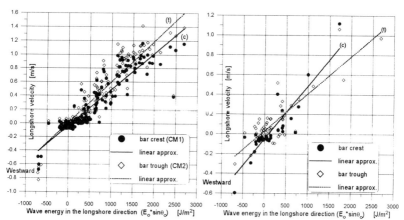

Figure 9. Longshore velocities measured at bar crest and trough vs. wave energy
('negative' to west) at mid-depth (left) and near bed (right)

Hence one can conclude that longshore velocities are almost identical at bar crests and troughs. Yet, in order to examine more closely the bar to crest relationships, one can enter details and present longshore velocities in terms of wave energy in the longshore direction, as compiled in Figure 9. Wave energy seems to be the most natural physical quantity encompassing the effects of both radiation stresses and the gradients of free surface elevation. For the sake of comparison, two water layers were selected at both troughs and crests – central upper (Fig.9a) and nearbed layer (Fig.9b). Seen in the former is a certain predominance of troughs over crests in the sense that for the same wave energy the velocities at troughs are generally higher than at crests. Note that „negative wave energy" is assumed for the westward velocities in order to distinguish between the two longshore orientations. On the other hand, the longshore velocities in the nearbed layer are higher at bar crests than at bar troughs (Fig.9b). Moreover, in the latter case there is a certain threshold velocity/energy about 0.2 and 500 respectively, where the two regression lines in Figure 9b intersect.

Figure 9 may contribute to our understanding of mechanisms of the generation of longshore currents but certainly requires a thorough examination of mathematical terms representing radiation stresses, gradiental components and perhaps some interactions intervening in the generation of longshore currents and their transformation between bar crests and troughs. At present it can only be postulated that the effects of transformations within the structure of longshore currents between bar crests and troughs become enhanced at higher wave energies. More detailed measurements in nearshore regions with multiple bars should shed light on this phenomenon.

Mathematical explanation of various terms intervening in the description of longshore currents can be based on a few approaches available in the pertinent literature (cf. i.a. *Proceedings of Coastal Dynamics '95*).

5. Closing remarks and conclusions

Tracer experiments (mostly bedload) point to the importance of longshore transport between bars, and exhibit the effect of transport parallelisation induced by bars. Confinement of transport between bars, even at high waves, suggests the dominance of `closed' circulation cells, especially in the transition zone inner/outer bars. Longshore velocities are almost identical at bar crests (c) and troughs (t) and display a fairly uniform vertical profile. Cross-shore velocities are vertically non-uniform but their *mean values* are close at (c) and (t). The cross-shore circulation, judged from mean water velocities, depends on wave energy and displays a vortex (whirl) between bars (Fig. 8). Free-surface elevation gradients may become as important as radiation stresses in generation of nearshore currents. Order-of-magnitude estimates of the terms intervening in mathematical models of 3-D circulation should help improve the modelling. Also, a reliable linkage between Lagrangian and Eulerian characteristics must be worked out to draw more conclusions on characteristics of nearshore velocities and sediment transport.

Acknowledgements
This paper results from EU-sponsored MAST III PACE project (MAS3-CT95-0002). The work was co-sponsored under ibwPAN-tem2/97 funded by the Polish Research Committee KBN, the support of which is gratefully acknowledged.

REFERENCES

Pruszak Z. & R.B. Zeidler (1995). Sediment transport in various time scales. *Proceedings 24th International Conf. Coastal Engineering* 2513-2526. ASCE et al., Kobe-New York.

Investigating Nearshore Circulation Using Inverse Methods

Falk Feddersen[1], R.T. Guza[1], and Steve Elgar[2]

Abstract

The mean (time-averaged) nearshore circulation is often forced by breaking waves. For longshore homogeneous waves and bathymetry, the depth-averaged longshore current is also longshore homogeneous, and is described by a one-dimensional momentum balance between forcing, mixing, and bottom stress. However, the performance of a one-dimensional longshore current model can be degraded if the neglected effects of longshore inhomogeneities are significant or if the forcing and mixing are not parameterized correctly. The two-dimensional equations of motion, coupled with observations of the two-dimensional circulation, and inverse methods, can be used to infer the forcing and mixing and to examine the effect of longshore inhomogeneities. An idealized example of the inverse approach to solving the equations of motion is shown to predict a rip current through a gap in a longshore inhomogeneous sandbar for the case of normally incident waves.

Introduction

The three-dimensional mean (time-averaged) nearshore circulation is driven by wave breaking-induced spatial gradients in the mean, wave-induced momentum flux (e.g. the radiation stress). Nearshore circulation models are often reduced to two dimensions (2-D) by depth-averaging (removing processes such as undertow). However, the physics of 2-D circulation are not well understood. The additional simplifying assumption of longshore homogeneity leads to a one-dimensional (1-D) longshore momentum equation that predicts accurately the longshore current on a planar beach. On a barred beach, the 1-D model performance is degraded, either because

---

[1]Scripps Inst. of Oceanography, 0209, La Jolla Ca, 92093, falk@coast.ucsd.edu
[2]Washington State University, Pullman Wa, 99164-2752

the neglected effects of longshore inhomogeneous bathymetry are significant, or because the forcing and mixing are not well parameterized.

Two-dimensional circulation models can be used to study the effects of inhomogeneous bathymetry, but the choice of model boundary conditions and parameterizations complicates interpreting model-data discrepancies. The two-dimensional dynamical equations, circulation observations, and inverse methods can be used to infer the forcing and mixing, and to examine the effect of longshore bathymetric inhomogeneities. An idealized example of the inverse approach is presented for the case where there is a symmetric gap in a sandbar and normally incident waves. The longshore variation in cross-shore wave forcing induces longshore variations in setup which drives a rip current through the gap. Application of inverse methods to the SandyDuck field experiment data set is planned.

Nearshore Circulation

The forced and dissipative shallow water equations for mass continuity and cross-(x) and longshore (y) momentum, widely used to model the steady, depth-integrated, wave-driven nearshore circulation are given, respectively, by

$$\frac{\partial}{\partial x}\left[h\overline{u}\right] + \frac{\partial}{\partial y}\left[h\overline{v}\right] = 0 \tag{1}$$

$$h\left(\overline{u}\frac{\partial\overline{u}}{\partial x} + \overline{v}\frac{\partial\overline{u}}{\partial y}\right) = -gh\frac{\partial\overline{\eta}}{\partial x} + \frac{1}{\rho}\left(\frac{\partial S_{xx}}{\partial x} + \frac{\partial S_{xy}}{\partial y} + \frac{\partial F_{xx}}{\partial x} + \frac{\partial F_{xy}}{\partial y}\right) - \tau_x^b \tag{2}$$

$$h\left(\overline{u}\frac{\partial\overline{v}}{\partial x} + \overline{v}\frac{\partial\overline{v}}{\partial y}\right) = -gh\frac{\partial\overline{\eta}}{\partial y} + \frac{1}{\rho}\left(\frac{\partial S_{xy}}{\partial x} + \frac{\partial S_{yy}}{\partial y} + \frac{\partial F_{xy}}{\partial x} + \frac{\partial F_{yy}}{\partial y}\right) - \tau_y^b \tag{3}$$

where $\overline{u}$ and $\overline{v}$ are the mean (depth and time-averaged over many wave periods) cross-shore and longshore velocities, $\overline{\eta}$ is the time-averaged deviation of the sea surface elevation from the still water depth h ($\overline{\eta} \ll h$ is assumed), g is gravity, ρ is the water density, S and F are the radiation and Reynold's stress tensors, respectively, and $\vec{\tau}^b$ is the bottom stress vector.

The assumption of longshore homogeneity ($\partial_y = 0$), coupled with the boundary condition of no cross-shore mass flux at the shoreline ($h\overline{u} = 0$ at $x = 0$) and continuity (1), yields $\overline{u} = 0$ everywhere. In this case, the cross-shore momentum equation simplifies to a balance between the cross-shore wave forcing (radiation stress gradient) and set-up, given by

$$\frac{\partial S_{xx}}{dx} = \rho gh\frac{\partial\overline{\eta}}{dx}. \tag{4}$$

The longshore momentum equation balances longshore wave forcing (from radiation stress gradients) and turbulent lateral mixing with the bottom stress (*Bowen*, 1969; *Longuet-Higgins*, 1970; *Thornton*, 1970),

$$\frac{\partial S_{xy}}{\partial x} + \frac{\partial F_{xy}}{\partial x} = \tau_y^b. \tag{5}$$

The radiation, Reynold's, and bottom stresses are difficult to measure directly and therefore must be parameterized in longshore current models. The radiation stress is given usually given by a wave transformation model based on either linear theory (*e.g. Thornton and Guza*, 1983) or more complex roller theories (*e.g. Dally and Brown*, 1995; and many others). The Reynolds stress F_{xy} is often related to the mean shear of the longshore current $\partial \bar{v}/\partial x$ through various mechanisms (*e.g. Battjes*, 1975; *Svendsen and Putrevu*, 1994). Assuming small wave angles and weak currents (*Longuet-Higgins*, 1970) the longshore bottom stress is parameterized typically as proportional to $u_{rms}\bar{v}$ where u_{rms} is the rms cross-shore orbital wave velocity. With these types of parameterizations, 1-D longshore current models reproduce $\bar{v}(x)$ observed on a longshore homogeneous planar beach with a small range of incident wave angles (*Thornton and Guza*, 1986).

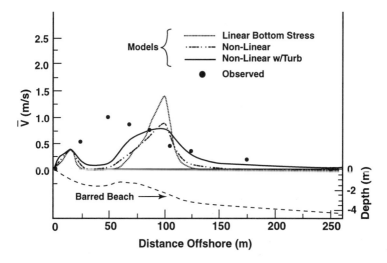

Figure 1: Longshore current (upper panel) and depth of the seafloor (lower panel) versus distance offshore. Filled circles are observed $\bar{v}(x)$ and curves are modeled $\bar{v}(x)$ with different bottom stress parameterizations (given in legend). (From *Church & Thornton* (1993)).

However, these models may not be applicable in general. A 1-D longshore current model with parameterizations similar to those for a planar beach predicts poorly $\bar{v}(x)$ observed at a barred beach near Duck, NC (*e.g. Church and Thornton*, 1993; *Feddersen et al.*, 1996). The modeled longshore current is maximum seaward of the bar crest and near the shoreline, whereas a single broad maximum is observed shorewards of the bar crest (*e.g.* Figure 1).

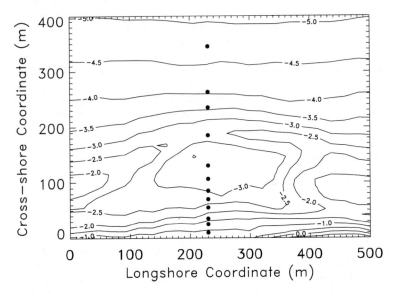

Figure 2: Contours (relative to mean-sea-level) of bathymetry at Duck on October 20, 1994. Longshore inhomogeneities are observed along the bar crest located about 150 m from the shoreline. The filled circles indicate flowmeter locations.

One possible cause for the poor model predictions is that the forcing and mixing parameterizations are inaccurate. Indeed, more sophisticated parameterizations (e.g. a wave roller model coupled with video observations of wave breaking) improve 1-D longshore current predictions (*Garcez-Faria*, 1997a) for cases with longshore homogeneity. However, the bathymetry at Duck can be quite complex (Figure 2), and nonlinear or pressure gradient terms may sometimes be important in the momentum balance. Two-dimensional model simulations with bathymetry similar to Figure 2 suggest that the neglected alongshore gradient terms can sometimes be important (*Sancho et al.,* 1995). The degree to which 2-D effects are important, and the extent to which the forcing and mixing parameterizations are correct, is difficult to assess from existing 1-D observations along on a single cross-shore transect (e.g. Figure 2).

Inverse Methods

Suppose that the 2-D shallow water equations (1-3) govern the flow, and that there are comprehensive 2-D circulation observations. A traditional modeling approach would be to solve (1-3) with prescribed boundary conditions and parameterizations, and compare model predictions with the observed circulation. However, the model solutions will depend explicitly on both the chosen boundary conditions and parameterizations, and therefore interpreting model-data discrepancies is difficult.

In principle, inverse solutions can be used to determine the extent to which the flow is 2-D and to infer the combined forcing and mixing. In the most general inverse problem, the flow ($\bar{u}$, $\bar{v}$, and $\bar{\eta}$), and combined forcing and mixing fields are found that are consistent with the shallow water equations (1-3) (*i.e.* small dynamical misfit), all available observations (*i.e.* small observation misfit), and make a global property of the solution (a norm) small. The norm replaces the need to satisfy boundary conditions at the longshore domain boundaries. Schematically, the cost function J is minimized, where

$$J = (\text{Dynamical Misfit})^2 + (\text{Observation Misfit})^2 + \text{Norm.} \qquad (6)$$

Comprehensive 2-D nearshore circulation observations that can be used in the general inverse problem do not yet exist. To demonstrate the approach, results from an idealized inverse problem that does not require any observations are presented. Assuming the forcing and mixing fields are known, the flow solution consistent with the shallow water equations (1-3) and also minimizing a norm is sought. Thus, the observation misfit term vanishes in (6) (because there are no observations) and the flow is the only unknown in the governing dynamical equations. The only specified boundary conditions are $\bar{u} = 0$ at the shoreline and the offshore boundary. The norm is somewhat arbitrary, and is chosen here to be the spatially integrated square deviation of the 2-D inverse solutions from the 1-D solutions (4 and 5). Solutions with this norm are equal to the 1-D solutions when conditions (bathymetry, forcing, and mixing) are longshore homogeneous, similar to the analytic solutions for 2-D circulation that are based upon a perturbation expansion around a 1-D base state (*Putrevu et al.*, 1995). The solution method is described in the Appendix.

The model bathymetry (Figure 3) is a barred beach with a symmetric gap, roughly similar to observed bathymetry (Figure 2) and bathymetry used by *Sancho et al.* (1995). Normally incident waves with offshore wave height $H_{rms} = 1$ m were transformed across the model domain using *Thornton and Guza* (1983). Refraction effects owing to longshore variations in bathymetry were ignored for simplicity, so that the longshore forcing ($\partial S_{xy}/\partial x$) is zero everywhere. The wave transformation model yields the cross-shore forcing (Figure 4a). Where the sandbar is pronounced (*e.g.* near the edges of the model domain) the cross-shore forcing is large over the bar and weak in the trough (cross-shore coordinates $x = 90$ m and $x = 40$ m respectively). Wave breaking does not occur over the gap in the bar (*e.g.* in the center of the domain), and the cross-shore forcing is weak there, but increases farther

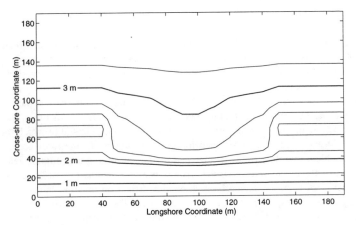

Figure 3: Contours (depth below mean sea level) of the model bathymetry. The shoreline is at the cross-shore coordinate origin. The sandbar has a gap in the center of the model domain. The bar crest is located at $x = 80$ m.

shoreward.

The solution for ($\bar{u}$, $\bar{v}$, and $\bar{\eta}$) indicates that the alongshore variation in cross-shore forcing creates alongshore variations in $\bar{\eta}$ (Figure 4b). At the edges of the model domain where the sandbar is pronounced, the setup is higher than at the center of the model domain at cross-shore locations $x = 30$ to $x = 80$ m. This difference in setup creates a longshore pressure gradient that drives a flow convergence shoreward of the sandbar gap, and a rip current through the gap (Figure 5). There is a weak return flow at each side of the rip current. For normally incident waves, the inverse (Figure 5) solution and that of *Sancho et al.* (1995) are qualitatively similar.

Summary

The effect of inhomogeneous bathymetry on the spatial variation of forcing, mixing, and the resulting 2-D nearshore circulation are not understood well. Interpreting observations of 2-D nearshore circulation through modeling is complex. The choices of boundary conditions and parameterizations strongly influence model solutions and complicate interpretation of model-data discrepancies. However, 2-D dynamical equations, 2-D circulation observations, and inverse methods might be used to infer the forcing and mixing, and to examine the effect of longshore bathymetric inhomogeneities. An idealized example of the approach on a barred beach predicts a rip current through a gap in the sandbar. This work is ongoing. Circulation data from a two-dimensional array of 36 near-bottom current meters are being obtained

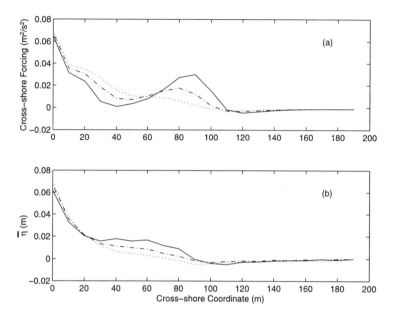

Figure 4: (a) Cross-shore wave forcing and (b) setup $\overline{\eta}$ from the model solution at three longshore locations versus cross-shore coordinate. The solid curve corresponds to the edge of the domain ($y = 10$ m), the dot-dashed curve to $y = 50$ m, and the dotted curve to the center of the domain (longshore coordinate $y = 90$ m). The bathymetry is shown in Figure 3.

as part of the SandyDuck field experiment, and application of the methods outlined here to this data set are planned.

Appendix

Here, the solution method for the idealized inverse problem where the forcing is assumed known and mixing neglected, is outlined. First, the governing equations (1-3) to be solved are discretized to the form

$$f_i(m_j) = F_i \tag{7}$$

where f_i represent the continuity and momentum equations and F_i represent the inhomogeneous forcing given at each grid point. With N and M grid points in the cross- and longshore respectively, there are $3(N-2)M$ equations. The m_j represent the unknowns ($\overline{u}$, $\overline{v}$, and $\overline{\eta}$ in this case). The boundary conditions are $\overline{u} = 0$ at the shoreline and $\overline{u} = 0$ and $\overline{\eta} = 0$ at the offshore boundary, resulting in $3(N - 1)M$

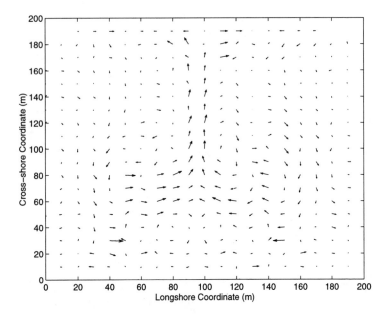

Figure 5: Mean currents predicted by the inverse model for normally incident waves on a barred beach. The bar is located at cross-shore coordinate 90 m, and a rip current flows seaward through the gap in the bar at longshore coordinate 100 m (Figure 3). The vectors point in the direction of the flow and their length is proportional to the magnitude of the current. The maximum current is 46 cm/s.

unknowns. Lacking longshore boundary conditions, there are more unknowns than equations so the problem is algebraically underdetermined. Instead, a solution is sought that is close to the 1-D solution (4 and 5). To find the model unknowns with the smallest (in least squares sense) deviation from the apriori 1-D solution, l_i, and from the observations of the model unknowns, o_i, as well as the smallest misfit to the dynamical equations (7), minimize the cost function

$$J = (m_i - l_i)W_{ij}^L(m_j - l_j) + e_i e_i + (m_i - o_i)W_{ij}^O(m_j - o_j), \qquad (8)$$

where e_i is the misfit to the dynamical equations, $e_i = f_i(m_j) - F_i$, and W_{ij}^L and W_{ij}^O are the weight matrices for the misfit of the apriori solution and the observations. The equations $f_i(m_j)$ are quadratically nonlinear so the minimum of J is found iteratively using the total inversion method of *Tarantola and Valette* (1982). Given a guess for the model unknowns $m_j^{(n)}$, Taylor expand the dynamical equations

$$f_i(m_j^{(n+1)}) \approx f_i(m_j^{(n)}) + \frac{\partial f_i^{(n)}}{\partial m_j^{(n)}}(m_j^{(n+1)} - m_j^{(n)})$$

where

$$\frac{\partial f_i^{(n)}}{\partial m_j} = \frac{\partial f_i(m_j^{(n)})}{\partial m_j^{(n)}},$$

and rewrite

$$\frac{\partial f_i^{(n)}}{\partial m_j} m_j^{(n+1)} \approx F_i - f_i(m_j^{(n)}) + \frac{\partial f_i^{(n)}}{\partial m_j} m_j^{(n)}.$$

Defining

$$G_{ij}^{(n)} = \frac{\partial f_i^{(n)}}{\partial m_j}$$

$$d_i^{(n)} = F_i - f_i(m_j^{(n)}) + \frac{\partial f_i^{(n)}}{\partial m_j} m_j^{(n)},$$

at each iteration step the misfit to the dynamical equations is written

$$e_i^{(n)} = G_{ij}^{(n-1)} m_j^{(n)} - d_i^{(n-1)}.$$

At each iteration step, minimize the cost function

$$J^{(n)} = (m_i^{(n)} - l_i) W_{ij}^L (m_j^{(n)} - l_j) + e_i^{(n)} e_i^{(n)} + (m_i^{(n)} - o_i) W_{ij}^O (m_j^{(n)} - o_j).$$

Iteratively minimizing this cost function yields the set of model solutions that correspond to a minimum of the cost function (8). Minimization is achieved by setting the derivative of the norm with respect to $m_l^{(n)}$ to zero

$$\frac{\partial J^{(n)}}{\partial m_l^{(n)}} = 0$$

yielding (the weight matrices are symmetrical)

$$(W_{lk}^L + W_{lk}^O + G_{il}^{(n-1)} G_{ik}^{(n-1)}) m_k^{(n)} = G_{il}^{(n-1)} d_i^{(n-1)} + W_{lk}^L l_k + W_{lk}^O o_k,$$

which can be solved for $m_k^{(n)}$ using standard methods.

Acknowledgements This research was supported by the Office of Naval Research, Coastal Sciences Program, and California Sea Grant.

References

[1] Battjes, J.A., Modeling of turbulence in the surfzone, *Proceedings of the Symposium on Modeling Techniques*, ASCE, 1050-1061, 1975.

[2] Bowen, A.J., The generation of longshore currents on a plane beach, *J. Mar. Res.*, **27**, 206-215, 1969.

[3] Church, J.C. and E.B. Thornton, Effects of breaking wave induced turbulence within a longshore current model, *Coastal Eng.*, **20**, 1-28, 1993.

[4] Dally, W.R. and C.A. Brown, A modeling investigation of the breaking wave roller with application to cross-shore currents, *J. Geophys. Res.*, **100**, 24,873-24,883, 1995.

[5] Feddersen, F., R.T. Guza, S. Elgar, and T.H.C. Herbers, Cross-shore structure of longshore currents during Duck94, *Proc. 25th Int. Coastal Engineering Conf.*, ASCE, 3666-3679, 1996.

[6] Garcez-Faria, A.F., PhD Thesis, Naval Postgraduate School, Monterey, CA, 1997a.

[7] Garcez-Faria, A.F., E.B. Thornton, T.P. Stanton, C.M.C.V. Soares, and T.C. Lippmann, Vertical profiles of longshore currents and related bed shear stress and bottom roughness, *J. Geophys. Res.*, (in press) 1997b.

[8] Longuet-Higgins, M.S., Longshore currents generated by obliquely incident sea waves 1, *J. Geophys. Res.*, **75**, 6790-6801, 1970.

[9] Putrevu, U., J. Oltman-Shay, and I.A. Svendsen, Effect of alongshore nonuniformities on longshore current predictions, *J. Geophys. Res.*, **100**, 16,119-16,130, 1995.

[10] Sancho, F.E., I.A. Svendsen, A.R. Van Dongeren, and U. Putrevu, Longshore nonuniformities of nearshore currents, *Coastal Dynamics 1995*, ASCE, 425-436, 1996.

[11] Svendsen, I.A. and U. Putrevu, Nearshore mixing and dispersion, *Proc. R. Soc. Lond. A* **445**, 561-576, 1994.

[12] Tarantola, A. and Valette B., Generalized nonlinear problems solved using the least squares criterion, *Rev. Geophys. Space Phys.*, **20**, 20,219-20,232, 1982.

[13] Thornton, E.B., Variation of longshore current across the surfzone, *Proc. 12th Int. Coastal Engineering Conf.*, New York, ASCE, 291-308, 1970.

[14] Thornton, E.B. and R.T. Guza, Transformations of wave height distribution, *J. Geophys. Res.*, **88**, 5925-5938, 1983.

[15] Thornton, E.B. and R.T. Guza, Surf zone longshore currents and random waves: field data and models, *J. Phys. Oceanogr.*, **16**, 1165-1178, 1986.

Hydrodynamics and Wave Transformation at Selected South Carolina Tidal Inlets

Paul A. Work[1] and Shams E. Riffat[2]

Abstract

A field study related to hydrodynamics, wave transformation, and sediment transport at tidal inlets is described, with emphasis on Price Inlet and Breach Inlet, South Carolina. The data indicate inlet morphology, and spatial and temporal variation in mean currents, wave heights, and energy flux. A significant gradient in wave energy flux is found along the Price Inlet channel centerline, inshore of the terminal lobe of the inlet. Numerical models of tidal hydrodynamics and wave transformation are employed to investigate sensitivity of wave transformation to tidal currents and other parameters. The wave transformation model is found to be weakly dependent on the calculated tidal currents.

Introduction

Prediction of the evolution of tidal inlets – particularly unstabilized tidal inlets – is a problem that has received attention because of increased development in coastal areas. Sediment transport at inlets is driven by astronomical tides, storms with associated winds, waves, and storm surge, and larger scale phenomena. Numerical modeling has emerged as a preferred tool for quantitative predictions of inlet hydrodynamics and sediment transport, due to the high costs of physical model studies and their accompanying scaling problems.

This paper discusses an ongoing field study in South Carolina (South Carolina Coastal Erosion Project) which is intended to describe long-term (years) evolution of untrained, natural tidal inlets (Work et al. 1995). Two-dimensional, numerical model predictions of tidal hydrodynamics are discussed. Measurements of non-directional

---
[1] Asst. Prof., Dept. of Civil Engineering, 110 Lowry Hall, Clemson University, Clemson, SC, 29634-0911. (864) 656-3000, pwork@eng.clemson.edu.
[2] Research Asst., Dept. of Civil Engineering, Clemson University.

wave energy spectra in an inlet channel are compared to numerical wave transformation model results. Additionally, the influence of the tidal currents on the wave conditions in the vicinity of the inlet is described.

Field Site and Measurements

Fieldwork completed to date includes bathymetric surveying and limited wave and current measurements (short-term; hours) within and near tidal inlets. Four inlets have been selected for study, all unstabilized, located northeast of Charleston Harbor (Figure 1). Mean and spring tidal ranges for the area are 1.6 and 1.8 m, respectively. The long-term mean, significant wave offshore has a height of 1 m and period of six seconds. Data availability is indicated in Table 1. The focus of this paper will be on Breach and Price inlets. The islands to either side of Price Inlet are essentially undeveloped, and the inlet appears to be relatively stable. Both sides of Breach Inlet have residential development, and the downdrift (south) side is the scene of periodic erosion problems.

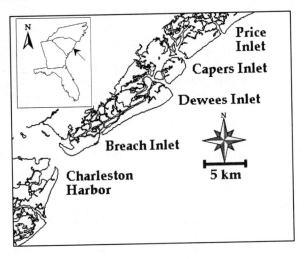

Figure 1. Surveyed tidal inlets in South Carolina.

Inlet	Bathymetry	Tidal Currents	Wave data
Breach	5/96, *5/98	5/96	
Dewees	8/94, 7/95, 5/97		
Capers	8/94, 7/95, 5/97	5/96	
Price	5/96, *5/98	5/96, 8/97	5/97

Table 1. Field data availability. * = Planned.

Bathymetric survey data are collected by boat, equipped with Global Positioning System (GPS) hardware, a digital heave-pitch-roll recorder, and an acoustic fathometer (Work et al., in review). Short-term (hours) records of tidal currents have been obtained using an acoustic doppler current profiler at different locations within the inlet regions. This instrument provides a three-dimensional record of the velocity profile over the water column. Positions of the measurements were established with GPS equipment. Figure 2 indicates the locations of the tidal current measurement stations at Breach Inlet, May 28, 1996. All stations at this inlet are located inland of a low bridge which restricts access.

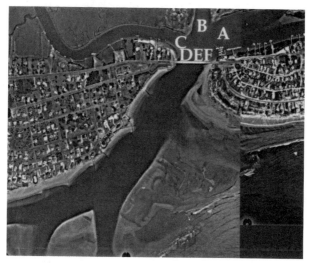

Figure 2. Composite, vertical, aerial photo of Breach Inlet, SC, 1993. Letters indicate approximate locations of 3-D tidal current measurements, 5/28/96.

Non-directional wave measurements were made during one day at several locations at Price Inlet. A pressure transducer was attached to an anchor and tethered to a small boat, where a data collection computer was located. The transducer was thrown over the side at points of interest and held in place by the anchor. The resulting pressure time series was analyzed to determine wave energy spectra, energy-based significant wave height, and modal and peak wave periods. Water depths at the measurement locations were 3.6 to 6.4 m. A floating buoy wave gage (National Data Buoy Center 41004, offshore of Charleston at a depth of 38 m) simultaneously recorded wind conditions and directional wave characteristics. Figure 3 indicates the locations of the wave measurement stations at Price Inlet, May 29, 1997.

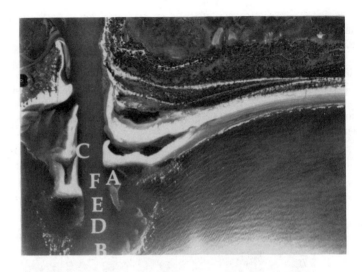

Figure 3. Vertical aerial photo of Price Inlet, 1993. Letters indicate approximate locations of non-directional wave spectra measurements, 5/29/97.

Hydrodynamic Modeling

A 2-D, finite element model was used to model tidal hydrodynamics associated with maximum ebb and flood tides, respectively, at Price Inlet (Norton et al. 1973). The short-term tidal current data set from May, 1996, was used for calibration purposes.

Figure 5 indicates the velocity fields corresponding to maximum ebb and flood scenarios. The bulk of the ebb flow occurs through the main inlet channel, whereas during flood tide, significant flow is shown to pass through the marginal flood channels, in agreement with observations by the authors and others (FitzGerald 1977, Hayes 1994). The model boundary conditions were established using predicted spring tides and a phase lag based on linear long wave theory. Maximum velocities, per the model, are 71/81 cm/s (ebb/flood), compared to observations of 71 cm/s during flood tide. The velocity fields shown in Figure 5 were used to assess qualitatively the influence of tidal currents on nearshore wave conditions, in conjunction with a numerical wave transformation model.

Measured Wave Conditions

Figure 6 indicates offshore wave conditions during the measurement period. Winds were relatively strong and steady; significant wave height, mean period, and mean direction were also nearly steady. In a sense, these were optimum conditions for the

nearshore wave measurements, because all temporal variation is then a result of the
change in tidal stage and tidal currents only.

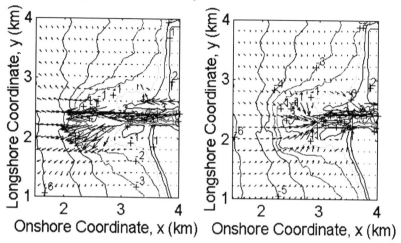

Figure 4. Hydrodynamic model (RMA-2, Norton et al. 1973) results for maximum ebb
(left) and flood (right) currents at Price Inlet. Contours indicate bathymetry at 1 m
intervals. Maximum currents: 71 cm/s (ebb), 81 cm/s (flood).

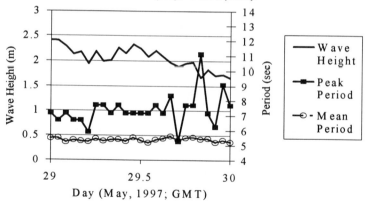

Figure 5. Wave conditions (NOAA 41004 Buoy) during measurement period.

Figure 6 provides an indication of the temporal variation in the energy spectrum at
a nearshore point within the zone of influence of the inlet. The two spectra were
measured at the same location at different times. The latter plot shows the spectrum
closer to high tide, when more energy reaches the site after propagating across the ebb
tidal shoals. The wave height increases from 50 to 80 cm despite the nearly constant
offshore conditions.

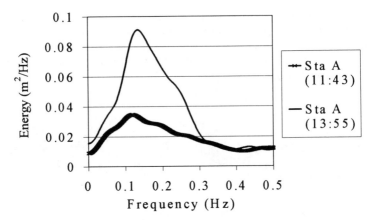

Figure 6. Pressure spectra from Station A (see Figure 3) at different times as tide was rising.

Figure 7 indicates the spatial dependence in the spectra. The two spectra were obtained at different locations, within 30 minutes of one another. The wave height at the more sheltered site (Station A, see Figure 3) is 50 cm, whereas the more exposed Station B yields a wave height of 140 cm, with similar tide and wave conditions.

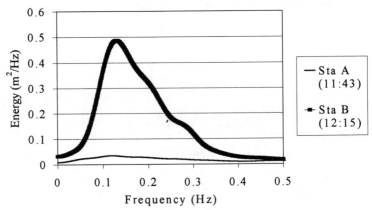

Figure 7. Pressure spectra from two locations (see Figure 3) measured 30 minutes apart.

Figure 8 indicates the variation in wave height along the centerline of Price Inlet. Four observations are shown over a 3-1/2 hour period during a rising tide. Despite the increase in tidal stage during the measurement period, and therefore a reduction in the wave energy dissipation on the shoals, a marked decrease in wave height is observed as one proceeds into the inlet from the seaward side. It is interesting to note the similarity to the measurements of Melo and Guza (1991) for a conceptually similar scenario. The reduction in energy flux, a better indicator of dissipation, is even more pronounced. Some hypothetical numerical wave model simulations of this scenario are discussed below.

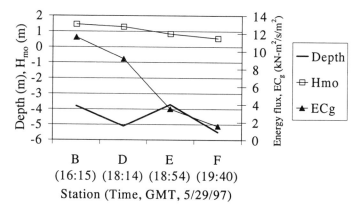

Figure 8. Measured wave heights, instantaneous water depths, and energy flux along Price Inlet centerline, 5/29/97. Slack high tide ~19:00.

<u>Wave Model Results</u>

The REF/DIF S numerical wave propagation model of Kirby and Ozkan (1994) was employed to investigate wave model sensitivity. This is a spectral wave transformation model, employing weakly nonlinear wave theory and a parabolic approximation to the mild slope equation. It allows a priori specification of a mean velocity field (thus neglecting wave-induced currents) and incorporates wave breaking via the statistical approach of Thornton and Guza (1983).

Hypothetical wave conditions were constructed to represent nominal ($H_{mo} = 1$ m, $T_p = 6$ sec) and swell ($H_{mo} = 3$ m, $T_p = 12$ sec) sea states. A series of wave model runs was executed, systematically altering tidal stage, tidal currents, and wave angle of incidence, height, and period.

Figure 9 shows the variation of wave height along the Price Inlet channel centerline and the dependence on tidal currents. The currents are seen to decay rapidly with distance from the inlet throat. Because of this, little difference in wave height is evident when comparing the slack tide case to the maximum ebb and flood cases. The gradient in wave height along the inlet channel, inshore of the terminal shoal, is much less than was observed in the field, although it should be noted that the field case involved waves having slightly oblique incidence.

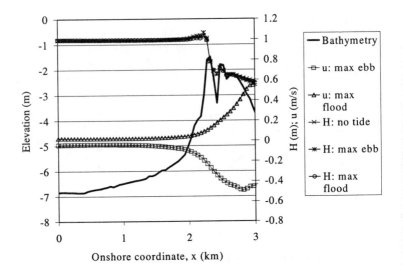

Figure 9. Influence of tidal current on wave heights along Price Inlet channel centerline, for nominal offshore wave condition (H_{mo} = 1 m, T_p = 6 sec, normal incidence, zero tide stage).

Tide stage has a much more pronounced effect on the model results. Figure 10 presents results for low, mid-, and high tide along the Price Inlet channel centerline. A larger incident wave (H_{mo} = 3 m, T_p = 12 sec) was used for this example, resulting in wave energy dissipation even well offshore of the inlet shoals. The large relative change in depth over the shoals as the tide stage changes modifies the energy dissipation significantly.

A similar trend, more pronounced, is evident at Breach Inlet. Inspection of Figure 2 reveals that the inlet channel does not follow a shore-normal line; thus waves must pass over sizeable shoals before reaching the inlet throat. At low tide, with shore-normal incident waves, the shoals dissipate almost all of the incident wave energy.

Conclusions

A South Carolina field study related to tidal inlet morphology, hydrodynamics, and wave transformation was described. The study includes bathymetric surveying and measurement of waves and tidal currents at selected inlets. The data set will be useful for future validation of numerical models of tidal hydrodynamics, wave transformation, and, ultimately, prediction of bathymetric evolution. A number of hypothetical scenarios have been explored through the use of numerical models for tidal hydrodynamics and wave transformation.

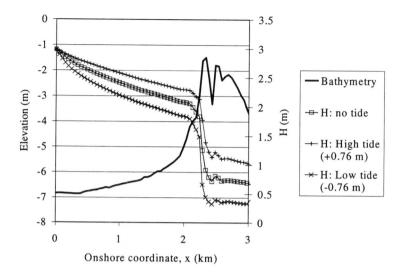

Figure 10. Influence of tidal stage on nearshore wave heights, Price Inlet channel centerline. No currents, normal incidence, H_{mo} = 3 m, T_p = 12 sec.

Shoals and tidal currents at the surveyed field sites quite obviously influence wave transformation. Accurate predictions of the velocity field and the spatial variation in the wave energy flux vector at tidal inlets are necessary for realistic models of sediment transport and inlet evolution. Laboratory studies of inlet wave-current interaction are quite difficult to perform because of the open, periodic boundary conditions, and more difficult still if one wishes to allow for sediment transport. Field data are required for numerical model validation, but suitable data sets are few.

Although the field data presented in this paper have not yet been directly compared to numerical model results, some qualitative conclusions based on the model results are possible. The hydrodynamic model yielded a velocity field that agrees with qualitative observations. The wave transformation model results were only slightly affected by the presence of a maximum flood or ebb current. The wave model also appears to underestimate the rate of energy dissipation within the inlet channel, landward of the inlet terminal lobe at Price Inlet.

There are several possible reasons for the minor influence of the tidal currents on the wave model results and for the underestimation of wave energy dissipation within the inlet channel. One possibility is that the hydrodynamic model is underestimating the velocities. Another is that the wave model, only weakly nonlinear, does not employ an adequate description of wave-current interaction. The dissipation in the bottom boundary layer may be underestimated. And, lastly, the model of Thornton and Guza (1983), employed to describe wave breaking, was developed for the case of waves breaking on a beach. It may not be adequate to apply this model to the case of wave breaking due to the combined influence of tidal currents and bathymetry, accounting only for the change in apparent wave frequency due to the presence of the current. More extensive study of this problem will be necessary for development of reliable models of tidal inlet evolution.

Acknowledgement

This work was supported as part of the South Carolina Coastal Erosion Study, funded by the U.S. Geological Survey, Coastal Geology group, and the South Carolina Sea Grant Consortium. Their support is gratefully acknowledged.

References

FitzGerald, D.M., 1977. Hydraulics, morphology, and sediment transport at Price Inlet, SC. Ph.D. diss., Dept. Geol., Univ. of South Carolina, Columbia, SC, 84 pp.

Hayes, M.O., 1994. Georgia bight. Ch. 9 in *Geology of the Holocene Barrier Island System*, Springer-Verlag, Berlin, 233-304.

Kirby, J.T., and Ozkan, H.T., 1994. Combined Refraction/Diffraction Model for Spectral Wave Conditions, REF/DIF S, version 1.1. CACR Rpt. No. 94-4, Center for Applied Coastal Research, Dept. of Civil Eng., Univ. of Delaware, Newark, DE.

Melo, E., and Guza, R.T., 1991. Wave propagation in jettied entrance channels. II. Observations. J. Waterway, Port, Coastal, and Ocean Eng., 117(5), ASCE, 493-510.

Norton, W.R., King, I.P., and Orlob, G.T., 1973. A finite element model for Lower Granite Reservoir. Water Resources Engineers, Inc., Walnut Creek, CA.

Thornton, E.B., and Guza, R.T., 1983. Transformation of wave height distribution. J. Geophys. Res., Vol. 88, No. C10, 5925-5938.

Work, P.A., Rogers, W.E., and Hayter, E.J., 1995, South Carolina Coastal Erosion Study: Inlet Morphodynamics and Sediment Transport. *Proc. Coastal Dynamics '95*, Gdansk, Poland, ASCE, New York, NY, 1047-1058.

Work, P.A., Hansen, M., and Rogers, W.E., in review. Bathymetric surveying with shipborne GPS and heave, pitch, and roll compensation.

CURRENT-INDUCED BREAKING AT AN IDEALIZED INLET

Jane McKee Smith[1], Donald T. Resio[2], and Charles L. Vincent[2]

ABSTRACT: Laboratory measurements of wave shoaling and breaking on an ebb current were made in an idealized inlet in a wave basin. The wave and ebb current measurements showed increased shoaling and breaking (compared to no current), shifting of the spectral peak to lower frequencies, and energy dissipation concentrated at the spectral peak and higher frequencies. The measurements were used to evaluate and develop formulations of wave dissipation on a current for application to a spectral wave model. White-capping dissipation formulations underpredicted breaking on a current in shallow water, whereas the bore-based wave dissipation relationship of Battjes and Janssen and a relationship developed in this study provided good estimates of wave-height decay on an ebb current.

INTRODUCTION

Waves in tidal inlets steepen and break on strong ebb currents. Although wave breaking at inlet entrances impacts navigation, sediment transport, and wave penetration into the inlet, the wave breaking process in the presence of a current is poorly understood. In this paper, wave breaking on a current is examined through physical-model measurements in an idealized inlet with a steady ebb current. Wave and current measurements are used to evaluate wave dissipation models. The goal

---

[1]Res. Hyd. Engr., U.S. Army Engineer Waterways Experiment Station, Coastal and Hydraulics Laboratory, 3909 Halls Ferry Rd., Vicksburg, MS 39180-6199, USA.
[2]Sr. Scientist, U.S. Army Engineer Waterways Experiment Station, Coastal and Hydraulics Laboratory, 3909 Halls Ferry Rd., Vicksburg, MS 39180-6199, USA.

of the study is to determine a dissipation function for wave breaking on a current that is based on integrated wave parameters, is applicable for arbitrary water depths, and is robust.   The remainder of the Introduction describes candidate wave dissipation formulations and selected previous studies of wave breaking on a current. Next, the laboratory experiments are described.   Under Results, trends in the measured spectra, evaluation and development of dissipation formulations, and application of the formulations to calculate wave height decay are presented.   Last, conclusions and future work are discussed.

Dissipation Formulations.   There is little information on breaking criteria for wave breaking on a current.   Most nearshore breaking criteria neglect current and are based on relative water depth, defined as the ratio of wave height to water depth, but existing criteria that include wave steepness are good candidates for application on a current, e.g., Miche's criterion (1951) given by

$$H_{max} = 0.142 \ L \ \tanh kd \tag{1}$$

where H_{max} is the limiting regular wave height, L is wavelength, k is wave number, and d is water depth.   The strength of this relationship is that it reduces to a steepness limit in deep water and a depth limit in shallow water, thus incorporating both limiting factors in a simple form.   Battjes and Janssen (1978) applied the Miche criterion with the concept of energy dissipation in a bore (LeMehaute 1962) for irregular waves in the following form

$$D = -0.25 \ Q_b \ f_m \ \frac{H_{max}^3}{d} \tag{2}$$

where D is the wave energy dissipation rate, Q_b is the percentage of waves breaking (function of the ratio of root-mean-square wave height to H_{max}), and f_m is the mean frequency.   Another class of breaking relationships with possible application with current are the whitecapping formulations used in spectral wave generation models (Hasselmann 1974).   Two such relationships are:

$$D = -a \ S^4 \ \omega_m \ \frac{k}{k_m} \ E(\omega) \tag{3}$$

given by Komen et al. (1984, 1994), where a is a coefficient, S is an integrated relative steepness parameter, ω_m is the mean angular frequency, k_m is the mean wave number, and E is the energy density spectrum, ω is angular frequency; and

$$D = -\frac{\varepsilon\ g^{0.5}\ E(f)^3\ k_m^{4.5}}{\tanh(k_m d)^{0.75}} \tag{4}$$

given by Resio (1987), where ε is a coefficient and g is the acceleration due to gravity. These dissipation relationships were developed for waves in the absence of current, but they are applied in this study of breaking on a current to give insight about the processes.

Previous Studies. Previous laboratory studies of wave breaking on a current included Hedges et al. (1985), Lai et al. (1989), and Suh et al. (1994). Hedges et al. developed a limiting spectral shape for waves breaking on a current in deep water and tested it with four spectra in a wave-current flume. Suh et al. extended the Hedges et al. formula to finite water depths and tested it with nine spectra in a flume. In both studies, little of the data is presented, and results are given in the form of limiting spectra. Lai et al. performed a detailed flume experiment of wave-current interaction kinematics in deep water. They observed that linear theory predicted kinematics well, if the Doppler shift is included; they confirmed deep-water blocking of waves if the ratio of ebb current velocity to wave celerity exceeded 0.25; and they observed a downshifting of the peak wave frequency for breaking on a strong current. Ris and Holtuijsen (1996) used Lai et al.'s deepwater breaking data to evaluate breaking criteria and found that the whitecapping formulation of Komen et al. (1984) underestimated dissipation, but supplementing this formulation with the Battjes and Janssen (1978) breaking formulation gave significantly better agreement with the data. In this study, we will evaluate dissipation formulations with a new data set that includes shallow to intermediate relative water depths.

LABORATORY MEASUREMENTS

A laboratory study of wave breaking on an ebb current was conducted in a 45-by-110-m wave basin. The model configuration was an idealized inlet, formed through a straight barrier island with shore-parallel offshore bathymetry contours and no ebb shoal. The model was symmetric about the inlet and included parallel rubble jetties. The inlet width and jetty separation were approximately 3.7 m, and the jetties were 5.5 m long. The unidirectional, spectral wave generator was located 22 m from the inlet, in a region with negligible flow. Waves were measured at the generator and in a six-gauge cross-shore array, with capacitance gauges evenly spaced to cover a region from 3 m offshore to 3 m inshore of the jetty heads. The water depths in the measurement section were 12.5 to 8.5 cm and were held constant. Currents were measured at approximately mid depth with acoustic Doppler velocimeters. The current meters were centered between the wave gauges at five locations.

Wave conditions for the study were zero-moment wave height $H_{mo}=3.7$ and 5.5 cm, peak spectral period $T_p=0.7$ and 1.4 s, and incident wave direction perpendicular to the jetties. The current velocities were 0, 12, and 24 cm/s. The current decreased approximately 20 percent seaward of the jetties. Twelve runs were analyzed for the work presented here (two periods, two heights, and three current speeds), but additional experiments with regular waves and oblique wave angles were run. A summary of wave parameters considered in this study is given in Table 1, where U is the average current magnitude, and C is the wave celerity. The wave and current parameters cover a wide range of values, which makes the data useful to evaluate the wave dissipation formulations for current-induced wave breaking. Each experiment run was repeated three times, first with the wave gauge and current meter array centered between the jetties and then offset to the left and to the right of the center line by 1 m. Waves and currents at the three positions across the channel were similar and were averaged.

Table 1. Laboratory Wave Parameters							
H_{mo} (cm)	T_p (s)	U (cm/s)	H_{mo}/L	U/C	H_{mo}/d	kd	$k\,H_{mo}/2$
3.7, 5.5	0.7, 1.4	0, 14, 24	0.025-0.11	0-0.45	0.25-0.63	0.4-1.4	0.07-0.3

RESULTS

The results are discussed in three categories: measured spectra, energy dissipation, and modeling of wave heights.

Measured Spectra. Selected spectra from two runs are shown in Figures 1 and 2. The peak periods for the runs were 0.7 and 1.4 s, respectively, and the incident height for both cases was 5.5 cm and the current was 24 cm/s. The curves are labeled for gauges 0, 1, 3, and 6; with gauge 0 offshore (near the wave generator), gauge 1 approximately 3 m offshore of the jetty heads, gauge 3 near the jetty heads, and gauge 6 approximately 3 m inshore of the jetty heads. The spectra show interesting trends. First, Figure 1 ($T_p = 0.7$ s) shows a significant downshifting of the peak frequency from the offshore (gauge 0) toward the inlet throat (gauge 6). The peak period increased up to 16 percent for $U = 24$ cm/s, 9 percent for $U = 14$ cm/s, and varied by 0-6 percent for no current. Lai et al. (1989) reported a similar trend which they attribute to nonlinear side band instabilities. The energy dissipated through breaking was extracted at the peak frequency and higher, with the slope of the high-frequency tail of the spectra remaining fairly constant. This implies that dissipation is related to energy at a given frequency or energy is nonlinearly redistributed to maintain the high-frequency slope. The energy in the low-frequency end of the spectra also increased, most noticeably in the cases with longer peak periods (e.g., Figure 2), with and without current.

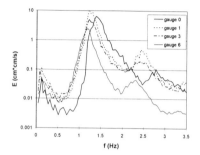

Figure 1. Spectra for $T_p = 0.7$ s, $U = 24$ cm/s, and $H_{mo} = 5.5$ cm

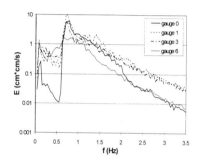

Figure 2. Spectra for $T_p = 1.4$ s, $U = 24$ cm/s, and $H_{mo} = 5.5$ cm

<u>Dissipation Rates</u>. The motivation for these laboratory experiments was to find a wave dissipation formulation for wave breaking on a current that can be applied in numerical wave transformation models. Two likely parameters to correlate with dissipation are wave steepness (as in deep-water breaking and whitecapping relationships) and wave height (as in bore dissipation models). Also, existing dissipation formulations discussed in the introduction are evaluated with dissipation calculated from the measurements.

Dissipation was calculated from the laboratory measurements by applying the action balance equation,

$$\frac{\partial}{\partial x}\left(E \ \frac{(C_{gr} + U)}{\omega_r} \right) = \frac{D}{\omega_r} \tag{5}$$

where C_{gr} is the wave group celerity, ω is the angular frequency, the subscript r denotes parameters in a reference frame moving with the current, and x is the direction of wave propagation. The action balance equation was applied between two gauges to solve for the dissipation, D. This one-dimensional formulation was developed under the assumption of no refraction or diffraction, which is a reasonable assumption for the idealized inlet, away from the jetties. Figures 3 and 4 show the calculated dissipation as a function of steepness and wave height, respectively. Figure 3 shows that the steepness parameterization segregates the data by peak wave period. Although dissipation increases with steepness, for a given steepness, dissipation is higher for longer peak periods. This result foreshadows that the whitecapping dissipation formulations, which are strongly a function of wave steepness, will not provide good estimates of dissipation in this data set. The calculated dissipation is highly correlated with wave height, and wave period does not seem to be a factor (Figure 4). This wave-height dependence implies that bore-type dissipation formulations, which are functions of wave height, are good candidates for estimating dissipation for this data set. Figure 5 shows the Miche limit (Equation 1) in terms of maximum wave steepness as a function of relative depth (solid line). The lab data, some of which are breaking and some not, are also plotted. The Miche criterion serves as a conservative upper limit to the data. The conservatism is not surprising because the formulation is for regular waves, and the data correspond to irregular waves.

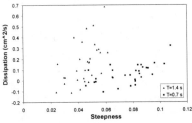

Figure 3. Dissipation versus steepness

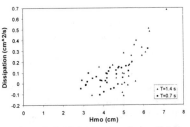

Figure 4. Dissipation versus wave height

Three dissipation formulations were evaluated with the laboratory data: Komen et al. (Equation 3), Resio (Equation 4), and Battjes and Janssen (Equation 2). Although current does not enter explicitly in any of these formulations, we have included the current effect in the calculation of wavelength and wave number using conservation of waves and linear theory (see, e.g., Jonsson 1990). The Resio formulation is given in Figure 6 as the solid line. This formulation slightly overpredicts dissipation for the 0.7-s waves, but significantly underpredicts dissipation for the 1.4-s waves. The correlation coefficient (r^2) for the Resio formulation (both periods) is 0.46. Likewise, the Komen et al. formulation (not shown) significantly underpredicts the dissipation for all wave conditions ($r^2 = 0$). Because these two formulations are strongly dependent on wave steepness, it is expected that they correlate poorly with the data set (see Figure 3). It is not surprising that these white-capping expressions do not represent wave breaking on a current in shallow water, because the formulations were developed to balance

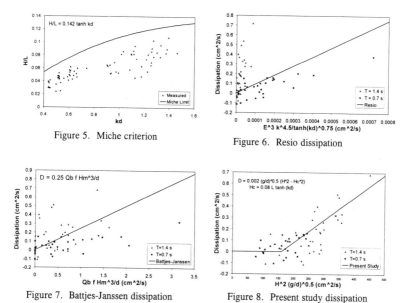

Figure 5. Miche criterion

Figure 6. Resio dissipation

Figure 7. Battjes-Janssen dissipation

Figure 8. Present study dissipation

excess wind input in saturated spectra without a current. The formulation of Battjes and Janssen is plotted in Figure 7 with the laboratory measurements and gives the correct trend in the data, but produces wide scatter (large deviation from the formulation), with a correlation coefficient of 0.25.

The poor performance of the dissipation formulations led to development of an alternative relationship. Figure 4 shows good correlation of dissipation to wave height, so a form consistent with dissipation in a bore was assumed, and the following relationship was developed using linear regression:

$$D \; = \; -0.002 \left(\frac{g}{d} \right)^{\frac{1}{2}} (H_{mo}^2 \; - \; H_c^2) \quad \text{for} \quad H_{mo} > H_c \tag{6}$$

where H_c is a critical wave height below which no dissipation occurs,

$$H_c \; = \; 0.08 \; L \; \tanh(kd) \tag{7}$$

The wavelength and wave number in Equation 7 include modification by a current. Equation 7 was chosen to be of the form of Miche's criterion. Equation 6 is plotted against the laboratory data in Figure 8 and shows reasonable agreement ($r^2 = 0.78$).

Wave Heights. Applying the action balance equation (Equation 5) between the wave gauges, together with the various dissipation models, gives a simple one-dimensional shoaling and decay model. The Miche criterion also was applied as a dissipation function by limiting the wave energy based on the maximum wave height given by Equation 1. This transformation technique was used to calculate wave shoaling and breaking through the gauge array for each of the twelve runs. Cases without current had little or no wave height decay through the gauge array. Example cases with the greatest dissipation are presented in Figures 9-11. Note in the figures that the second point ($x = 120$ cm) is consistently lower than the nearest points, probably due to a gauge calibration problem. This gauge also accounts for some of the scatter in the dissipation calculations.

The results generally cluster into two categories. The first category includes the formulations of Komen et al., Miche, and Resio (for $T_p = 1.4$ s), which show little or no dissipation for almost all runs and thus significantly overpredict the wave height through the gauge array. In Figures 9 and 11, the Komen et al. and Miche curves generally overlay each other and predict no dissipation. Even in the cases with the highest wave height and strongest current (Figures 10 and 12), the Komen et al. formulation predicts little or no dissipation, whereas the Miche criterion predicts some wave height decay. The second category includes the formulations of Battjes-Janssen, the present study (Equations 6 and 7), and Resio (for $T_p = 0.7$ s). These formulations fall on the lower portions of the figures and agree well with the measured wave heights for the twelve cases, with the exception of cases with $T_p = 0.7$ s and $U = 14$ cm/s (Figure 11). The reason for the poor agreement with this case is unclear. The correlation coefficient for the Resio formulation (both peak periods) was 0.71, for the Battjes and Janssen formulation 0.86, and for the present study formulation 0.87. Therefore, in spite of the large scatter between the Battjes-Janssen dissipation relationship and the data, the prediction of wave height is good, essentially equivalent to the expression developed in this study.

CONCLUSIONS

Results from a laboratory study of wave breaking on an ebb current at an idealized inlet are presented. The measurements show increased wave shoaling and breaking in the presence of the ebb current, as is expected from the conservation of waves and linear wave theory. The measurements also show a down shifting of the peak frequency (peak period became longer by up to 16 percent). Energy was dissipated at the peak and higher frequencies of the spectra. Dissipation rates calculated from the measurements were proportional to wave height.

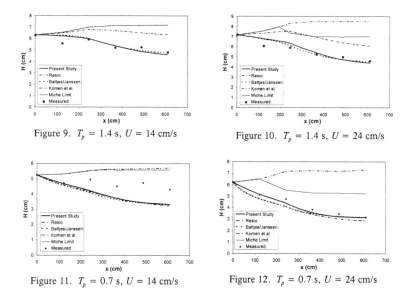

Figure 9. $T_p = 1.4$ s, $U = 14$ cm/s

Figure 10. $T_p = 1.4$ s, $U = 24$ cm/s

Figure 11. $T_p = 0.7$ s, $U = 14$ cm/s

Figure 12. $T_p = 0.7$ s, $U = 24$ cm/s

Dissipation algorithms were tested with the data. Whitecapping formulations (Komen et al. 1984, 1994, and Resio 1987), which are strongly dependent on wave steepness, generally underpredicted dissipation. Application of Resio's whitecapping formulation gave a correlation coefficient of 0.46 for predicting dissipation through the idealized inlet. The Battjes and Janssen breaking algorithm worked well for predicting wave height through the idealized inlet, in spite of the fact that there was considerable scatter in the dissipation prediction ($r^2 = 0.25$). A relationship for dissipation as a function of wave height squared gave improved agreement between calculated and predicted dissipations ($r^2 = 0.78$), but no substantial improvement over the Battjes and Janssen formulation for modeling the wave height. A dissipation function or breaking criterion applied at a coastal inlet must include relative depth and wave steepness, as well as wave-current interaction. Depth effects are more important for longer period waves and steepness for shorter period waves.

The next phase of experiments in the idealized inlet includes construction of an ebb shoal and extension of the measurement array farther offshore. Breaking over the ebb shoal, wave blocking, scaling relationships, and oblique wave incidence will be studied.

ACKNOWLEDGMENTS

Permission was granted by the Office, Chief of Engineers, U.S. Army Corps of Engineers, to publish this information. This research was supported by the U.S. Army Corps of Engineers, Coastal Inlets Research Program. Laboratory experiments were planned and executed by Gordon Harkins, Michael Briggs, Cecil Dorrell, and William Seabergh. Beneficial technical reviews were provided by Mary Cialone, Bruce Ebersole, Nicholas Kraus, William Seabergh, and Ernest Smith.

REFERENCES

Battjes, J.A., and J.P.F.M. Janssen. 1978. "Energy loss and set-up due to breaking of random waves," *Proc. 16th Coast. Engrg. Conf.*, ASCE, 569-587.

Hasselmann, 1974. "On the spectral dissipation of ocean waves due to whitecapping," *Boundary-Layer Meteorology*, 6(1-2), 107-127.

Hedges, T.S., K. Anastasiou, and D. Gabriel. 1985. "Interaction of random waves and currents," *J. Wtrway., Port, Coast., and Oc. Engrg.*, 111(2), 275-288.

Jonsson, I.G. 1990. "Wave-current interactions," *The Sea*, ed. B. LeMehaute and D.M. Hanes, John Wiley & Sons, Inc., New York, 65-120.

Komen, G.J., S. Hasselmann, and K. Hasselmann. 1984. "On the existence of a fully developed wind-sea spectrum," *J. Phys. Oceanogr.*, 14, 1271-1285.

Komen, G.J., L. Cavaleri, M. Donelan, K. Hasselmann, S. Hasselmann, and P.A.E.M. Janssen. 1994. *Dynamics and Modelling of Ocean Waves*, Cambridge, 300 pp.

Lai, R.J., S.R. Long, and N.E. Huang. 1989. "Laboratory studies of wave-current interaction: kinematics of the strong interaction," *J. Geophys. Res.*, 97(C11), 16,201-16,214.

LeMehaute, B. 1962. "On non-saturated breakers and the wave run-up," *Proc. 8th Coast. Engrg. Conf.*, ASCE, 77-92.

Miche, M. 1951. "Le Pouvoir Reflechissant des Ouvrages Maritimes Exposes a l' Action de la Houle," *Annals des Ponts et Chaussess*, 121e Annee, 285-319 (translated by Lincoln and Chevron, University of California, Berkeley, Wave Research Laboratory, Series 3, Issue 363, June 1954).

Resio, D.T. 1987. "Shallow-water waves. I: Theory," *J. Wtrway., Port, Coast., and Oc. Engrg.*, 113(3), 264-281.

Ris, R.C., and L.H. Holthuijsen. 1996. "Spectral modelling of current induced wave-blocking," *Proc. 25th Coast. Engrg. Conf.*, ASCE, 1247-1254.

Suh, K.D., Y.-Y. Kim, D.Y. Lee. 1994. "Equilibrium-range spectrum of waves propagating on currents," *J. Wtrway., Port, Coast., and Oc. Engrg.*, 120(5), 434-450.

SEASONAL CLOSURE OF TIDAL INLETS

Roshanka Ranasinghe[1] and Charitha Pattiaratchi[1]

Abstract

Seasonally open tidal inlets which occur mainly on parabolic shaped equilibrium beaches close to the ocean every year due to the formation of a sand bar. Field data at a typical seasonally open inlet indicate that cross-shore sediment transport processes may play an important role in the closure of this type of inlet. A Quasi-3D morphodynamic model, which takes cross-shore and longshore processes both into account is developed and implemented. The model validation tests yield impressive results. Application to two ideal cases indicate that the inclusion of cross-shore processes is crucial to determine the seasonal closure inlets.

Introduction

Seasonally open tidal inlets are commonly found in microtidal, wave-dominated coastal environments where strong seasonal variations of streamflow (into the estuary/lagoon connected to the inlet) and wave climate are experienced (Hodgkin and Clark, 1988; Largier et al., 1992; Cooper, 1994). This type of inlets are common in Australia, India, South Africa and Sri Lanka (Bruun, 1986; Hodgkin and Clark, 1988; Gordon, 1990; Largier et al., 1992; Cooper, 1994; Lanka Hydraulic Institute, 1997). These inlets close to the ocean every year due to the formation of a sand bar across the entrance, usually during summer when the streamflows are low and long-period swell waves dominate, or during monsoonal conditions when longshore transport rates are high. Because of their sheltered nature, most of these inlets are used as small fishing boat harbours and as recreational facilities. Hence, the seasonal closure of the inlet presents a two fold problem. Firstly, ocean access for boats is limited to when the inlet is open, and secondly the degradation of the water quality in the estuary/lagoon during the months of inlet closure.

---

[1] Centre for Water Research, University of Western Australia, Nedlands 6907, Western Australia

Inlet closure is generally attributed to the formation and progradation of an updrift spit due to strong longshore transport. However, many seasonally open tidal inlets are located on parabolic shaped equilibrium beaches, defined by Silvester and Hsu (1993), where longshore transport is negligible (Hodgkin and Clark, 1988; Gordon, 1990; Largier et al., 1992; Cooper, 1994, Lanka Hydraulic Institute, 1997). Therefore, it is unlikely that longshore transport is the dominant process leading to the seasonal closure of these inlets. The aim of this study was to gain insight into the governing mechanisms of seasonal inlet closure. The study comprised of field investigation and numerical modeling components.

Field Study

Wilson Inlet, Western Australia (Fig. 1) was selected as a case study for this investigation. The small oceanic tidal range (0.8m), strong seasonality of streamflow, and the parabolic shaped equilibrium beach on which the inlet is situated make Wilson Inlet an ideal case study. The inlet entrance is closed for about 6-7 months during summer due to the formation of a sandbar and is artificially opened during winter.

Offshore wave climate and inlet currents were recorded using S4 current meters. A Marsh McBirny current meter was used to measure nearshore currents along a number of cross-shore transects. Nearshore bed level changes and inlet bed level changes were recorded every 3 weeks. All measurements were made during summer 1995. The offshore wave data indicate predominant southerly swell wave conditions of 1m wave height (rms) and 12s period. Several on-off shore transport criteria predict on-shore transport under these wave conditions. The inlet currents measured were slightly flood dominant with maximum flood and ebb velocities of $1ms^{-1}$ and $0.8ms^{-1}$ respectively. The maximum longshore currents measured were rather weak (< 50cms^{-1}). The morphological measurements along channel transects do not indicate a gradual filling up of the inlet channel or the formation of an updrift spit as would be the case if longshore transport was causing inlet closure. Typical inlet bed level changes are shown in Fig. 2a. Nearshore bed level measurements along all cross-shore transects indicate a gradual build-up of the beach face and onshore migration of a shore-parallel bar during summer. Figure 2b shows nearshore bed level changes along a typical cross-shore transect. These results indicate that rather than longshore sediment transport blocking the inlet, cross-shore processes too may be critical for the seasonal closure of the entrance.

Numerical Study

Currently available state-of-the-art coastal area models (DeVriend, 1987; Anderson et al., 1988; O'Connor and Nicholson, 1992) are all two dimensional in the horizontal (2DH), and therein lie their main weakness. These models generally fail where, (a) the 3-D flow field is important, eg. in the vicinity of tidal inlets and coastal structures, and (b) cross-shore sediment transport processes are important (DeVriend et al, 1993).

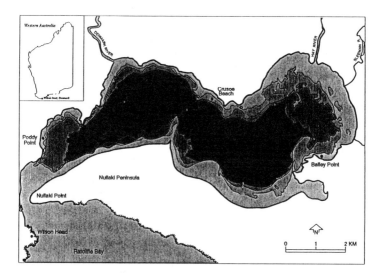

Figure 1. Wilson Inlet, Denmark, Western Australia.

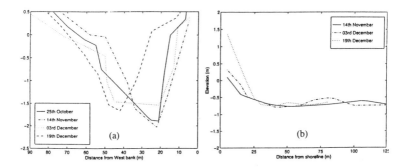

Figure 2. Bed-level changes through summer '95: (a) Inlet; (b) Nearshore.

Therefore, it is obvious that the currently available 2DH Medium Term Model (2DH MTM) concept is inappropriate to study the seasonal closure of tidal inlets. In this study, it was attempted to overcome the two main short comings of the MTM concept given above by (a) using a 3D hydrodynamic model to estimate near bottom tidal currents, and (b) including cross-shore processes such as undertow, wave asymmetry and bottom boundary layer flow which govern cross-shore sediment transport.

Model Formulation:
A three-dimensional hydrodynamic model, the HAMburg Shelf Ocean Model (HAMSOM), is used to obtain the near-bottom tidal current velocities and surface elevations that are required for the sediment transport calculations. The model is described by Backhaus (1985). HAMSOM has been successfully applied to several coastal systems since 1985.

A monochromatic wave transformation model, RCPWAVE, is used to obtain the wave heights and directions, wave numbers, and breaker positions that are required for the sediment transport calculations. The model development and validation is described in Ebersole et al. (1986). This model too has been used in many engineering projects.

The morphological model developed in this study consisted of, (a) a wave-driven currents module, and (b) a sediment transport module.

Wave-Driven Currents Module
To estimate the wave-driven longshore current and its distribution across the surf zone, Longuet-Higgins (1970) formulation is employed in this model. Rather than estimating the longshore current at mid-surf zone and then using some distribution pattern to estimate the current at different locations across the surf zone (Martinez and Harbaugh, 1993), it is more appropriate to use a formulation like that of Longuet-Higgins (1970), which estimates the distribution of the longshore current across the surf zone based on the radiation stress gradients. The two main Wave-driven cross-shore currents, undertow and bottom boundary layer flow (Stokes drift), which are neglected in 2DH models, are evaluated in the present model according to DeVriend and Stive (1987).

Sediment Transport Module
Both cross-shore and longshore transport are estimated using Bailard (1981) type energetics transport equations. Cross-shore and longshore transport rates estimated at each grid point are then resolved into x and y directions on the finite difference grid by performing a simple axes transformation. The bed elevations are then updated by solving the 2D sediment continuity equation using a Lax-Wendroff scheme. The stability of this scheme is dependent on a CFL type criterion for bed form motion (for a detailed description see Nairn and Southgate, 1993).

Model validation:
As it is imperative that in this case cross-shore *and* longshore processes are
represented accurately, model validation was undertaken separately for these two
modes of transport. No attempt was made to validate the hydrodynamic and wave
models as these models have already been validated and used successfully in a
number of studies.

Model simulations of onshore and offshore transport cases were compared with small
scale and prototype scale laboratory results, and field results. The comparisons are
shown in Figure 3 and indicate that model predictions of both on-shore and off-shore
transport are very close to the observations.

To validate longshore transport processes, a number of model simulations were
undertaken with different conditions and the model results were compared with the
values predicted by the Shore Protection Manual(SPM) (1984) equation for longshore
transport under the same conditions. This was considered to be a more reasonable
check as the SPM equation has been calibrated using a large data base. The forcing
conditions, model results, and SPM equation predictions for a mild slope (m = 0.01)
and a steep slope (m = 0.05) are given in Table 1. Although the model appears to be
consistently under-predicting longshore transport the model results are never more
than a factor of 2 away from the SPM equation predictions.

Table No. 1: Model predicted and calculated longshore transport rates

			m=0.01		m=0.05	
H_o(m)	T(s)	α_o(deg)	I_l(N/s) - SPM	I_l(N/s) - Energetics	I_l(N/s) - SPM	I_l(N/s) - Energetics
1.0	12	10	246	133	371	127
1.0	12	50	1004	832	1614	967
2.5	6	10	2270	1680	3149	2809
2.5	6	50	8559	6487	9805	7485

where, m = bed slope,H_o = offshore wave height,T = wave period, I_l-SPM and I_l-
Energetics = longshore transport rate given by SPM equation and model respectively.

From the results discussed above, it can be concluded that the model appears to be
capable of simulating the cross-shore and longshore processes reasonably well.

Model Application:
To determine the effect of the improvements made to the 2DH MTM concept in this
model, two ideal scenarios were considered:
(a) an inlet on a straight beach; and
(b) an inlet on a semi-circular beach

Small Scale Laboratory Tests

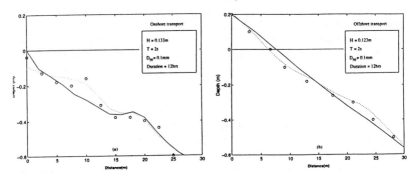

Prototype Scale Laboratory Tests

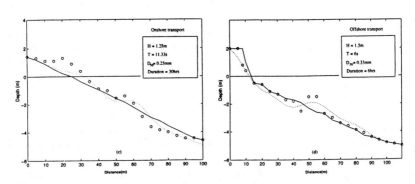

Field Experiments

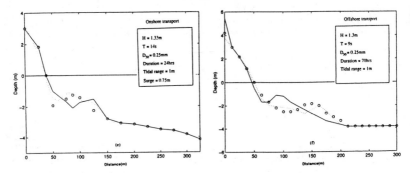

Figure 3. Model validation tests; Cross-shore transport (— = initial profile, o = final mesured profile, ... = final modelled profile)

The erosive/accretive pattern predicted by the model with only longshore transport (this would be the case with the usual 2DH MTM concept) was compared with that of when both longshore transport (LT) and cross-shore transport (CST) were considered. For all test cases, a diurnal tide with a range of 1m was specified at the offshore boundary. The offshore wave conditions with which the model has been forced are indicated in each figure. Figures 4 to 7 show the accretion and in meters after 5 days of simulation for the various cases.

Inlet on a straight beach:
(a) Erosive wave conditions
With LT only, a large shoal builds up on the updrift side of the inlet, while a much smaller shoal builds up on the downdrift side (Fig. 4a). The ebb channel is shifted downstream due to the build-up of the upstream shoal. This formation of two channel margin bars and the downstream shifting of the ebb channel has also been observed in the field. Under these conditions, the eventual growth of the upstream shoal would cause the closure of the inlet. Figure 4b indicates that when LT and CST are both included, still the eventual closure of the inlet would be due to the growth of the updrift shoal caused mainly by LT. However, in Fig. 4b, in addition to features that can be seen in Fig. 4a, the nearshore erosion and the growth of a shore-parallel offshore shoal (near the breaker position), which is very common under erosive wave conditions, can be seen clearly.

(b) Accretive wave conditions
Under these conditions, when only LT is considered, as in the previous case, two shoals can be seen forming updrift and downdrift of the inlet (Fig. 4c). However, as the LT is smaller than before, the shoals are much smaller than under the erosive case. Also, there is only a slight shifting of the ebb channel when compared with the previous case. Therefore, the inlet current appears to be strong enough to keep the inlet open. In contrast, when CST is included, offshore erosion and nearshore accretion is indicated along the entire shore line (Fig. 4d) which is common under prolonged accretionary swell wave conditions. The effect of LT can be seen on the updrift side of the inlet where the shoals build up more than in other areas. In this case then, in contrast to when only LT is taken into account (where the inlet shows a tendency to remain open (Fig. 4c)), the inlet shows a tendency to be closed due to the combined effect of LT and CST. Thus, under accretionary wave conditions, the inclusion of cross-shore processes into the model is critical, and CST is an important process leading to the closure of the inlet.

Inlet on a semi-circular beach:
(a) Erosive wave conditions
In this case, due to the curvature of the bottom contours, the offshore waves refract considerably before breaking, and approach the beach with angles close to 90^0. Therefore, the longshore currents are smaller than that of the straight beach case discussed above. Fig. 5a shows the erosion/accretion when only LT is taken into account. The erosive patches are where the incident wave angles depart from 90^0.

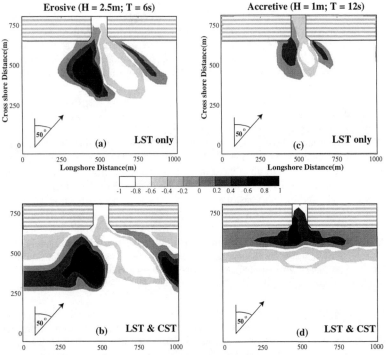

Figure 4. Model Predictions: Inlet on a straight beach; (a) Erosive waves - without cross-shore transport; (b) Erosive waves - with cross-shore transport; (c) Accretive waves - without cross -shore transport; and (d) Accretive waves - with cross-shore transport (negative = erosion, positive = accretion).

Again, two shoals can be seen developing on either side of the inlet and their orientation indicates that there is a tendency for the inlet to close eventually due to LT. When CST is included (Fig. 5b), as before, nearshore erosion and the formation of an offshore shoal is clearly indicated. However, the shoals attached to either side of the inlet is due to the action of the LT and the growth of these shoals will be the main process that will eventually close the inlet.

(b) Accretive wave conditions
Under accretive wave conditions when only LT is considered, again, two shoals can be seen forming on either side of the inlet (Fig. 5c). However, these shoals are much smaller than in the erosive case, and also their orientation indicates that the inlet currents are strong enough to keep the inlet open. When CST is included (Fig. 5d), there is a clear indication of offshore erosion and nearshore accretion along the entire beach. The accretion opposite the inlet is greater than away from the inlet due to combined effect of LT caused shoals and CST caused shoals. Thus, the inclusion of

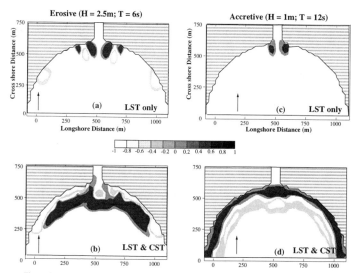

Figure 5. Model Predictions: Inlet on a semi-circular beach; (a) Erosive waves - without cross-shore transport; (b) Erosive waves - with cross-shore transport; (c) Accretive waves - without cross-shore transport; and (d) Accretive waves - with cross-shore transport (negative = erosion, positive = accretion).

CST in the model appears to be crucial under accretionary wave conditions, so much so, that the result of including CST would be to change the conclusion from 'a tendency for the inlet to remain open (with LT only)' to 'a tendency for the inlet to close'.

Conclusions

A Quasi-3D morphodynamic model which simulates longshore and cross-shore processes has been developed and implemented successfully. Application of the model to two ideal situations indicate that the inclusion of cross-shore processes results in a much more detailed representation of nearshore morphodynamics than in a conventional 2DH MTM model. Under reduced longshore transport conditions, such as that would occur on stable parabolic shaped beaches where many seasonally open tidal inlets are located, the inclusion of cross-shore processes is crucial to determine the behaviour of small tidal inlets. Model applications suggest that although longshore transport could close an inlet under energetic wave conditions, onshore sediment transport is the primary mechanism responsible for inlet closure under accretionary swell wave conditions.

References

Andersen, O. H., Hedegaard, I. B., Deigaard, R., de Girolamo, P., and Madsen, P. 1988. Model for morphological changes under waves and current. *Proc. IAHR symp. Math. Mod. Sed. Transp. Coastal Zone,* Copenhagen, DHI, Horsholm, 310-319.

Backhaus, J. O. 1985. A three dimensional model for the simulation of shelf sea dynamics. *Deutsche Hydrographische Zeitschrift.* Vol. 38: 165-187.

Bruun, P. 1986. Morphological and Navigational Aspects of Tidal Inlets on Littoral Drift Shores. *Journal of Coastal Research.* Vol. 2(2): 123-145.

Cooper, J. A. G. 1994. Sedimentary processes in the river-dominated Mvoti estuary, South Africa. *Geomorphology.* Vol. 9: 271-300.

DeVriend, H. J. 1987. 2DH mathematical modelling of morphological evolutions in shallow water. *Coastal Engineering.* Vol. 11(1): 1-27.

DeVriend, H. J. and Stive, M. J. F. 1987. Quasi-3D modelling of nearshore currents. *Coastal Engineering.* Vol. 11(5/6): 565-601.

DeVriend, H. J., Zyserman, J., Nicholson, J., Roelevink, J. A., Pechon, P., and Southgate, H. N. 1993. Medium-term 2DH coastal area modelling. *Coastal Engineering.* Vol. 21: 193-224.

Ebersole, B. A., Cialone, M. A., and Prater, M. D. 1986. Regional Coastal Processes Numerical Modelling System. US Army Corps of Engineers-CERC. Technical Report, CERC-86-4.

Gordon, A. D. 1990. Coastal Lagoon Entrance Dynamics. *Proc.International Conference on Coastal Engineering,* Delft, Netherlands, American Society of Civil Engineers, Vol. 3: 2880-2893.

Hodgkin, E. P. and Clark, R. 1988. Wilson, Irwin and Parry Inlets: the estuaries of the Denmark Shire. Environmental Protection Authority. Vol. 3.

Silvester, R. and Hsu, J. R. C. 1993. Coastal Stabilization. Prentice - Hall Inc., New Jersey, USA.

Lanka Hydraulic Institute. 1996. Chilaw Anchorage flushing study, Interim Report. LHI-96-2.

Largier, J. L., Springer, J. H. AND Taljaard, S. 1992. The stratified hydrodynamics of the Palmiet - a prototype bar-built estuary. *In* Prandle, D. (ed.), *Dynamics and Exchanges in Estuaries and the Coastal Zone.* American Geophysical Union, pp. 135-154.

Longuet-Higgins, M. S. 1970. Longshore Currents generated by obliquely incident sea waves. *Journal of Geophysical Research.* Vol. 75: 6778-6801.

Martinez, P. A. and Harbaugh, J. W. 1993. Simulating Nearshore Environments. Pergamon Press.

Nairn, R. B. and Southgate, H. N. 1993. Deterministic profile modelling of nearshore processes. Part 2. Sediment transport and beach profile development. *Coastal Engineering.* Vol. 19: 57-96.

O' Connnor, B. A. and Nicholson, J. 1992. An estuarine and coastal sediment transport model. *In* Prandle, D. (ed.), *Dynamics and Exchanges in Estuaries and the Coastal Zone.* American Geophysical Union, pp. 507-528.

Wave Groups from Deep to Shallow Water

C.C.Bird [†] and D.H.Peregrine [‡]

Abstract

Numerical computations using a full irrotational flow solver have been carried out to study the evolution of wave groups and the low frequency waves (currents) associated with them. In deep water we show the formation of steep wave groups from a slightly modulated uniform wave train. There is evidence of bound low frequency waves (currents) beneath these groups. In the intermediate to shallow water regimes, we study an isolated wave group, and show that this disperses as it progresses. The group emits long free waves which can be detected; this indicates the formation of a bound wave beneath the group. As shallow water is approached, Stokes' theory fails and the waves break.

Introduction

In water of intermediate or greater depth, wave groups generate long, low frequency waves (currents) beneath them. These travel with the group velocity and are known as bound waves. They create a depression of the mean surface level, especially under high waves, and hence a small backflow in the water underneath. If they are not accounted for in a given initial condition, then they will form. In doing so, the fluid repelled from underneath the group may then result in the generation of more low frequency waves. These are known as long free waves as they travel at the phase velocity appropriate for their wavelength and are not bound to the wave group.

Longuet-Higgins & Stewart (1962) derived the bound waves analytically, using a perturbation expansion, in water of intermediate depth. The bound waves detected in deep water are of a significantly smaller magnitude than those found in intermediate water. However, little is known about the influence of highly nonlinear (very steep) waves on the long wave properties; the following deep water study is part of an investigation of this.

Although the expansion for deriving bound waves is not valid in shallow water (it becomes singular as the group velocity approaches the phase velocity, ie. as Stokes' theory becomes inadequate), low frequency waves (LFW) exist in this region. In recent experiments and numerical studies on LFW in the surf zone (Barnes, Peregrine & Watson 1994 and Watson, Barnes & Peregrine 1994), it

---

[†]Research Student, School of Mathematics, University of Bristol, University Walk, Bristol, BS8 1TW, UK (*C.C.Bird@bristol.ac.uk*)

[‡]Professor of Applied Mathematics, School of Mathematics, University of Bristol (*D.H.Peregrine@bristol.ac.uk*)

was not clear how to initialise the LFW, as their precise form depends on the beach topography, on the earlier evolution of the waves. Computations by Barnes (Barnes & Peregrine 1995) provide some insight into this. They show a single wave group moving up a plane sloping beach and onto a level shallow water shelf. A number of interesting features were noted there, for example, the group spread out significantly, indicating that the effects of dispersion may negate those of shoaling; low frequency bound and free waves were detected, and the bound wave (or set down) develops significantly in shallow water. The following work on intermediate water is an attempt to clarify some of these features.

The numerical model

The computations were carried out using a fully accurate nonlinear irrotational flow solver. This is able to solve from deep water up to the first occurrence of wave breaking. The spatial domain may be either periodic or of infinite length; for the computations on deep water we chose the former and on intermediate-depth water the latter.

The fluid is considered to be two-dimensional, inviscid, incompressible and irrotational, and surface tension is neglected. Defining points (x, y) on the free surface by $\mathbf{R}(X(s,t), Y(s,t)) = \mathbf{R}(s,t)$, where s is a particle-following (Lagrangian) coordinate, we have:

$$\nabla^2 \phi = 0 \quad \text{in fluid body} \tag{1}$$

$$\frac{D\mathbf{R}}{Dt} = \nabla\phi \quad \text{on free surface} \tag{2}$$

$$\frac{D\phi}{Dt} = \frac{1}{2}|\nabla\phi|^2 - gy \quad \text{on free surface} \tag{3}$$

with the appropriate boundary and bed conditions. The velocity potential is represented by $\phi(x, y, t)$, h is the water depth in the intermediate depth situation and g is the acceleration due to gravity.

The model solves this system and hence computes the velocity at the free surface using a Cauchy theorem boundary integral. The surface is then stepped in time using a Taylor series truncated at the sixth order. Further details may be found in Dold (1992) for the periodic domain, and Tanaka *et al.* (1987) for the infinite domain.

Deep Water

Initial Conditions

The periodic model is initialised with an accurate steady Stokes' wave form of steepness $ak = 0.08$ and containing 25 waves, modulated first by 5 short 10% modulations each of 5 waves, and then by 1 long 5% modulation over all 25 waves. The progress of this wave was followed over 1000 time periods of the corresponding linear wave.

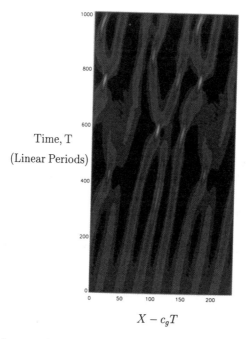

$$X - c_g T$$

Figure 1: A space-time plot of modulation amplitude in a frame of reference moving with the linear group velocity

Results

Figure 1 shows a plot of modulation, or envelope amplitude*. The lighter regions represent higher waves and vice versa. The spatial axis (x-axis) has been shifted with the group velocity, and there are $1\frac{1}{2}$ periods of the surface in space, hence $7\frac{1}{2}$ small groups may be detected initially instead of 5. Time in periods is indicated on the y-axis.

It is possible to see that the short modulations become steeper as they progress. This is consistent with the Benjamin-Feir instability (Benjamin & Feir 1967), which causes a uniform wave train subject to an infinitesimal disturbance to be unstable, and develop into a series of wave groups, ie. the energy self-focusses in these groups. Similarly, the nonlinear Schrödinger Equation (NLS+),

$$i\frac{\partial u}{\partial t} + \frac{\partial^2 u}{\partial x^2} + |u|^2 u = 0 \qquad (4)$$

which is an approximate governing equation in deep water ($kh > 1.363$) valid to third order in the wave steepness ak, also predicts this self-focussing of the energy.

---

*Courtesy of K. Henderson

Here, u may represent the modulation amplitude of the velocity potential or the surface elevation, and variables are scaled to give the simple form of equation 4.

Whereas in the case of a simple modulation, the wave train will reform (although not exactly to the initial form), in this situation the steepening groups interact further to form some very steep groups, at a time close to 600 periods.

To illustrate details, a time when there exists a particularly steep group is chosen. More precisely, we choose the time when there is a crest at the centre of this short, steep group (t=588.6 linear periods).

Analysis

The wave surface at this time consists of approximately three wave groups: one short and very steep group $(ak \approx 0.3)$ centred close to $x = 2.25$, another slightly less steep group $(ak \approx 0.2)$ at around $x = 4.0$, and a long relatively flat group.

In order to determine the form of the long wave current underneath this surface, the horizontal velocity $u = \partial\phi/\partial x$ along a line just beneath the wave surface was filtered using a low pass Fourier filter. This is illustrated, with the wave surface, in figure 2. In the plot of the current, we chose to retain just 9 modes, as the length of variation is then of approximately the same length as the steep groups.

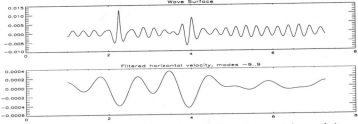

Figure 2: Wave surface and the Fourier filtered current beneath it

It is clear that the low frequency current is negative underneath the steep groups. This is consistent with a bound wave forming underneath each group. However, there is a lot of excess noise present in the form of extra oscillations. An explanation for this is that as the bound wave forms underneath a group, it leaves a significant amount of free energy in the system. This cannot escape since periodic boundary conditions have been imposed, and hence it bounces around the system and creates this noise.

Figure 3 shows a plot of filtered stream function in space and time, again with 9 modes retained. Two spatial periods have been included for clarity. This indicates the presence of the bound wave through the trace of negative current moving with the steep groups.

Intermediate towards Shallow Water, $kh < 1.363$

Initial Conditions

The model was initialised with a linear wavetrain modulated by the deep water envelope soliton solution of the nonlinear Schrödinger equation (NLS+)

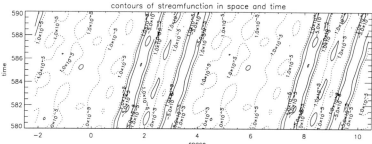

Figure 3: The 9-mode Fourier filtered stream function over 10 linear time periods

(Peregrine 1983). The surface elevation and velocity potential of the envelope soliton are given (at time $t = 0$) by

$$\eta \;=\; \frac{1}{2}\left(a\,\text{sech}\left(\sqrt{2}ak^2 x\right)e^{ikx} + c.c.\right) \tag{5}$$

$$\phi \;=\; -\frac{g}{2\omega}\left(ia\,\text{sech}\left(\sqrt{2}ak^2 x\right)e^{ikx} + c.c.\right). \tag{6}$$

Second order approximations to the surface elevation and velocity potential (eg. Brinch-Nielsen 1985) were added to this; these include terms of the appropriate Stokes' second harmonic.

The frequency ω comes from the linear dispersion relation

$$\omega^2 = gk\tanh(kh). \tag{7}$$

The value of this was chosen to be appropriate to the depth on which the computation was carried out, although strictly (for the purposes of the exact analytical solution of the NLS+) it should take the deep water value given by $\omega^2 = gk$.

By imposing a deep water group on water of constant depth and only adjusting the second harmonic, these initial conditions simulate a bed topography stepping abruptly from infinitely deep to intermediate depth water.

Results for $kh = 1.1$, $ak = 0.1$

Figure 4 shows the development computed by the (infinite domain) model of a such an envelope soliton, on a depth of $kh = 1.1$. It is clear that the group spreads out as it progresses. Looking at a selection of these profiles (figure 5), it can be seen that the lengthened group has a very flat topped envelope. It is also possible to detect that the waves at the front of the lengthened group are slightly longer than those at the back of the group. This is consistent with the dispersion relation given above (equation 7), which indicates that longer waves travel faster than shorter ones.

Figure 6 corresponds to the surface elevation at various times, with the region outside the primary group exaggerated and overplotted with the dotted line. As in the previous figure, the spatial coordinate has been shifted with the linear group velocity. One noticeable feature indicated by the exaggerated surface is

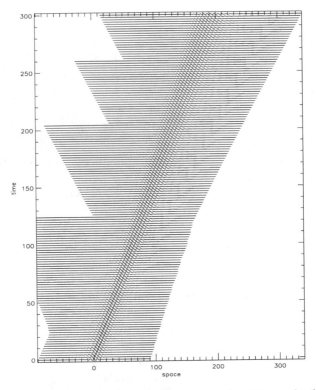

Figure 4: Stack plot of wave group of steepness $ak = 0.1$ on a depth $kh = 1.1$

the long wave of elevation radiated at the front of this group, and the smaller wave of depression moving away from the tail of the group. These result from the omission of the bound wave from underneath the group in the initial conditions. The long free waves are emitted as compensation for the bound wave which forms as the group progresses. The exaggerated profile also highlights a group of short waves, travelling slower than the group. These correspond to parts of the Stokes' second harmonic which occur in the third and higher order approximations to the surface. These were omitted from the initial condition, and so are emitted by the group as a 'correction'.

Comparisons with approximate theories

Whereas the NLS+ describes the behaviour of wave groups in deep water upto the third order in wave steepness, the relevant governing equations in intermediate depth are the Davey-Stewartson equations (Davey & Stewartson 1974). These are two coupled nonlinear equations which include terms representing the low frequency waves, or mean flow, associated with the group, as they are now of significant magnitude. They model three-dimensional wave groups; however, if

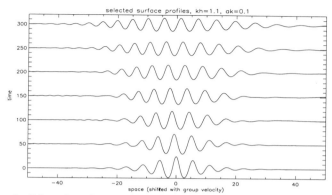

Figure 5: Selected surface profiles for wave group of steepness $ak = 0.1$ on a depth $kh = 1.1$

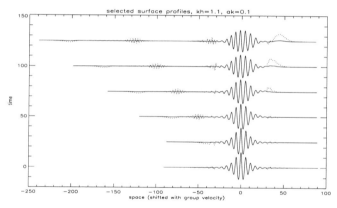

Figure 6: Selected surface profiles for wave group of steepness $ak = 0.1$ on a depth $kh = 1.1$

propagation is only in one space dimension, it is possible to eliminate the mean flow potential term ϕ_{01} and reduce the equations to a form of the nonlinear Schrödinger equation, this time known as the NLS− (or defocussing form) of the equation due to the sign in front of the nonlinear term. Appropriate scalings give

$$i\frac{\partial u}{\partial t} + \frac{\partial^2 u}{\partial x^2} - |u|^2 u = 0. \tag{8}$$

Computed solutions of the NLS− (not shown here) for an equivalent initial condition demonstrate that the group spreads out (disperses) at roughly the same rate as the computed full solution. The envelopes of lengthened groups are very noticeably flat topped.

The second of the Davey-Stewartson equations represents a long wave generated (or forced) by a short wave modulation. It may be written in an alternative form (not moving with the group) as

$$\frac{\partial^2 \phi_{10}}{\partial t^2} - gh \frac{\partial^2 \phi_{10}}{\partial x^2} = \left(\frac{\omega^3}{2k \tanh^2(kh)} + \frac{c_g \omega^2}{4 \sinh^2(kh)} \right) \frac{\partial \mid A \mid^2}{\partial x}. \tag{9}$$

Since the initial expression for A is known, it is possible to solve for ϕ_{01} and thus form an expression for the horizontal current. Writing κ_1 as the coefficient of the forcing term (on the right hand side of equation 9) gives

$$\frac{\partial \phi_{10}}{\partial x} = \frac{\kappa_1 a^2}{2(c_g^2 - c_p^2)} \left(1 - \frac{3c_g}{c_p} \right) \operatorname{sech}^2 \left(\kappa_2 (x - c_p t) \right) +$$
$$\frac{\kappa_1 a^2}{2(c_g^2 - c_p^2)} \left(1 - \frac{c_g}{c_p} \right) \operatorname{sech}^2 \left(\kappa_2 (x + c_p t) \right) + \frac{\kappa_1 a^2}{(c_g^2 - c_p^2)} \operatorname{sech}^2 \left(\kappa_2 (x - c_g t) \right). \tag{10}$$

Here the first term on the right hand side represents the long free wave which travels ahead of the group, the second is the long free wave behind the group, and the third is the bound long wave. κ_2 is a constant term. In order to assess the magnitudes of these waves, appropriate values may be inserted for the case of $kh = 1.1$, giving

$$\begin{aligned}
\frac{\partial \phi_{10}}{\partial x} &= 0.00704 \ \operatorname{sech}^2 \left(\kappa_2 (x - c_p t) \right) - 0.00144 \ \operatorname{sech}^2 \left(\kappa_2 (x + c_p t) \right) \\
&\quad - 0.01135 \ \operatorname{sech}^2 \left(\kappa_2 (x - c_g t) \right).
\end{aligned} \tag{11}$$

It is clear from these values that the wave behind the group and the bound wave are waves of depression, whereas the forward wave is one of elevation. It is also possible to see that the free wave ahead of the group is roughly 5 times larger than the free wave behind the group in the early evolution of the computation, and this is consistent with what is observed.

Further Results

For a wave of initial steepness $ak = 0.1$, changes in the behaviour of the wave group as it progresses are evident at a depths corresponding to $kh = 0.7$ or less. Figure 7 shows selected surface profiles from the computation on depth $kh = 0.6$. A strong indication that Stokes' theory fails at this depth is the presence of small 'blips' in the troughs around the centre wave, due to the addition of the Stokes' second harmonic terms. This is the same theory which is used to form the Davey-Stewartson equations, so this too becomes inappropriate at this depth. However, the emission of the higher harmonics is still evident as the group progresses, as is the presence of the long wave in front of the group. At this depth, the group velocity is much closer in value to the phase velocity and the long wave velocity, and so the long wave of elevation takes much longer to emerge from the front of the group. It is now much larger than before, and elevates the individual waves in the group as it emerges.

The shapes of the waves undergo a significant change as the group progresses. Even by a time of 100 (in non-dimensional units) the waves have developed long, flat troughs and short well defined crests; they appear very much like solitary

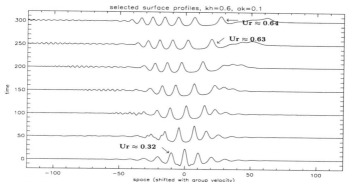

Figure 7: Selected surface profiles for wave group of steepness $ak = 0.1$ on a depth $kh = 0.6$

waves. This is particularly noticeable by the end of the computation, where the largest solitary wave is at the front of the group and is progressing faster than the ones behind it. It is moving independently of the rest of the group, a behaviour typical of the shallow water regime. Solitary waves are also forming from the long wave itself, as is to be expected.

Approximate Ursell numbers are indicated for some of the waves in this computation. The Ursell number is defined as $Ur = a/k^2h^3$, where a is the individual wave amplitude, k the wave number and h the depth of the water. The wave which is at the front of the group at $t = 300$ has doubled its value of Ur, from ~ 0.32 to ~ 0.64, and clearly does not satisfy the condition $Ur \ll 1$ required for Stokes' theory to hold.

Similar effects, although more pronounced, may be observed for lower water depths. However, by a depth of $kh = 0.4$, the waves rapidly steepen and break, indicating that we are too far into the shallow water regime for this choice of initial conditions.

Conclusions

In deep water, a modulated wave train is unstable and decomposes into a series of wave groups (the Benjamin-Feir instability). In this way, the wave energy self-focusses, as predicted by the NLS+, an approximate governing equation for this regime ($kh > 1.363$). In some cases, the energy may focus enough to form very nonlinear, steep groups. Using an accurate nonlinear flow solver to compute the evolution of a modulated wave train, it is observed that small bound wave currents form underneath these steep groups. The energy which is freed in the process of forming the bound waves is not able to escape from the imposed periodic spatial domain of the computation, and so creates excess noise in the low frequency current away from the steep groups.

In intermediate depth water, $kh < 1.363$, the Davey-Stewartson equations are the appropriate governing equations, valid to third order in the wave steepness. These reduce to the NLS− in one space dimension. The expected defocussing of wave energy is observed for the case of a single wave group on an infinite spatial domain, as the group spreads out (disperses). As a bound wave forms underneath the group, a long free wave of elevation is emitted ahead of the group, and also a smaller long free wave of depression which travels away from the back of the group. As shallow water is approached, Stokes' theory fails (and hence the Davey-Stewartson equations are no longer appropriate), and the evolved waves appear much like solitary waves, with long flat troughs and short well-defined crests. They appear to act independently of the rest of the group. A depth of $kh = 0.7$, which implies that the wavelength is approximately 10 times the depth of the water, is an approximate limit for this behaviour. Once shallower water is reached, the waves quickly steepen and break.

Acknowledgements

We are grateful for support from an EPSRC studentship, and from the Marine Technology Directorate Ltd. of EPSRC and industrial contributors to the programme on Uncertainties in Loads of Offshore Structures under contract GR/J23624.

References

T.C.D. BARNES and D.H. PEREGRINE. Wave groups approaching a beach: Full irrotational flow computations. *Proc. of Coastal Dynamics, 116-127,* (1995).

T.C.D. BARNES, D.H. PEREGRINE and G. WATSON. Low frequency wave generation by a single wave group. *Proc. Internat. Sympos. Waves* - Physical and Numerical Modelling, *280-286,* (1994). Eds. M. Isaacson and M.Quick.

T.B. BENJAMIN and J.E. FEIR. The disintegration of waves of deep water; Part 1, Theory. *J. Fluid Mech.,* **27:** *417-430,* (1967).

U. BRINCH-NIELSEN. Slowly modulated, weakly nonlinear gravity waves - fourth order evolution equations and stability analysis. Master's thesis, Technical University of Denmark, (1985).

A. DAVEY and K. STEWARTSON. On three-dimensional packets of surface waves. *Proc. R. Soc. Lond. A,* **338:** *101-110,* (1974).

J.W. DOLD. An Efficient Surface-Integral Algorithm Applied to Unsteady Gravity Waves. *J. Comp. Phys.,* **103:** *90-115,* (1992).

M.S. LONGUET-HIGGINS and R.W. STEWART. Radiation stress and mass transport in gravity waves, with application to 'surf beats'. *J. Fluid Mech.,* **13:** *481-504,* (1962).

D.H. PEREGRINE. Water waves, nonlinear Schrödinger equations and their solutions. *J. Austral. Math. Soc. Ser. B,* **25:** *16-43,* (1983).

M. TANAKA, J.W. DOLD, M. LEWY and D.H. PEREGRINE. Instability and breaking of a solitary wave. *J. Fluid Mech.,* **185:** *235-248,* (1987).

G. WATSON, T.C.D. BARNES and D.H. PEREGRINE. The generation of low-frequency waves by a single wave group incident on a beach. "*Proc. 24th Internat. Conf. on Coastal Engng.* Kobe, ASCE, **1:** *776-790,* (1994).

Cross-shore Variability of Infragravity Wave Pressure and Velocities
on a Barred Beach

T. C. Lippmann[1], T. H. C. Herbers[2], and E. B. Thornton[2]

Abstract

Field measurements of infragravity wave motions were collected on a natural
barred beach with a cross-shore array of co-located pressure sensors and bi-directional
current meters extending from the beach face to 5 m depth. Cross-shore variations of
infragravity pressure and velocity variances are examined in relation to the incident
wave energy and the mean alongshore current. During low-energy incident wave
conditions, infragravity pressure variances decrease with increasing depth qualitatively
consistent with the theoretically predicted unshoaling and trapping of gravity waves.
However, during high-energy incident wave conditions, the observed infragravity
pressure variances are nearly uniform across the surf zone, suggesting strong
scattering effects in a wide surf zone. Strong tidal modulations of both infragravity
velocity variances and the mean longshore current indicate that shear instabilities of the
longshore current contribute significantly to the infragravity velocity field.

Introduction

Observations from many natural beaches have shown that energy levels of infragravity waves (frequencies 0.004-0.05 Hz) are strongly correlated with incident wave
energy levels and tend to increase towards the shoreline, while the incident waves are
diminished by wave breaking (*e.g.*, Huntley, 1976; Huntley *et al.*, 1981; Holman,
1981; Guza and Thornton, 1982; Holman and Sallenger, 1985; Sallenger and
Holman, 1987; and many others). Much of the infragravity energy is contained in
gravity waves reflected at the shoreline with maximum energy at the shoreline and a
complex cross-shore pattern of nodes and anti-nodes. These standing waves can

[1]Center for Coastal Studies, Scripps Institution of Oceanography, University of California, San
Diego, 9500 Gilman Dr., La Jolla, CA, 92093-0209, USA
[2]Department of Oceanography, Naval Postgraduate School, Monterey, CA, 93943-5100, USA

either be trapped on the sloping beach and continental shelf through reflection from an offshore turning point (edge waves) or radiate energy to open ocean basins (leaky waves). Whereas the energy of leaky gravity waves is proportional to $h^{-1/2}$ (Guza and Thornton, 1985), edge waves decay more rapidly with increasing water depth, h, owing to the combined effects of unshoaling and trapping. On a planar beach the energy of an edge wave spectrum with a broad distribution of energy across all possible wavenumbers is proportional to h^{-1} (Herbers, et al, 1995; Lippmann, et al, 1997).

Observations have also shown energetic velocity fluctuations at infragravity- and lower-frequencies that have been attributed to instabilities of the longshore current, commonly referred to in the literature as shear waves (Bowen and Holman, 1989; Oltman-Shay, et al., 1989; and others). Shear waves have slow alongshore phase speeds (relative to gravity waves), and are distinguished by a dispersion relationship that depends on the mean current velocity (Bowen and Holman, 1989).

Resolving the detailed mode mix of gravity and shear waves at infragravity frequencies requires large aperture alongshore arrays of spatially lagged sensors (Howd, et al., 1991). During the Delilah experiment (discussed later) two alongshore arrays of current meters were deployed on a natural beach. Although these arrays were capable of identifying shear waves and resolving the mode mix of edge and leaky waves, little information could be extracted about the strong cross-shore variability of these motions. In this paper we present preliminary results of a study of the cross-shore variability of gross properties of infragravity wave motions using a dense cross-shore array of co-located pressure and bi-directional current meters that was also deployed in the Delilah experiment.

Observations

The field data examined in this study were obtained on a sandy ocean beach during the Delilah experiment, held at the Army Corps of Engineers Field Research Facility in Duck, NC in the fall of 1990. For details of the experiment see Birkemeier (1991) and Thornton and Kim (1993). Bi-directional velocity and near-bottom pressure measurements were collected at 9 locations along a cross-shore transect that spans the surf zone (Figure 1). The instruments were sampled continuously at 8 Hz. The present analysis is focused on an 11 day period from 6-16 October. In the shallow water that the data were collected (less than 5 m), depth attenuation by hydrodynamic filtering was negligible for the infragravity frequencies of interest. Although, wave pressures were converted to equivalent sea surface excursions using linear theory (e.g., Guza and Thornton, 1980), they will be referred to as pressure signals throughout the paper.

The beach at Duck runs approximately north-south and faces the Atlantic Ocean

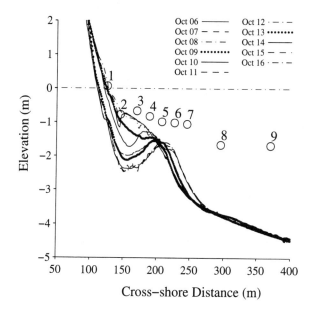

Figure 1. Daily bathymetric profiles measured along the instrumented transect from 6-16 October during the Delilah experiment. The locations of the current meters are indicated with circles (the pressure sensors are lower in the water column at the same cross-shore positions).

on the central eastern sea board of the U. S. Offshore bottom contours are nearly straight and parallel with a gentle slope of 0.0064. During Delilah, there was a single alongshore sandbar that varied substantially over the course of the experiment. The cross-shore bottom profiles measured daily along the instrumented transect are shown in Figure 1. The wave climate during Delilah was variable with offshore significant wave heights ranging from 0.4 m to 2.5 m in 8 m depth. Two extratropical storms passed during the experiment (on 11 and 13 October) radiating long-crested swell that approached the coast at between 40 and 20 degrees CW from shore normal (southeast). Consequently, wave driven longshore currents were predominantly towards the north. A weak northeaster (on 16 October) generated higher frequency waves from the northeast (approaching the coast at about 20 degrees CCW from shore normal) that caused a brief reversal of the longshore current towards the south. A more detailed discussion of the wave, current, and wind conditions during Delilah can be found in Thornton and Kim (1993).

Infragravity velocity and pressure variances were estimated from 34.1 minute data records by integrating the energy spectra over the frequency range 0.004-0.05 Hz. The estimated pressure and velocity variances over the duration of the experiment at each sensor location are shown in Figures 2-4. The high correlation between infra-gravity pressure variances and the incident wave variance E_{inc} (upper panel, based on integration of the spectra of the most seaward sensor 9 over the frequency range 0.05-0.5 Hz) is consistent with many previous observations (e.g., Guza and Thornton, 1982), and suggests that the infragravity motions are low-frequency gravity waves forced by nonlinear interactions of the incident waves in shallow water (e.g., Longuet-Higgins and Stewart, 1962; Gallagher, 1971). During the benign conditions of 6-11 October, the observed pressure variances decay seaward owing to unshoaling and trapping effects (Fig. 5). The observed decay between two sensors varies with tidal stage because the water depths at the instruments are affected by tidal sea level fluctuations. The observed decay is stronger than the theoretical $h^{-1/2}$ decay of leaky gravity waves (bottom panel) suggesting significant trapping of infragravity waves within the instrumented transect. However, during the larger wave events of October 11 and 13, the pressure variances do not decrease with increasing depth but are nearly constant across the surf zone with a slight increase over the bar. This homogenization of the infragravity pressure field suggests that long-wavelength gravity waves are scattered in a wide surf zone, potentially from the hydrodynamic irregularities associated with shorter wavelength broken incident wave bores.

The observed infragravity velocity variances (Figures 3 and 4) show a weaker correlation with E_{inc} and larger cross-shore variations than the pressure variances (Figure 2). During benign wave conditions, observed velocity variances are strongly tidally modulated, coincident with tidal variations in the mean longshore current V_{mean}

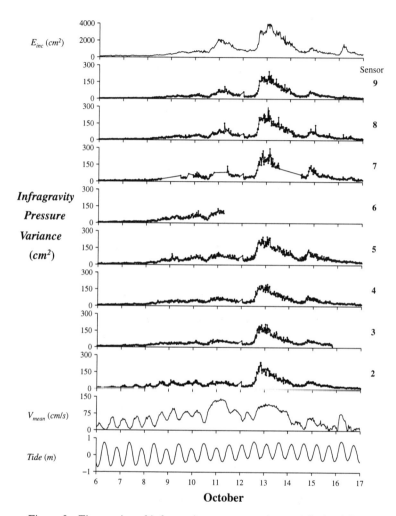

Figure 2. Time series of infragravity pressure variances (obtained from 34.1 minute records) at all 8 instrumented locations (the sensor positions relative to the bathymetry are shown in Figure 1). For reference, the incident (sea and swell) wave variance, E_{inc}, at the most seaward sensor is shown in the top panel, the magnitude of the mean longshore current, V_{mean}, at mid-array sensor 4 is shown in the 2nd panel from the bottom, and the tidal sea level variation about NGVD is shown in the bottom panel.

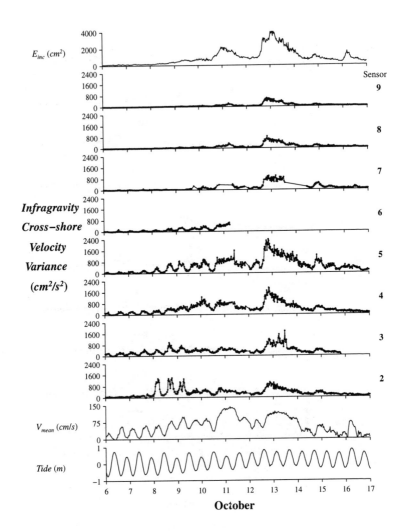

Figure 3. Same as Figure 2 for the infragravity cross-shore velocity variance.
Note that current meter 2 was not submerged at low tides prior to the 8th of October.

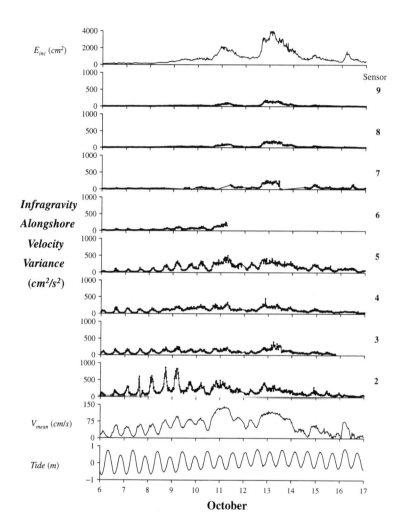

Figure 4. Same as Figure 2 for the infragravity alongshore velocity variance. Note that current meter 2 was not submerged at low tides prior to the 8th of October.

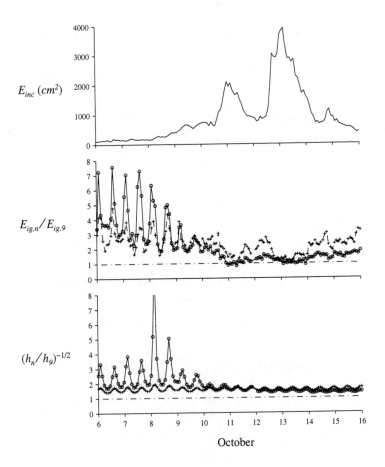

Figure 5. Time series of observed infragravity pressure variances $E_{ig,n}$ relative to the variance $E_{ig,9}$ at the most offshore sensor, for sensors $n = 2$ located near the shoreline (o) and $n = 5$ located on the bar crest (+). The corresponding theoretical variance ratios for leaky gravity waves are shown in the bottom panel, and the incident wave variance is shown for reference in the top panel.

(Figures 2-4, estimated from sensor 4 in the middle of the array). Thornton and Kim (1993) discuss tidal modulations of the mean longshore current in the same Delilah data set, and show that the current distribution moves on and offshore in response to the varying position of the breaking patterns relative to the shoreline (which also varies with tidal stage) and the offshore bar. Thornton and Kim (1993) observed the local V_{mean} and rms incident wave height to be in-phase or 180 degrees out of phase depending on the position of the sensor relative to the cross-shore shear in V_{mean} The observed infragravity velocity variances behave similarly, qualitatively consistent with shear wave contributions that are confined to the region of seaward shear in the longshore current profile (see Lippmann *et al.*, 1997 for further discussion).

Conclusions

The cross-shore variability of infragravity-frequency (0.004-.05 Hz) motions on a naturally barred beach is briefly examined with field measurements from a cross-shore array of co-located pressure sensors and bi-directional current meters. The data were obtained continuously over a three week period during the Delilah experiment, held at the Army Corps of Engineers Field Research Facility on the Outer Banks of North Carolina in the fall of 1990. Infragravity pressure and velocity variances (based on 34. 1 minute records) are compared at 9 positions spanning the width of the surf zone from very near the beach face to 4.5 m depth.

As reported in many previous studies, infragravity pressure variances are highly correlated with incident wave energy variances. During benign conditions the infragravity pressure variances decrease with increasing water depth consistent with the theoretically expected unshoaling and trapping of gravity waves. However, during large wave events the observed infragravity pressure variances are approximately uniform across the surf zone, suggesting that scattering effects are important in a wide turbulent surf zone.

The observed infragravity velocity variances are more variable than the pressure variances and show strong tidal variations that closely resemble the tidal modulations in mean longshore currents observed in the same data by Thornton and Kim (1993). These observations suggest that shear instabilities of the mean longshore current contribute significantly to the infragravity velocity field. The relative contributions of shear instabilities and gravity (*i.e.*, edge and leaky) waves to nearshore infragravity motions is examined in more detail in a subsequent paper (Lippmann, *et al.*, 1997).

Acknowledgments

This work was sponsored by the Office of Navel Research, Coastal Sciences Program. We are indebted to the capable staff of the FRF headed by Bill Birkemeier, and the many people who contributed to the Delilah experiment, in particular Tim Stanton and Rob Wyland of the Naval Postgraduate School who collected the pressure and velocity data use in this study.

References

Birkemeier, W. A., DELILAH nearshore processes experiment: Data summary, miscellaneous reports, Coastal Eng. Res. Cent., Field Res. Facil., U. S. Army Eng. Waterw. Exp. Sta., Vicksburg, Miss., 1991.

Bowen, A. J., and R. A. Holman, Shear instabilities of the mean longshore current 1. Theory, *J. Geophys. Res.*, 94(C12), 18023-18030, 1989.

Gallagher, B., Generation of surf beat by non-linear wave interactions. *J. Fluid Mech.*, 49, 1-20, 1971.

Guza, R. T., and E. B. Thornton, Local and shoaled comparisons of sea surface elevation, pressures, and velocities, *J. Geophys. Res.*, 85, 1524-1530, 1980.

Guza, R. T. and E. B. Thornton, Swash oscillations on a natural beach, *J. Geophys. Res.*, 87(C1), 483-491, 1982.

Guza, R. T., and E. B. Thornton, Observations of surf beat, *J. Geophys. Res.*, 90, 3161-3172, 1985.

Herbers, T. H. C., S. Elgar, R. T. Guza, and W. C. O'Reilly, Infragravity-frequency (0.005-0.05 Hz) motions on the shelf. Part II: Free waves, *J. Phys. Oceanogra.*, 25(6), 1063-1079, 1995.

Holman, R. A., Infragravity energy in the surf zone, *J. Geophys. Res.*, 86(C7), 6442-6450, 1981.

Holman, R. A. and A. H. Sallenger, Setup and swash on a natural beach, *J. Geophys. Res.*, 90(C1), 945-953, 1985.

Howd, P. A., J. M. Oltman-Shay and R. A. Holman, Wave variance partitioning in the trough of a barred beach, *J. Geophys. Res.*, 96, 12781-12795, 1991.

Huntley, D. A., Long period waves on a natural beach, *J. Geophys. Res.*, 8, 6441-6449, 1976.

Huntley, D. A., R. T. Guza and E. B. Thornton, Field observations of surf beats. 1. Progressive edge waves, *J. Geophys. Res.*, 86(C7). 6451-6466, 1981.

Lippmann, T. C., T. H. C. Herbers, and E. B. Thornton, Gravity and shear wave contributions to nearshore infragravity motions, *J. Phys. Ocean*, sub judice, 1997.

Longuet-Higgins, M. S., and R. W. Stewart, Radiation stress and mass transport in surface gravity waves with application to surf beats, *J. Fluid Mech.*, 13, 481-504, 1962.

Oltman-Shay, J. M., P. A. Howd, and W. A. Birkemeier, Shear Instabilities of the mean longshore current 2. Field observations, *J. Geophys. Res.*, 94(C12), 18031-18042, 1989.

Sallenger, A. H., and R. A. Holman, Infragravity waves over a natural barred profile, *J. Geophys. Res.*, 92(C9), 9531-9540, 1987.

Thornton, E. B., and C. S. Kim, Longshore current and wave height modulation at tidal frequency inside the surf zone, *J. Geophys. Res.*, 98, 16509-15519, 1993.

ANALYSIS OF INFRAGRAVITY WAVE-DRIVEN SEDIMENT TRANSPORT ON MACROTIDAL BEACHES

Andrew N. Saulter, Tim J. O'Hare, Paul E. Russell[1]

ABSTRACT

Previous works have noted that under high energy conditions the inner surf zone is dominated by infragravity wave components. Time-series analysis is made on data from Llangennith beach (S.Wales, U.K.) with the aim of discerning a mechanism for sediment transport at these frequencies. In this situation the major mechanism for suspension of sediment is deemed to be an offshore directed velocity associated with the troughs of incoming long waves. Analysis of third order velocity moments yields a similar prediction, and points toward velocity skewness as a significant factor in associating this hydrodynamic component with the observed suspended sediment concentrations.

INTRODUCTION

Under storm conditions, the surf zone becomes one of the most dynamic areas of the nearshore in terms of sediment mobilization and transport. Nonetheless, this region remains one of the least well understood as far as the actual mechanisms leading to spatial and temporal variations in morphodynamics are concerned.

Looking simply at the hydrodynamics, three major components can be seen making up the instantaneous velocity field - a mean flow, and two oscillatory components relating to gravity and infragravity waves. For high energy conditions field studies have found the inner surf zone to be dominated by energy attributed to the latter of these (Wright *et al.* 1982, Russell 1990), suggesting an increasing importance of the infragravity component in the sediment transport problem near to shore. Studies by Hanes and Huntley (1986), Beach and Sternberg (1987), Osborne and Greenwood (1993), and Russell (1993) all observe significant components of sediment suspension at infragravity frequencies, but with differing suggested mechanisms underlying these events.

This paper makes a further examination of the mechanisms lying behind a set of infragravity frequency events observed by Russell (1990) in a particularly high

---

[1] Institute of Marine Studies, University of Plymouth, Drake's Circus, Plymouth, PL4 8AA

energy situation (observed breaker height 3m) in the inner surf zone, where these events are seen to dominate sediment transport processes. The primary aim is to see if a particular component of the instantaneous velocity field may be associated with these largest suspension events, although the discussion is taken further by making an examination of the velocity moments generated from the data set in order to obtain a comparison between the observed sediment transport, and a predictive scheme using hydrodynamic conditions only.

FIELD DATA AND PROCESSING

The data herein comes from a study conducted on Llangennith beach (S. Wales, U.K.) over two tides during a storm event on November 17th 1988, and forms part of the B-BAND (British Beach and Nearshore Dynamics) data set (Russell *et al.* 1991). The entire programme aimed to gather data from a variety of macrotidal beaches, of which Llangennith provides a dissipative example. During the experiment the beach was virtually planar in profile, with a slope of 0.0145, and made up of non-cohesive moderately well sorted medium-fine quartz sand (D_{50}=0.23mm).

Because the beach is situated in the Bristol Channel, which is a highly macrotidal region (spring range approximately 8.5m), any fixed sensor experiences a change in hydrodynamic regime with time. For this experiment a single rig of colocated sensors, consisting of a pressure transducer, electromagnetic current meter (EMCM, positioned 12cm above bed level) and optical backscatter sensor (OBS, elevation 4cm), was positioned such that data could be logged approximately 90 minutes either side of high water. Each run consisted of 1024 points sampled at 3Hz (i.e. time-series length 5.68 minutes). For the present analysis a time-series record length of 2048 points, 11.36 mins, could be used whilst still retaining good levels of resolution and confidence for spectral estimates. Such a record length means that the maximum depth change observed over one 11.36 min record is approximately 0.1m, giving the highest amount of resolution which may reasonably be employed if sets of consecutive records are used to examine the data in terms of cross-shore distribution. Here a predicted cross-shore position for the sensor rig is determined for each record by applying a tidal curve, fitted to mean depths as calculated from the pressure transducer time-series, to the beach profile, which for this experiment was assumed planar. Whilst this technique has a number of errors inherent to it, for example not including effects of set-up, it allows for at least a qualitative idea of the cross-shore distribution of processes to be gained. The 4 channels of data recorded - sea-surface elevation, cross-shore and long-shore velocity, and suspended sediment concentration - are here reduced to 3 channels as the normally incident swell experienced during the storm event led to a comparatively small long-shore component.

HYDRODYNAMICS

Of primary interest to the mobilization of sediment at infragravity frequencies are the behaviour of the oscillatory components, i.e. waves of gravity and infragravity frequencies. Figure 1 indicates the cross-shore distribution of values of significant wave height divided by depth ($H_{sig}/h=\gamma$) for two suites of records 'Storm Day' and 'Storm Night'. Significant wave height is determined following Guza and Thornton (1980) using

$$H_{sig}=4\sigma$$

where σ is the standard deviation of the sea surface elevation, time-series calculated for each record using the area under the sea surface elevation spectrum. This allows for easy further decomposition of H_{sig} into gravity and infragravity components, here using a cut-off at 0.05 Hz. At gravity frequencies γ is of a relatively constant value throughout the region observed for both Storm Day and Storm Night (Figure 1), indicating that wave height is depth limited, and therefore that the observations come from the mid to inner surf zone. Thus at gravity frequencies a diminishing amount of energy is observed as the sensors move closer to the shoreline.

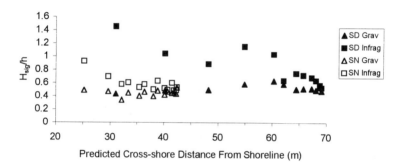

Figure 1: (Qualitative) Cross-shore distribution of Storm Day and Storm Night γ

γ at infragravity frequencies is seen to behave rather differently, showing a general increase in the shoreward direction. Here however Storm Day and Storm Night show marked differences in magnitude. This is attributable to the time at which each tide occurred relative to the storm event, in that the Storm Day records were made at the storm's peak, whilst the Storm Night records come from when the swell was beginning to diminish, with a corresponding decrease in the long wave energy present. Both sets of records however show a qualitatively similar shoreward rate of increase of wave height versus depth, indicating that the infragravity waves present undergo a shoaling process in the shoreward direction, and hence a corresponding increase in wave energy at these frequencies. This is manifested in sea surface elevation spectra as an increasingly dominant peak at (bin) frequencies of 0.00586 Hz and 0.01172 Hz from records made at increasing times before or after high water.

Since the long wave energy is not being dissipated through the surf zone reflection of long waves occurs near to the shoreline leading to the possible formation of standing or partially standing long wave structures through the cross-shore. This is investigated through the calculation of frequency based reflection coefficients (R) using a principal component analysis as detailed by Tatavarti *et al* (1988). R is the ratio of the outgoing wave amplitude to the incoming wave amplitude so that a value of 1 indicates complete reflection. Figure 2 shows the temporal distribution of values of R for the lowest three infragravity frequency bins used in spectral estimates for the Storm Night data. At 0.00586 Hz and 0.01172 Hz a reasonably consistent value of approximately 0.75 is obtained for R, showing that much of the energy at the dominant low frequencies is reflected. The behaviour of the 0.01758 Hz bin is illustrative of that of the other frequencies higher in the infragravity range in that whilst a certain amount of energy appears to be reflected very near to shore, little is actually retained in the offshore propagating wave signal.

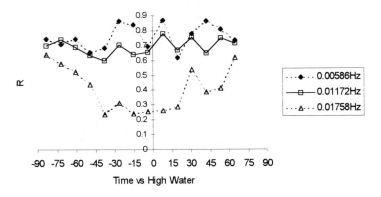

Figure 2: Temporal distribution of R for Storm Night at lowest three infragravity spectral (bin) frequencies

SEDIMENT DYNAMICS

The total flux of suspended sediment (Figure 3) was calculated using point measurements and following the method of Jaffe *et al.* (1984). The flux is directed offshore for both suites of records as an effect of coupling between suspended sediments and offshore directed currents due to both infragravity and undertow components of velocity (Russell 1990). Furthermore, in the Storm day data, a set of records occurring approximately 1 hour after high water were found to yield significantly larger (by up to an order of magnitude) fluxes than those records taken prior to or at high water. On inspection of the suspended sediment concentration (SSC) time-series it was noted that these largest fluxes were in turn dominated by a set of large sediment suspension events, as can be seen when comparing time-series from two records taken just prior to high water, and approximately 1 hour into the ebb phase (Figures 3a and 3b). Whilst Figure 3a shows some elements of infragravity suspension of sediment (at approximately 0.3Hz, 30 secs period), in Figure 3b the major suspension events are not only of up to an order of magnitude larger, but occur at periods of approximately 100 seconds, thus associating themselves with the frequencies dominating the oscillatory components and associated with infragravity waves.

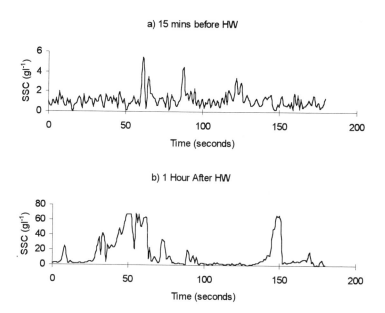

a) 15 mins before HW

b) 1 Hour After HW

Figure 3: SSC time-series from a) 15 minutes before high water, b) 1 hour after high water.

Suspension of sediment is due to instantaneous velocity shear and at such low frequencies has two likely major forcing mechanisms, firstly shear due to long wave velocities, or secondly stirring by the largest gravity frequency waves such that suspension occurs at wave group frequencies (e.g. Hanes and Huntley 1986, Osborne and Greenwood 1993). To investigate which of these mechanisms is appropriate in this instance Figure 4 shows an example of correlation between signals for the low frequency changes in sea-surface elevation, and the gravity frequency wave group envelope (determined following List 1991), versus the SSC time-series for a record containing some of the largest suspension events. Of note is that some gravity wave groupiness is retained in this region, possibly as an effect of low frequency changes in water depth through which the gravity waves propagate (c.f. Nelson and Gonsalves 1992). For the correlation analysis a 95% significance level was determined following Garrett and Toulany (1981), and is marked on the figure as the two tramlines. For these records significant correlation is only found between SSCs and the low frequency signal, occurring with a lag of approximately -10 seconds, i.e. a trough in sea surface elevation at infragravity frequencies is observed approximately 10

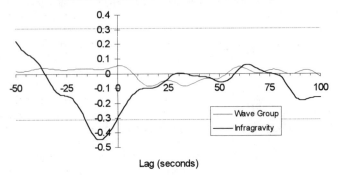

Figure 4: Correlations between SSCs and wave group envelope/low frequency fluctuations of sea surface elevation for a record 1 hour after high water. Significant (95%) correlation is indicated by the dotted tramlines.

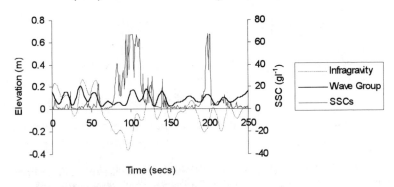

Figure 5: Time-series of SSC, wave group envelope, and low frequency fluctuation of sea surface elevation for record 1 hour after high water.

seconds prior to a peak in SSC. In this situation quantifying the lag is somewhat anomalous in that the behaviour of the SSC time-series is non-sinusoidal, and peaks in events are not well defined, however the lag is small enough to suggest a physical process whereby velocity drives local suspension.

This is further illustrated in the time-series shown in Figure 5, where the largest suspension events are clearly seen to coincide with the largest troughs in sea surface elevation at low frequencies. Peaks in the gravity wave envelope, representing the largest waves in each group, do appear to drive smaller more frequent suspension events seen in the SSC records, but no significant change in wave size is seen between the smallest and largest events, suggesting that the gravity waves do not provide the critical driving force for the largest suspension events.

Further examination of the infragravity wave contribution to suspension of sediment was made by performing a similar time domain analysis comparing SSCs with velocities associated with the incoming and outgoing components of the infragravity wave field, these signals having been obtained using the technique of Guza *et al* (1984). Correlations for each are shown in Figures 6a and 6b. Figure 6a is seen to mirror the behaviour of the low frequency component in Figure 4, with a significant trough at a lag of about -10 seconds (in fact the correlation in Figure 6a is improved), now indicating a correlation between offshore directed velocities associated with the incoming infragravity waves and peaks in SSC. Significant (although smaller) correlations are also observed between outgoing wave velocities and SSCs (Figure 6b), but at rather larger lags than would be expected for any physical link to be associated with the two processes. In fact the significant peak in Figure 6b sees a correlation between the velocity associated with the (reflected) outgoing wave troughs, which now give rise to an onshore directed velocity, and the suspension events these waves had initiated during their incoming phase. Thus the major oscillatory component attributable to the largest infragravity frequency suspension events is seen in this analysis to a velocity associated with troughs in the incoming long wave component.

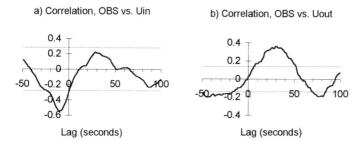

Figure 6: Correlations between SSCs and a) velocity associated with the incoming long wave component, b)) velocity associated with the outgoing long wave component.

DISCUSSION

The analysis in the previous section has, almost through a process of elimination, identified offshore directed velocities associated with incoming infragravity waves as being the key factor in some very large sediment suspension events. This does not however dismiss the role played by other hydrodynamic components; for example the fact that these events are associated with offshore velocities only may imply a level of coupling between these velocities and the undertow, whilst velocities associated with gravity frequency waves will further affect the instantaneous velocity field. Nevertheless a mechanism has been identified which dominates the instantaneous velocity field driving for the main infragravity frequency suspension events.

In this instance, the velocities driving suspension are seen to be associated with the incoming waves in particular. Further analysis shows this result to be mirrored within the behaviour of third order velocity moments derived from the field data. For the Llangennith dataset four terms associated with the hydrodynamic components are seen to dominate predicted sediment flux (c.f. Foote *et al.* 1994, Russell and Huntley, in review); the skewness (cubic terms) of mean and long wave velocity, and the two interaction terms between gravity and infragravity wave stirring (quadratic terms) coupled with the mean velocity. Here, as an illustrative example, the long wave skewness (u_L^3) term is further decomposed into its incoming and outgoing wave associated signals, leading to four further moment terms

$\langle u_{Lin}^3 \rangle$ incoming long wave skewness,

$\langle u_{Lout}^3 \rangle$ outgoing long wave skewness,

$3.\langle u_{Lin}^2 . u_{Lout} \rangle$ incoming-outgoing component velocity interaction,

$3.\langle u_{Lout}^2 . u_{Lin} \rangle$ outgoing-incoming component velocity interaction.

Figure 7 shows a (qualitative) distribution of the first two of these terms, examining the skewness of the incoming and outgoing components, through the cross-shore. Here a clear difference in the contribution made by the incoming and outgoing long waves is illustrated. Whilst outgoing wave skewness is negative (i.e. offshore directed) it also remains small and fairly constant throughout this region of the surf zone. The incoming waves however show a marked increase in magnitude of negative velocity skewness toward shore, likely due to shoaling effects, and it is with the most skewed records that the largest suspension events are associated. Though not shown here, the interaction terms virtually mirror each other in behaviour, with $3.<u_{Lin}^2.u_{Lout}>$ providing a negative moment, and $3.<u_{Lout}^2.u_{Lin}>$ a positive moment. However the incoming wave shoaling leads to a rather larger $3.<u_{Lin}^2.u_{Lout}>$ term, and means that the resultant of the interaction terms nearest to the shoreline is an offshore directed moment of approximately the same magnitude as that observed for the velocity skewness of the incoming wave velocities. Overall, infragravity wave skewness is seen to have a marked increase, due to shoaling in the incoming infragravity waves, which corresponds with the largest sediment suspension events.

Similar comparisons may be made between the time-series analysis and the other appropriate moments generated from the data. Thus in this high energy situation, with sediment readily available for suspension, and reasonably clear and long lived hydrodynamic and sediment suspension events dominant, velocity moments can be linked to observations of sediment suspension. Here skewness of the incoming and outgoing long wave components and their respective phase are the important characteristics in causing the larger negative long wave moments associated with the largest bursts of sediment. This dominance of the incoming long wave component in this situation is somewhat surprising when it is borne in mind that the reflection coefficient for the dominant long wave frequencies is reasonably high

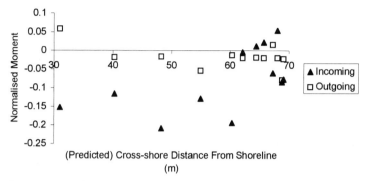

Figure 7: Cross-shore distribution of Storm Day skewness terms for incoming and outgoing long waves.

(approximately 0.75), and it may well be that some change in wave shape occurs through the reflective processes such that the outgoing waves are less skewed than their incoming counterparts. Consideration of such effects should be necessary in the accurate modelling of sediment transport in high energy nearshore situations.

CONCLUSIONS

Analysis of a high energy dissipative inner surf zone shows sediment flux to be dominated by a set of large suspended sediment events occurring at infragravity frequencies.

These events are seen to be driven by infragravity waves rather than any groupiness of gravity frequency waves. Further analysis shows the dominant component driving these suspension events to be an offshore directed velocity associated with the incoming infragravity waves.

Third order velocity moments are found to (qualitatively) retain the trends seen in terms of observed SSCs and their corresponding fluxes, and point to long wave skewness as being an important factor in associating this component with the largest sediment suspension events.

REFERENCES

Beach, R.A., Sternberg, R.W., 1987. Suspended sediment transport in the surf zone: response to cross-shore infragravity motion. Marine Geology, 80, 61-79.

Foote, Y., Huntley, D.A., O'Hare, T.J., 1994. Sand transport on macrotidal beaches. Proc. Euromech 310 Colloquium (Le Havre), 360-374.

Garrett, C.J.R., Toulany, B., 1991. Variability of the flow through the Strait of Belle Isle. Journal of Marine Research, 39, 163-189.

Guza, R.T., Thornton, E.B., 1980. Local and shoaled comparisons of sea surface elevations, pressures and velocities. Journal of Geophysical Research, 85, 1524-1530.

Guza, R.T., Thornton E.B., Holman, R.A., 1984. Swash on steep and shallow beaches. Proc. 19th International Conference on Coastal Engineering, ASCE, 708-723.

Hanes, D.M., Huntley, D.A., 1986. Continuous measurement of suspended sand concentrations in a wave dominated nearshore environment. Continental Shelf Research, 6(4), 585-596.

Jaffe, B.E., Sternberg R.W., Sallenger, A.H., 1984. The role of suspended sediment in shore-normal beach profile change. Proc. 19th International Conference on Coastal Engineering, ASCE, 1983-1986.

List, J.H., 1991. Wave groupiness variations in the nearshore. Coastal Engineering, 15, 475-496.

Nelson, R.C., Gonsalves, J., 1992. Surf zone transformation of wave height to water depth ratios. Coastal Engineering, 17, 49-70.

Osborne, P.D., Greenwood, B., 1993. Sediment suspension under waves and currents: time scales and vertical structure. Sedimentology, 40, 599-622.

Russell, P.E., 1990. Field studies of suspended sand transport on a high energy dissipative beach. Ph.D. Thesis, Univ. of Wales, 318pp (Unpubl.).

Russell, P.E., 1993. Mechanisms for beach erosion during storms. Continental Shelf Research, 13(11), 1243-1265.

Russell, P.E., Davidson, M., Huntley, D.A., Cramp, A., Hardisty, J., Lloyd, G., 1991. The British Beach and Nearshore Dynamics (B-BAND) programme. Proc. Coastal Sediments '91, 371-384.

Russell, P.E., Huntley, D.A. In Review. The spatial distribution of cross-shore transport on high energy beaches.

Wright, D., Guza, R.T., Short, A.D., 1982. Dynamics of a high energy dissipative surf zone. Marine Geology, 45, 41-62.

ACKNOWLEDGEMENTS

        This research was carried out with the financial assistance of the Natural Environment Research Council. The authors would also like to thank David Huntley and Mark Davidson for their ongoing advice and input into this project.

RIPPLE CHARACTERISTICS AND BED ROUGHNESS UNDER TIDAL FLOW

Richard Whitehouse[1], Helen Mitchener and Richard Soulsby

Abstract

This paper presents the results of laboratory experiments on the development of sand ripples in a simulated tidal flow. The longitudinal profile of the bed is measured every half-hour through the experiment and simultaneous measurements of the vertical profile of the flow velocity are obtained. Analysis of the measurements shows how the ripple morphology varies through 6-half tidal cycles and the flow data gives the corresponding information for the hydraulic roughness of the bed. The relevance of the findings to observations made in the sea is discussed and the measurements are compared with the predictions made from an existing hydraulic roughness model.

Introduction

In shallow coastal waters the flow and sediment transport are largely controlled by seabed friction. Consequently, computations of these two quantities are found to depend strongly on the bed roughness, parameterised by the seabed drag coefficient C_D, the Nikuradse equivalent roughness k_s, or the bed roughness length z_0 ($k_s/30$). The specification of the roughness is, however, a major source of uncertainty and there is some evidence that it can vary through a tidal cycle by up to two orders of magnitude (Whitehouse and Chesher, 1994). The exact behaviour is sensitive to the tidal current velocity, the sediment grain size, the size and orientation of the bedforms with respect to the flow direction, and the amount of sediment in motion as bedload or suspended load. This paper describes the results of a laboratory experiment to investigate the link between tidal ripples and their hydraulic roughness. The dimensions of ripples are of the order 2cm height and 20 to 30cm wavelength.

Field observations of rippled bed roughness in tidal flow

Based upon field measurements of the velocity profile above rippled sand Soulsby (1997) has suggested a typical value for z_0 is 0.6cm. However, when the seabed sediment is mobile the value of z_0 can vary in a systematic fashion through the tide. Whitehouse and Chesher (1994) have reviewed 15 field datasets for the variation in z_0 over gravel and sand beds. The following types of behaviour have been measured over mobile sand beds:

(a) a monotonic increase in z_0 through the period of a half tidal cycle (e.g. Dyer, 1980; Wilkinson. 1986). In the data of Dyer measured over a rippled bed of 300 micron sand z_0 increased from 0.01 to 5cm during the tide.

[1]Marine Sediments Group, HR Wallingford, Howbery Park, Wallingford, Oxon OX10 8BA, UK

(b) an increase of z_0 in proportion to the increasing shear velocity as ripples grow (e.g. Dyer, 1980). The range of z_0 was 0.1 to 3cm.

(c) a reduction in z_0 at periods of peak tidal flow as ripples become washed out (e.g. Dyer, 1980; McLean, 1983; Soulsby et al., 1983). In the data of McLean z_0 was 0.1cm during the peak of the tidal flow as opposed to 0.6cm at other times.

(d) under an open tidal ellipse z_0 varies with time as the direction of flow changes with respect to the orientation of the ripple crests (Soulsby, 1990). The range of z_0 values was 0.1 to 1.5cm.

These observations have shown that apparently large variations in the value of z_0 can occur through a tidal cycle arising from the complex interactions of flow, bedforms and sediment transport. In contrast, over an immobile bed the value of z_0 is approximately constant during most of the tide. Wilkinson (1986) made a systematic attempt to explain the behaviour observed in (a). No wholly satisfactory explanation was forthcoming in terms of either (i) flow acceleration due to tidal variation or topography (producing apparent changes in roughness length), (ii) sediment transport (saltation or density stratification) or (iii) changes in ripple geometry. Dyer (1980) attributed 50% of the change in (a) to be due to changes in the size and shape of the ripples. Soulsby et al. (1983) observed that although the value of z_0 can increase by two orders of magnitude during the tide the ripple height or wavelength does not increase by this amount.

In the absence of measurements values of z_0 appropriate to current ripples are often estimated using semi-empirical expressions based upon ripple height and steepness, height H divided by wavelength L. Typical of these expressions is the one given by Soulsby (1997).

$$z_{or} = a_r \frac{H^2}{L} \qquad (1)$$

This type of expression has been calibrated with roughness measurements made above asymmetrically shaped, equilibrium morphology ripples and Soulsby quotes values of the constant a_r varying between 0.3 and 3 taking a typical value of 1. Van Rijn (1993) uses $a_r = 0.66$ but finds from steady flow data that k_s/H due to ripple roughness is a much steeper function of H/L in the range 0.1 to 0.2 than is predicted by Equation 1. The variability in a_r arises at least partly due to natural variations in ripple cross-section and plan shape but also due to the varying definitions of ripple height and wavelength adopted by different authors. The appropriateness of Equation 1 in the marine environment, where the ripples are constantly changing their shape, remains untested due to the lack of data with simultaneous observations of hydraulic roughness and ripple morphology. This lack of data led to the laboratory experiment described below from which data was generated to explain some of the observations made in the sea and to test the performance of Equation 1 under tidal conditions.

The approach taken was to simulate the bottom 0.2m of a natural tidal flow at prototype period and velocity amplitude (12.5 hours, 0.5ms^{-1}) in the laboratory. The shallow water depth was used to reduce significantly the influence of non-stationarity and secondary circulation (in a flume) on the flow profiles. A natural quartz sand was used ($d_{50} = 0.510$mm, $d_{84} = 0.715$mm, $d_{16} = 0.326$mm) which, with typical near-bottom flow velocities acting on it, produced ripples with similar height, wavelength, asymmetry, migration rate, and plan-shape characteristics as would be encountered on the seabed. A coarse sand was chosen to restrict the sediment transport to bedload and to thereby remove the complicating influence of suspended sediments on the form of the velocity profile.

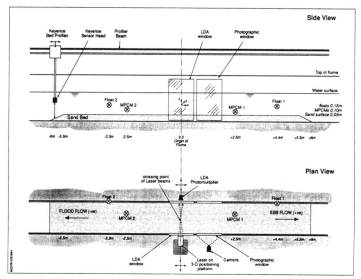

Figure 1 Positioning of the measurement equipment with respect to the sand bed

Test facility

The tests were performed in a large reversing flume (length of straight channel 23m, width 0.62m and overall depth 0.4m). The facility is described in Whitehouse et al. (1994) and has been used previously to examine the variation of turbulence quantities through a tidal cycle (Anwar and Atkins, 1980). The channel has smooth boundaries and guide vanes at each end of the test section to provide uniform and symmetrical flow in either direction. Water is circulated through pipework adjacent to the flume and the pump can be programmed to reproduce a prescribed time variation in flow speed and direction. The central window in the flume was used for measurements of the flow velocity with a Laser Doppler Anemometer (LDA) and an adjacent window was used for photography of the sand bed during tests (Figure 1). The sediment bed for the tests was formed from a 0.05m thick sand sheet placed centrally in the length of the flume, extending 5.5m either side of the centreline. The ends of the sand sheet were ramped over a length of 0.5m (slope approximately 1:10) to allow a gradual progression of the flow from the smooth bed of the flume onto the sand surface. The nominal water depth over the sand bed was 0.2m and the mean value of the kinematic viscosity was $1.27 \ 10^{-6} \mathrm{m}^2\mathrm{s}^{-1}$.

Description of experiments

The main aim of the experiments was to investigate the variation in shear velocity u_* and roughness z_0 through a tidal cycle. In addition estimates of the threshold of motion were made for flat and rippled beds and the development of ripples from a flat bed was measured, plus their response to step changes in discharge (Whitehouse et al., 1994). For steady flow over equilibrium ripples the value of k_s/H was about 2.

Velocity measurements were made over the crest and trough of an immobile rippled bed to verify that the spatially averaged flow field could be sampled by measuring the vertical profile of current velocity. The results (Figure 2) demonstrated that when the profiles were plotted with respect to the spatially averaged (centreline) bed level they merged at a height greater than about 0.08m above the rippled bed (i.e. about 4 ripple heights). The velocity profile measured above this height was taken to indicate the spatially averaged total drag (Chriss and Caldwell, 1982; Kapdasli and Dyer, 1986). This will be similar to the parameters measured in most field experiments where the lowest current meter is typically at about 0.10m above the bed (e.g. Dyer, 1980; Wilkinson, 1986).

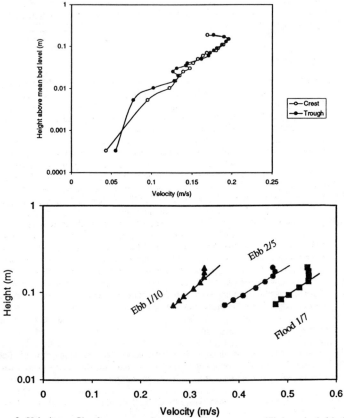

Figure 2 Velocity profiles from crest and trough positions on an equilibrium rippled bed (top; curved line fitted) and from the tidal flow tests (bottom; with log profile fits)

The magnitude and direction of the flow was varied in a sinusoidal, rectilinear reversing fashion and with an approximately prototype tidal period (see Figure 3). A series of 6 half-tidal cycles were simulated in the experiments and the bed characteristics and vertical velocity profiles were measured at regular intervals. The tests were denoted Flood 1, Ebb 1, Flood 2, Ebb 2, Flood 3 and Ebb 3. Test Flood 1 included the growth of ripples from a flat bed. The duration of the half-cycles was not exactly the same and the 'tide' was slightly asymmetric as can be seen in Figure 3. The first four tests were taken on near consecutive days but there was a period of one week between tests Ebb 2 and Flood 3.

The data collection included flow discharge, water surface slope and reference current speeds at two locations. Sequential vertical profiling of the flow above the sediment bed was carried out with the LDA (at 8 heights above flume floor level - 0.120m, 0.130m, 0.140m, 0.160m, 0.180m, 0.200m, 0.220m, 0.240m). Each profile took about 15 minutes to collect in the times between the traverse of the bed profiler. The velocity measurements were averaged over 60 seconds at each height and interpolated in time using a linear scheme to provide quasi-instantaneous velocity profiles at times corresponding to the measured bed profiles. The velocity profiles were reduced to the spatially averaged bed level and the semi-logarithmic velocity profile (Equation 2) was fit to the data between heights of 0.08m and 0.15m, using the method of Wilkinson (1984, see Figure 2 for example data and log profile fits).

$$U(z) = \frac{u_*}{0.40} \ln\left(\frac{z}{z_0}\right) \tag{2}$$

A computerised traversing carriage with a laser displacement sensor fixed at 0.12m above the bed was used to obtain centreline bed profiles every 30 minutes. The vertical accuracy of the sensor was 50 microns and a reading was obtained every 15mm over 4.7m either side of the mid-point of the flume (Figure 1). Time-lapse photographs of bed morphology were also obtained to coincide with the times that the profiler had completed its traverse.

Although the bedforms in the flume looked like the ripples found on the seabed, and there was no discernible interaction of the morphology with the free surface, an irregular low amplitude undulation of the bed was identified. This was subtracted from the bed trace using a 'box car' filter of (arbitrary) length equal to 3 times the water depth (Whitehouse et al., 1994). The de-trended (ripple) traces were analysed to determine the ripple height, wavelength and mean slopes of the ripple flanks. The mean bed level obtained from the original unfiltered bed signal did not vary by more than ± 2mm during the experiment. The ripple height was defined as $2\sqrt{2}\sigma_m$ where σ_m is the standard deviation of the ripple trace about the mean level. This definition corresponds to the equivalent height of a sinusoidal bed undulation. The ripple wavelength was obtained from a zero up-crossing (counting) analysis of the trace with regard to the mean level. The mean positive and negative slopes were determined by calculating the bedslope over each 15mm segment of the ripple trace. Positive and negative slope values were stored separately and an average value of each obtained for every trace. The time variation of the bed quantities and $U_{0.10}$ the flow velocity at 0.10m above the bed are shown in Figure 3.

Results and discussion

The results of the threshold of motion tests were broadly in agreement with published data (Soulsby, 1997) and the current speeds $U_{0.10}$ relating to the initiation of continuous sediment motion were about $0.4 \mathrm{ms}^{-1}$ on the flat bed and $0.3 \mathrm{ms}^{-1}$ on the rippled bed.

The bedform data in Figure 3 shows how the ripple wavelength and height varied through the experiment, with a data point obtained every half-hour. During Flood 1 the ripples grew and became well formed with a 3-dimensional plan shape after about 4 hours. After the reversing of the flow direction in Ebb 1 the ripples remained more or less the same shape until about 1.5 hours into the test when the crests started to reverse their orientation and after 2 hours the ripple asymmetry had become well defined in the flow direction. The switching of the ripple orientation is evident in the bottom panel of Figure 3 where the downstream facing slope on Flood 1 and the downstream facing slope on Ebb 1 are reversed until the ripple form has adjusted to the new flow direction. This switching of ripple orientation was evident through the rest of the experiment. The bedform height and wavelength continued to grow during Ebb 1 until the bed became immobile during the last hour and a half of the tide. The estimates of slope in the bottom part of Figure 3 are derived from around 700 individual measurements taken at 15mm intervals along the bed. Because the avalanche slopes of the ripples only have a plan length of 3 to 4cm the traverse-averaged slope will be much less than the angle of repose.   A typical value for the upstream facing slope is 0.15 which corresponds to about 8.5 degrees.   On Flood 2 the wavelength was larger during the peak of the tide and the height was lowered as the ripples were partially washed out to produce low flat forms with very short avalanche slopes.

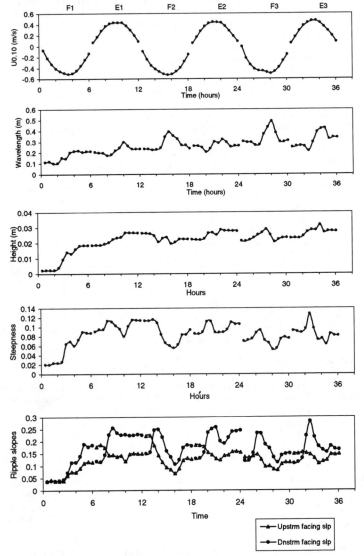

Figure 3 Time series (from top to bottom) of the flow velocity at 0.10m above the bed, ripple wavelength L and height H, ripple steepness H/L and average ripple slopes (see text)

As the flow decreased over the last part of the tide the wavelength and height nearly reverted back to the values at the beginning of the tide and the ripples became more 3 dimensional again. The ripple steepness was significantly reduced under peak flow. During Ebb 2 the crests of the ripples were reversed after 2 hours and the height increased through the tide in a similar fashion to Ebb 1 and Ebb 3. The overall pattern of variation in wavelength and height during Ebb 1, 2 and 3 is similar as are the variations observed during Flood 2 and 3. The typical wavelength (0.3m) corresponds to about $600d_{50}$ and the ripple migration rate at the peak of Ebb 2 was about $4\ 10^{-4}\mathrm{ms}^{-1}$. It is interesting to note that the pattern of variation in the ripple steepness is broadly similar to the variation in the value of the downstream facing slope.

The hydraulic data corresponding to the bedform data plotted in Figure 3 are shown in Figure 4. As with the bedform plots a value is plotted every half an hour. The full dataset is reported in Whitehouse et al. (1994) but some screening of the data was applied here to remove those estimates where the semi-logarithmic velocity profile assumption was invalidated by a poor least-squares fit of Equation 2 to the measurements. For the purposes of these plots an arbitrary cut off value of $r^2 = 0.9$ has been chosen. The error bars on the individual estimates of $u*$ and z_0 are shown for the three velocity profiles in Figure 2 (95% confidence limits). Some of the other estimates will have larger error bounds but if we assume that each estimate is statistically independent, and because there appears to be a repeating pattern of behaviour between the

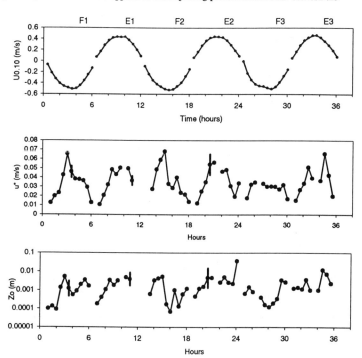

Figure 4   Time series corresponding to Figure 3 of (from top to bottom) the flow velocity measured at 0.10m above the bed and the spatially averaged shear velocity $u*$ and roughness length z_0

consecutive flood and ebb tides, it appears that the results are meaningful.

The value of z_0 at the start of Flood 1 was about 1×10^{-4}m, i.e. twice the grain related roughness of 4.25×10^{-5}m ($2.5d_{50}/30$), but this is not surprising as the prepared bed was not totally smooth. As the ripples developed during Flood 1 the value of z_0 increased to around the value of 0.6cm quoted for sand ripples by Soulsby (1997).

The pattern of z_0 variation in Figure 4 displays the first three types of behaviour listed in the Introduction. All the ebb tests indicated an overall increase in z_0 through the tide by at least an order of magnitude, similar to type (a). On test Flood 1 z_0 increased as the ripples grew (type b). During tests Flood 2 and Flood 3 the ripples were partially washed out during the peak tidal flow and z_0 was reduced during this period by an order of magnitude (type c), recovering at the end of the tide as the ripples were reformed. The reduction in z_0 during the peak of these tides also coincided with the reduction in ripple steepness (Figure 3).

Another interesting feature of the z_0 time series is the systematic way in which z_0 is lower at the start of a half cycle than at the end of the preceding one. This behaviour was observed in the sea by Wilkinson (1986). As noted above the ripple asymmetry is effectively reversed at the start of the tide and the flow passes over the ripple in the 'wrong' direction until the ripple geometry changes as a result of sediment transport. Laboratory measurements made over fixed bedforms by Lau (1988) have shown the friction factor for reverse flow is about 50% of the value obtained when flow is passing over the bedforms in the normal direction (shallow slope facing upstream). Converting his measured difference in friction factor f to a difference in drag coefficient ($C_D = f/8$) and back calculating to give z_0 assuming a semi-logarithmic velocity profile indicates a factor of 10 reduction on the reverse flow values of z_0. Thus reverse flow effects would appear to account for the lowering of z_0 observed at the start of the tide in our data and in the sea.

Comparison of ripple roughness predictor with data

Finally it is worthwhile to compare the performance of the z_0 predictor (Equation 1) with the observed values of z_0. The predictor was driven with the measured time series of ripple height and wavelength, and thus steepness, depicted in Figure 3 and using a value of $a_r = 1.0$ as recommended by Soulsby (1997). The predicted and observed values are shown in Figure 5 where the total value of z_0 (SUM) is the sum of the grain related part ($z_0 1 = 2.5d_{50}/30$) and $z_0 2$ is the ripple roughness given by Equation 1. The predicted values of z_0 follow reasonably well the upper part of the data which is to be expected as this is the region where the ripples are nearest to an equilibrium shape. Some of the details are not picked out, such as the drop in z_0 at the start of each half cycle due to reverse flow asymmetry, the high values of z_0 at the end of Ebb 2 and Ebb 3, and the low values during the peak of Flood 2 and Flood 3. Also during Flood 1 the roughness appears to increase more quickly than is suggested by the ripple steepness and height; the predictions lag the observations. The influence of reverse flow asymmetry on z_0 could be included in the predictions by using a reduced value of $a_r = 0.1$ during the periods when the ripple asymmetry is in the reverse orientation.

In the lower panel of Figure 5 the range in total roughness predictions with values of $a_r = 0.3$ and 3 has been plotted. Based upon this comparison it appears that a model based upon ripple height and steepness H/L (Equation 1) can be used to predict the likely range of z_0 during a tidal cycle using the upper and lower values for a_r suggested by Soulsby (1997) from steady flow over ripples.

A fully comprehensive roughness predictor for tidal flow will require the prediction of the time development of the ripples as a function of the flow speed and sediment characteristics, perhaps

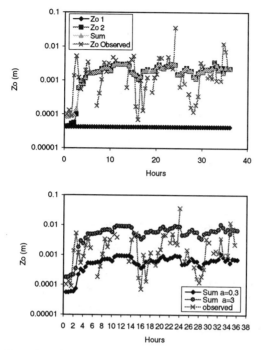

Figure 5 Comparison of the total bed roughness length predictions with measurements, $a_r = 1$ (top) and $a_r = 0.3$ and $a_r = 3$ (bottom)

in a similar approach to that suggested by Baas (1993). This bedform model can then be linked to a hydraulic roughness model to predict the time variation in z_0. Some additional terms will be required to obtain a closer fit with the observations such as specifying the effect of ripple plan shape and crest orientation on the roughness.

Conclusions

- Laboratory data for the variation of the bed roughness length z_0 of a rippled sand bed under simulated tidal flow have been obtained.   The influences of density stratification due to suspended sediment and non-stationarity on the velocity profiles were precluded in the experiment. Simultaneous high resolution surveys of the ripple bed morphology were made and the relationships between the flow and bedform characteristics have been examined. The behaviour of z_0 observed in these experiments displays broadly similar characteristics to the time variation in z_0 that has been observed at a number of seabed locations.

- The ripple height and steepness measurements have been used with an expression for steady flow ripple roughness to predict the time variation of z_0. The predictions have been compared with the observations. The overall trend in z_0 is predicted quite well and the variation in the calibration constant for steady flow conditions accounts for the majority of the range of z_0 values observed under tidal conditions.

Acknowledgements

The funding for parts of this work was provided by the UK Ministry of Agriculture, Fisheries and Food (Flood and Coastal Defence with Emergencies Division) and for other parts by the EC MAST contract MAS2-CT92-0027 (G8M Coastal Morphodynamics). The authors are grateful to Les Smith, Ian Payne, Philip Houghton, Liz Stevenson and Robert Adams for assistance with the equipment, testing and data analysis. Kate Shaw (industrial placement student from the University of Cardiff) assisted with the analysis presented in this paper.

References

Anwar, H O and Atkins, R (1980). Turbulence measurements in simulated tidal flow. Journal of Hydraulic Engineering, ASCE, 106, 1273-1289.

Baas, J H (1993). Dimensional analysis of current ripples in recent and ancient depositional environments. PhD Thesis, University of Utrecht, Netherlands, 199pp.

Chriss, T M and Caldwell, D R (1982). Evidence for the influence of form drag on bottom boundary layer flow. J. Geophys. Res. vol. 87, no. C6, 4148-4154.

Dyer, K R (1980). Velocity profiles over a rippled bed and the threshold of movement of sand. Estuarine and Coastal and Marine Science, vol. 10, 181-199.

Kapdasli, M S and Dyer, K R (1986). Threshold conditions for sand movement on a rippled bed. Geo-Marine Letters, vol. 6, 161-164.

Lau, Y L (1988). Roughness of reverse flow over dunes and its application to the modelling of the Pitt River. Canadian Journal of Civil Engineering, vol. 15, 547-552.

McLean, S R (1983). Turbulence and sediment transport measurements in a North Sea tidal inlet (The Jade). In: North Sea Dynamics. Eds. J Sundermann and W Lenz, 436-452. Springer-Verlag.

Soulsby, R L, Davies, A G and Wilkinson R H (1983). The detailed processes of sediment transport by tidal currents and by surface waves. Inst. of Oceanographic Sci. Report 152, 80pp.

Soulsby, R L (1997). Dynamics of Marine Sands. A Manual for Practical Applications. Thomas Telford Publications.

Van Rijn, L C (1993). Principles of Sediment Transport in Rivers, Estuaries and Coastal Seas. Aqua Publications, Amsterdam.

Wilkinson, R H (1984). A method for evaluating statistical errors associated with logarithmic velocity profiles. Geo-Marine Letters, vol. 3, 49-52.

Wilkinson, R H (1986). Variation of roughness length of a mobile sand bed in a tidal flow. Geo-Marine Letters, vol. 5, 231-239.

Whitehouse, R J S, Williamson, H J and Soulsby R L (1994). Laboratory experiments on ripple development and bed roughness in tidal flow. HR Wallingford Report SR 385.

Whitehouse, R J S and Chesher, T J (1994). Seabed Roughness in Tidal Flows: A Review of Existing Measurements. HR Wallingford Report SR360.

Large-Eddy Computation of Horizontal Fluidized Bed by a New Multiphase LES

Yasuo Nihei[1], Kazuo Nadaoka[2] and Hiroshi Yagi[3]

ABSTRACT

For extending the applicability of a new multiphase LES (GAL-LES model) recently developed by the authors to multiphase flow with high particle concentration, it has been attempted to improve the GAL-LES model by incorporating a new modeling of the particle collision. To confirm the validity of the present model, a large-eddy computation of a horizontal fluidized bed as a typical multiphase flow with high particle concentration is performed. The present model has succeeded in simulating the formation of an instability in the flow field, giving good agreement with the existing experimental data.

1. INTRODUCTION

Sheet-flow-type sediment transport has significant importance in coastal sedimentary process. Although numerous attempts of laboratory experiment have been made to clarify the sheet-flow dynamics, difficulty in the measurements of velocity and particle concentration in the sheet-flow field has shown increasing demands to develop a numerical model as an alternative efficient tool for the sheet-flow study.

The authors have recently developed a new framework for LES modeling of multiphase turbulent flow with high accuracy (Nadaoka et al., 1995; 1996; 1997). The framework is based on a new formulation of solid-particle motion, referred to as GAL (Grid-Averaged Lagrangian) model. Combining the GAL model with a SGS (Sub-Grid Scale) model for fluid-phase turbulence, a new multiphase LES has been formulated (**GAL-LES model**). The preliminary version of the GAL-LES model by Nadaoka et al. (1995; 1996; 1997) has been applied to a particle plume as a typical

[1]Research Associate and [2]Professor, Graduate School of Information Science and Eng., Tokyo Institute of Technology, O-okayama, Meguro-ku, Tokyo 152, Japan
[3]Associate Professor, Dept. of Civil Eng., Tokyo Institute of Technology

two-phase turbulent flows. The GAL-LES model has succeeded in reproducing a characteristic large eddy structure observed in the previous experiment of particle plume, demonstrating the fundamental performance of the GAL-LES model. However this version has not taken account of particle collision and hence may not be applied to a multiphase flow with high particle concentration such as sheet-flow.

For this reason, in the present study, it has been attempted to extend the applicability of the GAL-LES model to multiphase flow with high particle concentration by incorporating a new modeling of the particle collision. To confirm the fundamental performance of the GAL-LES model with new particle collision model, a large-eddy computation of a horizontal fluidized bed with high particle concentration as a simplified computation of sheet-flow was performed and compared with the experimental results by Turian & Yuan (1977).

2. OUTLINE OF THE GAL-LES MODEL

2.1 Basic concept of the GAL-LES model

For the treatment of the fluid-phase motion, a usual LES formulation is applied, in which turbulence is separated into GS (Grid Scale) and SGS components by a spatial filtering over a grid volume. The GS component in the fluid phase is directly obtained by the time integration of the grid-averaged momentum equation and the SGS component is modeled with the SGS turbulence kinetic energy. The equations of the fluid-phase motion are therefore composed of the continuity, momentum and SGS turbulence kinetic energy equations in which the particle concentration and the interaction forces between two phases are taken into account.

On the other hand, the solid-particle motion is represented by the new formulation, named GAL model, developed by the authors. The GAL model is based on an averaging operation for a Lagrangian-type equation of a particle over all the particles within a grid volume and a procedure of reallocation of solid-particle clouds to each grid. The model is therefore a mixed Eulerian-Lagrangian model which can effectively reduce computational time compared with the existing Lagrangian-type models, without losing the advantage of Lagrangian-type models that they can properly describe the dynamical evolution of particles. Due to the grid-averaging operation, the solid-particle motion in the GAL model is represented with the grid-averaged value and the variance of the particle fluctuation around it. Since the GAL model is derived with a grid-averaging operation, it is easily combined with the fluid-phase LES formulation. For details of the equations of the GAL-LES model, one can refer to Nadaoka *et al.* (1995; 1996; 1997).

2.2 New Modeling of Particle Collision

For extending the applicability of the GAL-LES model to multiphase flow with high particle concentration, it is necessary to incorporate new modeling of the particle collision into the GAL model. The effect of the particle collision is, in the present study, estimated by the following seven steps:

(1) One representative particle and four neighboring particles are assumed within a grid, as shown **Fig. 1**.

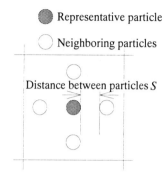

● Representative particle

○ Neighboring particles

Distance between particles S

Fig. 1 Position of each particle
in a grid

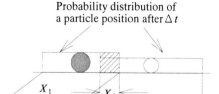

Probability distribution of
a particle position after Δt

X_1 X_2

X_1'

Fig. 2 Estimation of a particle-collision
probability P_k

(2) The distance S between the representative and its neighboring particle is defined
with a function of the particle diameter d, the particle concentration C and the
maximum possible solid-particle concentration C_{MAX} (Bagnold , 1954) :

$$S = d \cdot \frac{C_{MAX}^{1/3} - C^{1/3}}{C^{1/3}}. \tag{1}$$

(3) All the quantities for the representative particle are defined to be equal to the
values at the grid, while those of four neighboring particles are obtained by
interpolating the values at the grid. For example, the grid-averaged particle
velocity for the neighboring particle k in the i direction, $U_{si}[k]$, is given as

$$U_{si}[k] = U_{si} + \frac{\partial U_{si}}{\partial X_i} \cdot S . \tag{2}$$

(4) After each particle is advected and diffused during the computational time
interval Δt under the assumption of no collision between particles, the
probability distribution of a particle position for each particle is obtained like
Fig. 2, in which the distribution shape of the particle position is assumed to be
uniform.

(5) The collision probability P_k between the representative particle and the
neighboring particle k is given as

$$P_k = \frac{X_2^2}{X_1 X_1'} , \tag{3}$$

where X_1, X_1' and X_2 are defined in **Fig. 2**.

(6) The grid-averaged velocity and velocity variance of the particle after the particle
collision are estimated from the impulsive equations between two particles,

which are written as

$$U_{si_{(k)}}^{(1)} = U_{si}^{(0)} + J_{i_{(k)}}^{(0)} / m, \tag{4}$$

$$J_{i_{(k)}}^{(0)} = J_{n_{(k)}}^{(0)} n_{i_{(k)}} = M(1+e)\Big(G_j[k]^{(0)} n_{j_{(k)}}\Big) n_{i_{(k)}}, \tag{5}$$

where the superscript 0 and 1 represent the pre- and post-collision quantities, the subscript k denotes the number of neighboring particle, n_i the unit normal vector at the collision surface, J_i the impulsive force in the n_i direction, e the restitution coefficient, m the mass of the particle and $G_i[k]$ the relative velocity between the representative particle and the neighboring particle k. Also, M is defined as $m/2$ in case of particle-particle collision and m for particle-wall collision. With eqs. (4) and (5), the grid-averaged velocity and velocity variance after particle collision, $U_{si_{(k)}}^{(1)}$ and $\overline{u_{si}^2}_{(k)}^{(1)}$, are formulated as follows;

$$U_{si_{(k)}}^{(1)} = U_{si}^{(0)} + A\Big\{ G_j^{(0)}[k] n_{j_{(k)}} \Big\} n_{i_{(k)}}, \tag{6}$$

$$A = \frac{M}{m}(1+e), \tag{7}$$

$$\overline{u_{si}^2}_{(k)}^{(1)} = \overline{u_{si}^{2(0)}} + \beta A^2 \Big(\overline{u_{sj}^{2(0)}} + \overline{u_{sj}^2[k]^{(0)}} \Big), \tag{8}$$

$$n_{i_{(k)}} = G_{i_{(k)}}^{(0)} \Big/ \Big| G_{i_{(k)}}^{(0)} \Big|, \tag{9}$$

where β is the parameter to describe the change in the velocity variance due to particle collision and is given here to be 0.01.

(7) The grid-averaged velocity and velocity variance of the particle including the effect of the particle collision, $U_{si}^{(2)}$ and $\overline{u^2}_{si}^{(2)}$, are finally represented with the collision probability P_k between the representative particle and the neighboring particle k;

$$U_{si}^{(2)} = \left(1 - \sum_{k=1}^{k=4} P_k\right) U_{si}^{(0)} + \sum_{k=1}^{k=4} P_k U_{si_{(k)}}^{(1)}, \tag{10}$$

$$\overline{u_{si}^{2(2)}} = \left(1 - \sum_{k=1}^{k=4} P_k\right) \overline{u_{si}^{2(0)}} + \sum_{k=1}^{k=4} P_k \overline{u_{si}^2}_{(k)}^{(1)}. \tag{11}$$

3. COMPUTATION OF A HORIZONTAL FLUIDIZED BED

3.1 Computational Conditions

To check the fundamental performance of the GAL-LES model with the new particle collision model for the multiphase flow with high particle concentration, a large-eddy computation was conducted for a horizontal fluidized bed as a

Flow direction

Flow direction

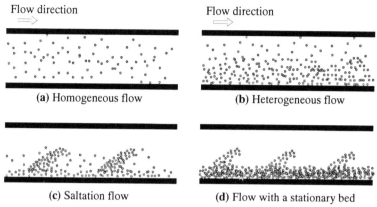

(a) Homogeneous flow

(b) Heterogeneous flow

(c) Saltation flow

(d) Flow with a stationary bed

Figure 3 Flow pattern in a horizontal fluidized bed (Turian & Yuan, 1977).

simplified computation of a sheet-flow. The conditions for the computation are the same as those for the experiment conducted by Turian & Yuan (1977); i.e., the diameter and specific gravity of a particle are respectively 1mm and 2.97, the cross-sectional mean particle concentration C_m is 5%. Two cases of the cross-sectional mean horizontal fluid velocity U_m, 200cm/s (case1) and 350cm/s (case2), were adopted. As shown in **Fig. 3**, Turian & Yuan (1977) classified a horizontal fluidized bed into the four typical flow patterns; *homogenous flow, heterogeneous flow, saltation flow* and *flow with a stationary bed*. According to this experimental classification, the flow conditions of case1 and case2 in the present computation correspond to those of the saltation flow and heterogeneous flow, respectively. Two-dimensional computation was done for the domain of 160cm (80 grid points) in the horizontal direction and 5.25cm (30 grid points) in the vertical direction. The periodic boundary condition was applied at the up-stream and down-stream boundaries, while the logarithmic law of the wall was employed both at the upper and lower boundaries. As the initial conditions, the particle concentration was set to be constant at all grid points (=0.05) and the velocity was assumed to be based on the logarithmic wall law.

3.2 Results and Discussion

(a) Characteristics of Instantaneous Flow Pattern
Figure 4 shows the instantaneous spatial distribution of the fluid-phase velocity vectors and the particle concentration in case of relatively lower fluid velocity (case1). For understanding the detailed structure of the fluid-phase motion, 150cm/s is subtracted from the horizontal component of the velocity vectors in the figure and the vertical component is exaggerated ten times larger. At the beginning stage of the computation, $t=1.2$s, the particle concentration near the bottom boundary is higher than that near the upper boundary because the specific gravity of the particle is more than unity. The corresponding velocity vectors of the fluid phase

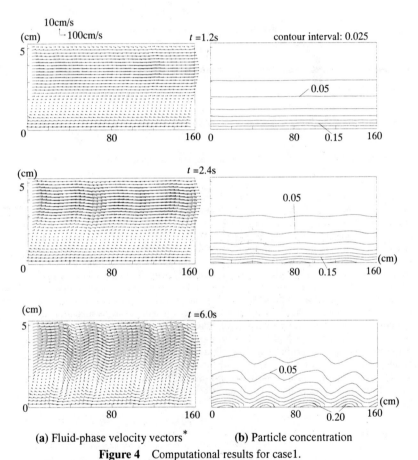

(a) Fluid-phase velocity vectors[*]   (b) Particle concentration

Figure 4   Computational results for case1.
([*]Indicated by subtracting 150cm/s from the horizontal component of
the velocity vectors)

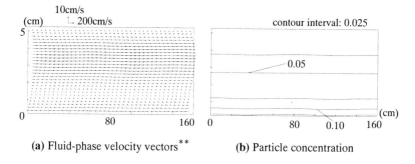

Figure 5   Computational results for case2 at $t=6.0$s.
([**] Indicated by subtracting 250cm/s from the horizontal component of the velocity vectors)

near the bottom boundary are retarded because of the relatively higher particle concentration. It is also noted that both the distribution are horizontally uniform. On the other hand, at $t=2.4$s, the spatial fluctuation of the fluid-phase velocity vectors and the particle concentration occurs in a horizontal alignment. And then, at $t=6.0$s, a series of large eddies in the fluid-phase motion are developed further and the spatial fluctuation of the corresponding particle concentration appears more noticeably by the large eddy structure in the fluid-phase motion. The overall characteristic of the flow pattern after $t=6.0$s is almost similar to that at $t=6.0$s. Comparing with the experimental classification of horizontal fluidized bed as shown in **Fig. 3**, one can find the flow pattern of case1 is similar to the saltation flow.

On the other hand, the computational results in case of relatively higher fluid velocity (case2) at $t=6.0$s is shown in **Fig. 5**. After this time stage, the overall characteristic of the flow pattern was confirmed to hold. **Figure 5** indicates that the particle concentration and horizontal fluid velocity close to the bottom boundary are respectively higher and lower than those close to the upper boundary. Also the results of case2 represent no appearance of the large eddies like those observed in case1. Hence the computational results of case2 are found to be similar to the heterogeneous flow.

The fact that this difference in the development of large eddies according to the variation in the mean fluid velocity is also observed in the previous experiment by Turian & Yuan (1977) demonstrates the validity of the present model. The similar 2-D large-scale eddies is observed in a flow acceleration phase of sheet-flow development (Conley & Inman, 1992). Therefore the present attempt of the large-eddy computation of a horizontal fluidized bed may become a significant step to develop a more general LES tool for the sheet-flow simulation.

(b) Vertical profile of statistic quantities

Figures 6(a) and **(b)** represent the vertical distributions of the horizontal fluid velocity and particle concentration in each case averaged over the horizontal

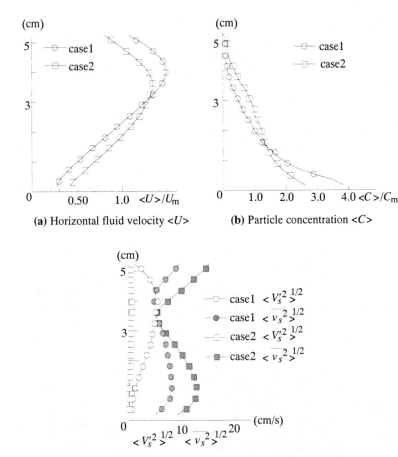

(a) Horizontal fluid velocity <U>

(b) Particle concentration <C>

(c) Turbulence intensity of the particle velocity in the vertical direction
Figure 6 Vertical distribution of statistical quantities.

computational domain. In these figures, the vertical coordinate represents the height from the bottom boundary, while the horizontal coordinates indicate the normalized velocity $<U>/U_m$ and particle concentration $<C>/C_m$, respectively. The fluid-phase velocity near the bottom boundary is lower than that near the upper boundary and the opposite is found for the particle concentration. These characteristics are more marked in case1 than case2. The relatively larger retardation of the fluid velocity near the bottom boundary observed in case1 is mainly due to the higher particle concentration.

The cause of the difference in the particle concentration between case1 and 2 may be found by examining the vertical distribution of the turbulence intensity of the particle velocity in the vertical direction, because it is closely related with the vertical distribution of the particle concentration. Since the solid-particle motion in the GAL model is represented with the grid-averaged velocity and velocity variance components, the particle velocity turbulence can be expressed separately with the large-scale fluctuation of the grid-averaged velocity V_s' and the small-scale turbulence corresponding to the velocity variance. The intensity of the former turbulence component $<V_s'^2>^{1/2}$ was estimated from the r.m.s. value of the deviation of the grid-averaged particle velocity from its horizontally averaged value, while the latter was obtained from the horizontally averaged value of the particle velocity variance $<\overline{v_s^2}>^{1/2}$. **Figure 6(c)** shows the vertical distribution of these two components of the particle turbulence intensity. Although the large-scale turbulence intensity is appreciable only in case1, the total intensity in the lower layer is smaller in case1 than case2. This may be the main cause to produce higher particle concentration and hence larger velocity retardation near the bottom boundary in case1.

CONCLUSION

1) The new modeling of the particle collision suitable to the GAL-LES model is presented for extending the applicability of the GAL-LES model to multiphase flow with high particle concentration.
2) The present GAL-LES model has been applied to a horizontal fluidized bed, demonstrating the validity of the GAL-LES model with the new particle collision model for the multiphase turbulent flow with high particle concentration through the comparison with the experimental results by Turian & Yuan (1977) .

REFERENCES

Bagnold, R.A. (1954): Experiments on a gravity-free dispersion of large solid spheres in a Newtonian fluid under shear, *Proc. R. Soc. Lond. A.*, Vol.225, pp.49-63.
Turian, R. M. and Yuan, T. (1977): Flow of slurries in pipelines, *AIChE Journal*, Vol.23, No.3, pp.232-243.
Conley, D. C. and Inman, D. L. (1992): Field observations of the fluid-granular boundary layer under near-breaking waves, *J. Geophys. Res.*, Vol.97, No.C6, pp.9631-9643.
Nadaoka, K., Nihei, Y. and Yagi, H. (1995): A new framework for LES modeling of solid-fluid phase turbulent flow —A mixed Euler-Lagrangian approach—, *Proc. 2nd Int. Conf. Multiphase Flow '95-KYOTO*, Vol. 2, pp.p3-71-74.
Nadaoka, K., Nihei, Y. and Yagi, H. (1996): An LES modeling for solid-liquid phase flow based on new formulation of solid-particle motion, *J. Hydraulic, Coastal and Environmental Eng.*, JSCE, No. 533, pp.61-73. (*in Japanese*)
Nadaoka, K., Nihei, Y. and Yagi, H. (1997) : Grid-Averaged Lagrangian LES model for multiphase turbulent flow, *Int. J. Multiphase Flow,* (to be submitted).

Subject Index

Page number refers to the first page of paper

Author Index

Page number refers to the first page of paper